Molecular Methods in Plant Pathology

Edited by

Rudra P. Singh

Uma S. Singh

CRC LEWIS PUBLISHERS

Boca Raton London Tokyo

Library of Congress Cataloging-in-Publication Data

Molecular methods in plant pathology / edited by Rudra P. Singh, Uma S. Singh.
p. cm.
Includes bibliographical references (p.) and index.
ISBN 0-87371-877-1
1. Plant diseases—Molecular aspects—Research—Methodology.
2. Phytopathogenic microorganisms—Molecular aspects—Research—Methodology.
I. Singh, Rudra P. II. Singh, Uma S.
SB732.65.M64 1994
581.2′9—dc20 94-10126
CIP

Lewis Publishers is an imprint of CRC Press

International Standard Book Number 0-87371-877-1
Library of Congress Card Number 94-10126
Printed in the United States of America 1 2 3 4 5 6 7 8 9 0
Printed on acid-free paper

PREFACE

In recent years, we have witnessed the introduction and/or extensive use of several powerful molecular, biological, immunological, tissue culture, biochemical, histochemical, and microscopic techniques in plant pathology. Some of these techniques have been described recently, while others are the result of conceptual and instrumentational advancements of the comparatively older techniques. Increasing applications of these methods in plant pathology are bringing about the information explosion in our understanding of the complex phenomena like host-pathogen compatibility/specificity, host resistance/susceptibility, pathogen avirulence/virulence and pathogenic variability. Some of these techniques have found wide applicability in disease diagnosis and quarantine. They are also being used to develop disease-resistant plants against diseases where natural sources of resistance are not available.

Most of the techniques covered in this publication have much broader application in biology, and a good amount of the published information, although scattered widely, is available on methodology. Considering this fact, in most of the articles major emphasis has been placed on actual and potential applications of different techniques in plant pathology and their advantages and limitations in solving the plant pathological problems in addition to brief methodology and basic principles on which these techniques are based. Detailed protocols are included for only those techniques which are either more specific to or have wider application in plant pathology. For detailed protocols of other techniques, authors are referred to the reference and further reading sections.

All the chapters are contributed by the scientists who are using these techniques in their research. The authors were requested to write not merely a review but a thought-provoking article giving due consideration to their own experiences and perceptions. For the benefit of readers, chapters have been grouped into four sections and each chapter includes a list of critical references and suggestions for further reading.

Section I — Biophysical Methods deals with light and electron microscopic and electrophoretic techniques. There has been renewed interest in some biochemical characteristics of plant host as a marker for disease resistance and isozyme patterns in fungal taxonomy/variability. These methods are listed in **Section II — Biochemical Methods.** The use of molecular biological techniques is now commonplace in many laboratories and a considerable arsenal of techniques has been developed to tackle a variety of plant pathological problems. All these techniques are described in **Section III — Biomolecular Methods.** In spite of a number of limitations, monoclonal antibodies and plant tissue culture have found quite extensive use in plant pathology. Articles on these techniques and one article on the present status of conventional breeding for disease resistance are included in **Section IV — Biotechnological Methods.**

We fully realize that inclusion of a few more articles on recent advances in light microscopy, scanning-tunneling and atomic force electron microscopy, population genetics, histochemistry, RFLPs, etc. would have enhanced usefulness of the work. However, in this kind of publication, compromises have to be made with regard to the availability of a suitable author, time, and space. There is always scope to improve future editions. As editors, our major aim has been to produce a cohesive volume that could be used effectively by students, teachers, and researchers and not fall into the category of either an encyclopedia or a ragbag of specialized chapters. How far we have succeeded in our aim is left to the reader to judge.

We extend our thanks to *bel spirit* members of the "Editorial Advisory Committee", Drs. Hei-Ti Hsu, Wilford M. Hess, Noel T. Keen, and Barbara Baker, for their invaluable help at various stages of the project. Our thanks also go to all the contributors for their willing participation in this project and tolerance to editorial interferences. We hope that their imaginative and thought-provoking articles will stimulate plant pathologists and plant biotechnologists to use newer and more accurate techniques in solving their research problems.

We are indebted to Jon Lewis, Publisher, and the skillful staff at Lewis Publishers for their unexcelled cooperation. Our sincere thanks to Agriculture and Agri-Food Canada and G.B. Pant University of Agriculture and Technology for permitting us to undertake this project.

Rudra P. Singh

Uma S. Singh

Rudra P. Singh, Ph.D., is Principal Research Scientist at Agriculture Canada, Research Centre, Fredericton, New Brunswick, Canada.

Dr. Singh, a native of Sariya, India, graduated from the Agricultural College, Kanpur, U.P., India, and obtained his Ph.D. in 1966 from North Dakota State University, Fargo, ND. His scientific career began at the Research Station, Fredericton, where he worked as a Postdoctoral Fellow of the National Research Council of Canada (1966 to 1967). Later, he became a Research Scientist in Potato Virology at the same Agriculture Canada Research Station, became Senior Research Scientist in 1978, and since 1992 has been the Principal Research Scientist.

Dr. Singh's continuing research on spindle tuber disease of potatoes, which began in 1962, culminated in 1971 in the discovery of the ribonucleic acids of a low-molecular weight nature (viroid) as the causal agent of the disease. In addition to this discovery, he also demonstrated seed and pollen transmission of viroids, a general method of viroid detection using reverse polyacrylamide gel electrophoresis, and was instrumental in eradicating the spindle tuber disease from Canada. Recently, Dr. Singh has been researching various virus and viroid diseases of potatoes.

Dr. Singh has been involved in the Canadian International Development Agency (CIDA) projects in Brazil, China, and Tunisia; in the Inter-American Institute for Cooperation on Agriculture project in Brazil and in Food & Agriculture Organization projects in India. He teaches virology component at Nova Scotia Agricultural College, Truro, N.S. and is Adjunct Professor or Honorary Research Associate of N.S. Agricultural College and the University of New Brunswick, respectively.

Dr. Singh is a member of the American Phytopathological Society, the Potato Association of America, the Canadian Phytopathological Society, the European Association for Potato Research, the Indian Potato Association, and the Indian Virological Society. Dr. Singh has served on the editorial committees of the American Phytopathological Society, the Canadian Phytopathological Society, and the Potato Association of America. Dr. Singh has been honored by the New Brunswick Seed Potato Growers Association (1990) in recognition of outstanding service in the development of practices for improving seed potato quality in Canada. He was later honored by the "Friends of Potato Industry" (1991) for his leadership in handling the PVY^N outbreaks in Canada. He has been the organizer, contributor, and co-editor of "Potato Pest Management in Canada" and served as the first chairman of the International Viroid Working Group.

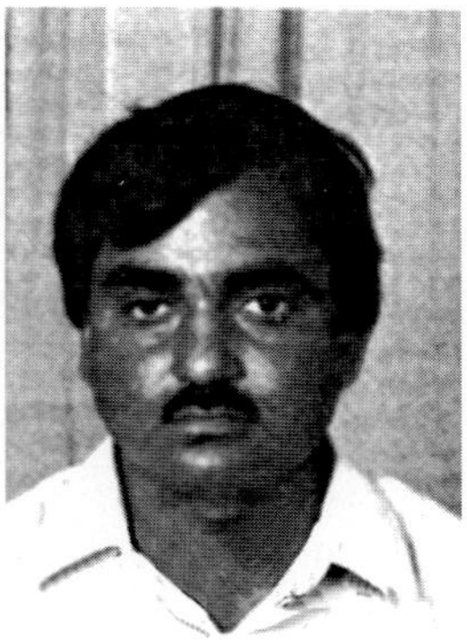

Uma S. Singh, Ph.D., is Associate Professor at G. B. Pant University of Agriculture and Technology, Pantnagar, India.

Dr. Singh, a graduate of Delhi University, obtained his Ph.D. from G. B. Pant University of Agriculture and Technology, Pantnagar in 1983 and immediately after completion of the degree joined the faculty of the same university as an Assistant Professor. He served at the International Rice Research Institute, Los Banos, Philippines from June 1990 to October 1992 as a postdoctoral scientist.

The major research areas of Dr. Singh are systemicity of fungicides in plants and host-pathogen interactions in fungal diseases. He demonstrated the existence of the nonsymplastic route, in addition to the symplastic route, for the downward translocation and role of plant species in the uptake, translocation, and distribution of fungicides in plants. He also showed that leaf epicuticular waxes play an active role in genetic and ontogenic resistance of rice blast. Recently Dr. Singh has been working on the mode of action of *Alternaria brassicae* toxins and the generation of toxin-resistant rape seed and mustard plants.

At Pantnagar Dr. Singh is associated with the departments of Plant Pathology and Molecular Biology and Genetic Engineering. He teaches "Phytopathological Techniques" and advanced courses related to host-parasite interaction in plant diseases.

For his research contributions Dr. Singh was awarded the Prof. M. J. Narasimhan Academic Merit Award of the Indian Phytopathological Society. Dr. Singh and students who worked under his guidance have been awarded the Pesticide India Award of the Society of Mycology and Plant Pathology four times for their research contributions on fungicides.

Dr. Singh has co-authored/co-edited four books, namely *Experimental and Conceptual Plant Pathology* (3 volumes; Gordon and Breach Science Publishers, New York and Oxford & IBH, New Delhi), *Perspectives in Phytopathology* (Today's & Tomorrow, New Delhi), *Plant Disease Management: Principles and Practices* (CRC Press, Boca Raton, FL), and *Plant Diseases of International Importance* (four volumes; Prentice Hall, Englewood Cliffs, NJ).

Dr. Singh is a member of the Indian Phytopathological Society, the Society of Mycology and Plant Pathology, the Society of Biological Chemists, and the Association of Microbiologists of India.

EDITORIAL ADVISORY COMMITTEE

CONTRIBUTORS

Jacqueline Agnan
Department of Botany and Plant Pathology, Oregon State University, Cordley Hall 2082, Corvallis, OR 97331-2902
FAX: (503) 737-3573
Phone: (503) 737-5303

Richard Belanger
Department de phytologie, Faculte des Sciences, de l'Agriculture et de l'Alimentation, Universite Laval, Ste-Foy, Quebec, G1K 7P4, Canada
FAX: (418) 656-7856
Phone: (418) 656-7517

Nicole Benhamou
Recherche en Sciences de la vie et de la Santé, Pavillon C. E. Marchand, Universite Laval, Ste-Foy, Quebec, G1K 7P4, Canada
FAX: (418) 656-7176
Phone: (418) 656-7517/2165

Madan K. Bhattacharyya
Plant Biology Division, The Samuel Roberts Noble Foundation, Inc., P. O. Box 2180, 2510 Highway 199 East, Ardmore, OK 73402
FAX: (405) 221-7380
Phone: (405) 223-5810

M. R. Bonde
USDA-ARS, Foreign Disease—Weed Science Research, Fort Detrick, Building 1301, Frederick, MD 21702
FAX: (301) 619-2880
Phone: (301) 619-2860

Laurence M. Brill
Department of Biology, University of California, Los Angeles, 405 Hilgard Avenue, Los Angeles, CA 90024-1606
FAX: (310) 206-5413

Yang-Sun Chan
Institute of Molecular and Cell Biology, National University of Singapore, 10 Kent Ridge Crescent, Singapore 0511, Republic of Singapore
FAX: (65) 7791117

Tet-Fatt Chia
Institute of Molecular and Cell Biology, National University of Singapore, 10 Kent Ridge Crescent, Singapore 0511, Republic of Singapore
FAX: (65) 7791117

R. G. Christie
Department of Agronomy, University of Florida, Gainesville, FL 32611

Nam-Hai Chua
Laboratory of Plant Molecular Biology, The Rockefeller University, 1230 York Avenue, New York, NY 10021

James N. Culver
Department of Plant Pathology, University of California, Riverside, CA 92521
FAX: (301) 314-9075

Richard A. Dixon
Plant Biology Division, The Samuel Roberts Noble Foundation, Inc., P. O. Box 2180, 2510 Highway 199 East, Ardmore, OK 73402
FAX: (405) 221-7380
Phone: (405) 223-5810

J. R. Edwardson
Department of Agronomy, University of Florida, Gainesville, FL 32611

John M. S. Forrest
Scottish Crop Research Institute, Invergowrie, Dundee DD2 5DA, Scotland, U.K.
FAX: (382) 562-426
Phone: (382) 562-731
Telecom Gold/CGNET: 10087 : NQQ003

Abed Gera
Department of Virology, Agricultural Research Organization, Institute of Plant Protection, The Volcani Center, Bet Dagan, Israel
FAX: (972) 3-9604180

A. Hadidi
National Germplasm Resources Laboratory, USDA-ARS, Beltsville, MD 20705-2350
FAX: (301) 504-5551
Phone: (301) 504-6460

Robert M. Hansch
Botanical Institute, Technical University of Braunschweig, Humboldstreet 1, 38106 Braunschweig, Germany

Wilford M. Hess
Microscopy Laboratory, Department of Botany and Range Science, 401 WIDB, Brigham Young University, Provo, UT 84602
FAX: (801) 378-7499
Phone: (801) 378-2582

C. Z. Hess
Microscopy Laboratory, Department of Botany and Range Science, 401 WIDB, Brigham Young University, Provo, UT 84602
FAX: (801) 378-7499
Phone: (801) 378-2582

Hei-Ti Hsu
Florist and Nursery Crops Laboratory, 10300 Baltimore Avenue, Room 238, Bldg. 010A, BARC-West, USDA-ARS, Beltsville, MD 20705-2350
FAX: (301) 504-5096
Phone: (301) 504-6570

Yau-Heiu Hsu
Agricultural Biotechnology Laboratories, National Chung Hsing University, Taichung, Taiwan 40227, Republic of China
FAX: (886) 4-286-1905
Phone: (886) 4-287-4754

Roger Innes
Department of Biology, Indiana University, Jordon Hall 142, Bloomington, IN 47405
FAX: (812) 855-6705

Hideo Ishii
Fruit Tree Research Station, Ministry of Agriculture, Forestry & Fisheries, 2-1 Fujimoto, Tsukuba, Ibaraki, 305 Japan
FAX: (0298) 38-6541
Phone: (0298) 38-6544

Ramon Jordan
Research Plant Pathologist, USDA-ARS, USNA, Floral and Nursery Plants Research Unit, Bldg. 004, Room 208, BARC-West, 10300 Baltimore Avenue, Beltsville, MD 20705-2350
FAX: (301) 504-5096
Phone: (301) 504-5646/6570

Ravinder K. Khetarpal
National Bureau of Plant Genetic Resources, Pusa Campus, New Delhi-110012, India
Phone: (91, 11) 5785619, 5784835

C. A. Kumar
4/27, The Avenue, Balaclava, Melbourne, VIC-3183, Australia
Phone: (03) 5278407

J. Kumar
Department of Plant Pathology, Hill Campus of G. B. Pant University, Ranichauri, P. O. Ranichauri 249 199, Distt. Tehri (U.P.), India

George H. Lacy
Laboratory for Molecular Biology of Plant Stress: Phytobacteriology Section, Department of Plant Pathology, Physiology & Weed Science, College of Agriculture and Life Sciences, Virginia Polytechnic Institute and State University, Blacksburg, VA 24061-0330
FAX: (703) 231-7477
Phone: (703) 231-6361

Roger H. Lawson
Florist and Nursery Crops Laboratory, 10300 Baltimore Avenue, Room 238, Bldg. 010A, BARC-West, USDA-ARS, Beltsville, MD 20705-2350
FAX: (301) 504-5096
Phone: (301) 504-6570

L. Levy
National Germplasm Resources Laboratory, USDA-ARS, Beltsville, MD 20705-2350
FAX: (301) 504-5551
Phone: (301) 504-5437

Na-Sheng Lin
Institute of Botany, Academia Sinica, Nankang, Taipei, Taiwan 11529, Republic of China
FAX: (886) 2-782-7954
Phone: (886) 2-782-3050

Kevin McCluskey
Department of Plant Pathology, University of Arizona, Tucson, AZ 85721
FAX: (602) 621-9290
Phone: (602) 621-1828

Ralf R. Mendel
Botanical Institute, Technical University of Braunschweig, Humboldstreet 1, 38106 Braunschweig, Germany

Ray Mernaugh
Pharmacia P-L Biochemicals, 2202 North Bartlett Avenue, Milwaukee, WI 53202
FAX: (414) 227-3759/278-0688
Phone: (414) 227-3662

G. Mernaugh
Pharmacia P-L Biochemicals, 2202 North Bartlett Avenue, Milwaukee, WI 53202
FAX: (414) 227-3759
Phone: (414) 227-3662

J. A. Micales
USDA-Forest Service, Forest Products Laboratory, One Gifford Pinchot Drive, Madison, WI 53705
FAX: (608) 231-9592
Phone: (608) 231-9215

Sally A. Miller
Department of Plant Pathology, Ohio Agricultural Research and Development Center, The Ohio State University, 1680 Madison Avenue, Wooster, OH 44691-4096
FAX: (216) 263-3841
Phone: (216) 263-3678

Dallice Mills
Department of Botany and Plant Pathology and Genetics Program, Oregon State University, Cordley Hall 2082, Corvallis, OR 97331-2902
FAX: (503) 737-3573
Phone: (503) 737-5303

Olga V. Nikolaeva
Department of Virology, Moscow State University, Moscow 119899, Russia
FAX: 7 (095) 292-6511 for VIROLOGY BOX 8121
Present Address
Department of Plant Pathology, University of California, Riverside, CA 92521-0122
FAX: (909) 787-4294
E-mail: BITNET:PLPATH@UCRVMS

Kazusato Ohshima
Department of Applied Biological Sciences, Faculty of Agriculture, Saga University, 1-banchi, Honjomachi, Saga 840, Japan
FAX: (0952) 22-6274
Phone: (0952) 24-5191

Nancy L. Paiva
Plant Biology Division, The Samuel Roberts Noble Foundation, Inc., P. O. Box 2180, 2510 Highway 199 East, Ardmore, OK 73402
FAX: (405) 221-7380
Phone: (405) 223-5810

J. E. Parlevliet
Department of Plant Breeding, Agricultural University, P.O.B. 386, 6700 AJ, Wageningen, The Netherlands
FAX: (31) 8370-8-34-57
Phone: (31) 8370-8-28-36

E. V. Podleckis
USDA, APHIS, PPQ Biological Assessment and Taxonomic Support, Room 625, 6505 Belcrest Road, Hyattsville, MD 20705
FAX: (301) 436-4300
Phone: (301) 436-5014
E-mail: A348EPODLECK@ATTMAIL.COM

Reuven Reuveni
Division of Plant Pathology, Agricultural Research Organization, Newe Ya'ar Research Center, P.O. Box 9000 Haifa, 31-900, Israel
FAX: (972) 4-9836936
Phone: (972) 4-9833186

Brian Russell
Department of Botany and Plant Pathology, Oregon State University, Cordley Hall 2082, Corvallis, OR 97331-2902
FAX: (503) 737-3573
Phone: (503) 737-5303

Maarten H. Ryder
CSIRO Division of Soils and Cooperative Research Centre for Soil and Land Management, Glen Osmond, SA 5064, Australia
Postal Address:
Private Bag No. 2, Glen Osmond, SA 5064, Australia
FAX: (08) 303-8550
Phone: (08) 303-8534

Amita Sachan
Department of Plant Pathology, College of Agriculture, G. B. Pant University of Agriculture and Technology, Pantnagar 263145, Distt. Nainital, U.P., India
FAX: (5948) 33473

G. W. Simone
Plant Pathology Department, University of Florida, Gainesville, FL 32611

Pratibha Singh
Department of Molecular Biology and Genetic Engineering, College of Basic Sciences and Humanities, G. B. Pant University of Agriculture and Technology, Pantnagar 263145, Distt. Nainital, U.P., India
FAX: (5948) 33473

Rudra P. Singh
Research Centre, Agriculture & Agri-Food Canada, P. O. Box 20280, Fredericton, New Brunswick, E3B 4Z7, Canada
FAX: (506) 452-3316
Phone: (506) 452-3260

Uma S. Singh
Department of Plant Pathology, College of Agriculture, G. B. Pant University of Agriculture and Technology, Pantnagar 263145, Distt. Nainital, U.P., India
FAX: (5948) 33473
Phone: (5948) 33076

O. K. Sinha
Indian Institute of Sugarcane Research, Lucknow 226002, India
FAX: (0522) 51028
Phone: (0522) 51504

Michael D. Standing
Microscopy Laboratory, P.O. Box 25181, Brigham Young University, Provo, Utah, 84602
FAX: (801) 378-7499
Phone: (801) 378-4011

Verlyn K. Stromberg
Laboratory for Molecular Biology of Plant Stress: Phytobacteriology Section, Department of Plant Pathology, Physiology & Weed Science, College of Agriculture Life Sciences, Virginia Polytechnic Institute and State University, Blacksburg, VA 24061-0330
FAX: (703) 231-7477
Phone: (703) 231-6361

Tsung-Che Tseng
Academia Sinica, Institute of Botany, Nankang, Taipei, Taiwan 11529, Republic of China
FAX: (886) 2-782-7954
Phone: (886) 782-3050 ext. 424

Soledad Verdejo-Lucas
Departmento de Patologia Vegetal, Institut de Recerca i Tecnologia Agroalimentaries (IRTA), Centre de Cabrils, Crta. de Cabrils s/n., 08348 Cabrils (Barcelona), Spain
FAX: 1 (9) 3-753 39 54
Phone: 1 (9) 3-750 75 11

Norman R. M. Watts
Research Centre, Agriculture & Agri-Food Canada, P. O. Box 20280, Fredericton, New Brunswick, E3B 4Z7, Canada

Keith L. Wycoff
Department of Biology, University of California, Los Angeles, 405 Hilgard Avenue, Los Angeles, CA 90024-1606
FAX: (310) 206-5413

Nevin Dale Young
Department of Plant Pathology, University of Minnesota, Twin Cities, 495 Borlaug Hall, 1991 Upper Buford Circle, St. Paul, MN 55108
FAX: (612) 625-9728
Phone: (612) 625-8200
E-mail: Bitnet:neviny@puccini.crl.umn.edu

CONTENTS

Section 1
BIOPHYSICAL METHODS

Chapter 1

X-Ray Microanalysis

W. M. Hess, C. Z. Hess, and M. D. Standing

TABLE OF CONTENTS

I. INTRODUCTION

The use of X-ray microanalysis to study plant-disease associations has many potential applications. Since the technology is relatively new and the equipment is not readily available in all laboratories, the applications to plant-disease associations have been limited. However, with the background knowledge now known and the increasing availability of equipment in laboratories, it is hoped that X-ray microanalysis will be used more extensively for investigations of plant-disease associations. The purpose of this short review is to briefly discuss the technology and limitations and provide limited examples of applications. Therefore, the discussion is not limited to plant-disease associations. The types of potential applications to plant-disease associations are diverse and, in some cases, may be quite specific. Additional applications of X-ray microanalysis to a variety of plant-disease associations may be suggested or apparent after the brief overview that follows.

Since the early 1960s, X-ray microanalysis has provided a method of elemental analysis at the microscopic level and has been used at the subcellular level to determine the location of major inorganic ions for both plant and animal systems.[1] Characteristic X-rays are emitted from samples which have been excited from some energy source. The source of excitement may be either X-rays or a high-energy electron beam. The most common energy source for biological samples is the high-energy electron beam because of the ability to excite very small areas of a sample. As primary electrons strike a solid specimen, they interact with the specimen to create several types of signals that provide different types of information. The signals are[2] (1) secondary electrons, which provide morphological information; these electrons generally have energies less than 50 eV. (2) Backscatter electrons, which provide "atomic number contrast"; backscatter shows where various elements are because of the atomic number. Elements with higher atomic numbers emit more backscatter electrons and, therefore, appear brighter. Backscatter electrons have energies greater than 50 eV. (3) Auger electrons, which have extremely low energy, provide information about the top few angstroms of the surface of a specimen. (4) Characteristic X-rays provide very exact information about the elemental composition of the specimen. (5) An additional signal which is produced is continuum X-rays. These X-rays are also known as background X-rays and serve to interfere with analysis. It is because of these X-rays that it is not possible to detect very small quantities of elements in a sample with energy dispersive X-ray analysis (EDX).

The production of X-rays makes it possible to gather information about the elements within the specimen in the region being excited. This provides a means of correlating morphological information at the ultrastructural level with elemental analysis of specimens and regions in specimens. It is possible to use X-rays to identify the elements and, with proper sample preparation, to quantify the elements present within a specimen.[3] The principles and instrumentation have been discussed by several authors.[2–8]

Absolute sensitivity and resolution were discussed by Läuchli and Boursier.[1] The absolute sensitivity with X-ray microanalysis is approximately 10^{-19} g. The sensitivity limits for frozen hydrated plant specimens is about 1 mol m^{-3} for K^+ and Cl^- and about 5 mol m^{-3} for Na^+.[9] Spatial resolution is influenced by several factors, which include the diameter of the electron beam at the specimen surface,

0-87371-877-1/95/$0.00+$.50

the depth of electron penetration, and lateral electron spread within the specimen.[1] As an example, it was estimated that the accelerating voltage of 10 kV, frequently used for bulk frozen hydrated specimens, would normally provide a spatial resolution of about 2 to 3 μm for biologically important elements.[9] "Thus, the resolution of X-ray microanalysis provides the opportunity for analyses in plant cell compartments, i.e., cytoplasm, vacuole, and chloroplast, but measurements at the level of the cell wall are at the limits of resolution."[1]

II. EQUIPMENT

There are many types of equipment which can be used for X-ray microanalysis of biological samples. These instruments include the following: (1) X-ray fluorescence (XRF), (2) electron microprobe or "probe", (3) scanning electron microscope (SEM), (4) transmission electron microscope (TEM), and (5) scanning transmission electron microscope (STEM). XRF and the electron probe generally are not used for biological samples. XRF is not generally used because it requires a relatively large sample and does not provide very good spatial resolution. The electron probe uses very large beam currents which can easily damage a specimen and, because of the large spot size of the electron beam, it has relatively poor resolution. The SEM is the instrument of choice when examining bulk samples. Its rastering beam makes it ideal for obtaining morphological information followed by precisely positioning the beam on a desired spot or area for subsequent X-ray analysis. The TEM and STEM are used for observing and analyzing thin-sectioned material. The TEM is limited in its analytical capability because the beam does not raster across the sample. It is therefore necessary to increase the magnification of an area of interest in order to obtain X-ray information. Even then, some of the surrounding area may be excited by the beam and emit X-rays. The STEM combines the rastering beam, which can be focused to a very fine spot, possible with a SEM, with the higher-energy emissions and resolution of the TEM. Therefore, the STEM is the ideal instrument for X-ray analysis on thin-sectioned material.

Although there are several types of equipment which are used to generate characteristic X-rays from a specimen, there are only two types of X-ray spectrometers which are used. One is the EDX spectrometer, and the other is the wavelength dispersive X-ray spectrometer (WDX). The EDX system measures the energies of X-rays coming into the detector and the WDX system detects X-rays of a particular wavelength. Each system has its advantages and disadvantages.

The EDX system is able to detect all energies of X-rays at the same time. With this capability, all detectable elements can be analyzed simultaneously. This makes it possible to determine what the major elements are within a sample. With proper sample preparation, quantitative X-ray analysis to within a relative error of about 10% is possible with the EDX spectrometer. However, the sensitivity of this type of detector is limited. If the elements being analyzed constitute less than 1% of the area being analyzed, then they probably will not be detected. Also, if the elements produce very soft X-rays (X-rays less than 1 keV), then the percent concentration must be higher in order for detection to be possible. EDX detectors will also generally have a window which protects the actual detector from contamination. Contamination is an important consideration since the detector is kept at liquid nitrogen temperatures. If the window being used is made of beryllium, then the lowest energy of X-rays which can be detected is about 1 keV, which means that elements of atomic number lower than 11 cannot be detected. There are some thin windows available which allow detection of elements down to atomic number 4. Windowless detectors also allow for detection down to atomic number 4, with better detection of the low numbers since there is no energy-absorbing window. In practical application it is difficult to accurately detect elements below carbon with an atomic number of 6. It has been pointed out that low-energy X-rays may be more susceptible to absorption due to contamination, which necessitates an evaluation of the effects of contamination on the intensities of characteristic X-ray lines,[10] and X-ray absorption corrections may be necessary.[11] The EDX spectrometer can be attached to all of the types of equipment mentioned above, and for biological applications it is generally the preferred system.

The WDX system is limited to analysis of only one wavelength at a time, and therefore only one element can be analyzed at a time. Precise geometry between the primary beam, the sample, the reflecting crystal, and the detector is essential because of how the wavelengths of the desired X-rays are selected. Due to the tedious nature of setup and analysis required with this type of system, general elemental identification is usually carried out with an EDX spectrometer prior to using a WDX spectrometer. WDX is most commonly used with XRF and electron probe systems, and occasionally with SEM systems. WDX is not readily available for TEM and STEM systems. WDX systems also require high beam currents in order to obtain high enough count rates to provide data in a reasonable amount of time. Because of these limitations, spatial resolution becomes poorer and the possibility of beam damage to the specimen increases. The advantage of WDX analysis is that trace elements can be detected in significantly lower concentration ratios. Quantitative analysis can also be carried out with greater accuracy, and light-element detection is more accurate since no window is used in this system. WDX can be a powerful tool when analyzing for trace elements, but for most biological work its applications may be limited.

The microscopes and the X-ray systems both have some fantastic capabilities. When combining these technologies, care should be taken to be sure all of the desired capabilities are incorporated into the analytical system. It is possible to combine the rastering of a SEM or STEM with the X-ray analysis of the EDX spectrometer to localize elements in various areas of a specimen simultaneously to create an X-ray map. By using this capability, several previously identified elements can be localized and correlated with a morphological image. When several elements are mapped at a time each may be displayed in a different color. X-ray maps can then be overlaid to show how the various elemental distributions overlap. X-ray line scans can be used in a similar manner to show intensities of X-ray emission of various elements across a scanned line over a portion of a sample. X-ray maps and line scans, in conjunction with backscatter and/or secondary electron images, can provide much useful information about the elemental composition correlated with the morphology of a specimen.

III. THE SPECIMEN

Specimen preparation is very important. It is difficult to maintain cellular and subcellular structure and prevent loss and redistribution of diffusible elements while preparing the specimen for the harsh environment of the microscope. All of the electron microscopes mentioned in the previous section require a very high vacuum in order to allow electrons to flow freely at the required energy levels. Sample preparation must therefore preserve the specimen structure, remove or solidify the water within the sample, preserve the diffusible elements within the sample, and, for X-ray work, not add elements which would interfere with the analysis to be performed. If the water-soluble ionic elements are the primary target of interest, then some type of cryo work would be a desirable method to obtain meaningful data.

A commonly used cryo method is frozen hydrated bulk specimens. It is necessary to rapidly freeze-fix a specimen and to maintain it in the frozen hydrated state. This necessitates the use of specialized cryo equipment and a cold stage on the microscope. One of the important factors in obtaining good spectra is the topography of the specimen. If samples have excessive topography, X-rays may not be detectable from various areas of interest. X-rays travel in straight lines, and if a structure lies in the path of the X-rays, then the X-rays may be blocked from reaching the detector. Freeze-hydrated, fractured specimens normally have relatively rough surfaces. The deep penetration of the probe into the specimen limits the resolution to 4 to 8 μm and the probe may cause some redistribution of diffusible ions due to localized melting.[2]

It is possible to obtain quantitative information with frozen hydrated bulk plant specimens. If specimens are not relatively flat, quantitative microanalysis may be restricted. Läuchli and Boursier[1] pointed out that data expressed as peak/background ratios give relative quantitative information about elemental distribution in specimens. They stated that standards which resemble biological specimens in chemical, physical, and structural properties can be used to provide quantitative data in units of concentration.

Very thin samples, such as small particulate, extracts, or thin sections may be examined with better spatial resolution using a TEM, or preferably a STEM.[2] As an example, samples of plant sap can be freeze-dried on thin films and may be compared with droplets of standards.[12] Samples which have been previously prepared should be placed on carbon or beryllium grids (other grids may produce X-rays which would interfere with elements of interest) which have a thin carbon support film on them. If morphologic information is desired as well as X-ray information, then sample thickness should not be greater than 100 nM. If the specimens are thin, qualitative and semiquantitative data can be obtained;[2] if morphologic information is not critical, then sections may be used up to 1000 nM. As with frozen hydrated bulk samples, frozen hydrated sections may be ideal when water-soluble elements are of interest. This technology is not new; it has been used since the 1970s.[13] A TEM or STEM equipped with a cryo stage, to keep the samples frozen, would provide means to analyze elements without mobilization, although some movement of ions may occur due to localized melting of the section caused by heating from the electron beam. The vitreous water in a frozen hydrated biological specimen contributes about 97% of the continuum X-radiation.[14] Lazof and Läuchli reported that using their instrumentation and procedures, P, S, Cl, and K are detectable down to 24 mM, and Na and Mg are detectable down to 38 mM in frozen hydrated tissue. However, in the presence of 100 mM K, detection of calcium is possible down to 22 and 2 mM in frozen hydrated and freeze-dried tissue, respectively, suggesting an improvement in minimally detectable elemental concentrations of some elements in freeze-dried tissues. However, precautions should be taken to reduce ion beam damage and element loss when biological tissues are analyzed.[15]

With most nonbiological specimens there is not as much concern with diffusible elements. Examples of nonbiological applications are analysis of asbestos fibers,[16,17] identification of metal-containing reactive dyes,[18] structure and chemical composition of superalloys,[19] and distribution of heavy metals in activated sludge.[20]

When cryo systems are not available for observing frozen sections, another way to prepare samples for TEM or STEM studies would be to use a freeze-substitution method of sample preparation. With this method, the sample is quick-frozen in various solvents at liquid nitrogen temperatures as mentioned above, but then the frozen samples are put into cold (about −70°C) acetone and kept cold for several days, changing the acetone periodically. By doing this, dehydration of the sample occurs as is required for standard TEM sample preparation, but the washes in a water-based solution are eliminated, reducing the chances of losing the soluble ions. The sample is then warmed to room temperature in a closed system to prevent water contamination. Embedding in resin is then carried out using a variety of resins which do not contain elements which interfere with the elements of interest.

Cryo work can be used for both bulk samples and thin samples. Freeze substitution can also be used for bulk samples. Following the substitution, samples can be processed through any of the standard drying techniques used for SEM preparation. Freeze-drying of bulk or thin samples may be suitable for preparation for X-ray microanalysis. Several other sample preparation methods can be used, depending on the specific nature of the experiment being performed.

Freeze substitution and freeze-drying can be effectively used with plant tissues for localization and analysis of nondiffusible elements. In some instances diffusible elements are retained *in situ* when these procedures are used.[21] Hodson et al.[22] used conventional preparation procedures, TEM, and EDX to study silica deposits in lemma and glumes of *Phalaris canariensis* and freeze substitution with sections in a TEM.[23] Both procedures were effectively used to analyze immobile deposits of silicon, but were inadequate to study the soluble silicon. Although cryo-SEM and freeze substitution with TEM are preferred, freeze-drying and resin embedment have also been used for mineral partitioning in plant tissues.[24,25] For localization of heavy metals in plant roots, Wasserman et al.[26] sectioned tissues with a cryostat, dehydrated in an ethanol series, and critical point-dried the tissues. Kunoh et al.[27] used conductive staining to aid with microdissection with SEM observations. This helped to clarify relationships between the surface of the host tissue, the plant pathogen on the leaf surface, and the interiors of biological materials. An early compilation of general methodology was published in 1983.[28] A more recent treatment was published in 1992.[5]

IV. INCLUSION IDENTIFICATIONS

Higher plants, algae, and fungi often have inclusions. Since the inclusions normally appear to be relatively stable structures, embedding and sectioning have been a common procedure to use for elemental analysis investigations. During the 1970s, investigators commonly embedded tissues in resin and examined inclusions with TEM and EDX. Examples include the elemental composition of gamma particles in a phycomycete fungus,[29] cellular inclusions in algal cells,[30] plant statolith analysis,[31] elemental localization in chloroplasts,[32] and the analysis of phosphorus-containing inclusions in cyanobacteria.[33]

During the 1980s, resin sections were still commonly used with TEM and EDX to evaluate elemental composition of electron-dense bodies in mycorrhizas[34] and other fungi,[35] although osmium tetroxide was often omitted[36] or freeze substitution was used.[37] Various other procedures have also been used, including critical point drying and freeze-fracturing at liquid nitrogen temperatures followed by freeze-drying.[38] As was mentioned above, the unavailability of equipment to study tissues in the frozen hydrated state in many laboratories imposes unfortunate limitations. Nevertheless, for the examination of relatively stable cellular structures in higher plant, algal, and fungal cells, elemental composition studies can be conducted with equipment which is more readily available.

V. SALT AND ELEMENTAL CONCENTRATION

During the mid-1970s, silver chloride precipitation was used to attempt to determine where chloride ions were located in plant tissues of the halophyte *Salicornia pacifica* var. *utahensis*.[39]

Then frozen sections were fractured under liquid nitrogen and scanned for Na^+, K^+, and Cl^- with WDX.[40] Both procedures demonstrated that the salt appears to be excluded from photosynthetic cells. However, it would be desirable to use X-ray microanalysis of freeze-hydrated tissues for similar studies. Although the examination of freeze-hydrated tissues appears to provide the most reliable data when compared to other procedures, the freezing process is a very important step. Single cells are relatively easy to rapid freeze.[41] Fungal cells require more precautions than single cells because the most rapid freezing procedures only freeze a few layers of cells well without formation of ice crystals in the tissues.[42] Much better freezing can be attained with high-pressure freezing.[43] Various attempts have been used to overcome the freezing problems.[44–53]

In their studies of cell lines of *Nicotiana* sensitive and resistant to sodium chloride, Dix et al.[54] used cryosectioning and freeze-drying to determine that both cell lines rapidly accumulated Na^+ from the culture medium, and X-ray microanalysis did not help them elucidate the differences in the cell lines. However, in studies with salt tolerance in *Plantago* using freeze substitution and hydrated specimens,[55,56] it was possible to determine that the differences in salinity tolerance between the species was related to the efficiency of ion transport to the shoot and compartmentation of ions in root cortical cells.

Similar studies were conducted with root cells of maize. Differences in salinity concentrations were demonstrated using X-ray microanalysis of freeze-substituted tissue.[57] In subsequent studies[58] X-ray microanalysis, compartmental analysis, and longitudinal ion profiles were used. In the salt-sensitive variety, X-ray microanalysis, compartmental analysis, and longitudinal profiles yielded approximately the same data for cytoplasmic K^+, but the methods disagreed for cytoplasmic Cl^- where compartmental analysis was about four times that for X-ray microanalysis and longitudinal profiles.

Freeze substitution with dry thin sections and X-ray microanalysis were used to demonstrate that sodium and chloride concentrations increased with salinity in the cytoplasm, vacuole, and cell wall of the marine fungus *Dendryphiella,* although potassium concentrations decreased with salinity.[59] Freeze substitution and X-ray microanalysis were also used to demonstrate that an important factor with salt damage in rice is dehydration caused by extracellular accumulation of salt. Excised leaf tissues were frozen in 8% (v/v) methylcyclohexane in 2-methylbutane cooled with liquid nitrogen to about −170°C. Leaves were fractured and freeze substituted in acetone at −70°C. Flat, dry thin sections, coated with carbon, were used for X-ray microanalysis. Using these procedures it was concluded that extracellular salt accumulation can be the factor which initiates salinity damage.[60] When freeze substitution and thin sectioning are used, frozen hydrated sections can be transferred to the stage of a STEM or TEM and freeze-dried in the microscope before X-ray microanalysis. Unfortunately, few laboratories have these capabilities.

Examples of studies which involved X-ray microanalysis of elemental accumulation in plant tissues include the use of freeze substitution and dry sections to study zinc concentrations in ectomycorrhizal fungal mycelium,[61] standard sectioning to study metal accumulation in a heavy metal-tolerant fungus,[62] the use of standard sectioning and lyophilized cryosections to study platinum localization in corn roots,[63] the use of embedded tissue and bulk specimens from powdered-compressed tissue to study calcium and potassium accumulation in corn bran and oat hulls,[64] and Epon®-araldite-embedded sections to study accumulation of aluminum in cell walls and organelles in aluminum accumulator plants.[65]

VI. ELEMENTAL LOCALIZATION AND DISTRIBUTION

The uses of X-ray microanalysis to help solve biological problems are diverse. This technology has been used to identify elements in cotton, corn, and soybean dusts which may cause lung dysfunction.[66] Mineralogical particles, thought to be of soil origin, were identified. The corn and soybean dusts were different from the cotton dust. Mineralized metal powders were used for intrafloral pollen tracking. Backscatter SEM and X-ray microanalysis were used to detect the micronized metal dusts zinc and tin.[67] Backscatter SEM analysis requires an atomic number contrast between the elements of interest to differentiate the metals. The number of backscatter electrons emitted from a sample is directly proportional to the atomic number of the elements present. In this instance the tin particles appeared brighter than the zinc particles, and both were brighter than the carbon background of the biological sample.

Ethidium bromide was used to provoke potassium efflux in yeast cells followed by WDX analysis for potassium content. This helped to demonstrate that cationic dyes can induce an almost complete loss of potassium from yeast cells.[68] In a study with higher plants, lanthanum was used to confirm that solutes enter the apoplasm of root meristems and move to the stele of the root and to the shoots of several species of intact plants studied. The presence of lanthanum was confirmed by X-ray microanalysis with unstained thin sections.[69]

Treeby et al.[70] discussed the importance of Pi concentrations and localization in leaves because of the role Pi plays in cytoplasmic and vacuolar pH control, stomatal activity, movement of pulvini, and carbohydrate activity. They used WDX with hydrated, bulk-frozen samples to demonstrate that inorganic phosphate contents of mesophyl cells were shown to be highly dependent on phosphate nutrition. For localization of hemicellulose in birch wood decayed by fungi, a STEM equipped with EDX was used to localize thiocarbohydrazide-silver proteinate in sections.[71] These procedures made it possible to determine the patterns of wall degradation which did not necessarily correspond with electron density in micrographs. In a similar study, bromination was used to selectively bind bromine to lignin. Brominated wood pieces were embedded in resin and sections were examined in a STEM equipped with EDX. Based upon these studies it was concluded that all white rot basidiomycetes do not remove lignin from the cell wall in the same manner.[72]

Other examples of elemental localization with X-ray microanalysis are antimonate precipitation to localize biological elements,[73,74] mercuric chloride for cytochemical localization of mercury in *Saccharomyces,*[75] concentration of ATPase activity in plant cells,[76,77] DNA localization,[78] ion localization in root meristem cells,[79] effects of nitrogen compounds upon potassium distribution in plant roots and the rhizosphere,[80] soluble ion localization in fungal hyphal cells,[81] and changes in texture, cell wall structure, and composition during storage of calcium-treated and untreated apples.[82] Although a variety of techniques and procedures have been used for many different applications to localize elements in biological tissues, it is important to emphasize that the use of frozen hydrated samples has important advantages for analysis of plant and fungal tissues. In 1986, Echlin and Taylor[83] stated, "the X-ray microanalysis of bulk, frozen-hydrated samples is now an accepted method for making quantitative measurements of local elemental concentrations in biological material."

VII. ENVIRONMENTAL APPLICATIONS

There are many environmental applications of X-ray microanalysis to plant and fungal tissues. Only a few studies are mentioned here as examples of applications. Sela et al.[84] shock-froze fronds of *Azolla* in isopentane at $-176°C$, freeze-dried at $-45°C$, and fixed tissues with paraformaldehyde vapors. The freeze-dried tissues were embedded in Lowicryl® resin and sections were examined in a TEM with EDX. The authors studied the storage and distribution of copper, cadmium, and uranium for their effects upon the ionic contents of the shoots and roots. Similar studies were conducted with shoots of an aquatic liverwort to demonstrate lead accumulation in plant tissues.[85] The lead was from an abandoned lead mine. In this instance the authors used standard fixation and dehydration procedures, and embedment was in Spurr resin. Unstained sections were analyzed in a TEM with EDX.

Nitrogen form and solution pH were studied to determine the effects upon growth and nutrition of *Vaccinium.*[86] Frozen root and shoot samples were freeze-dried and mounted onto aluminum stubs covered with double-stick tape. Carbon paint was used to ground the samples and nickel was evaporated onto samples to make them conductive. The tissues were analyzed in a SEM with EDX.

The concern with aluminum toxicity and localization and mineral element distribution in Norway spruce roots led to studies with X-ray microanalysis.[87] Root and fungal material was freeze substituted in acetone and embedded in Taab Transmit EM resin. Dry sections were carbon coated in hinged copper grids. Analysis was in a STEM equipped with EDX. Seven elements were studied: aluminum, silicon, phosphorus, sulfur, chlorine, potassium, and calcium. Aluminum was confined to the cortical cell walls and was not in the endodermis. The presence of a fungus significantly increased aluminum concentrations in cortical cell walls.

Other examples of environmental applications of X-ray microanalysis include studies of aluminum accumulation in birch leaves and twigs, soil, and grass due to atmospheric pollution from a smelter,[88] inhibition of tomato fruit ripening by silver,[89] changes in elemental concentrations in tobacco leaf cells during growth and senescence,[90] mercury-induced loss of potassium from yeast cells,[91] and elemental changes in fresh-water algal cells during blooms.[92]

VIII. PLANT DISEASE AND X-RAY MICROANALYSIS

Agrios[93] defined plant disease as "malfunctioning of host cells and tissues that results from their continuous irritation by a pathogenic agent or environmental factor and leads to the development of symptoms. Disease is a condition involving abnormal changes in the form, physiology, integrity, or behavior of the plant. Such changes may result in partial impairment or death of the plant or its parts." By this definition most abnormal changes in plant growth and development could be called plant disease.

Tissue preparation procedures are very important in all studies involving living biological materials and X-ray microanalysis. When dynamic living systems are studied, including host-pathogen interfaces, rapid freeze-fixation without the addition of chemical treatments may be the first important step in obtaining reliable information. Unfortunately, even a thin layer of fungal cells on a support membrane will often not freeze adequately.[42] The development of high-pressure freezing, although the equipment is expensive, has made it possible to freeze up to approximately 500 μm.[43] Therefore, even under the most ideal conditions and with the best equipment available, there are significant limitations with rapid freeze-fixation. For freeze-fixation of spores or single cells the equipment may be relatively inexpensive and the freezing procedures may be easily accomplished.[94] However, for many, and perhaps most, applications with host-pathogen associations it would be desirable to work with multicellular systems which are very difficult to adequately freeze-fix. For this reason it is necessary to use alternative approaches. Some examples of approaches have been discussed above.

In addition to the approaches discussed above, electron spectroscopic imaging (ESI) and electron energy loss spectroscopy techniques can be used with equipment such as a Zeiss® TEM for elemental analysis of plant materials.[95] Even with these approaches the preferred procedures were to use cryofixation followed by freeze substitution. Atomic absorption spectrometry and neutron activation analysis have also been used with X-ray microanalysis.[96] These procedures are particularly useful when elements are of interest which are not detectable with X-ray microanalysis.

Freeze-fracture cytoimmunochemistry and the use of colloidal gold[97,98] or other markers make it possible to label specific sites in the specimen before it is frozen, after it has been fractured, after it has been sectioned, or after it has been platinum shadowed and/or carbon coated. Visualization of the labeled cellular structures can be achieved by a variety of different methodologies. If there is a question about the identity of the markers, X-ray microanalysis can be used with SEM,[99] TEM, STEM, or with ESI as with a Zeiss® TEM.[100] Examples of applications of this technology to plant-disease associations include localization of glycoproteins in necrotic *Nicotiana* tissue infected with tobacco mosaic virus[101] and studies of lignin peroxidates in wood degraded by a white rot fungus.[102]

Other examples of applications of X-ray microanalysis of plant diseases are analysis of antimonate precipitates in healthy and virus-infected tobacco leaves;[103] elemental composition of barley coleoptile papillae related to *Erysiphe* leaf penetration;[104] the use of SEM and WDX for analysis of fungal tissues[105–107] and SEM, TEM, and SEM with WDS to study virus-induced leaf tumors from maize plants;[108] a comparison of laser microscopy and X-ray microanalysis to evaluate chemical composition of the fungus *Trichothecium;*[109] and the use of low-temperature SEM of frozen hydrated *Penicillium* cultures.[110] A useful general reference was published by Flegler and Baker.[111]

In this brief overview we have attempted to discuss applications of X-ray microanalysis to plant and fungal tissues with emphasis upon plant-disease associations. It is obvious that X-ray microanalysis has been used for a wide variety of applications, and diverse methods and procedures have been used. In the future the methodology and equipment will continue to improve and change. We will be able to more carefully characterize the chemical associations in biological tissues and more particularly in plant-disease associations. Unfortunately, the availability of equipment for most laboratories, the cost and maintenance of the equipment, and the unavailability of funds for personnel and supplies are all limiting factors. Even with these limitations significant progress has been made in attempts to better understand chemical associations, and X-ray microanalysis will continue to help make significant contributions to biological knowledge, including plant-disease interfaces.

REFERENCES

1. Läuchli, A. and Boursier, P. J., Compartmentation in root cells and tissues: X-ray microanalysis of specific ions, *Methods Enzymol.,* 174, 267, 1989.
2. Bozzola, J. J. and Russell, L. D., *Electron Microscopy: Principles and Techniques for Biologists,* Jones and Bartlett, Boston, 1992, chap. 15.
3. Chandler, J. A., *X-ray Microanalysis in the Electron Microscope,* North-Holland, Amsterdam, 1977.
4. Heinrich, K. F. J., *Electron Beam X-ray Microanalysis,* Van Nostrand Reinhold, New York, 1981.
5. Goldstein, J. I., Newbury, D. E., Echlin, P., Joy, D. C., Romig, A. D., Jr., Lyman, C. E., Fiori, C., and Lifshin, E., *Scanning Electron Microscopy and X-Ray Microanalysis,* Plenum Press, New York, 1992.
6. Lechene, C. P. and Warner, R. R., *Microbeam Analysis in Biology,* Academic Press, New York, 1979.
7. Newbury, D. E., Joy, D. C., Echlin, P., Fiori, C. E., and Goldstein, J. I., *Advanced Scanning Electron Microscopy and X-Ray Microanalysis,* Plenum Press, New York, 1986.
8. Woldseth, R., *X-Ray Energy Spectrometry,* Kevex Corporation, Burlingame, CA, 1973.
9. Pitman, M. G., Läuchli, A., and Stelzer, R., Ion distribution in roots of barley seedlings measured by electron probe X-ray microanalysis, *Plant Physiol.,* 68, 673, 1981.
10. Horita, Z., Sano, T., and Nemoto, M., Effect of contamination on quantitative X-ray microanalysis in the analytical electron microscope, *Philos. Mag.* B, 63, 965, 1991.
11. Horita, Z., Sano, T., and Nemoto, M., A new form of the extrapolation method for absorption correction in quantitative X-ray microanalysis with the analytical electron microscope, *Ultramicroscopy,* 35, 27, 1991.
12. Malone, M., Leigh, R. A., and Tomos, A. D., Concentrations of vacuolar inorganic ions in individual cells of intact wheat leaf epidermis, *J. Exp. Bot.,* 42, 305, 1991.
13. Saubermann, A. J. and Echlin, P., The preparation, examination and analysis of frozen hydrated tissue sections by scanning transmission electron microscopy and X-ray microanalysis, *J. Microsc.,* 105, 155, 1975.
14. Lazof, D. and Läuchli, A., Complementary analysis of freeze-dried and frozen-hydrated plant tissue by electron-probe X-ray microanalysis: spectral resolution and analysis of calcium, *Planta,* 184, 327, 1991.
15. Kirby, B. J. and Legge, G. J. F., Ion beam induced damage and element loss during a microanalysis of biological tissue, *Nucl. Inst. Met. Phy. Res.,* B54, 98, 1991.
16. Altree-Williams, S. and Preston, J. S., Asbestos and other fibre levels in buildings, *Ann. Occup. Hyg.,* 29, 357, 1985.
17. Gylseth, B., Mowe, G., and Wannag, A., Fibre type and concentration in the lungs of workers in an asbestos cement factory, *Br. J. Ind. Med.,* 40, 375, 1983.
18. Sakata, A., Mamba, K., Kitahama, M., and Seta, S., Application of the energy dispersive X-ray microanalyzer equipped with the scanning electron microscope to the criminal identification of the metal-containing reactive dyes, *Jpn. J. Legal Med.,* 33, 358, 1979.
19. Link, T. and Osterle, W., Röntgenmikroanalyse im elektronenmikroskop Teil IV: Superlegierungen, *Pract. Met.,* 28, 101, 1991.
20. Schulze, G. and Gunkel, G., Verteilung und umsetzungen von schwermetallen in der biologischen stufe einer kommunalen kläranlage, *Vom Wasser,* 70, 209, 1988.
21. Marshall, A. T., Freeze-substitution as a preparation technique for biological X-ray microanalysis, *Scanning Electron Microsc.,* II, 395, 1980.
22. Hodson, M. J., Sangster, A. G., and Parry, D. W., An ultrastructural study on the developmental phases and silicification of the glumes of *Phalaris canariensis* L., *Ann. Bot.,* 55, 649, 1985.
23. Hodson, M. J. and Sangster, A. G., Techniques for the microanalysis of higher plants with particular reference to silicon in cryofixed wheat tissues, *Scanning Microsc.,* 4, 407, 1990.
24. Eschrich, W., Fromm, J., and Essiamah, S., Mineral partitioning in the phloem during autumn senescence of beech leaves, *Trees,* 2, 73, 1988.
25. Fritz, E., X-Ray microanalysis of diffusible elements in plant cells after freeze-drying, pressure infiltration with ether and embedding in plastic, *Scanning Microsc.,* 3, 517, 1989.
26. Wasserman, J. L., Mineo, L., Majumdar, S. K., and Van Tyne, C., Detection of heavy metals in oak mycorrhizae of northeastern Pennsylvania forests, using X-ray microanalysis, *Can. J. Bot.,* 65, 2622, 1987.

27. Kunoh, H., Watanabe, T., Yamada, M., and Nagatani, T., The conductive staining method suitable for X-ray microanalysis and micromanipulation of plant materials, *J. Electron Microsc.*, 24, 301, 1975.
28. Roomans, G. M. and Shelburne, J. D., *Basic Methods in Biological X-Ray Microanalysis*, Scanning Electron Microscopy, Inc., AMF O'Hare, IL, 1983.
29. Hutchinson, T. E., Cantino, M. E., and Cantino, E. C., Calcium is a prominent constituent of the gamma particle in the zoospore of *Blastocladiella emersonii* as revealed by X-ray microanalysis, *Biochem. Biophys. Res. Commun.*, 74, 336, 1977.
30. Bochem, H. P. and Sprey, B., Laser microprobe analysis of inclusions in *Dunaliella salina*, *Z. Pflanzenphysiol.*, 95, 179, 1979.
31. Steudle, E., Läuchli, A., and Sievers, A., X-Ray microanalysis of barium and calcium in plant material: significance for the analysis of statoliths, *Z. Naturforsch.*, 33, 444, 1978.
32. Jánossy, A. G. S., Mustárdy, L. A., and Faludi-Dániel, A., X-Ray microanalytical study of Mn and Fe compartmentation in maize chloroplasts, *Acta Histochem.*, 58, 317, 1977.
33. Kessel, M., Identification of a phosphorus-containing storage granule in the cyanobacterium *Plectonema boryanum* by electron microscope X-ray microanalysis, *J. Bacteriol.*, 129, 1502, 1977.
34. Cox, G., Moran, K. J., Sanders, F., Nockolds, C., and Tinker, P. B., Translocation and transfer of nutrients in vesicular-arbuscular mycorrhizas. III. Polyphosphate granules and phosphorus translocation, *New Phytol.*, 84, 649, 1980.
35. Hobbs, G., Griffiths, A. J., and Hann, A. O. C., X-ray microanalysis of electron-dense bodies in *Cladosporium resinae*, *FEMS Microbiol. Lett.*, 33, 285, 1986.
36. Grellier, B., Strullu, D. G., Martin, F., and Renaudin, S., Synthesis *in vitro*, microanalysis and ^{31}P-NMR study of metachromatic granules in birch mycorrhizas, *New Phytol.*, 112, 49, 1989.
37. Orlovich, D. A., Ashford, A. E., and Cox, G. C., A reassessment of polyphosphate granule composition in the ectomycorrhizal fungus *Pisolithus tinctorius*, *Aust. J. Plant Physiol.*, 16, 107, 1989.
38. Benson, D. R., Greenhut, V. A., and Eveleigh, D. E., Scanning electron microscopy of the nitrogen-fixing actinorhiza of *Myrica pensylvanica* L. (bayberry), *New Phytol.*, 83, 755, 1979.
39. Hess, W. M., Hansen, D. J., and Weber, D. J., Light and electron microscopy localization of chloride ions in cells of *Salicornia pacifica* var. *utahensis*, *Can. J. Bot.*, 53, 1176, 1975.
40. Weber, D. J., Rasmussen, H. P., and Hess, W. M., Electron microprobe analyses of salt distribution in the halophyte *Salicornia pacifica* var. *utahensis*, *Can. J. Bot.*, 55, 1516, 1977.
41. Ridge, R. W., A simple apparatus and technique for the rapid-freeze and freeze-substitution of single-cell algae, *J. Electron Microsc.*, 39, 120, 1990.
42. Hoch, H. C., Freeze-substitution of fungi, in *Ultrastructure Techniques for Microorganisms*, Aldrich, H. C. and Todd, W. J., Eds., Plenum Press, New York, 1986, chap. 7.
43. Studer, D., Michel, M., and Müller, M., High pressure freezing comes of age, *Scanning Microsc. (Suppl.)* 3, 253, 1989.
44. Hagler, H. K., Artifacts in cryoelectron microscopy, in *Artifacts in Biological Electron Microscopy*, Crang, R. F. E. and Klomparens, K. L., Eds., Plenum Press, New York, 1988, chap. 10.
45. Jeffree, C. E., Read, N. D., Smith, J. A. C., and Dale, J. E., Water droplets and ice deposits in leaf intercellular spaces: redistribution of water during cryofixation for scanning electron microscopy, *Planta*, 172, 20, 1987.
46. Dahl, R. and Staehelin, L. A., High-pressure freezing for the preservation of biological structure: theory and practice, *J. Elect. Microsc. Tech.*, 13, 165, 1989.
47. Fujikawa, S., Suzuki, T., Ishikawa, T., Sakurai, S., and Hasegawa, Y., Continuous observation of frozen biological materials with cryo-scanning electron microscope and freeze-replica by a new cryo-system, *J. Electron Microsc.*, 37, 315, 1988.
48. Kessel, R. G. and Beams, H. W., Freeze fracture and scanning electron microscope studies on the nuclear envelope and perinuclear cytomembranes (parabasal apparatus) in the protozoan, *Lophomonas blattarum*, *J. Submicrosc. Cytol. Pathol.*, 22, 367, 1990.
49. Castejón, O. J., Freeze-fracture scanning electron microscopy and comparative freeze-etching study of parallel fiber-Purkinje spine synapses of vertebrate cerebellar cortex, *J. Submicrosc. Cytol. Pathol.*, 22, 281, 1990.
50. Diller, K. R., A note on optimal techniques of rapid cooling for low temperature preparation of electron microscopy specimens, *Cryo-Letters*, 11, 75, 1990.
51. Cole, R., Matuszek, G., See, C., and Rieder, C. L., A simple pneumatic device for plunge-freezing cells grown on electron microscopy grids, *J. Elect. Microsc. Tech.*, 16, 167, 1990.

52. Macfarlane, D. R. and Forsyth, M., Recent insights on the role of cryoprotective agents in vitrification, *Cryobiology,* 27, 345, 1990.
53. Thom, F. and Matthes, G., Remarks on ice formation in binary aqueous solutions of ethylene glycol and dimethylsulfoxide. II. Dimethylsulfoxide, *Cryo-Letters,* 11, 13, 1990.
54. Dix, P. J., Plunkett, A., and Toth, I., The use of tissue culture for investigating the physiology of salt stress, in *Plant Tissue Culture in Relation to Biotechnology,* 1982, 95–104.
55. Harvey, D. M. R. and Brandtner, R., The regulation of ion levels within *Plantago maritima* roots: an examination of freeze-substituted and frozen hydrated specimens, *Acta Histochem. Cytochem.,* 20, 527, 1987.
56. Harvey, D. M. R., Salinity tolerance in *Plantago* species: an ultrastructural and X-ray microanalytical investigation of differences in ion transport in the roots, *Scanning Microsc.,* 3, 527, 1989.
57. Hajibagheri, M. A., Harvey, D. M. R., and Flowers, T. J., Quantitative ion distribution within root cells of salt-sensitive and salt-tolerant maize varieties, *New Phytol.,* 105, 367, 1987.
58. Hajibagheri, M. A., Flowers, T. J., Collins, J. C., and Yeo, A. R., A comparison of the methods of X-ray microanalysis, compartmental analysis and longitudinal ion profiles to estimate cytoplasmic ion concentrations in two maize varieties, *J. Exp. Bot.,* 39, 279, 1988.
59. Clipson, N. J. W., Hajibagheri, M. A., and Jennings, D. H., Ion compartmentation in the marine fungus *Dendryphiella salina* in response to salinity: X-ray microanalysis, *J. Exp. Bot.,* 41, 199, 1990.
60. Flowers, T. J., Hajibagheri, M. A., and Yeo, A. R., Ion accumulation in the cell walls of rice plants growing under saline conditions: evidence for the Oertli hypothesis, *Plant Cell Environ.,* 14, 319, 1991.
61. Denny, H. J. and Wilkins, D. A., Zinc tolerance in *Betula* spp. IV. The mechanism of ectomycorrhizal amelioration of zinc toxicity, *New Phytol.,* 106, 545, 1987.
62. Fukami, M., Nakano, Y., Hasegawa, Y., and Toda, S., Intensified copper signal observed in the osmium-fixed cell of *P. ochro-chloron* by X-ray microanalysis, *Spectrochim. Acta,* 36B, 655, 1981.
63. Filippenko, V. N., Ivanov, V. B., Sanygin, V. P., Kvardakov, A. M., Pogorelov, A. G., and Allakhverdov, B. L., X-Ray microanalysis of the location of platinum absorbed from the solutions of cisplatin in the meristematic cells of corn roots, *Tsitologiya,* 32, 1140, 1990.
64. Dintzis, F. R., Baker, F. L., and Nelsen, T. C., X-Ray microanalysis of Ca and K in corn bran and oat hulls, *Scanning Microsc.,* 4, 399, 1990.
65. Cuenca, G., Herrera, R., and Mérida, T., Distribution of aluminum in accumulator plants by X-ray microanalysis in *Richeria grandis* Vahl leaves from a cloud forest in Venezuela, *Plant Cell Environ.,* 14, 437, 1991.
66. Goynes, W. R., Ingber, B. F., and Palmgren, M. S., Microscopical comparison of cotton, corn, and soybean dusts, *Environ. Health Perspect.,* 66, 125, 1986.
67. Wolfe, A. D., Estes, J. R., and Chissoe, W. F., III, Tracking pollen flow of *Solanum rostratum* (Solanaceae) using backscatter scanning electron microscopy and X-ray microanalysis, *Am. J. Bot.,* 78, 1503, 1991.
68. Theuvenet, A. P. R., Bindels, R. J. M., Van Amelsvoort, J. M. M., Borst-Pauwels, G. W. F. H., and Stols, A. L. H., Interaction of ethidium bromide with yeast cells investigated by electron probe X-ray microanalysis, *J. Membr. Biol.,* 73, 131, 1983.
69. Peterson, T. A., Swanson, E. S., and Hull, R. J., Use of lanthanum to trace apoplastic solute transport in intact plants, *J. Exp. Bot.,* 37, 807, 1986.
70. Treeby, M. T., Van Steveninck, R. F. M., and De Vries, H. M., Quantitative estimates of phosphorus concentrations within *Lupinus luteus* leaflets by means of electron probe X-ray microanalysis, *Plant Physiol.,* 85, 331, 1987.
71. Blanchette, R. A. and Abad, A. R., Ultrastructural localization of hemicellulose in birch wood (*Betula papyrifera*) decayed by brown and white rot fungi, *Holzforschung,* 42, 393, 1988.
72. Otjen, L., Blanchette, R. A., and Leatham, G. F., Lignin distribution in wood delignified by white-rot fungi: X-ray microanalysis of decayed wood treated with bromine, *Holzforschung,* 42, 281, 1988.
73. Luke, K. E., Hess, W. M., and Smith, B. N., Effects of mineral deficiencies upon fungus infected plants: energy dispersive X-ray microanalysis of antimony precipitation products in *Cucurbita pepo* infected by *Sclerotinia sclerotiorum, J. Plant Nutr.,* 3, 93, 1981.
74. Van Iren, F., Van Essen-Joolen, L., Van der Duyn Schouten, P., Boers-Van der Sluijs, P., and de Bruijn, W. C., Sodium and calcium localization in cells and tissues by precipitation with antimonate: a quantitative study, *Histochemistry,* 63, 273, 1979.

75. Whittaker, S. G., Smith, D. G., Foster, J. R., and Rowland, I. R., Cytochemical localization of mercury in *Saccharomyces cerevisiae* treated with mercuric chloride, *J. Histochem. Cytochem.*, 38, 823, 1990.
76. Van Steveninck, R. F. M., The verification of cytochemical tests for ATP-ase activity in plant cells using X-ray microanalysis, *Protoplasma,* 99, 211, 1979.
77. Weber, D. J., Hess, W. M., and Kim, C. K., Distribution of ATPase in cells of *Salicornia pacifica* var. *utahensis* as determined by lead phosphate precipitation and X-ray microanalysis, *New Phytol.*, 84, 285, 1980.
78. Testillano, P. S., Sanchez-Pina, M. A., Olmedilla, A., Ollacarizqueta, M. A., Tandler, C. J., and Risueño, M. C., A specific ultrastructural method to reveal DNA: the NAMA-Ur, *J. Histochem. Cytochem.*, 39, 1427, 1991.
79. Cameron, I. L., Hunter, K. E., and Smith, N. K. R., The subcellular concentration of ions and elements in thin cryosections of onion root meristem cells: an electron-probe EDS study, *J. Cell Sci.*, 72, 295, 1984.
80. Weiming, S. and Zhiyu, L., Effects of NH_3 and urea on K distribution in root and rhizosphere of maize, *Fert. Res.*, 14, 235, 1987.
81. Galpin, M. F. J., Jennings, D. H., Oates, K., and Hobot, J. A., Localization by X-ray microanalysis of soluble ions, particularly potassium and sodium, in fungal hyphae, *Exp. Mycol.*, 2, 258, 1978.
82. Glenn, G. M. and Poovaiah, B. W., Calcium-mediated postharvest changes in texture and cell wall structure and composition in 'Golden Delicious' apples, *J. Am. Soc. Hortic. Sci.*, 115, 962, 1990.
83. Echlin, P. and Taylor, S. E., The preparation and X-ray microanalysis of bulk frozen hydrated vacuolate plant tissue, *J. Microsc.*, 141, 329, 1986.
84. Sela, M., Tel-Or, E., Fritz, E., and Huttermann, A., Localization and toxic effects of cadmium, copper, and uranium in *Azolla, Plant Physiol.*, 88, 30, 1988.
85. Satake, K., Takamatsu, T., Soma, M., Shibata, K., Nishikawa, M., Say, P. J., and Whitton, B. A., Lead accumulation and location in the shoots of the aquatic liverwort *Scapania undulata* (L.) Dum. in stream water at Greenside Mine, England, *Aquat. Bot.*, 33, 111, 1989.
86. Rosen, C. J., Allan, D. L., and Luby, J. J., Nitrogen form and solution pH influence growth and nutrition of two *Vaccinium* clones, *J. Am. Soc. Hortic. Sci.*, 115, 83, 1990.
87. Hodson, M. J. and Wilkins, D. A., Localization of aluminum in the roots of Norway spruce [*Picea abies* (L.) Karst.] inoculated with *Paxillus involutus* Fr., *New Phytol.*, 118, 273, 1991.
88. Pyatt, F. B. and Lacy, D., An appraisal of atmospheric pollution by aluminum fumes emanating from smelter works in western Norway, *Environ. Int.*, 14, 407, 1988.
89. Hobson, G. E., Nichols, R., Davies, J. N., and Atkey, P. T., The inhibition of tomato fruit ripening by silver, *J. Plant Physiol.*, 116, 21, 1984.
90. Echlin, P., Changes in the cellular concentrations of elements in tobacco leaf tissues during growth and senescence, *Beitr. Tabakforsch. Int.*, 14, 297, 1989.
91. Kuypers, G. A. J. and Roomans, G. M., Mercury-induced loss of K^+ from yeast cells investigated by electron probe X-ray microanalysis, *J. Gen. Microbiol.*, 115, 13, 1979.
92. Clay, S., Sigee, D. C., and Bellinger, E., X-Ray microanalytical studies of freshwater biota: changes in the elemental composition of *Anabaena spiroides* during blooms of 1988 and 1989, *Scanning Microsc.*, 5, 207, 1991.
93. Agrios, G. N., *Plant Pathology,* Academic Press, New York, 1988, 6.
94. Ridge, R. W., A simple apparatus and technique for the rapid-freeze and freeze-substitution of single-cell algae, *J. Electron Microsc.*, 39, 120, 1990.
95. Lehmann, H., Kramer, A., and Schulz, D., Preparation of plant material for elemental analysis using ESI and EELS techniques, *Ultramicroscopy,* 32, 26, 1990.
96. Wróblewski, R., Wroblewski, J., Lundström, H., Edström, L., and Jansson, E., Energy dispersive X-ray microanalysis, neutron activation analysis and atomic absorption spectrometry—comparison using biological specimens, *Scanning Microsc.*, 3, 467, 1989.
97. Severs, N. J., Freeze-fracture cytochemistry: review of methods, *J. Elect. Microsc. Tech.*, 13, 175, 1989.
98. Nakata, T., Sobue, K., and Hirokawa, N., Conformational change and localization of calpactin I complex involved in exocytosis as revealed by quick-freeze, deep-etch electron microscopy and immunocytochemistry, *J. Cell Biol.*, 110, 13, 1990.
99. Peschke, T., Augsten, K., Sasama, F., and Gabert, A., Determination of gold-labelled surface receptors on single cells by X-ray microanalysis, *J. Microsc.*, 156, 191, 1989.

100. Özel, M., Pauli, G., and Gelderblom, H. R., Electron spectroscopic imaging (ESI) of viruses using thin-section and immunolabelling preparations, *Ultramicrosc.*, 32, 35, 1990.
101. Benhamou, N., Mazau, D., Esquerré-Tugayé, M., and Asselin, A., Immunogold localization of hydroxyproline-rich glycoproteins in necrotic tissue of *Nicotiana tabacum* L. cv. Xanthi-nc infected by tobacco mosaic virus, *Physiol. Mol. Plant Pathol.*, 36, 129, 1990.
102. Daniel, G., Pettersson, B., Nilsson, T., and Volc, J., Use of immunogold cytochemistry to detect Mn(II)-dependent and lignin peroxidases in wood degraded by the white rot fungi *Phanerochaete chrysosporium* and *Lentinula edodes, Can. J. Bot.*, 68, 920, 1990.
103. Favali, M. A., Barbieri, N., and Bonecchi, R., Electron microscopy and X-ray microanalysis of antimonate precipitates in healthy and virus infected leaves of tobacco, *Caryologia*, 31, 331, 1978.
104. Kunoh, H., Aist, J. R., and Israel, H. W., Elemental composition of barley coleoptile papillae in relation to their ability to prevent penetration of *Erysiphe graminis, Physiol. Mol. Plant Pathol.*, 29, 69, 1986.
105. Thibaut, M., Ansel, M., and de Azevedo Carneiro, J., X-Ray microanalysis as applied to fungal tissues, *Br. J. Exp. Pathol.*, 58, 209, 1977.
106. Thibaut, M., Ansel, M., and de Azevedo Carneiro, J., A microanalysis approach to investigate problems encountered in mycology, *Am. J. Pathol.*, 90, 23, 1978.
107. Thibaut, M. and Ansel, M., Mycological applications of X-ray microanalysis, *J. Bacteriol.*, 116, 1181, 1973.
108. Favali, M. A., Barbieri, N., Bianchi, A., Bonecchi, R., and Conti, M., X-Ray microanalysis of leaf tumors from maize plants experimentally infected with maize rough dwarf virus: scanning and transmission electron microscopic study, *Virology*, 103, 357, 1980.
109. Thibaut, M., Observations on the chemical composition of *Trichothecium roseum* Link determined by X-ray microanalysis and by laser microscopy, *Microbios*, 38, 139, 1983.
110. Allan-Wojtas, P. and Yang, A. F., Solutions to difficulties encountered in low-temperature SEM of some frozen hydrated specimens: examination of *Penicillium nalgiovense* cultures, *J. Elect. Microsc. Tech.*, 6, 325, 1987.
111. Flegler, S. L. and Baker, K. K., The use of scanning electron microscopy in plant pathology, *Scanning Microsc.*, 4, 1707, 1983.

FURTHER READING

Bozzola, J. J. and Russell, L. D., *Electron Microscopy: Principle and Techniques for Biologists*, Jones and Bartlett, Boston, 1992.

Goldstein, J. I., Newbury, D. E., Echlin, P., Joy, D. C., Romig, A. D., Jr., Lyman, C. E., Fiori, C., and Lifshin, E., *Scanning Electron Microscopy and X-Ray Microanalysis*, Plenum Press, New York, 1992.

Zeyen, R. J., Applications of *in situ* microanalysis in understanding disease: X-ray microanalysis, *Annu. Rev. Phytopathol.*, 20, 119, 1982.

Chapter 2

Immunoelectron Microscopy

Nicole Benhamou and Richard Bélanger

TABLE OF CONTENTS

I. INTRODUCTION

In recent years, considerable progress has been made in elucidating the structural and molecular organization of cell compartments in plants. It has become increasingly apparent that *in situ* localization of molecules could lead to a better understanding of the functional activity of the plant cell during various biological processes such as metabolic and hormonal regulation, seed storage, transport, and recognition.[1,2] A growing body of evidence from numerous reports indicates that visual determination of the precise location and distribution of all constituents is also of crucial importance in studies dealing with the mechanisms underlying microbial pathogenesis.[36] Indeed, current knowledge about pathogen modes of action and plant reactions to microbial attack has greatly benefitted from cytological investigations of the spatiotemporal distribution of a wide range of molecules including carbohydrates, glycoproteins, and enzymes.[6–8] Taken together, cyto- and immunocytochemical techniques applied to plant tissues or cells can provide unique information on various topics such as: (1) chemical composition of cell structures; (2) vulnerability of wall-bound polymers to microbial enzymes; (3) spatiotemporal changes in protein distribution during hypersensitivity; (4) reinforcement of cell walls as a response to stress (i.e., accumulation of callose, lignin, and hydroxyproline-rich glycoproteins [HRGPs]); (5) accumulation sites of newly synthesized gene products; (6) antimicrobial activity of specific molecules such as hydrolytic enzymes and secondary metabolites, and (7) expression of "foreign" genes in transgenic plants. Immunocytochemical innovations appear with increasing frequency, and it is expected that new developments will extend the applicability of the method to more and more research areas in plant pathology.

Since the first introduction by Coons et al.[9] of fluorescent antibodies to identify sites of antigen-antibody reaction, significant advance in the study of the molecular structure of cells and tissues has been

0-87371-877-1/95/$0.00+$.50

made through the discovery of new reagents and the development of novel approaches of investigation at the electron microscope level.[10–12] Because immunoglobulins are not electron opaque, antibody binding sites in a tissue can only be visualized with a detectable marker. A number of electron-dense markers (i.e., peroxidase, ferritin) have been used over the years, each providing improvement in terms of specificity, sensitivity, and resolution. However, the most recent advancement in this field was the introduction of colloidal gold particles as a tracer for antisera.[10] The increasing interest in colloidal gold as an immunocytochemical marker of choice was related to its specific properties, such as electron opacity, stability, and particulate nature.[13] The use of this powerful marker has been extended to other cytochemical approaches using lectins,[14] enzymes,[15] or other proteins[5] with specific binding affinity for a given molecule. Gold-labeled probes presently provide the highest resolution in cytochemistry and have become increasingly popular during the last 10 years.

The ability to localize a molecule cytochemically depends on both its immobilization and retention of its biological activity during tissue processing. These requirements impose obvious criteria on fixative and resin selection for sample preparation. The ideal situation would be a procedure that, at the same time, could preserve ultrastructure and reactivity, and provide optimal access of the gold-labeled probe.[16] However, such a situation is quite unrealistic, and most often the choice of tissue preparation method must be a compromise between obtaining cell structures as close as possible to the native state, and preserving the molecules under study in correct proportions to yield a true image of their *in situ* distribution. The problem is even more accute in the case of low-molecular weight (MW) soluble proteins such as pathogenesis-related (PR) proteins which can easily diffuse from one cell compartment to another during tissue fixation with aldehydes. Rapid-freeze fixation[17] or microwave energy fixation[18] appear to be the most appropriate procedures for retaining optimal levels of diffusible antigens.

Despite some technical limitations, cytochemical techniques have acquired increasing applicability in plant pathology. It is not our purpose in this chapter to review the wide range of applications that have been reported in the literature. Instead, our objectives are to outline the methods that are currently used in cyto- and immunocytochemistry, to discuss alternative approaches, and to present a few selected examples in which gold labeling has proven to be a powerful tool for elucidating some aspects of the interaction between a host plant and a pathogen. Although lectins are of nonimmune origin,[19] we will consider here the possibility of using these reagents in addition to antibodies for detecting carbohydrate-containing molecules in infected plant tissues.

II. PRINCIPLES

Cyto- and immunocytochemical techniques are based on the affinity properties existing between macromolecules. Thus, several substances, once tagged directly or indirectly with an electron-opaque marker, enable the ultrastructural localization of their target molecules, provided they have sufficient access to the intracellular structures. With the development of post-embedding labeling of thin sections, pre-embedding has been rarely used. The rationale for such an interest in using post-embedding techniques was that not only did the probes have direct access to the cell structures at the surface of the section, but also the procedure did not require permeabilization of membranes and did not encounter problems of restricted diffusion related to cell walls as with pre-embedding techniques.[2] However, conditions which allow good ultrastructural preservation and retain the biological activity of the macromolecules have to be worked out in each case for obtaining optimal and precise labeling.

A. PROPERTIES OF THE COLLOIDAL GOLD MARKER

Colloidal gold is a negatively charged hydrophobic sol, composed of electron-opaque, metallic particles which are capable of strong emission of secondary electrons and can absorb macromolecules under specific conditions of pH and concentration.[20] The particulate nature of colloidal gold allows a precise identification of the labeled structures and is easily amenable to quantitation. Monodisperse gold sols with particle sizes ranging from 3 to 150 nm can be prepared and used in transmission, scanning, or light microscopy.

Colloidal gold is formed by reducing tetrachloroauric acid ($HAuCl_4$) with organic agents such as white phosphorus, formaldehyde, tannic acid, ascorbic acid, and sodium citrate. The most popular approach is the sodium citrate method[21] to produce monodisperse gold particles ranging from 12 to 150 nm in diameter, depending on the amount of sodium citrate added. As an example, gold particles with a uniform size of 12 nm are obtained by adding 4 ml of 1% (w/v) aqueous sodium citrate to a

100-ml, boiling solution of 0.01% (w/v) aqueous $HAuCl_4$. The reduction process is completed when the suspension turns red-orange. In recent years, a large number of reviews have been devoted to the preparation and stabilization of colloidal gold sols,[12,22–24] and the reader is invited to consult these reports for practical details of the techniques.

The principle underlying the absorption of proteins to gold particles is a still incompletely understood phenomenon. However, it is generally thought that protein absorption results from an electrostatic interaction between the negatively charged surface of gold particles and positively charged groups of proteins.[12] A number of physicochemical factors, such as pH of the colloidal gold, salt concentration of the protein solution, and protein concentration, have been shown to influence the absorption process.[12–22]

B. LECTIN CYTOCHEMISTRY

Lectins can be defined as a group of carbohydrate-binding proteins (usually glycoproteins) of nonimmune origin that occur predominantly in plants and invertebrates.[25] Because of their specific binding properties, lectins have become essential tools in carbohydrate cytochemistry.[19] Originally applied for the investigation of cell surface architecture,[26] lectins have found since then wide application in the study of intracellular carbohydrate-containing molecules.[14] A large number of lectins have been purified from plant and animal sources, and are readily available from several commercial companies (for more details, see Reference 14). Among the lectins that have been widely applied to plant tissues, one can cite: (1) wheat germ agglutinin (WGA), specific for *N*-acetylglucosamine residues and used for the localization of fungal chitin in fungus-infected plant tissues;[27,28] (2) *Ricinus communis* agglutinin, specific for galactose residues;[29] (3) *Helix pomatia* agglutinin, specific for *N*-acetylgalactosamine residues;[14] and (4) *Aplysia* gonad lectin (AGL), a lectin isolated from the sea mollusc *Aplysia depilans* and found to specifically bind to polygalacturonic acids.[30] Unfortunately, to date no lectins are known for the identification of polysaccharides with linear β-1-4- linkages which are, by far, the most important components in plant cells walls.

Lectins as tools in cytochemistry have proved extremely useful in studies of surface-related biological phenomena such as recognition, degradation, and attachment. In addition, they may be of potential value for the *in situ* identification of molecules inducing plant defense responses. These molecules, called elicitors, are known to be oligosaccharides released from plant or pathogen cell walls.[5] Lectins with high MW ($<$15 kDa) can be complexed to colloidal gold and directly applied to tissue sections.[14] Usually, lectins with low MW ($<$15 kDa) cannot be conjugated to gold. In such cases, an indirect labeling method where the marker is complexed to a secondary reagent that has affinity for the lectin is used.[14] These secondary compounds include glycoproteins such as ovomucoid (for WGA), polysaccharides with appropriate sugar binding sites, or antilectin antibodies.

C. IMMUNOCYTOCHEMISTRY

Immunocytochemistry in use for the localization of plant antigens is identical in principle to lectin cytochemistry. However, success of the method is contingent upon highly specific antibodies, appropriate preservation of protein antigenicity in plant tissues, and sufficient access of immunoglobulins to their corresponding molecules.[6] Both monoclonal and polyclonal antibodies have been used successfully for immunocytochemistry in plant cells.[2,6] In all cases, careful production and screening of the antibody probes prior to immunolabeling are required in order to avoid nonspecific interactions between the immunological probe and the tissue section.

1. Screening of Antibody Probes

The potential value of immunolabeling techniques relies in the use of highly specific antibodies. Polyclonal antisera are usually raised in rabbits against purified molecules. The immunogen may be an antigen alone, or a chemically coupled conjugate of antigen and carrier molecules.[31] Polyclonal antibodies contain a mixture of immunoglobulins that react with particular epitopes of the immunogen. Monoclonal antibodies derive from antibody-secreting hybridoma cells lines, and bind to one single epitope of an antigen molecule. For immunocytochemical purposes, polyclonal antisera raised against highly purified antigens are currently used. They offer the advantage of being easy to prepare and cheaper than monoclonal antibodies which are time-consuming and require expensive procedures. Usually, a sample of preimmune serum, collected just before the immunization process, is used as a control to assess the restricted reaction of the antiserum towards the antigen against which it has been produced.

Antigen purity is often determined by polyacrylamide gel electrophoresis (PAGE) followed by Coomassie blue or silver staining.[32] Analysis by PAGE may be performed under native or denaturing (sodium dodecyl sulfate) conditions, using one-dimensional (D), or two-D gel systems. The specificity of the antiserum raised against a purified antigen has to be verified not only against the antigen itself, but also against whole tissue components.[6] The most popular approach is the western blotting (or immunoblotting) of proteins transferred on nitrocellulose membranes.[33]

Cross reactivity of the antiserum may arise if the antigen itself carries epitopes that occur in other molecules (i.e., polysaccharide moiety of glycoproteins). This problem may be circumvented by carefully deglycosylating the molecule prior to rabbit immunization[34] or by using affinity chromatography to remove from the antiserum those immunoglobulins that react with the sugar portion.[35]

2. Principles of Immunogold Labeling

Antigen-antibody reactions are usually visualized under the electron microscope through the use of gold-conjugated secondary reagents such as protein A, or goat antiserum to rabbit immunoglobulins (rabbit antiserum to mouse immunoglobulins in the case of monoclonal antibodies). Protein A, a cell wall protein produced by most strains of *Staphylococcus aureus,* displays the ability to interact with immunoglobulins, notably immunoglobulin G (IgG) in the Fc region.[36] One molecule of protein A contains four homologous Fc region binding sites and is able to react with two IgG molecules, thus giving a strong signal. However, there is some variation in protein A reactivity of the IGs from different animal species. Recently, Bendayan[37] introduced the protein G as an alternative to protein A for high-resolution immunocytochemistry. The author reported that the avidity of protein G (from *S. aureus*) for IGs was greater than that of protein A, especially for monoclonal antibodies from mice or rats.

Secondary antibodies conjugated to colloidal gold may be used in place of protein A or protein G. Such labeled antibodies are commercially available in various gold particle sizes. They are highly stable at -4°C and can be stored for several months before use.

D. TISSUE PREPARATION

Post-embedding techniques are the most flexible and widely used approaches in immunocytochemistry. Although there is no standard protocol for the preparation of tissue for on-grid immunocytochemistry, the main objective remains the preservation of both the ultrastructure and antigenic immunoreactivity. Thus, conditions for tissue fixation and embedding have to be worked out in order to obtain optimal degrees of structural preservation and immunoreactivity.

1. Fixation

Tissue fixation is essential to restrict diffusion of compounds into and out of cells, and to strengthen the plant structures against the effect of other reagents during tissue processing. However, tissue fixation leading to satisfactory morphological preservation often precludes sufficient retention of antigenicity. Fixation should therefore be efficient enough to retain antigenic sites and ultrastructure without preventing immunoreagent accessibility. Several excellent reviews have been published on the subject and should be consulted for practical details.[16,38]

Glutaraldehyde, a dialdehyde that efficiently cross links protein molecules, is probably the most widely used electron microscopical fixative. It can be used alone at a concentration ranging from 1 to 3%, or in combination with formaldehyde. Conventional fixation procedures usually recommend tissue post-fixation with osmium tetroxide. However, this fixative, which acts as an excellent membrane stabilizer and contrasting agent, can also mask or destroy protein antigens.[38] It is thus suggested to avoid the use of this fixative for immunocytochemistry. By contrast, there is evidence that most lectin binding sites are not altered by osmium tetroxide.[14] Thus, for each system, conditions that yield optimal labeling and best ultrastructural preservation have to be worked out.

Various procedures have been developed to circumvent problems of antigenic deterioration. Among them, freeze-substitution and freeze-drying techniques have obtained some success. These techniques rely on a rapid cooling of tissue in liquid nitrogen followed by dehydration with slow warming at room temperature. The tissue is then exposed to osmium tetroxide vapor and embedded. Another procedure that has recently been introduced in the field of plant immunocytochemistry is microwave energy fixation.[18] This procedure, based on the use of microwave irradiation during aldehyde fixation, was found effective for preserving cellular structures and maintaining soluble proteins in their cell compartments.[18] Maximal tissue preservation is usually obtained with 15 to 20 s of microwave irradiation and a final fixative solution temperature of 37 to 40°C. It is likely that diffusion of the fixative is enhanced by the thermal effect of microwave irradiation. This approach has the advantage of being inexpensive.

2. Embedding Media

Several types of resin are available and have been successfully used for immunocytochemical purposes.[40] Two broad types of resin are currently used: the epoxy resins and the hydrophylic cross-linked acrylics.[40] Epoxy resins such as Epon®, Spurr®, and Araldites® exhibit low water absorption and are hydrophobic. They offer the advantage of yielding good ultrastructural preservation and high beam stability. However, their hydrophobic nature may affect retention of antigenicity, thus resulting in reduction of labeling. It is noteworthy that successful results of tissue preservation and immunoreactivity have been obtained with Epon®.[3–5] Hydrophylic resins including LR white and Lowicryl® K4M are gaining in popularity. Usually these resins exhibit excellent retention of antigenicity, but lower degrees of morphological preservation.

E. SECTION PREPARATION

Grids of gold or other nonoxidizable metals such as nickel have to be used to collect ultrathin tissue sections (60 to 90 nm in thickness). Because copper may react with the buffer solution used during cyto- and immunocytochemical procedures, copper grids are not recommended. Stability of the sections can be increased by previous grid coating with Formvar® or collodion support film.

III. METHODOLOGY

A. LECTIN LABELING PROCEDURES

Both direct and indirect labeling with lectins can be applied to ultrathin tissue sections for localizing specific sugar residues.[14] As a general rule, all experiments are performed in a moist chamber to avoid dessication.

1. Direct Labeling

A typical protocol for direct labeling with lectins is outlined in Table 1. Tissue sections are incubated with the gold-complexed lectin and at the electron microscope level the lectin-sugar interactions are readily visualized by the gold particles.

2. Indirect Labeling

Table 2 illustrates a standard protocol for indirect labeling of sugar molecules with lectins. The sugar-lectin interaction is detected through the use of a gold-complexed secondary reagent, the latter being chosen for its high affinity for the lectin.

3. Cytochemical Controls

Proper controls are necessary in determining the specificity of labeling. These controls include: (1) incubation with the lectin-gold complex to which was previously added its corresponding sugar; (2) incubation with the uncomplexed lectin, followed by incubation with the gold-complexed lectin; (3) for the indirect labeling, incubation with the lectin previously absorbed with its corresponding sugar,

Table 1 **Direct labeling with lectins**

Step	Procedure	Duration
1. Pre-incubation	Grids are floated on a drop of phosphate-buffered saline (PBS) containing 0.01% (w/v) on polyethylene glycol (PEG 20000); the pH is adjusted according to the pH of optimal activity of the lectin	5–10 min
2. Incubation	Grids are transferred onto a drop of gold-complexed lectin at the appropriate dilution in PBS-PEG	30 min
3. Rinsing	Grids are thoroughly washed with PBS and rinsed with distilled water and air dried	15 min
4. Staining	Grids are contrasted with uranyl acetate and lead citrate	

Table 2 **Indirect labeling with lectins**

Step	Procedure	Duration
1. Pre-incubation	Grids are pre-incubated on a drop of PBS, pH 7.2	5–10 min
2. Incubation	Grids are transferred onto a drop of the uncomplexed lectin at the appropriate dilution in PBS	30 min
3. Rinsing	Grids are washed with PBS, pH 7.2; the excess of buffer is removed with filter paper	10 min
4. Incubation	Grids are incubated on the gold-complexed secondary reagent at the appropriate dilution	30 min
5. Rinsing	Grids are washed with PBS and rinsed with distilled water	15 min
6. Staining	Grids are stained with uranyl acetate and lead citrate	

Table 3 **Indirect immunogold labeling with protein A**

Step	Procedure	Duration
1. Pre-incubation	Grids are pre-incubated on a drop of blocking buffer (PBS containing bovine serum albumin or ovalbumin).	10 min
2. Incubation	Grids are transferred onto a drop of primary antibody diluted in blocking buffer	2 h at 37°C
3. Rinsing	Grids are washed with the blocking buffer	15 min
4. Incubation	Grids are incubated on a drop of protein A-gold diluted in PBS-PEG, pH 7.2	30 min
5. Rinsing	Grids are washed with PBS and rinsed with distilled water	15 min
6. Staining	Grids are stained with uranyl acetate and lead citrate	

Table 4 **Indirect immunogold labeling with a second antibody**

Step	Procedure	Duration
1. Pre-incubation	Grids are incubated on a drop of blocking buffer	10 min
2. Incubation	Grids are transferred onto a drop of normal goat serum diluted in blocking buffer	30 min
3. Incubation	Grids are incubated on a drop of primary antibody (raised in rabbit), diluted in blocking buffer	2 h at 37°C
4. Rinsing	Grids are washed with Tris-buffered saline (TBS), pH 8.2, containing 1% (w/v) BSA	15 min
5. Incubation	Grids are incubated onto a drop of goat antiserum to rabbit immunoglobulin complexed to gold, diluted in TBS-BSA, pH 8.2	1 h
6. Rinsing	Grids are washed with TBS and rinsed with distilled water	15 min
7. Staining	Grids are stained with uranyl acetate and lead citrate	

followed by incubation with the secondary reagent complexed to gold; and (4) incubation with the gold-complexed secondary reagent alone.

B. IMMUNOGOLD LABELING PROCEDURE

In general, indirect methods for antigen localization at the electron microscope level are used. These methods are based on the use of a secondary reagent such as protein A (or G) and a second antibody. Typical protocols for immunogold labeling are outlined in Tables 3 and 4.

Immunocytochemical controls include: (1) use of pre-immune serum in place of primary antibody, (2) omission of the primary antibody step, and (3) pre-incubation of the primary antibody with its corresponding antigen prior to section labeling.

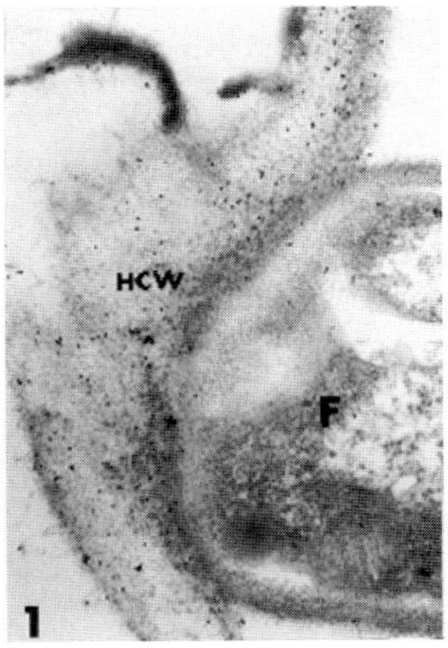

Figure 1 Cytochemical localization of pectin in bean leaf cell walls following infection by *Colletotrichum lindemuthianum.* The section was incubated with the gold-complexed *Aplysia* gonad lectin (AGL). A few gold particles are associated with the highly altered host cell wall. F, fungus; HCW, host cell wall. (Magnification × 36,000.)

IV. APPLICATIONS IN PLANT PATHOLOGY

A. APPLICATIONS OF THE LECTIN-GOLD TECHNIQUE FOR POST-EMBEDDING LOCALIZATION OF SUGAR RESIDUES

The reliability of the lectin-gold approach has been abundantly demonstrated by the successful localization of different classes of sugar residues, such as wall-bound carbohydrate-containing molecules, notably chitin and pectin, in infected plant tissues.[8,27,28] Several other glycoconjugates including galactose,[29] mannose/glucose, *N*-acetylgalactosamine, and fucose[41] have also been identified *in planta,* allowing a better characterization of the chemical composition of diverse structures, such as papillae formed in response to fungal attack. Collectively, the data obtained from the use of lectin-gold complexes have brought new insights into various biological events occurring during host-pathogen interactions, notably cell attachment, cell-cell recognition, cell wall degradation, and host reactions to pathogen attack.

The AGL was recently introduced in the field of plant cytochemistry[30] and found useful for the *in situ* localization of galacturonic acid-rich molecules.[5,8] Studies dealing with the use of this lectin provided new information on both the vulnerability of pectic compounds to fungal pectinolytic enzymes and the involvement of pectic fragments in disease resistance.[5,8,42,43] The marked alteration of pectin in bean leaf cell walls following infection by *Colletotrichum lindemuthianum* is illustrated in Figure 1. It is interesting to note that pectin breakdown was found to occur at a distance from the fungus pathway (arrow), thus indicating that fungal pectinases have the ability to diffuse extracellularly. Pectin degradation was associated with the release of fragments (Figure 2) that, in turn, could be involved in the induction of plant defense responses as suggested by others.[42] In addition, pectin was found to be associated with physical barriers (Figure 3), thus confirming that this structural polymer plays important functions in plant resistance to fungal ingress.

The application of the lectin-gold method has also been useful in delineating the mode of action of biocontrol agents. Indeed, in the development of these agents, it has become increasingly essential to determine whether a mycoparasite uses enzymes, antibiotics, or both against its fungal host. For instance, a large number of *Trichoderma* isolates have been shown to excrete hydrolytic enzymes such as chitinases

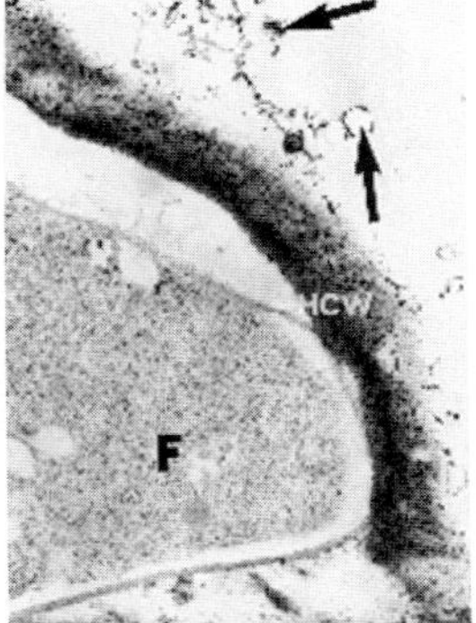

Figure 2 Cytochemical labeling of pectin in bean leaf cell walls following infection by *C. lindemuthianum.* Labeling with the AGL-gold complex. Labeled fragments are released from the host cell wall (arrows). F, fungus; HCW, host cell wall. (Magnification × 18,000.)

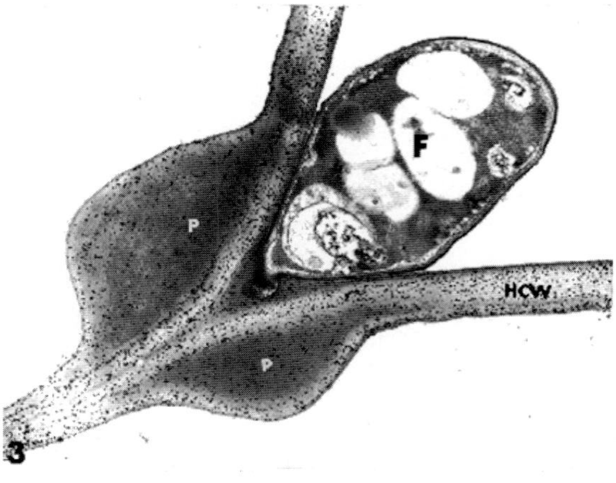

Figure 3 Cytochemical labeling of pectin in bean leaf cell walls following inoculation by *C. lindemuthianum.* Labeling with the AGL-gold complex. Wall appositions, termed papillae, are significantly labeled. F, fungus; HCW, host cell wall; P, papilla. (Magnification × 21,500.)

and β-1,3-glucanases *in vitro* when grown in media supplemented with chitin or laminarin.[45,46] However, few studies have been able to correlate the production of hydrolases *in vitro* with true antifungal activity *in vivo.* In fact, it is well known that the capacity of fungi or bacteria to produce enzymes *in vitro* does not necessarily indicate an effective enzymatic activity in the mode of action of biocontrol agents.[47] To this end, the use of the lectin-gold method offers a unique and powerful tool to visualize whether or not the cell wall structure is affected in the presence of an antagonist *in situ.*

The reliability of this approach was recently demonstrated by Hajlaoui et al.[28] Using the WGA/ovomucoid-gold complex for chitin labeling, it was clearly shown that *Stephanoascus flocculosus* induced a rapid collapse of the cytoplasm in the pathogen, *Sphaerotheca pannosa* var. *rosae,* while the cell walls remained intact even at a late stage of infection (Figure 4). This study provided indirect evidence that antibiosis rather than enzymatic processes could be a major determinant of *Stephanoascus flocculosus* antagonism. Considering the growing interest in the study of antagonist-pathogen interactions for the development of biocontrol agents, it is quite evident that gold labeling procedures will find increasing applications in the field of plant pathology.

B. APPLICATIONS OF THE IMMUNOGOLD TECHNIQUE FOR POST-EMBEDDING LOCALIZATION OF ANTIGENS

In recent years, several studies have taken advantage of the remarkable capacity of antibodies for binding specifically to certain antigens on tissue sections.[6] A large number of antibodies have been produced against a variety of proteins and found useful for elucidating some physiological processes involved in host-pathogen interactions.

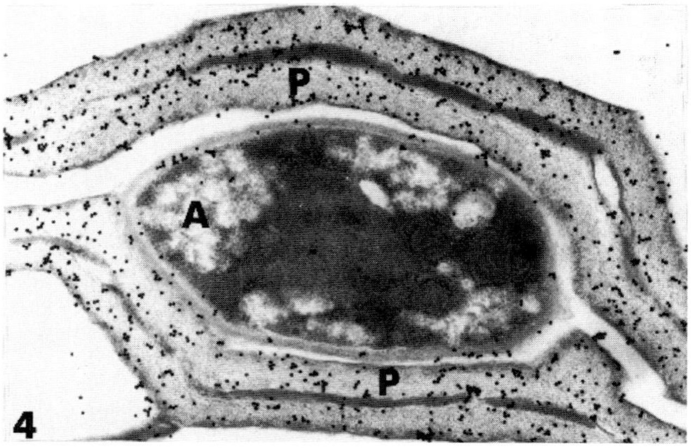

Figure 4 Cytochemical labeling of *N*-acetylglucosamine (chitin) during the interaction between *Sphaerotheca pannosa* var. *rosae* and the antagonist *Stephanoascus flocculosus.* Labeling with the WGA/ovomucoid-gold complex. Host cytoplasm has completely leaked out while cell walls are still evenly labeled. A, antagonist; P, pathogen. (Magnification × 54,000.)

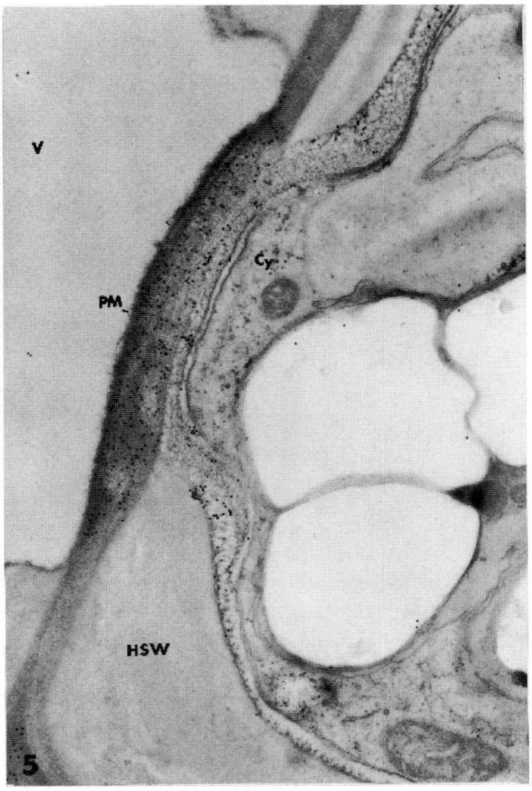

Figure 5 Immunogold labeling of a toxic glycopeptide produced by *Ophiostoma ulmi* in elm wood tissues. The section was incubated with a monoclonal antibody followed by gold-complexed protein A. Labeling is mainly associated with the pit membrane. Cy, cytoplasm; HSW, host secondary wall; PM, pit membrane; V, vessel. (Magnification × 45,000.)

Monoclonal antibodies have been successfully used in evaluating the role and mode of action of fungal metabolites during the infection process. One area of particular interest has been the use of monoclonal antibodies in conjunction with gold-complexed protein A for the *in situ* detection of a toxic glycopeptide produced by *Ophiostoma ulmi,* the Dutch elm disease pathogen.[48] Observations at various intervals after inoculation indicated that the toxin was obviously capable of diffusing rapidly from cell to cell through pits and plasmodesmata, causing marked cell alterations in advance of pathogen penetration (Figure 5). This study provided valuable information on the mode of action of fungal toxins in the Dutch elm disease. An additional study dealing with the use of monoclonal antibodies raised against a synthetic dsRNA (poly I: poly C) suggested that the virulence of some *O. ulmi* strains could be influenced by the presence of dsRNA-rich mycoviruses.[49]

More recently, the immunocytolocalization of plant proteins has been an exciting area that has provided new insights into the complex mechanisms of disease resistance. Post-embedding localization of proteins such as HRGPs, PR proteins, and enzymes has been accomplished using polyclonal antibodies raised against highly purified antigens.[3,4,50]

HRGPs are structural glycoproteins present in low amounts in the cell walls of green plants. It has been convincingly shown that the level of wall-bound HRGPs markedly increases upon pathogenic infection.[51] Immunocytolocalization of HRGPs in tomato root cells infected by *Fusarium oxysporum* f. sp. *radicis-lycopersici* (FORL) is depicted in Figure 6. In this study, root samples were fixed in glutaraldehyde and embedded in Epon®. Sections were incubated with polyclonal antibodies raised against deglycosylated HRGPs, followed by goat anti-rabbit gold as outlined in Table 4. Clearly, HRGPs were found to accumulate in host cell walls as well as in wall appositions formed in response to infection. A time-course study of HRGP accumulation revealed that these molecules were deposited earlier and to a higher extent in resistant than susceptible plants.[50] Their occurrence in physical barriers provided support to their implication in the protection against fungal invasion.[50] A similar conclusion was reached by O'Connell et al.,[52] who reported the localization of HRGPs in bean leaf cells infected by bacteria and fungi.

Among the proteins newly synthesized in plants infected by pathogens, the so-called PR proteins have received particular attention in terms of physicochemical properties and gene expression at the

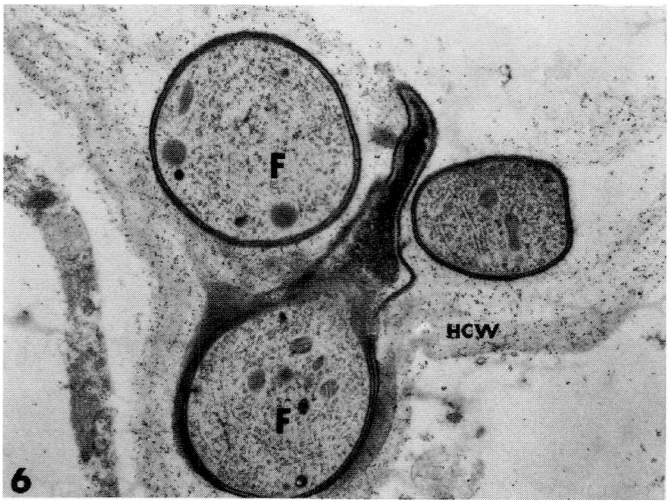

Figure 6 Immunogold localization of hydroxyproline-rich glycoproteins ($HRGP_s$) in tomato root tissues infected by *Fusarium oxysporum* f. sp. *radicis-lycopersici.* The section was incubated with a polyclonal antibody raised against deglycosylated $HRGP_s$, followed by goat anti-rabbit-gold (10 nm).$HRGP_s$ accumulate heavily in host cell walls. F, fungus; HCW, host cell wall. (Magnification × 21,500.)

mRNA level.[53] These proteins have been grouped into five families. Understandably, much interest has been devoted to PR proteins exhibiting β-1,3-glucanase and chitinase activites because of their believed antimicrobial potential, as suggested by *in vitro* investigations. Obviously, localizing these enzymes in infected plant tissues could provide more conclusive evidence on their function *in vivo.* Using polyclonal antisera, Benhamou et al.[3,4] investigated the subcellular localization of both β-1,3 glucanase and chitinase in tomato root tissues fixed with glutaraldehyde and embedded in Epon®. Time-course studies of enzyme accumulation in resistant and susceptible tomato plants generated key information on the spatiotemporal distribution of these PR proteins. As an example, localization of β-1,3 glucanase in susceptible tomato root tissues is illustrated in Figure 7. Taken together, these studies revealed that: (1) both enzymes occurred at the fungal cell surface, thus supporting the view of an antifungal activity; (2) chitinase activity was likely preceded by the action of β-1,3-glucanase; (3) both enzymes accumulated earlier in resistant than in susceptible plants; (4) induction of β-1,3 glucanase was an early event likely associated with the protection against fungal invasion; and (5) production of chitinase was a punctual response possibly triggered by β-1,3-glucan fragments released from fungal cell walls through the action of β-1,3 glucanase.[3,4]

Another group of PR proteins that has received much attention is the PR-1 (or PR P14 in tomato) group.[53] However, the biological function of these proteins is still unknown. The immunocytolocalization of this group of proteins in Epon®-embedded tomato root tissues (infected by FORL) revealed their predominant association with host cell walls and intercellular spaces and their absence at the fungal cell surface[54] (Figure 8). These observations supported the view that PR-1 proteins were likely involved in restricting fungal invasion through their association with the plant cell wall and the physical barriers.

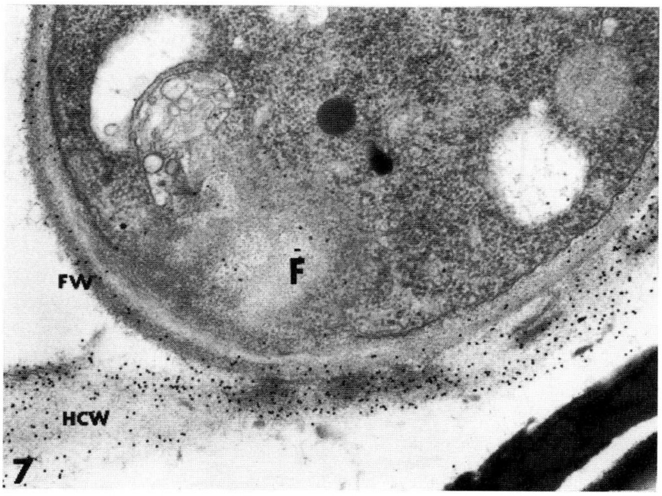

Figure 7 Immunogold localization of β-1,3-glucanase in tomato root tissues infected by *F. oxysporum* f. sp. *radicis-lycopersici.* The section was labeled with a polyclonal antibody raised against a tobacco β-1,3-glucanase, followed by goat anti-rabbit-gold (10 nm). Gold particles occur over the fungus cell wall, and accumulate over the host cell wall. F, fungus; FW, fungus wall; HCW, host cell wall.

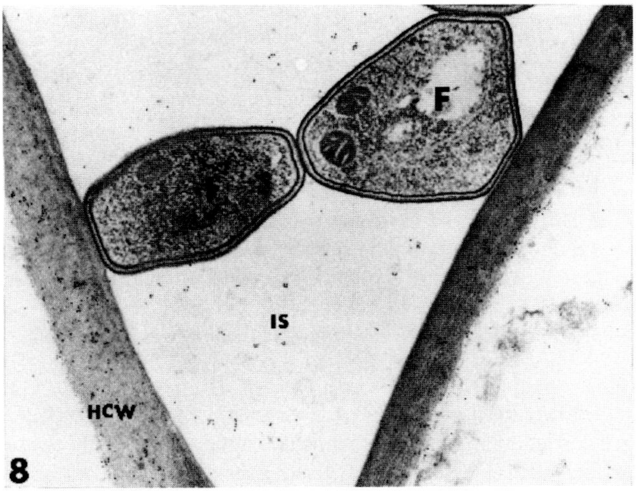

Figure 8 Immunogold localization of PR P14 in tomato root tissues infected by *F. oxysporum* f. sp. *radicis-lycopersici.* The section was labeled with a polyclonal antibody raised against a tomato PR P14, followed by goat anti-rabbit-gold. Gold particles are present in the intercellular space and in the host cell walls. F, fungus; HCW, host cell wall; IS, intercellular space. (Magnification × 36,000.)

Unlike chitinase and β-1,3-glucanase, these newly synthesized proteins appeared to be free of direct antimicrobial activity.

Invertase, the enzyme that hydrolyzes sucrose into glucose and fructose, was also localized in FORL-infected tomato root tissue.[55] (Figure 9). The enzyme was found to markedly increase upon fungal infection, especially in resistant plants, and to accumulate predominantly in host cell walls. It is speculated that induction of invertase (β-fructosidase) can be a signal that converts infected cells into sinks where sucrose is unloaded and carbohydrates rapidly mobilize to provide carbon sources required for the establishment of defense responses.[55]

There is no doubt that applications of immunogold labeling to specific topics in plant pathology are increasing each year. In the near future, immunocytochemistry will refine even more our understanding of the complex cellular and molecular events occurring during host-pathogen interactions. The information derived from such studies will undoubtedly be of great help in genetic transformation of plants and microorganisms as well as in biological control of virulent pathogens.

V. ADVANTAGES AND LIMITATIONS

The potential value of immunocytochemical techniques in plant pathology has been convincingly demonstrated through the *in situ* localization of various molecules that could not be detected by other means. These approaches offer several advantages over other methods such as: (1) the possibility of obtaining labelings of high specificity and high resolution, (2) the possibility of quantifying the labeling,

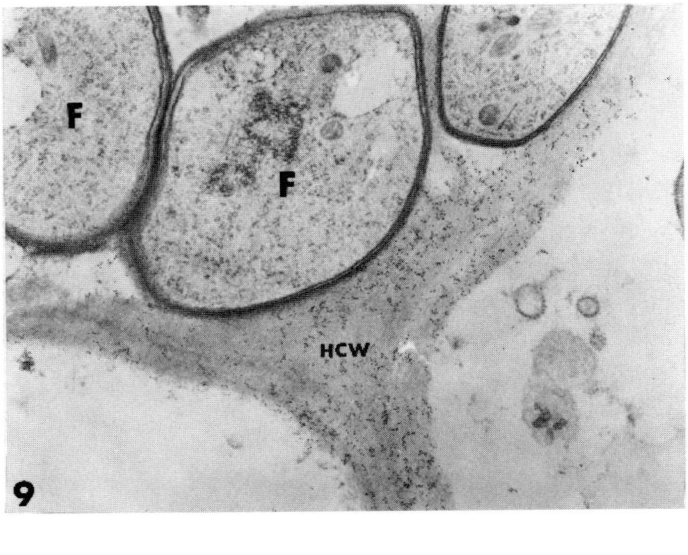

Figure 9 Immunogold localization of invertase in tomato root tissues infected by *F. oxysporum* f. sp. *radicis-lycopersici.* The section was labeled with a polyclonal antibody raised against a deglycosylated carrot invertase, followed by goat anti-rabbit-gold. Labeling is associated with the host cell wall. F, fungus; HCW, host cell wall. (Magnification × 27,000.)

and (3) the possibility of detecting two types of molecules over the same tissue section by using gold particles of different sizes. In addition, these techniques are easy to perform and do not require costly instrumentation.

However, one should keep in mind that immunocytochemical techniques may have some problems and limitations. Among the difficulties that can be encountered are the following:

1. Purity of the antigen used for animal immunization
2. Specificity of the antibodies raised against the purified antigen
3. Preservation of antigenic sites in tissue sections, which may be limited by the nature of the fixatives
4. Accessibility of the probe to cellular structures, which may be limited by the nature of the embedding resin
5. Specificity of the labeling pattern observed on tissue sections, which implies rigorous control tests to assess the validity of the results[6,14]

Enhancing antigenic preservation while maintaining satisfactory ultrastructure is undoubtedly the greatest challenge that immunocytochemistry has to meet in the future. Improvements in methods such as rapid-freeze fixation and freeze substitution, as well as the use of acrylic resin at low temperature, will yield to the true localization of various molecules that until now were difficult to preserve due to their solubility, their presence in low amounts, and/or their low MW.

REFERENCES

1. Herman, E. M., Immunocytochemical localization of macromolecules with the electron microscope, *Annu. Rev. Plant Physiol. Plant Mol. Biol.*, 39, 139, 1988.
2. Vandenbosh, K. A., Immunogold labeling, in *Electron Microscopy of Plant Cells,* Hall, J. L. and Hawes, C., Eds., Academic Press, New York, 1991, 181.
3. Benhamou, N., Grenier, J., Asselin, A., and Legrand, M., Immunogold localization of β-1,3-glucanases in two plants infected by vascular wilt fungi, *Plant Cell,* 1, 1209, 1989.
4. Benhamou, N., Joosten, M. H. A. J., and DeWit, P. J. G. M., Subcellular localization of chitinase and of its potential substrate in tomato root cells infected by *Fusarium oxysporum.* f.sp. *radicis-lycopersici, Plant Physiol.,* 92, 1108, 1990.
5. Benhamou, N., Lafitte, C., Barthe, J. P., and Esquerré-Tugayé, M. T., Cell surface interactions between bean leaf cells and *Colletotrichum lindemuthianum:* cytochemical aspects of pectin breakdown and fungal endopolygalacturonase accumulation, *Plant Physiol.,* 97, 234, 1991.
6. Benhamou, N. and Asselin, A., *In situ* immunocytochemical localization of plant proteins, in *Methods in Plant Molecular Biology and Biotechnology,* Thompson, J. E. and Glick, B. R., Eds., CRC Press, Boca Raton FL, 1993, 225.
7. Benhamou, N., Chamberland, H., Ouellette, G. B., and Pauzé, F. J., Ultrastructural localization of β-(1 → 4) glucans in two pathogenic fungi and in their host tissues by means of an exoglucanase-gold complex, *Can. J. Microbiol.,* 33, 405, 1987.
8. Benhamou, N., Chamberland, H., and Pauzé, F. J., Implication of pectic components in cell surface interactions between tomato root cells and *Fusarium oxysporum* f.sp. *radicis-lycopersici:* a cytochemical study by means of a lectin with polygalacturonic acid-binding specificity, *Plant Physiol.,* 92, 995, 1990.
9. Coons, A. H., Creech, H. J., and Jones, R. N., Immunological properties of an antibody containing a fluorescent group, *Proc. Soc. Exp. Biol.,* 47, 200, 1941.
10. Faulk, N. P. and Taylor, G. M., An immunocolloid method for the electron microscope, *Immunocytochemistry,* 8, 1081, 1971.
11. Bendayan, M., Protein A-gold electron microscopic immunocytochemistry: methods, applications, and limitations, *J. Electr. Microsc. Tech.,* 1, 243, 1984.
12. Roth, J., Application of lectin-gold complexes for electron microscopic localization of glycoconjugates on thin sections, *J. Histochem. Cytochem.,* 31, 987, 1983.
13. De Mey, J., The preparation of gold probes, in *Immunocytochemistry: Modern Techniques and Applications,* Polak, J. M. and Van Noorden, S., Eds., John Wright & Sons, Bristol, 1986, 3.
14. Benhamou, N., Preparation and application of lectin-gold complexes, in *Colloidal Gold: Principles, Methods, and Applications.* Hayat, M. A., Ed., Academic Press, New York, 1989, 95.

15. Bendayan, M., Enzyme-gold microscopic cytochemistry: a new affinity approach for the ultrastructural localization of macromolecules, *J. Electr. Microsc. Tech.*, 1, 349, 1984.
16. Brandtzaeg, P., Tissue preparation methods for immunocytochemistry, in *Techniques in Immunocytochemistry*, Vol. 1, Bullock, G. R. and Petrusz, P., Eds., Academic Press, New York, 1982, 2.
17. Craig, S. and Staehelin, L. A., High pressure freezing of intact plant tissues. Evaluation and characterization of novel features of the endoplasmic reticulum and associated membrane systems, *Eur. J. Cell Biol.*, 46, 80, 1988.
18. Benhamou, N., Noël, S., Grenier, J., and Asselin, A., Microwave energy fixation of plant tissues: an alternative approach that provides excellent preservation of ultrastructure and antigenicity, *J. Electr. Microsc. Tech.*, 17, 18, 1990.
19. Horisberger, M., Lectin cytochemistry, in *Immunolabelling for Electron Microscopy*, Polak, J. M. and Varndell, I. M., Eds., Elsevier Press, New York, 1984, 249.
20. Roth, J., The colloidal gold marker system for light and electron microscopic cytochemistry, in *Techniques in Immunocytochemistry*, Vol. 2, Bullock, G. R. and Petrusz, P., Eds., Academic Press, New York, 1983, 217.
21. Frens, G., Controlled nucleation for the regulation of the particle size in monodisperse gold solutions, *Nature Phys. Sci.*, 241, 20, 1973.
22. Horisberger, M., The gold method applied to lectin cytochemistry in transmission and scanning electron microscopy, in *Techniques in Immunocytochemistry*, Vol. 3, Bullock, G. R. and Petrusz, P., Eds., Academic Press, New York, 1985, 155.
23. De Mey, J., Colloidal gold probes in immunocytochemistry, in *Immunocytochemistry: Practical Applications in Pathology and Biology*, Polak, J. M. and Van Noorden, S., Eds., Wright-PSG, Bristol, 1983, 82.
24. Smit, J. and Todd, N. J., Colloidal gold labels for immunocytochemical analysis of microbes, in *Ultrastructure Techniques for Microorganisms*, Aldrich, H. C. and Todd, W. J., Eds., Plenum Press, New York, 1986, 469.
25. Kocourek, J. and Horejsi, V., A note on the recent discussion on definition of the term "lectin", in *Lectins*, Vol. 3, Bog-Hansen, T. C. and Spengler, G. M., Eds., Walter de Gruyter, Berlin, 1983, 3.
26. Lotan, R., Qualitative and quantitative aspects of labeling cell surface carbohydrates using lectins as probes, in *Scanning Electron Microscopy*, Johari, O., Ed., S.E.M., Chicago, 1979, 549.
27. Nicole, M. and Benhamou, N., Ultrastructural localization of chitin in cell walls of *Rigidoporus lignosus*, the white rot fungus of rubber tree roots, *Physiol. Mol. Plant Pathol.*, 29, 415, 1991.
28. Hajlaoui, M. R., Benhamou, N., and Bélanger, R., Cytochemical study of the antagonistic activity of *Sporothix flocculosa* on rose powdery mildew, *Sphaerotheca pannosa* var. *rosae*, *Phytopathology*, 82:583–589.
29. Benhamou, N., Chamberland, H., Ouellette, G. B., and Pauzé, F. J., Detection of galactose in two fungi causing wilt diseases and in their host tissue by means of gold-complexed *Ricinus communis* agglutinin I., *Physiol. Mol. Plant Pathol.*, 32, 249, 1987.
30. Benhamou, N., Gilboa-Garber, N., Trudel, J., and Asselin, A., A new lectin-gold complex for the ultrastructural localization of galacturonic acids, *J. Histochem. Cytochem.*, 36, 1403, 1988.
31. De Mey, J. and Moeremans, M., Raising and testing polyclonal antibodies for immunocytochemistry, in *Immunocytochemistry: Modern Methods and Applications*, Polak, J. M. and Van Noorden, S., Eds., Wright & Dons, Bristol, 1986, 3.
32. Morrissey, J. H., Silver stain for proteins in polyacrylamide gels: a modified procedure with enhanced uniform sensitivity, *Anal. Biochem.*, 117, 307, 1981.
33. Towbin, H., Staehelin, T., and Gordon, J., Electrophoretic transfer of proteins from polyacrylamide gels to nitrocellulose sheets: procedure and some applications, *Proc. Natl. Acad. Sci. U.S.A.*, 76, 4350, 1979.
34. Edge, A. S. B., Faltynek, D. R., Hof, L., Reichert, L. E., Jr., and Weber, P., Deglycosylation of glycoproteins by trifluoromethanesulfonic acid, *Anal. Biochem.*, 118, 131, 1981.
35. Greenwood, J. S., Keller, G. A., and Chrispeels, M. J., Localization of phytohemaglutinin in the embryonic axis of *Phaseolus vulgaris* with ultrathin cryosections embedded in plastic after indirect immunolabelling, *Planta*, 162, 548, 1984.
36. Slot, J. W. and Geuze, H. J., Gold markers for single and double immunolabelling of ultrathin cryosections, in *Immunolabelling for Electron Microscopy*, Polak, J. M. and Varndell, J. M., Eds., Elsevier Press, New York, 1984, 129.

37. Bendayan, M., Introduction of the protein G-gold complex for high-resolution immunocytochemistry, *J. Elect. Microsc. Tech.,* 6, 7, 1987.
38. Roland, J. C. and Vian, B., General preparation and staining of thin sections, in *Electron Microscopy of Plant Cells,* Hall, J. L. and Haines, C., Eds., Academic Press, New York, 1991, 1.
39. Frens, G., Controlled nucleation for the regulation of particle size in monodisperse gold solutions, *Nature Phys. Sci.,* 241, 10, 1973.
40. Causton, B. E., The choice of resins for immunocytochemistry, in *Immunolabelling for Electron Microscopy,* Polak, J. M. and Varndell, I. M., Eds., Elsevier Press, New York, 1984, 29.
41. Chamberland, H., Benhamou, N., Ouellette, G. B., and Pauzé, F. J., Cytochemical detection of saccharide residues in paramural bodies formed in tomato root cells infected by *Fusarium oxysporum* f.sp. *radicis-lycopersici, Physiol. Mol. Plant Pathol.,* 34, 131, 1989.
42. Chérif, M., Benhamou, N., and Bélanger, R. R., Ultrastructural and cytochemical studies of fungal development in cucumber plants infected by *Pythium ultimum, Physiol. Mol. Plant Pathol.,* 39, 353, 1991.
43. Hajlaoui, M., Benhamou, N., and Bélanger R. R., Cytochemical aspects of fungal penetration, haustorium formation and interfacial material in rose leaves infected by *Sphaerotheca pannosa* var. *rosae, Physiol. Mol. Plant Pathol.,* 39, 341, 1991.
44. West, C. A., Moesta, P., Jin, D. F., Lois, A. F., and Wickham, K. A., The role of pectic fragments of the plant cell wall in the response to biological stresses, in *Cellular and Molecular Biology of Plant Stress,* Key, J. L. and Kosuge, T., Eds., Alan R. Liss, New York, 1985, 335.
45. Chet, I., *Trichoderma*—application, mode of action, and potential as biocontrol agent of soilborne plant pathogenic fungi, in *Innovative Approaches to Plant Disease,* Chet, I., Ed., John Wiley & Sons, New York, 1987, 137.
46. Sivan, C. J. and Chet I., Degradation of fungal cell walls by lytic enzymes of *Trichoderma harzianum, J. Gen. Microbiol.,* 135, 675, 1989.
47. Semal, J., Traité de pathologie végétale, Les Presses Agronomiques de Gembloux, 621 pp., 1989.
48. Benhamou, N., Lafontaine, J. G., Joly, J. R., and Ouellette, G. B., Ultrastructural localization in host tissues of a toxic glycopeptide produced by *Ophiostoma ulmi,* using monoclonal antibodies, *Can. J. Bot.,* 63, 1185, 1985.
49. Benhamou, N., Parent, J. G., Garzon, S., Asselin A., and Ouellette, G. B., Use of a monoclonal antibody against poly I. poly C for detecting mycoviruses and potential applications to potato spindle tuber viroid and animal reoviruses, *Can. J. Plant Pathol.,* 9, 106, 1987.
50. Benhamou, N., Mazau, D., Grenier, J., and Esquerré-Tugayé, M. T., Time-course study of the accumulation of hydroxyproline-rich glycoproteins in root cells of susceptible and resistant tomato plants infected by *Fusarium oxysporum* f. sp. *radicis-lycopersici, Planta,* 184, 196, 1991.
51. Mazau, D. and Esquerré-Tugayé, M. T., Hydroxyproline-rich glycoprotein accumulation in the cell walls of plants infected by various pathogens, *Physiol. Mol. Plant Pathol.,* 86, 540, 1986.
52. O'Connell, R. J., Brown, I. R., Mansfield, J. W., Bailey, J. A., Mazau, D., Rumeau D., and Esquerré-Tugayé, M. T., Immunocytochemical localization of hydroxyproline-rich glycoproteins accumulating in melon and bean at sites of resistance to bacteria and fungi, *Mol. Plant-Microbe Interact.,* 3, 33, 1990.
53. Carr, J. P. and Klessig, D. F., The pathogenesis-related proteins in plants, in *Genetic Engineering, Principles and Methods,* Vol. 11, Setlow, J. K., Ed., Plenum Press, New York, 1989, 65.
54. Benhamou, N., Grenier, J., and Asselin, A., Immunogold localization of pathogenesis-related protein P14 in tomato root cells infected by *Fusarium oxysporum* f.sp. *radicis-lycopersici, Physiol. Mol. Plant Pathol.,* 38, 237, 1991.
55. Benhamou, N., Grenier, J., and Chrispeels, M. J., Accumulation of β-fructosidase in the cell walls of tomato roots following infection by a fungal wilt pathogen, *Plant Physiol.,* 97, 739, 1991.

FURTHER READING

Aldrich, H. C. and Todd, W. J., *Ultrastructure Techniques for Microorganisms,* Plenum Press, New York, 1986, 533.

Bullock, G. R. and Petruz, P., *Techniques in Immunocytochemistry,* Academic Press, New York, Vol. 1, 1982, pp. 1–306; Vol. 2, 1983, pp. 1–290; Vol. 3, 1985, 241.

Glauert, A. M., *Practical Methods in Electron Microscopy,* North Holland, Amsterdam, 1977, 312.

Hall, J. L. and Hawes, C., *Electron Microscopy of Plant Cells,* Academic Press, New York, 1991, 466.
Hayat, M. A., *Colloidal Gold: Principles, Methods and Applications,* Vol. 1, Academic Press, New York, 536.
Polak, J. M. and Varndell, I. M., *Immunolabelling for Electron Microscopy,* Elsevier Press, New York, 1984, 370.
Sommerville, J. and Scheer, V., *Electron Microscopy in Molecular Biology: a Practical Approach,* IRL Press, Oxford, 1989, 248.

Chapter 3

Diagnosing Plant Virus Diseases by Light Microscopy*

R. G. Christie, J. R. Edwardson, and G. W. Simone

TABLE OF CONTENTS

I. INTRODUCTION

Diagnosis of plant virus infections has been greatly assisted by the classification of viruses into groups. Viruses within groups have similar properties, many of which are not shared by viruses in other groups. Such properties are often referred to as the "main characteristics" of the group. Particle morphology, serological relationships, and mode of transmission, among others, represent such characteristics. When a virus collected from the field matches certain of the main characteristics, it can be tentatively assigned to a group. When this is accomplished the diagnostician can predict a number of additional properties that can be useful in control strategies even though the virus has not been completely described.

A number of methods have been developed for the detection and diagnosis of virus diseases. The three methods most commonly used are bioassay, electron microscopy, and serology. Bioassay is probably the most widely used approach, because specialized skills are not required to perform the test. Electron microscopy is useful for the detection of a number of viruses, but this instrument is expensive and its availability is limited. Although serological techniques have proved to be valuable diagnostic tools, their use in detecting a broad spectrum of viruses is limited by the availability of antisera. In recent years, cytological techniques have been developed for the detection of virus-induced inclusions. These intracellular structures are characteristic for the virus inducing them and have proved to be valuable agents in the diagnosis of plant virus diseases.

Plant virus inclusions are direct intracellular evidence of virus infection. They may consist of aggregated virus particles, aggregated coat protein, virus-directed nonstructural proteins, and, in some cases, mixtures of these. They may also be made up of altered host constituents. Inclusions differ from surrounding cytoplasm and organelles in structure and staining reactions. Virus inclusions have been induced by all plant viruses studied cytologically. Inclusions induced by a specific virus maintain a characteristic appearance over a host range. When properly stained, most inclusions can be readily detected with a light microscope. Light microscopic recognition of inclusion types offers a reliable, practical, and economical method for identifying virus diseases at the group level and can often lead to a specific diagnosis when the virus host range is considered.

* Mention in this paper of a trademark or proprietary product does not constitute a guarantee or warranty of the product by the University of Florida and does not imply its approval to the exclusion of other products that may also be available.

0-87371-877-1/95/$0.00+$.50

Cytological studies with the electron microscope have resolved the distinctive structure and composition of many inclusions. Once these inclusion features were described at the ultrastructural level, stains were designed which were capable of detecting and differentiating many of the same features in the light microscope. The ability to identify a particular inclusion type with both the light and electron microscope has enabled inclusions to be described in terms common to both levels of microscopy. For instance, an inclusion shown to consist of virus particles with electron microscopy can be similarly identified in the light microscope as a virus aggregate, even though individual particles cannot be resolved by light microscopy. Although in this article we deal exclusively with how inclusions appear in the light microscope, the descriptions have their basis in electron microscopy as well. Simple, rapid light microscopic techniques designed to select and differentiate inclusions induced by a wide range of viruses infecting many host species have been described in detail previously.[1]

II. QUALITY AND SELECTION OF TISSUES

The quality of the plant sample will often determine the choice of diagnostic method, the amount of replication, and the probability of success in problem definition. Sample quality and quantity are often determined by a second party—not by the individual responsible for plant disease determination. In the examination for viral inclusions, the concerns of sample quality and quantity are very important in deriving maximum utility from this technique.

The relative economic importance of plant virus groups has recently been reviewed.[28] The viral groups of greatest importance primarily represent those possessing inclusions generally distributed throughout plant tissues. Quality plant samples, therefore, must consist of sufficient symptomatic plant canopy in a turgid state to allow processing by epidermal strips or sections through leaf mesophyll areas. Although most diagnostic cycles result in the definition of "known" viruses in documented hosts, unknown viruses or undocumented virus/host combinations do exist in the field. For this reason, complete symptomatic plants or subsampled plants (i.e., new growth, old growth, root system) should be submitted for processing. Additionally, the submission of "healthy" control plant material is very useful for individuals just implementing viral inclusion methods. The efficiency of this method is highly correlated with knowledge of plant anatomy, cytology, and light microscopy. Processing healthy plant material will familiarize the investigator with normal plant cytology, such as the presence of microbodies, plant crystals, plastid morphology, chromatin distribution in the nuclei, nucleolus number, etc. Being familiar with the normal cell will allow faster focusing on the viral-induced inclusion(s) amid the diversity present at the plant cell level.

The choice of infected plant material can be critical to detection of inclusions, since symptom expression may not always be correlated with the presence of inclusions. Inclusions may be difficult to detect when chlorosis or necrosis is severe because they may not have reached the size or stage of development necessary for recognition or they may have begun to disintegrate in dying cells. Often, inclusions are prominent just before symptom expression or in tissues with mild or undetectable symptoms. Therefore, samples should be taken not only from areas with symptoms, but also from tissues of varying ages, regardless of symptom expression.

The epidermal cells of leaves, flowers, and fruit are often easy to prepare for observation and can be rich sources of inclusions induced by viruses of many groups. These tissues should be the starting point on the search for inclusions. If inclusions are not found in the epidermis, then the vascular tissues of the veins, stems, and roots should be examined. In some cases, it is necessary to examine such tissues as apical meristems and lateral buds. Specific directions for preparing all these tissues have been presented in a previous study.[1]

III. STAINING INCLUSIONS

When appropriate tissue pieces have been prepared, they need to be stained to enhance the viral inclusions. Differential stains have been developed in our laboratory to provide rapid penetration and uniform distribution, even into relatively thick tissue pieces. One stain is a combination of two dyes, one orange (Calcomine Orange 2RS) and one green (Luxol Brilliant Green BL), that will here be subsequently referred to as the O-G combination. This combination differentially stains plant organelles and inclusions containing protein. Another staining solution, Azure A, is designed to detect inclusions containing nucleic acids. When used under the conditions described,[1] Azure A is metachromatic, rendering virus inclusions containing ribonucleic acid (RNA) red-violet and deoxyribonucleic acid (DNA) blue. The two stains and their sources and methods for preparing them have been described in detail in the Appendix at the end of this chapter. The staining reactions of both the O-G combination and the Azure A stain in healthy tissues are summarized in Table 1. The materials and the protocol for applying the stains are given in the Appendix.

When the O-G combination is used, stained plastids often obscure small inclusions. The plastids can be dissolved by floating the tissue on a 2% solution of Triton® X-100 (Rohm & Haas Co., Philadelphia, PA 19105) for 5 min prior to staining (see the Appendix). This treatment is especially useful for detecting the characteristic cylindrical inclusions of the potyvirus group, especially during early stages of infection when these inclusions are very small and often located at the cell periphery. When Triton® X-100 treatment is used, untreated controls should be included, since some inclusions are dissociated by this procedure (Table 2).

IV. DIAGNOSIS WITH VIRUS INCLUSIONS

Diagnosis of plant viral diseases does not differ from that conducted with any other pathogen group. This diagnostic process is a deductive one that logically proceeds in the following manner:

a) identification of the host species
b) perception of plant symptoms that imply viral etiology
c) access to a relevant plant disease index to focus the direction of investigation
d) choice of investigatory techniques to define pathogen etiology
e) literature confirmation for a "known" viral pathogen
f) application of Koch's postulates for investigation of an unreported virus or virus/host combination

Selection of plant inclusion methodology offers a strength above all other viral diagnostic technologies. This method is the only unbiased one available to answer the fundamental diagnostic hypothesis: "Is there a virus present in this sample?" Plant viral inclusions define viral etiology regardless of viral particle morphology, nucleic acid composition, or transmissibility requirements.

The presence of a particular viral-induced inclusion can establish that a virus is present in a particular sample and thus eliminate from consideration other conditions that may mimic viral symptoms, e.g., pesticide damage. The next step is to compare the types of inclusion present with those characteristic of different virus groups. If an unknown virus is found to induce inclusion types with similar characteristics to those of a particular group, it can be assumed that the virus belongs to that group. Placing a virus within a group eliminates from consideration all viruses outside the group and at the same time allows inferences to be made about properties that the virus may have in common with group members. This is especially important in cases where the virus in question is undescribed and information on its properties is lacking.

When using inclusions for diagnosis, five distinctive inclusion features need to be considered in describing them. These are (1) structure; (2) composition, e.g., protein or nucleoprotein; (3) intracellular location; (4) tissue location; and (5) reaction to differential stains. Inclusions can be distinguished from one another based on differences in one or more of these criteria.

V. INCLUSION DEVELOPMENT

Inclusions are not static, but go through developmental stages. In the initial stages they may be small and difficult to detect. However, over time they increase in size and, in some cases, complexity. When conditions are favorable, they reach a "mature" state where they display their most characteristic appearance. It is usually at this stage that they are most valuable for diagnosis. For this reason it is of utmost importance that different stages of infected tissue be examined to assure that such stages are included in the sample.

Certain environmental conditions, such as temperature and light, can affect the rate at which the inclusions develop. The relative tolerance or resistance of the host genome is also important. In tolerant hosts inclusions may mature and reach their characteristic forms even though symptoms are not evident. On the other hand, resistant hosts can delay or even arrest inclusion development. In such cases there are also few or no symptoms evident, and the inclusions are harder to detect. However, proper sampling

Table 1 **Staining reactions of host cell constituents present in both healthy and virus diseased tissue**

Stain	Chromatin	Nucleoplasm	Nucleolus	Cell wall	Cytoplasm	Plastids	Microbodies & microcrystals	P-protein (phloem)	Inorganic crystals (druses, raphides, etc.)	Starch granules
Azure A	Blue	Clear	Red-violet	Colorless	Colorless	Colorless[a]	Colorless	Colorless	Colorless	Colorless
O-G Stain	Green	Orange	Green	Yellow-green	Yellow-green	Yellow-green	Green	Green	Colorless	Colorless

[a]In diseased cells the cytoplasm and plastids may stain reddish.

Table 2 **Important virus groups including inclusions that are of diagnostic value**

A) Virus groups with inclusions generally distributed throughout plant tissues

Virus group	Inclusions	Diagnostic inclusions	Triton® X-100[a]	Color reaction[b] O-G	Color reaction[b] A	Comments
Carlavirus	Cytoplasmic					
	1) Vacuolate-vesiculate	1,2	R	Brownish-Green	Red/violet	These inclusions are common to all carlaviruses.
	2) Paracrystals[c]		D	Green	Red/violet	Some paracrystals are banded. Paracrystals can reach lengths where they double back upon themselves.
	Additional					
	3) Tonoplast aggregates[c]		D	Green	Red/violet	Carlaviruses have a propensity for virus particles to mass near the tonoplast and project into the central vacuole. These inclusions are similar in appearance to the vacuolate-vesiculate ones, but stain much more intensely.
Caulimovirus	Cytoplasmic					
	1) Rounded or elongated bodies with small vesicles	1	R	Green	Red/violet	Thin section studies reveal the presence of DNA containing virus particles within the vesicles. *These vesicles will stain blue.*
	Additional					
	2) Nuclear		R	Green	Red/violet	Spherical inclusions have been associated with several caulimovirus infections. Information on these inclusions is very limited and their value in diagnosis undetermined.
Comovirus[e]	Cytoplasmic					
	1) Vacuolate-vesiculate	1,2,3	R	Brownish-green	Red/violet	The vacuolate inclusions appear at early stages of infection. They become large and dense, often exceeding the nucleus in size.
	2) Virus crystals and aggregates[c]		(R)	Green	Red/violet	These inclusions follow the appearance of the vacuolate ones. They persist longer, especially in glandular hairs, guard cells and phloem-associated parenchyma. They are not as evenly distributed as the vacuolate inclusions and may be widely scattered in older infections.

Table 2—*Continued*

A) Virus groups with inclusions generally distributed throughout plant tissues

Virus group	Inclusions	Diagnostic inclusions	Triton® X-100[a]	Color reaction[b] O-G	Color reaction[b] A	Comments
	3) Xylem blockage and crystals[c]		R	Green	Red/violet	Xylem blockages have been detected in all comoviruses studied by light microscopy. *This characteristic considered along with the vacuolate-vesiculate and the virus crystals located in specialized cells is diagnostic for this group and the sobemoviruses.*
Cucumovirus	Cytoplasmic					
	1) Vacuolate-vesiculate		R	Brownish-green	Red/violet	These inclusions are often similar in size and shape to plastids. They can be differentiated from the latter since they are resistant to Triton® and stain in Azure A, while the plastids are dissociated by Triton® and fail to stain in Azure A.
	2) Virus crystals, aggregates[c]		(R)	Green	Red/violet	These inclusions are present both in the cytoplasm and the central vacuole. In the case of cucumber mosaic virus (CMV) they are often hollow, *a characteristic that appears diagnostic for many CMV isolates.*
Furovirus	Cytoplasmic					
	1) Virus aggregates and paracrystals[c]		(R)	Green	Red/violet	Wheat soilborne mosaic the type member of this group induces both 1 and 2 types cytoplasm inclusions. *The paracrystals, unlike those of the tobamoviruses, stain in Azure A without heat.*
	Additional					
	2) Vacuolate		R	Brownish-green	Red/violet	
Nepovirus	Cytoplasmic					
	1) Vacuolate-vesiculate		R	Brownish-green	Red/violet	The inclusions appear during early stages of infection. They may grow to a large size, but are not dense and stain only lightly.
	2) Virus crystals and aggregates[c]		(R)	Green	Red/violet	Located both in the cytoplasm and the central vacuole. These inclusions are especially evident in meristems, even when symptoms are absent. *This characteristic may be of diagnostic value for nepovirus infections.*

Pea enation mosaic (monotypic)	Cytoplasmic					
	1) Irregular-shaped, often near nucleus	1,2,3	D	Green	Red/violet	Both cytoplasmic irregular-shaped and the nuclear inclusions appear to be confined to the area of the lesions (enations).
	2) Crystalline					
	Nuclear[c]					
	3) Diffuse nuclear[c]		R	Green	Red/violet	
Potexvirus	Cytoplasmic					
	1) Banded bodies[c]	1,2	D	Green	Red/violet	*The banded bodies induced by potexviruses differ sufficiently in structure to be of diagnostic value.* These inclusions are delicate and can be easily destroyed by certain fixations and solvents. It is useful to study these inclusions in the uninjured epidermis of a thick tissue mount.
	2) Thick fusiform[c] (pillow-shaped or spindle-shaped)		D	Green	Red/violet	The fusiform inclusions like the banded bodies are viral aggregates. The lack of banding may be due to the angle of observation, damage resulting from tissue preparation, or the reaction to solvents. However, in some cases these structures (e.g., cactus virus X) may represent a distinctive inclusion type.
	Additional					
	3) Vacuolate-vesiculate		(R)	Green	Red/violet	These inclusions appear early during infection. They may be small and diffuse in staining reaction. However, in certain potexvirus infections (e.g., papaya mosaic) they can be quite large and often contain small virus paracrystals.
	4) Laminate inclusion components		R	Green	Red/violet	*Laminate inclusion components are unique and diagnostic of Potato virus X isolates.*[25] They are not characteristic for the potexvirus group.

Table 2—*Continued*

A) Virus groups with inclusions generally distributed throughout plant tissues

Virus group	Inclusions	Diagnostic inclusions	Triton® X-100[a]	Color reaction[b]		Comments
				O-G	A	
Potyvirus	Cytoplasmic					
	1) cylindrical (protein) containing tubes. These inclusions are described as scrolls in cross section in the electron microscope. (Subdivision, I and IV)[d]	1 or 2 or 3	R	Green	Colorless	*The potyviruses induce characteristic, proteinaceous cylindrical inclusions that are diagnostic at the group level.* The structural nature of the tubes and plates can be determined by changing the focus of the microscope. *In addition, they induce a variety of other cytoplasmic as well as nuclear inclusions that are useful for diagnosis when considered along with the cylindrical inclusions.* This table separates the potyviruses into three subdivisions based on cylindrical inclusion structure.
	2) Cylindrical (protein) containing plates. These inclusions are described as laminated aggregates in the electron microscope. (Subdivision II)[d]		R	Green	Colorless	
	3) Cylindrical (protein) containing both tubes and plates. (Subdivision III)[d]		R	Green	Colorless	
Rhabdovirus	Cytoplasmic					
	1) Irregular-shaped viroplasms	2,3				A number of rhabdoviruses induce viroplasm-like structures in the cytoplasm. However, none of these inclusions have been tested with either the O-G or Azure A stains.
	Nuclear					
	2) Roundish virus aggregates[c] (often more than one)		R	Green	Faint red	During late stages of infection, these inclusions may be diffused or absent, leaving a nucleus that appears empty.

	3) Viroplasm-like structures; irregular (similar in color to the nucleus)		R	Green	Red/violet	In thin sections, these inclusions contain many partially enveloped virus particles embedded in a dense staining matrix. They are usually observed during earlier stages of infection.
Sobemovirus[e]	Cytoplasmic					
	1) Vacuolate-vesiculate	1,2,3,	R	Brownish-green	Red/violet	The vacuolate inclusions appear at early stages of infection. They become large and dense, often exceeding the nucleus in size.
	2) Virus crystals and aggregates[c]		(R)	Green	Red/violet	These inclusions follow the appearance of the vacuolate ones. They persist longer, especially in glandular hairs, guard cells, and phloem-associated parenchyma. They are not as evenly distributed as the vacuolate inclusions and may be widely scattered in older infections.
	3) Xylem blockage and crystals[c]		R	Green	Red/violet	Xylem blockages have been detected in all sobemoviruses studied by light microscopy. *This characteristic, considered along with the vacuolate-vesiculate and the virus crystals located in specialized cells is diagnostic for this group and the comoviruses.*
Tenuivirus	Cytoplasmic					
	1) Masses of thread-like material	1	R	Green	Faint red	These inclusions are readily detected in epidermal tissues. Since they are lightly stained in Azure A, they appear to be primarily proteinaceous. They have been found in all members of this group that have been studied.[26,29] *Their unique appearance and staining reaction make them diagnostic for infections by members of the tenuivirus group.*
Tobamovirus	Cytoplasmic					
	1) Virus crystals and aggregates[c] a) hexagonial in face view, rectangular in side view	1 or 2 or 3	D	Green	No heat colorless +Heat red/violet	A number of tobamoviruses can be distinguished based on differences in the structure of the virus aggregates.[27] *The requirement that heat be used in order for these inclusions to stain in Azure A is diagnostic for the tobamovirus group.*[1]

Table 2—*Continued*

A) Virus groups with inclusions generally distributed throughout plant tissues

Virus group	Inclusions	Diagnostic inclusions	Triton® X-100[a]	Color reaction[b] O-G	Color reaction[b] A	Comments
	b) stacked plates rounded in face view, varying in lengths in side view		D	Green	No heat colorless +Heat red/violet	
	c) Angled-layer aggregates (appear as fibrous masses in light microscope)		D	Green	No heat colorless +Heat red/violet	
	Additional					
	2) Paracrystals[c]		(R)	Green	No heat colorless +Heat red/violet	These virus aggregates are most prominent in older infections and occur for the most part in the central vacuoles.
	3) Vacuolate-vesiculate (X bodies)		R	Brownish-green	Red/violet	The material contained in these inclusions aggregates in some tobamovirus infections, but remains dispersed in others. In the latter case many small cytoplasmic granules can be observed. The X-bodies are more numerous during early stages of infection.
Tospovirus	Cytoplasmic					
	1) Dense, irregular with projections[c]	1	(R)	Green	Red/violet	Thin section studies of these inclusions indicate that the finger-like projections are proliferated, swollen endoplasmic reticulum, and/or dictyosomes containing masses of virus particles. The finger-like projections can be resolved in the light microscope by changing the focus.
	Additional					
	2) Irregular-shaped protein bodies		R	Green	Colorless	Present in certain tospovirus infections.

Tombusvirus	Cytoplasmic					
	1) Virus crystals and aggregates.[c] Variable in size and shape.	1,2,3	(R)	Green	Red/violet	Cytoplasmic crystalline virus aggregates are frequently found in cells infected by tombusviruses. *Aggregates occurring both in the cytoplasm and central vacuole appear to be a distinctive feature, when considered along with multivesicular bodies.*
	2) Multivesicular bodies		D	Green	Faint red	Multivesicular bodies are derived from peroxisomes (microbodies), mitochondria, and chloroplasts.[17–19] These inclusions stain green with O-G combination, but only faintly with Azure A. *They appear to be unique and diagnostic for the tombusvirus group.*[18]
	Nuclear					
	3) Spherical[c]		R	Green	Red/violet	The nuclear inclusions of the tombusviruses vary considerably in size and are often diffuse in staining reaction, making them more difficult to detect than the cytoplasmic inclusions.
Tymovirus	Cytoplasmic					
	1) Clumping of altered chloroplasts with deep staining cytoplasmic matter		(R)	Green (matrix)	Red/violet (matrix)	Although the plastids are dissociated by Triton®, the matrix appears to be fairly stable. *The massive clumping of the plastids within the matrix is diagnostic for tymoviruses.*[15]
	Additional					
	2) Virus crystals and aggregates[c]		(R)	Green	Red/violet	These inclusions are located both in the cytoplasm and the central vacuoles.
	Nuclear					
	3) Diffuse nuclear (protein)		R	Green	Colorless (nuclei appear empty)	The tymovirus nuclear inclusions are aggregated viral coat protein without nucleic acid.

Table 2—*Continued*

B) Virus groups with inclusions associated predominantly with vascular tissues

Virus group	Inclusions	Diagnostic inclusions	Triton® X-100[a]	Color reaction[b] O-G	Color reaction[b] A	Comments
Closterovirus	Cytoplasmic					
	1) Paracrystals, banded bodies[c]	1,2	D	Green	Red/violet	*Because of the structure and tissue location of the closterovirus inclusions, they can be considered diagnostic at the group level.*
	2) Densely staining phloem cells, often with many vacuoles		R	Green	Red/violet	
Geminivirus	Nuclear					
	1) Rounded dense bodies[c] (one or more may be present)	1	R	Green	Blue	*The occurrence of these virus aggregates in the nuclei of phloem parenchyma cells along with their distinctive blue color in Azure A make them diagnostic for this group.*[20]
	Additional					
	2) Ring-shaped			Green	Blue/green	These inclusions appear ring-shaped in the light microscope.[20] They are very small and difficult to detect. *However, when detected, their presence is diagnostic for members of the geminivirus group.*
Luteovirus	Cytoplasmic					
	1) Virus crystals and aggregates[c]		(R)	Green	Red/violet	These inclusions are often irregular in shape. Phloem necrosis is usually associated with the inclusions, especially during later stages of infection. These inclusions may be confused with the virus aggregates induced by the reoviruses. Since the virus particles induced by these two groups differ considerably in size, electron microscopy can be used to separate them.

Reovirus (subgroups Fijivirus, Phytoreovirus)	Cytoplasmic				
	1) Virus crystals and aggregates[c]	(R)	Green	Red/violet	These inclusions can be both angular or irregular in outline.[21]
	2) Rounded and/or elongate	R	Green	Red/violet	The presence of phloem-associated elongate viroplasms, spherical in the case of Fijiviruses and elongate for the Phytoreoviruses,[21] appears to be a unique feature of this group and *may be of diagnostic value,* especially when considered together with the cytoplasmically located phloem virus crystals and aggregates. Reovirus-induced viroplasms may also be useful in distinguishing reovirus and luteovirus infections.

[a] The effects of a 5-min treatment by 2% Triton® X-100 at room temperature. D = dissociated. R = resistant. (R) = virus inclusions and crystals that may be dissociated by longer treatments.

[b] Inclusions containing ribonucleic acid (RNA) stain varying shades of red, depending on the density of the inclusion. The smaller inclusions are red, while the larger, denser ones tend toward purple and violet. Inclusions containing desoxyribonucleic acid (DNA) stain blue.

[c] Inclusions consisting of virus particles as determined by electron microscopy.

[d] The cylindrical inclusions of the potyvirus group have been divided into four subdivisions based on electron microscopy.[9] However, in the light microscope subdivisions I and IV appear similar and are difficult to distinguish; therefore, they are grouped together under I.

[e] Inclusions induced by the comoviruses and sobemoviruses are similar.

will still reveal the characteristic inclusions, although they are reduced in number and limited in distribution. Such information can be very valuable to programs breeding for resistance to viral diseases.

VI. INCLUSION CHARACTERISTICS FOR A VIRUS GROUP

No viruses are known that do not induce inclusions. Inclusions are now considered as a main characteristic of most currently recognized plant virus groups.[2–4] Virus groups are defined as a collection of viruses and virus strains, each of which shares with the type member all, or nearly all, the main characteristics of the group.[5] An inclusion is characteristic for a group when it is induced by most group members and is similar to those induced by other members in structure, composition, intracellular location, tissue location, and staining reactions. The inclusions induced by some viruses are so unique that they are diagnostic for a particular virus.

Members of the potyvirus group (the largest of the plant virus groups) are among the most commonly encountered viruses in crops and weeds. Therefore, it is very important to become familiar with the different types of inclusions induced by viruses of this group.

The most distinctive inclusion types induced by the potyviruses are the cytoplasmic cylindrical inclusions.[8] These inclusions are coded for by the viral genome[6] and are considered as diagnostic at the group level.[7] At early stages of infection they appear first at the cell periphery where they increase in number and in size.[10] As infection progresses, they begin to accumulate in groups in the central portions of the cell. Eventually, they are found only in large masses. In some cases, this process is arrested and the inclusions never reach the massing stage, but instead remain at the cell periphery. The recognition of the inclusions at the peripheral stage can be important in the diagnosis of potyvirus infections.

The potyviruses have been subdivided based on differences in cylindrical inclusion structures as seen in thin section.[8,9] Two different structural components of potyvirus cylindrical inclusions can be recognized in the light microscope. The first of these appears tubular in shape. In the electron microscope these inclusions are described as scrolls in cross section. The tubular structures can be likened to a group of needles laid side by side. If the needles are oriented so that the tips point toward the viewer, then they would appear as a group of dots. When the focus of the microscope is changed, the dots remain visible as dots, while the needles oriented on their sides leave the field of view. The second type recognizable in the light microscope consists of plate-like structures. These structures are described as laminated aggregates in terms of the electron microscope. When a group of plates belonging to an individual cylindrical inclusion are viewed from the side, they appear as a group of parallel lines. When such an inclusion is seen from the end, it would look like an asterisk. As the focus of the microscope is changed the plates shift position, but they still appear as lines. Therefore, simply by changing the focus of the microscope, it is possible to distinguish between the tubular and plate-containing cylindrical inclusions. Certain potyviruses induce only the tubular components, while others induce only the plate-like structures. In addition, there are those that induce both types, as in the case of viruses such as turnip mosaic, a member of Subdivision III.[9] Many potyviruses can be distinguished in the light microscope based solely on the differences in cylindrical inclusion structure.

The tubular types of cylindrical inclusions can in some respects resemble groups of virus paracrystals. Paracrystalline virus aggregates are induced by a number of potyviruses, as well as by viruses of other groups. The cylindrical inclusions can be distinguished from paracrystals based on differences in their staining reactions in Azure A. The virus paracrystals, which contain RNA, will stain red with Azure A, while the proteinaceous cylindrical inclusions, which lack RNA, will not stain.

VII. VIRAL AGGREGATES

Many virus aggregates can be characteristic for a group, although not necessarily diagnostic. Aggregation of virus particles is apparently a common phenomenon among plant viruses. Aggregates can occur in the cytoplasm, vacuoles, and nuclei. They may vary considerably in size, but are usually sufficiently large to be detected in the light microscope. Virus aggregates, like all virus inclusions, go through developmental stages. During early stages of infection, they may be small and ill defined, while at later stages they may disintegrate or be present only in isolated cells or tissues. Therefore, it is important that the aggregates be detected at a stage of infection where they are plentiful, and where they display their most characteristic appearance. The Azure A technique is well suited for this purpose, because it stains viral aggregates vividly and allows extensive areas of epidermis, mesophyll, and vascular tissue to be searched for their presence.

Two virus groups, the tobamoviruses and the potexviruses, both characterized by elongate virus particles, induce virus aggregates that are characteristic for their respective groups. The aggregates induced by certain viruses within each of these groups differ sufficiently in structure that they can be used to distinguish the virus inducing them from other group members. Polyhedral viruses also induce virus aggregates that can be useful for diagnosis, especially when their structure and location are considered along with the presence of additional inclusion types (Table 2).

The aggregates of polyhedral virus particles are not as easy to detect as those induced by the elongate viruses. This is because they are not as uniformly distributed in the tissues and do not persist as long. This is especially true of virus aggregates located in the cell vacuoles, as in the case of cucumber mosaic virus (CMV). During early stages of infection, CMV crystals are abundant and easy to detect. In older tissues with long-standing infections the inclusions are fewer and widely dispersed, although in such cases, abundant inclusions can still be found in very young leaves of the same plant. A distinctive feature of CMV inclusions is that they often appear as hollow shells.[10] This feature has proved to be of diagnostic value.

VIII. INCLUSIONS INDUCED IN ADDITION TO THOSE CHARACTERISTIC FOR A VIRUS GROUP

A. ADDITIONAL CYTOPLASMIC INCLUSIONS

Besides the inclusions that are characteristic for the group, i.e., those in common with other group members, many viruses induce additional inclusions that can be useful in separating groups and in some instances viruses within groups. Additional inclusions may be unique and diagnostic themselves, as in the case of the nuclear inclusions induced by tobacco etch virus isolates[11] and the cytoplasmic laminate inclusion components (LIC) induced by potato virus X.[25] Additional inclusions may also differ in some respect from inclusions induced by other groups, and may be used to differentiate viruses whose characteristic inclusions may be similar to those in other groups. For instance, both the carlaviruses and the nepoviruses induce vacuolate, irregular inclusions that are characteristic for their groups. However, the carlaviruses induce additional paracrystalline inclusions, while the nepoviruses induce additional crystalloid virus aggregates.[10] Differences between the two additional inclusion types are readily apparent and allow separation of members of these two groups.

Certain potyviruses induce irregular, cytoplasmic inclusions that are proteinaceous, but also have an RNA associated with them. These inclusions are also products of the viral genome. Their constituent protein has been demonstrated to be involved with aphid transmission of the virus.[12] This protein does not aggregate into inclusions in all potyvirus infections. The presence of such inclusions in addition to the characteristic cylindrical inclusions can be used to separate many potyviruses.

Inclusions that are induced in addition to those characteristic for the group are included in Table 2. When describing these inclusions, the same five criteria should be used as was the case in describing the characteristic inclusions; namely, their structures, compositions, cellular locations, tissue locations, and staining reactions.

B. ADDITIONAL NUCLEAR INCLUSIONS

A number of plant viruses induce nuclear inclusions. Such inclusions may be virus aggregates, virus-directed nonstructural proteins, coat protein shells, or membranous structures. Certain potyviruses, such as tobacco etch virus, induce nuclear inclusions so distinctive that their presence can even be used to distinguish among closely related strains.[11] These inclusions are products of the viral genome and represent aggregations of two proteins that have putatively been designated as a viral polymerase[13] and a protease.[14] They are usually distinctive and well defined in shape. A number of viruses closely related to bean yellow mosaic virus also induce distinctive nuclear inclusions that differ in structure sufficiently to be useful for diagnosis.

Viruses in several groups induce distinctive nuclear inclusions that consist of virus aggregates. Descriptions of some of these inclusions are contained in Table 2. Nuclei containing these inclusions are often swollen and distorted. The inclusions themselves often have no distinctive shape, but can be distinguished based on their staining reactions. In the light microscope nuclear aggregates containing RNA will stain red to violet in Azure A. Nuclear aggregates induced by the geminiviruses will stain blue in Azure A, since they contain DNA. The distinct color of these nuclear inclusions coupled with the fact that they are associated with vascular tissues and not generally distributed in other plant tissues make them diagnostic for the geminivirus group.

Table 2 describes a number of nuclear inclusions. Nuclear inclusions, taken either alone or when considered in conjunction with other inclusion types that may be present, have proven extremely valuable for virus diagnostic purposes.

IX. ALTERED HOST CONSTITUENTS

A number of plant viruses induce inclusions that contain altered organelles and other cytoplasmic constituents. Some of the vacuolate-vesiculate types of inclusions referred to in Table 2 are probably of this nature. Such inclusions can be very useful for diagnosis both by themselves and when considered along with other characteristic inclusions. The distinctive clumping of plastids induced by the tymoviruses is an example of altered plant organelles themselves being diagnostic for infections at the group level.[15] Multivesicular bodies, derived from pre-existing cellular components, such as peroxisomes,[16,17] mitochondria,[18] and plastids,[19] have been suggested to be of diagnostic value for the tombusvirus infections.[18] In healthy tissues, microbodies (peroxisomes) contain a protein that stains green with the O-G combination, but remains unstained in Azure A, indicating the presence of protein and the absence of RNA. In tomato bushy stunt-infected tissue the microbodies increase considerably in volume, often reaching the size of small plastids. In addition to the protein present, these inclusions are also found to stain a light red with Azure A, indicating the presence of RNA. Such inclusions appear to be unique to the tombusvirus group. The tombusviruses also induce both cytoplasmic and nuclear virus aggregates, both of which stain red with Azure A (Table 2). The presence of these latter two inclusion types, taken together with the multivesicular bodies, whatever their particular origin, appears to be a diagnostic feature of the tombusviruses.

X. VIRUSES OF VASCULAR TISSUES

Section B of Table 2 includes four virus groups: the closteroviruses, geminiviruses, luteoviruses, and the plant reoviruses. Each of these viruses induce inclusions associated principally with the living cells of the vascular system. Therefore, techniques designed for exposing these tissues, such as sectioning or abrading, are necessary.[1] Inclusions induced by viruses in all of the above mentioned groups are best detected using the Azure A staining procedure.

The closteroviruses can be distinguished from the other vascular-inhabiting viruses based on the presence of characteristic paracrystalline aggregates located in the cell cytoplasm. In addition, many large, vacuolate, intensely red-violet staining cells are also present. The geminiviruses are characterized by their distinctive blue staining nuclear inclusions, which are aggregates of DNA containing virus particles. These inclusions are diagnostic for infections by viruses in the geminivirus group.[20] Both the luteoviruses and the reoviruses induce virus aggregates in the cytoplasm. In addition to the virus aggregates, reoviruses also induce dense, spherical viroplasms in the leafhopper-borne subgroup (Fijiviruses) and amorphous, elongate viroplasms in the planthopper-transmitted Phytoreovirus subgroup.[21] It can be readily seen that by using the five aspects for describing inclusions (see Section IV) we can distinguish the vascular-inhabiting virus groups from each other as well as the virus groups in Section A of Table 2.

Two virus groups, the comoviruses and sobemoviruses, form large blockages of xylem elements. These inclusions have been demonstrated through electron microscopy to consist of masses of virus particles. They stain red-violet in Azure A and can be seen at relatively low magnifications in the light microscope. These characteristic xylem-located inclusions, when considered with the other inclusions, make it possible to separate the comoviruses and sobemoviruses from viruses in all other groups.

It should be mentioned that certain other pathogens of the vascular system also stain with Azure A. Mycoplasmas appear granular in the light microscope when viewed in longitudinal sections and are located in the sieve elements where masses of them often block the sieve plates. These organisms stain red-violet in Azure A. Fastidious plant bacteria, such as *Xylella fastidiosa,* which also stain red-violet, are limited to the xylem elements and are easily discerned by their bacilliform shape in transverse sections. Both the mycoplasma and the fastidious bacteria stain green with the O-G method.

XI. MIXED INFECTIONS

Experience has demonstrated that mixed virus infections are commonly present in field samples. Such infections are often difficult to detect, since diagnostic host ranges may overlap, may be depressed, or the viruses involved may not have been characterized.

One of the most important features of the light microscopic techniques for plant virus identification is the ability to easily detect multiple infections. Inclusions induced by many different viruses are distinctive enough to be distinguished when they occur within the same host and even the same cell.

XII. USE OF TABLE 2

The ideal way to demonstrate the distinctive characteristic inclusions displayed by the light microscope is by depicting them with color light micrographs. Unfortunately, space restrictions and cost considerations of this article prevent the inclusion of such material. However, the reader is referred to a number of articles that contain such micrographs.[1,10,22–24,26,28] It is possible, however, to present much valuable information on inclusions in tabular form. This has been done in Table 2. Table 2 contains information on the structure, composition, intracellular location, tissue location, and staining reactions of inclusions induced by viruses in 20 groups. The reaction of certain inclusion types to the plastid-solubilizing detergent Triton® X-100 is also included. In Table 2 the virus groups are listed alphabetically under two sections. Section A contains those virus groups whose inclusions are found generally distributed in the plant tissues. Section B includes those viruses that induce inclusions associated predominately with vascular tissues. Several virus groups that are included in Table 2 have been studied to only a limited degree by light microscopy. Certain members of these groups have been investigated extensively in ultrastructural studies which have demonstrated the presence of a number of characteristic inclusions. Since these inclusions are similar in structure and location to those that have been studied by light microscopy, their staining reactions have been predicted and included in Table 2.

It will be noted in Table 2 that certain inclusion types, such as cytoplasmic, cylindrical inclusions induced by the potyvirus group, are unique and therefore diagnostic for the group. These inclusions are so designated. Inclusions that are induced in addition to those characteristic of the group are also included in Table 2. The presence of these additional inclusions can be very useful in separating viruses within the group and in certain cases viruses of other groups whose characteristic inclusion(s) may appear similar.

When an inclusion or inclusions associated with an unidentified virus are found to be similar to the characteristic and additional inclusions of a virus group as described in Table 2, the virus can be tentatively assigned to that particular group. Since viruses within a group have most characteristics in common, it would be expected that the unidentified virus would also possess most of those characteristics. Because mode of transmission is a main group characteristic, possible control measures could be immediately recommended even though the unknown in question has not been characterized as to its particular relationship to the group.

XIII. CONCLUSIONS

The identification of inclusions by light microscopy, utilizing the O-G combination protein stain and the Azure A nucleic acid stain, offers a reliable, practical, and economical method for the diagnosis of many plant viral diseases. With this method it is possible to diagnose virus infections at the group level and sometimes at the specific level. Determining that a virus belongs to a particular group based on the presence of characteristic inclusions can predict many properties that this virus has in common with the group, whether the virus has been previously described or not. This information may suggest possible control measures for a particular crop situation, although the exact identity of the virus remains undetermined.

Designating the virus group also enhances the effectiveness of other diagnostic probes by narrowing the choice of viruses that need to be considered as possible causal agents. This step can be especially helpful to clinics that do not have the extensive facilities needed for indexing or have access to a broad spectrum of antisera. In addition, the presence of distinctive inclusion types can be used to diagnose multiple infections. This attribute of the technique is especially important, since mixed infections of viruses of the same group and/or different groups are of common occurrence.

REFERENCES

1. Christie, R. G. and Edwardson, J. R., Light microscope techniques for detection of plant virus inclusions, *Plant Dis.,* 70, 273, 1986.
2. Fenner, F. F., Classification and nomenclature of viruses, *Intervirology,* 7, 4, 1976.
3. Matthews, R. E. F., Classification and nomenclature of viruses, *Intervirology,* 17, 4, 1982.
4. Francki, R. I. B., Fauquet, C. M., Knudson, D. L., and Brown, F., Classification and nomenclature of viruses, Fifth Report International Committee on Taxonomy of viruses, *Arch. Virol.* (Suppl. 2), 3, 1991.
5. Harrison, B. D., Finch, J. T., Gibbs, A. J., Hollings, M., Shepherd, R. J., Valenta, V., and Wetter, C., Sixteen groups of plant viruses, *Virology,* 45, 356, 1971.
6. Dougherty, W. G. and Hiebert, E., Translation of potyvirus RNA in a rabbit reticulocyte lysate: identification of nuclear inclusion proteins as products of tobacco etch virus RNA translation and cylindrical inclusion protein as a product of the potyvirus genome, *Virology,* 104, 174, 1980.
7. Edwardson, J. R., *Some Properties of the Potato Virus-Y Group,* Fla. Agric. Exp. Stn. Monogr., No. 4, 1, 1974.
8. Edwardson, J. R. and Christie, R. G., Use of virus-induced inclusions in classification and diagnosis, *Annu. Rev. Phytopathol.,* 16, 31, 1978.
9. Edwardson, J. R., Christie, R. G., and Ko, N. J., Potyvirus cylindrical inclusions—subdivision IV, *Phytopathology,* 74, 1111, 1984.
10. Christie, R. G. and Edwardson, J. R., *Light and Electron Microscopy of Plant Virus Inclusions,* Fla. Agric. Exp. Stn. Monogr., No. 9, 1, 1977.
11. Christie, R. G. and Edwardson, J. R., Biological variants of tobacco etch virus that induce morphologically distinct nuclear inclusions, *Arch. Virol.,* Suppl. 5, 223, 1992.
12. DeMejia, M. V. G., Hiebert, E., Purcifull, D. E., Thornberry, D. W., and Pirone, T. P., Identification of potyviral amorphous inclusion protein as a nonstructural, virus specific protein related to helper component, *Virology,* 142, 34, 1985.
13. Dougherty, W. G. and Carrington, J. G., Expression and function of potyviral gene products, *Annu. Rev. Phytopathol.,* 26, 123, 1988.
14. Carrington, J. G. and Dougherty, W. G., Small nuclear inclusion protein encoded by a plant potyvirus genome is a protease, *J. Gen. Virol.,* 61, 2540, 1987.
15. Lesemann, D. E., Virus group-specific and virus-specific cytological alterations induced by members of the tymovirus group, *Phytopathol. Z.,* 90, 315, 1977.
16. Martelli, G. P. and Russo, M., Use of thin sectioning for visualization and identification of plant viruses, *Methods Virol.,* 8, 143, 1984.
17. Martelli, G. P., diFranco, A., and Russo, M., The origin of multivesicular bodies in tomato bushy stunt virus-infected Gomphrena globosa plants, *J. Ultrastruct. Res.,* 88, 275, 1984.
18. Martelli, G. P., Gallitelli, D., and Russo, M., Tombusvirus group, in *Atlas of Plant Viruses,* Vol. 1, CRC Press, Boca Raton, FL, 181, 1985.
19. Bassi, M., Barbieri, N., Appiano, A., and D'Agostino, G., Origin and function of tomato bushy stunt virus-induced bodies. I. Differential staining, cytochemistry, and serial sectioning, *J. Ultrastruct. Mol. Struct. Res.,* 96, 194, 1986.
20. Christie, R. G., Ko, N. J., Falk, B. W., Hiebert, E., Lastra, R., Bird, J., and Kim, K. S., Light microscopy of geminivirus induced nuclear inclusion bodies, *Phytopathology,* 76, 124, 1986.
21. Shikata, E., Reoviruses, in *Handbook of Plant Virus Infections and Comparative Diagnosis,* Kurstak, E., Ed., Elsevier/North-Holland, Biomedical Press, New York, 423, 1981.
22. Edwardson, J. R. and Christie, R. G., *Viruses Infecting Forage Legumes,* Vol. 3, Fla. Agric. Exp. Stn. Monogr., No. 14, Appendix E, 1986.
23. Edwardson, J. R. and Christie, R. G., *The Potyvirus Group,* Vol. 4, Fla. Agric. Exp. Stn. Monogr., No. 16, 1991.
24. Edwardson, J. R. and Christie, R. G., *Handbook of Viruses Infecting Legumes,* CRC Press, Boca Raton, FL. 1991.
25. Shalla, T. A. and Shepard, J. F., The structure and antigenic analysis of amorphous inclusion bodies induced by potato virus X, *Virology,* 49, 654, 1972.
26. Overman, M. A., Ko, N. J., and Tsai, J. H., Identification of viruses and mycoplasmas in maize by use of light microscopy, *Plant Dis.,* 76, 318, 1992.

27. Edwardson, J. R. and Christie, R. G., Tobacco mosaic virus: cytopathological effects, in *The Plant Viruses* 2, Van Regenmortel, M. H. V. and Fraenkel-Conrat, H., Eds., Plenum Press, New York. 1986, 153.
28. Edwardson, J. R., Christie, R. G., Purcifull, D. E., and Petersen, M., Inclusions in diagnosing plant virus diseases, in *Diagnosis of Plant Virus Diseases,* Mathews, R. E. F., Ed., CRC Press, Boca Raton, FL, 1993, 101.
29. Hiebert, E., Purcifull, D. E., and Christie, R. G., Purification and immunological analyses of plant viral inclusion bodies., *Methods Virol.,* 8, 225, 1984.

APPENDIX

I. PREPARATION OF VIRUS INCLUSION STAINS

A. Preparation of the orange-green (O-G) protein stain. Stain powders should be prepared separately as follows:

a. Add 1 g of Calcomine Orange to 100 ml of 2-methoxyethanol, stir thoroughly, and filter.

b. Add 1 g of Luxol Brilliant Green BL to 100 ml of 2-methoxyethanol, stir thoroughly, and filter. The stains should be stored in brown bottles and will keep indefinitely if tightly capped. Prepare the final staining solution by mixing one part distilled water, one part of the orange dye, and eight parts of the green dye. This solution is stable and can be used as needed. In the staining procedures this solution is referred to as the Orange-Green stain.

B. Preparation of the Azure A nucleic acid and nucleoprotein stain:

a. Azure A powder should be stirred into 100 ml of 2-methoxyethanol to achieve a 0.1% dye (g/100 ml) content (Azure A powders vary in dye content). This stain will keep indefinitely if capped and stored in a brown bottle.

b. Prepare a 0.2 *M* solution of dibasic sodium phosphate ($Na_2HPO_4 \cdot 7H_2O$). It is important to use a hydrated (not anhydrous) form. Prepare the final staining solution by adding one part of the phosphate solution to nine parts of Azure A. This solution must be prepared fresh with each staining sequence. Do not reuse it.

II. SOURCE OF STAINS, SOLVENTS, MOUNTING MEDIA, AND FORCEPS

Aldrich Chemical Co., Inc., P.O. Box 14508, St. Louis, MO 63178-9916
- Luxol Brilliant Green BL—cat. #27,726-6
- Azure A—cat. #86,104-9
- 2-Methoxyethyl acetate—cat. #30,826-9
- Triton® X-100—cat. #23,472-9

Fisher Scientific, 711 Forbes Ave., Pittsburgh, PA 15219-8300
- 2-Methoxyethanol=Ethylene Glycol Monomethyl Ether=Methyl Cellosolve cat. #E182
- 9 cavity plate 85–100 mm—cat. #13748B

Carolina Biological Supply, Burlington, NC 27215
- Euparal—cat. #86,1890
- Euparal Vert.—cat. #86,1910

Ernest F. Fullam, Inc.
- Dumont #5 sharpened S.S. forceps—cat. #13020

Pylam Products Co. Inc., 1001 Stewart Ave., Garden City, NY 11530

CAUTIONARY NOTE:

Use of polyvinyl, single-use gloves and adequate ventilation are advised when handling the stains and solvents used in these procedures. Please consult the manufacturer's material safety data sheets for further information.

III. MATERIALS NEEDED FOR STAINING PROCEDURES

A. A compound light microscope with an oil immersion objective and oculars capable of achieving 1000× or greater.

B. Stains: Orange-Green stain
Azure A stain

C. Triton® X-100 plastid solubilizing agent
When the O-G combination is used, stained plastids often obscure small inclusions. The plastids can be dissolved by treating tissue pieces with a 2% (2 ml concentrate: 98 ml H_2O) solution of Triton® X-100 (Rohm and Haas Co., Philadelphia, PA 19105) for 5 min at room temperature or 10–15 s in a microwave oven before staining. This treatment is especially useful for detecting the cylindrical inclusions of potyviruses, particularly during their early stages of development when these small inclusions are located at the cell periphery.

D. Rinsing solution: 70 ml of 2-methoxyethyl acetate: 30 ml of 95% EtOH (70:30; MeA:EtOH Rinse). If 2-methoxyethyl acetate is not available, the tissues can be rinsed in 95% ethanol. However, the tissues should not remain in the EtOH longer than 1 min, since it will remove the stains.

E. Holding solution (optional): Pure 2-methoxyethyl acetate.
This step can be used following the 70:30 MeA:EtOH. Since the stains are insoluble in 2-methoxyethyl acetate, the tissue can remain for extended periods of time. This step is useful when a break is needed in the schedule. More permanent preservation of color is achieved by use of this step.

F. Mounting agents: Euparal (straw color) and
Euparal "Vert" (green color)

G. Storage of slides: store slides in a cool, dark place.

H. Materials for the mechanics of the technique:
glass slides, cover slips, fine-tipped forceps,
staining dishes, disposable pipettes, razor blades.

IV. STAINING PROCEDURES

A. Pull epidermal strips (or cut appropriate sections) and place into:
 a. Orange-Green stain (Sample A)
 b. 2% Triton® X-100 (Sample B)
 c. Azure A (Sample C)
 See Procedure Flow Chart

B. Microwave 10–15 s at full power. Enclose a beaker of water during use. If a microwave is not available, stain for 10–15 min at room temperature.

C. Remove Orange-Green stain from Sample A and Azure A stain from Sample C with disposable pipettes and replace with 70:30 MeA:EtOH mix.

D. Rinse once or twice more with 70:30 MeA:EtOH mix until excess stain is removed.

E. Mount tissue pieces from Sample A and Sample C onto a clean slide in Euparal "Vert" (green color) and Euparal (straw color), respectively.

F. Remove Triton® X-100 from Sample B with pipette. Replace with Orange-Green stain. Microwave 10–15 s or stain for 10–15 min at room temperature.

G. Remove the Orange-Green stain from Sample B with disposable pipette and replace with 70:30 MeA:EtOH.

H. Rinse once or twice more with 70:30 MeA:EtOH, mix until excess stain is removed.

I. Mount tissue pieces from Sample B into Euparal "Vert" (green color).

J. Examine for inclusions.

V. PROCEDURE FLOW CHART

1) Sample A	2) Sample B	3) Sample C
↓	↓	↓
Orange-Green stain	2% Triton® X-100	Azure A stain combined with dibasic sodium phosphate
↓	↓	↓
Microwave 10–15 s or stain 10 min at room temperature	Microwave 10–15 s or treat 10 min at room temperature	Microwave 10–15 s or stain 10 min at room
↓	↓	

Rinse 2 or 3 times until clear with 70:30; 2MeA:EtOH
↓
Place in 100% 2MeA (optional) for 1–2 min at room temperature
↓
Mount in Euparal "Vert" (green color)
↓
Examine slide for inclusions

Remove Triton® X-100 Replace with the Orange-Green stain
↓
Microwave 10–15 s or stain 10 min at room temperature
↓
Rinse 2 or 3 times until clear with 70:30; 2MeA:EtOH
↓
Place in 100% 2MeA (optional) for 1–2 min at room temperature
↓
Mount in Euparal "Vert" (green color)
↓
Examine slide for inclusions

temperature
↓
Rinse 2 or 3 times until clear with 70:30; 2MeA:EtOH
↓
Place in 100% 2MeA (optional) for 1–2 min at room temperature
↓
Mount in Euparal (straw color)
↓
Examine slide for inclusions

VI. AZURE A METHOD FOR THE DETECTION OF TOBAMOVIRUSES

Azure A stains most inclusions at room temperature, but does *not* stain the crystalline, paracrystalline, or angled-layer aggregate inclusions induced by the tobamoviruses unless heat is applied during staining.

VII. PROCEDURAL FLOW CHART FOR TOBAMOVIRUSES

Sample 1
↓
Azure A combined with dibasic sodium phosphate (room temperature)
↓
Rinse 2 or 3 times until clear with 70:30; 2MeA:EtOH
↓
Place in 100% 2MeA (optional) for 1–2 min at room temperature
↓
Mount in Euparal (straw color)
↓
Examine for clear crystalline inclusions

Sample 2
↓
Azure A combined with dibasic sodium phosphate (heat at about 60°C in the stain for 1–2 min)
↓
Rinse 2 or 3 times until clear with 70:30; 2MeA:EtOH
↓
Place in 100% 2MeA (optional) for 1–2 min at room temperature
↓
Mount in Euparal (straw color)
↓
Examine for stained (red/violet) crystalline, paracrystalline, angled-layer aggregate inclusions

Chapter 4

Use of DNA-Binding Fluorochromes for the Nuclear Staining in Fungi

Uma S. Singh, J. Kumar, Amita Sachan, and Pratibha Singh

TABLE OF CONTENTS

I. INTRODUCTION

Nuclear staining in fungi is essential not only for the study of genetics, nuclear division, ploidy level (i.e., haploid, diploid, or polyploid), and to ascertain number of chromosomes per nucleus, but also for the determination of number of nuclei per cell[1] which is an important taxonomic criterion in certain fungi (e.g., *Rhizoctonia solani* and *Ceratobasidium* sp. are distinguished from each other based on the number of nuclei per cell[2]). The number of nuclei in spores is an important consideration where protoplast fusion or plasmid mediated recombination experiments are conducted since it is necessary to isolate uninucleate cells in such studies.[3]

Nuclear staining in fungi has been difficult because of their low DNA content.[4] Commonly used nuclear stains like iron-hematoxylin, Geimsa, acetocarmin, acetoorcein, etc. are difficult to be employed for the routine nuclear staining in fungi because of their low specificity for nucleus and cumbersome staining process involving critically timed hydrolysis of RNA. Only an experienced person can use these stains since sub-optimal acid hydrolysis may result in poor staining and/or artefact leading to erroneous conclusions.[37]

There are a number of DNA-intercalating fluorescent compounds such as acridine orange, DAPI, DIPI, Hoechst 33258, Hoechst 33342, quinacrine mustard, acriflavin, mithramycin, auramine O, olivomycin A, and ethidium bromide, which bind very specifically to the nucleus. They can be used directly on viable (vital dye) or fixed (post-vital dye) fungal mycelium, spores or fruiting bodies for rapid, one-step, and highly reproducible nuclear staining. Because of the direct relationship between DNA content and intensity of the fluorescence, some of the fluorochromes could even be used for the quantitative estimation of the DNA in fungal nuclei and also for deciding ploidy level and duplication cycle of the nucleus.[5–8]

0-87371-877-1/95/$0.00+$.50

II. FLUOROCHROMES

A. ACRIDINE ORANGE

Acridine orange is a 3, 6 *bis* (dimethylamino) acridine zinc chloride double salt.[9] The dye binds to both DNA and RNA. The acridine orange bound to DNA fluoresces green whereas its complex with RNA gives brick red fluorescence.[10] The excitation wavelength peak for acridine orange is 365 nm whereas cut off wavelengths for dichroic mirror and barrier filters are 450 and 440 nm, respectively.[11] Malachite or methyl green can be used to quench excessive fluorescence.[10]

Acridine orange has been used extensively in animal systems and medicine, as well as in cell physiology, virology, and bacteriology. It is most commonly employed in the evaluation of malignancy or to elucidate viral infections.[10] Probably because of its ability to stain fungal lysosomes brighter than the nuclei, acridine orange has limited application in staining of nuclei in fungi. Yamamoto and Uchida[12] used acridine orange for staining nuclei in hyphae of *R. solani,* mycelium and young gametangia of *Phytophthora capsici,* uredospores of *Puccinia oxalidis,* conidia of *Phytostricta capitalensis* and sporidia of *Ustilago maydis.* Wilson et al.[13] observed that when used as vital dye, as in animal cells, acridine orange could serve as a lysosomal marker in *Ceratocystis ulmi, Cryptococcus neoforenans,* and *Botrytis cinerea.*

Major disadvantages with the use of acridine orange are (1) it is not very specific to nucleus, (2) it does not stain DNA quantitatively, and (3) careful control of pH (4 to 5) is required for DNA staining.

B. AURAMINE O

Auramine O is 4, 4-(imidocarbomyl) *bis* (N, N-dimethyl aniline) monohydrochloride.[9] It has binding specificity for DNA and is commonly used for the detection of acid fast bacilli, particularly *Mycobacterium tuberculosis* and *M. Leprae*[10] which stain orange or yellow.

Raju[14] reported that auramine O was good for quick and reliable nuclear counts in *Neurospora* but was not useful for staining meiotic chromosomes in the ascus. Using auramine O, Bonfante-Fasolo et al.[15] successfully studied the distribution of nuclei in the different fungal structures (i.e., inter- and intracellular hyphae, arbuscules, and vesicles) formed by vesicular-arbuscular mycorrhizae endophyte during host root colonization.

C. DIPI

DIPI [4′, 6-*bis* (2′-imidazolinyl-4H, 5H)-2-phenylindole][9] specifically binds to A=T rich region of double stranded DNA.[16] The combination of UG1 365 as excitation filter, FT 420 as chromatic beam splitter, and GG 435 as barrier filter has been used for the visualization of DIPI-DNA complex. Since performance of DIPI is not as good as other DNA-intercalating agents including DAPI, it is no longer used as a nuclear stain.

D. OLIVOMYCIN A

Olivomycin A is 3-0 {2, 6 dideoxy-3-C-methyl-4-O-(2-methyl-1-oxypropyl)-α-L-arabino-hexapyranosyl}-olivomycin D.[9] It is soluble in alcohol, ether, and chloroform but insoluble in water. Olivomycin A binds with DNA and fluoresces yellow when observed under fluorescent microscope using KP 400 as excitation filter, 455 nm dichroic mirror, and K 460 as barrier filter.[14] Raju[14] found olivomycin A quite satisfactory as post-vital nuclear stain for counting nuclei per cell in *Neurospora* but was not suitable for the chromosome analysis during meiosis. It is not often used for the nuclear staining in fungi.

E. HOECHST 33258

Hoechst 33258 or H-stain is 2-(2-(4-hydroxyphenyl)-6-(benzimidazolyl)-6-(1-methyl-4-piperazyl)-benzimidazol-trichloride.[9] It preferentially binds to A=T rich region of DNA. Wavelength peak for excitation filter and cutoff wavelengths for dichroic mirror and barrier filter are 365, 420, and 410 nm, respectively.[11] Hoechst 33258 is both heat and light sensitive and because of this its stock solution needs storage at 4°C in dark.

Hough et al.[11] used Hoechst 33258, both as a vital and post-vital dye, for the detection of pollen grain and pollen tube nuclei. When used in combination with callose stain, sirofluor, Hoechst 33258 enabled differentiation of generative and vegetative nuclei. Otto and Tsou[16] found it at par with Hoechst 33342, DAPI and DIPI for the staining of DNA in Chinese hamster. In another comparative study, Hoechst 33258 was found as good as any other fluorochrome for quick and reliable nuclear count in spores and mycelium of *Neurospora* but was not suitable for the chromosomal analysis during meiotic division.[14]

F. HOECHST 33342

Like Hoechst 33258, Hoechst 33342 is also a bisbenzimidazole derivative and binds with A=T rich region of the DNA.[9] Staining characteristics of both these fluorochromes are quite similar,[16] though Hoechst 33258 is more popular as a DNA stain.

G. QUINACRINE MUSTARD

Combination of IF 490 excitation and O515 as barrier filter are used for observation. Geimsa stained specimens can be destained by immersing in acidic methanol then restained with quinacrine mustard. Quinacrine-HCl can also be used in place of quinacrine mustard.[10]

Quinacrine mustard is widely used in clinical laboratory staining. It is specific for the chromosomes and convenient for the observation of genetic abnormalities, demonstration of genetic mutation of malignant cells, and also for chromosome mapping.[10] However, we are not aware of its use as a nuclear or chromosomal stain in fungi.

H. MITHRAMYCIN

Mithramycin (= mitramycin or antibiotic 7017) or aureolic acid (= aurelic acid) is a member of a group of antitumour antibiotics produced by *Streptomyces plicatus.*[18] It is similar to chromomycins and olivomycins.[9] Mithramycin is inert towards RNA and protein but fluoresces yellow when bound to double stranded DNA in the presence of Mg.[24] For observation under fluorescent microscope KP 400 excitation filter, 455 nm dichroic mirror, and K 460 barrier filter can be used. Initial fluorescent is brighter with excitation filter KP 500, a 510 nm dichroic mirror, and K 515 barrier filter.[18]

Slater[18] was probably the first to use mithramycin for staining nuclei in yeast (*Saccharomyces cerevisiae*). He found it quite suitable for monitoring yeast cell cycle or sporulation as they occur rather than after they are completed. Coleman and Goff[19] found mithramycin extremely useful for the study of pollen development and growth. It stained nuclei brilliantly both in living and fixed pollen thereby permitting rapid scanning for pollen abnormalities and easy observation of nuclear details. Panwar et al.[1] used mithramycin for the single step rapid staining of nuclei in unfixed mycelium and spores of filamentous fungi like *R. solani, Ceratobasidium* sp., *Ascochyta rabiei* and *Phytophthora nicotianae* var. *parasitica.* They observed that mithramycin was so specific for nucleus that counter stains like fluorescamine, fluorescein isothiocyanate, or acridine orange were needed to visualize cell boundaries. Mithramycin binds quantitatively with DNA and thus may be used to quantitate ploidy changes and to study nuclear cycle. The major disadvantages of mithramycin are its high cost, poor stability in solution, and fading of fluorescence during observation.[4]

I. ACRIFLAVIN

Acriflavin is 3, 6-diamino-10 methyl acridinium chloride mixture with 3, 6-acridinediamine.[9] It specifically binds to DNA. Excitation occurs at 450 nm and emission at 540 nm.[14]

Levinston et al.[20] used acriflavin for staining and Tanke and van Ingen[21] used it for the quantitative DNA measurement of animal nuclei. It proved to be at par with other DNA-intercalating fluorochromes like DAPI, Hoechst 33258, olivomycin, etc. for quick and reliable nuclear staining in *Neurospora*[14] and *Colletotrichum* spp.[3] However, it was superior to others for detailed chromosomal analysis and observation on nucleolar organization region.[14] Acriflavin is specific for DNA, the chromosomes stain well and ribosomal RNA rich nucleolus appears as a ghost. Since spindles and spindle pole bodies do not fluoresce at all, the fluorochrome is very useful for determining chromosome number during division stages in the *Neurospora* ascus.[14] The major problem with acriflavin staining is that unlike other fluorochromes it involves acid hydrolysis.

J. ETHIDIUM BROMIDE

Ethidium bromide is 2, 7-diamino-10-ethyl-9-phenyl phenanthridium bromide or 3, 8-diamino-5-ethyl-6-phenyl phenanthridium bromide.[9] It intercalates between adjacent base pairs, specifically G-C base pairs in both RNA and DNA.[22] It fluoresces bright brick red (<590 nm) under G excitation 465 to 500 nm) and orange (<420 nm) under U excitation (330 to 380 nm).[4] However, considering the brightness of the fluorescence, observations are recorded under G excitation with the DM 580 dichroic mirror, B 545 exciter filter, and O 590 barrier filter, supplementary barrier filter R 610 may be used to enhance the contrast of nuclear fluorescence by minimizing the background fluorescence.[4]

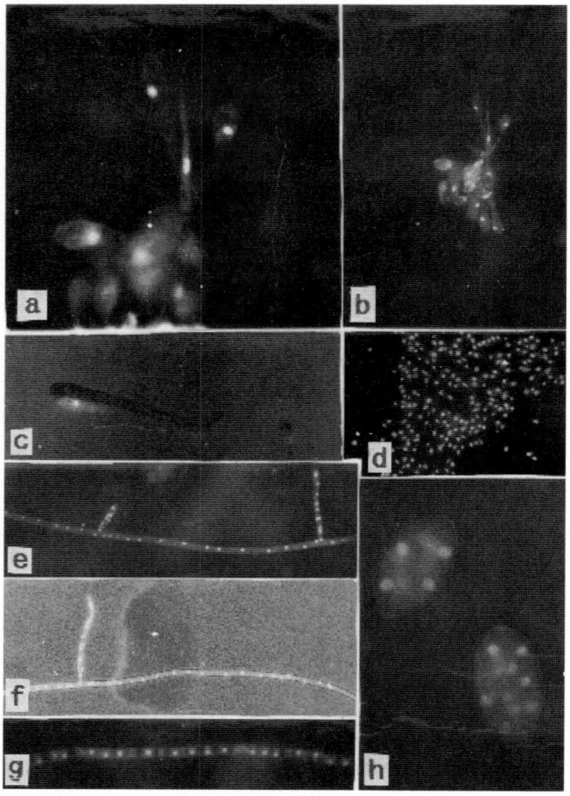

Figure 1 Fungal nuclei stained with ethidium bromide and observed under fluorescent light (green excitation region). **(a).** Conidia of *Venturia inaequalis.* × 312.5. **(b).** Conidia of *V. inaequalis.* × 62.5. **(c).** Germinating conidia of *V. inaequalis,* double exposure (once under normal and once under fluorescent light). × 125. **(d).** Spores of *Fusarium oxysporum.* × 62.5. **(e).** Hyphae of *Rhizoctonia solani.* × 62.5. **(f).** Hyphae of *R. solani,* normal light superimposed on fluorescent light to visualize septum (arrow). × 62.5. **(g).** Hyphae of *R. solani.* × 125. **(h).** Sporangium of *Phytophthora infestans.* × 312.5.

Ethidium bromide is commonly employed for DNA and RNA staining in electrophoresis.[23] Hough et al.[17] used it as a vital stain or following fixation for the study of nuclei during pollen development and growth. Singh and Kumar[4] used it for the first time to stain nuclei in several fungi like *R. solani, Fusarium oxysporum, Phytophthora infestans, Venturia inaequalis,* etc. Excellent nuclear staining was obtained in all of the test fungi. Nuclear fluorescence was bright and very stable. Even if the stained hyphae/spores dried out because of evaporation of mounting medium, they were remoistened and could be observed without loss of nuclear fluorescence.[4] The stock solution of ethidium bromide was stable for more than a year. Because of the high efficiency and specificity, better stability of stain and fluorescence, and low cost and ready availability, ethidium bromide is considered better than any other fluorochrome for the staining of nuclei in fungi. Now it is being used as a routine fungal nuclear stain in our laboratory. In spite of the bright and stable fluorescence and positive correlation between nucleic acid content and fluorescence intensity, ethidium bromide may not be suitable for the quantitative estimation of DNA content in nuclei as it intercalates with both double stranded RNA and DNA.

Like mithramycin, ethidium bromide cannot stain cell boundaries. Thus far, we have not been able to identify a suitable counter-stain. However, to visualize septa and cell walls, as may be desired in some studies, normal transmitted light could be superimposed on reflected fluorescent light (see Figure 1f). Alternatively, if possible, one can go for double exposure (one under fluorescent and one under normal light) of the object without disturbing its position (see Figure 1c).

Ethidium bromide is a powerful mutagen and is moderately toxic.[33] Therefore, adequate caution should be taken. Gloves should be worn while working and after use ethidium bromide solution and glassware may be decontaminated by using slurries of activated charcoal or Amberlite XAD-16.[23]

K. DAPI

DAPI is 4′, 6-diamidino-2-phenyl indole. It specifically binds to A=T rich region of DNA and exhibit intense brownish fluorescence. Its excitation and emission frequencies are 365 and 450 nm, respectively.[24] Filters often used for DAPI-DNA fluorescence are 365 nm excitation, 420 nm dichroic mirror, and 410 nm barrier.

DAPI was first synthesized by the Otto Dann's laboratory at Erlangen in 1971.[24] It binds preferentially to AT-rich dsDNA. Williamson and Fennel[24] used DAPI for the staining and separation of mitochondrial DNA. Since then, it has been used most widely as a vital and post-vital stain for the staining of nuclei in fungi,[3,8,14,25–29] because it is specific, highly reproducible, quick, convenient, and produces sharper images of nuclei with little background. However, for the staining of meiotic chromosomes in *Neurospora* and *Coprinus,* it was inferior to acriflavin.[14]

Since DAPI is specific to DNA, DAPI-DNA complex fluoresces at 15 to 20 times the intensity of background DAPI alone. Fluorescence fades only comparatively slowly under excitation and the intensity of fluorescence emitted is directly proportional to amount of DNA. The DNA content of individual nuclei can be estimated by measuring the emission from fluorescent nuclei photometrically. DAPI has been widely used for measuring DNA content in nuclei of several fungal species.[3,8,30–35] The DAPI stain enables the detection of a heterokaryon in which one nucleus had approximately twice as much DNA as the other.[28] Using this fluorochrome, Whittaker et al.[6] demonstrated that progeny of sexual crosses of *P. infestans* involving isolates of different DNA contents produced individuals of diverse DNA contents.

III. METHODOLOGY

Among the fluorochromes described in Section II, acridine orange, mithramycin, acriflavin, ethidium bromide, and DAPI are excellent for routine nuclear staining in fungi; acriflavin and DAPI are very good for staining of meiotic chromosomes, and DAPI has been used most extensively for the estimation of DNA content in fungal nuclei.

Usually high speed films (≥400 ASA) are used for the photography of fluorescent nuclei.

Basic protocols for the staining of fungal nuclei using common fluorochromes are as follows:

A. ACRIDINE ORANGE

For the staining of nuclei, acridine orange can be used either as vital or post-vital dye. However, for the staining of lysosomes it can be used only on unfixed fungal mycelium and spores as a vital dye. Of crucial importance is the accurate adjustment of pH of the staining solution. The following protocol is modified from Yamamoto and Uchida.[12]

1. Take fresh or fixed (with alcohol-acetic acid) mycelium or spores on glass slide.
2. Add a few drops of acridine orange solution (25 μg ml^{-1}) in veronal acetate buffer (pH 4.5).
3. Spread mycelium with needle and mount.
4. Observe under fluorescent microscope using filter combination as described in Section II.

Veronal acetate solution	
Sodium acetate	971 mg
Soluble veronal (sodium diethyl barbiturate)	1471 mg
Distilled water	50 ml
Veronal acetate buffer	
Veronal acetate solution	50 ml
HCl (0.01 M)	74 ml
Distilled water	100 ml
Adjust pH to 4.5	
Acridine orange solution	
Acridine orange	2.5 mg
Veronal acetate buffer	100 ml

B. MITHRAMYCIN

It can be used on living or fixed fungal mycelium or spores or as a vital dye can be added into culture or germination medium. However, results are best when it is used on fresh mycelium or spores. Fluorescamine, fluorescein isothiocyanate, or acridine orange are used as counter-stains, alone with mithramycin to visualize cell wall (cell boundaries and septa).

The specific binding and fluorescent properties of mithramycin make it possible to stain fungal nuclei in a rapid one step procedure. The following procedure is based on Panwar et al.[1] and Slater.[18]

1. Place fresh mycelium or spores on a glass slide.
2. Add 2 drops of mithramycin solution.

3. If staining of cell boundaries and septa is desired, add two drops of aqueous solution of fluorescamine (30 μg ml^{-1}), fluorescein isothiocyanate (100 μg ml^{-1}), or acridine orange (1.0 μg ml^{-1}). In case of acridine orange, the mycelia/spores should be stained with mithramycin until the nuclei are visible before adding counter-stain while two other counter-stains can be added simultaneously with the fluorochrome.
4. Place cover slip and observe under fluorescent microscope using filter combination as described in Section II.

Mithramycin solution	
Mithramycin	2.0 mg
Aqueous ethanol (25%, v/v) containing 15 mM $MgCl_2$	10 ml

C. ACRIFLAVIN

The following protocol is based on Raju.[14]

1. Hydrolyse unfixed, intact mycelia/spores/perithecia in 4N HCl for 20 to 30 minutes at 30°C.
2. Rinse once in water and stain in a solution containing acriflavin (100 to 200 μg ml^{-1}) and $K_2S_2O_5$ (5 mg ml^{-1}) for 20 to 30 minutes at 30°C.
3. Wash stained mycelia/spores/perithecia thrice (3 to 5 minutes each) in concentrated HCl and in 70% ethanol mixture (2:98, v/v) at 30°C to remove non-covalently bound stain from cells.
4. Wash three times in distilled water.
5. Mount in 25% glycerol. In case of perithecia, dissect in a drop of 25% glycerol and squash asci under cover glass.
6. Observe under fluorescent microscope using filter combination as described in Section II.

D. ETHIDIUM BROMIDE

The following protocol is based on Singh and Kumar.[4]

For mycelium:

1. Immerse unfixed fungal mycelia for 5 minutes in 0.1% solution of ethidium bromide in ethanol-water (1:3, v/v) on a glass slide.
2. Decant stain by tilting slide.
3. Wash twice in distilled water.
4. Mount in water.

For spores:

1. Mount fungal spores in 0.1% solution of ethidium bromide in ethanol-water (1:3, v/v) on a glass slide.
2. After 5 minutes, gradually replace ethidium bromide with distilled water by carefully absorbing the stain with blotting paper from one end of the cover slip and simultaneously adding water from the other end.
3. Observe under fluorescent microscope using filter combination as described in Section II.

E. DAPI

The following protocol is based on Hooley et al.[8] and Gu et al.[34]

1. Fixed (in 5% glutaraldehyde in 0.05 M-Tris/HCl buffer; pH 7.0) or unfixed fungal mycelium/spores are dipped in DAPI solution (1 μg ml^{-1} in 0.1 M sodium phosphate; pH 7.0) for 5 minutes.
2. Excess stain is then removed by washing three times in distilled water.
3. Mount in water or stain.

To help prevent photobleaching of the DAPI fluorochrome, 0.05% n-propyl gallate in 75% glycerine buffered with a monobasic-bibasic sodium phosphate solution (0.1 M; pH 9.0) is added just prior to placing the cover slip on the slide.

IV. CONCLUSION

DNA intercalating fluorochromes offer a quick, specific, and highly reproducible method for the staining of nuclei both in living and fixed fungal mycelium and spores. Some of these can also be used for the staining of meiotic chromosomes and also for the quantitative estimation of nuclear DNA content thereby determining the ploidy level of nuclei. With the advancement of con-focal microscopy,[36] use of these fluorochromes in fungal cytogenetics is likely to become more common in the future. Major disadvantages with flurochromes is the requirement of fluorescent microscope and mutagenic nature of some of these fluorochromes. Therefore, appropriate precautions should be taken to avoid contact of the fluorochromes with the skin.

REFERENCES

1. Panwar, R., Singh, U. S., and Singh, R. S., Fluorescent staining of nuclei in filamentous fungi, *Stain Technol.,* 62, 205, 1987.
2. Parmeter, J. R., Jr., *Rhizoctnia solani: Biology and Pathology,* University of California Press, Berkeley, 1970.
3. TeBeest, D. O., Shilling, C. W., Riley, L. H., and Weidemann, G. J., Number of nuclei in spores of three species of *Collectotrichum, Mycologia,* 81, 147, 1989.
4. Singh, U. S. and Kumar, J., Staining of nuclei in fungi by ethidium bromide, *Biotechnic Histochem.,* 66, 266, 1991.
5. Bianciotto, V. and Bonfante, P., Quantification of the nuclear DNA of two arbuscular mycorrhizal fungi, *Mycol. Res.,* 96, 1071, 1992.
6. Whittaker, S. L., Shattock, R. C., and Shaw, D. S., Inheritance of DNA contents in sexual progenies of *Phytophthora infestans, Mycol. Res.,* 95, 1094, 1992.
7. Whittaker, S. L., Shattock, R. C., and Shaw, D. S., The duplication cycle and DAPI-DNA contents in nuclei of germinating zoospore cysts of *Phytophthora infestans, Mycol. Res.,* 96, 355, 1992.
8. Hooley, P., Fyfe, A. M., Maltese, C. E., and Shaw, D. S., Duplication cycle in nuclei of germinating zoospores of *Phytophthora drechsleri* as revealed by DAPI staining, *Trans. Brit. Mycol. Soc.,* 79, 563, 1982.
9. Windholz, M. and Budavari, S., *The Merck Index: An Encyclopaedia of Chemicals, Drugs, and Biologicals,* 10th Ed., Merck & Co., Rahway, N.J. 1983.
10. *The Use of Olympus Fluorescence Microscope,* Olympus Optical, Tokyo, Japan, 1982.
11. McGovein, R. J., Horst, R. K., and Israel, H. W., Chrysanthemum phloem necrosis detection by epifluorescence microscopy, *Can. J. Bot.,* 67, 1024, 1989.
12. Yamamoto, D. T. and Uchida, J. Y., Rapid nuclear staining of *Rhizoctonia solani* and related fungi with acridine orange and with safranin O, *Mycologia,* 74, 145, 1982.
13. Wilson, C. L., Jumper, G. A., and Mason, D. L., Acridine orange as a lysosome marker in fungal spores, *Phytopathology,* 68, 1564, 1978.
14. Raju, N. B., A simple fluorescent staining method for meiotic chromosomes of *Neurospora, Mycologia,* 78, 901, 1986.
15. Bonfante-Fasolo, P., Berta, G., and Fusconi, A., Distribution of nuclei in a VAM fungus during its symbiotic phase, *Trans. Brit. Mycol. Soc.,* 88, 263, 1987.
16. Otto, F. and Tsou, K. C., A comparative study of DAPI, DIPI and Hoechst 33258 and 33342 as chromosomal DNA stains, *Stain Technol.,* 60, 7, 1985.
17. Hough, T., Bernhardt, P., Knox, R. B., and Williams, E. G., Applications of fluorochromes to pollen biology. The DNA probes ethidium bromide and Hoechst 33258 in conjugation with the callose-specific aniline blue fluorochrom, *Stain Technol.,* 60, 155, 1985.
18. Slater, M. L., Staining fungal nuclei with mithramycin, *Methods Cell Biol.,* 20, 135, 1978.
19. Coleman, A. W. and Goff, L. J., Application of fluorochromes in pollen biology. Mithramycin and DAPI as vital stains and for quantitation of nuclear DNA, *Stain Technol.,* 60, 145, 1985.
20. Levinston, J. W., Retzel, S., and McCormick, J. J., An improved acriflavin-Feulgen method, *J. Histochem. Cytochem.,* 25, 355, 1977.
21. Tanke, H. J. and van Ingen, E. M., A reliable Feulgen-acriflavin-SO_2 staining for quantitative DNA measurements, *J. Histochem. Cytochem.,* 28, 1007, 1980.
22. LePecq, J. B. and Paoletti, C., A fluorescent complex between ethidium bromide and nucleic acids, *J. Molec. Biol.,* 12, 87, 1967.
23. Sambroock, J., Fritsch, E. F., and Maniatis, T., *Molecular Cloning: A Laboratory Manual,* Vol. I, Cold Spring Harbor Laboratory Press, New York, 1989, 616.

24. Williamson, D. H. and Fennel, D. J., The use of fluorescent DNA binding agent for detecting and separating yeast mitochondrial DNA, *Methods Cell Biol.,* 12, 335, 1975.
25. Martigani, E. and Trezzi, F., Fluorescent staining of *Neurospora* nuclei with DAPI, *Neurospora Newsl.,* 26, 20, 1979.
26. Lemke, P. A., Kugelman, B., Morimoto, H., Jacobs, E. C., and Ellison, J. R., Fluorescent staining of fungal nuclei with a benzimidazole derivative, *J. Cell Sci.,* 29, 77, 1978.
27. Wagstaff, J. E., Klapholz, S., and Esposoto, R. E., Meiosis in haploid yeast, *Proc. Natl. Acad. Sci. U.S.A.,* 79, 2986, 1982.
28. Whittakar, S. L., Shattock, R. C., and Shaw, D. S., Variation in DNA content of nuclei of *Phytophthora infestans* as measured by microfluorometric method using the fluorochrome DAPI, *Mycol. Res.,* 95, 602, 1991.
29. Hyde, G. J. and Hardham, A. K., Microtubules regulate the generation of polarity in zoospores of *Phytophthora cinnamoni, Eur. J. Cell Biol.,* in press, 1993.
30. Butt, T. M., Hoch, H. C., Staples, R. C., and St. Leger, R. J., Use of fluorochromes in the study of fungal cytology and differentiation, *Exp. Mycol.* 3, 303, 1989.
31. Motta, J. J., Peabody, D. E., and Peadbody, R. B., Quantitative differences in nuclear DNA content between *Armillaria mellea* and *Armillaria bulbosa, Mycologia,* 78, 963, 1986.
32. Bresinsky, A., Fischer, M., Meixner, B., and Paulus, W., Speciation in *Pleurotus, Mycologia,* 79, 234, 1987.
33. Talbot, N. J., Coddington, A., Roberts, I. N., and Oliver, R. R., Diploid construction by protoplast fusion in *Fulvia fulva* (syn. *Cladosporium fulvum*): genetic analysis of an imperfect fungal plant pathogen, *Curr. Genet.,* 14, 567, 1988.
34. Gu, W., Spielman, L. J., Matuszak, J. M., Aist, J. R., Bayles, C. J., and Fry, W. E., Measurement of nuclear DNA content of Mexican isolates of *Phytophthora infestans, Mycol. Res.,* 97, 857, 1993.
35. Rutherford, F. S. and Ward, E. W. B., Estimation of relative DNA content in nuclei of *Phytophthora megasperma* var. *glycinea* by quantitative fluorescent microscopy, *Can. J. Gen. Cytol.,* 27, 614, 1985.
36. Kwon, Y. H., Wells, K. S., and Hoch, H. C., Fluorescence confocal microscopy: Applications in fungal cytology, *Mycologia,* 85, 721, 1993.
37. Singh, U. S., personal observation.

Chapter 5

Virion Gel Electrophoresis

Rudra P. Singh and Norman R. M. Watts

TABLE OF CONTENTS

I. INTRODUCTION

The detection and diagnosis of viruses in plants have classically involved the observation of gross symptoms, such as lesion appearance and leaf morphology, and the determination of transmissibility, host range, and vector relationships. The utility of these methods is limited in part due to environmental influences. Thus, more recently, the detection and identification of the viral genome by hybridization and sequencing approaches, and of the capsid protein by serological means, have gained in importance. Serology in particular has become widely implemented due to its inherent simplicity, sensitivity, and, especially with monoclonal antibodies, specificity. However, these two methodologies share the drawback of requiring the production of a probe (an involved procedure in both cases) and the inherent failure to detect nontarget viruses. The application of techniques based solely on the electrophoretic properties of viruses, or their constituents, not only overcomes these drawbacks, but also has some additional practical advantages. The diagnosis of plant viruses by the electrophoretic analysis of double-stranded (ds) RNA and of intact virions are notable examples.

Of these two methods the analysis of plant viral dsRNA has seen somewhat more applications and will not be discussed at length here. An early review of the subject[1] provides a general introduction to the applications and methodology and gives a number of additional practical advantages of the technique. A more recent review by Valverdre[2] provides a detailed protocol for all aspects of the technique and also a discussion of some of the limitations. There are also numerous recent applications.[3–11]

II. VIRION ELECTROPHORESIS

A. METHODOLOGY

1. Electrophoresis in Free Solution

Electrophoresis of virions in free solution predominated in early studies, with the 2-ml Tselius cell (Perkin-Elmer® model 38-A) fitted with schlieren optics being most prevalent.[12–18] Typically, only brief

0-87371-877-1/95/$0.00+$.50

methodological descriptions were provided. A more detailed description involving density gradients in a laboratory-made U-tube apparatus may be found in Ball.[19] Other electrophoretic separations in density gradients have been employed.[20–23]

2. Electrophoresis in Gels

The majority of virion separations have been performed in stabilizing media such as polyacrylamide and agarose, with the latter now being used almost exclusively.

a. Polyacrylamide

Polyacrylamide was first used for electrophoresis by Tselius et al.[24] and has since seen numerous applications for resolution of intact virions. Separations typically involve low-concentration gels (typically 2.4 or 3%) and slightly alkaline buffers (frequently Tris-glycine).[15,17,18,21,22,25–29]

b. Agarose

Although agarose gel electrophoresis of intact virions has been used for a long time,[30–44] a review of the literature indicates considerable refinement in the technique in recent years and it appears to have become the method of choice. In particular, the work by Serwer and group[31] has contributed significantly to optimizing this technology and for this reason Serwer et al.[31] are frequently cited. The reader is directed particularly towards Serwer[37] for detailed descriptions for optimizing apparatus, gel preparation, sample preparation, sample application, electrophoresis, staining, and documentation, as well as for a discussion of the theory. This latter work is centered around the preparation of multiple-concentration agarose gels embedded within a supporting frame gel. Considerations of the sieving of rod-shaped vs. isometric virions in agarose may be found in Griess et al.[45]

c. Polyacrylamide/Agarose

The incorporation of agarose into low-percentage acrylamide gels improves their strength and elasticity. Polyacrylamide/agarose gels have been used to purify turnip yellow mosaic virus,[24] tobacco mosaic virus (TMV),[46] and to analyze TMV disassembly intermediates.[33]

3. Isoelectric Focusing

Isoelectric focusing (IEF) has proven to be a very powerful technique useful in resolving, characterizing (determination of isoelectric point [pI]), and evaluating the purity (and microheterogeneity) of proteins. However, IEF of intact virions from plants has seen relatively little application. For detailed reviews of all aspects of IEF methodology see Righetti.[47]

B. APPLICATIONS

1. Typing

a. Electrophoresis

The predominant application of virion electrophoresis, whether in solution or agarose or acrylamide gels, has been for purposes of characterizing a virus, or simply discriminating between closely related, and even serologically indistinguishable, isolates. Electrophoretic diversity between various isolates has been studied in southern bean mosaic virus,[16] cucumber mosaic virus (CMV),[26–28] peanut stunt virus (PSV),[26,27] andean potato latent virus,[30] grapevine fanleaf virus,[46] sharka virus,[49] cucumoviruses (CMV) and soybean stunt virus, chrysanthemum mild mottle virus,[29] red clover mottle virus,[50,51] selected tobamoviruses,[19,34] hibiscus chlorotic ringspot virus,[36] various tombusviruses,[35,52] sweet clover necrotic mosaic virus (SCNMV),[53] tomato aspermy virus,[54] various dianthoviruses (SCNMV), carnation ringspot virus (CRSV), and red clover necrotic mosaic virus.[39]

b. pH-Mobility Curves

An interesting application of virion typing by electrophoresis was described by Hurtt et al.[41] Building on a two-dimensional electrophoretic technique originally devised by Rosengren et al.[55] to select optimal pH conditions for gel electrophoresis, and later advanced by Righetti and group,[56–59] Hurtt et al.[41] first established a pH gradient in an agarose gel containing 2.5% carrier ampholytes and then, after applying virus in a continuous band across the pH gradient, applied an electric field perpendicular to the first field. Virus at each pH along the gradient migrated with a particular mobility, giving a pH-mobility curve characteristic for each virus.

Aside from the ratio of titratable acidic and basic groups on the surface of the virion particle, swelling (discussed below in Section B.2.c) may also play a role, both in terms of binding ampholytes and possible interactions with the gel matrix.[41] In addition to determination of the pI and net surface charge, the curves also reveal virion pH instabilities and particle heterogeneity.[41] Other possible applications for this technique indicated by the authors include the study of the effect of pH on the susceptibility of virions to degrading agents, mutations, and host passage.

c. *Genetic Reassortants*

The study of typing and assortment of genetic determinants in divided genome viruses involves the construction of the genetic reassortants and the subsequent characterization of the hybrids, typically by infectivity, serological specificity, and hybridization analysis. Pappu and Hiruki[39] have devised a simple and reliable criterion for studying the reassortants of SCNMV based on virion mobility during agarose gel electrophoresis.

SCNMV consists of 30–35 nm isometric particles, each containing two single-stranded genomic RNAs (RNA-1, 4.3 kb and RNA-2, 1.4 kb).[60] The coat protein is coded on RNA-1.[61] Virion electrophoresis of particles from plants infected by homologous and heterologous mixtures of the RNAs 1 and 2 purified from electrophoretically distinct strains revealed the hybrid nature of the assortants. In each case the reassortant migrated with the mobility of the parent contributing the RNA-1 moiety.[39] Virion electrophoresis may similarly be applied to other divided genome viruses such as CMV, red clover mosaic virus, CRSV, where the location of the coat protein gene is known, and possibly even rod-shaped viruses.[39]

d. *Isoelectric Focusing*

IEF of the plant viruses CCMV, BMV, TYMV, TMV, STNV and the bacteriophage Qb has been described by Rice and Horst.[62] IEF of plant viruses has also been described more recently.[53]

2. Physical Characterization

a. *Purification*

The isolation and maintenance of pure virus isolates is essential and is usually achieved by sequential single-lesion passage. The use of virion electrophoresis to obtain pure preparations has been described for turnip yellow mosaic virus,[24] TMV,[46] broad bean mottle virus, and alfalfa mosaic virus (AMV).[64] A method for purifying cacao swollen shoot virus by electrophoresis in a saccharose gradient has also been reported.[65]

b. *Physical Properties*

The physical characterization of viruses including their electrophoretic properties is less common than the simple differentiation of isolates such as described above (Section B.1.a). The early observation[12] of host passage effects in TMV was facilitated by the application of virion gel electrophoresis. The physical characteristics of saguaro virus,[23] prune dwarf virus,[18] beet yellows virus and beet mild yellowing virus,[66] citrus ringspot virus,[38] and eggplant mosaic virus[67] have been described. Virion gel electrophoresis has also been employed to demonstrate that particle size heterogeneity and not density is the cause of centrifugal heterogeneity in tobacco streak virus.[17]

c. *Virion Conformational Changes*

Agarose gel electrophoresis has also been used to monitor conformational states of intact spherical virions.[44] It is known that several small spherical plant viruses undergo conformational changes (swelling in some instances up to a 20% increase in diameter) when divalent cations, principally calcium, are removed either by chelation or by slightly alkaline pH (e.g., 8.5). Many subunit bonds depend on the presence of divalent cations, and several plant viruses have binding sites for such ions and are stabilized by them. When the cations are replaced the virions typically contract again, though apparently by a different pathway. The conformational changes involved in swelling have previously been monitored by involved techniques such as hydrogen ion titration, photon correlation spectroscopy, analytical ultracentrifugation, X-ray crystallography, and neutron scattering, among others.[44]

In this study[44] it was shown that these conformational changes can be observed as a marked change in electrophoretic mobility in agarose gels. Turnip crinkle carmovirus (TCV) and tomato bushy stunt tombusvirus (TBSV) migrated more slowly during agarose gel electrophoresis and stained more intensely with ethidium bromide when in the swollen state. TCV could be stabilized to pH-induced changes by the addition of 5.0 m*M* CaCl, but not by 5.0 m*M* MgCl. By comparison TBSV, being more stable, required the presence of chelators in addition to an alkaline pH before swelling. Thus agarose gel electrophoresis provides a simple and rapid method for assaying pH and chelator-induced changes in conformation and permeability in virions.

d. Electrophorotype Conversion

Two electrophoretic forms have been reported for the comoviruses bean pod mottle virus (BPMV)[13] and cowpea mosaic virus (CPMV).[68] The origin of the alternative forms, slow in the case of BPMV and fast for CPMV, was thought to be due to heterogeneity of the structural proteins[15] and shown *in vitro* to be due to proteolytic digestion.[20] Geelen et al.[21,22] later showed that in the case of CPMV the removal of a 22- to 25-amino acid fragment from the carboxyl end of the viral S protein correlated with the conversion of the slow to fast form. Smith[69] was able to isolate the two forms by fast protein liquid chromatography. Interestingly, Kartaatmadja and Sehgal[70] have reported that, while there is a correlation between changes in the electrophoretic mobility of BPMV virions and the well-established progressive decline in specific infectivity (SI) with increasing duration of infection,[14] this was not causal. Rather, SI decline was due to the selective *in situ* degradation of the RNA-1 component which was thought to be physically less stable than RNA-2.

Langham et al.[71] have demonstrated the *in vivo* and *in vitro* conversion of the comoviruses BPMV, CPMV, and squash mosaic virus by both trypsin and the protease-containing regurgitant of various leaf-feeding beetles, using agarose gel electrophoresis. While converted and unconverted viral forms existed in infected tissue, only the converted forms were found in beetle regurgitant. Conversion was both stable (giving no intermediates in agarose gel electrophoresis, even upon prolonged digestion) and reproducible, and did not impair infectivity. It was considered likely that only converted forms of the virus initiate infection when transmitted by leaf-feeding beetles.

Recent studies of the TCV have utilized agarose gel electrophoresis to follow changes in the domains of the capsid protein induced by site-directed mutagenesis[42] and also to monitor encapsidation in these virions.[43]

e. Nucleoprotein Electrophoresis

Hogue and Asselin[32,33] have studied alkali-induced disassembly intermediates of TMV known as partially stripped virus. These nucleoproteins were studied in polyacrylamide/agarose gels. It was found that even distantly related strains of TMV (e.q., U1 vs. U6) displayed a common pattern of polar and sequential disassembly. The technique holds promise for guiding experiments directed at elucidating RNA-protein interactions, disassembly kinetics, and the mechanism of *in vivo* disassembly.[33] Theriault and Asselin[72] have also used agarose gel electrophoresis to follow the *in vitro* disassembly of native and reconstituted hybrid particles of cowpea strain TMV. These authors demonstrated distinct differences in sensitivity of the L and S forms to alkaline and urea treatments, which may provide useful information for studying the strong and labile protein-RNA interactions.

Nucleoprotein electrophoresis of prunus necrotic ringspot virus (PNRSV) in agarose gels[40] enabled the resolution of serologically indistinguishable isolates. Electrophoresis of PNRSV nucleoproteins by agarose gel electrophoresis and electrophoresis on cellulose acetate sheets[73] suggested that net surface charge is primarily responsible for differences in the relative mobility among isolates separated on cellulose acetate sheets, while differences in size and charge give rise to the band patterns seen in agarose gels.

The complicated relationship between the protein and RNA species of AMV, involving 4 major and at least 13 minor nucleoproteins, has been investigated by Bol and Lak-Kaashoek.[25]

III. CONCLUSION

Virion electrophoresis is a simple, rapid, and inexpensive technique useful for the purification, resolution, characterization, and analysis of viruses. Agarose gel electrophoresis in particular holds promise for the determination of the biophysical properties of virions.

ACKNOWLEDGMENTS

Norman R. M. Watts was supported by a Natural Sciences and Engineering Research Council (NSERC) Visiting Researcher Fellowship.

REFERENCES

1. Dodds, J. A., Worris, T. J., and Jordan, R. L., Plant viral double-stranded RNA, *Annu. Rev. Phytopathol.*, 22, 151, 1984.
2. Valverdre, R. A., Analysis of double-stranded RNA for plant virus diagnosis, *Plant Dis.*, 74, 255, 1990.
3. Yoshikawa, N. and Converse, R. H., Strawberry pallidosis disease: distinctive dsRNA species associated with latent infections in indicators and in diseased strawberry cultivars, *Phytopathology*, 80, 543, 1990.
4. Moreno, P., Guerri, J., and Munoz, N., Identification of spanish strains of citrus tristeza virus by analysis of double-stranded RNA, *Phytopathology*, 80, 477, 1990.
5. Smith, O. P., Hunst, P. L., Hewings, A. D., Stone, A. L., Tolin, S. A., and Damsteegt, V. D., Identification of dsRNAs associated with soybean dwarf virus-infected soybean, *Phytopathology*, 81, 131, 1991.
6. Zhang, L., Zitter, T. A., and Palukaitis, P., Maize white line mosaic virus double-stranded RNA, replicative structure, and in vitro translation product analysis, *Phytopathology*, 81, 1253, 1991.
7. Romaine, C. P. and Schlagnhaufer, B., Hybridization analysis of the single-stranded RNA bacilliform virus associated with la France disease of Agaricus bisporus, *Phytopathology*, 81, 1336, 1991.
8. Rezian, M. A., Krake, L. R., Cunying, Q., and Hazzalin, C. A., Detection of virus-associated dsRNA from leafroll infected grapevines, *J. Virol. Meth.*, 31, 325, 1991.
9. James, D., Isolation of a high molecular weight dsRNA associated with twisted leaf disease in cherry, *Can. J. Plant Pathol.*, 14, 281, 1992.
10. Pares, R. D., Gillings, M. R., and Gunn, L. V., Differentiation of biologically distinct cucumber mosaic virus isolates by PAGE of double-stranded RNA, *Intervirology*, 34, 23, 1992.
11. Daniels, J., Characterization of cucumber mozaic virus isolates from California, *Plant Dis.*, 76, 1245, 1992.
12. Bawden, F. C., Reversible changes of tobacco mosaic virus from leguminous plants, *J. Gen. Microbiol.*, 18, 751, 1958.
13. Bancroft, J. B., Purification and properties of bean pod mottle virus and associated centrifugal and electrophoretic components, *Virology*, 16, 419, 1962.
14. Gillaspie, A. G. and Bancroft, J. B., The rate of accumulation, specific infectivity, and electrophoretic characteristics of bean pod mottle virus in bean and soybean, *Phytophathology*, 55, 906, 1965.
15. Semancik, J. S., Studies on electrophoretic heterogeneity in isometric plant viruses, *Virology*, 30, 698, 1966.
16. Magdoff-Fairchild, B. S., Electrophoretic and buoyant density variants of southern bean mosaic virus, *Virology*, 31, 142, 1967.
17. Lister, R. M., Ghabrial, S. A., and Saksena, K. N., Evidence that particle size heterogeneity is the cause of centrifugal heterogeneity in tobacco streak virus, *Virology*, 49, 290, 1972.
18. Halk, E. L. and Fulton, R. W., Stabilization and particle morphology of prune dwarf virus, *Virology*, 91, 434, 1978.
19. Ball, E. M., A technique for comparing the electrophoretic mobility rates of viruses or virus strains, *Arch. Biochem. Biophys.*, 114, 547, 1966.
20. Niblett, C. L. and Semancik, J. B., Conversion of the electrophoretic forms of cowpea mosaic virus in vivo and in vitro, *Virology*, 38, 685, 1969.
21. Geelen, J. L. M. C., van Kammen, A., and Verduin, B. J. M., Structure of the capsid of cowpea mosaic virus. The chemical subunit: molecular weight and number of subunits per particle, *Virology*, 49, 205, 1972.
22. Geelen, J. L. M. C., Rezelman, G., and Van Kammen, A., The infectivity of the two electrophoretic forms of cowpea mosaic virus, *Virology*, 51, 279, 1973.
23. Nelson, M. R. and Tremaine, J. H., Physicochemical and serological properties of a virus from saguaro cactus, *Virology*, 654, 309, 1975.
24. Tsalius, A., Hjerten, S., and Jerstedt, S., Particle-sieve electrophoresis of virus in polyacrylamide gels exemplified by purification of turnip yellow mosaic virus, *Arch. Gesamte Virusforsch.*, 17, 512, 1965.
25. Bol, J. F. and Lak-Kaashoek, M., Composition of alfalfa mosaic virus nucleoproteins, *Virology*, 60, 476, 1974.
26. Lot, H. and Kaper, J. M., Physical and chemical differentiation of three strains of cucumber mosaic virus and peanut stunt virus, *Virology*, 74, 209, 1976.

27. Diaz-Ruiz, J. R., Kaper, J. M., Waterworth, H. E., and Devergne, J. C., Isolation and characterization of peanut stunt virus from alfalfa in Spain, *Phytopathology,* 69, 504, 1979.
28. Dodds, J. A., Cross-protection and interference between electro-phoretically distinct strains of cucumber mosaic virus in tomato, *Virology,* 118, 235, 1982.
29. Hanada, K., Electrophoretic analysis of particles of fourteen cucumovirus isolates, *Ann. Phytopathol. Soc. Jpn.,* 50, 361, 1984.
30. Koenig, R., Fribourg, C. E., and Jones, R. A. C., Symptomological, serological, and electrophoretic diversity of isolates of andean potato latent virus from different regions of the Andes, *Phytopathology,* 69, 748, 1979.
31. Serwer, P., Moreno, E. T., Hayes, S. J., Berger, P., Langham, M., and Toler, R. W., Rapid detection and characterization of plant viruses by agarose gel electrophoresis: size, surface charge, heterogeneity of panicum mosaic and related viruses, *Electrophoresis,* 5, 202, 1984.
32. Hogue, R. and Asselin, A., Study of tobacco mosaic virus in vitro disassembly by sucrose density gradient centrifugation and agarose gel electrophoresis, *Can. J. Bot.,* 62, 457, 1984.
33. Hogue, R. and Asselin, A., Polyacrylamide-agarose gel electrophoretic analysis of tobacco mosaic virus disassembly intermediates, *Can. J. Bot.,* 62, 2336, 1984.
34. Asselin, A. and Grenier, J., Detection of tobamoviruses from purified or crude extracts after agarose gel electrophoresis, *Can. J. Plant Pathol.,* 7, 223, 1985.
35. Koenig, R. and Gibbs, A., Serological relationships among tombusviruses, *J. Gen. Virol.,* 67, 75, 1986.
36. Hurtt, S. S., Detection and comparison of electrophorotypes of hibiscus chlorotic ringspot virus, *Phytopathology,* 77, 845, 1987.
37. Serwer, P., Use of gel electrophoresis to characterize multimolecular aggregates, *Methods Enzymol.,* 130, 116, 1986.
38. Derrick, K. S., Bralansky, R. H., Lee, R. F., Timmer, L. W., and Nguyen, T. K., Partial characterization of a virus associated with citrus ringspot, *Phytopathology,* 78, 1298, 1988.
39. Pappu, H. R. and Hiruki, C., Electrophoretic variability among dianthoviruses, *Phytopathology,* 79, 1253, 1989.
40. Ong, C.-A. and Wink, G. I., Evaluation of agarose gel electrophoresis for resolving nucleoprotein components of prunus necrotic ringspot virus, *Phytopathology,* 79, 613, 1989.
41. Hurtt, S. S., Tietz, D., Fawcett, J. S., and Chrambach, A., Agarose gel electrophoretic pH-mobility (titration) curves of isometric plant viruses, *Phytopathology,* 79, 661, 1989.
42. Heaton, L. A., Lee, T. C., Wei, N., and Morris, T. J., Point mutations in the turnip crinkle virus capsid protein affect the symptoms expressed by Nicotiana benthamiana, *Virology,* 183, 143, 1991.
43. Hacker, D. L., Petty, I. T. D., Wei, N., and Morris, T. J., Turnip crinkle virus genes required for RNA replication and virus movement, *Virology,* 186, 1, 1992.
44. Heaton, L. A., Use of agarose gel electrophoresis to monitor conformational changes of some small spherical plant viruses, *Phytopathology,* 82, 803, 1992.
45. Griess, G. A., Moreno, E. T., Hermann, R., and Serwer, P., The sieving of rod-shaped viruses during agarose gel electrophoresis. I. Comparison of the sieving of spheres, *Biopolymers,* 29, 1277, 1990.
46. Asselin, A., Sevigny, P., and Hogue, R., A note on the preparation of electrophoretically purified isolates of tobacco mosaic virus, *Phytoprotection,* 65, 84, 1984.
47. Righetti, P. G., Isoelectric focusing: theory, methodology and applications, in *Laboratory Techniques in Biochemistry and Molecular Biology,* Elsevier Biomedical Press, Amsterdam, 1983, 386.
48. Vrdoljak, M. and Saric, A., Electrophoretic investigation of some grapevine fanleaf virus isolates, *Acta Bot. Croat.,* 40, 51, 1981.
49. Rankovic, M. and Velickovic, D., Possibility of sharka virus strain differentiation by electrophoresis, *Zast. Bilja Plant Prot.,* 34, 47, 1983.
50. Lapchik, L. G. and Kuznetsova, L. L., Characteristics of electrophoretic forms and capsid protein of red clover mottle virus, *Mikrobiol. Zh. (Kiev),* 46, 53, 1984.
51. Musil, M. and Gallo, J., Serological and electrophoretic typing of red clover mottle virus isolates, *Acta Virol.,* 28, 515, 1984.
52. Hollings, M. and Stone, O. M., Serological and immunoelectrophoretic relationships among viruses in the tombusvirus group, *Ann. Appl. Biol.,* 80, 37, 1975.
53. Pappu, H. R., Tittle, F. L., and Hiruki, C., IEF of plant viruses with mini IEF cell, *Bio Radiat.,* 68, 6, 1988.

54. Plazibat M. and Stefanac, Z., Differences in particle stability and electrophoretic behavior among tomato aspermy virus isolates, *Acta-Hortic.* Wageningen, International Society for Horticultural Science, 234, 469, 1988.
55. Rosengren, A., Bjellqvist, B., and Gasparic, V., A simple method for choosing optimum pH-conditions for electrophoresis, in *Electrofocusing and Isotachophoresis,* Radola, B. J. and Graesslin, D., Eds., W. de Gruyter, Berlin, 1977, 608.
56. Krishnamoorthy, R., Bosisio, A. B., Labis, D., and Righetti, P. G., Titration curves of liganded haemoglobins by combined isoelectric focusing-electrophoresis, *FEBS Lett.,* 94, 319, 1978.
57. Righetti, P. G., Krishnamoorty, R., Gianzza, E., and Labis, D., Protein titration curves by combined isoelectric focusing-electrophoresis with haemoglobin mutants as models, *J. Chromatogr.,* 166, 455, 1978.
58. Righetti, P. G. and Gianazza, E., pH-Mobility curves of proteins by isoelectric focusing combined with electrophoresis at right angles, in *Electrophoresis '79,* Radola, B. J., Ed., W. de Gruyter, Berlin, 1979, 858.
59. Valentini, L., Gianzza, E., and Righetti, P. G., pK Determinations via pH-mobility curves obtained by isoelectric focusing-electrophoresis: theory and experimental verification, *J. Biochem. Biophys. Methods,* 3, 323, 1980.
60. Hiruki, C., Sweet clover necrotic mosaic virus, *Descriptions of Plant Viruses,* No. 321, Commonwealth Mycological Institute/Association of Applied Biologists, 1987.
61. Okuno, T., Hiruki, C., Rao, D. V., and Figueiredo, G. C., Genetic determinants distributed in two genomic RNAs of sweet clover necrotic mosaic, red clover mosaic, and clover primary leaf necrosis viruses, *J. Gen. Virol.,* 64, 1907, 1983.
62. Rice, R. H. and Horst, J., Isoelectric focusing of viruses in polyacrylamide gels, *Virology,* 49, 602, 1972.
63. Pappu, H. R., Hiruki, C., and Inuoye, N., A new type of sweet clover necrotic mosaic virus, *Phytopathology,* 78, 1343, 1988.
64. Makkouk, K. M., Katul, L., and Rizkallah, A., Electrophoretic separation: an alternative simple procedure for the purification of broad bean mottle mosaic virus, *F.A.B.I.S. Newsl. Faba Bean Inf. Serv.,* 19, 12, 1987.
65. Partiot, M. and Djiekpor, E. K., Cacao swollen shoot virus: development of a method for purifying the virus by electrophoresis on a sacharrose gradient, Proceedings, 9th International Cocoa Research Conference, Lagos, Nigeria: Cocoa Producers Alliance, 1985.
66. Chevalier, D., Engel, A., Wurtz, M., and Putz, C., The structure and characterization of a closterovirus, beet yellows virus, and a leuteovirus, beet mild yellowing virus, by scanning transmission electron microscopy, optical diffraction of electron images and acrylamide gel electrophoresis, *J. Gen. Virol.,* 64, 2289, 1983.
67. de Souza, V. B. V., Gergerich, R. C., Kim, K. S., and Langham, M. A. C., Properties and cytopathology of a tymovirus isolated from eggplant, *Phytopathology,* 80, 1092, 1990.
68. Agrawal, H. O., Identification of cowpea mosaic virus isolates, *Mededel,* Landbouwhogeschool Wageningen, Neth., 64–5, 3–53, 1964.
69. Smith, T. J., The isolation of the two electrophoretic forms of cowpea mosaic virus using fast protein liquid chromatography, *J. Virol. Methods,* 16, 263, 1987.
70. Kartaatmadja, S. and Sehgal, O. P., Decline of bean pod mottle virus specific infectivity in vivo correlates with degradation of encapsidated RNA-1, *Phytopathology,* 80, 1182, 1990.
71. Langham, M. A. C., Gergerich, R. C., and Scott, H. A., Conversion of comovirus electrophoretic forms by leaf-feeding beetles, *Phytopathology,* 80, 900, 1990.
72. Theriault, C. and Asselin, A., In vitro disassembly of native and reconstituted hybrid particles of cowpea strain tobacco mosaic virus, *Can. J. Plant Pathol.,* 8, 249, 1986.
73. Crosslin, J. M. and Mink, G. I., Biophysical differences among prunus necrotic ringspot ilarviruses, *Phytopathology,* 82, 200, 1992.

FURTHER SELECTED READINGS

Dillon, J. R., Benzanson, G. S., and Yeung, K.-H., Basic techniques, in *Recombinant DNA Methodology,* Dillon, J. R., Nasim, A., and Nestman, E. R., Eds., John Wiley & Sons, New York, 1985, chap. 1.

Maniatis, T. I., Fritsch, E. F., and Sambrook, J., *Molecular Cloning: a Laboratory Manual,* Cold Spring Harbor Laboratory, Cold Spring Harbor, NY, 1982.

Rickwood, D. and Hames, B. D., Eds., *Gel Electrophoresis of Nucleic Acids: a Practical Approach,* IRL Press, Oxford, 1982, 242.

Righetti, P. G., Immobilized pH gradients: theory and methodology, in *Laboratory Techniques in Biochemistry and Molecular Biology,* Elsevier Biomedical Press, Amsterdam, 1990.

Serwer, P., Agarose gel electrophoresis of bacteriophages and related particles, *J. Chromatogr.,* 418, 345, 1987.

Voytas, D., Resolution and recovery of large DNA fragments: agarose gel electrophoresis, in *Current Protocols in Molecular Biology,* Ausubel, F. M., Brent, R., Kingston, R. E., Moore, D. D., Siedman, J. G., Smith, J. A., and Struhl, K., Eds., John Wiley & Sons, New York, 1989, chap. 2.

Chapter 6

On the Mechanism of Viroid Strain Separation in Gel Electrophoresis

Norman R. M. Watts and Rudra P. Singh

TABLE OF CONTENTS

I. INTRODUCTION

A. THE NATURE OF VIROIDS

Viroids cause serious diseases in several economically important crops including potatoes, tomatoes, grapes, citrus, and oilpalm, among others. Therefore it is of considerable importance to be able to determine the presence of viroids during plant quarantine and introduction, germ plasm collection, tissue-culture multiplication, and crop certification. Typically this has been accomplished by time-consuming bioassays involving cross-protection studies in which indicator plants are inoculated with test material and subsequently challenged with a more virulent form of the viroid.[1–4] A more rapid means for both early detection of viroids and the further discrimination between mild and severe strains was therefore important. Viroids may now be detected by molecular hybridization and electrophoresis of isolated nucleic acids,[1] and strains can be resolved by either return-polyacrylamide gel electrophoresis (R-PAGE)[5] or temperature-gradient gel electrophoresis (TGGE) (reviewed in Riesner et al.).[6,7] The objective of this paper is to review the mechanisms of viroid strain separation by R-PAGE and TGGE in view of viroid and RNA secondary structure.

Viroid structure and biology have been extensively reviewed.[2,8–13] Viroids are low-molecular weight,[14,15] covalently closed, single-stranded, circular RNA molecules having a highly base-paired, rod-like secondary structure[16] consisting of a serial arrangement of double-helical sections and internal loops.[17]

Viroids are localized in the nucleolus[18] and differ from viruses in that they consist solely of RNA, have no translocation protein, and have no known mRNA activity. They are therefore completely dependent on host enzyme systems for their replication[3] which is thought to occur by a rolling circle mechanism.[19–21] For a recent modification of this model see Singh.[1] Self-cleavage of the linear transcripts resulting during replication may proceed to a greater[22–24] or lesser[25,26] degree and by RNase T1.[27,28] For reviews of viroid replication see Singh[3] and Tabler and Tsargis.[29]

Viroids have been recognized only recently[3,14,15] and may be of recent origin.[8,30] However, it now seems more likely that they may be escaped introns[31,32] that were introduced from the wild[30] by intensive cultivation methods, especially monoculture. Viroids are transmitted by plant propagules, some insect vectors, and mechanically by horticultural implements (reviewed in Singh).[3]

B. R-PAGE AND TGGE

R-PAGE is a bidirectional electrophoretic procedure used to isolate viroids by exploiting their unique circular nature.[33] The subsequent modification of this technique to discriminate between mild and severe

0-87371-877-1/95/$0.00+$.50

strains of potato spindle tuber viroid (PSTVd) on the basis of their native secondary structure[5] has provided a rapid and sensitive alternative to the bioassay. An overview of R-PAGE methodology may be found in Singh,[4] and for a detailed discussion of a possible mechanism of PSTVd strain separation see Singh et al.[34]

TGGE is a recently developed technique which, like R-PAGE, is capable of resolving viroid strains, but unlike R-PAGE, does so by exploiting their thermally inducible transitions and is particularly suitable for the analysis of coexisting secondary structures. For a general review of TGGE methodology see Riesner et al.,[7] and for a detailed description of a recent application, Steger et al.[28]

II. VIROID AND RNA STRUCTURE

A. VIROID PRIMARY AND SECONDARY STRUCTURE

The nucleotide sequence of several viroids and their naturally occurring isolates has been extensively reviewed in Singh.[3] There are four generally recognized groups: PSTVd, hop stunt viroid, CCVd, and avocado sunblotch viroid (ASBVd). The sequence homology within a group is high and considerably lower between groups. Isolated exchanges, insertions, and deletions account for most of the differences in sequence homology between strains. The number of nucleotides in viroids varies from 246 in coconut cadang-cadang viroid (CCCVd), the smallest, to 371 to 375 in citrus exocortis viroid (CEVd). In the case of CCCVd, analysis of several isolates reveals that they contain the entire sequence plus varying amounts of additional duplicated sequence.[35] Herold et al.[35] have described five new field isolates of PSTVd of different virulence in which the prototypical number of 359 nucleotides has not been conserved, but varies between 356 and 360 nucleotides. Most of the changes are located within the virulence-modulating and the variable regions.

The elucidation of viroid secondary structure, from both experimental results and theoretical considerations, has been reviewed.[3,9,10] The rod-like structure of viroids observed at low resolution by electron microscopy[37] was shown by chemical modification and dye binding,[36,39] oligonucleotide binding and susceptibility to enzymatic attack,[9,39] and theoretical calculations[9,40] to consist of an unbranched series of double-helical sections alternating with single-stranded regions and a hairpin loop on each end. The viroids known to date have 18 to 32 single-stranded and 19 to 33 double-stranded regions.[3] In the case of ASBVd a low percentage (0.2%) of the molecules exist in a bifurcated form under native conditions.[40] Branched structures have also been proposed for the carnation stunt associated viroids[41] and pear blister canker viroid (and possibly several others).[42] From hydrodynamic studies it is known that viroids are relatively stiff molecules with a persistence length of 30 nm, limiting bending to a quarter of a circle,[9,43] and in the case of PSTVd have an axial ratio of approximately 20.[16,44]

B. FUNCTIONAL DOMAINS OF VIROIDS

Five functional domains have been proposed in native viroids. These include the conserved central region, the flanking pathogenicity and variable domains, and left and right terminal loops,[45–47] which correspond in part to physical domains of the native viroid molecule (see Riesner).[11] The UCCR (Upper Conserved Central Region) is characteristically a region of 20 to 30 conserved nucleotides in the center of the molecule with an 18-base imperfect direct repeat bordered by inverted direct repeats and a run of 11 to 18 purines at the left end. As well, there is a C-A base pair at the 3′ end of the left portion of the inverted repeat and a U-G at the 5′ end of the right portion (see Singh).[3] These elements bear a resemblance to similar features in integratable RNA elements,[48] suggesting a derivatory relationship. A speculative integration mechanism based on these observations has been proposed.[49]

C. SOLUTION STRUCTURE OF RNA

Very little detailed knowledge is available on RNA structure.[50] Transfer RNA is still the only biologically active RNA molecule completely solved by crystallography.[51,52] Consequently, indirect experimental data from chemical and enzymatic probes,[53,54] nuclear magnetic resonance,[55] random and site-directed mutagenesis,[46] and theoretical considerations (such as phylogenetic analysis[50,56,57] and energy minimization)[28] are employed to determine the solution structure of RNA.

Structure prediction based solely on energy minimization considerations is complicated by the flexibility of RNA about seven intra- and internucleotide bonds.[50] Recourse is thus often taken to interactive modeling which, however, introduces a considerable subjective component. To date, no algorithm exists which can produce an objective three-dimensional structure for a reasonably large RNA (for reviews of the extensive modeling literature see Zuker[57] and Gautheret and Cedergren).[50]

The three-dimensional structure of RNA molecules is often complex, but can be broken down into a limited set of basic structural elements which include sections of double-stranded helix, hairpin loops, bulge loops, internal loops, helical junctions, and pseudoknots.[58] The latter two elements, however, are not found in native viroids.

In RNA, as in DNA, single-stranded regions stack in a right-handed helix. Stacking propensity is in the order C > A > C > U; thus, if a U occurs between two others the U will bulge out of stack.[59] The ribose conformations are 3′-endo in single-stranded stacked RNA compared to 2′-endo in DNA, except in tertiary interactions.[58]

Double-stranded A-form RNA (the predominant form in natural RNAs) has a deep and narrow major groove and a shallow minor grove, the ribose moieties remain 3′endo, the glycosidic bonds are anti, the helix remains right handed, and base pairs determine stacking.[58] Stacking is not, however, limited to Watson-Crick base pairs as in DNA.[60–63] The presence of noncanonical base pairs (common in viroids) does not seem to significantly distort the helix.[58] Double-stranded RNA helices are characterized by the following parameters: the number of base pairs per turn, the axial rise per base pair, and base pair propeller twist, axial displacement, and tilt and roll (where the tilt axis is perpendicular to the base pair axis and the roll axis is coincident with the base pair axis).[58] The contribution of the base pair tilt and roll components to axial bending in RNA helices can be found in Koo and Crothers.[64] The range of values for all these parameters is limited, since they are interdependent.[58,65] Hydrated double-stranded RNA has a diameter of 29 Å.[43]

In addition to the double-stranded helices discussed above, RNA frequently contains hairpin loops, bulge loops, and internal loops. Little is known about these. In general, RNA hairpin loops seem to involve stacking at the 5′ end of the loop and an abrupt turn at the 3′ end.[58,66] Nuclear magnetic resonance studies of tetraloops[67,68] suggest that hairpin loops are more complex than the tRNA anticodon loop.[50] Tetraloops have not been found in native viroids so far, but may exist in the trihelical form of PSTVd.[28]

Bulge loops are due to unpaired bases occurring on only one strand of a double-stranded helix. In DNA a single base may[69] or may not[70] swing out of the helix, depending on stacking energy parameters.[71] Bulge loops cause the helix to bend.[72–74] The magnitude of the bend depends on the nature of the base and may be due to the time the base spends stacked in, or bulged out of, the helix.[75] The bulge bend induced by an adenine in DNA involves 20° of tilt and 8° of roll for a total bend angle of 21° in the direction away from the bulged base. A bulge also reduces the helical advance to a degree corresponding to an unwinding of 25°.[75] The bending of the RNA helix axis is also dependent on the number and type of bases in the bulge loop,[72,76] as well as the phasing (helical distance between two successive bulge loops), and sequence context of the bulge.[74] The origin of the contextual contribution to kinking is unclear, but may be due to strand asymmetry of flanking purines or the possibility of breakage of base pairing on one side of the bulge.[74] The magnitude of kinking is less in RNA than in DNA of the same sequence and size.[72]

Interior loops contain two or more mismatched bases. There is very little structural information on these elements.[58,77] Nuclear magnetic resonance studies suggest that interior loops are not open and that extensive stacking is conserved. Any perturbations do not extend beyond the limits of the loop. The main differences between A-form RNA and the internal loops are thought to be dynamic ("breathing") and not structural.[77] Generally, in modeling, interior loops are considered to maintain double-helical integrity by optimizing base pairing and stacking unless suggested otherwise by data.[50] Asymmetrical interior loops have much less effect on helix bending than bulge loops.[74]

Tertiary interactions such as base triples[58] and pseudoknots[78] are not found in native viroids, though possibly in hairpin I of partially denatured ones.[79]

III. METHODS OF VIROID STRAIN SEPARATION: R-PAGE AND TGGE

A. ELECTROPHORETIC THEORY

Separation of nucleic acids in both agarose and polyacrylamide media has become increasingly important. The mobility of nucleic acids during PAGE is determined largely by the porosity of the gel, the potential gradient, the nature and ionic strength of the buffer, the presence or absence of divalent cations and denaturants, the temperature, the size of the molecules, and in the case of single-stranded nucleic acids such as viroids, their conformation.[34] The differential mobility of viroid strains is probably due to differences in their physical interaction with the gel matrix as a result of their intrinsic conformations and flexibilities.

Polyacrylamide gels are thought to be a random meshwork of fibers. The pore "size" is determined by the concentrations of acrylamide and bisacrylamide used, as well as the polymerization conditions, and has a skewed size distribution. In 5% polyacrylamide gels such as used in R-PAGE and TGGE, pore size may be in the range of 3 nm (see Figure 3 in Chrambach and Rodbard).[80] The gel fibers themselves are thought to have a hydrated diameter of about 1 nm.[80]

Theoretical work on the motion of nucleic acids through gels has been reviewed by Zimm and Levine.[81] de Gennes[82] assumed that a DNA chain moves in a snake-like fashion through a tightly restrictive "tube" defined by the dense mesh of the gel matrix and the limits of its own flexibility. This concept in a more relaxed form, in which the chain could fold back on itself within the "tube", was further developed by Doi and Edwards.[83] Later theoretical advances based on reptation theory[84–86] have been concerned mainly with the folding and draping motions of nucleic acid chains. These predictions have largely been born out experimentally when individual fluorescently labeled DNA molecules were observed during agarose gel electrophoresis by light microscopy.[87,88] However, these models usually involved chains longer and more flexible than viroids and, in particular, were mostly applicable to agarose gel electrophoresis where the pore size is considerably larger than in polyacrylamide,[81] suggesting that other factors are involved in strain resolution.

B. METHODOLOGY OF R-PAGE

A method of two-dimensional gel electrophoresis for the separation of viroids from contaminating host nucleic acids was developed by Schumacher et al.[89] In this procedure viroids are first separated from high-molecular weight contaminants during a nondenaturing run and subsequently separated from comigrating linear nucleic acids during a denaturing run in the reverse direction.[33] During this "return" separation under denaturing conditions (changes in ionic strength and temperature of the buffer) the open circular viroids migrate with drastically reduced mobility.[89] Modification of the technique has enabled the separation of viroid strains.[5] Detailed descriptions of this methodology and applications can be found in Singh[4] and a proposed mechanism for viroid strain separation in Singh et al.[34] Typically, plant tissues are homogenized with an extraction buffer (containing ammonium hydroxide, EDTA, Tris, LiCl, and bentonite), phenol extracted, and ethanol precipitated. Dried samples are dissolved in an 89-m*M* Tris/borate pH 8.3 sample buffer and resolved in 5% acrylamide gels, first under nondenaturing conditions (89 m*M* Tris/borate, pH 8.3, 25°C), and then, with reversed polarity, under denaturing conditions (10 m*M* Tris/borate, 70°C). Gels are visualized with silver.

C. MECHANISM OF STRAIN SEPARATION IN R-PAGE

Experiments with gel-purified viroid preparations[34] showed that strains exhibited differential mobilities in unidirectional electrophoresis under nondenaturing conditions (Figure 1), but that this was abolished under denaturing conditions (Figure 2). Under the conditions of the unidirectional denaturing gels and the return run of R-PAGE, circular viroid molecules are resolved from linear forms as expected.[90] However, essentially no differences were observed between circular forms, or linear forms, of different strains. Any difference was attributed to residual base pairing due to incomplete denaturation.[34] These experiments showed that viroid strain separation actually occurred during the initial nondenaturing run in R-PAGE. Thus, these authors proposed, the differential mobilities under nondenaturing conditions were due to conformational differences between the viroid strains and not due to the minor (1- to 3-nucleotide) differences in their chain lengths.

D. INFLUENCE OF NUCLEIC ACID BENDING ON ELECTROPHORETIC MOBILITY

A marked reduction in mobility of nucleic acids during electrophoresis originally reported for intrinsically bent DNA[91] and RNA[72,74] was found to be a nonlinear function of the curvature[64] induced by bulge loops. The magnitude of the bulge effect on mobility is influenced by the number and nature of the bases in the bulge loop, and its sequence context[92] and phasing.[74] The bulge effect was recently integrated into the reptation model.[93]

The image of a viroid that emerges from the discussions above is one of a multiply kinked rod-like structure with a diameter on the order of 3 nm (perhaps somewhat greater at interior loops) and approximately 60 nm long in which the degree of kinking varies with the base composition of the bulges and in which the inter-kink sections project at varying rotational angles as a function of the helical distances between the bulges. Moreover, one might expect various degrees of flexibility at various points along the rod due to interior loops, between different strains, since the persistence length

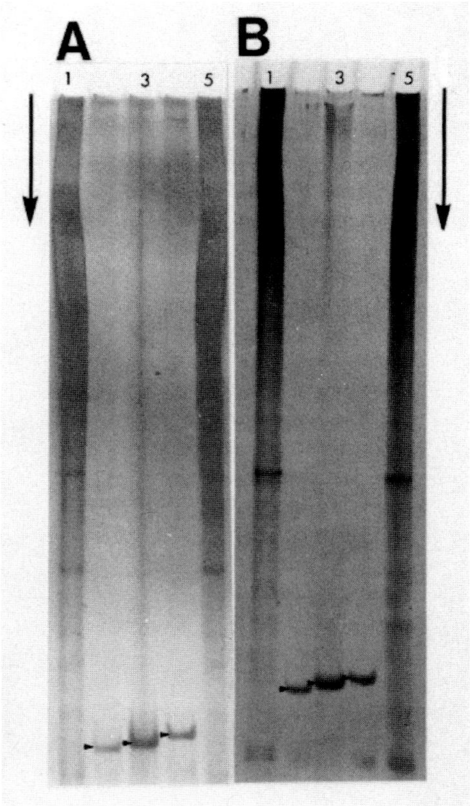

Figure 1 Differential migration of PSTVd strains during nondenaturing polyacrylamide gel electrophoresis. Lanes 1 and 5 contain nucleic acids from healthy tomato plants; lane 2, gel-purified PSTVd-FS (severe); lane 3, gel-purified PSTVd-FSL (lethal); and lane 4, gel-purified PSTVd-FM (mild). Migration is from top to bottom. Arrowheads indicate viroid bands. Under the nondenaturing conditions, circular and linear molecules co-migrate. Strains of different virulence exhibit differential mobility. (From Singh et al., *Can. J. Plant Pathol.,* 13, 202, 1991. With permission.)

of 30 nm is only an average value.[9,43] Envisioning such a viroid molecule tightly enmeshed within a gel matrix with approximately 3-nm pores suggests that while the reptation model applies in polyacrylamide to DNA molecules longer than about 50 base pairs,[81] and therefore presumably to viroids such as PSTVd which has about 360 base pairs,[36] the draping and coiling motions attributed to long chains by the reptation theory do not account for differential strain mobility. Recent advances in the theoretical model incorporating the elastic energy of the chain[93] (in this case the unique combinations of curvature, leverage, and flexibility of the viroid molecule) and the flexibility[94] and randomness of the gel (reviewed by Zimm and Levine)[81] may be required.

E. METHODOLOGY OF TGGE

In TGGE a linear temperature gradient (e.g., 20 to 60°C) is established across an acrylamide slab gel perpendicular to the electric field by means of a thermostated platten. Gels are typically 5% acrylamide and buffered with 17.8 m*M* Tris/borate. Nucleic acid samples are applied in a broad slot at one end of the gel and parallel to the temperature gradient. As a consequence, any given molecule will migrate at a constant temperature and characteristic mobility during the electrophoretic run. Staining the gels after the run reveals the thermal transition curve. Riesner et al.[7] provide an overview of the theory, instrumentation, and methodology of TGGE while a more detailed integration of the qualitative and quantitative theories of PSTVd conformational transitions and experimental results can be found in Steger et al.[28]

F. MECHANISM OF STRAIN SEPARATION IN TGGE

Viroids undergo distinct structural transitions during thermal denaturation. Typically there are one sharp and one or two broader thermal transitions.[44,95] The midpoint temperature (Tm) of the main transition is at approximately 50°C, about 20°C lower than the Tm of double-stranded DNA, and about 30°C lower than the Tm of double-stranded RNA of similar G-C content. The broader transitions occur at temperatures 10 to 20°C higher. The main transition is highly cooperative and is due to the ready ability of the circular viroid molecule to form alternative helices by pairing of distal palindromic sequence

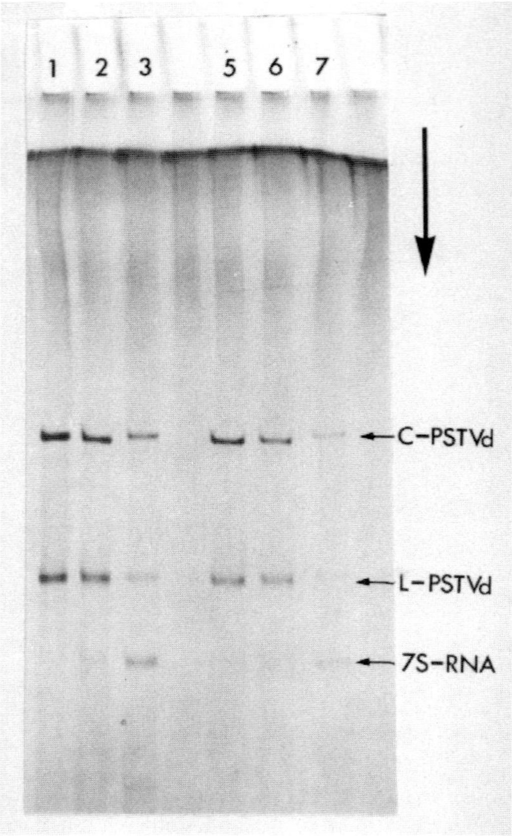

Figure 2 Uniform migration of PSTVd strains during denaturing polyacrylamide gel electrophoresis. Lanes 1 and 5, PSTVd-FM; lanes 2 and 6, PSTVd-FSL; lanes 3 and 7, PSTVd-FS. Under denaturing conditions, linear and circular forms no longer co-migrate and the migration rates of the strains are essentially identical. (From Singh, et al., *Can. J. Plant Pathol.* 13, 202, 1991. With permission.)

elements.[48] The main transition begins in the left half of the native secondary structure and results in two (CEVd, CSVd) or three (PSTVd) secondary helices. In PSTVd hairpins I, II, and III are formed by nucleotides (79 to 110), (227 to 328), and (127 to 168), respectively (reviewed in Singh).[3] These helices can be observed in electron micrographs.[44,96] Viroids differ significantly from random double-stranded RNA in this cooperativity and sharp transition. It is thought to represent an optimal compromise of stability, allowing replication and affording self-protection.[44] For reviews see Riesner and Gross[9] and Singh.[3]

The transition from the native structure is influenced by the presence of denaturants and the ionic strength and temperature of the buffer. Low ionic strength and low temperature favor extended native structure.[28] Conversely, high ionic strength provides counter ions to backbone charges and therefore, together with elevated temperatures, facilitates the formation of branched and eventually open single-stranded structures. During TGGE the cooperative transition of a molecule such as PSTVd from a rod-like structure to a partially denatured state involving three alternative hairpins, and the separate thermal transitions in turn of these helices as discussed above, leads to different frictional coefficients at the various temperatures. Molecules therefore migrate at a rate determined by their degree of thermal denaturation, with the branched structures and large loops migrating more slowly. The mobility of the spectrum of structures across the temperature gradient provides the transition curve(s). The transition curves of viroid isolates should therefore vary with the changes in stability of their various helices.[7,28] This has indeed been observed. Studies with PSTVd of different virulence showed that TGGE can distinguish between some, but not all strains[7] (Figure 3).

IV. CONCLUSION

At present, the large degree of conformational freedom of RNA molecules and the considerable influence of subjective input into the interactive modeling process in the absence of specific long-range constraints make it unlikely that these programs will reveal a nucleic acid conformation with a true energy minimum.[50]

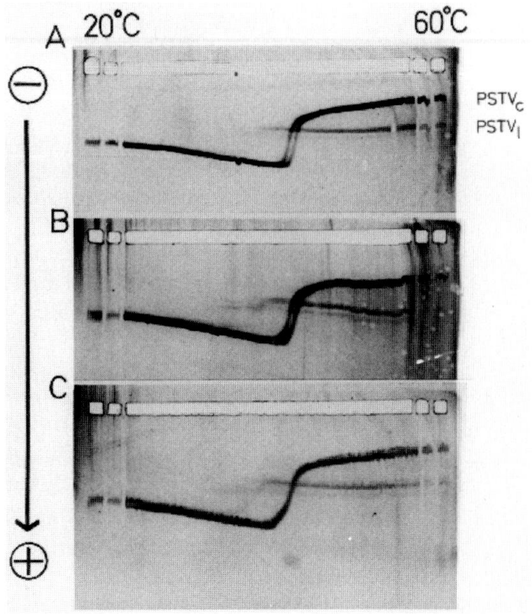

Figure 3 Resolution of PSTVd strains by TGGE. Panel A: PSTVd lethal (KF440-2) and PSTVd intermediate (DI); Panel B: PSTVd lethal (RG-1) and PSTVd intermediate (DI); Panel C: PSTVd lethal (KF440-2) and PSTVd lethal (RG-1). Electrophoresis is from top to bottom with the temperature gradient as indicated. Note that the lethal and intermediate forms exhibit different thermal denaturation curves (Panels A and B) and can thereby be resolved from one another while the two lethal forms cannot. (From Zimmat et al., in *Proceedings of the 6th Conversation in Biomolecular Stereodynamics,* Vol. 3, Sarma, R. and Sarma, M., Eds., Adenine Press, Schenectady, NY, 1990, 339. With permission.)

Attempts to deal with suboptimal structures with very similar energy minima have been described. However, PSTVd (and probably viroids in general) is unusual in that minor changes in energy parameters are unlikely to significantly perturb the predicted folding.[57] This should facilitate obtaining reliable three-dimensional structure predictions and comparisons. Viroid structure can be further probed by chemical and enzymatic means[53,54] and site-directed mutagenesis.[46] More subtle probes of RNA structure involving incorporation of modified bases, ribose moieties, and phosphate groups have also been developed.[98] R-PAGE, and particularly TGGE, are well suited for analyzing such modified viroid molecules. This in turn may facilitate understanding the structure-based pathogenicity of viroids.

With regard to screening applications of R-PAGE and TGGE, it should be borne in mind that all natural viroid infections may contain more than one sequence variant.[99]

REFERENCES

1. Singh, R. P., Diagnosis, characteristics and management of viroid diseases, in *Perspectives in Plant Virology 1985,* Gupta, B. M., Singh, B. P., Verma, H. N., and Srivastava, K. M., Eds., Print House, Lucknow, India, 1986, 1.
2. Singh, R. P., Techniques in the study of viroid diseases of tropical and subtropical plants, *Rev. Trop. Plant Pathol.,* 6, 81, 1989.
3. Singh, R. P., Plant viroids: a biochemical novelty, in *Plant Viruses.,* Vol. I: *Structure and Replication,* Mandahar, C. L., Ed., CRC Press, Boca Raton, FL, 1989, 259.
4. Singh, R. P., Return-polyacrylamide gel electrophoresis for the detection of viroids, in *Viroids and Satelites: Molecular Parasites at the Frontier of Life,* Maramarosch, K., Ed., CRC Press, Boca Raton, FL, 1991, 89.
5. Singh, R. P. and Boucher, A., Electrophoretic separation of a severe from mild strain of potato spindle tuber viroid, *Phytopathology,* 77, 1588, 1987.
6. Riesner, D., Steger, G., Zimmat, R., Owens, R. A., Wagenhöfer, M., Hillen, W., Vollbach, S., and Henco, K., Temperature-gradient gel electrophoresis: analysis of conformational transitions, sequence variations, and protein nucleic acid interactions, *Electrophoresis,* 10, 377, 1989.
7. Riesner, D., Henco, K., and Steger, G., Temperature-gradient gel electrophoresis: a method for the analysis of conformational transitions and mutations in nucleic acids and proteins, in *Advances in Electrophoresis,* Vol. 4, Chrambach, A., Dunn, M. J., and Radola, B. J., Eds., VCH, Verlagsgesellschaft, Weinheim, 1991, 169.

8. Sänger, H. L., Minimal infectious agents: the viroids, in *The Microbe 1984: Part I Viruses,* Mahy, B. W. J. and Pattison, J. R., Eds., Cambridge University Press, Cambridge, 1984, 28.
9. Riesner, D. and Gross, H. J., Viroids, *Annu. Rev. Biochem.,* 54, 531, 1985.
10. Diener, T. O., Ed., *The Viroids,* Plenum Press, New York, 1987.
11. Riesner, D., Structure of viroids and their replication intermediates. Are thermodynamic domains also functional domains?, *Semin. Virol.,* 1, 83, 1990.
12. Riesner, D., Viroids: from thermodynamics to cellular structure and function, *Mol. Plant-Microbe Interact.,* 4, 122, 1991.
13. Symons, R. H., The intriguing viroids and virusoids: what is their information content and how did they evolve?, *Mol. Plant-Microbe Interact.,* 4, 111, 1991.
14. Diener, T. O., Potato spindle tuber "virus". IV. A replicating, low-molecular-weight RNA, *Virology,* 45, 411, 1971.
15. Singh, R. P. and Clark, M. C., Infectious low-molecular-weight ribonucleic acid from tomato, *Biochem. Biophys. Res. Commun.,* 44, 1077, 1971.
16. Sänger, H. L., Klotz, G., Riesner, D., Gross, H. J., and Kleinschmidt, A. K., Viroids are single-stranded covalently closed circular RNA molecules existing as highly base paired rod-like structures, *Proc. Natl. Acad. Sci. U.S.A.,* 73, 3852, 1976.
17. Langowski, J., Henco, K., Riesner, D., and Sänger, H. L., Common structural features of different viroids: serial arrangement of double helical sections and internal loops, *Nucleic Acids Res.,* 5, 1589, 1978.
18. Riesner, D., Physical-chemical properties. Structure Formation, in *The Viroids,* Diener, T. O., Ed., Plenum Press, New York, 1987, chap. 3.
19. Branch, A. D. and Robertson, H. D., A replication cycle for viroids and other small infectious RNAs, *Science,* 223, 450, 1984.
20. Ishikawa, M., Meshi, T., Ohno, T., Okada, Y., Sano, T., Ueda, I., and Shikata, E., A revised replication cycle for viroids: the role of longer than unit length RNA in viroid replication, *Mol. Gen. Genet.,* 196, 421, 1984.
21. Hutchins, C. J., Keese, P., Visvader, J. E., Rathjen, P. D., McInnes, J. I., and Symons, R. H., Comparison of multimeric plus and minus forms of viroids and virusoids, *Plant Mol. Biol.,* 4, 293, 1985.
22. Hutchins, C. J., Rathjen, P. D., Forster, A. C., and Symons, R. H., Self-cleavage of plus and minus RNA transcripts of avocado sunblotch viroid, *Nucleic Acids Res.,* 14, 3627, 1986.
23. Forster, A. C. and Symons, R. H., Self-cleavage of virusoid RNA is performed by the proposed 55-nucleotide activity site, *Cell,* 50, 9, 1987.
24. Symons, R. H., Self-cleavage of RNA in the replication of viroids and virusoids, *Semin. Virol.,* 1, 117, 1990.
25. Robertson, H. D., Rosen, D. L., and Branch, A. D., Cell-free synthesis and processing of an infectious dimeric transcript of potato spindle tuber viroid RNA, *Virology,* 142, 441, 1985.
26. Tsargis, M., Tabler, M., and Sänger, H. L., Oligomeric potato spindle tuber viroid (PSTV) RNA does not process autocatalytically under conditions where other RNAs do, *Virology,* 157, 227, 1987.
27. Tsagris, M., Tabler, M., and Sänger, H. L., Ribonuclease T1 generates circular RNA molecules from viroid-specific RNA transcripts by cleavage and intramolecular ligation, *Nucleic Acids Res.,* 19, 1605, 1991.
28. Steger, G., Baumstark, M., Mörchen, M., Tabler, M., Sänger, H. L., and Riesner, D., Structural requirements for viroid processing by RNase T1, *J. Mol. Biol.,* 227, 719, 1992.
29. Tabler, M. and Tsargis, M., Viroid replication mechanisms, in *Recognition and Response in Plant-Virus Interactions,* Vol. H41, Fraser, R. S. S., Ed., NATO ASI series, Springer-Verlag, Berlin, 1990, 185.
30. Diener, T. O., Viroids, *Adv. Virus Res.,* 28, 241, 1983.
31. Diener, T. O., Are viroids escaped introns?, *Proc. Natl. Acad. Sci. U.S.A.,* 78, 5014, 1981.
32. Dinter-Gottlieb, G., Viroids and virusoids are related to group I introns, *Proc. Natl. Acad. Sci. U.S.A.,* 83, 6250, 1986.
33. Schumacher, J., Meyer, N., Riesner, D., and Weidemann, H.-L., Diagnostic procedure for detection of viroids and viruses with circular RNAs by "Return"-gel electrophoresis, *J. Phytopathol.,* 115, 332, 1986.
34. Singh, R. P., Boucher, A., Haas, B., and Sänger, H. L., Differential migration during polyacrylamide gel electrophoresis suggests conformational differences among strains of potato spindle tuber viroid, *Can. J. Plant Pathol.,* 13, 202, 1991.

35. Haselhoff, J., Mohamed, N. A., and Symons, R. H., Viroid RNA of cadang-cadang disease of coconuts, *Nature (London)*, 299, 316, 1982.
36. Herold, T., Haas, B., Singh, R. P., Boucher, A., and Sänger, H. L., Sequence analysis of five new field isolates demonstrates that the chain length of potato spindle tuber viroid (PSTVd) is not strictly conserved but as variable as in other viroids, *Plant Mol. Biol.*, 19, 329, 1992.
37. Sogo, J. M., Koller, T. H., and Diener, T. O., Potato spindle tuber viroid X. Visualization and size determination by electron microscopy, *Virology*, 55, 70, 1973.
38. Singh, R. P. and Williams, R. E., Potato spindle tuber viroid: circular dichroism spectrum and physical chemical studies of its interaction with ethidium bromide, *Can. J. Biochem.*, 56, 934, 1978.
39. Gross, H. J. and Riesner, D., Viroids: a class of subviral pathogens, *Agnew. Chem. Int. Ed. Engl.*, 19, 231, 1980.
40. Steger, G., Hofmann, H., Förtsch, J., Gross, H. J., Randles, J. W., Sänger, H. L., and Riesner, D., Conformational transitions in viroids and virusoids: comparison of results from energy minimization algorithm and from experimental data, *J. Biomol. Struct. Dyn.*, 2, 543, 1984.
41. Hernandez, C., Daros, J. A., Elena, S. F., Moya, A., and Flores, R., The strands of both polarities of a small circular RNA from carnation self-cleave in vitro through alternative double- and single-hammerhead structures, *Nucleic Acids Res.*, 20, 6323, 1992.
42. Hernandez, C., Elena, S. F., Moya, A., and Flores, R., Pear blister canker viroid is a member of the apple scar skin subgroup (apscaviroids) and also has sequence homology with viroids from other subgroups, *J. Gen. Virol.*, 73, 2503, 1992.
43. Riesner, D., Kaper, J. M., and Randles, J. W., Stiffness of viroids in solution, *Nucleic Acids Res.*, 10, 5587, 1982.
44. Riesner, D., Henco, K., Rokohl, U., Klotz, G., Klienshmidt, A. K., Domdey, H., Jank, P., Gross, H. J., and Sänger, H. L., Structure and structure formation of viroids, *J. Mol. Biol.*, 133, 85, 1979.
45. Keese, P. and Symons, R. H., Domains in viroids: evidence of intermolecular RNA rearrangements and their contribution to viroid evolution, *Proc. Natl. Acad. Sci. U.S.A.*, 82, 4582, 1985.
46. Owens, R. A., Thompson, S. M., and Steger, G., Effects of random mutagenesis upon potato spindle tuber viroid replication and symptom expression, *Virology*, 185, 18, 1991.
47. Sano, T., Candresse, T., Hammond, R. W., Diener, T. O., and Owens, R. A., Identification of multiple structural domains regulating viroid pathogenicity, *Proc. Natl. Acad. Sci. U.S.A.*, 89, 10104, 1992.
48. Kiefer, M. C., Owens, R. A., and Diener, T. O., Structural similarities between viroids and transposable genetic elements, *Proc. Natl. Acad. Sci. U.S.A.*, 80, 6234, 1983.
49. Zimmern, D., Do viroids and RNA viruses derive from a system that exchange genetic information between eukaryotic cells, *Trends Biochem. Sci.*, 7, 205, 1982.
50. Gautheret, D. and Cedergren, R., Modelling the three-dimensional structure of RNA, *FASEB J.*, 7, 97, 1993.
51. Quigley, G. J., Seeman, N. C., Wang, A. H.-J., Suddath, F. L., and Rich, A., Yeast phenylalanine transfer RNA. Atomic coordinates and torsion angles, *Nucleic Acids Res.*, 2, 2329, 1975.
52. Jack, A., Ladner, J. E., and Klug, A., Crystallographic refinement of yeast phenylalanine transfer RNA at 2.5 angstroms resolution, *J. Mol. Biol.*, 108, 619, 1976.
53. Ehresmann, C., Baudin, F., Hougel, M., Rumby, P., Ebel, J.-P., and Ehresmann, B., Probing the structure of RNAs in solution, *Nucleic Acids Res.*, 22, 9109, 1987.
54. Latham, J. A. and Chech, T. R., Defining the inside and outside of a catalytic RNA molecule, *Science*, 245, 276, 1989.
55. Gabriele, V., Wimberly, B., and Tinoco, I., Jr., Conformation and dynamics of an RNA internal loop, *Biochemistry*, 28, 7760, 1989.
56. Gutell, R. R., Power, A., Hertz, T. Z., Putz, E. J., and Stormo, G. D., Identifying constraints on the higher-order structure of RNA: continued development and application of comparative sequence analysis methods, *Nucleic Acids Res.*, 20, 5785, 1992.
57. Zuker, M., On finding all suboptimal foldings of an RNA molecule, *Science*, 244, 48, 1989.
58. Tinoco, I., Jr., Davis, P. W., Hardin, C. C., Puglisi, J. D., Walker, C. T., and Wyatt, J., RNA structure from A to Z, *Cold Spring Harbor Symp. Quant. Biol.*, 52, 135, 1987.
59. Lee, C.-H. and Tinoco, I., Jr., Conformation studies of 13 trinucleoside diphosphates by 360 MHz NMR spectroscopy. A bulged base conformation. I. Base protons and H1′ protons, *Biophys. Chem.*, 11, 283, 1980.

60. Patel, D. J., Kozlowski, S. A., Marky, L. A., Rice, J. A., Broka, C., Dallas, J., Itakura, K., and Breslauer, K. J., Structure, dynamics, and energetics of deoxyguanosine thymidine wobble base pair formation in the self-complementary d(CGTGAATTCGCG) duplex in solution, *Biochemistry,* 21, 437, 1982.
61. Patel, D. J., Kozlowski, S. A., Ikuta, S., and Itakura, K., Deoxyguanosine-deoxyadenosine pairing in the d(CGAGAATTCGCG) duplex. Conformation and dynamics at and adjacent to the dG-dA mismatch site, *Biochemistry,* 23, 3207, 1984.
62. Hunter, W. N., Brown, T., Anand, N. N., and Kennard, O., Structure of an adenine-cytosine base pair in DNA and its implications for mismatch repair, *Nature,* 320, 552, 1986.
63. Arnold, F. H., Wolk, S., Cruz, P., and Tinoco, I., Jr., Structure, dynamics, and thermodynamics of mismatched DNA oligonucleotide duplexes $d(CCCAGGG)_2$ and $d(CCCTGGG)_2$, *Biochemistry,* 26, 4068, 1987.
64. Koo, H.-S. and Crothers, D. W., Calibration of DNA curvature and a unified description of sequence-directed bending, *Proc. Natl. Acad. Sci. U.S.A.,* 85, 1763, 1988.
65. Zhurkin, V. B., Lysov, V. P., and Ivanov, V. I., Different families of double-stranded conformations of DNA as revealed by computer calculations, *Biopolymers,* 17, 377, 1978.
66. Haasnoot, C. A. G., Hilbers, C. W., van der Marel, G. A., van Boom, J. H., Singh, U. C., Pattebiraman, H., and Kollman, P. A., On loop folding in nucleic acid hairpin-type structures, *J. Biomol. Struct. Dyn.,* 3, 843, 1986.
67. Cheong, C., Varini, G., and Tinoco, I., Jr., Solution structure of an unusually stable RNA hairpin, 5′GGAC(UUCG)GUCC, *Nature (London),* 346, 680, 1990.
68. Heus, H. A. and Pardi, A., Nuclear magnetic resonance studies of the hammerhead ribozyme domain. Secondary structure formation and magnesium ion dependence, *J. Mol. Biol.,* 217, 113, 1991.
69. Morden, K. M., Chu, Y. G., Martin, F. H., and Tinoco, I., Jr., The unpaired cytosine in the deoxyoligonucleotide duplex $dCA_3CA_3G\text{-}dCT_6G$ is outside of the helix, *Biochemistry,* 22, 5557, 1983.
70. Patel, D. J., Kozlowski, S. A., Marky, L. A., Rice, J. A., Broka, C., Itakura, K., and Breslauer, K. J., Extra adenosine stacks into the self-complementary d(CGCAGAATTCGCG) in solution, *Biochemistry,* 21, 445, 1982.
71. Turner, D. E., Sugimoto, N., and Frier, S. M., RNA structure prediction, *Annu. Rev. Biophys. Biophys. Chem.,* 17, 167, 1988.
72. Bhattacharyya, A., Murchie, A. I. H., and Lilley, D. M. J., RNA bulges and the helical periodicity of double-stranded RNA, *Nature,* 343, 484, 1990.
73. Puglisi, J. D., Tan, R., Calnan, B. J., Frankel, A. D., and Williamson, J. R., Conformation of the TAR RNA-arginine complex by NMR spectroscopy, *Science,* 257, 77, 1992.
74. Riordan, F. A., Bhattacharyya, A., McAteer, S., and Lilley, D. M. J., Kinking of RNA helices by bulged bases, and the structure of the human immunodeficiency virus transactivator response element, *J. Mol. Biol.,* 226, 305, 1992.
75. Rice, J. A. and Crothers, D. M., DNA bending by the bulge defect, *Biochemistry,* 28, 4512, 1989.
76. Tang, R. S. and Draper, D. E., Bulge loops used to measure the helical twist of RNA in solution, *Biochemistry,* 29, 5232, 1990.
77. Varani, G., Wimberly, B., and Tinoco, I., Conformation and dynamics of an RNA internal loop, *Biochemistry,* 28, 7760, 1989.
78. Studnicka, G. M., Rahn, G. M., Cummings, I. W., and Salzer, W. A., Computer method of predicting the secondary structure of single-stranded RNA, *Nucleic Acids Res.,* 5, 3365, 1978.
79. Branch, A. D., Benenfeld, B. J., and Robertson, H. D., Ultraviolet light-induced crosslinking reveals a unique region of local tertiary structure in potato spindle tuber viroid and HeLa 5S RNA, *Proc. Natl. Acad. Sci. U.S.A.,* 82, 6590, 1985.
80. Chrambach, A. and Rodbard, D., Polyacrylamide gel electrophoresis, *Science,* 172, 440, 1971.
81. Zimm, B. H. and Levine, S. D., Problems and prospects in the theory of gel electrophoresis of DNA, *Q. Rev. Biophys.,* 25, 171, 1992.
82. de Gennes, P.-G., Reptation of a polymer chain in the presence of fixed obstacles, *J. Chem. Phys.,* 55, 572, 1971.
83. Doi, M. and Edwards, S. F., Dynamics of concentrated polymer solutions, *J. Chem. Soc. Faraday Trans.,* 274, 1789, 1978.
84. Deutsch, J. M., Dynamics of pulsed-field electrophoresis, *Phys. Rev. Lett.,* 59, 1255, 1987.

85. Zimm, B. H., Size fluctuations can explain anamalous mobility in field-inversion electrophoresis of DNA, *Phys. Rev. Lett.*, 61, 2965, 1988.
86. Deutsch, J. M., Theoretical studies of DNA during gel electrophoresis, *Science,* 240, 922, 1988.
87. Smith, S. B., Aldridge, P. K., and Callis, J. B., Observation of individual DNA molecules undergoing gel electrophoresis. *Science,* 243, 203, 1989.
88. Schwartz, D. C. and Koval, M., Conformational dynamics of individual DNA molecules during gel electrophoresis, *Nature,* 338, 520, 1989.
89. Schumacher, J., Randles, J. W., and Riesner, D., A two-dimensional electrophoretic technique for the detection of circular viroids and virusoids, *Anal. Biochem.,* 135, 288, 1988.
90. Sänger, H. L., Ramm, K., Domdey, H., Gross, H. J., Henco, K., and Riesner, D., Conversion of circular viroid molecules to linear strands, *FEBS Lett.,* 99, 117, 1979.
91. Marini, J. C., Levene, S. D., Crothers, D. M., and Englund, P. T., Bent helical structure in kinetoplast DNA, *Proc. Natl. Acad. Sci. U.S.A.*, 79, 7664, 1982.
92. Bhattacharyya, A. and Lilley, D. M. J., The contrasting structures of mismatched DNA sequences containing looped-out bases (bulges) and multiple mismatches (bubbles), *Nucleic Acids Res.,* 17, 6821, 1989.
93. Levene, S. D. and Zimm, B. H., Understanding the anamalous electrophoresis of bent DNA molecules: a reptation model, *Science,* 245, 396, 1989.
94. Calladine, C. R., Collis, C. W., Drew, H. R., and Mott, M. R., A study of electrophoretic mobility of DNA in agarose and polyacrylamide gels, *J. Mol. Biol.,* 221, 981, 1991.
95. Henco, K., Sänger, H. L., and Riesner, D., Fine structure melting of viroids as studied by kinetic methods, *Nucleic Acids Res.,* 6, 3041, 1979.
96. Klotz, G. and Sänger, H. L., Electron microscopic evidence for viroid conformers, *Eur. J. Cell. Biol.,* 25, 5, 1981.
97. Zimmat, R., Gruner, R., Hecker, R., Steger, G., and Riesner, D., Analysis of mutations in viroid RNA by non-denaturing and temperature-gradient gel electrophoresis, in *Proceedings of the 6th Conversation in Biomolecular Stereodynamics,* Vol. 3, Sarma, R. and Sarma, M., Eds., Adenine Press, Schenectady, NY, 1990, 339.
98. Usman, N. and Cedergren, R., Exploiting the chemical synthesis of RNA, *Trends Biol. Sci.,* 17, 334, 1992.
99. Visvader, J. E. and Symons, R. H., Eleven new sequence variants of citrus exocortis viroid and the correlation of sequence with pathogenicity, *Nuceic Acids Res.,* 13, 2907, 1985.

Chapter 7

Electrophoretic Karyotyping: Method and Applications

Dallice Mills, Kevin McCluskey, Brian W. Russell, and Jacqueline Agnan

TABLE OF CONTENTS

I. INTRODUCTION

Electrophoretic karyotyping is a relatively new experimental technique whereby the chromosomes of lower eukaryotes are physically separated in an agarose gel matrix by pulsed field gel electrophoresis (PFGE), stained with ethidium bromide, and visualized when irradiated with ultraviolet light. The electrophoretic karyotype consists of a pattern of DNA bands that correspond to intact chromosomes which are generally resolved according to their size. Because the velocity of migration in the gel is a function of the size of the DNA molecule, and the distance migrated is inversely proportional to the size of the chromosome, the chromosomes are usually arrayed from top to bottom in the gel in decreasing order of size.[1,2] Hence, this technique provides an estimate of the number of chromosomes, except for those fortuitously of similar size, and an accurate measurement of their sizes. Moreover, the summation of the sizes of individual bands, with care taken to account for presumed doublet bands, will provide an estimate of the size of the genome.

0-87371-877-1/95/$0.00+$.50

The ability to obtain a clearly resolved electrophoretic karyotype for an organism is dependent upon several factors. Probably most important is the ability to make chromosome preparations that are amenable to electrophoresis. The number of relatively large chromosomes, and the number of chromosomes of similar size, also impact on the ease with which useful karyotypes are generated. The recent development of new instrumentation along with new methodology for preparing chromosome samples have greatly alleviated problems initially encountered in obtaining reproducible electrophoretic karyotypes.

Unlike traditional cytological procedures or ultrastructural reconstructions[3] which provided no opportunity to recover any of the chromosomes visualized with the microscope, electrophoretically separated chromosomes are amenable to a multitude of procedures for the analysis of DNA that has been resolved in a gel. The resolution of intact chromosomes makes possible (i) the rapid assignment of fragments and genes to specific linkage groups by Southern blot hybridization, (ii) the construction of chromosome-specific libraries for the analysis of chromosome organization, (iii) the rapid development of physical and genetic linkage maps, and, perhaps most promising, (iv) the molecular analysis of genomes of organisms that have either proven recalcitrant to genetic analysis or have no known sexual stage.

II. METHODS OF RESOLVING DNA BY PFGE

A. GENERAL PRINCIPLES

The separation of DNA molecules by conventional gel electrophoresis is based on their size, charge, and conformation.[4,5] DNA molecules less than approximately 50 kb in length are fractionated in agarose matrices by the sieving action of the pores.[6] The smaller molecules pass through the pores and travel in a linear fashion, whereas the larger unresolved DNA molecules are not sieved and, consequently, their velocity is not proportional to their length. PFGE separates DNA molecules in agarose matrices by an electric field that alternates between two directions.[1] The time required for a DNA molecule to reorient in response to a change in the direction of the electric field is size dependent. In general, DNA molecules of increasing size require more time to reorient and spend less time migrating relative to smaller DNA molecules. Therefore, the resolution of DNA molecules of a targeted size is dependent upon the switching interval, and pulse times are selected so that the larger DNA molecules of a targeted size spend more time reorienting during the pulse than the smaller molecules, and the larger molecules move more slowly through the agarose matrix than smaller molecules. For most organisms, the chromosomes are sufficiently different in size that all of them cannot be resolved in a single gel, and two or more sets of switching parameters are required.

B. INSTRUMENTATION

Following the first reported electrophoretic karyotypes in 1984,[1,2] the instrumentation has evolved to alleviate problems associated with the resolution and migration of DNA molecules larger than 20 kb pairs in agarose matrices. The chronological order in which various systems emerged is discussed below. Schwartz and Cantor[1] used a PFGE apparatus consisting of two transverse alternating electric fields at approximately 90°, one of which was inhomogeneous, to obtain a partial electrophoretic karyotype of the budding yeast, *Saccharomyces cerevisiae.* This initial karyotype was distorted and consisted of approximately 11 DNA bands that ranged up to 2000 kb in size. However, because of variation in the field strength across the gel, DNA molecules in different lanes migrated at different velocities, and comparison of the sizes of DNA molecules from different lanes could not be accomplished with accuracy. Carle and Olson[2,7] developed a modification of PFGE, designated orthogonal-field-alternation gel electrophesis (OFAGE), which resolved 15 of the 16 yeast chromosomes in a single gel. This system utilized two alternating inhomogeneous electric fields, but because the angle between the electric fields at the top and bottom of the gel varied, the DNA molecules migrated at different rates at different positions in the gel, resulting in a curved migration outward at the bottom of the gel.

Carle et al.[8] later used field-inversion gel electrophoresis to fractionate large DNA molecules. This approach differs from PFGE and OFAGE in that a single electric field is applied and periodically reversed. By briefly reversing the current, the DNA molecules reorient and migrate backwards, but because the forward cycle is of longer duration, the molecules have a net forward migration through the gel and are fractionated. The fractionation of any size range can be maximized by selecting the appropriate pulse intervals for the forward and reverse cycles, and by a continuous ramp of the pulses from one interval to the other.[8] The system has been reported to be popular for separation of smaller molecules ranging up to approximately 800 kb.[9]

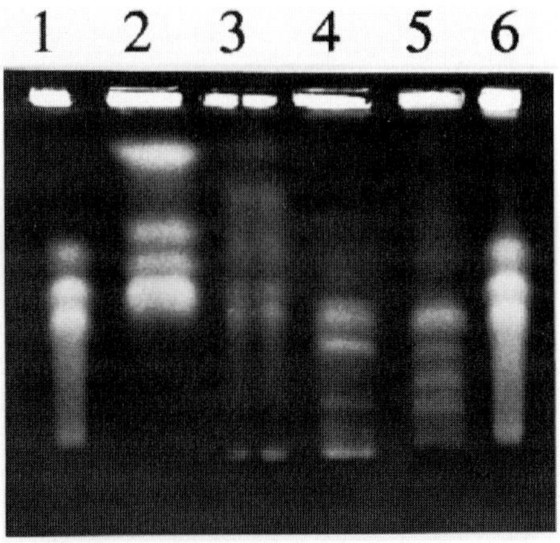

Figure 1 Fractionation of large fungal chromosomes by PACE PFGE. Lanes 1 and 6, *Schizosaccharomyces pombe;* lane 2, *Neurospora crassa* 74A; lane 3, *Magnaporthe grisea* cpa83; lane 4, *Nectria haematococca* 230-25-7; lane 5, *N. haematococca* T8. Samples were electrophoresced in a 0.7% agarose gel (low EEO agarose, Gibco BRL) buffered with 0.5 × TBE at 11°C initially for 75 h at 1.1 V/Cm with a 106° orientation angle and 90-min switch interval, followed by 125 h at 1.0 V/Cm with a 106° reorientation angle and 140-min switch interval. Additional parameters for the resolution of *N. haematococca* chromosomes are described in Miao et al.[12] (Photograph courtesy of M. Orbach and H. Van Etten.)

The transverse alternating field electrophoresis (TAFE) system has the gel oriented vertically with the electrodes oriented parallel to the gel faces.[10] The DNA molecules transverse in a zigzag manner through the gel and, because all lanes experience similar field strengths, the molecules move in straight lanes.

The contour-clamped homogeneous electric field (CHEF) system utilizes 24 electrodes arranged in a closed hexagonal contour with orientation angles of 120°, and separates large and small DNA molecules in straight lanes in pulsed fields.[11] This system has gained great popularity in recent years, and has been used to obtain karyotypes of a wide variety of fungi. Large numbers of samples may be analyzed in a single gel, allowing direct comparisons of the karyotypes of at least 40 individuals when two gels are stacked. A variation of this system, designated programmable autonomously controlled electrode (PACE), allows each of the 24 electrodes to be independently controlled. This flexibility enables the system to have a variable reorientation angle, as well as variable field strengths and pulse parameters. It is especially promising for resolving very large DNA molecules in a shorter period of time. Presented in Figure 1 is the resolution of very large chromosomes attainable by this technique.

For the investigator who is willing to construct a PFGE system at considerable savings, the electrophoresis device (ED) system has been described by Schwartz et al.[13] which provides flexibility for both field shape and reorientation angle.

C. PREPARATION OF CHROMOSOMES FOR PFGE

The key to obtaining high-resolution karyotypes resides in the identification of a successful method of preparing chromosomes for PFGE. Methods that yield excellent karyotypes of one fungus may be completely ineffectual with others. In preparing samples, it is also imperative that the large, fragile DNA molecules not be broken in the process.

1. Protoplasts

The initial method of preparation of *S. cerevisiae* chromosomes served as a model for similar studies with plant pathogens. This method involved embedding cells in agarose plugs and diffusing cell wall degrading enzymes and various reagents into the agarose plugs to remove the cell walls and disrupt cell membranes.[1,2] The advantages of this protocol were its ease and the protection from shearing it afforded the large DNA molecules. A major disadvantage of this system was the realization that enzymes that efficiently degrade yeast cell walls were frequently ineffectual with other fungi. Moreover, the ratio of nuclear mass to cell mass for some filamentous fungi may be greatly reduced relative to yeast, resulting in very faint banding patterns under ideal conditions. To circumvent the problem of having too much cellular mass, filamentous fungi are typically treated enzymatically to first produce protoplasts, which are then embedded in agarose and treated with detergents and proteolytic enzymes to disrupt the cellular membranes and degrade proteins. This approach gained general acceptance for nonfilamentous fungi as well, because the efficiency of the cell wall-degrading enzymes could be monitored by microscopic observation. Protocols that result in a high percentage of protoplasts offer a better opportunity to develop good karyotypes. The protoplasting step is carried out in the presence of an osmoticum to maintain integrity of the cells. The steps used to lyse the embedded protoplasts are carried out at elevated temperatures and in the presence of ethylenediamine tetraacetic acid (EDTA), which protects the DNA from nucleases.

As previously stated, many phytopathogenic fungi are recalcitrant to the enzymatic treatment, and procedures that were essentially 100% efficient in generating protoplasts from yeast cells may not produce any protoplasts from other fungi.[14] For many early studies, it was necessary to produce the enzymes in one's laboratory because they were either commercially unavailable on a large scale, or too expensive. Moreover, specific enzymes when used alone are frequently ineffective, and an enzyme cocktail may be required to produce protoplasts.

Many fungi of interest contain chitin and glucan as major cell wall components, which prompted a study to evaluate the effect of several commercially available polysaccharases on the release of protoplasts from fungi.[14] It was found that these enzyme activities were important for efficient protoplast formation in some fungi and, moreover, β-D-glucanase and chitinase activities were associated with commercial enzymes such as Novozym 234 (CalBiochem) and Cellulase CP (Sigma), which gave the highest yields of protoplasts.

When an effective method of efficiently producing protoplasts is identified, the karyotypes may sometimes be characterized by diffuse bands with evidence of chromosomal fragments. Furthermore, strains are frequently encountered that appear unresponsive to a method that works efficiently with other related strains. Therefore, to suggest a general procedure which might be appropriate for a variety of fungi may be foolhardy. However, most of the techniques which have been adopted are essentially modifications of existing procedures. Therefore, as a starting point in karyotype analysis, a procedure is outlined below which has been used to make protoplasts from a variety of smut fungi.[15,16] Care is taken to point out steps where modifications may result in more efficient protoplast formation.

Preparation of Protoplasts of Tilletia *and* Ustilago *Species*

1. Grow cultures in liquid medium until early to mid-log phase. Typically, 100 to 125 ml quantities are grown of both filamentous and sporidial cultures.
2. Harvest the culture by centrifugation at 2000 × g for 10 min.
3. Wash the cells by suspending the cell pellet (ideally 1.5 to 2 g wet wt) in 10 ml of TE (25 m*M* Tris-HCl pH 7.5, 25 m*M* EDTA) which also contains 10 m*M* dithiothreitol. Incubate at room temperature for 15 min.
4. Collect cells by centrifugation at 2000 × g.
5. Suspend the cell pellet in 10 ml of digestion buffer (50 m*M* NaH_2PO_4, 50 m*M* Na_2HPO_4, pH 5.8, 1 *M* sorbitol) containing 32 U/ml (about 1 mg/ml) Novozym 234 and 4000 U/ml β-glucuronidase. Incubate with constant shaking (100 rpm) at room temperature for up to 3 h. The concentration and type of enzymes which are effective for protoplast formation as well as the incubation times are factors that must be determined for each species. Visually inspect the samples at intervals using a hemacytometer and a light microscope to ascertain the extent of protoplast formation over the course of 3 h. Significantly longer incubation would suggest that other enzymes should be tested or the concentrations modified. Other osmotica (e.g., MgSO4, NaCl, KCl, manitol) may be effectively substituted for sorbitol.
6. For some filamentous fungi, it may be necessary to filter the digestion mixture through a 20μm nylon screen to remove undigested cells and debris. Otherwise, this step can be omitted.
7. Collect protoplasts from step 5 or from the filtrate (step 6) by low-speed centrifugation (about 1000 × g), taking care not to disrupt the protoplasts.
8. Suspend the pellet in one volume of TE containing 1*M* sorbitol as the osmoticum, mix with one volume of 2.5% molten (50°C) low gelling temperature agarose dissolved in osmotically stabilized buffer to avoid premature lysis of the protoplasts, and cast into plugs.
9. Incubate the plugs in 0.45 *M* EDTA, 1% SDS, and 1 mg/ml protease (type XIV, Sigma) at 50°C for 3 to 24 h.
10. Rinse with 0.5 *M* EDTA and store plugs in EDTA (0.125 to 0.5 *M*) at 4°C.

2. Nonprotoplast

It has been possible to circumvent problems of poor protoplast formation in some fungi by employing a technique which eliminates the need to use cell wall-degrading enzymes.[17] High-quality karyotypes of *Ustilago* and *Tilletia* species,[15–17] as well as *Leptosphaeria maculans*[18] and *Phymatotrichum omni-*

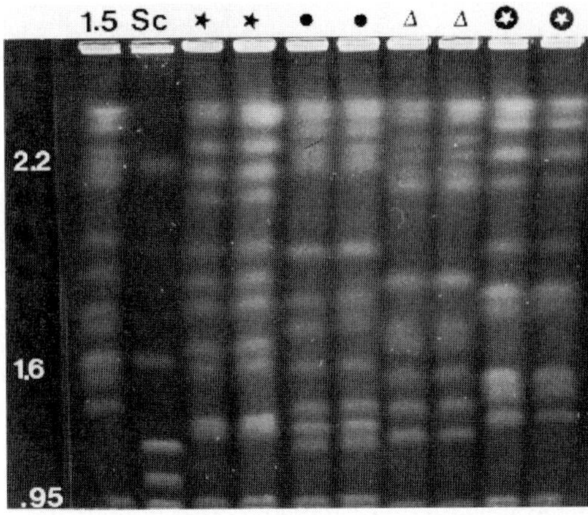

Figure 2 Electrophoretic karyotypes of a tetrad from *Leptosphaeria maculans* obtained by the nonprotoplast method.[17] Lanes 1 and 2 contain chromosomes used as molecular size standards. Lanes with like symbols above represent identical spore pairs of the eight-spored ascus. Consult Plummer and Howlett[18] for additional details. (Photograph courtesy of K. M. Plummer and B. J. Howlett.)

vorum[19] were obtained by embedding slightly ground mycelia or intact sporidia in agarose and treating them with EDTA, SDS, and protease type XIV (Sigma) at elevated temperatures (50 to 65°C) for periods up to 24 h. The karyotypes produced with this procedure are typically sharper, with less evidence of degraded DNA than karyotypes obtained from protoplasts. Moreover, karyotypes have been produced from strains that cannot be readily proplasted.[17–19] An additional advantage of this technique is that the karyotypes have been obtained from individual colonies removed from agar plates and grown in broth or spread onto agar plates to obtain sufficient numbers of cells.[20] This extends the analysis of karyotypes to the population level because upwards of 40 colonies may be readily examined using small-toothed combs and stacking two gels for each run, much as one would prepare miniprep plasmid DNA.

The protocol for the nonprotoplast method of preparing material for electrophoretic karyotyping is presented below. Again, it is important to establish the types and concentrations of enzymes and reagents, as well as the optimum time of incubation that produces the sharpest karyotypes. The karyotypes of strains of *L. maculans* which have been obtained with this procedure are presented in Figure 2.

Nonprotoplast Method for Generating Karyotypes of Smut Fungi

1. Grow and harvest cultures from liquid medium according to the method described for protoplasting cells. For single-celled fungi, the contents of a 100-ml flask of cells are repeatedly centrifuged in a 15-ml graduated centrifuge tube to obtain a pellet (proceed to step 3). This step can be modified for fungi which grow as single colonies, and are easily removed from an agar surface. Single colonies of *Ustilago* are removed from petri dishes with a sterile toothpick and spread on another plate. After overnight growth, the colony is scraped from the agar surface into a microfuge tube.
2. For filamentous fungi, suspend the cell pellet (about 1 g) in 10 ml of TE buffer (10 m*M* Tris-HCl, 1 m*M* EDTA, pH 8.0) which contains 1 *M* sorbitol, and gently macerate the mycelia in a glass tissue grinder at room temperature to break up large mats.
3. For single-celled fungi (*Ustilago*), wash the pellets with TE buffer. For filamentous fungi, resuspend each pellet in 1 ml of TE buffer containing 1 *M* sorbitol and repeat centrifugation at 5000 × g for 5 min.
4. For single-celled fungi (*Ustilago*), suspend the pellet in an equal volume of molten (50°C) 2.5% low gelling temperature agarose and cast plugs in a plug-casting apparatus. For filamentous fungi, suspend each pellet in an equal volume of molten (50°C) 1.5% low gelling temperature agarose which was dissolved in a solution containing 125 m*M* EDTA and 500 m*M* sorbitol, and cast plugs as described above.
5. Incubate five to ten plugs in 20 ml of 0.45 *M* EDTA, 1% SDS, and 1 mg/ml protease (type XIV) at 50°C for 24 h. Change the incubation buffer after 12 h. For some samples it has been necessary to incubate at temperatures up to 65°C.
6. Rinse the plugs with 5 ml of 0.5 *M* EDTA and store them in 0.5 *M* EDTA at 4°C.

3. Agarose Bead Encapsulation Method

When a comparison of karyotypes is being made among different strains or related species, it is essential that equal amounts of DNA are loaded in each well. Typically, trial PFGE is performed to visualize the karyotypes and ascertain the amount of each plug necessary to produce karyotypes of uniform intensity across all lanes. This is particularly important if the strains are suspected to be aneuploid and meaningful quantitation of the band intensities is desired. Although it is possible to carefully slice the agarose plugs and obtain a sliver of the precise size, the results are frequently disappointing. This problem can be essentially alleviated by preparing the cells by the agarose bead encapsulation method described by Overhauser and Radic[21] and modified by Bakalinsky.[22] The cells are harvested and suspended in 10 ml of SE (75 m*M* NaCl, 25 m*M* Na_2EDTA, pH 8.0), washed twice, and resuspended in 4 ml of SE. The cells are then mixed with 5 ml of 1% low gelling temperature agarose in a 45°C water bath prior to the addition of 20 ml temperature-equilibrated mineral oil. The resultant emulsion is vigorously mixed for 30 s before being poured into a beaker that contains 100 ml of ice-cold SE and a stir bar that is stirring at medium speed. Upon contact with ice-cold SE, the cells are encapsulated in tiny beads of solidified agarose. After several minutes, the contents are transferred to graduated polypropylene tubes and the beads are pelleted by centrifugation. The beads are then treated with cell wall-degrading enzymes and other reagents routinely used to obtain protoplasts, or they can be treated as described for the nonprotoplast method. After treatment, the beads are stored as a slurry in EDTA solution at 4°C, and these preparations have produced reproducible karyotypes of *S. cerevisiae* after storage for 2 years.[22] The beads are pipetted into the wells of the gel for PFGE, which provides a more accurate method of arriving at precisely the amount required to obtain a uniform karyotype for all lanes. Modifications to this technique involve the use of substituted reagents and enzymes, reduced incubation times for enzymatic digestion and centrifugations, and procedures for examining individual colonies from petri plates.

D. ELECTROPHORESIS CONDITIONS

The electrophoresis conditions are ultimately determined by the size of the chromosomes. However, without prior knowledge of their size, the initial conditions most conveniently adopted are those that resolve the chromosomes of *S. cerevisiae*. Under these conditions chromosomes ranging from approximately 2 megabase pairs (Mb) down to about 0.2 Mb are resolved in approximately 24 h. In the CHEF apparatus, the chromosomes of *U. hordei* in this size range are resolved in 1% agarose with 0.5 × TBE buffer[23] at 14°C, using a 70-s switch interval for the initial 15 h at 200 V, followed by a 120-s switch interval for the final 11 h. These conditions resolved all but the largest three to five chromosome bands detected in various strains of *U. hordei*. However, under identical conditions only about 8 to 12 of the approximately 20 chromosome bands of *T. caries* and *T. controversa* are resolved because of their larger size. For chromosomes larger than 2 Mb, extended periods of electrophoresis are required at low voltage, with longer switching periods, and perhaps a lower concentration (0.65 to 1%) of agarose. For example, the larger chromosomes of *Tilletia* spp., which range from 2 to 4.5 Mb, were resolved in 0.9% agarose gels in 4 d using ramped pulses from 900 to 480 s and beginning at 75 V with increases of 5 V at 24-h intervals.[16] Much longer periods of time are required to resolve the largest chromosomes found in other plant pathogens. For example, the largest chromosome of *Magnaporthe grisea*, which exceeds 10 Mb in length, is resolved using other parameters[24] (and Figure 1).

E. AGAROSE

Little information is available, other than from vendors, about the advantages of using different types of agarose for PFGE. Agarose is commercially available in ultra pure and reagent grade. When cost is a factor, it is prudent to compare the performance of various brands and grades of agarose. Some vendors have made available a special PFGE grade of agarose that has high tensile strength to accommodate low gel concentrations. This is typically very expensive, and low-percentage gels of conventional agarose cast on a support of a higher concentration agarose serve the same purpose. A gel showing the resolution of *Tilletia* chromosomes in two different grades of agarose (low or medium EEO) is presented in Figure 3. The results of this comparison indicated that the time required to resolve *Tilletia* chromosomes could be shortened by approximately 1 d using low EEO ($-Mr = 0.12$) agarose.

III. APPLICATIONS IN PLANT PATHOLOGY

A. FUNGAL KARYOTYPE ANALYSIS

1. Karyotype Variability

The number of fungi for which electrophoretic karyotypes have been obtained continues to grow at a rapid pace. Many recent karyotypes are of plant pathogens which have not previously been the subject

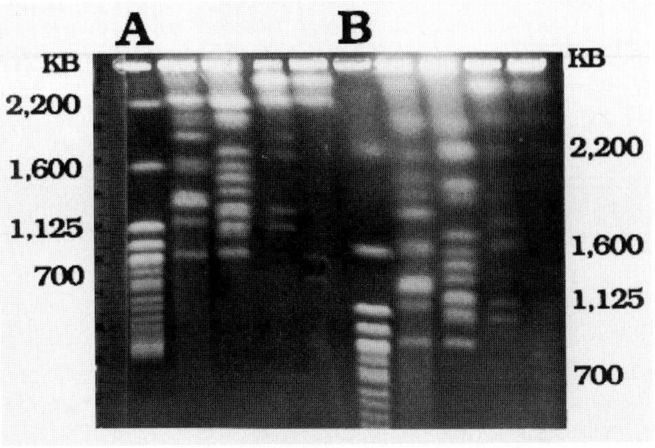

Figure 3 The effect of medium and low EEO grade agarose on the migration of fungal chromosomes. The gel consists of two kinds of agarose; Panel A. Medium EEO (−0.16), Sigma; Panel B. Low EEO (−0.12), International Biotechnologies, Inc. For each panel, lane 1, *Sacharomyces cerevisiae;* lane 2, *Tilletia caries;* lane 3, *Tilletia* spp.; lane 4, *N. haematococca* T57; lane 5, *N. haematococca* 8A-59. (Photograph provided by L. Cuiffetti and B. W. Russell.)

of molecular genetic analysis. These results are very exciting because it should be possible to use the resolved chromosomes to conduct a multitude of experiments that are possible with any isolated DNA molecule. Information about the number of chromosomes, their size range, and the instruments used to resolve the chromosomes of numerous fungi is presented in Table 1, and the references may serve as a source of other valuable information concerning the preparation of chromosomes.

One of the unexpected results of several of these studies has been the frequent observation that strains of many species appear to have variable chromosome numbers.[42] Several explanations may be advanced to explain these results. First, some of the bands in the gel could be fragmented chromosomes that have broken at specific fragile sites, resulting in fragments of a precise size. Secondly, specific chromosome bands may be absent from the karyotypes of some strains because of chromosome-length polymorphisms that result from deletions, insertions, and translocations. Following such events, a chromosome whose size was previously dissimilar from all other chromosomes may coincidentally

Table 1 **Chromosome number and size range of selected phytopathogenic fungi**

Organism	Number	Size range	Technique	Ref.
Absidia glauca	10	1.2–7.0	RAGE	25
Cephalosporium acremonium	8	1.7–4.0	TAFE	26
Cladosporium fulvum	11	1.9–5.4	CHEF	27
Cochliobolus heterostrophus	16	1.3–3.7	TAFE, CHEF	28
Colletotrichum gloeosporioides	6–15	0.27–6.0	CHEF	29
Curvularia lunata	12	1.4–4.0	CHEF	30
Fusarium oxysporum				
f.sp. *cubense*	8	1.3–4.0	CHEF	31
F. oxysporum				
f.sp. *conglutinans*	8	2.2–6.2	CHEF	32
f.sp. *raphani*	11	0.63–6.4	CHEF	32
Leptosphaeria maculans	6–14	0.44–2.2	TAFE	18, 33
Magnaporthe grisea	5–10	0.5–12.0	CHEF	24, 34
Nectria haematococca	10–15	0.4–4.0	CHEF	12, 35, 36
Phanerochaete chrysosporium	7	0.5–5.0	CHEF	37
Phoma tracheiphila	12	0.75–2	OFAGE	38
Phymatotrichum omnivorum	4	2.0–6.0	CHEF	19
Phytophthora megasperma	9–14	1.4–4	CHEF	39
Septoria tritici	17–18	0.33–3.5	TAFE	40
Tilletia caries	12–15	0.8–4.0	CHEF	16
T. controversa	12–16	0.8–4.0	CHEF	16
Ustilago hordei	16–21	0.17–3.15	CHEF	15
U. maydis	20	0.3–>2.0	OFAGE	41

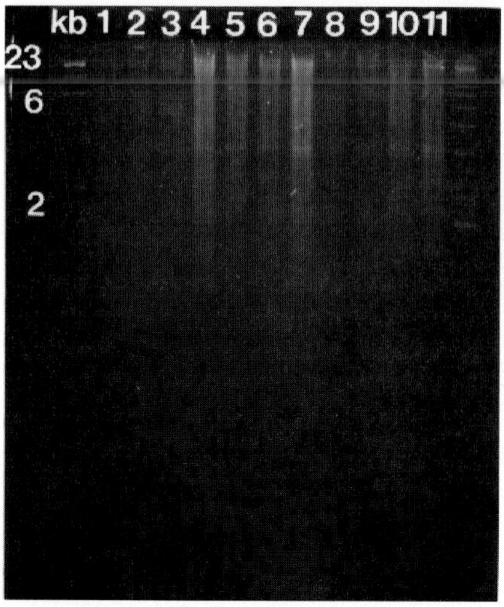

Figure 4 Fractionation of *Bgl*II restriction fragments of chromosomes of *Ustilago hordei* isolated by the freeze-thaw method. Lanes 1 to 11 contain DNA fragments from individual chromosome bands (smallest to largest) resolved by conventional gel electrophoresis. Extreme lanes contain molecular size markers.

assume the size of another chromosome, and a doublet band in some karyotypes may be comprised of nonhomologous chromosomes. Thirdly, many strains may be aneuploid and have two or more copies of some chromosomes that are resolved because of chromosome length polymorphisms. Finally, some species may carry dispensible B chromosomes which are present only in some strains. It is possible without any prior knowledge of the genetics of the fungus in question to determine which of these pertains for any karyotype.

2. Ascertaining the Number of Linkage Groups

The number of linkage groups is equal to the haploid (N) number of chromosomes, barring the occurrence of dispensible B chromosomes. However, when two or more presumptive haploid strains have varying chromosome numbers, it is necessary to determine the basis of the discrepancy. One frequently used approach is to react hybridization probes made of cloned DNA fragments with Southern-blotted chromosomes. The hybridization profile will reveal whether the cloned fragment is single-copy DNA, which will hybridize with only one chromosome, or repetitive DNA, which will hybridize with more than one chromosome if the repeated sequence is dispersed throughout the genome. Single-copy probes which are linked will hybridize with a single chromosome and define that linkage group. If the chromosome is present in multiple copies in aneuploid strains, and the chromosomes are polymorphic in length, additional bands will be detected. Hybridization probes made of conserved genes from other fungi often have sufficient homology with cognate genes that it is possible to assign them to a linkage group when probed onto Southern-blotted chromosomes. For example, using the rDNA genes from *Neurospora crassa* and the actin gene from *Aspergillus nidulans* as molecular probes, the respective genes were assigned to specific chromosomes of *T. caries,*[16] *T. controversa,*[16] and *U. hordei.*[20] It was of interest that the rDNA genes in different strains of *Tilletia* spp. were present on from one to three chromosomes, suggesting that the strains were either aneuploid or that these genes were translocated in some strains. This powerful technique is particularly useful for assigning genes to linkage groups of fungi that either do not have a sexual cycle or are difficult to use for classical genetic analysis.

Some bands appear to be unresolved doublets or higher orders of chromosomes. To resolve the number of chromosomes contained within a band, it has been useful to construct a hybridization probe of a telomere repeat, which is usually isolated from another fungus, and probe it onto Southern-blotted restriction fragments of total genomic DNA, or fragments from individual chromosomes. Because each chromosome has the telomere repeat at its termini, two restriction fragments from each chromosome will hybridize to the probe. When the blot of a gel of *U. hordei* chromosomal fragments as shown in Figure 4 is probed with the telomere repeat $[TTAGGG]_{18}$ from *Fusarium oxysporum,*[43] the CHEF-fractionated chromosome bands could be shown to contain one, two, or additional chromosomes (Figure 5). Procedures for isolating individual chromosomes for this analysis are described in the following sections.

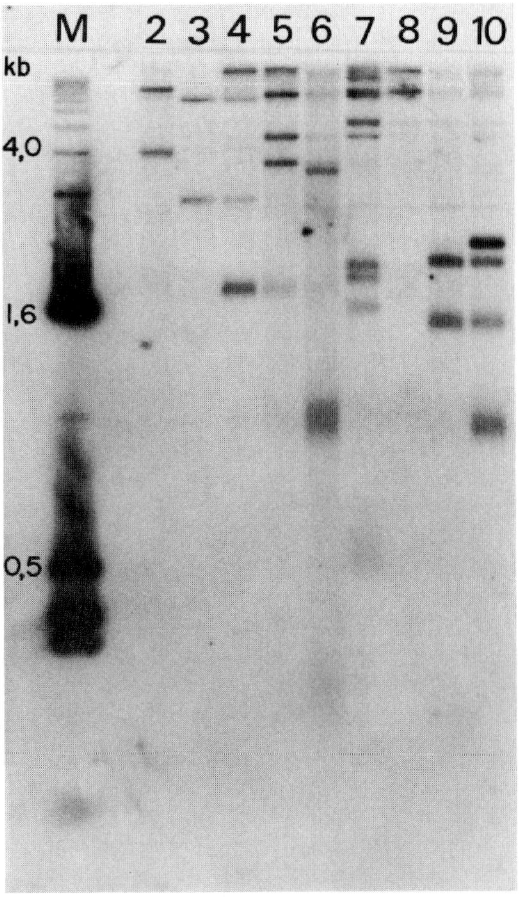

Figure 5 The utility of a telomere hybridization probe to ascertain the number of chromosomes present in an excised chromosome band. *U. hordei* chromosome bands resolved by CHEF were excised, digested with *Bam*HI, separated by electrophoresis Southern blotted and probed with the telomere repeat from *F. oxysporium.* Lanes 2 to 10 contain chromosome bands 2 to 10, respectively; lane M, molecular size markers. Some bands (e.g., 5 and 7) were presumed to be doublets. Note that bands 4 and 10 contain contaminating chromosomal DNA from chromosomes 3 and 9, respectively. Consult text for discussion.

B. CONSTRUCTING CHROMOSOME-SPECIFIC PROBES AND LIBRARIES

Procedures have been described in detail for making hybridization probes of randomly cloned genomic DNA fragments,[23,44] and assigning the fragment to a particular chromosome by Southern-blot hybridization analysis.[15,16] However, when it is necessary to assign several fragments to a linkage group, it is often very desirable and more efficient to make the DNA probes from specific chromosomes rather than using random fragments. Several techniques have been described for the recovery of chromosome-specific DNA from gels. A freeze-thaw method is inexpensive and consistently gives high yields of chromosomal DNA.[12,20] Other methods utilize β-agarase I or sodium iodide to degrade the agarose, which may be more desirable if large inserts are important as, for example, in the construction of a cosmid library.

1. Recovery of Chromosomal DNA by the Freeze-Thaw Method

The chromosomes of a single strain are resolved in several lanes of a low gelling temperature agarose gel to obtain a uniform karyotype across all lanes. The band of interest is identified by ethidium bromide staining and an agarose slab is excised, taking care to limit exposure of the DNA to ultraviolet light to 1 min or less. It is important that all skin be protected from exposure to the UV rays. The agarose slab containing a single band from as many as eight lanes is then suspended in a 5-ml solution of 25 m*M* Tris-HCl pH 7.5, 1 m*M* EDTA, and 1 *M* NaCl in a 15-ml polypropylene centrifuge tube. The agarose is melted by incubation at 70°C for 20 min, brought to room temperature, and then solidified at −20°C (1 to 24 h). The sample is then incubated at 37°C for 1 h, and the tube is frequently and gently inverted to uniformly distribute the melted agarose throughout the sample. This process is then repeated for a total of two freeze and thaw cycles which releases the DNA. After the final thaw, the agarose is pelleted by centrifugation for 30 min at 10,000 × g, and the supernatant containing the DNA is precipitated with an equal volume of isopropanol at −20°C for at least 1 h. The DNA is then recovered by centrifugation at 10,000 × g (30 min), 400 μl of TE buffer is added to each pellet, and the sample is extracted twice with an equal volume of phenol and once with chloroform. Finally, 50 μl of 3M sodium acetate (pH 5.2) and two volumes of 95% ethanol are added to the aqueous phase, and the DNA is precipitated at −20°C (1 to 20 h). The DNA is pelleted by centrifugation, washed with ethanol, dried, and dissolved in 20 μl of TE, or the buffer used for restriction endonuclease digestion. A typical yield of restriction fragments of individual chromsomes from *U. hordei* is presented in Figure 4. In some lanes, the chromosomes contain a heavy band which is suggestive that a repeated sequence is present on several chromosomes. The fragments from each chromosome may be cloned into pUC vectors for ease in manipulation, or into a vector of choice.

2. Recovery of Chromosomal DNA by Agarose Digestion

a. Enzymatic Digestion

Alternative methods of releasing DNA molecules from agarose include both enzymatic and chemical digestion of the agarose. β-Agarase I is an agarose-degrading enzyme that is encoded by a gene cloned from *Pseudomonas atlantica* and commercially available. The enzyme acts by cleaving carbohydrate bonds, thereby freeing the trapped DNA, and the residual carbohydrate molecules are incapable of gelling. Neither the remaining gel material nor the β-agarase interferes with restriction digestion, ligation, and transformation. This technique should be especially suited for recovery of large DNA fragments for making chromosome-specific cosmid libraries. Detailed instructions are provided by the manufacturers for use in recovering DNA.

b. Chemical Digestion

Agarose also may be solubilized in sodium iodide and the chromosomal DNA may be adsorbed to glass powder or to various silica gel suspensions that are available in kit form and marketed as Qiaex (Qiagen Inc., Chatsworth, CA), Geneclean™ (Bio 101, La Jolla, CA) and Elu-Quik (Schleicher & Schuell, Keene, NH). These kits offer ease in the recovery of DNA, and some kits are claimed to effectively adsorb DNA in the range of 50 bp to greater than 200 kb. DNA recovered by these procedures may be digested with enzymes that are sensitive to polysaccharide contamination.

3. Composition and Efficacy of Chromosome-Specific DNA Libraries

a. Composition

Theoretically, all members of a truly chromosome-specific library will hybridize with the targeted chromosome. However, it should be noted that a library made of a particular chromosome is, in reality, only enriched for DNA of that chromosome, and it often contains fragments of other chromosomes. This is especially true if the library is made of chromosome bands that are not sufficiently resolved, or made of a smaller chromosome. Note that fragments that hybridized with the telomeric probe in some libraries were from chromosomes in adjacent bands (Figure 5). The gel slabs of lower bands also may contain fragments of larger chromosomes which move at a similar velocity. However, it is a relatively easy task to sort out these anomolous fragments if they hybridize with a single chromosome. If the library is made of fragments resulting from complete digestion of the chromosome, and an insert hybridizes with two chromosome bands, it will be necessary to ensure that the insert is a single fragment rather than two noncontiguous fragments from heterologous chromosomes. However, because some of the library clones may contain repeated DNA, they also could hybridize with various other chromosomes. Finally, some library clones fail to hybridize with the chromosomes of related strains either because they are deleted for this region of the chromosome[40] or the insert was cloned from a dispensible chromosome which is present only in some strains.[12]

b. Efficacy

Chromosome-specific cosmid libraries should be extremely valuable for cloning genes by function or complementation, provided the fungus is amenable to transformation. If genetic data indicate that a gene for a mutant phenotype, e.g., auxotrophy, morphology, and drug resistance, is linked to another gene of interest, e.g., a dominant avirulence gene, the mutant gene could be cloned by complementation using a genomic library. Subsequently, a hybridization probe made of this cloned gene could be hybridized with Southern-blotted chromosomes to identify the appropriate linkage group. Finally, a cosmid library of that chromosome could be used to transform a race lacking the specific avirulence gene, and individual transformants could be analyzed for acquisition of the new phenotype. This approach dramatically reduces the number of transformants that have to be screened for the avirulence phenotype.

The development of yeast artificial chromosomes (YAC)-like vectors for plant pathogenic fungi would be most desirable. Typically, these vectors encode drug resistance which is expressed both by *Escherichia coli* and by the fungus, and contain a centromere, a fungal sequence for autonomous replication (ARS), a bacterial origin of replication (ori), telomeres, and a cloning site. These vectors accommodate large inserts (>50 kb) and greatly reduce the number of library clones needed to contain either the entire genome or a single chromosome. Moreover, they would greatly enhance studies of the organization of complex genetic loci, as well as the organization of specific chromosomes.

C. MAPPING GENETIC MARKERS BY PFGE AND SOUTHERN ANALYSIS

1. Use of Restriction Enzymes that Infrequently Cut DNA

When the linkage of two or more cloned markers has been established, the physical distance separating these markers on a chromosome can be determined without the need for genetic crosses. The large fragments resulting from digestion of total genomic DNA with an enzyme that cuts infrequently (e.g., *Not*I, *Sfi*I) can be resolved by PFGE and analyzed by Southern blot hybridization. Hybridization probes made of linked markers often hybridize to a single fragment which immediately reveals the maximum distance between the markers. When this technique is repeated using other infrequent-cutting restriction enzymes, or combinations of these enzymes, the physical distances between the markers can be more precisely established. Moreover, a rough restriction map can be generated of chromosomes that have multiple sites for cutting by these enzymes. Subsequently, a refined restriction map can be generated of the larger fragments following their excision from the gel. By these approaches, an accurate measurement of the relationship of the physical map to the genetic map is feasible. This procedure has been used to obtain a physical map of *Schizosaccharomyces pombe*.[45] Similarly, the relationship of the physical and genetic maps of chromosome III of *Saccharomyces cerevisiae* was established and refined by the development of the entire nucleotide sequence of that chromosome.[46] Its utility may be extremely valuable for studies of genome evolution and the organization of specific chromosomes.

2. Mapping with Novel Chromosome-Breaking Vectors

PFGE led to the discovery that two pathogenicity genes, *Pda*6 and *Mak*1 of *Nectria hematococca,* are linked to a 1.6-Mb dispensible chromosome[12] whose presence or absence can lead to karyotypic variability, as well as different virulence phenotypes. That the number of chromosomes among field isolates of other fungal pathogens is variable (Table 1) is suggestive that aneuploidy may also play a significant role in karyotype variability. If either phenomenon pertains for a pathogen, it is potentially possible to map genes that express mutant phenotypes to these chromosomes by deletion mutation without undue concern that the deletions will necessarily lead to concomitant lethality. For example, if an organism is disomic for a chromosome, and the homologs are heterozygous at a locus, removal of the wild-type allele by deletion would lead to the mutant phenotype. If the deletion extends into flanking DNA, the corresponding wild-type alleles at other loci on the nonmutated homolog would complement any genes that were deleted, regardless of the size of the deletion. Such events may be detected by PFGE and the deletion would not lead to lethality. The sequential deletion of either arm of a disomic chromosome, or a B chromosome, would therefore provide a method of mapping the position of markers along the chromosome.

The opportunity to develop an *in vivo* deletion mutagenesis system that could be used to identify extra chromosome copies, or B chromosomes, emerged somewhat unexpectedly from studies aimed at developing an efficient transformation system for *N. haematococca*.[43] Plasmid pFOLT4R4 replicates autonomously as a linear molecule in *F. oxysporum,* it has a repeated hexanucleotide telomere consensus sequence (TTAGGG) at its termini, and a selectable marker conferring resistance to hygromycin B.[43] The karyotypes of *N. haematococca* strains transformed with this vector were unchanged, and the vector appeared to replicate autonomously.[43] However, a derivative of pFOLT4R4, designated pLD, which lacks the telomere consensus sequence at one end, was shown to be integrated into the resident chromosomes of transformants.[36] Its integration into a 2.4-Mb chromosome resulted in a large deletion that presumably extends distal from the site of integration. The resulting karyotype revealed the loss of the 2.4-Mb band and the appearance of novel smaller bands which are mitotically stable. It is presumed that the 2.4-Mb chromosome is nonessential for growth because a hygromycin-sensitive revertant which had lost this entire chromosome was viable. This chromosome may be analogous to the B chromosomes of plants which are thought to be modified copies of other chromosomes, because

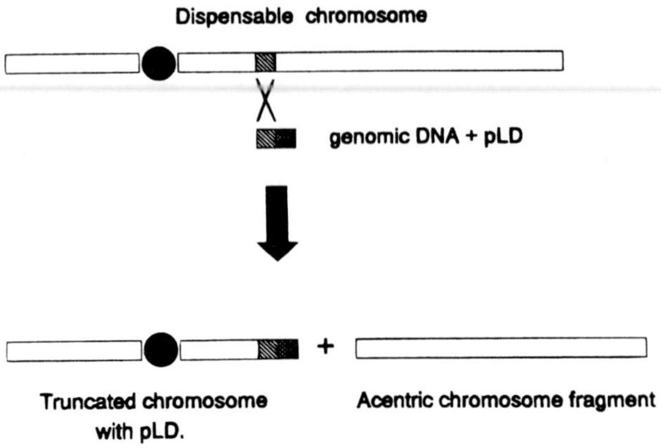

Figure 6 Proposed method by which chromosome-breaking vectors induce mutations. Consult text and Kistler and Miao[47] for further discussion of this strategy. (Illustration courtesy of C. Kistler.)

it shares significant homology with another larger (7-Mb) chromosome. Alternatively, it could be a homologous chromosome that has simply undergone rearrangement by deletion.

Chromosome-breaking vectors offer the possibility of constructing sequential overlapping deletions in B chromosomes, or along either arm of a chromosome that is known to be present in multiple copies. The strategy would be to target the integration of a chromosome-breaking vector (e.g., pLD) to such chromosomes by cloning fragments from the chromosome of interest into the vector to provide sites of homology for recombination (Figure 6). Hence, it should be possible to develop mutations and map them to specific chromosomes in fungal pathogens which do not have a sexual cycle, using other techniques that have emerged from PFGE technology.

D. ELECTROPHORETIC KARYOTYPING: AN ESSENTIAL TOOL FOR FUNGAL TAXONOMISTS

1. Utility of Fungal Electrophoretic Karyotypes

The ability to obtain electrophoretic karyotypes is regarded as a useful addendum to existing criteria for the proper classification of closely related fungi. For example, if two closely related organisms have greatly dissimilar karyotypes, it could be expected that chromosome pairing during meiosis would be impaired, which would present a barrier to interbreeding. Conversely, two organisms that have similar karyotypes and synonymy among linkage groups, as determined by Southern hybridization analysis, would constitute substantial and compelling evidence for conspecific status of the organisms of interest.

a. Karyotypes of T. controversa *and* T. caries *Suggest Conspecific Status*

T. controversa and *T. caries* are closely related filamentous basidiomycetes that incite dwarf bunt and common bunt diseases, respectively, of wheat. These pathogens readily mate and segregate genes in Mendelian ratios as expected for single unlinked genes among their viable offspring.[48] The morphological and physiological features which have been used to classify these pathogens as separate species can be shown on close inspection to be characteristics of either organism. Moreover, it has been argued that some traits, for example, teliospore morphology[49] and autofluorescence,[50] and the temperature at which the teliospores germinate[50] may be determined by one or a few genes.

Electrophoretic karyotyping provided a higher level of refinement than previously available to critically scrutinize the genomes of these pathogens and their hybrid progeny. In a recent study,[16] the karyotypes of four strains of each pathogen and five hybrid progeny were determined to be essentially identical, except for chromosome polymorphisms. The electrophoretic karyotypes of the progeny clearly showed that their chromosomes had gone through the reduction division step of meiosis. Moreover, preliminary Southern hybridization analysis with homologous and heterologous probes revealed respective linkage groups that were synonymous in all strains. These results, coupled with existing genetic and physiological data, provide strong evidence that these pathogens are not different species.

b. Karyotyping and Species Delimitation in the Genus Ustilago

Many species in the genus *Ustilago* parasitize cereal crop plants, and features used to distinguish these species include their host range, infection cycle, mating reactions, physiology, morphology, and development.[51] However, because some of these closely related species readily produce hybrid progeny,

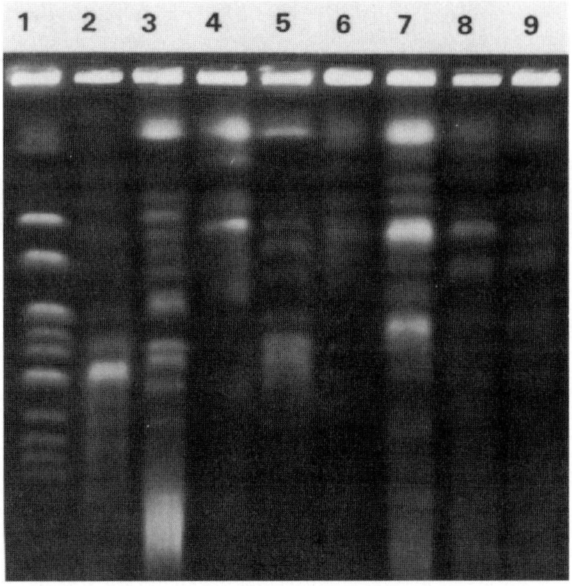

Figure 7 Electrophoretic karyotypes of smut fungi. Lane 1, *S. cerevisiae;* lane 2, *U. maydis;* lane 3, *U. hordei;* lane 4, *U. tritici;* lane 5, *U. bullata;* lane 6, *U. avenae;* lane 7, *U. nigra;* lane 8, *U. kolleri;* lane 9, *U. aegilopsidis.* (−1 Mb)

and information regarding the size and numbers of chromosomes is generally lacking, it is perhaps questionable whether or not some of them should be assigned species status. An example of the diversity of karyotypes may be visualized by the preliminary analysis of several *Ustilago* species presented in Figure 7. The electrophoresis parameters used in this analysis only resolved chromosomes up to approximately 2500 kb in size, but it appears that *U. avenae, U. kolleri,* and *U. tritici* have most chromosomes larger than 1 Mb in size, whereas *U. hordei, U. bullata,* and *U. nigra* have most chromosomes smaller than 1 Mb. The development of complete karyotypes of additional strains, coupled with Southern hybridization analysis to establish synonymy for linkage groups, will reveal whether the karyotypes of some of these fungi either support or refute their being separate species.

c. Karyotyping of other Fungi as an Addendum to Species Delimitation

Evidence in support of using electrophoretic karyotypes for the proper classification of fungal organisms has been presented for two additional fungal species. The highly virulent and weakly virulent isolates of *L. maculans* fail to mate and they have recently been shown to have very dissimilar electrophoretic karyotypes.[33] These results have been used to argue that they are different species. Among strains that do mate and produce viable progeny, chromosome length polymorphisms can be seen to segregate among members of the tetrad (Figure 2).

The closely related yeasts *Kluyveromyces marxianus* var. *marxianus* and *K. m.* var. *lactis*[52,53] also have very dissimilar karyotypes, although the two strains can mate. Interestingly, the progeny of these crosses do not show a reduced chromosome number, and apparently have complements of chromosomes from both parents. These results helped solidify arguments that support their being separate species. The ability to generate fungal electrophoretic karyotypes will undoubtely be an invaluable aid in the taxonomic classification of many fungal organisms and greatly reduce the emphasis presently placed on the morphologic species concept.

E. CONCLUDING REMARKS

The recent development of technology for resolving fungal chromosomes by PFGE has greatly advanced the opportunity to clone and map genes, develop an understanding of the basis of the genomic plasticity characteristic of some fungi, and have a more accurate basis for the taxonomic classification of closely related fungi. Additional unforeseen advances certain to emerge in the future, coupled with advances in fungal transformation, will greatly facilitate the analysis of fungal genomes, particularly the more recalcitrant fungi, such as the rust fungi that lack a sexual cycle. Finally, as these technologies evolve, molecular genetic analyses of fungal pathogens of plants may become a reality regardless of the organism of interest.

ACKNOWLEDGMENTS

We wish to thank those individuals identified in the legends to the figures for providing material for this chapter. A portion of the work described herein was supported by the Agricultural Research Station, from which this is technical paper No. 10,136, and by USDA NRICGP grant Nos. 91-37303-6462 and 94-37303-0617 and USDA-CSRS Special Res. Grant No. 88-34134-3489 to D.M.

REFERENCES

1. Schwartz, D. C. and Cantor, C. R., Separation of yeast chromosome-sized DNAs by pulsed field gradient gel electrophoresis, *Cell,* 37, 67, 1984.
2. Carle, G. F. and Olson, M., Separation of chromosomal DNA molecules from yeast by orthogonal-field-alteration gel electrophoresis, *Nucleic Acids Res.,* 12, 5647, 1984.
3. Boehm, E. W., Wenstrom, J. C., McLaughlin, D. J., Szabo, L. J., Roelfs, A. P., and Bushnell, W. R., An ultrstuctural pachytene karyotype for *Puccinia gramminis* f.sp. *tritici, Can. J. Bot.,* 70, 401, 1992.
4. Allen, R. C., Saravis, C. A., and Maurer, H. R., Gel electrophoresis and isoelectric focusing of proteins: selected techniques, in *Gel Electrophoresis of Nucleic Acids: a Practical Approach,* Richwood, D. and Hames, B. D., Eds., IRL Press, Oxford, 1982.
5. Clark, J. M., Lai, E., Birren, B. W., and Hood, L., A novel instrument for separating large DNA molecules with pulsed homogenous electric fields, *Science,* 241, 1203, 1988.
6. Cantor, C. R. and Schimmel, P. R., *Biophysical Chemistry,* Part 2, W. H. Freeman, San Fransisco, 1980, 676.
7. Carle, F. G. and Olson, M. V., An electrophoretic karyotype for yeast, *Proc. Natl. Acad. Sci. U.S.A.,* 82, 3756, 1985.
8. Carle, F. G., Frank, M., and Olson, M. V., Electrophoretic separations of large DNA molecules by periodic inversion of the electric field, *Science,* 232, 65, 1986.
9. Eby, M. J., Pulsed field separations: continued evolution, *Biotechnology,* 8, 243, 1990.
10. Gardener, K. and Patterson, D., Transverse alternating electrophoresis, *Nature* (*London*), 331, 371, 1988.
11. Chu, G., Vollrath, D., and Davis, R. W., Separation of large DNA molecules by contour-clamped homogeneous electric fields, *Science,* 234, 1582, 1986.
12. Miao, V. P. W., Covert, S. F., and VanEtten, H. D., A fungal gene for antibiotic resistance on a dispensable (B) chromosome, *Science,* 254, 1773, 1991.
13. Schwartz, D. C., Smith, C. L., Baker, M., and Hsu, M., ED: pulsed electrophoresis instrument, *Nature* (*London*), 342, 575, 1989.
14. Hamlyn, P. F., Bradshaw, R. E., Mellon, F. M., Santiago, J. M., and Peberdy, J. F., Efficient protoplast isolation from fungi using commercial enzymes, *Enzyme Microb. Technol.,* 3, 321, 1981.
15. McCluskey, K. and Mills, D., Identification and characterization of chromosome-length polymorphisms among strains representing fourteen races of *Ustilago hordei, Mol. Plant-Microbe Interact.,* 3, 366, 1990.
16. Russell, B. W. and Mills, D., Electrophoretic karyotypes of *Tilletia caries, T. controversa,* and their F_1 progeny: further evidence for conspecific status, *Mol. Plant-Microbe Interact.,* 6, 66, 1993.
17. McCluskey, K., Russell, B. W., and Mills, D., Electrophoretic karyotyping without the need for generating protoplasts, *Curr. Genet.,* 18, 385, 1991.
18. Plummer, K. M. and Howlett, B. J., Major chromosome length polymorphisms are evident after meiosis in the phytopathogenic fungus, *Leptosphaeria maculans, Curr. Genet.,* 24, 107, 1993.
19. Crouch, V. W., *Electrophoretic Karyotyping of Phymatotrichum omnivorum,* Masters thesis, Texas A&M University, College Station, TX, 1992.
20. McCluskey, K., *Molecular Karyotyping of Ustilago hordei and Related Smut Fungi,* Ph.D. thesis, Oregon State University, 1991.
21. Overhauser, J. and Radic, M. Z., A proprietary newsletter published by Bethseda Research Laboratories, *Focus,* 9, 8, 1987.
22. Bakalinsky, A. T., Electrophoretic karyotypes of wine strains of *Saccharomyces cerevisiae,* Bio-Rad Laboratories, Bulletin 1647, 1990.
23. Maniatis, T., Fritsch, E. F., and Sambrook, J., *Molecular Cloning: a Laboratory Manual,* Cold Spring Harbor Laboratory, Cold Spring Harbor, NY, 1982.

24. Skinner, D. Z., Budde, A. D., and Leong, S. A., Molecular karyotype analysis of fungi, in *More Gene Manipulations in Fungi,* Bennett, J. W. and Lasure, L. L., Eds., Harcourt Brace Jovanovich, San Diego, 1991, 87.
25. Kayser, T. and Wostemeyer, J., Electrophoretic karyotype of the zygomycete *Absidia glauca:* evidence for differences between mating types, *Curr. Genet.,* 19, 279, 1991.
26. Skatrud, P. L. and Queener, S. W., An electrophoretic karyotype for an industrial strain of *Cephalosporium acremonium, Gene,* 7, 331, 1989.
27. Talbot, N. J., Oliver, R. P., and Coddington, A., Pulsed field gel electrophoresis reveals chromosome length differences between strains of *Cladosporium fulvum* (syn. Fulvia *fulva*), *Mol. Gen. Genet.,* 229, 267, 1991.
28. Tzeng, T., Lyngholm, L. K., Ford, C. F., and Bronson, C. R., A restriction fragment length polymorphism map and electrophoretic karyotype of the fungal maize pathogen *Cochliobolus heterostrophus, Genetics,* 130, 81, 1992.
29. Masel, A. Braithwaite, K., Irwin, J., and Manners, J., Highly variable karyotypes in the plant pathogen *Colletotrichum gloeosporioides, Curr. Genet.,* 18, 81, 1990.
30. Osiewacz, H. D. and Ridder, R., Genome analysis of imperfect fungi: electrophoretic karyotyping and characterization of the nuclear gene coding for glyceraldehyde-3-phosphate dehydrogenase (gpd) of *Curvularia lunata, Curr. Genet.,* 20, 151, 1991.
31. Miao, V. P., Using karyotype variability to investigate the origins and relatedness of isolates of *Fusarium oxysporum* f. sp. *cubense,* in *Fusarium Wilt of Banana,* Ploetz, R. C. Eds., APS Press, St. Paul, MN, 1990.
32. Momol, E. A. and Kistler, H. C., Mitochondrial plasmids do not determine host range in crucifer-infecting strains of *Fusarium oxysporum, Plant Pathol.,* 41, 103, 1992.
33. Taylor, J. L., Borgmann, I., and Seguin-Swartz, G., Electrophoretic karyotyping of *Leptosphaeria maculans* differentiates highly virulent from weakly virulent isolates, *Curr. Genet.,* 19, 273, 1991.
34. Orbach, M. J., Electrophoretic karyotype of the *Magnaporthe grisea* genome, *Fung. Genet. Newslett.,* 36, 14, 1989.
35. Miao, V. P., Matthews, D. E., and VanEtten, H. D., Identification and chromosomal location of a family of cytochrome P-450 genes for pisatin detoxification in the fungus *Nectria haematococca, Mol. Gen. Genet.,* 226, 214, 1991.
36. Kistler, H. C. and Benny, U., Autonomously replicating plasmids and chromosome rearrangement during transformation of *Nectria haematococca, Gene,* 117, 81, 1992.
37. Covert, S. F., Bolduc, J., and Cullen, D., Genomic organization of a cellulase gene family in *Phanerochaele chrysosporium, Curr. Genet.,* 22, 407, 1992.
38. Rollo, F., Ferracuti, T., and Pacilli, A., Separation of chromosomal DNA molecules from *Phoma tracheiphila* by orthogonal-field-alternation gel electrophoresis, *Curr. Genet.,* 16, 477, 1989.
39. Howlett, B. J., An electrophoretic karyotype for *Phytophthora megasperma, Exp. Mycol.,* 13, 199, 1989.
40. McDonald, B. A. and Martinez, J. P., Chromosome length polymorphisms in a *Septoria tritici* population, *Curr. Genet.,* 19, 265, 1991.
41. Kinscherf, T. G. and Leong, S. A., Molecular analysis of the karyotype of *Ustilago maydis, Chromosoma (Berl),* 96, 427, 1988.
42. Mills, D. and McCluskey, K., Electrophoretic karyotypes of fungi: the new cytology, *Mol. Plant-Microbe Interact.,* 3, 366, 1990.
43. Powell, W. A. and Kistler, H. C., *In vivo* rearrangement of foreign DNA by *Fusarium oxysporum* produces linear self-replicating plasmids, *J. Bacteriol,* 3163, 1990.
44. Feinberg, A. P. and Vogelstein, B., A technique for radiolabelling DNA restriction endonuclease fragments to high specific activity, *Anal. Biochem.,* 132, 6, 1983.
45. Fan, J.-B., Chikashige, Y., Smith, C. L., Niwa, O., Yanagida, M., and Cantor, C. R., Construction of a *Not*1 restriction map of the fission yeast *Schizosaccharomyces pombe* genome, *Nucleic Acids Res.,* 17, 2801, 1989.
46. Oliver, S. G., Vander Aart, Q. J., Agostoni-corbone, M., *et al.,* The complete DNA sequence of yeast chromosome III, *Nature (London),* 357, 38, 1992.
47. Kistler, H. C. and Miao, V. P. W., New modes of genetic exchange, *Annu. Rev. Phytopathol.,* 30, 131, 1992.

48. Trail, F. and Mills, D., Growth of haploid *Tilletia* strains in planta and genetic analysis of a cross of *Tilletia caries* X *T. controversa, Phytopathology,* 8, 367, 1990.
49. Huang, H. Q. and Nielsen, J., Hybridization of the seedling-infecting *Ustilago* spp. pathogenic on barley and oats, and a study of the genotypes conditioning the morphology of their spore wall, *Can. J. Bot.,* 62, 603, 1984.
50. Russell, B. W. and Mills, D., Morphological, physiological, and genetic evidence in support of a conspecific status for *Tilletia caries, T. controversa* and *T. foetida, Phytopathology,* 84, 576, 1994.
51. Fischer, G. W. and Holton, C. S., *Biology and Control of the Smut Fungi,* The Ronald Press, New York, 1957.
52. Steensma, H. Y., de Jongh, F. C. M., and Linnekamp, M., The use of electrophoretic karyotypes in the classification of yeast: *Kluyveromyces marxianus* and *K. lactis, Curr. Genet.,* 14, 311, 1988.
53. Sor, F. and Fukuhara, H., Analysis of chromosomal DNA patterns of the genus *Kluyveromyces, Yeast,* 5, 1, 1989.

FURTHER READING

Birren, B. and Lai, E., *Pulsed Field Electrophoresis: a Practical Guide,* Academic Press, San Diego, 1993, 253 pp.

Debets, A. J. M., Holub, E. F., Swart, K., van den Broek, H. W. J., and Bos, C. J., An electrophoretic karyotype for *Aspergillus niger, Mol. Gen. Genet.,* 224, 264, 1990.

Horton, J. S. and Raper, C., Pulsed-field gel electrophoretic analysis of *Schizophyllum commune* chromosomal DNA, *Curr. Genet.,* 19, 77, 1991.

Keller, N. P., Cleveland, T. E., and Bhatnagar, D., Variable electrophoretic karyotypes of *Aspergillus* Section Flavi, *Curr. Genet.,* 21, 371, 1992.

Passoth, V., Hansen, M., Klinner, U., and Emeis, C. C., The electrophoretic banding pattern of the chromosomes of *Pichia stipitis* and *Candida shehatae, Curr. Genet.,* 22, 429, 1992.

Walz, M. and Kuck, U., Polymorphic karyotypes in related *Acremonium* strains, *Curr. Genet.,* 19, 73, 1991.

Section—II
BIOCHEMICAL METHODS

Chapter 8

Biochemical Markers for Disease Resistance*

Reuven Reuveni

TABLE OF CONTENTS

I. PREFACE

A broad spectrum of crop diseases occurring throughout the world can, under appropriate conditions, limit crop production or render the yield of grains or fruits unsatisfactory for use. Even major crops, throughout the developed and developing countries, rarely justify the use of expensive pesticides for disease control.

Long-term and immediate health and environmental considerations, which can have corresponding impact, also dictate the need for alternative methods of disease control. Breeding for resistance may be an achievable goal. Yield and quality of many crops were the main targets of breeding programs. However, resistance to pests and diseases was often not of major interest to the breeders. Therefore, repeated backcrosses are necessary today to reintroduce genes for resistance lost in the course of the search for higher yields and quality while growing millions of acres of crops over the many years of agriculture. The conventional selection process is long, difficult, and exposed to the environment, which has a negative effect on the expression of resistance. However, biochemical markers, which may or may not be part of the resistance mechanism, are a practical and reliable tool for predicting resistance to diseases. One of the problems with the traditional breeding strategy is the incorporation of single genes for resistance into germplasm as the basis of protection. The current approach is based on the "single gene for specific resistance" concept, whereas the biochemical markers that will be reviewed here can be used also to detect resistance based on the "multiple genes for widespread resistance" concept.

The literature chosen to illustrate herein the various details of the potential markers for resistance against plant pathogens is very much a personal selection. However, an attempt was made to emphasize the literature on particular topics by key data, and the selection is by no means exhaustive. It is hoped that this chapter will stimulate both plant pathologists and breeders as to the potential use of biochemical markers in breeding programs.

* Contribution from the Agricultural Research Organization, The Volcani Center, Bet Dagan, Israel. No. 1032-E, 1993 series.

0-87371-877-1/95/$0.00+$.50

Table 1 **Markers for disease resistance**

Type	Expression
Visual-morphological	Lesion size, shape and color; sporulation potential
Structural-mechanical	Leaf surface construction, cell wall composition—lignin, cutin, suberin, callose
Physiological-biochemical	Active enzymes in the defense, their substrates, and/or products; low-molecular weight antimicrobial compounds (phenols, hydroxamates, phytoalexins, anthocyanins)
Genetic-molecular	Structure of genomes and their products

II. INTRODUCTION

A long history of research has shown that single genes are often not durable and preclude the expression for resistance of multigenic factors that are involved in the resistance mechanisms available to the plant. Examples of this problem are many and include both abiotic (pesticide-resistant races of fungi) and biotic (phytopathogenic races of fungi) agents. The breakdown of resistance due to the evolution of new races or biotypes is determined as "genetic vulnerability.[1]" This means that resistance declines in a few years due to the occurrence of new races or biotypes of pathogens. The appearance of several new races of *Exserohilum turcicum,* the causal organism of northern leaf blight (NLB) of corn, has occurred since the introduction of the HtN gene family into commercial hybrids in the corn belt of the U.S.[2–4] Widespread use of the gene HtN applied selection pressure on the pathogen population, resulting in an increased frequency of biotypes that were virulent against that resistance.[4] In addition, highly virulent biotypes of *Helminthosporium carbonum,* an organism considered of no significance in corn, have appeared over the last 20 years.[5]

Marker identification, as evidence of the presence or an indication of the character, is the basic approach adopted to bring together a broad spectrum of possibilities to detect resistance by means of "biochemical" assays rather than by the conventional "biological" host-pathogen interactions, as expressed by artificial or natural inoculations.

At this stage, several terms regarding resistance or susceptibility need to be mentioned. *Resistance* is the ability of the host to suppress or restrict the activity of the pathogen. On the other hand, *nonhost resistance* is expressed when a pathogen of a particular host plant does not normally cause disease on other species. The organism therefore will be *pathogenic* or *nonpathogenic.* In case of *race-specific resistance,* which is the ability of the host cultivar to restrict the activity of one physiological variant of a pathogen, but not another, the organism will be *virulent* or *nonvirulent.* Such interaction between a virulent race and the susceptible cultivar is *compatible,* whereas *incompatible* is the interaction between a nonvirulent race and the resistant cultivar. Little is known about the structure and the regulation of genes for disease resistance. Data are limited as to how these genes interact biochemically with pathogen nonvirulent genes as their products. Furthermore, there is no evidence that both race-specific or non-specific resistance are determined by highly specific mechanisms. In most reports it seems that race specificity is determined by genes and the interaction of gene-regulated products, or processes which regulate the "when" and "how much" of a response.[6–8] The expression of resistance in plants to a variety of fungal pathogens has been shown to be associated with physical, physiological, biochemical, and molecular aspects of the host plants. These host-pathogen interactions eventually represent various aspects of markers for disease resistance (Table 1).

The data suggest, however, that the differences between susceptible plants in a compatible/virulent interaction and resistant plants in an incompatible/nonvirulent interaction, are determined by the rapidity and magnitude of the response rather than by the specificity of one or several possible defense mechanisms (see reviews).[6–7,9–14] A reliable determination is obligatory for all aspects of the resistance expression. Such a determination usually takes place after the inoculation of the host with the pathogen. This approach, however, is based on well-founded evidence that certain traits are consistently associated with the expression of resistance. These traits are each associated with aspects of physical, physiological, biochemical, and molecular markers that are inherent in the expression of the resistance.

Manipulation of disease reaction during the development of the pathogen may be a useful tool to exploit in studies of chemically based resistance. In most cases initiation of resistance expression starts with molecular changes as a result of the inoculation of the host with its compatible pathogen, followed by physiological and biochemical events and determined as physical traits of the host such as lignification (see reviews),[15,16] callose,[17] and papilla formation.[18-20] The papilla, the cell wall appositions—or lignitubers when lignified—are believed to provide an inducible barrier to cellular penetration by the potential pathogen (see reviews).[20-22] A papilla commonly comprises a callous matrix and various incorporating pectic materials—suberin, cellulose, gums, calcium, silicon, and protein, including peroxidase (POX).[21,22] Within this long and multifarious process, many stages can be used as evidence of resistance expression or as a marker.

In this review, several, but certainly not all, possible biochemical markers for disease resistance will be described. They will include: (a) enzymes involved in the defense of pathogenesis reactions of host plants against their potential pathogens such as POX and fungal cell wall-degrading enzymes and (b) low-molecular weight compounds such as phenols and hydroxamates which are involved in host-pathogen relations and other stress complexes.

Many have worked in this field and investigated enzymes and metabolites including phenolics, cyanogenic glycosides, alkaloids, terpenes, and the activity of phenoloxidases and peroxidases as candidate markers. The work gradually fell "out of fashion" in science, mainly due to inconclusive results, and probably also to a lack of conceptual information and technology. Studies in recent years in our laboratory,[23-32] and the interpretation of the data of others, strongly indicate the validity of the concept within the present pathological, physiological, and biochemical framework of knowledge and technology.

III. PEROXIDASE

For many years the role of oxidative enzymes and their metabolic products in the plant defense mechanism has been extensively studied. For an early review of the subject see Reference 51. The role of POX, phenoloxidase, and other enzymes was also investigated as markers for resistance.[33-38] The studies were carried out with the irrelevant, but perhaps justified simple scientific view that the enzymes were the resistance mechanism, or at least a direct part of the resistance mechanism, but not that they served as metabolic indicators or markers of a resistance potential, which is indeed the rapidity and magnitude of a resistance response to infection. Reviewing early studies, one can see that they often did not consider sufficiently the effects of environmental conditions, senescence, and the nature or stage of tissue development on the enzymes studied, and the data were clearly difficult to interpret. Age-[39-42] and temperature-related resistance[33-45] and other factors such as inoculum pressure on resistance expression[23,46] are documented in recent literature and their molecular mechanisms are currently being investigated.[47-50]

POX is one of the most investigated enzymes, mainly due to its involvement in so many molecular, physiological, and morphological events in the plant life cycle. POX activity is frequently increased in plants infected by pathogens, and the level of its activity is often closely correlated with disease resistance, as documented by Kosuge[51] more than 20 years ago. Enhanced POX activity is very often associated with resistance phenomena such as lignin production,[52-54] phenylalanine ammonia lyase activity, and phenol accumulation.[55-57] Furthermore, the direct role of POX in the defense reaction of plant resistance has been supported by the findings of Macko et al.,[58] Leherer,[59] Lovrekovich et al.,[60] and recently by Pang and Kuc.[61] However, little attention has been given to this enzyme in resistant plants before they are infected. The results of an investigation carried out by Reuveni and co-workers[24-30] and Shimoni et al.[62] strongly suggest that POX activity is a biochemical marker which may or may not be part of the resistance mechanism, but can be used to predict resistance to disease. A high correlation was found in noninfected plants of 12 tomato cultivars having or lacking the *Ve* gene for resistance against *Verticillium dahliae* (Table 2).[25]

Similar correlations were also found between high POX activity in noninfected muskmelon susceptible or resistant to *Sphaerotheca fuliginea*[24] and *Pseudoperonospora cubensis*.[26,27,30] In the latter case, POX activity and its distribution were investigated in populations derived from crosses of susceptible and resistant lines.[30] Such crosses—which were obtained from one resistant line in order to produce a reasonable number of segregating populations of a backcross of F_1 with the susceptible line and F_2—are common in breeding programs. Among the tested individuals, 100% of the resistant plants might be

Table 2 **Peroxidase activity in roots and wilt assessment of tomato seedlings 30 d after inoculation with *Verticillium dahliae***

Cultivar	Presence of *Ve* gene	Assessment of wilting (%)[a]	Peroxidase activity ($\times 10^a$)[b]		
			Infected	Noninfected	Ratio[c]
Flora Dade	+	20	90.6 ± 3.2[d]	112.8 ± 2.1	1.26
Robake	+	5	80.4 ± 1.9	112.2 ± 4.1	1.39
Rodade	+	40	93.6 ± 2.8	102.2 ± 3.1	1.09
R 296/52-4-34	+	5	82.8 ± 3.4	102.0 ± 4.1	1.23
RL 1/323-31-31	+	10	103.8 ± 2.3	196.0 ± 4.9	1.40
E (52/EC 02-9	+	25	113.4 ± 6.3	135.0 ± 3.8	1.19
MEL 2668170 G[e]	+	5	98.4 ± 8.3	131.0 ± 2.1	1.33
Red Kaki	−	90	58.8 ± 1.6	88.4 ± 2.2	1.50
Heinz 1370	−	100	63.0 ± 1.9	104.4 ± 3.0	1.65
L 2024	−	90	49.8 ± 2.4	84.0 ± 3.4	1.68
B W2	−	65	46.5 ± 0.9	97.2 ± 2.2	2.09
Karino[f]	−	90	129.6 ± 2.9	148.8 ± 4.8	1.14

[a]Percentage of wilted plants.
[b]Peroxidase activity expressed as changes in absorbance min^{-1} g^{-1} fresh weight.
[c]Infected to noninfected.
[d]Means and standard errors.
[e]Supplied by Prof. R. G. Gardner, North Carolina State University.
[f]Introduced from North Carolina in 1968.
Adapted from Reuveni and Ferreira, Reference 25.

predicted on the basis of POX values, detected in the second leaf of a three-leaf-stage seedling before inoculation with the pathogen. In practical terms, this information[30] shows that a primary selection to eliminate susceptible individuals among F_1 populations—either their backcrosses or "families", derived from a single fruit—from among the F_2 populations can be made reliable in an early stage (Figure 1). In this particular case, one favorable "family" originated from a single fruit (Group F, in Figure 1)—which represents one of three tested families with the highest mean POX activity and the largest number of resistant individuals—will be taken for further selection tests.

In lettuce (*Lactuca sativa*), a trend was apparent indicating that one component of field resistance to *Bremia lactucae* could be related to a high level of POX prior to infection[28] (Table 3). The highest activity was detected in the "Iceberg" cultivar (the one with the highest level of field resistance and a source of resistance being used in current breeding programs). "Santa Anna", a butterhead cultivar with a high level of field resistance, also exhibited high POX activity. An intermediate level of POX activity was detected in cvs "Grand Rapids" and "Lobjoits Cos", which exhibit moderate to low field resistance. In "Cobham Green" (a susceptible butterhead type) and "Ithaca" (a highly susceptible crisphead type), POX activity was low. A similar trend was found also between POX activity and the field resistance of three wild (*L. serriola*) lines.[28] Field resistance/susceptibility data for F_3 families from the cross Iceberg (resistant) × Vanguard 75 (moderately susceptible) were obtained, and eight of these families were chosen on the basis of contrasting responses. POX assays were carried out "blind" on 50 to 90 individuals from each of these eight families, and the same families were again assessed for field response to downy mildew. The data show that those families with the highest and lowest mean mildew resistance also exhibited the highest and lowest POX activity. The relatively high POX activity of "Vanguard 75" and the moderately high susceptibility of this cultivar emphasize, however, that POX levels alone are insufficient to explain field response to lettuce downy mildew.

To determine how the field performance of F_3 families correlated with the POX levels observed in their F_2 progenitors, an F_2 population from the cross Iceberg × Vanguard 75 (resistant × susceptible) was assayed for POX activity. The distributions of POX activity among individuals of "Iceberg", "Vanguard 75", and F_2 progeny of Iceberg × "Vanguard 75" are shown in Table 4. The data indicate that high POX activity equivalent to that observed for "Iceberg" was evident in the F_2 population, suggesting dominant inheritance. In practical terms, the data presented in Table 4 suggest that imposition of 50% selection at F_2 on the basis of high POX activity would result in retention of individuals yielding

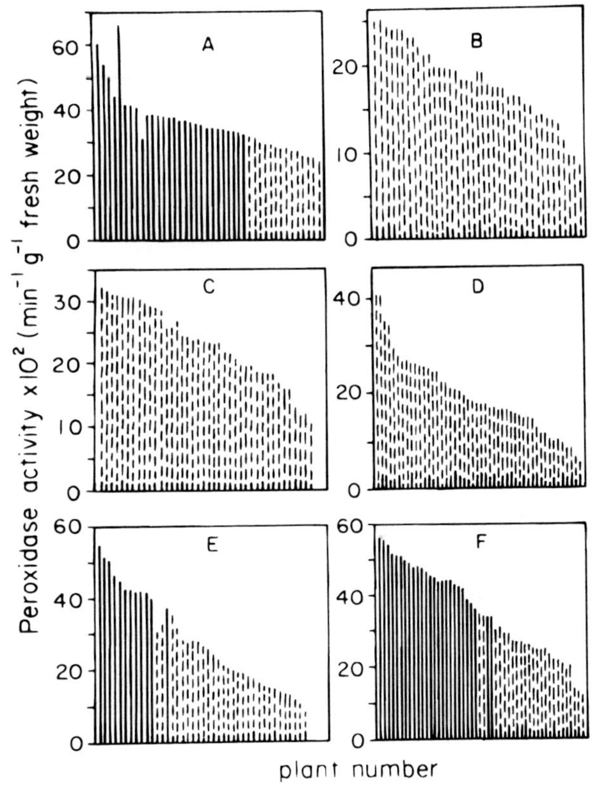

Figure 1 The relationship between peroxidase activity and resistance of individual muskmelon plants to *Pseudoperonospora cubensis* in the resistant parent, susceptible parent, and progenies from their crosses. Resistant plants are represented by solid lines and susceptible plants by dotted lines. A, Parent 1, resistant line 534; B, Parent 2, susceptible line CKR; C, backcross (BC), population from crossing F_1 (534 × CKR) with CKR; D,E,F, three families from self-breeding of the above cross and considered as F_2 populations. Peroxidase activities (change in absorbance at 470 nm × 100 min^{-1} g^{-1} fresh weight) are for individual plants and are the means of the activities obtained from two leaf disks from the second true leaf. The third leaf of each individual plant was inoculated and evaluated as resistant or susceptible. (Adapted from Reuveni et al., Reference 30.)

Table 3 **Relationship between peroxidase activity and field resistance/susceptibility of six lettuce cultivars to downy mildew**

Cultivar	Peroxidase activity[a]	Rank[b]	Resistance/susceptibility: Expected rank[c]	Number of infected leaves[d]	Actual rank
Iceberg	23.50	1	1	2.82	1
Santa Anna	18.44	2	2	3.70	2.5
Grand Rapids	14.50	4	3	3.70	2.5
Lobjoits Cos	16.04	3	4	7.28	6
Cobham Green	13.04	5	5	6.02	5
Ithaca	8.00	6	6	4.46[e]	4

[a]Mean of two experiments; peroxidase activity expressed as changes in absorbance min^{-1} g^{-1} fresh weight ($\times 10^3$).
[b]Based on peroxidase activity.
[c]Based on previous field trials.
[d]Resistance/susceptibility was assessed in 1985 in a trial comprising 5 replicates of 12 plants per plot.
[e]Low figure due to very heavy early infection which affected new leaf production.
Adapted from Reuveni et al., Reference 28.

80% of the most resistant F_3 progeny. This selection procedure could permit a 50% reduction in field trial size without drastically reducing the efficiency of selecting for resistance. Alternatively, if field trials of the same size were performed, a comparable increase in the efficiency of selection for other important traits would be possible. Based on the concept that the mechanisms for disease resistance may be specific or nonspecific, but the rapid recognition of a pathogen by the plant as virulent could be highly specific in all the above mentioned host-pathogen systems, total POX activity represents a marker for nonspecific response of the compatible host to the pathogen. Since all mechanisms for disease resistance reported to date are nonspecific,[6,7,10–14,63] total activity of POX might be a useful tool

Table 4 **Distribution of individuals with different levels of peroxidase activity among lettuce cultivars resistant (Iceberg) and susceptible (Vanguard) to downy mildew and the F_2 of a cross between them**

	% of total plants tested		
Peroxidase activity[a]	**Iceberg**	**Vanguard 75**	**F_2 Iceberg × Vanguard 75**
5.0–9.9	2.5	46.4	5.1
10.0–14.9	18.0	39.3	15.5
>15.0	79.5	14.3	79.4

[a]Peroxidase activity is expressed as change in absorbance min^{-1} g^{-1} fresh weight ($\times 10^3$). Number of plants tested was 78, 28, and 97, respectively, for Iceberg, Vanguard 75, and the F_2 of Iceberg × Vanguard 75.
Adapted from Reuveni et al., Reference 28.

to predict such a nonspecific resistance. On the other hand, quantitative differences in POX activity in noninoculated leaves between resistant and susceptible maize were found only in plants of the same genetic background, but differing in their susceptibility to *E. turcicum* (B37, B37HtN, B73, B73HtN).[62] Polyacrylamide gel electrophoresis (PAGE) separation has revealed different isoperoxidase banding patterns for resistant isolines containing the gene HtN (B37HtN, B73HtN), susceptible isolines, (B37, B73), or inbred "Jubilee". Additionally, such differences in the banding pattern of POX isozymes in the parental, F_1, F_2, and backcross populations from crosses between the inbred W64A and W22HtN (susceptible and resistant to *E. turcicum,* respectively) have been detected prior to inoculation. Distinct POX variant bands were identified not only for the susceptible and resistant inbred lines, but also a near perfect correlation was found between individual plant isozyme variant patterns and plant expression of resistance conferred by the *HtN* allele in the segregating BC_1 and F_2 populations.[29]

As has been found in maize isolines having the HtN gene for resistance against *E. turcicum,* total enhanced POX activity may not consistently reflect a potential resistance, but activity of specific isozyme(s) may reflect resistance potential for specific disease. Being highly aware that environmental and physiological factors affect POX activity, and since experiments have been conducted under closely controlled conditions, the results obtained by Reuveni and co-workers[24–30] suggest, indeed, that POX activity in noninfected plants can be used as a marker for resistance under such conditions. It should be emphasized again that reliability of resistance expression and correlation with POX activity should be prerequisites for each new tested host-parasite system. A rapid assay for monitoring POX activity as a nonspecific marker for primary selection, as suggested by Reuveni et al.,[27] can be used. In this assay, 96 leaf disks, 3 to 4 mm in diameter, can be sampled in a nondestructive assay and placed in a 96-well cassette containing reaction mixture to detect POX activity in an ELISA reader. Results regarding POX activity generated from each leaf disk in each well are available 3 min later and within 30 s of cassette reading (Figure 2).

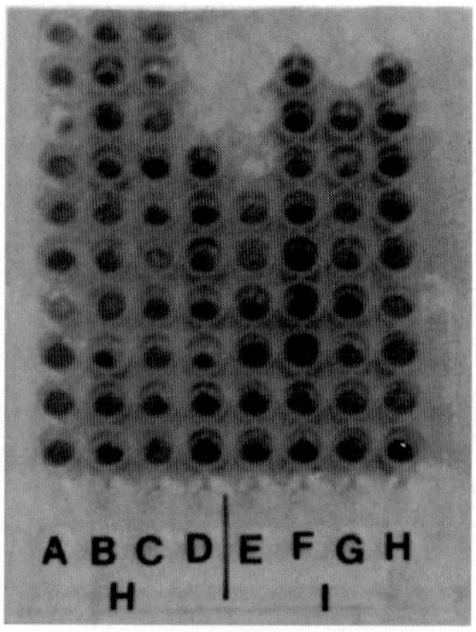

Figure 2 A wells assay for peroxidase activity in leaf disks of inoculated and noninoculated melons, with various levels of susceptiblity or resistance to *Pseudoperonospora cubensis.* Photograph was taken 5 min after one leaf disk was placed in each well containing a reaction mixture for peroxidase activity. H, Noninoculated; I, inoculated, 3 d after inoculation. A,E, 202, highly susceptible line; B,F, individuals from one of the F_2 families which were derived from self-breeding of FI1L (534 × CKR) × CKR. C,G, Noy Yizre'el, susceptible; D,H, 534, resistant line. (Adapted from Reuveni et al., Reference 27)

A. MATERIAL AND METHODS FOR PEROXIDASE

Preparation of enzyme extract and detection of total POX activity are described by Reuveni et al.[30] and that of native gels and electrophoresis is described by Hames[64] and Davis,[105] respectively. Detection, stabilization, and double staining of POX and proteins have been suggested by Shimoni and Reuveni,[66] as follows: after the electrophoretic separation, gels are soaked for 15 to 30 min in 15 m*M* sodium phosphate buffer, pH 6.0, containing 1 m*M* H_2O_2 and 0.1 m*M* *o*-methoxy phenol (guaiacol). The gels are rinsed with deionized water and transferred to a Coomassie blue solution (0.1% Coomassie blue R-250 in water:methanol:glacial acetic acid, 5:5:2 v/v) for 1.5 to 2.0 h. Gels are destained with a solution containing 25% methanol and 10% acetic acid. The destaining solution is replaced several times; it is not recommended to destain longer than 24 h, because Coomassie-stained bands lose their intensity.

IV. CELL WALL-DEGRADING ENZYMES

The degradative enzymes β-1,3-glucanase and chitinase have been reported to accumulate in higher plants following pathogen attack and environmental stress, which indicate their possible involvement in plant defense responses. These enzymes are particularly important for resistance of plants to fungal invasion, as chitin and glucans are major cell wall components of certain fungi[16,67] and their antifungal potential has been reported.[68–72] Individual assays of β-1,3-glucanase and chitinase have shown no growth inhibition of several fungi, but their combination caused the degradation of various fungal cell walls.[73] This kind of "defense compound", which would prevent fungal infection, can be used as a potential biochemical marker to detect resistance against fungal pathogens. Support for this approach was provided recently,[74] demonstrating a different isozymes pattern of β-1,3-glucanase isozymes between maize isolines susceptible and resistant to *E. turcicum.* Further evidence for this possibility was provided[42] by a demonstration that in tobacco plants, β-1,3-glucanase and chitinase total activity were about tenfold higher in leaf tissue from the main stalk (resistant to *Peronospora tabacina*) than from suckering stems (susceptible to *P. tabacina*). Isozyme patterns of both enzymes in all resistant tissues were typical of those of the tissues which had been systemically protected by either foliar inoculation with TMV or stem injection with *P. tabacina.*[42]

A. β-1,3-GLUCANASE

β-1,3-Glucanase was first reported in almond seeds in 1934.[75] Since then it has been documented as one of the most important pathogenesis-related (PR) proteins,[76] which accumulates in many host plant species upon infection by their pathogens.[77] This enzyme has been suggested as an important component of plant defense mechanisms against pathogens[42,74,76,78–82] in both dicotyledonous[83,84] and monocotyledonous[85–89] species of host plants. The local and systemic accumulation of these proteins is correlated with the induction of local and systemic resistance against various types of pathogens.[42,73,74,78–80,87,90]

The relationship between β-1,3-glucanase activity in leaf crude extract of maize and its resistance to *E. turcicum* has been described.[87] Based on the quantitative data regarding β-1,3-glucanase activity in isolines which differ in susceptibility to *E. turcicum,* the authors concluded that induction of the activity in inoculated leaves is a consequence of expression of resistance. Indeed, quantitative data regarding key enzymes in the defense mechanisms in plants can be used to express resistance of the host, as discussed here for the possible relationship between POX activity in noninoculated plants and the resistance to their pathogens. However, such data are not always sufficient to predict resistance in noninoculated plants, especially when the resistance is specific. Using the method described for direct detection of β-1,3-glucanase isozymes on PAGE and isoelectrofocusing,[91] one can efficiently investigate the possible use of this enzyme as a marker for resistance.

Positive or negative correlations between total activity of β-1,3-glucanase and resistance have been demonstrated in melon-*Fusarium oxysporum* interaction[83] or tomato-*Verticillium albo-atrum* interaction,[70,82] respectively. In pea pods, however, there was no difference in the time course or level of induction of total activity of β-1,3-glucanase upon inoculation with compatible strains of *F. solani.*[80] In maize, the induction in total activity of β-1,3-glucanase is hypothesized to be a consequence of the expression of resistance in response to *E. turcicum.*[87] The possibility of detecting β-1,3-glucanase isozymes in susceptible or resistant interactions, rather than the total activity of the enzyme, might be of more direct significance in the attempt to identify a marker for resistance. Recent data from our laboratory show that while using native PAGE and staining with 2,3,5,-triphenyltetrazolium chloride, a remarkable separation of β-1,3-glucanase isozymes from leaf extracts of maize was obtained.[74] In a set of two near isogenic lines of maize (B37, B37HtN), differing in a single gene for resistance to NLB, together with a susceptible sweet corn cultivar ("Jubilee"), different banding patterns of β-1,3-glucanase were visualized for susceptible and resistant maize to *E. turcicum* before and after the inoculation with the pathogen. In the susceptible line B37, one intense band appeared (Rf = 0.77). Bands with Rfs of 0.70 and 0.87 were detected in the resistant isoline B37HtN. After inoculation with *E. turcicum,* three isozymes of β-1,3-glucanase which had been absent in both the resistant and susceptible plants appeared only in the resistant plants. These differences were apparent only in native PAGE,[74] but not in SDS gel as reported by Nasser et al.,[92] who found one isozyme of β-1,3-glucanase in noninoculated maize plants.

Although the direct role of these isozymes in the resistance expression of maize to *E. turcicum* is not clear yet, the isoenzymes might be used as a reliable tool to detect an early response of the resistant maize genotypes to inoculation with *E. turcicum.* These markers will permit a rapid detection of the HtN gene in corn, and will overcome the problem of prolonged incubation periods needed for the detection of resistance. The possibility of using these isozymes as specific markers for resistance and subsequently in genetic and molecular manipulations is being investigated in several laboratories.

B. CHITINASE

Chitinase, a hydrolase with an antifungal potential,[72,79,93] is one of the PR proteins which is induced in cucumber seedlings[94] or tobacco plants in response to infections producing necrotic symptoms.[14,95] The localization of this enzyme is intriguing with regard to its possible antifungal function. In cucumber, for instance, where chitinase accumulates in the extracellular space, it has been hypothesized that the enzyme can directly "attack" incoming fungal hyphae.[96] Chitinase is detectable as a strongly staining band on electrophoretic SDS gels of acid-extractable proteins from infected cucumber[94,97] or tobacco[98] plants.

A recent report[99] provides a useful method to detect chitinase on native gels together with β-1,3-glucanase and protein patterns, as will be described in the Materials and Methods section. With this powerful technique, an overlay gel containing glycol chitin as a substrate for chitinase is incubated in close contact with the resolving gel, immediately after electrophoresis or isoelectrofocusing. This method reduces remarkably the variability that might result from using different separations of samples. In addition, the procedure facilitates research on the coordinated induction of chitinase and β-1,3-glucanase[98] and enables the possible detection of these enzymes as markers for resistance.

C. MATERIALS AND METHODS FOR β-1,3-GLUCANASE AND CHITINASE

1. Preparations of Enzyme Extract

Leaf samples (7 g each) are extracted by homogenizing them in 0.05 *M* sodium acetate buffer (pH 5.0) with a mortar and sea sand at 4°C. The extracts are dialyzed against 0.01 *M* sodium acetate buffer (pH 5.0) overnight at 4°C.

2. Enzyme Activity Determinations

a. β-1,3-Glucanase

Total activity is determined by measuring the rate of release of reducing sugar from laminarin as a substrate.[68] The assay mixture consisted of 0.4 ml of 0.1 *M* acetate buffer (pH 5.0) containing 1% laminarin and various volumes of enzyme extract. After a 30-min incubation, 1 ml of Somogyi reagent[101] is added at 37°C. After addition of 0.6 ml distilled water, the solution is heated in boiling water for 15 min and cooled to room temperature. Then 0.5 ml of Nelson's reagent[101] is added, mixed, and incubated 15 min at room temperature before measuring the absorbance at 660 nm.[102]

b. Chitinase

Total activity is assayed by measuring the hydrolysis of colloidal chitin to *N*-acetyl glucosamine, determined by the colorimetric assay of Boller and Mauch.[103] Alternatively, chitinase activity can be determined by radiochemical assay with [^{3}H] chitin as a substrate.[104] Different dilutions for each leaf homogenate should be tested to determine the appropriate dilutions for the assay. The reaction mixture consists of 0.8 mg of [^{3}H] chitin, 10 to 50 μl diluted tissue homogenate, and 5 μmol of sodium phosphate buffer (pH 6.4) in a final volume of 250 μl. After a 45-min incubation at 37°C, the reaction is stopped by the addition of 0.25 ml 10% aqueous trichloroacetic acid. The radioactivity of 0.2 ml of the supernatant is determined after centrifugation.

3. Detection of Isozymes on Native PAGE

Polyacrylamide resolving gels, 15%, 1.5 mm thick, are prepared according to Hames[64] and Davis[105] with slight modifications by Pan et al.[99] After electrophoresis, the gels remain attached to supporting glass plates and are incubated in 0.1 *M* sodium acetate (pH 5.0) for 5 min. They are then covered with a 7.5% (0.75 mm thick) polyacrylamide overlay gel which is attached to another supporting glass plate containing 0.04% glycol chitin in 0.1 *M* sodium acetate. The gels are incubated at 40°C for 15 h under moist conditions (pH 5.0). This overlay gel is for the detection of chitinase activity indicating isozymes existence in the original gel. Overlay gels are incubated in freshly prepared 0.01% (w/v) fluorescent brightener 28 in 500 m*M* Tris-HCl (pH 8.9) at room temperature for 5 min. Chitinase isozymes are visualized after 2 h incubation in water in the dark as cleared zones by placing the overlay gels on a UV transilluminator.

For staining β-1,3-glucanase isozymes, the gels are washed in water and incubated in 0.05 *M* sodium acetate (pH 5.0) for 5 min, and then incubated at 40°C for 30 min in a mixture containing 75 ml of 0.05 *M* sodium acetate (pH 5.0) and 1 g of laminarin dissolved in 75 ml of water (by heating in a boiling water bath). The gels are then incubated in a mixture of methanol, water, and acetic acid (5:5:2, v/v) for 5 min, and washed with water. Gels are stained with 0.3 g 2,3,5-triphenyltetrazolium chloride in 200 ml of 1.0 *M* NaOH in a boiling water bath until red bands appear. The stained gels can then be stained for proteins with Coomassie Brilliant Blue R250.

V. PHENOLS

The first evidence of the relation between phenolic compounds toxic to stem rust and their presence in resistant varieties in quantities sufficient to account for the resistance was presented by Newton and Anderson[106] in 1928. Several years later, Walker and Link[107] reported that catechol and protocatechuic acid were detectable in tissues of onion resistant to the fungus *Colletotrichum circinans,* but were absent in the susceptible cultivars.

A wide range of phenolic compounds are synthesized in plant tissues during normal growth and development via the phenylpropanoid biosynthetic pathway. The compounds are building blocks for cell wall structure[108] and plant pigment production[109] and serve as protection from ultra-violet light and as a defense against pathogens.[110] The synthesis of phenols is also hypothesized to modify hormonal activities,[111] and their production in plants might be induced in response to hormonal and environmental stimuli,[112] infection by biotic agents, and wounding.[50]

Many workers have examined phenolic acids in susceptible and resistant interactions with pathogens. A positive correlation was found between the phenol content of different potato cultivars resistant or susceptible to *V. albo-atrum*[113] or several horticultural crops resistant or susceptible to *Dematophora necatrix* Hartig.[114] Similarly, it was found[115] that "Little Club"—a cultivar of wheat (*Triticum* spp.) susceptible to most stem rust races caused by *Puccinia graminis Pers.* f. sp. *tritici*—had the lowest content of phenolic compounds, while "Khapli"—which is resistant to most races—had the highest content. However, in cultivars between these two extremes no relation was found between the content of phenolics and rust resistance. Interpretations of the results from each of these studies are difficult, mainly because the biology of the interactions was not well known at the time the early work was done, and due to the possible use of a different methodology.

In an investigation[44] of the levels of phenols in near-isogenic wheat possessing the Sr6 allele, conferring resistance to *P. graminis* f. sp. *tritici,* the first samples for analysis were taken 96 h after inoculation. Later histological work demonstrated that the first host reaction of the resistance expression to the stem rust fungus in wheat possessing the Sr6 allele was seen as early as 20 h after the inoculation.[32,116] Furthermore, by 60 h some colonies had ceased to grow. Therefore, when Seevers and Daly[44] noted no differences in total individual phenolic acids released by acid or alkaline hydrolysis, their findings may have been not related to the resistance reaction, which is considered to have been expressed several days earlier. Similarly, Rohringer et al.[31] and Fuchs et al.,[117] while using the same lines of wheat with the same resistance allele, reported greater incorporation of labeled precursors into bound phenolic esters in the resistant interaction than in the susceptible one. In this case, leaves were fed the radioactive compounds at day 6 after inoculation, and the level of incorporation was likely to follow the expression of resistance by several days. This apparent contradiction of the above reports[31,117] by that of Seevers and Daly[44] is an example of a possible mistake in the interpretation of data, derived from the contrast between studies of phenolics synthesis and total levels of the compounds in the respective reports.

Recent biochemical investigations have indicated a strong relation between the increased levels of a number of the intermediate compounds of the phenylpropanoid pathway through the pathogenesis and the expression of resistance in wheat to rust fungi.[43,44,118–120] Various steps in the synthesis of lignin via the phenylpropanoid pathway[17,120,121] and its phenolic precursors,[122] together with many other phenolic acids[123,124] which have been found to have strong antifungal activity, might be used as potential markers to detect resistance in the host plants at early stages of the screening.

Materials and methods for phenols—Preparation of plant tissue and methods for determination of phenols are described by Bray and Thorpe.[125]

VI. HYDROXAMATES

The glucoside of cyclic hydroxamate, 2,4-dihydroxy-7-methoxy-1,4-benzoxazin-3-one (DIMBOA), is present in most normal varieties of corn (*Zea mays*) and other cereals. After wounding, an enzymatic reaction releases the aglycon DIMBOA, which is toxic to several fungi including *H. turcicum* (=*Exserohilum turcicum*)—the causal organism of NLB of maize[126,127] and to the larva of the European corn borer, *Ostrinia nubilalis.*[128] The relation between DIMBOA and resistance has been reported in several host-pathogen systems such as wheat (*T. aestivum* L.) and resistance to stem rust caused by *P. graminis* var. *tritici* Erikss.[128,129] and maize and resistance to stalk rot caused by *Diplodia maydis* (Berk) Sacc.[130] Based on data[131] indicating that corn plants carrying the dominant gene HtN for resistance to NLB produced much smaller chlorotic lesions when inoculated with an avirulent race of *H. turcicum,* Couture et al.[126] investigated the role of cyclic hydroxamates in resistance to *E. turcicum.* They used a recessive corn mutant (bx) in which the glucosides are present only in low concentrations,[132] and compared the lesions area of leaves of four lines of corn inoculated with *E. turcicum:* with (BxBx) and without (bxbx) normal levels of cyclic hydroxamates, and with (HtHt) and without (htht) the gene for resistance. The presence of cyclic hydroxamate is readily noted when a blue coloration develops after crushing the mesocotyl of a 6-d-old seedling in 0.1 ml of $FeCl_3$. Using this simple method, Couture et al.[126] found a direct correlation between cyclic hydroxamate concentration and resistance to NLB. Later, Long et al.[127] confirmed similar correlations between resistance to NLB in the field and hydroxamate concentration. However, no apparent correlation was found between field resistance to either NLB or European corn borer feeding in extensive evaluations of several thousand corn genotypes in the field.[133]

Similarly, other conflicting reports, including the involvement of these compounds in resistance of corn to Erwinia stalk rot[130] and of wheat to stem rust, have been ultimately refuted.[65,129] Furthermore, in a recent publication[134] evidence based on *in vitro* as well as greenhouse experiments showed that the cyclic hydroxamate concentrations in 11 corn cultivars are not related to the resistance or susceptibility of the cultivars to anthracnose caused by *C. graminicola,* nor was there any relation to developmental differences such as leaf position and age, and the resistance of corn leaves to the disease. These findings show that, other than in juvenile leaves of corn, DIMBOA-glc does not persist in the leaf blade at substantial concentrations beyond the time of collaring. This means that the attempts by previous researchers to correlate cyclic hydroxamate—by measuring the aglycone DIMBOA concentration—with resistance of various foliar pathogens[65,126,127,129] may be irrelevant. In other words, despite their apparent antimicrobial properties and their "active" role in the defense mechanism, hydroxamates can be used as markers for resistance only in those host-pathogen systems where their relation to resistance expression is solidly based and regardless of their direct involvement in the defense. The report[135] that cyclic hydroxamates are distributed between the mesophyll cells and vascular bundles, but are not detectable in the epidermis of wheat, may explain their limited effectiveness in restricting fungal penetration. This is another example indicating that useful markers for disease resistance may be those which are correlated with the resistance phenomena and not necessarily part of its mechanism.

Materials and methods for hydroxamates—Three procedures are currently used to detect the concentration of DIMBOA: colorimetric,[136] spectrophotofluorometric,[137] and isotope dilution.[138] Although these procedures offer high sensitivity, a rapid procedure was suggested[139] for estimating cyclic hydroxamates in the large numbers required in a breeding program. Tissue samples (0.2 to 0.5 g) are frozen in vials overnight; subsequent thawing of the tissue enables hydrolysis of the glucosides by the *p*-glucosidase and release of DIMBOA. The homogenates of the tissues are prepared by mortar and pestle in 1.0 ml of 95% ethanol and 0.1 N HCl (1:1, v/v). After bringing the homogenates to a volume of 1.9 ml and centrifugation, 0.1 ml of $FeCl_3$ was added to the supernatant, followed by immediate reading of absorbance in a spectrophotometer at 570 nm.

VII. LIMITATIONS AND ADVANTAGES

As resistance expression is affected by biotic and abiotic factors, the assays for biochemical markers should be conducted under highly controlled environmental and physiological conditions. Standardized methods for each host-pathogen interaction will provide a reliable determination of the potential marker. Data from studies in many different host-pathogen interactions document quite conclusively the effectiveness of biochemical markers for identifying and investigating traits responsible for resistance against various soilborne as well as foliar pathogens. These markers, which may or may not be part of the resistance mechanism, are a practical and reliable tool to predict resistance to diseases and may also help in the understanding of genetic and metabolic factors determining resistance. The routine use of a non-destructive method to detect markers for resistance eliminates time-consuming extensive field trials and may also expedite the breeding program. Such markers provide the breeder with metabolic and eventually genetic markers to pinpoint efficiently the most advantageous direction of breeding programs and offer the possibility of an advanced approach to plant breeding.

REFERENCES

1. National Academy of Science, *Genetic Vulnerability in Major Crops,* Printing & Publication Office, N.A.S., Washington D.C., 1972.
2. Jordon, E. G., Perkins, J. M., Schall, R. A., and Pedersen, W. L., Occurrence of race 2 of *Exserohilum turcicum* on corn in the central and eastern United States, *Plant Dis.,* 67, 1163, 1983.
3. Leonard, K. L., Proposed nomenclature for pathogenic races of *Exserohilum turcicum* on corn, *Plant Dis.,* 73, 766, 1989.
4. Smith, D. R. and Kinsey, J. C., Further physiologic specialization in *Helminthosporium turcicum, Plant Dis.,* 64, 779, 1980.
5. Dodd, J. L. and Hooker, A. L., Previously undescribed pathotype of *Bipolaris zeicola* on corn, *Plant Dis.,* 74, 530, 1989.
6. Kuc, J., Plant immunization-mechanisms and practical implications, in *Active Defense Mechanisms in Plants,* Wood, R. K. S. and Tjamos, E., Eds., Plenum Press, New York, 1982, 157.
7. Kuc, J. and Rush, J., Phytoalexins, *Arch. Biochem. Biophys.,* 236, 455, 1985.
8. Salt, S. and J. Kuc, Elicitation of disease resistance in plants by the expression of latent genetic information, in *Bioregulators for Pest Control,* Hedin, P. A. Ed., American Chemical Society Press, Washington, D.C., 1985, 47.
9. Ye, X. S., Strobel, N., and Kuc, J., Induced systemic resistance (ISR): Activation of natural defense mechanism as part of integrated pest management (IPM), in *Novel Approaches to Integrated Pest Management,* Reuveni, R., ed., CRC Press, Boca Raton, FL, 1995, 95.
10. Kuc, J., Plant immunization, *Bioscience,* 32, 854, 1982.
11. Kuc, J. and Tuzun, S., Immunization for disease resistance in tobacco, *Rec. Adv. Tobacco Sci.,* 9, 179, 1983.
12. Kuc, J. and Preisig, C., Fungal regulation of disease resistance mechanisms in plants, *Mycologia,* 76, 767, 1984.
13. Kuc, J., Phytoalexins and disease resistance mechanisms from a perspective of evolution and adaptation, in *Origins and Development of Adaptation,* Ciba Foundation Symposium 102, Pitman, London, 1984, 100.
14. Kuc, J., Expression of latent genetic information for disease resistance in plants, in *Cellular and Molecular Biology of Plant Stress,* Key, J. and Kosuge, T., Eds., Alan R. Liss, New York, 1985, 303.
15. Friend, J., Plant phenolics, lignification and plant disease, in *Progress in Phytochemistry,* Vol. 7, Reinhold, L., Harbone, J. B., and Swain, T., Eds., Pergamon Press, Oxford, 1981, 197.
16. Wessels, J. G. H. and Sietsma, J. H., Fungal cell walls: a survey, in *Encyclopedia of Plant Physiology,* Vol. 13B: *Plant Carbohydrates* II, Tanner, W. and Loewus, F. A., Eds., Springer-Verlag, Berlin, 1981, 352.
17. Tiburzy, R. and Reisener, H. J., Resistance of wheat to *Puccinia graminis* f. sp. *tritici:* Association of the hypersensitive reaction with the cellular accumulation of lignin-like material and callose, *Physiol. Mol. Plant Pathol.,* 36, 109, 1990.
18. Edwards, M. C. and Ayres, P. G., Cell death and cell wall papillae in the resistance of oak species to powdery mildew disease, *New Phytol.,* 89, 411, 1981.

19. Ride, J. P. and Pearce, R. B., Lignification and papilla formation at sides of attempted penetration of wheat leaves by non-pathogenic fungi, *Physiol. Plant Pathol.*, 15, 79, 1979.
20. Ride, J. P., Cell walls and other structural barriers in defence, in *Biochemical Plant Pathology*, Callow, J. A., Ed., John Wiley & Sons, Chichester, 1983, 215.
21. Aist, J. R., Papillae and related wound plugs of plant cells, *Annu. Rev. Phytopathol.*, 14, 145, 1976.
22. Aist, J. R., Structural responses as resistance mechanisms, in *The Dynamics of Host Defense*, Bailey, J. A. and Deverall, B. J., Eds., Academic Press, New York, 1983, 33.
23. Reuveni, R., Resistance reaction of *Cucumis melo* to inoculation with *Pseudoperonospora cubensis*, *Ann. Appl. Biol.*, 102, 533, 1983.
24. Reuveni, R. and Bothma, G. C., The relationship between peroxidase activity and resistance of *Sphaerotheca fuliginea* in melons, *Phytopathol. Z.*, 114, 260, 1985.
25. Reuveni, R. and Ferreira, J. F., The relationship between peroxidase activity and the resistance of tomatoes (*Lycopersicum esculentum*) to *Verticillium dahliae*, *Phytopathol. Z.*, 112, 193, 1985.
26. Reuveni, R. and Karchi, Z., Peroxidase activity—a possible marker for resistance of melon against downy mildew, *Phytopathology*, 77 (Abstr.), 1724, 1987.
27. Reuveni, R., Shimoni, M., and Karchi, Z., A rapid assay for monitoring peroxidase activity in melons as a marker for resistance to *Pseudoperonospora cubansis*, *J. Phytopathol.*, 129, 333, 1990.
28. Reuveni, R., Shimoni, M., and Crute, I. R. An association between high peroxidase activity in lettuce (*Lactuca sativa*) and field resistance to downy mildew (*Bremia lactucae*), *J. Phytopathol.*, 132, 312, 1991.
29. Reuveni, R., Shimoni, M., Bar-Zur, A., and Juvik, J. A., The association of peroxidase isozymes with the resistance of maize to *Exserohilum turcicum*, *Phytoparasitica*, 20 (Abstr.), 257, 1992.
30. Reuveni, R., Shimoni, M., Karchi, Z., and Kuc, J., Peroxidase activity as a biochemical marker for resistance of muskmelon (*Cucumis melo*) against *Pseudoperonospora cubensis*, *Phytopathology*, 82, 749, 1992.
31. Rohringer, R., Fuchs, A., Lunderstaedt, J., and Samborski, D. J., Metabolism of aromatic compounds in healthy and rust-infected primary leaves of wheat. I. Studies with $^{14}CO_2$, quinate-U-^{14}C, and shikimate-U-^{14}C as precursors, *Can. J. Bot.*, 45, 863, 1967.
32. Rohringer, R., Kim, W. K., and Samborski, D. J., A histological study of interactions between avirulent races of stem rust and wheat containing resistance genes *Sr5, Sr6, Sr8*, or *Sr22*, *Can. J. Bot.*, 57, 324, 1979.
33. Fehrmann, H. and Dimond, A. E., Peroxidase activity and *Phytophthora* resistance in different organs of the potato plant, *Phytopathology*, 57, 69, 1967.
34. Hammerschmidt, E., Nuckles, E. M., and Kuc, J., Association of enhanced peroxidase activity with induced systemic resistance to *Colletotrichum lagenarium*, *Physiol. Plant Pathol.*, 20, 73, 1982.
35. Johnson, L. B. and Lee, R. F., Peroxidase changes in wheat isolines with compatible and incompatible leaf rust infection, *Physiol. Plant Pathol.*, 13, 173, 1978.
36. Maraite, H., Changes in polyphenoloxidase and peroxidase in muskmelon (*Cucumis melo* L.), infected by *Fusarium oxysporum* f. sp. *melonis*, *Physiol. Plant Pathol.*, 3, 29, 1973.
37. Maxwell, D. P. and Bateman, D. F., Changes in the activities of some oxidases in extracts of *Rhizoctonia*-infected bean hypocotyls in relation to lesion maturation, *Phytopathology*, 57, 132, 1967.
38. Seevers, P. M., Daly, J. M., and Catedral, F. F., The role of peroxidase isozymes in resistance of wheat stem rust disease, *Plant Physiol.*, 48, 353, 1971.
39. Bhattacharyya, M. K. and Ward, E. W. B., Expression of gene-specific and age-related resistance and the accumulation of glyceollin in soybean leaves infected with *Phytophthora megasperma* f. sp. *glycinea*, *Physiol. Mol. Plant Pathol.*, 29, 105, 1986.
40. Jeun, Y. C. and Hwang, B. K., Carbohydrate, amino acid, phenolic and mineral nutrient contents of pepper plants in relation to age-related resistance to *Phytophthora capsici*, *J. Phytopathol.*, 131, 40, 1991.
41. Ward, E. W. B., Susceptibility of immature soybean leaves to *Phytophthora* species, *Physiol. Mol. Plant Pathol.*, 34, 393, 1989.
42. Wyatt, S. E., Pan, S. Q., and Kuc, J., β-1,3-Glucanase, chitinase and peroxidase activity in tobacco tissues resistant and susceptible to blue mould as related to flowering, age and sucker development, *Physiol. Mol. Plant Pathol.*, 39, 433, 1991.
43. Moerschbacher, B. M., Witte, E., Koenigs, D., and Reisener, H.-J., Changes in the level of enzyme activities involved in lignin biosynthesis during the temperature sensitive resistant response of wheat (Sr6) to stem rust (P6), *Plant Sci.*, 65, 183, 1989.

44. Seevers, P. M. and Daly, J. M., Studies on wheat stem rust resistance controlled at the *Sr6* locus. I. The role of phenolic compounds, *Phytopathology,* 60, 1322, 1970.
45. Seevers, P. M. and Daly, J. M., Studies on wheat stem rust resistance controlled at the *Sr6* locus. II. Peroxidase activities, *Phytopathology,* 60, 1642, 1970.
46. Reuveni, R., Bar-Zur, A., and Shimoni, M., A rapid detection procedure of the *HtN* gene under controlled artificial inoculation of maize with *Exserohilum turcicum, Plant Dis.,* 77, 580, 1993.
47. Cramer, C., Ryder, T., Bell, J., and Lamb, C., Rapid switching of plant gene expression induced by fungal elicitor, *Science,* 227, 1240, 1985.
48. Lamb, C., Bell, J., Norman, P., Lawton, M., Dixon, R., Rowell, P., and Bailey, J., Early molecular events in the phytoalexin defense response, in *Structure and Function of the Plant Genome,* Dure, L. and Ciferri, O., Eds., NATO Adv. Study Institute, Plenum Press, New York, 1983.
49. Lawton, M. A., Dixon, R., Hahlbrook, K., and Lamb, C., Elicitor induction of m-RNA activity; rapid effects of elicitor on phenylalanine ammonia-lyase and chalcone synthetase m-RNA activities in bean cells, *Eur. J. Biochem.,* 130, 131, 1983.
50. Lawton, M. A. and Lamb, C. J., Transcriptional activation of plant defense genes by fungal elicitor, wounding and infection, *Mol. Cell Biol.,* 7, 335, 1987.
51. Kusuge, T., The role of phenolics in host response to infection, *Annu. Rev. Phytopathol.,* 7, 195, 1969.
52. Hammerschmidt, R. and Kuc, J., Lignification as a mechanism for induced systemic resistance in cucumber, *Physiol. Plant Pathol.,* 20, 61, 1982.
53. Ride, J. P., Lignification in wounded wheat leaves in response to fungi and its possible role in resistance, *Physiol. Plant Pathol.,* 5, 124, 1975.
54. Vance, C. P., Kirk, T. K., and Sherwood, R. T., Lignification as a mechanism of disease resistance, *Annu. Rev. Phytopathol.,* 18, 259, 1980.
55. Pollock, C. J. and Drysdale, R. B., The role of phenolic compounds in the resistance of tomato cultivars to *Verticillium alboatrum, Phytopathol. Z.,* 86, 56, 1976.
56. Rahtmell, W. G., Phenolic compounds and phenylalanine ammonia-lyase activity in relation to phytoalexin biosynthesis in infected hypocotyls of *Phaseolus vulgaris, Physiol. Plant Pathol.,* 3, 259, 1973.
57. Tena, M. and Valbuena, L. R., Increase in phenylalanine ammonia lyase activity caused by *Plasmopara palstedii* in sunflower seedlings resistant and susceptible to downy mildew, *Phytopathol. Z.,* 107, 47, 1983.
58. Macko, V., Woodbury, W., and Stahmann, M. A., The effect of peroxidase on the germination and growth of mycelium of *Puccinia graminis* f. sp. *tritici, Phytopathology,* 58, 1250, 1969.
59. Leherer, R. E., Antifungal effects of peroxidase systems, *J. Bacteriol.,* 99, 361, 1969.
60. Lovrekovich, L., Lovrekovich, H., and Stahmann, M. A., The importance of peroxidase in the wildfire disease, *Phytopathology,* 58, 193, 1968.
61. Peng, M. and Kuc, J., Peroxidase-generated hydrogen peroxide as a source of antifungal activity in vitro and on tobacco leaf disks, *Phytopathology,* 82, 696, 1992.
62. Shimoni, M., Bar-Zur, A., and Reuveni, R., The association of peroxidase activity and resistance of maize to *Exserohilum turcicum, J. Phytopathol.,* 131, 315, 1991.
63. Dean, R. and Kuc, J., Induced systemic protection in plants, *Trends Biotechnol.,* 3, 125, 1985.
64. Hames, B. D., An introduction to polyacrylamide gel electrophoresis, in *Gel Electrophoresis of Proteins: a Practical Approach,* Hames, B. D. and Rickwood, D., Eds., IRL Press, Oxford, 1981, 1.
65. Elnaghy, M. A. and Show, M., Correlation between resistance to stem rust and the concentration of a glucoside in wheat, *Nature,* 210, 417, 1966.
66. Shimoni, M. and Reuveni, R., A method for staining and stabilizing peroxidase activity in polyacrylamide gel electrophoresis, *Anal. Biochem.,* 175, 35, 1988.
67. Bartnicki-Garcia, S., Cell wall chemistry, morphogenesis and taxonomy of fungi, *Annu. Rev. Microbiol.,* 22, 87, 1969.
68. Abeles, F. B., Bosshart, R. P., Forrence, L. E., and Habig, W. H., Preparation and purification of glucanase and chitinase from bean leaves, *Plant Physiol.,* 47, 129, 1971.
69. Broekaert, W. F., Van Parijs, J., Allen, A. K., and Peumans, W. J., Comparison of some molecular, enzymatic and antifungal properties of chitinase from thorn-apple, tobacco and wheat, *Physiol. Mol. Plant Pathol.,* 33, 319, 1988.
70. Dixon, G. R. and Pegg, G. F., Hyphal lysis and tylose formation in tomato cultivars infected by *Verticillium albo-atrum, Trans. Br. Mycol. Soc.,* 53, 109, 1969.
71. Kean, N. T. and Yoshikawa, M., β-1,3-Endoglucanase from soybean releases elicitor-active carbohydrates from fungus cell walls, *Plant Physiol.,* 71, 460, 1983.

72. Schlumbaum, A., Mauch, F., Vogeli, U., and Boller, T., Plant chitinases are potent inhibitors of fungal growth, *Nature,* 324, 365, 1986.
73. Mauch, F., Mauch-Mani, B., and Boller, T., Antifungal hydrolases in pea tissue. II. Inhibition of fungal growth by combinations of chitinase and β-1,3-glucanase, *Plant Physiol.,* 88, 936, 1988.
74. Shimoni, M., Bar-Zur, A., and Reuveni, R., The relation between isozymes of β-1,3-glucanase and resistance of near-isogenic maize inbred lines to *Exserohilum turcicum, Can. J. Plant Pathol.,* 14, 285, 1992.
75. Grassmann, W., Zechmeister, L., Bender, R., and Toth, G., Ueber die Chitinspaltung durch Emulsin-praeparate. III. Mitteilungen ueber enzymatosche Spaltung von Polysacchariden, *Berichte,* 67, 1, 1934.
76. Kauffmann, S., Legrand, M., Geoffroy, P., and Fritig, B., Biological function of 'pathogenesis-related' proteins: four PR proteins of tobacco have 1,3-β-glucanase activity, *EMBO J.,* 6, 3209, 1987.
77. Loon, L. C. van, Pathogenesis-related proteins, *Plant Mol. Biol.,* 4, 111, 1985.
78. Boller, T., Induction of hydrolases as a defense reaction against pathogens, *UCLA Symp. Mol. Cell. Biol. New Ser.,* 22, 247, 1985.
79. Boller, T., Hydrolytic enzymes in plant disease resistance, in *Plant-Microbe Interactions, Molecular and Genetic Perspectives,* Vol. 2, Kosuge, T. and Nester, E. W., Eds, MacMillan, New York, 1987, 385.
80. Mauch, F., Hadwiger, I. A., and Boller, T., Antifungal hydrolases in pea tissue. I. Purification and characterization of two chitinases and two β-1,3-glucanases differentially regulated during development and in response to fungal infection, *Plant Physiol.,* 87, 325, 1988.
81. Wargo, P. M., Lysis of the cell wall of *Armillaria melles* by enzymes from forest trees, *Physiol. Plant Pathol.,* 19, 391, 1975.
82. Young, D. H. and Pegg, G. F., Purification and characterization of 1,3-β-glucano-hydrolases from healthy and *Verticillium albo-atrum* infected tomato plants, *Physiol. Plant Pathol.,* 19, 391, 1981.
83. Netzer, D., Kritzman, G., and Chet, I., β-(1,3)-Glucanase activity and quantity of fungus in relation to *Fusarium* wilt in resistant and susceptible near-isogenic lines of muskmelon, *Physiol. Plant Pathol.,* 14, 47, 1979.
84. Skuijns, J. J., Potgieter, H. J., and Alexander, M., Dissolution of fungal cell walls by a streptomycete chitinase and β-1,3-glucanase, *Arch. Biochem. Biophys.,* 111, 358, 1965.
85. Huber, D. J. and Nevins, D. J., β-D-Glucan hydrolase activity in *Zea* coleoptile cell walls, *Plant Physiol.,* 65, 768, 1980.
86. Huber, D. J. and Nevins, D. J., Partial purification of endo- and exo-β-D-glucanase enzymes from *Zea mays* L. seedlings and their involvement in cell wall autolysis, *Planta,* 151, 206, 1981.
87. Jondle, D. J., Coors, J. A., and Duke, S. H., Maize leaf β-1,3-glucanase activity in relation to resistance to *Exserohilum turcicum, Can. J. Bot.,* 67, 263, 1989.
88. Kivilaan, A., Bandurski, R. S., and Schulze, A., A partial characterization of autolytically solubilized cell wall glucan, *Plant Physiol.,* 48, 389, 1971.
89. Lee, S., Kivilaan, A., and Bandurski, R. S., *In vitro* autolysis of plant cell walls, *Plant Physiol.,* 42, 968, 1967.
90. Ye, Y. S., Pan, S. Q., and Kuc, S., Association of pathogenesis-related proteins and activities of peroxidase β-1,3-glucanase and chitinase with systemic induced resistance to blue mould of tobacco but not to systemic tobacco mosaic virus, *Physiol. Mol. Plant Pathol.,* 36, 523, 1990.
91. Pan, S. Q., Ye, Y. S., and Kuc, J., Direct detection of β-1,3-glucanase isozymes on poly-acrylamide electrophoresis and isoelectro-focusing gels, *Anal. Biochem.,* 182, 136, 1989.
92. Nasser, W., Tapia, M. de, and Burkard, G., Maize pathogenesis-related proteins: characterization and cellular distribution of 1,3-β-glucanase and chitinases induced by brome mosaic virus infection or mercuric chloride treatment, *Physiol. Mol. Plant Pathol.,* 36, 1 1990.
93. Boller, T. and Kende, H., Hydrolytic enzymes in the central vacuole of plant cells, *Plant Physiol.,* 83, 1123, 1979.
94. Metraux, J. P. and Boller, T., Local and systemic induction of chitinase in cucumber plants in response to viral, bacterial and fungal infections, *Physiol. Mol. Plant Pathol.,* 28, 161, 1986.
95. Tuzun, S., Rao, M. N., Vogelie, U., Schardl, C. L., and Kuc, J., Induced systemic resistance to blue mold: early induction and accumulation of β-1,3-glucanase, chitinase and other pathogenesis-related proteins (b-proteins) in immunized tobacco, *Phytopathology,* 79, 979, 1989.
96. Boller, T. and Metraux, J. P., Extracellular localization of chitinase in cucumber, *Physiol. Mol. Plant Pathol.,* 33, 11, 1988.
97. Metraux, J. P., Streit, L., and Staub, T., A pathogenesis-related protein in cucumber is chitinase, *Physiol. Mol. Plant Pathol.,* 33, 1, 1988.

98. Pan, S. Q., Ye, Y. S., Tusun, J., and Kuc, J., Chitinase isozyme pattern and its coordinated induction with β-1,3-glucanase in tobacco plants immunized by *Peronospora tabacina* and tobacco mosaic virus, *Phytopathology,* 80, 1024, 1990.
99. Pan, S. Q., Ye, Y. S., and Kuc, J., A technique for detection of chitinase; β-1,3-glucanase and protein patterns after a single separation using polyacrylamide gel electrophoresis or isoelectro-focusing, *Phytopathology,* 81, 970, 1991.
100. Somogyi, M., Notes on sugar determination, *J. Biol. Chem.,* 195, 19, 1952.
101. Nelson, N., A photometric adaptation of the Somogyi method for the determination of glucose, *J. Biol. Chem.,* 153, 275, 1944.
102. Ashwell, G., Colorimetric analysis of sugars, *Methods Enzymol.,* 3, 73, 1971.
103. Boller, T. and Mauch, F., Colorimetric assay for chitinase, *Methods Enzymol.,* 161, 430, 1988.
104. Boller, T., Gehri, A., Kauch, F., and Vogeli, V., Chitinase in bean leaves: induction by ethylene, purification, properties and possible function. *Planta,* 157, 22, 1983.
105. Davis, B. J., Disc-electrophoresis; method and application to human serum proteins, *Ann. N.Y. Acad. Sci.,* 2, 404, 1964.
106. Newton, R. and Anderson, J. A., Studies on the nature of rust resistance in wheat. IV. Phenole compounds of the wheat plant, *Can. J. Res.,* 1, 86, 1928.
107. Walker, J. C. and Link, K. P., Toxicity of phenolic compounds to certain onion bulb parasites, *Bot. Gaz.,* 96, 468, 1935.
108. Hahlbrock, K. and Grisebach, H., Enzymic controls in the biosynthesis of lignin and flavonoids, *Annu. Rev. Plant Physiol.,* 30, 105, 1979.
109. Ebel, J. and Hahlbrock, K., Biosynthesis, in *The Flavonoids, Advances in Research,* Harborne, J. B. and Mabry, T. J., Eds., Chapman & Hall, New York, 1982, 641.
110. Dixon, R. A., Dey, P. M., and Lamb, C. J., Phytoalexins: enzymology and molecular biology, *Adv. Enzymol. Relat. Areas Mol. Biol.,* 55, 1, 1983.
111. Jacobs, M. and Rubery, P. H., Naturally occurring auxin transport regulators, *Science,* 241, 346, 1988.
112. Ecker, J. R. and Davis, R. W., Plant defense genes are regulated by ethylene, *Proc. Natl. Acad. Sci. U.S.A.,* 84, 5202, 1987.
113. Mclean, J. G., Le Tourneau, D. J., and Guthrie, J. W., Relation of histochemical tests for phenols to *Verticillium* wilt resistance of potatoes, *Phytopathology,* 51, 84, 1961.
114. Sztejenberg, A., Azaizia, H., and Chet, I., The possible role of phenolic compounds in resistance of horticultural crops to *Dematophora necatrix* Hartig, *Phytopathol. Z.,* 107. 318, 1983.
115. Kiraly, Z. and Farkas, G. L., Relation between phenols' metabolism and stem rust resistance in wheat, *Phytopathology,* 52, 657, 1962.
116. Skipp, R. A. and Samborski, D. J., The effect of the *Sr6* gene for host resistance on histological events during the development of stem rust in near-isogenic wheat lines, *Can. J. Bot.,* 52, 1107, 1974.
117. Fuchs, A., Rohringer, R., and Samborski, D. J., Metabolism of aromatic compounds in healthy and rust-infected primary leaves of wheat. II. Studies with L-phenylalanine-U-^{14}C, L-tyrosine-U-^{14}C, and ferulate-U-^{14}C, *Can. J. Bot.,* 45, 2137, 1967.
118. Moerschbacher, B. M., Noll, U. M., Flott, B. E., and Reisener, H.-J., Lignin biosynthetic enzymes in stem rust infected, resistant and susceptible near-isogenic wheat lines, *Physiol. Mol. Plant Pathol.,* 33, 33, 1988.
119. Southerton, S. G. and Deverall, B. J., Changes in phenylalanine ammonia-lyase and peroxidase activities in wheat expressing resistance to the leaf rust fungus, *Plant Pathol.,* 39, 223, 1990.
120. Southerton, S. G. and Deverall, B. J., Histochemical and chemical evidence for lignin accumulation during the expression of resistance to leaf rust fungi in wheat, *Physiol. Mol. Plant Pathol.,* 36, 483, 1990.
121. Beardmore, J., Ride, J. P., and Granger, J. W., Cellular lignification as a factor in the hypersensitive resistance of wheat to stem rust, *Physiol. Plant Pathol.,* 22, 209, 1983.
122. Keen, N. T. and Littlefield, L. J., The possible association of phytoalexins with resistance gene expression in flax to *Melampsora lini, Physiol. Plant Pathol.,* 14, 265, 1979.
123. Mayama, S., Tani, T., Matsuura, Y., Ueno, T., and Fukami, H., The production of phytoalexins by oat in response to crown rust, *Puccinia coronata* f. sp. *avenae, Physiol. Plant Pathol.,* 19, 217, 1981.
124. Morris, P. F., Savard, M. E., and Ward, E. W. B., Identification and accumulation of isoflavonoids and isoflavone glucosides in soybean leaves and hypocotyls in resistance responses to *Phytophthora megasperma* f.sp. *glycinea, Physiol. Mol. Plant Pathol.,* 39, 229, 1991.

125. Bray, H. C. and Thorpe, W. W., Analysis of phenolic compounds of interest in metabolism, *Methods Biochem. Anal.,* 1, 27, 1954.
126. Couture, R. M., Routley, D. G. R., and Dunn, G. M., Role of cyclic hydroxamic acids in monogenic resistance of maize to *Helminthosporium turcicum, Physiol. Plant Pathol.,* 1, 515, 1971.
127. Long, B. J., Dunn, G. M., and Routhe, D. G., Relationship of hydroxamate concentration in maize and field reaction to *Helminthosporium turcicum, Crop Sci.,* 18, 573, 1987.
128. Anglade, P. and Molot, P. M., Mise en evidence d'une relation entre les sensibilites de lignes de mais la pyrale (*Ostrinia nubilalis* Hbn.) et a l'helminthosporiose (*Helminthosporium turcicum* Pass.),. *Ann. Epiphyt. (Paris),* 18, 279, 1967.
129. Elnaghy, M. A. and Linko, P., The role of 4-*O*-glucosyl-2,4-dihydroxy-7-methoxy-1,4-benzoxazin-3-one in resistance of wheat to stem rust, *Physiol. Plant.,* 15, 764, 1962.
130. BeMiller, J. N. and Pappelis, A. J., 2,4-Dihydroxy-1,4-benzoxazin-3-one glucoside in corn. I. Relation of water-soluble, 1-butanol-soluble glycoside fraction content of pith cores and stalk rot resistance, *Phytopathology,* 55, 1237, 1965.
131. Hilu, A. M. and Hooker, A. L., Monogenic chlorotic lesion resistance to *Helminthosporium turcicum* in corn seedlings, *Phytopathology,* 53, 909, 1963.
132. Hamilton, R. H., A corn mutant deficient in 2,4-dihydroxy-7-methoxy-1,4-benzoxazin-3-one with an altered tolerance of atrazin, *Weeds,* 12, 27, 1964.
133. Guthrie, W. D., Barry, B. D., Roosman, E. C., and Jarvis, S. L., Correlation between leaf feeding resistance to European corn borer (*Lepidoptera:* Pyralidae) and resistance to Northern leaf blight, *J. Econ. Entomol.,* 78, 811, 1985.
134. Lyons, C. P. and Nicholson, R. L., Evidence that cyclic hydroxamate concentrations are not related to resistance of corn leaves to anthracnose, *Can. J. Plant Pathol.,* 11, 215, 1989.
135. Argandona, V. H., Niemayer, H. M., and Corcuera, L. J., Effect of content and distribution of hydroxamic acids in wheat on infestation by the aphid *Schizaphis graminum, Phytochemistry,* 20, 673, 1981.
136. Hamilton, R. H., Tolerance of several grass species to 2-chloro-s-triazine herbicides in relation to degradation and content of benzoxzinon derivatives, *J. Agric. Food Chem.,* 12, 14, 1973.
137. Bowman, M. C., Morton, B., and Kalum, J. A., Photo-fluorometric determination of 6-methoxy-2-linone, an indicator of resistance to European borer in *Zey mays, J. Econ. Entomol.,* 120, 1969.
138. Klun, J. A. and Brindley, T. A., Role of 6-methoxybenzoxazolinone in inbred resistance of host plant (maize) to first-brood larvae of European corn borer, *J. Econ. Entomol.,* 59, 711, 1966.
139. Long, B. J., Dunn, G. M., and Routhy, D. G., Rapid procedure for estimating cyclic hydroxamate (DIMBOA) concentration in maize, *Crop Sci.,* 14, 601, 1974.

FURTHER READING

Bailey, J. A. and Deverall, B. J., Eds., *The Dynamics of Host Defense,* Academic Press, New York, 1983.
Boller, T., Hydrolytic enzymes in plant disease resistance, in *Plant-Microbe Interactions, Molecular and Genetic Perspectives,* Vol. 2, Kosuge, T. and Nestor, E. W., Eds., MacMillan, New York, 1987.
Hames, B. D. and Rickwood, D., Eds., *Gel Electrophoresis of Proteins: a Practical Approach,* IRL Press, Oxford, 1981.
Singh, R. S., Singh, U. S., Hess, W. M., and Weber, D. J., Eds., *Experimental and Conceptual Plant Pathology,* Vol. III, *Defence,* Gordon & Breach Science Publishers, New York, 1988.

Chapter 9

Isozymes: Methods and Applications*

J. A. Micales and M. R. Bonde

TABLE OF CONTENTS

Keywords—Isozymes, plant pathology, genetics, taxonomy

I. INTRODUCTION

Isozyme analysis is a powerful biochemical technique with numerous applications in plant pathology. It has long been used by geneticists to study the population genetics of fish, mammals, insects, nematodes, and higher plants. Mycologists and plant pathologists more recently adopted the procedure, and it is now being used routinely to settle taxonomic disputes, identify "unknown" cultures, "fingerprint" patentable fungal lines and plant cultivars, analyze genetic variability, trace pathogen spread, follow the segregation of genetic loci, and determine ploidy levels of fungi and other plant pathogens. These topics have been recently reviewed.[1,2] The large number of publications in this field each year indicates the widespread interest in isozyme analysis.

In this paper, we discuss some major applications of isozyme analysis in basic and applied plant pathology. The technique is particularly useful with fungi; the greatest advances have been mostly with fungal pathogens. Isozyme banding patterns obtained from fungi are usually relatively uncomplicated and easy to interpret. Isozyme analysis can be readily performed in most laboratories with relatively little expense. With the development of computer programs that enable large numbers of comparisons at the gene level, much information can be obtained about the population genetics and life cycle of the organism. Isozyme analysis has proven particularly useful in situations where it is necessary to differentiate among two or more morphologically similar fungi. These and other uses for isozyme analysis will be discussed, along with advantages and disadvantages as compared to alternative techniques.

* Forest Products Laboratory is maintained in cooperation with the University of Wisconsin. This article was written and prepared by U.S. Government employees on official time, and it is therefore in the public domain and not subject to copyright.

0-87371-877-1/95/$0.00+$.50

II. PRINCIPLES

Isozymes are defined as multiple molecular forms of a single enzyme. These forms usually have similar, if not identical, enzymatic properties, but slightly different amino acid compositions due to differences in the nucleotide sequence of the DNA that codes for the protein. Often the only difference among isozymes is the substitution of one to several amino acids.

Only those isozymes that have large variations in size or shape or that differ in net charge can be separated by electrophoresis. Differences in net charge can occur when a basic amino acid, such as lysine, is substituted for an acidic amino acid, such as aspartic acid. Only 28.7% of all amino acid substitutions will change the net charge of a protein.[3] Some amino acid substitutions that do not involve charge differences can also affect the electrophoretic mobility of a protein, presumably by altering the tertiary structure of the enzyme.[4] Thus, about one third of all single amino acid substitutions will be electrophoretically detectable, and several simultaneous substitutions can cancel out the effect. Isozyme analysis therefore provides a very conservative estimate of the extent of genetic variability within a population.[3]

Detectable isozymes can arise from three different genetic and biochemical conditions: (1) multiple alleles at a single locus, (2) single or multiple alleles at multiple loci, and (3) secondary isozymes, usually arising from post-translational processing.

A. MULTIPLE ALLELES AT A SINGLE LOCUS

In a fungal population, any given genetic locus can be monomorphic (i.e., expresses a single allele in 99% or more of the population) or polymorphic (i.e., expresses more than one allele in 99% of the population). When a genetic locus is polymorphic, the isozymes formed by the expression of the different alleles are termed "allozymes". Each allele codes for a structurally distinct version of a particular polypeptide chain. The primary structure of an allozyme therefore depends on the number of alleles present and their nucleotide sequences. The number of alleles in any organism varies with its nuclear condition (monokaryotic, dikaryotic), ploidy number (haploid, diploid, polyploid), and genetic makeup (homozygous, heterozygous). The allozymes of individuals that are haploid or homozygous produce simple electrophoretic banding patterns due to the expression of a single allele. Allozymes of organisms that are diploid or dikaryotic and heterozygous produce more complex banding patterns due to the expression of two separate alleles.

Enzymes can consist of one or more polypeptide chains. Monomeric enzymes consist of a single polypeptide chain; multimeric (or oligomeric) enzymes are comprised of two or more polypeptide chains. Most multimeric enzymes are either dimeric (two chains) or tetrameric (four chains). The electrophoretic banding pattern obtained for monomeric enzymes is usually simple and easy to interpret, even if the organism is heterozygous (Figure 1). Each allele is expressed as a single polypeptide band. The heterozygous condition appears as a mixture of isozymes produced by the two corresponding homozygotes.

More complicated patterns are formed in heterozygotes when the enzymes are multimeric due to the formation of intermediate, "heteromeric" (or hybrid) bands (Figure 1). These are in addition to the two "homomeric" forms associated with each homozygote. For example, if allele A codes for polypeptide a, and allele A′ codes for polypeptide a′, the following allozymes will be formed for a tetrametic enzyme: aaaa (homomeric), a′a′a′a′ (homomeric), aaaa′ (heteromeric), aaa′a′ (heteromeric), and aa′a′a′ (heteromeric). Other examples of heteromeric band formation for a single locus are detailed in Figures 1 and 2.

The frequency of occurrence of the forms of multimeric enzymes, assuming completely random combination of the polypeptide chains, should follow Mendelian ratios. The ratio of all possible isozymes formed in heterozygotes (assuming two alleles in the population) is 1:1 for monomers, 1:**2**:1 for dimers, 1:**3**:**3**:1 for trimers, and 1:**4**:**6**:**4**:1 for tetramers (heteromeric bands are boldfaced) (Figure 1). The relative quantities of the different isomers can often be recognized by differences in staining intensity; the heteromeric bands, which have a higher probability of being formed and are therefore present in larger quantities, should stain darker or more intensely. In some instances, a particular polypeptide will not contribute equally to the activity of the enzyme due to a slower rate of synthesis, low stability, or a tendency to break down before it can be assembled into the final enzyme. Certain polypeptide chains may reduce the activity of an enzyme by decreasing its stability or by reducing its catalytic ability. In such cases, the enzymes would not be detected in the expected ratios of staining intensity on the gel.

Extremely complex banding patterns may be obtained if three or more alleles are present in a population (Figure 2). Numerous heteromeric bands may be resolved. In most cases, the interpretation of such complex banding patterns should be confirmed by comparison to crosses of known genotypes; otherwise, readers and editors will be skeptical that the data were interpreted properly. The polypeptide composition of many enzymes is often conserved among organisms.[6,7] The genetic interpretation of banding patterns is much simpler when the polymeric composition of the enzyme is known.

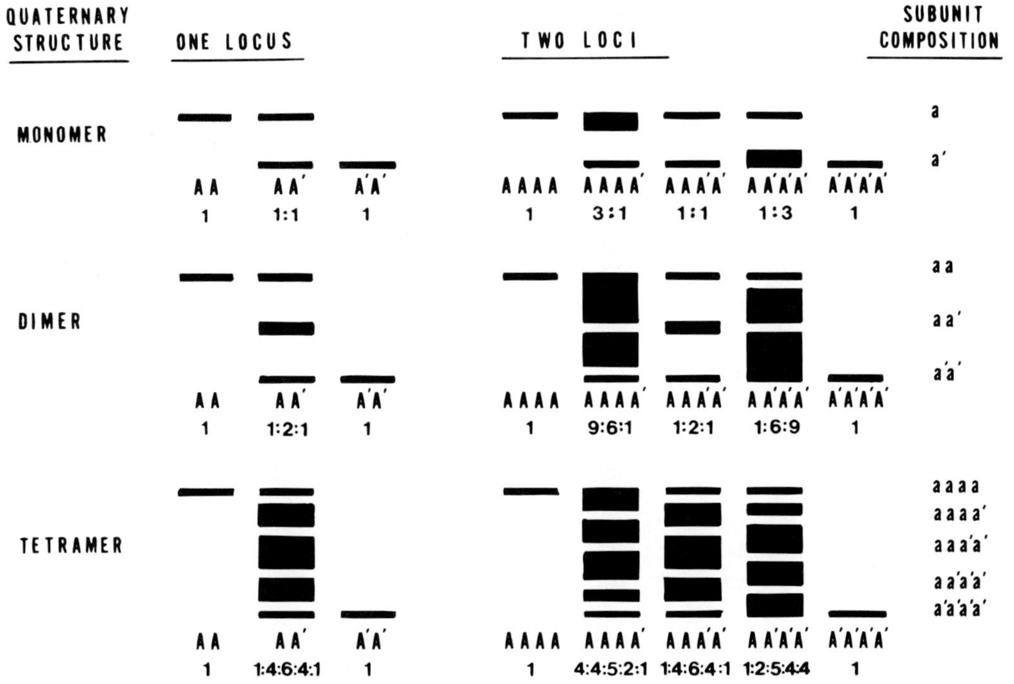

Figure 1 Predicted banding patterns for one locus with two alleles (A and A′) and two segregating loci that share the same alleles for monomeric, dimeric, and tetrameric enzymes of a diploid or dikaryotic organism. Genotypes (in capital letters) are listed below each banding pattern. Subunit composition of each protein band is shown on right; lowercase letters represent subunit designations. The expected ratios of banding intensity for each phenotype is presented beneath the genotype. (Adapted from May, Reference 5.)

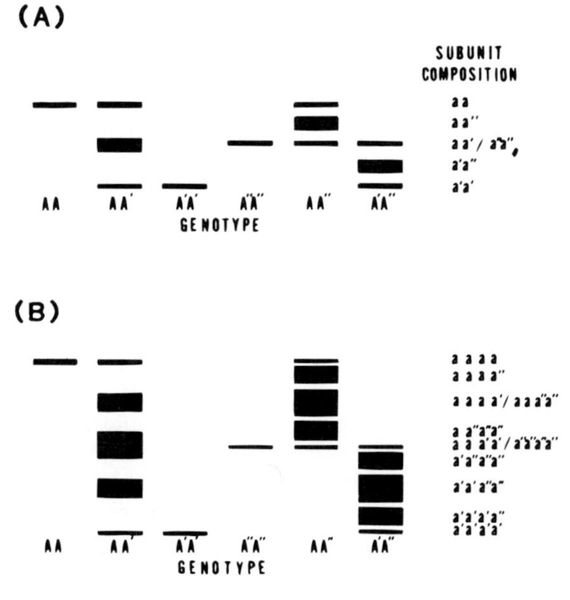

Figure 2 Predicted banding patterns of dimeric (A) and tetrameric (B) enzymes coded by a single locus and three electrophoretically distinct alleles (A, A′, and A″) in a diploid or dikaryotic organism. Abbreviations as in Figure 1. (Adapted from May, Reference 5.)

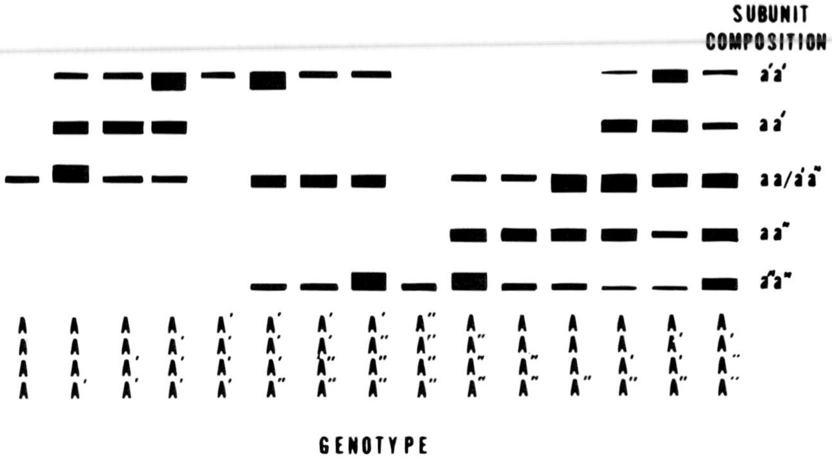

Figure 3 Predicted banding patterns of a dimeric enzyme coded by two loci that share the same electrophoretically distinct alleles (A, A′, and A″). Abbreviations as in Figure 1. (Adapted from May, Reference 5.)

B. SINGLE OR MULTIPLE ALLELES AT MULTIPLE LOCI

Multiple loci may also code for a series of isozymes.[6] Some stains, such as esterase and acid/alkaline phosphatase, are not very specific and detect broad classes of isozymes, often at multiple loci. Different loci can also be expressed in different tissues of an organism or are compartmentalized in different areas of the cell. Malate dehydrogenase, for example, is often expressed by two different loci for cytoplasmic and mitochondrial forms.[6] The distribution of enzymes within an organism is usually constant for a species. Isozymes coded by different loci are often detected in separate regions of the electrophoretic gel due to their greater differences in charge and conformation than usually associated with multiple alleles at a single locus. Heteromeric bands can form from polypeptides coded by different loci. Banding patterns may be quite complex when a number of different loci and alleles are expressed (Figures 1 and 3). It is often difficult to provide a genetic interpretation of such banding patterns.

C. SECONDARY ISOZYMES

Electrophoretic bands may not appear to follow expected genetic patterns due to post-translational processing and other events that form secondary isozymes. Common modifications include deamidation, acetylation, oxidation of sulfhydryl groups, additions and removals of carbohydrate and phosphate moieties, cleavage by proteases, and aggregation or polymerization of protein. The formation of secondary isozymes is usually uniform within a species or group and can often be recognized by the production of a series of closely migrating bands for each allele.[6]

Glycoproteins, which can have large amounts of carbohydrate covalently attached to the protein backbone, will often display a series of electrophoretic bands. Alternatively, the isozymes may all migrate electrophoretically as a large complex, aggregated together by the carbohydrate, and fail to resolve into a tight band. Glycoproteins can be treated with carbohydrate-degrading enzymes to remove the associated carbohydrate. Procedures for working with glycoproteins are given by Beeley.[8]

Conformational isomerism may also generate secondary isozymes. Some enzymes may have several stable configurations that vary in tertiary or quaternary structure. Such forms frequently have different electrophoretic mobilities. A single preparation will usually contain all possible configurations. Conformational isomers will appear as a closely migrating series of bands for each allele.

Enzymes that require cofactors, such as flavins or B vitamins, may vary in their electrophoretic mobility, depending on the degree of saturation of the enzyme with the cofactor. Cofactors and substrates should not be limiting in staining solutions, or inconsistent results may occur. Cofactors and substrates can be incorporated into the gel or sample buffer before electrophoresis to help maintain the activity and stability of the enzyme.[6]

Proteolysis during extraction and storage may also be responsible for artifactual, secondary bands. Samples should be kept cold (below 4°C) during extraction. Proteinase inhibitors, such as phenylmethylsulfonylflouride (PMSF), also can be added to the sample buffer to prevent proteolysis (Section III).

Proteins may aggregate in a sample, especially if the pH or the ionic strength of the buffer is incorrect. Protein aggregation will result in poor resolution and uninterpretable banding patterns. Several different buffers should be tried during a preliminary "screening" run to determine which buffers prevent aggregation and provide the best resolution for a particular enzyme.

III. METHODOLOGY

A. SAMPLE SELECTION AND PREPARATION

Sample selection and preparation are the most critical steps in isozyme analysis. The quality of genetic information obtained from an experiment is only as good as the sampling of isolates from which the data are derived. The number of isolates and their geographic and host range will all affect data interpretation. Care in sample preparation is also essential for a successful study. Poor resolution, faint staining or absence of bands, and irregular banding patterns can be caused by the incorrect choice of sample buffer or improper extraction techniques. Sample preparation has been thoroughly discussed elsewhere.[1,2]

The specific activity stains used in isozyme analysis detect only active enzymes; denaturation must be prevented during and after sample preparation. Samples should be kept cold (below 4°C) during preparation and storage. They can often be frozen at −80°C for up to 1 year, but this can vary for different organisms. Repeated freezing and thawing will result in denaturation. Enzymes may aggregate or precipitate if they are in concentrations that are too high or low or if the sample buffer is of the wrong pH or ionic strength. A commonly used sample buffer is Tris-HCl (0.01 to 0.1 *M* pH 6.8 to 7.5). Chelating agents, protease inhibitors, and enzyme stabilizers, such as EDTA, polyvinylpyrrolidone, PMSF, and dithiothreitol, can be added to the sample buffer to increase resolution. *Note: many of these reagents are extremely toxic. Use proper safety precautions to avoid contact with these compounds.* 2-Mercaptoethanol (20 μl/100 ml sample buffer) and bovine serum albumin (4 mg/ml) can also improve resolution by reducing the effects of resins, phenolics, and free fatty acids. Such contaminants are usually more of a problem when plant tissue is being extracted, although some fungal pigments fall into this category. The addition of small quantities of substrate (20 mg/100 ml sample buffer) may also help to stabilize some enzymes. Several substrates can be incorporated into a single sample buffer as long as they do not interact with each other. Alternatively, substrates can be added to the gel or electrode buffer, although this is generally not as effective.[9]

B. ELECTROPHORETIC TECHNIQUES

Different electrophoretic techniques can be used to separate isozymes, including starch gel electrophoresis, polyacrylamide gel electrophoresis (PAGE), isoelectric focusing, and two-dimensional electrophoresis. Advantages, disadvantages, and protocol references for different electrophoretic procedures have been summarized.[2] Traditionally, isozyme analysis was performed with starch or PAGE, but isoelectric focusing is now being used more commonly. Isoelectric focusing and two-dimensional electrophoresis resolve larger numbers of isozymes than do the other techniques, but the electrophoretic banding patterns may not lend themselves to a genetic interpretation due to their complexity.

The identification and visualization of individual enzymes using specific activity stains requires the presence of active, nondenatured enzymes. Dissociating procedures, such as SDS-PAGE (polyacrylamide gel electrophoresis in which the proteins are denatured with the detergent sodium dodecyl sulfate), cannot be used for isozyme analysis.

C. STAINING

Isozymes are visualized on the electrophoretic gel by reaction with specific activity stains. Detection of specific enzymes is possible because the appropriate substrates and cofactors required for activity are provided in the staining solution. The enzymatic reaction forms a colored product, either through direct activity with a dye or by involving other enzymes in a series of reactions with the generation of a colored product as a final result. For some enzymes, such as superoxide dismutase, isozymes are seen as white bands on a dark background. Fluorescent products can be detected with ultraviolet light. Conversely, nonfluorescent products can be visualized as "negatively stained" by reacting the starch with a fluorescent compound. The biochemistry of the different staining reactions has been discussed,[6,10] and stain "recipes" for many different enzymes have been described.[6,9,11–15]

The selection of enzymes to study is an important part of isozyme analysis and can have a dramatic impact on the results and genetic interpretation of the data. Some stains, such as esterases, phosphatases, and peroxidases, are not substrate specific and detect entire groups of enzymes. Isozymes that are coded by multiple loci are usually resolved by these stains, so banding patterns are often very complex and difficult to interpret. In addition, esterases, phosphatases, and other nonregulatory enzymes usually display more genetic variation than do regulatory enzymes involved with energy metabolism.[16] Utilization of only nonregulatory enzymes may detect a disproportionately high level of intraspecific variation and greatly overestimate the amount of genetic diversity in a population. On the other hand, the variability associated with nonregulatory enzymes may be desirable for identifying or "fingerprinting" subspecific taxa such as races or *formae speciales.*

D. GENETIC INTERPRETATION

Electrophoretic banding patterns of isozymes can be interpreted in terms of the alleles and loci that code for the polypeptides. Specific banding patterns are associated with certain genetic conditions, as described in Section II, and can be easily recognized. Journals often accept such genetic interpretations of the data without concurrent crossing experiments between isolates as long as the banding patterns are clear and always consistent with the genetic interpretation.

Electrophoretic data can be presented in many different forms. Statistical methods have been developed for population genetics, numerical taxonomy, and cladistics (i.e., systematics based on phylogenetic relationships) to express relatedness among samples. Each band on a gel can be assigned a descriptive value based either on the net migration of the band from the origin (an Rf value) or its position relative to that of the band coded by the most common allele.[11] Data can be analyzed using a variety of tests. If no genetic interpretation is planned, each pair of isolates can be compared using a variety of simple matching coefficients.[17]

More complex comparisons can be made when the data are interpreted in terms of loci and allele frequency. The data can be expressed in terms of genetic similarity (which describes the closeness of the relationship of two individuals or populations) or genetic distance (which indicates the amount of dissimilarity between two individuals or populations). Similarity values vary from 0 to 1.0; closely related organisms have values close to 1.0. Distance values vary from 0 to infinity; closely related organisms have distance coefficients close to 0. Different formulas can be used for calculating genetic distance and similarity.[17–21] Theoretical aspects of the use of some of these different statistical values are discussed by Buth.[22] Examples of calculations are presented by Ferguson.[23]

Matching, similarity, and distance coefficients can then be subjected to cluster analysis using multivariate analysis or other clustering procedures[17] to group together the different individuals or populations that resemble each other and to identify causes of variability (i.e., geographic location, subspecific groupings, etc.). Several cluster analysis programs are available that use slightly different parameters to analyze the data. Two of the most commonly used clustering procedures are single-linkage cluster analysis and unweighted pair group mean average cluster analysis.[17,24] More complex cladistic procedures can also be used.[25] The relationships of individual isolates or entire populations can be summarized in the form of clusters or dendrograms.[23,26] Most standard statistical software packages will perform the calculations necessary for isozyme analysis. More specialized programs are also available (e.g., "Allozyme," R. Struss, University of Arizona, Tucson).

IV. APPLICATIONS OF ISOZYME ANALYSIS

Many aspects of plant pathology, both applied and basic, can be studied with isozyme analysis.[2,27] Most applications have involved fungal pathogens, but the technique has also been used for nematodes[28] and bacteria.[29,30] Isozymes are frequently used by plant geneticists and breeders as genetic markers for resistance. This application is very important for plant pathology, but it is beyond the scope of this paper. Readers interested in using isozyme analysis to study host plants are referred to Conkle et al.,[9] Cheliak and Pitel,[31] Conkle,[32] and Tanksley and Orton.[33]

A. TAXONOMY

Isozyme analysis is frequently used for taxonomic purposes, especially when a taxon is morphologically diverse or plastic. In most cases, fungal species are easily differentiated by electrophoresis. The technique is commonly used to make recommendations on the separation or combination of species.[34–39] Subspecies, varieties, and intersterility groups have also been separated.[40–44] Simple band-counting procedures can be used to distinguish taxa, although cladistic and phylogenetic information can be derived from the allelic frequencies and ratios derived from a genetic interpretation of the data. Isozyme analysis is most successful in distinguishing species and subspecies when the amount of intraspecific genetic variation is limited within a population. Otherwise, intraspecific variability will obscure interspecific differences. The selection of enzyme systems is very important in taxonomic applications (Section III, C). Any study that uses only nonregulatory enzymes or stains that visualize broad classes of enzymes (such as esterases, alkaline/acid phosphatases, and peroxidases) will display disproportionally high levels of intraspecific variation. Such a study would probably not be able to resolve taxonomic issues. Exaggerated levels of intraspecific variability can often be avoided by using both regulatory and nonregulatory enzymes and by using specific stains that react with single enzymes.

One important taxonomic question is how much variability can be allowed within a taxon before it should be split into a new species or subspecies. Thorpe[45] and Ayala[46] have provided guidelines for the separation of populations, subspecies, and species based on statistical interpretations of isozyme data derived from vertebrates, invertebrates, and plants. Such guidelines should be applied to fungi cautiously. Many genera, in which species are clearly defined, may fall into such discrete categories.[47,48] Other genera consist of poorly delineated species that exist in a continuum or "complex". Using restrictive statistical cutoff values may eliminate outgroups that really belong within the continuum. Other fungal attributes, including differences or plasticity in morphology, cultural characteristics, and host preference, must be taken into account when deciding whether or not an organism needs to be reclassified.

B. IDENTIFICATION OF UNKNOWN ORGANISMS

The ability of isozyme analysis to differentiate species and subspecies leads to its application in the identification of plant pathogens. Isozyme analysis can be used both to identify unknown pathogens and to "fingerprint" commercially important strains. The correct, rapid identification of an unknown pathogen may allow early implementation of control measures that will prevent large economic loss. State and federal agencies also need to be able to identify pathogens of regulatory significance, often from very small samples. Industry must be able to identify commercial strains that have been developed and patented. Of all the applications of isozyme analysis, pathogen identification is the one most important economically. This topic has been recently reviewed.[1]

The identification of unknown pathogens is dependent on the identification of monomorphic loci; i.e., loci that are invariable within a species (or subspecies). This must be determined by screening large numbers of isolates from a broad geographic range for many different enzyme systems and selecting those loci that do not demonstrate intraspecific (or intra-subspecific) variability. Subsequent electrophoretic runs should include a standardized strain of the suspected pathogen for comparison. Enzyme preparations of the standard strain can usually be prepared in large quantities and stored in liquid nitrogen for 1 year.

Isozyme analysis can also be used to identify the various pathogens present in a mixed infection. This has been especially useful in identifying mycorrhizal fungi.[49–51]

The "fingerprinting" of specific strains is dependent on the presence of polymorphic loci within the species. An allele, or combination of several alleles, must be identified that is unique to that particular strain. Often, enzymes that express high degrees of variability, such as esterases, phosphatases, and peroxidases, are useful for this application. Roux and Labarere,[52] for example, found that even closely related strains of *Agaricus bitorquis* (Quel.) Sacc. could be differentiated by their banding patterns for alcohol dehydrogenase, phenoloxidase, esterase, and peroxidase. Strains of commercial mushrooms,[52–56] biological control agents,[57–59] and mycorrhizal fungi[60] have been successfully "fingerprinted" with this technique. Isozyme analysis is easier and considerably less expensive than comparable molecular biological techniques that are used for strain identification, such as restriction fragment length polymorphism.

C. GENETICS

Genetic information about a pathogen can be derived from isozyme analysis, including the amount of genetic variability (i.e., the percent polymorphism) of a species or population, the amount of heterozygosity, the linkage of specific loci, and genetic maps of the chromosomes. As genetic markers, isozymes are useful for studying population structure, tracing epidemics, establishing the origins of new pathogenic forms, and analyzing crosses.[61]

Isozymes and virulence are the most common markers used in fungal population genetics. Isozymes are generally more selectively neutral than virulence genes and usually demonstrate less variability.[62,63]

The inheritance of virulence may be quite complicated, involving dominance and recessiveness. The genetic patterns associated with isozymes are usually more simple; isozymes are usually expressed as codominant alleles at one locus or a few loci.[64] Virulence studies are also quite labor intensive, involving large numbers of different hosts, thus restricting sample size.[65] Isozyme tests can easily accommodate large numbers of samples.

The sample size, number of loci studied, and accuracy of species definition are all essential to obtaining valid estimates of genetic diversity. The type of enzyme selected is also important, since some enzymes are known to be more variable than others. Isozyme analyses that look only at esterases, phosphatases, and polyphenoloxidases, for example, would greatly overestimate the amount of variability in the genome.

The amount of genetic variability of a population has important implications for plant pathologists. Pathogens with a large amount of genetic diversity are more likely to become rapidly resistant to fungicides or virulent to resistant hosts. The amount of variability in a species (or population) is often related to the pathogenicity of an organism (Table 1). Obligate pathogens, which are highly specialized and have a relatively uniform substrate and environment, often are very uniform genetically and have low levels of polymorphism, even for neutral markers.[61,76] *Erysiphe graminis* f. sp. *hordei,* for example, is entirely monomorphic for over 50 different loci.[88] Certain hosts, such as barley, may impose strict biochemical requirements on pathogens and prevent the survival of recombinant forms.[88] Facultative pathogens and saprophytes, which find themselves in much more diverse environments and broader host ranges, are usually more genetically variable than obligate pathogens.[61,76] There are exceptions, however. For example, *Uromyces appendiculatus* (Pers.: Pers.) Unger, the causal agent of bean rust, is an obligate, autoecious, macrocyclic pathogen, yet 67% of its loci are polymorphic.[66] In contrast, *Fusarium oxysporum,* a pathogen with a very broad host range, displays only 24% polymorphism.[47]

The amount of polymorphism is also dependent on the amount of sexual reproduction in a population or species. Low levels of genetic diversity are often associated with species that are maintained asexually, such as *Phakopsora pachyrhizi* Sydow[89] and *Puccinia striiformis.*[87] There was no variation among *P. graminis* f. sp. *tritici* collections from Australia, where the fungus is maintained asexually.[86] Collections from the U.S., where sexual reproduction was common until the eradication of the barberry in the 1920s to 1930s, contain 38% polymorphic loci.[78] In contrast, sexual populations of *U. appendiculatus* displayed less genetic diversity than did asexual populations.[66] The authors concluded that mutation and selection would lead to greater divergence and higher levels of polymorphism in an asexual population since there is no exchange of genes. Fungi with a very high reproductive potential, including those that form massive quantities of asexual spores, may possess tremendous genetic diversity due to mutation alone.[70] Lack of genetic variation may also indicate that a pathogen has developed from a limited number of recent introductions and that insufficient time has elapsed for variation to develop.[70,82] Low variability may also suggest that electrophoretic variants are unfit for survival or that coadapted isozyme complexes, which have similar electrophoretic migration rates, may exist.[64] Clearly, genetic diversity (or uniformity) can result from several different factors.

Genetic diversity can also be measured as the average number of alleles per locus or by percentage of heterozygosity in a population or species. Again, the selection of enzymes is very important (see Section III) since the loci must be representative of the genome. Heterozygous banding patterns are often readily recognizable (Section II). The frequency of heterozygous loci appears to be quite variable in fungi (Table 2). In most instances, estimates of heterozygosity are extremely conservative, since electrophoresis only detects one third of the heterozygosity that actually exists.[3] Overestimates of heterozygosity are a danger when working with heterokaryotic species, since the organism may be expressing the gene products of different homozygous nuclei rather than one heterozygous nucleus.[82]

Isozymes can also be used as markers to trace hybridizations that occur naturally or are induced. Burdon et al.[91] used isozyme analysis to show that a common Australian race of *P. graminis* f. sp. *tritici* originated as a somatic hybrid of other races. Another study with *P. graminis* in Australia demonstrated that collections made from the grass *Agropyron scabrum* (Labill.) Beauv. originated as a somatic hybrid of *P. graminis* f. sp. *tritici* and *P. graminis* f. sp. *secalis.* Linde et al.[63] used phosphoglucomutase as a marker to examine selfing and crossing in the common bean rust fungus *U. appendiculatus.* They concluded that pathogens artificially increased in the greenhouse may not represent original populations in the field. Isozyme markers were also used to demonstrate that urediniospores could act as spermatia in isolates of *U. appendiculatus* that fail to initiate a sexual cycle.[92]

Table 1 **Polymorphic loci in fungal populations and species as estimated by isozyme analysis**

Species	Polymorphism (%)	Ref.
Pleurotus ostreatus (Jacq. ex Fr.) Quel.	94	66
Agaricus campestris Fr.	87	67
Atkinsonella hypoxylon (Peck) Diehl	85	48
Suillus variegatus (Fr.) O. Kuntze	71	50, 51
S. plorans (Roll.) Sing.	67	60
Uromyces appendiculatus (Pers.) Unger	67	68
Puccinia graminis Pers. f. sp. *tritici* (worldwide)	61	69
Pyricularia oryzae Cavara	55	70
Lentinula edodes (Berk.) Pegler	55	71
Phytophthora infestans (Mont.) deBary	54	72
Tilletia indica Mitra	44, 52	73, 74
S. bovinus (Fr.) O. Kuntze	50	50, 51
S. tomentosus (Kaufm.) Snell, Singer & Dick	47	50, 51
S. placidus (Bon.) Sing.	47	60
Agaricus brunnescens Peck	43	54
Heterobasidion annosum (Fr.) Bref.	40	75
Intersterility group "spruce"		
Ustilago zeae (Beckm.) Ung.	40	76
Rhizoctonia solani Kuehn	38	77
Rhynchosporium secalis (Oudem.) J. J. Davis	38	76
Cryphonectria cubensis (Bruner) Hodges	38	37
Puccinia graminis f. sp. *tritici* (U.S.)	38	78
Endocronartium harknessii (J. P. Moore) Y. Hiratsuka	38, 12	79, 80
Ceratocystiopsis ranuculosis J. R. Bridges & T. J. Perry	36	36
Leptographium wageneri (Kendrick) Wingfield	30, 48	42, 75
Volvariella volvacea (Bull. ex Fr.) Sing.	29	81
Cronartium quercuum (Berk.) Miyabe ex Shirai f. sp. *banksianae*	25	80
Fusarium oxysporum Schlect, emend Snyd. & Hans.	24	47
Peronosclerospora sorghi (Weston & Uppal) C. G. Shaw	23	34
Phytophthora cinnamoni Rands	23	82
Heterobasidion annosum		
Intersterility group "pine"	20	43, 75
Erysiphe graminis D.C. f. sp. *tritici*	20	83
E. graminis f. sp. *secalis*	20	83
Stagonospora nodorum (Berk.) Castellano & E. G. Germano	11	76
U. spinificis Ludw.	11	84
Pyricularia oryzae (rice isolates)	11	70
U. bullata Berk.	11	85
Puccinia recondita Robs. ex Desm. f. sp. *tritici*	9	86
P. striiformis West. f. sp. *tritici*	0	87
P. striiformis f. sp. *hordei*	0	87
P. graminis f. sp. *tritici*	0	86
P. hordei Otth.	0	87
E. graminis f. sp. *hordei*	0	88

Careful analysis of isozyme patterns can be used to determine whether individual loci are inherited independently or are linked. This information can be used to form genetic maps of chromosomes. Nine different loci were inherited independently in *U. appendiculatus;* there was no evidence for linkage.[64] Two of six genetic loci were linked in *Lentinula edodes.*[71] The frequency of crossing over was used to estimate the distance of two linked loci from the centromere in *Agaricus brunnescens.*[93] Similar linkage studies have been done with *A. campestris,*[65] *Volvariella volvacea,*[81] *P. graminis* f. sp. *tritici,*[69] and *Ustilago bullata.*[94]

Table 2 **Heterozygosity in different fungal species as estimated by isozyme analysis**

Species	Heterozygosity (%)	Ref.
Tilletia indica Mitra	7.2	73
Ustilago bullata Berk.	4.8	85
Puccinia graminis Pers. f. sp. *tritici*	33.0	86
P. recondita Robs. ex Desm. f. sp. *tritici*	15.0	86
Agaricus campestris Fr.	28.0	65
Scytinostroma galactinum (Fr.) Donk	0.0	90
S. protrusum (Burt) Nakas. subsp. *protrusum* Nakas.	12.5	90
S. protrusum subsp. *septentrionale* Nakas.	37.5	90
Heterobasidion annosum (Fr.) Bref. intersterility group S	10.0	43
H. annosum intersterility group P	2.0	43

The ploidy level (haploid, diploid or dikayotic, and polyploid) of a fungus can often be determined from isozyme data (Section II). Studies of the life cycle of the organism can thus be performed. Isozyme analysis was used to show that mating patterns of *Phytophthora infestans* were random in Mexico, where the sexual stage of the pathogen exists.[72] Asexual populations were identified by lack of recombination in the U.S., Canada, and Europe. In England and Wales, 10% of *P. infestans* isolates were of the uncommon A2 mating type (usually found only in Mexico).[95] Additional circumstantial evidence attested to low levels of sexual reproduction in the U.K.[95] Low levels of recombination were also demonstrated in *Rhynchosporium secalis,* an organism that has no known sexual phase.[65]

Atypical meiosis has been detected in some organisms by isozyme analysis. Basidiospores arising from germinated teliospores of *Tilletia indica* did not inherit alleles with equal frequency, and some basidiospores appeared to inherit both alleles.[35] The authors proposed that some basidiospores may receive two haploid nuclei from the promycelium or that the spores are actually aneuploids. This interpretation has been supported by cytological evidence.[96] Atypical meiosis has also been demonstrated in homokaryotic lines of *A. brunnescens.*[93]

D. EPIDEMIOLOGY

In many cases, fungi of a single species from different geographic sources can be differentiated from each other by isozyme analysis. This usually occurs due to genetic isolation. The differentiation of electrophoretic patterns in isozyme analysis is often the first indication that organisms are beginning to evolve into different species. Such information can be used to identify the origin of pathogens and to document their movement.

Cluster analysis was used to place isolates of *Endocronartium harknessii,* causal agent of western gall rust, into two distinct groups that corresponded to their geographic source.[79] This confirmed an earlier premise that *E. harknessii* is a western rust and that a different species occurs in the eastern U.S. Local populations of *Morchella deliciosa* Fr., *M. esculenta* (L.) Pers.,[97] *Neurospora intermedia* Tai,[98] and *Suillus* spp.[50,51,60] were also shown to be genetically distinct. Only 7% of alleles were shared among isolates of *Phakopsora pachyrhizi* from the Eastern and Western hemispheres.[35] When genetic distances are so large, the original classification of the organisms as a single species must be questioned. Isozyme patterns have been used to trace independent introductions of *Puccinia recondita* Robs. ex Desm. to the U.S.[99] and *P. graminis* f. sp. *tritici* to Australia.[100]

Geographic localities in which large amounts of genetic variability are encountered often represent the place of origin of a species. Kerrigan and Ross[101] thus speculated that *A. bisporus* (Lange) Imbach is indigenous to North America, just as Tooley et al.[72,102] traced the evolutionary origin of *Phytophthora infestans* to Mexico.

E. PATHOGENICITY AND VIRULENCE

There have been mixed results in using isozyme analysis to differentiate races, *formae speciales,* and other subgroups differing in host preference, pathogenicity, and virulence. Such separations have been made for some fungi, including *Cronartium quercuum,*[103] *F. oxysporum,*[47,104] *Erysiphe graminis,*[83] *Hetero-*

basidion annosum,[43] *Atkinsonella hypoxylon*,[48] *Cochliobolus carbonum* R. R. Nels.,[105] *Cryphonectria cubensis*,[106] *Phytophthora megasperma* Drechs.,[107] *Puccinia graminis*,[87] *P. sorghi* Schwein.,[108] and *Phyllotopsis nidulans* (Pers. ex Fr.) Sing. [= *Pleurotus nebrodensis* (Pers. ex Fr.) Kummer].[109] Races of *P. graminis* f. sp. *tritici* have been separated, but only in asexual populations.[78] Sexual populations of the organism had much higher levels of genetic variability, which obscured differences associated with virulence. In many cases, the genes responsible for virulence are quite distinct and are different from those assayed in isozyme analysis. There is usually more genetic variability among virulence genes than is detected by isozyme analysis due to the strong selection pressures placed upon pathogens to infect and colonize resistant hosts.

V. ADVANTAGES AND DISAVANTAGES

Isozyme analysis, as any technique, has its strengths and weaknesses. With starch gel electrophoresis, the technique is relatively inexpensive and results in less exposure to toxic chemicals (except for certain stain components). More staining systems can be used with starch gel electrophoresis than PAGE. This allows the researcher to compare large numbers of enzymes from many different metabolic pathways and to obtain information about many different genetic loci. In most cases with fungal pathogens, good resolution and successful genetic interpretations can be obtained with a panel of 15 to 25 enzymes. Such an isozyme test can be completed within a day. Most stains used in isozyme analysis are specific for a single enzyme. This greatly simplifies data interpretation since only a limited number of bands are visualized for each sample. In contrast, general protein stains usually detect large numbers of bands, which makes data collection and interpretation (e.g., the calculation of simple matching coefficients or other statistical values) very difficult.

The greatest disadvantage to isozyme analysis is the relatively large quantities of an organism that are often required for extracting sufficient enzyme. This is usually not a problem with facultative fungi that can be cultured on artificial media. Obligate pathogens may require considerable effort to obtain the necessary quantities of fungal tissue. For example, with maize downy mildew fungi, conidia can be washed directly from the plant surface and concentrated by centrifugation.[34,38] Similar difficulties arise when analyzing urediniospores of rust fungi. In general, 50 to 100 mg (wet weight) of mycelium, 50 mg (wet weight) of downy mildew condia, or 30 to 50 mg (pregermination dry weight) of germinating rust urediniospores are needed for isozyme analysis.

Time requirements may be another disadvantage to isozyme analysis, depending upon the application. Although electrophoresis can be conducted rapidly, several days or even weeks are often needed to isolate and grow the organisms. For example, the germination of viable teliospores of *T. controversa* Kuhn, the causal agent of dwarf bunt of wheat, can require 6 to 8 weeks, and continued growth of mycelium is then necessary for another week to obtain sufficient fungal material for the test.[1] Such time requirements are unacceptable in situations where identification is required within hours. In such cases, alternative procedures, such as gene probes, may be more satisfactory.

Using isozyme analysis with bacteria presents another problem. The slimy polysaccharides of the capsule must be removed, using ultracentrifugation or enzymatic treatments.[1,29] Even without excess carbohydrate, the electrophoretic migration rates of bacterial proteins are often quite similar, and only small differences in banding patterns may be observed. It is usually necessary to repeat the analysis several times to ensure that the bands are scored correctly. Alternative methods, including various nucleic acid procedures, may be more effective in separating and analyzing bacterial plant pathogens.

VI. CONCLUSIONS

Isozyme analysis is a simple, efficient, and inexpensive technique for evaluating the taxonomy, genetics, virulence, and epidemiology of plant pathogens, especially fungi. The technique also has practical applications for pathogen detection and identification. Recently, there has been an explosion in the number of publications describing the application of isozyme analysis to phytopathological questions. Objections raised by "classical" geneticists have subsided as genetic interpretations of banding patterns have been confirmed by crossing experiments. Isozyme analysis is becoming a standard technique for the study of plant pathogens.

REFERENCES

1. Bonde, M. R., Micales, J. A., and Peterson, G. L., The use of isozyme analysis for the identification of plant pathogenic fungi, *Plant Dis.*, 77, 961, 1993.
2. Micales, J. A., Bonde, M. R., and Peterson, G. L., Isozyme analysis in fungal taxonomy and molecular genetics, in *Handbook of Applied Mycology: Fungal Biotechnology,* Vol. 4, Arora, D. K., Elander, P., and Mukerji, K. G., Eds., Marcel Dekker, New York, 1992, 57.
3. Shaw, C. R., How many genes evolve? *Biochem. Genet.*, 4, 275, 1970.
4. Bonaventura, J. and Riggs, A., Hemoglobin Kansas, a human hemoglobin with a neutral amino acid substitution and an abnormal oxygen equilibrium, *J. Biol. Chem.*, 243, 980, 1968.
5. May, B., Salmonid genome: evolutionary restructuring following a tetraploid event, Ph.D. thesis, Pennsylvania State University, 1980, 199 pp.
6. Harris, H. and Hopkinson, D. A., *Handbook of Enzyme Electrophoresis in Human Genetics* [with supplements], Oxford American Elsevier, New York, 1976.
7. Darnall, D. W. and Klotz, I. M., Protein subunits: a table (revised edition), *Arch. Biochem. Biophys.*, 149, 1, 1972.
8. Beeley, J. G., *Glycoprotein and Proteoglycan Techniques,* Elsevier, New York, 1985.
9. Conkle, M. T., Hodgskiss, P. D., Nunnally, L. B., and Hunter, S. C., Starch gel electrophoresis of conifer seeds: a laboratory manual, *Gen. Tech. Rep. PSW-64,* Pacific Southwest Forest and Range Exp. Stn., U.S. Forest Service, U.S. Department of Agriculture, Berkeley, CA, 1982.
10. Gabriel, O., Locating enzymes on gels, in *Methods of Enzymology,* Vol. 2, Colowick, S. P. and Kaplan, N. O., Eds., Academic Press, New York, 1971, 578.
11. Micales, J. A., Bonde, M. R., and Peterson, G. L., The use of isozyme analysis in fungal taxonomy and genetics, *Mycotaxon,* 27, 405, 1986.
12. Shaw, C. R. and Prasad, R., Starch gel electrophoresis of enzymes—a compilation of recipes, *Biochem. Genet.*, 4, 297, 1970.
13. Siciliano, M. J. and Shaw, C. R., Separation and visualization of enzymes on gels, in *Chromatographic and Electrophoretic Techniques: Zone Electrophoresis,* Vol. 2, Smith, I., Ed., A. W. Heinemann, London, 1976, 185.
14. Vallejos, C. E., Enzyme activity staining, in *Isozymes in Plant Genetics and Breeding,* Part A, Tanksley, S. D. and Orton, T. J., Eds. Elsevier, Amsterdam, 1983, 469.
15. Werth, C. R., Implementing an isozyme laboratory at a field station, *Va. J. Sci.*, 36, 53, 1985.
16. Johnson, G. B., Enzyme polymorphism and metabolism, *Science,* 184, 28, 1974.
17. Sneath, P. H. A. and Sokol, R. R., *Numerical Taxonomy,* W. H. Freeman, San Francisco, 1973.
18. Nei, M., Genetic distance between populations, *Am. Nat.*, 106, 283, 1972.
19. Nei, M., Estimation of average heterozygosity and genetic distance from a small sample of individuals, *Genetics,* 89, 583, 1978.
20. Nei, M. and Li, W. H., Mathematical models for studying genetic variation in terms of restriction endonucleases, *Proc. Natl. Acad. Sci. U.S.A.*, 76, 5269, 1979.
21. Rogers, J. S., IV, Measures of genetic similarity and genetic distance, *Stud. Genet.*, 7, 145, 1972.
22. Buth, D. G., The application of electrophoretic data in systematic studies, *Annu. Rev. Ecol. Syst.*, 15, 501, 1984.
23. Ferguson, A., *Biochemical Systematics and Evolution,* John Wiley & Sons, New York, 1980.
24. Sokal, R. R. and Sneath, P. H. A., *Principles of Numerical Taxonomy,* W. H. Freeman, San Francisco, 1963.
25. Liu, Z. L. and Sinclair, J. B., Genetic diversity of *Rhizoctonia solani* anastomosis group 2, *Phytopathology,* 82, 778, 1992.
26. Dunn, G. and Everitt, B. S., *An Introduction to Mathematical Taxonomy,* Cambridge University Press, New York, 1982.
27. Burdon, J. J. and Marshall, D. R., The use of isozymes in plant disease research, in *Isozymes in Plant Genetics and Breeding,* Tanksley, S. D. and Orton, T. J., Eds., Elsevier, Amsterdam, 1983, 401.
28. Huettel, R. N., Dickson, D. W., and Kaplan, D. T., Biochemical identification of the two races of *Radopholus similis* by starch gel electrophoresis, *J. Nematol.*, 15, 338, 1983.
29. Bonde, M. R., Peterson, G. L., and Dowler, W. M., Comparative isozyme analysis of species of *Xanthomonas* attacking rice, *Phytopathology,* in press.

30. Kubicek, Q. B., Civerolo, E. L., Bonde, M. R., Hartung, J. S., and Peterson, G. L., Isozyme analysis of *Xanthomonas campestris* pv. *citri*, *Phytopathology*, 79, 297, 1989.
31. Cheliak, W. M. and Pitel, J. A., Techniques for starch gel electrophoresis of enzymes from forest trees, *Information Report PI-X-42*. Petawawa National Forestry Institute, Canadian Forest Service, 1984.
32. Conkle, M. T., (technical coordinator), Proceedings of the symposium on isozymes of North American forest trees and forest insects, *Gen. Tech. Rep. PSW-48*, Pacific Southwest Forest and Range Exp. Stn., U.S. Forest Service, U.S. Department of Agriculture, Berkeley, CA, 1981.
33. Tanksley, S. D. and Orton, T. J., Eds., *Isozymes in Plant Genetics and Breeding*, Part A, Elsevier, New York, 1983.
34. Bonde, M. R., Peterson, G. L., Dowler, W. M., and May, B., Isozyme analysis to differentiate species of *Peronosclerospora* causing downy mildews of maize, *Phytopathology*, 74, 1278, 1984.
35. Bonde, M. R., Peterson, G. L., and Royer, M. H., Inheritance of isozymes in the smut pathogen *Tilletia indica*, *Phytopathology*, 78, 1276, 1988.
36. Harrington, T. C. and Zambino, P. J., *Ceratocystiopsis ranaculosus*, not *Ceratocystis minor* var. *barrasii*, is the mycangial fungus of the southern pine beetle, *Mycotaxon*, 38, 103, 1990.
37. Micales, J. A., Stipes, R. J., and Bonde, M. R., On the conspecificity of *Endothia eugeniae* and *Cryphonectria cubensis*, *Mycologia*, 79, 707, 1987.
38. Micales, J. A., Bonde, M. R., and Peterson, G. L., Isozyme analysis and aminopeptidase assays of the genus *Peronosclerospora*, *Phytopathology*, 78, 1396, 1988.
39. Powers, H. R., Jr., Snow, G., Lin, D., and Hubbes, M., Isozyme analysis as an indicator of synonymy of the causal agents of gall rust on sand and Virginia pine, *Plant Dis.*, 75, 1225, 1991.
40. Bielenin, A., Jeffers, S. N., Wilcox, W. F., and Jones, A. L., Separation by protein electrophoresis of six species of *Phytophthora* associated with deciduous fruit crops, *Phytopathology*, 78, 1402, 1988.
41. Faris, M. A., Sabo, F. E., and Cloutier, Y., Intraspecific variation in gel electrophoresis patterns of soluble mycelial proteins of *Phytophthora megasperma* isolated from alfalfa, *Can. J. Bot.*, 64, 262, 1986.
42. Otrosina, W. J. and Cobbs, F. W., Jr., Analysis of allozymes of three distinct variants of *Verticicladiella wageneri* in Western North America, *Phytopathology*, 77, 1360, 1987.
43. Otrosina, W. J., Chase, T. E., and Cobb, F. W., Jr., Allozyme differentiation of intersterility groups of *Heterobasidion annosum* isolated from conifers in the western United States, *Phytopathology*, 82, 540, 1992.
44. Zambino, P. J. and Harrington, T. C., Isozyme variation within and among host-specialized varieties of *Leptographium wageneri*, *Mycologia*, 81, 122, 1989.
45. Thorpe, J. P., Enzyme variation, genetic distance and evolutionary divergence in relation to levels of taxonomic separation, in *Protein Polymorphism: Adaptive and Taxonomic Significance*, Oxford, G. S. and Rollinson, D., Eds., Academic Press, New York, 1983, 131.
46. Ayala, F. J., Genetic differentiation during the speciation process, in *Evolutionary Biology*, Vol. 8, Dobzhansky, T., Hecht, M. K., and Steere, W. C., Eds., Plenum Press, New York, 1975, chap. 1.
47. Bosland, P. W. and Williams, P. H., An evaluation of *Fusarium oxysporum* from crucifers based on pathogenicity, isozyme polymorphism, vegetative compatibility, and geographic origin, *Can. J. Bot.*, 65, 2067, 1987.
48. Leuchtmann, A. and Clay, K., Isozyme variation in the fungus *Atkinsonella hypoxylon* within and among populations of its host grasses, *Can. J. Bot.*, 67, 2600, 1989.
49. Rosendahl, S., Sen, R., Hepper, C. M., and Azcon-Aguilar, Quantification of three vesicular-arbuscular mycorrhizal fungi (*Glomus* spp) in roots of leek (*Allium porrum*) on the basis of activity of diagnostic enzymes after polyacrylamide gel electrophoresis, *Soil Biol. Biochem.*, 21, 519, 1989.
50. Sen, R., Intraspecific variation in two species of *Suillus* from Scots pine (*Pinus sylvestris* L.) forests based on somatic incompatibility and isozyme analyses, *New Phytol.*, 114, 607, 1990.
51. Sen, R., Isozymic identification of individual ectomycorrhizas synthesized between Scots pine (*Pinus sylvestris* L.) and isolates of two species of *Suillus*, *New Phytol.*, 114, 617, 1990.
52. Roux, P. and Labarere, J., Isozyme characterization of dikaryotic strains of the edible basidiomycete *Agaricus bitorquis* (Quel.) Sacc. (Syn. *Agaricus edulis*), *Exp. Mycol.*, 14, 101, 1990.
53. Royse, D. J. and May, B., Use of isozyme variation to identify genotypic classes of *Agaricus brunnescens*, *Mycologia*, 74, 93, 1982.

54. Royse, D. J. and May, B., Genetic relatedness and its application in selective breeding of *Agaricus brunnescens, Mycologia,* 74, 569, 1982.
55. Royse, D. J., Jodon, M. H., Antoun, G. G., and May, B. P., Confirmation of intraspecific crossing and single and joint segregation of biochemical loci of *Volvariella volvacea, Exp. Mycol.,* 11, 11, 1987.
56. Royse, D. J., Spear, M. C., and May, B., Cell line authentication and genetic relatedness of lines of the shiitake mushroom, *Lentinus edodes, J. Gen. Appl. Microbiol.,* 29, 205, 1983.
57. Stasz, T. E., Weeden, N. F., and Harman, G. E., Methods of isozyme electrophoresis for *Trichoderma* and *Gliocladium* species, *Mycologia,* 80, 870, 1988.
58. Stasz, T. E., Nixon, K., Harman, G. E., Weeden, N. F., and Kuter, G. A., Evaluation of phenetic species and phylogenetic relationships in the genus *Trichoderma* by cladistic analysis of isozyme polymorphism, *Mycologia,* 81, 391, 1989.
59. Zamir, D. and Chet, I., Application of enzyme electrophoresis for identification of isolates of *Trichoderma harzianum, Can. J. Microbiol.,* 31, 578, 1985.
60. Keller, G., Isozymes in isolates of *Suillus* species from *Pinus cembra L., New Phytol.,* 120, 351, 1992.
61. Newton, A. C., Markers in pathogen populations, in *Genetics and Plant Pathogenesis,* Day, P. R. and Jellis, G. J., Eds., Blackwell Scientific, Oxford, 1987, 187.
62. Lenne, J. M. and Burdon, J. J., Preliminary study of virulence and isozymic variation in natural populations of *Colletotrichum gloeosporioides* from *Stylosanthes guianensis, Phytopathology,* 80, 728, 1990.
63. Linde, D. C., Groth, J. V., and Roelfs, A. P., Inheritance of two phosphoglucomutase isozymes in *Uromyces appendiculatus* and comparison of crossing selfing methods, *J. Hered.,* 80, 337, 1989.
64. Linde, D. C., Groth, J. V., and Roelfs, A. P., Comparison of isozyme and virulence diversity patterns in the bean rust fungus *Uromyces appendiculatus, Phytopathology,* 80, 141, 1990.
65. McDermott, J. M., McDonald, B. A., Allard, R. W., and Webster, R. K., Genetic variability for pathogenicity, isozyme, ribosomal DNA, and colony color variants in populations of *Rhynchosporium secalis, Genetics,* 122, 561, 1989.
66. Kulkarni, R. K., Kamerath, C. D., and Allred, K. L., Genetic diversity between isolates of *Pleurotus ostreatus* as revealed by enzyme analysis, *Proc. Sym. Scientific and Technical Aspects of Cultivating Edible Fungi,* Pennsylvania State University, University Park, PA, 1986, 171.
67. May, B. and Royse, D. J., Genetic variation and joint segregation of biochemical loci in the common meadow mushroom, *Agaricus campestris. Biochem. Genet.,* 20, 1165, 1982.
68. Lu, T. H. and Groth, J. V., Isozyme detection and variation in *Uromyces appendiculatus, Can. J. Bot.,* 66, 885, 1987.
69. Burdon, J. J., Roelfs, A. P., and Brown, A. H. D., The genetic basis of isozyme variation in the wheat stem rust fungus (*Puccinia graminis tritici*), *Can. J. Genet. Cytol.,* 28, 171, 1986.
70. Leung, H. and Williams, P. H., Enzyme polymorphism and genetic differentiation among geographic isolates of the rice blast fungus, *Phytopathology,* 76, 778, 1986.
71. Royse, D. J., Spear, M. C., and May, B., Single and joint segregation of marker loci in the shiitake mushroom, *Lentinus edodes, J. Gen. Appl. Microbiol.,* 29, 217, 1983.
72. Tooley, P. W., Fry, W. E., and Villarreal Gonzalez, M. J., Isozyme characterization of sexual and asexual *Phytophthora infestans* populations, *J. Hered.,* 76, 431, 1985.
73. Bonde, M. R., Peterson, G. L., and Matsumoto, T. T., The use of isozymes to identify teliospores of *Tilletia indica, Phytopathology,* 79, 596, 1989.
74. Bonde, M. R., Peterson, G. L., Dowler, W. M., and May, B., Comparison of *Tilletia indica* isolates from India and Mexico by isozyme analysis (Abstr.), *Phytopathology,* 75, 1309, 1985.
75. Otrosina, W. J., Chase, T. E., Cobb, F. W., Jr., and Taylor, J. W., Isozyme structure of *Heterobasidion annosum* isolates relating to intersterility groups, in *Proc. Seventh Int. Conf. on Root and Butt Rots of Forest Trees,* Morrison, D. J., Ed., Forestry Canada, Victoria, BC, 1989, 406.
76. Newton, A. C., Isozyme variability in isolates of some facultative phytopathogenic fungi, *J. Phytopathol.,* 131, 199, 1991.
77. Liu, Z., Nickrent, D. L., and Sinclair, J. B., Genetic relationships among isolates of *Rhizoctonia solani* anastomosis group-2 based on isozyme analysis, *Can. J. Plant Pathol.,* 12, 376, 1990.
78. Burdon, J. J. and Roelfs, A. P., The effect of sexual and asexual reproduction on the isozyme structure of populations of *Puccinia graminis, Phytopathology,* 75, 1068, 1985.

79. Tuskan, G. A., Walla, J. A., and Lundquist, J. E., Genetic-geographic variation in *Peridermium harknessii* in the north-central United States, *Phytopathology,* 80, 857, 1990.
80. Tuskan, G. A. and Walla, J. A., Isozyme characterization of *Peridermium harknessii* and *Cronartium quercuum* f. sp. *banksianae* with starch gel electrophoresis, *Phytopathology,* 79, 444, 1989.
81. Royse, D. J. and May, B., Identification of shiitake genotypes by multilocus enzyme electrophoresis: catalog of lines, *Biochem. Genet.,* 25, 705, 1987.
82. Old, K. M., Morgan, G. F., and Bell, J. C., Isozyme variability among isolates of *Phytophthora cinnamomi* from Australia and Papua New Guinea, *Can. J. Bot.,* 62, 2016, 1984.
83. Koch, G. and Kohler, W., Isozyme variation and genetic distances of *Erysiphe graminis* DC. Formae Specialis, *J. Phytopathol.,* 129, 89, 1990.
84. Andrews, R., Kirby, G. C., and Adams, M., The population biology of a smut fungus, *Ustilago spinificis* Ludw. II. Isozyme polymorphism and geographic differentiation, *Aust. J. Bot.,* 36, 347, 1988.
85. Kirby, G. C., Variation in *Ustilago bullata* on *Agropyron scabrum* in Australia and New Zealand, *Australas. Plant Pathol.,* 16, 10, 1987.
86. Burdon, J. J., Luig, N. H., and Marshall, D. R., Isozyme uniformity and virulence variation in *Puccinia graminis* f. sp. *tritici* and *P. recondita* f. sp. *tritici* in Australia, *Aust. J. Biol. Sci.,* 36, 403, 1983.
87. Newton, A. C., Caten, C. E., and Johnson, R., Variation for isozymes and double-stranded RNA among isolates of *Puccinia striiformis* and two other cereal rusts, *Plant Pathol.,* 34, 235, 1985.
88. Clark, J., Butters, J., Brent, K. J., and Hollomon, D. W., Isozyme uniformity in *Erysiphe graminis* f. sp. *hordei, Mycol. Res.,* 92, 404, 1989.
89. Bonde, M. R., Peterson, G. L., and Dowler, W. M., A comparison of isozymes of *Phakopsora pachyrhizi* from the eastern hemisphere and the new world, *Phytopathology,* 78, 1491, 1988.
90. Nakasone, K. K. and Micales, J. A., *Scytinostroma galactinum* species complex in the United States, *Mycologia,* 80, 546, 1988.
91. Burdon, J. J., Marshall, D. R., and Luig, N. H., Isozyme analysis indicates that a virulent cereal rust pathogen is a somatic hybrid, *Nature,* 293, 565, 1981.
92. McCain, J. W. and Groth, J. V., Isozyme evidence of pycnial fertilization by urediniospores of the bean rust fungus, *Can. J. Bot.,* 67, 1844, 1989.
93. Spear, M. C., Royse, D. J., and May, B., Atypical meiosis and joint segregation of biochemical loci in *Agaricus brunnescens, J. Hered.,* 74, 417, 1983.
94. Kirby, G. C. and Mulley, A. M., Inheritance of isozyme mobility differences between two races of the head smut of grasses, *Ustilago bullata* Berk., *Bot. Gaz. (Chicago),* 143, 243, 1982.
95. Shattock, R. C., Shaw, D. S., Fyfe, A. M., Dunn, J. R., Loney, K. H., and Shattock, J. A., Phenotypes of *Phytophthora infestans* collected in England and Wales from 1985 to 1988: mating type, response to metalaxyl and isoenzyme analysis, *Plant Pathol.,* 39, 242, 1990.
96. Goates, B. J. and Hoffmann, J. A., Nuclear behavior during teliospore germination and sporidial development in *Tiletia caries, T. fostida,* and *T. controversa, Can. J. Bot.,* 65, 512, 1986.
97. Gessner, R. V., Romano, M. A., and Schultz, R. W., Allelic variation and segregation in *Morchella deliciosa* and *M. esculenta, Mycologia,* 79, 683, 1987.
98. Spieth, P. T., Population genetics of allozyme variation in *Neurospora intermedia, Genetics,* 80, 785, 1975.
99. Burdon, J. J. and Roelfs, A. P., Isozyme and virulence variation in asexually reproducing populations of *Puccinia graminis* and *P. recondita* on wheat, *Phytopathology,* 75, 907, 1985.
100. Burdon, J. J., Marshall, D. R., Luig, N. H., and Gow, D. J. S., Isozyme studies on the origin and evolution of *Puccinia graminis* f. sp. *tritici* in Australia, *Aust. J. Biol. Sci.,* 35, 231, 1982.
101. Kerrigan, R. W. and Ross, I. K., Allozymes of a wild *Agaricus bisporus* population: new alleles, new genotypes, *Mycologia,* 81, 433, 1989.
102. Tooley, P. W., Therrien, C. D., and Ritch, D. L., Mating type, race composition, nuclear DNA content, and isozyme analysis of Peruvian isolates of *Phytophthora infestans, Phytopathology,* 79, 478, 1989.
103. Powers, H. R., Jr., Lin, D., and Hubbes, M., Interspecific and intraspecific differentiation within the genus *Cronartium* by isozyme and protein pattern analysis, *Plant Dis.,* 73, 691, 1989.
104. Katan, T., Zamir, D., Sarfatti, M., and Katan, J., Vegetative compatibility groups and subgroups in *Fusarium oxysporum* f. sp. *radicis-lycopersici, Phytopathology,* 81, 255, 1991.
105. Simcox, K. D., Nickrent, D., and Pedersen, W. L., Comparison of isozyme polymorphism in races of *Cochliobolus carbonum, Phytopathology,* 82, 621, 1992.

106. Alfenas, A. C., Jeng, R., and Hubbes, M., Isoenzyme and protein patterns of isolates of *Cryphonectria cubensis* differing in virulence, *Can. J. Bot.*, 62, 1756, 1984.
107. Nygaard, S. L., Elliott, C. K., Cannon, S. J., and Maxwell, D. P., Isozyme variability among isolates of *Phytophthora megasperma, Phytopathology,* 79, 773, 1989.
108. Bergquist, R. R. and Pryor, A. J., Virulence and isozyme differences for establishing racial identity in rusts of maize, *Plant Dis.*, 68, 281, 1984.
109. Boisselier-Dubayle, M. C., Taxonomic significance of enzyme polymorphism among isolates of *Pleurotus* (Basidiomycetes) from umbellifers, *Trans. Br. Mycol. Soc.*, 81, 121, 1983.

RECOMMENDED READING

Bonde, M. R., Micales, J. A., and Peterson, G. L., The use of isozyme analysis for the identification of plant pathogenic fungi, *Plant Dis.*, 77, 961, 1993.

Burdon, J. J. and Marshall, D. R., The use of isozymes in plant disease research, in *Isozymes in Plant Genetics and Breeding,* Tanksley, S. D. and Orton, T. J., Eds., Elsevier, Amsterdam, 1983, 401.

Buth, D. G., The application of electrophoretic data in systematic studies, *Ann. Rev. Ecol. Syst.*, 15, 501, 1984.

Ferguson, A., *Biochemical Systematics and Evolution,* John Wiley & Sons, New York, 1980.

Hames, B. D., An introduction of polyacrylamide gel electrophoresis, in *Gel Electrophoresis of Proteins: a Practical Approach,* Hames, B. D. and Rickwood, D., Eds., IRL Press, Washington, D.C., 1981, chap. 1.

Harris, H. and Hopkinson, D. A., *Handbook of Enzyme Electrophoresis in Human Genetics* [with supplements], Oxford American Elsevier, New York, 1976.

Micales, J. A., Bonde, M. R., and Peterson, G. L., The use of isozyme analysis in fungal taxonomy and genetics, *Mycotaxon,* 27, 405, 1986.

Micales, J. A., Bonde, M. R., and Peterson, G. L., Isozyme analysis in fungal taxonomy and molecular genetics, in *Handbook of Applied Mycology: Fungal Biotechnology,* Vol. 4, Arora, D. K., Elander, P., and Hukarji, K. G., Eds., Marcel Dekker, New York, 1992, 57.

Newton, A. C., Markers in pathogen populations, in *Genetics and Plant Pathogenesis,* Day, P. R. and Jellis, G. J., Eds., Blackwell Scientific, Oxford, 1987, 187.

Warth, C. R., Implementing an isozyme laboratory at a field station, *Va. J. Sci.*, 36, 53, 1985.

Section—III

BIOMOLECULAR METHODS

Chapter 10

Nucleic Acid Hybridization Methods in Diagnosis of Plant Viruses and Viroids

Olga V. Nikolaeva

TABLE OF CONTENTS

I. INTRODUCTION

Plant virus and viroid diseases can be traditionally detected by bioassay on suitable plant cultivars. This assay is very sensitive, but unfortunately it is laborious, expensive, and time consuming. Modern express methods of plant virus and viroid detection are based on the identification of a specific molecular component(s) of the causal agent in tested samples. Virus-specific protein can be detected by any appropriate serological technique (immunofluorescence, radioimmunoassay, enzyme-linked immunosorbent assay). The genetic material of the pathogen (nucleic acid) can be detected by nucleic acid hybridization assay. This nonimmunological detection technique was initially used in phytopathology practice for viroid detection.[1] Later this technique was adopted for the detection of a number of plant viruses.[2] The sensitivity of the assay is of the same order as that of ELISA. The nucleic acid hybridization assay is useful for the detection of viroids, which are plant pathogens consisting solely of a short RNA molecule and hence are not detectable by any serological assay. The similar situation is known for some plant virus infections, when virus coat protein is not produced and such infections cannot be identified with serological techniques.[3,4] The nucleic acid hybridization assay is also a powerful tool for the detection of virus satellite RNAs, which are not detectable by serological methods.[5,6] In the nucleic acid hybridization assay the whole genome of the plant pathogen can be probed, compared with 2 to 5% of the viral genome encoding antigenic determinants of the virus coat protein. Due to this reason the nucleic acid hybridization assay is widely used for differentiation of virus strains, which have the similar coat proteins, but produce significant differences in pathogenecity or vector transmissibility and cannot be discriminated serologically.[7–12] Moreover, high-quality virus-specific antisera are not always readily available because of difficulties in virus purification. In these cases, the nonimmunological nucleic acid hybridization assay may be a valuable approach in plant virus detection.

0-87371-877-1/95/$0.00+$.50

II. PRINCIPLES OF THE NUCLEIC ACID HYBRIDIZATION ASSAY

The nucleic acid hybridization assay is based on the formation of a duplex "target-probe" between the nucleic acid of a pathogen (target sequence) and a pathogen-specific complementary nucleic acid (probe). The duplex formation process is termed the hybridization reaction. As a rule, the probe molecules are modified by a so-called "reporter group" or "label", which can be detected in the hybridization product (duplex) by an appropriate method. The hybridization reaction may be carried out in solution.[13] The presence of hybridization products in this case is estimated by S1 nuclease digestion separately for each sample. This labor-intensive solution hybridization technique is suitable for testing only of a small number of samples. Usually, a lot of samples must be handled simultaneously in phytopathological practice. For these purposes, the mixed-phase hybridization technique on solid supports is considered to be a more convenient tool for rapid screening. Two forms of solid support are often used in the hybridization assay, either nitrocellulose or nylon membranes (filters). The nucleic acid hybridization assay on membrane support is termed as a dot-blot nucleic acid hybridization assay and includes the following steps: (1) sample preparation; (2) sample application and immobilization of the target sequence; (3) prehybridization; (4) hybridization with the complementary nucleic acid probe; (5) removing of the excess probe (washing); and (6) detection of hybridization products.

III. NUCLEIC ACID HYBRIDIZATION ASSAY STEPS

A. SAMPLE PREPARATION

1. Sap Treatment Procedures

The first step of the nucleic acid hybridization assay is sample preparation. Usually, a few grams (or less amounts) of suspect tissue are homogenized with the special buffer solution in 1:2 (w/v, g/ml) proportion.[2] Sometimes, squashing or squeezing of the suspect material onto a membrane can be used.[9,14] Samples prepared by these procedures are termed "crude sap samples" and may be directly used in the assay.

Nevertheless, many researchers have found that crude sap component(s) may interfere with the assay. Because of this phenomenon, the false negative signals, on the one hand, and false positive signals, on the other hand, have been often detected. The main reason of the false signals is the influence of different cell substances which occur naturally in tested tissues. Some kinds of tissues may contain small amounts of these cell substances. In such cases the sap interfering effect is not crucial and crude sap samples may be used without any special treatment.[1,2] More frequently, interfering cell substances are present in suspect tissues in large amounts and a strong sap effect is clearly observed. In the case of a strong sap inhibition effect a signal of the assay is very low and a special treatment of crude sap sample is needed. This inhibition may be eliminated by treating the crude plant tissue homogenate either with chloroform[15,16] or with sodium dodecyl sulfate (SDS) and phenol-chloroform.[17] Sometimes a strong sap inhibition effect may take place for one kind of host plant tissue, while an inhibition effect is not observed in the sap sample from another kind of tissue of the same plant. For example, purified tobacco rattle virus (TRV) added to a homogenate of healthy tulip bulbs (not leaves) gave a much lower signal upon nucleic acid hybridization assay than virus diluted in buffer, indicating that inhibitory substances in the bulb homogenate interfered with the assay.[17] Similarly, in the detection of arabis mosaic virus (AMV) there was a considerable inhibition effect of strawberry sap on the assay when compared with the extracts of petunia and cucumber.[18] The nonspecific false positive signals also may take place in dot-blot hybridization assay because of nonspecific binding of the labeled probe.[19–22] Sometimes the false positive signals are clearly pronounced for sap samples from one plant species and are not observed for sap samples from another plant species, as it has been shown for the detection of tomato spotted wilt virus (TSWV) in *Lycopersicon esculentum* plants and in *Nicotiana rustica* plants, respectively.[19] Strong false positive reactions can be eliminated by partial purification of crude sap samples. As a rule, the method used is a combination of phenol-chloroform extraction and nucleic acid precipitation.[19–21] In some cases, the additional procedures are needed to eliminate nonspecific signals.[21–23]

2. Nucleic Acid Denaturation Procedures

An important step in the dot-blot nucleic acid hybridization assay is denaturation of nucleic acids, because for successful binding to nitrocellulose, the nucleic acid should not have a secondary structure. Double-stranded RNA or DNA should be denatured by heating or alkali treatment before sample spotting onto the membrane.[2] There are numerous reports on the effect of denaturation of single-stranded nucleic

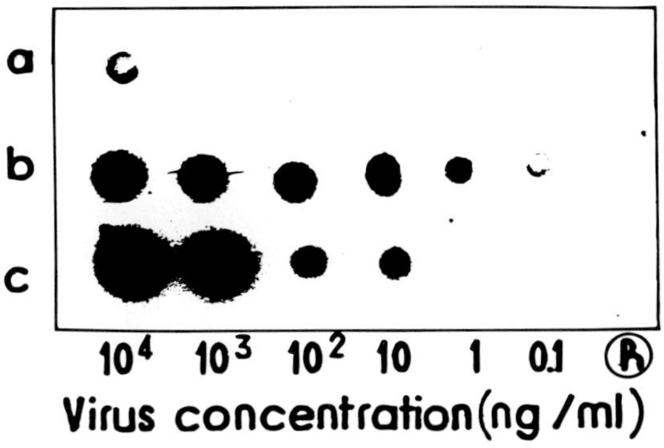

Figure 1 Effect of the membrane treatment technique on potato virus S (PVS) detection by dot-blot hybridization assay. Lane a, conventional hybridization assay; lanes b and c, modified hybridization assay. Lanes a and b, dilutions of PVS in buffer; lane c, dilutions of PVS in healthy potato tuber extract. The last spot in lane c: healthy potato tuber extract (h). The probe used was ^{32}P nick-translated pBR322-derivated recombinant cDNA (10^6 cpm/ml). The probe synthesized was divided into three equal portions for independent hybridization with the a, b, and c strips.

acid on its binding to nitrocellulose membrane. As reported by Thomas,[24] single-stranded RNA did not need to be denaturated for effective binding to nitrocellulose. Nevertheless, in some cases single-stranded nucleic acid binds to nitrocellulose much more effectively after denaturation: treatment of samples with formaldehyde enhanced the signal for barley yellow dwarf virus (BYDV)-infected samples and reduced false positive signals for healthy tissue samples.[25] Formaldehyde treatment was successfully used for detection of apple chlorotic leaf spot virus in apple trees.[21] It has been shown for some RNA plant viruses that certain denaturants as glyoxal or alkali decreased sensitivity of viral RNA detection.[2] At the same time, glyoxylation increased the sensitivity of papaya mosaic virus RNA detection by severalfold and eliminated nonspecific hybridization signals from tested samples, as compared to formaldehyde treatment.[20] Glyoxal treatment was also successfully used for detection of AMV in strawberry samples.[18] Although viroids are small, circular, single-stranded RNAs, they have a strong secondary structure because of internal self-complementary sites. So it is desirable for viroid-infected sap samples to be denatured to increase the sensitivity of the assay. For these purposes the formaldehyde treatment is often used.[23,26–28] Also, heat denaturation of viroid RNA was found to increase the sensitivity of the assay for potato spindle tuber viroid (PSTV) and citrus exocortis viroid (CEV) by five- to tenfold.[29] Denaturation of PSTV RNA with glyoxal prior to spotting onto nitrocellulose filters improved the sensitivity of viroid detection considerably, but did not improve the signal-to-noise ratio.[30]

B. SAMPLE APPLICATION AND TARGET NUCLEIC ACID IMMOBILIZATION

A single spot of each sap sample is applied by pipette (1 to 2 μl) or by gentle suction through vacuum Manifold apparatus (up to 200 μl) to nitrocellulose filters which must be presoaked in 6–20xSSC solution (1xSSC = 0.15 *M* NaCl and 0.0015 *M* sodium citrate). Nitrocellulose filters may be cut to any appropriate size. For example, an 11.5 × 16.5 cm nitrocellulose sheet can hold up to 150 spots, each 2 μl in volume.[31] After sample application the filters must be air dried and baked in vacuo at 80°C for 2 h to immobilize the target nucleic sequence. Before baking, filters may be treated additionally: in our experiments we have developed a rapid and simple technique to enhance the sensitivity of the conventional dot-blot hybridization assay (Reference 32 and Figure 1). This procedure follows that of Baulcombe et al.,[9] but includes short-time (15 to 20 min) heating of the nitrocellulose filter in 10x SSC, 0.5% SDS solution at 55°C after sample application. After baking, filter sheets can be stored until required for probing.

C. PREHYBRIDIZATION

Prior to the hybridization step, free binding sites on the nitrocellulose must be blocked with nonhomologous DNA (usually, briefly sonicated and denatured salmon sperm or calf thymus DNA) and a protein (bovine serum albumin or nonfat dried milk) for 2 to 5 h at 65°C. The composition of the prehybridization solution may slightly vary in different experiments. For RNA plant virus detection the following buffer may be used: 3xSSC, 4xDenhardt mixture (1xDenhardt mixture = 0.02% each of bovine serum albumin, Ficoll 400, and polyvinylpyrrolidone), 250 μg/ml of sonicated and denatured salmon sperm DNA.[2]

D. HYBRIDIZATION

The quantitative aspects of thermodynamics and kinetics of nucleic acid hybridization are discussed in detail elsewhere.[33–35] Here it should be mentioned that the two complementary strands of nucleic acid anneal at specific temperature and salt concentrations. At higher temperature the strands of the duplex will separate (melting temperature, T_m). T_m is affected by salt concentration and by the length and composition of the nucleic acid. Empirically, the maximum rate of hybridization has been found to be about 25°C below the T_m at 1.0 *M* NaCl; hybridization rate is not affected significantly at 0.4 to 1.0 *M* NaCl. For detection of plant pathogens hybridization is often carried out at 65°C in 0.3 to 0.7 *M* NaCl. Addition of formamide to the hybridization mix reduces the temperature of optimum hybridization by 0.7°C for each percent of formamide in solution. So, prehybridization and hybridization may be carried out at 42°C in the presence of 50% formamide. Usually, the composition of the hybridization solution is the same as the composition of prehybridization one, except that the denatured probe is included. Usually, hybridization is carried out for 18 to 20 h.

E. WASHING PROCEDURE

It has been shown that the degree of homology between the RNAs of different viruses may be influenced by stringency during the washing procedure.[36] Thus, when the washing temperature was raised from 50 to 65°C, mainly heterologous, but not homologous binding of cDNA was reduced. To enhance the stringency conditions of the washing procedure, a low salt concentration may also be used (0.1xSSC solution). In some cases strong washing conditions are needed to reduce (or to eliminate) the false positive signals with cDNA.[20]

F. DETECTION

After the washing procedure, products of the nucleic acid hybridization assay (duplex "target probe") can be detected using an appropriate method. Duplex developing procedures may vary, depending on the kind of label attached to the probe molecule (see below). Radioactive areas on the filter paper may be visualized through autoradiography at −70°C using an intensifying screen for 48 h. For quantification, radioactive areas may be excised and scintillation counted. The autoradiographic results of a conventional dot-blot hybridization assay using radiolabeled nucleic acid probes are presented in Figure 1.

IV. NUCLEIC ACID PROBES

Different kinds of nucleic acid probes may be used in nucleic acid hybridization assay. The nucleic acid probe can be single-stranded or double-stranded RNA.[37–39] A single-stranded DNA probe (cDNA) can be prepared *in vitro* by reverse transcription of viral RNA (most viruses have a plus-strand RNA genome) using RNA-dependent DNA-polymerase (reverse transcriptase, RT). Complementary DNA probes can be synthesized *in vitro* for each special experiment. Usually, for screening purposes cDNA is inserted into a bacterial plasmid DNA or phage DNA for its amplification in recombinant DNAs to give a practically unlimited amount of uniform, well-characterized DNA probes. So, the repeated isolation and purification of plant pathogen for cDNA production is not needed. The basic principles and detailed protocols for cloning of DNA are given in several well-known laboratory manuals.[35,40]

Nucleic acid probes must be carefully checked for their specificity before use. The specificity of the hybridization assay may vary, depending on the portion of the genome cloned. Thus, nucleic acid probes from the coat-protein coding region of MAV isolate of BYDV hybridized well only to the homologous MAV isolate (27-fold greater than for PAV isolate of BYDV), while those from elsewhere hybridized effectively also with the PAV isolate.[16,41] Probes representing nonstructural viral protein genes were equally sensitive in detecting both serotypes D and M of plum pox virus (PPV).[39]

The sensitivity of the dot-blot nucleic acid hybridization assay using recombinant DNAs may be related to the size of the complementary DNA insert: for purified BYDV the signal of the assay was proportional to the size of the cDNA insert.[41] Similarly, a full-size PSTV-specific probe was on average three orders of magnitude more efficient in dot-blot hybridization compared with the short (19 to 20 nucleotides) synthetic probes.[42] At the same time, the nucleic acid probe with an 87-bp insert was almost as sensitive as the one carrying the full genome PSTV copy.[26] The concatameric insert nucleic acid probe (6.2 copies of a full-length PSTV) cut out from a plasmid was four times more sensitive than the monomeric one for viroid detection in dormant potato tubers and allows detection of 0.5 pg of viroid RNA.[27]

In some cases the synthetic probes may be used in hybridization reactions. Oligonucleotides have the ability to discriminate precisely between the target nucleic acids even if the difference between target and probe molecules is only in one base. The synthetic nucleic acid probes directed toward the variable region of the plant pathogen genome are useful for identification of different strains of the

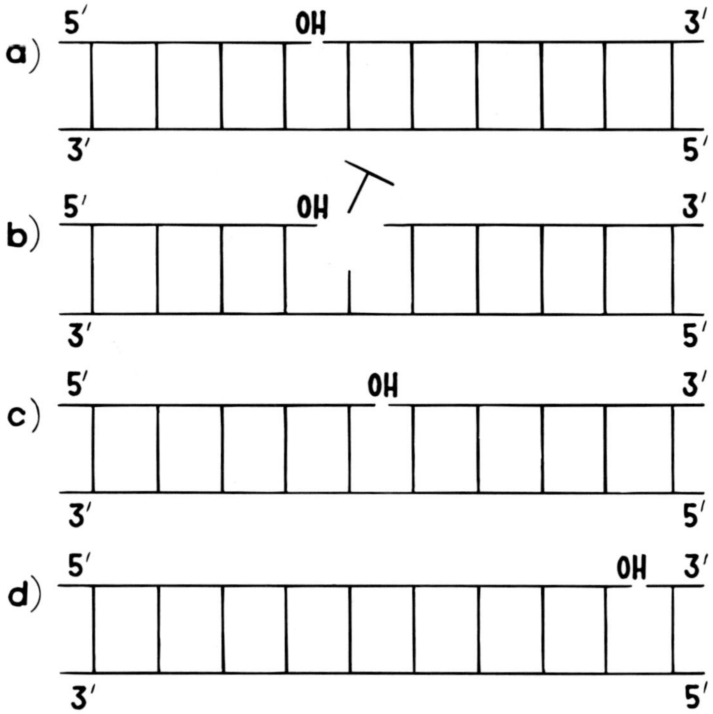

Figure 2 Nick-translation procedure. (a) Using DNAse 1, a nick with a 3-hydroxyl terminus is introduced into a double-stranded DNA probe; using *Escherichia coli* DNA polymerase 1; (b) the first nucleotide on the 5′ phosphate side of the nick is removed; (c) a labeled nucleotide is inserted instead of the unlabeled one; (d) the nick has been translated along the DNA chain, in a 5′ to 3′ direction.

same pathogen, which are indistinguishable by the conventional diagnosis technique.[43] Thus, hop stunt viroid (HSV) and its closely related strains (HSV-grapevine, HSV-cucumber, HSV-citrus) cannot be differentiated by bioassay on cucumber and by hybridization assay using full-size cDNA probes, because they induce the similar symptoms on cucumber plants and have up to 99% sequence homology.[44] The synthetic probes with specifically programmed sequences, complementary toward the conservative region of the pathogen genome will be useful for simultaneous detection of different strains of pathogen.[26,42–44]

V. LABELING OF NUCLEIC ACID PROBES

A. RADIOACTIVE LABELING

During the last decades the most common nucleic acid labels are radioisotopes, which are easily incorporated into nucleic acid probes by a variety of enzymatic techniques using radioactive NTPs or dNTPs as a substrate and can be directly detected by autoradiographic or scintillation-counting methods. Radiolabeled single-stranded RNA probes can be prepared by *in vitro* transcription by SP6 or T7 RNA polymerases utilizing as a template a plasmid DNA carrying a pathogen-specific cDNA insert.[30,45–48] Single-stranded radiolabeled DNA probes can be prepared by reverse transcription of viral RNA or by primer extension from a single-stranded M13 DNA template or by assymmetric polymerase chain reaction (PCR) using Taq polymerase. A uniform way of labeling recombinant double-stranded DNA probes is provided by the nick-translation enzymatic procedure,[49] using labeled dNTPs. The basic principle of the nick-translation reaction is schematically represented on Figure 2. After melting recombinant nick-translated double-stranded DNA probes are able to hybridize with both virus RNA target sequence and plasmid DNA sequence, forming a so-called net-structure. Unfortunately, radioisotope-labeled probes are unsuitable for routine use in detection of plant pathogens because of isotope instability, health hazards, disposal problems, and extended time for autoradiography (up to 48 h).

B. NONRADIOACTIVE LABELING

In the course of the past decade, nonradioactive detection systems have been developed. These systems are based on modification of the probe molecule by any nonradioactive reporter group to be detectable. The reporter systems may vary and can be divided into two different types. The first one includes direct chemical modifications of bases in the nucleic acid chain. Using a 2-acetylaminofluorene (AAF)-labeled cDNA probe, such a direct system has been applied for the routine detection of beet necrotic yellow vein virus (BNYVV) in crude beet root extracts and allow detection of 0.2 ng of purified virus per spot.[11] It was reported that a 150-pg target membrane-bound sequence can be detected using an ultraviolet (UV)-labeled DNA probe, followed by detection of the UV-labeled DNA by enzyme-conjugated antibodies raised against UV-irradiated DNA.[50] Unfortunately, these systems need an individual modification of the probe, coupled for each type of target sequence. The second type of nonradioactive detection system includes the systems where reporter groups are not directly linked to the probe, but linked either to NTPs, which can be incorporated into the nucleic acid probe (for example, by nick-translation) or to a polynucleotide chain using cross-linking agents. The indirect systems involve more steps, but are more convenient due to available universal signal-generating detection systems.

1. Biotinylated Probes

The most widely used nonradioactive indirect detection system includes labeling with biotin (vitamin H), wich binds very tightly ($K_{dis} = 10^{-15}$ *M*) to avidin (a 68−kDa glycoprotein, isolated from egg white) and streptavidin (isolated from *Streptomyces avidinii*). Each avidin molecule has four biotin-specific binding sites. Using avidin coupled to enzyme (usually alkaline phosphatase or horseradish peroxidase) it is possible to detect the biotin in hybrid "target probe" by measuring enzymatic activity with chromogenic or fluorogenic or chemiluminogenic substrates. Biotin can also be detected by unlabeled avidin with the addition of a biotin-enzyme conjugate. The sensitivity of target nucleic acid detection may be enhanced up to 50-fold using a biotinylated enzyme polymer and from 1 to 2 up to 0.1 pg of membrane-bound target sequence can be easily detected.[51,52] Biotin-specific antibodies can also be used to detect a biotinylated probe in hybrid "target probe".

There are several ways by which biotin can be linked to the nucleic acid probe. The first way is direct incorporation of biotinylated dNTPs or NTPs into the probe, because biotinylated nucleoside derivates are good substrates for DNA and RNA polymerases.[30,45,51–55] So, biotinylated nucleic acid probes can be prepared by the same ways as radioactive ones. The sensitivity of the hybridization assay may depend upon the length of the arm attached to the biotin molecule.[51,55] For example, the sensitivity of detection was approximately 1000-fold greater when biotin-14-dATP was used instead of biotin-7-dATP.[53] The melting temperature of biotinylated probes is slightly reduced, but the labeling does not affect the efficiency of hybridization.[56] The second way of labeling is by attaching biotin to the 5′ end or 3′ end of the nucleic acid probes.[57,58] But only one biotin molecule per probe molecule may be joined this way. Another way includes cross linking to the nucleic acids (DNA) of the reporter group which is a long-chain biotin (polybiotin) chemically linked to the basic molecule. This provides a very sensitive probe by which 10 to 50 fg of target DNA can be easily visualized using an avidin-enzyme conjugate.[59] Nucleic acid probes may be also labeled by cross linking the polynucleotide chain with biotin derivate, termed photobiotin, which attaches to double-stranded and single-stranded nucleic acids under intense light.[60] Detection of PAV-BYDV in plant sap using a photobiotin-labeled DNA probe was as sensitive as the radioactive ^{32}P assay.[25] In the dot-blot hybridization assay using purified double-stranded CARNA5 labeled with photobiotin as a probe, as little as 2 pg of dsCARNA5 was detected.[61]

Biotinylated nucleic acid probes can be stored at −20°C for long periods of time (1 to 2 years) without losing activity and can be reused several times without losing sensitivity of detection.[45,51] Unfortunately, an endogeneous biotin (vitamin H), which plays an important role in metabolic processes within the cells, induces problems regarding specificity and background. On the other hand, both avidin and streptavidin bind nonspecificically with proteins of the membrane-blocking agents. To improve the signal-to-noise ratio, some additional procedures may be used (preincubating the blotting membranes with high-ionic strength buffer, a special blocking procedure).

2. Digoxigenin-Labelling of Nucleic Acid Probes (DIG Probes)

Recently, an alternative nonradioactive labeling system was developed, based on the specific interaction between the cardenolide-steroid digoxigenin (DIG) from *Digitalis* plants and a high-affinity DIG-specific antibody coupled with a reporter group (i.e., hapten-antihapten interaction).[62] This system does not encounter the problem of nonspecific reaction, because the cardenolide DIG occurs naturally exclusively in *Digitalis* plants. The labeling of nucleic acid probes with digoxigenin can be performed by chemical, photochemical, and enzymatic reactions. Chemical digoxigenin incorporation is possible using several reactive groups of nucleic acids, such as the N6 position of the cytosine base, the C-5 position of the uracil base, 5′ end, 3′ end, and an internal phosphodiester bond. Photochemical digoxigenin labeling of DNA and RNA can be performed using a photoreactive azide compound as a photoactive reagent under shortwave light (λ_{max} = 254 nm). In this reaction the photoactive digoxigenin derivate (photo-DIG) is transferred into a reactive nitrene intermediate which reacts with the nucleic acid in a random manner. The photochemically labeled nucleic acids contain an average of one digoxigenin modification per 200 to 400 nucleotides. Digoxigenin labeling can be performed enzymatically by the incorporation of digoxigenin-modified deoxyribo- or ribonucleoside triphosphates (for example, by random-primed synthesis, nick-translation, PCR, or reverse transcription). Using these reactions, DNA is labeled with an average of one digoxigenin modification per 25 to 36 nucleotides. The digoxigenin labeling system allows detection of up to 0.1 pg of membrane-bound target sequence and may be successfully used for plant pathogen detection: for PSTV detection, the digoxigenin-labeled probe is as sensitive as the radioactive one, detecting as little as 2.5 pg of PSTV RNA.[63]

3. Enzyme Labeling of Nucleic Acid Probes

Another approach in nonradioactive labeling of nucleic acid probes is to couple the enzyme molecule (alkaline phosphatase or horseradish peroxidase) to polyethyleneimine to supply the enzyme with a positively charged tail,[64] which binds electrostatically to the nucleic acid probe and may be covalently linked with glutaraldehyde. The DNA probe labeled in this way can be added directly to a hybridization mixture without the probe purification procedure. After washing the enzyme-bound probe may be made visible using a dye-precipitation substrate, or the nucleic acid duplex may be detected by another appropriate procedure. The detection limit of this method is in the 0.4 to 5 pg range of target sequence[64,65] a level comparable to conventional hybridization techniques using radioactive probes. Unfortunately, when crude sap samples are used in the assay instead of purified nucleic acids, specific and nonspecific background signals may be undistinguishable (own results, unpublished). Probably, this high-speed, nonradioactive technique may be used in plant pathogen detection using a special sample preparation procedure.

4. Advantages and Limitations of Nonradioactive Systems

Nonradioactive nucleic acid probes are attractive alternatives to radioactive ones due to their safety, long period of life, and short time for label detection. Nevertheless, it was reported by many authors that nonspecificity is usually the main drawback in nonradioactive labeling systems. While nonspecificity may be negligible using purified nucleic acids as a sample, this problem becomes important when crude sap samples must be tested.[52,54–56] In some cases, nonspecific signals may be significantly reduced or eliminated with special procedures for sample preparation, blocking, and washing steps. Another way to overcome the nonspecificity is use of the so-called "sandwich" hybridization technique. In this method a pair of probes complementary to different regions of target sequence is used (Figure 3a). One of them is immobilized on the solid support and is used to capture the target, while the other probe carries the nonradioactive detectable label.[61,66,67] The hybridization between capture probe, target, and labeled probe is carried out in solution, so all the material in the extract which induces nonspecific signals can be washed away. Sandwich hybridization may be performed in solution with significantly improved kinetic parameters, compared with sandwich hybridization using an immobilized capture probe.[69] With this technique hybrids may be detected using an affinity-based procedure: after hybridization, sulfonated capture DNA is collected in polystyrene microtitre wells precoated with antibodies to sulfone-modified DNA (Figure 3b). The sandwich hybridization technique is now under investigation to be adopted for plant virus and viroid detection. It has been found in preliminary tests that the sandwich hybridization technique shows considerable promise.[70]

VI. CONCLUSION

The repertoir of modern diagnostic methods is large. In the past decade considerable progress has been made in the nucleic acid hybridization assay, which seems to be a good alternative to the ELISA technique, when virus-specific antisera are not available or pathogen-specific protein is not produced in host plants and, hence, such infections are not detectable serologically. The nucleic acid hybridization method is a simple, sensitive, and flexible approach in plant virus infection diagnosis and studying of relationships between viruses or viroids, and which is able to detect precisely any part of the plant pathogen genome. Applicability of nucleic acid hybridization assay will be extended in the future by developments in reliable nonradioactive detection systems and in studying of viral genome sequences. It is known that infection symptoms in host plants often vary in distribution, both spatially and temporally. These peculiarities may affect the success of any diagnostic method, including the nucleic acid hybridiza-

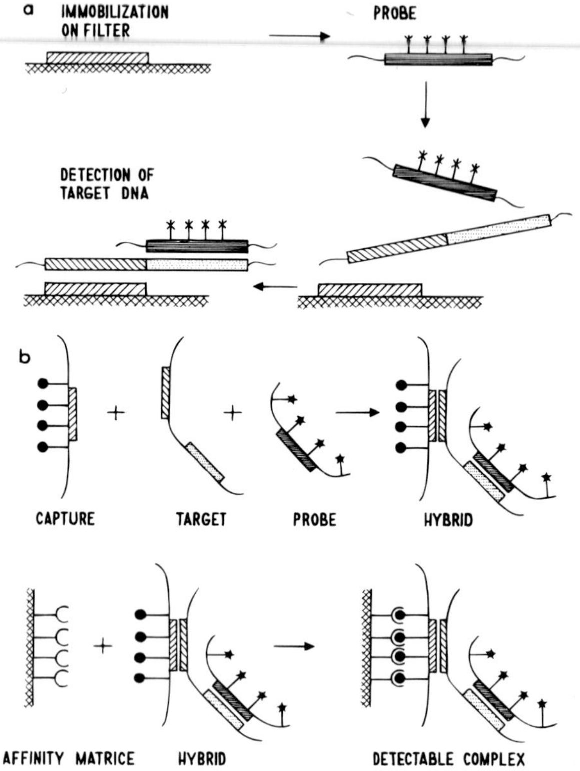

Figure 3 Sandwich hybridization procedure. (a) The capture sequence is affixed to a solid matrix and is used to immobilize the hybrid between the probe and target sequences. (b) In a modified technique,[69] the capture, target, and probe sequences are hybridized in liquid. The capture sequence contains a hapten (e.g., protein or biotin) which can be "captured" by an affinity matrix (e.g., antiserum to the protein or avidin).

tion assay. Detection of viroids and some viruses may be crucial if the concentration of these pathogens is below detection levels of conventional methods, both ELISA and the nucleic acid hybridization assay. This problem may be overcome by using samples with an amplified target sequence, as in the case of plant pathogens with known primary structure of the genome. This target sequence amplification may be done in a cyclic PCR using specific oligonucleotides as primers and Taq-polymerase. PCR is a highly efficient and specific method, theoretically capable of synthesizing 10^6 and more copies of product from a single DNA target sequence. After the target sequence, amplification samples can be analyzed by electrophoresis in agarose gel.[71,72] In this electrophoretic assay, detectable amounts of plant virus genome in infected plant tissue are approximately three to four orders of magnitude lower as compared with those detectable by ELISA and direct nucleic acid hybridization assay. Sometimes, PCR may be inhibited in nucleic acid extracts from some kinds of plant tissue, and a special sample purification procedure is needed to eliminate the inhibitory effect and enable PCR amplification of the virus-specific sequence in a tissue sample.[73] Combining PCR with molecular hybridization further increases the sensitivity of detection to a gain of four to five orders of magnitude as compared with direct molecular hybridization and enables the detection of up to a few molecules of plant pathogen genome.[22,74] The combination of PCR and the nucleic acid hybridization assay allows detection with the highest level of sensitivity and should be important in the future.

ACKNOWLEDGMENT

I am grateful to Prof. D. J. Gumpf for encouragement and helpful discussions. This work was funded in part by USDA/ARS cooperative grant to D. J. Gumpf.

REFERENCES

1. Owens, R. A. and Diener, T. O., Sensitive and rapid diagnosis of potato spindle tuber viroid disease by nucleic acid hybridization, *Science,* 213, 670, 1981.

2. Maule, A. J., Hull, R., and Donson, J., The application of spot hybridization to the detection of DNA and RNA viruses in plant tissues, *J. Virol. Meth.,* 6, 215, 1983.
3. Harrison, B. D. and Robinson, D. J., Genome reconstitution and nucleic acid hybridization as method of identifying particle-deficient isolates of tobacco rattle virus in potato plants with stem-mottle disease, *J. Virol. Meth.,* 5, 255, 1982.
4. Harrison, B. D., Robinson, D. J., Mowat, W. P., and Duncan, G. H., Comparison of nucleic acid hybridization and other tests for detecting tobacco rattle virus in narcissus plants and potato tubers, *Ann. Appl. Biol.,* 102, 331, 1983.
5. Francki, R. I. B., Plant viral satellites, *Annu. Rev. Microbiol.,* 39, 151, 1985.
6. Kaper, J. M. and Waterworth, H. E., Cucumber mosaic virus-associated RNA 5: causal agent for tomato necrosis, *Science,* 196, 429, 1977.
7. Rosner, A. and Bar-Joseph, M., Diversity of citrus tristeza virus strains indicated by hybridization with cloned cDNA sequences, *Virology,* 139, 189, 1984.
8. Rosner, A., Lee, R. F., and Bar-Joseph, M., Differential hybridization with cloned cDNA sequences for detecting a specific isolate of citrus tristeza virus, *Phytopathology,* 76, 820, 1986.
9. Baulcombe, D., Flavell, R. B., Boulton, R. E., and Jellis, G. J., The sensitivity and specificity of a rapid nucleic acid hybridization method for the detection of potato virus X in crude sap samples, *Plant Pathol.,* 33, 361, 1984.
10. Burgermeister, W., Koenig, R., Weich, H., Sebald, W., and Lesemann D.-E., Diversity of the RNAs in thirteen isolates of beet necrotic yellow vein virus in *Chenopodium quinoa* detected by means of cloned cDNAs, *J. Phytopathol.,* 115, 229, 1986.
11. Sakamoto, H., Lemaire, O., Merdinoglu, D., and Guesdon, J.-L., Comparison of enzyme-linked immunosorbent assay (ELISA) with dot hybridization using ^{32}P- or 2-acetylaminofluorene (AAF)-labelled cDNA probes for the detection and characterization of beet necrotic yellow vein virus, *Mol. Cell. Probes,* 3, 159, 1989.
12. Weidemann, H. L. and Koenig, R., Differentiation of isolates of potato virus S which infect *Chenopodium quinoa* systematically by means of quantitative cDNA hybridization test, *Z. Pflanzenkr. Pflanzenschutz,* 97, 323, 1990.
13. Jayasena, K. W., Randles, J. W., and Barnett, O. W., Synthesis of complementary DNA probe specific for detecting subterranean clover red leaf virus in plants and aphids, *J. Gen. Virol.,* 65, 109, 1984.
14. Czosnek, H. and Navot, N., Virus detection in squash-blots of plants and insects: applications in diagnostics, epidemiology, and breeding, in *Biotechnology in Agriculture,* Alan R. Liss, New York, 1988, 83.
15. Waterhouse, P. M., Gerlach, W. L., and Miller, W. A., Serotype-specific and general luteovirus probes from cloned cDNA sequences of barley yellow dwarf virus, *J. Gen. Virol.,* 67, 1273, 1986.
16. Fattouh, F. A., Ueng, P. P., Kawata, E. E., Barbara, D. J., Larkins, B. A., and Lister, R. M., Luteovirus relationships assessed by cDNA clones from barley yellow dwarf viruses, *Phytophatology,* 80, 913, 1990.
17. Van Der Vlugt, C. I. M., Linthorst, H. J. M., Asjes, C. J., Van Schadewijk and Bol, J. F., Detection of tobacco rattle virus in different parts of tulip by ELISA and cDNA hybridization assays, *Neth. J. Plant Pathol.,* 94, 149, 1988.
18. Jelkmann, W., Maiss, E., Breyel, E., and Casper, R., Production and use of cDNA clones from arabis mosaic virus, *Ann. Appl. Biol.,* 113, 483, 1988.
19. Ronco, A. E., Dal Bo, E., Ghiringhelli, P. D., Medrano, C., Romanowski, V., Sarachu, A. N., and Grau, O., Cloned cDNA probes for the detection of tomato spotted wilt virus, *Phytopathology,* 79, 1309, 1989.
20. Roy, B. P., Aboul Haidar, M. G., Sit, T. L., and Alexander, A., Construction and use of cloned cDNA biotin and ^{32}P-labeled probes for the detection of papaya mosaic potexvirus RNA in plants, *Phytopathology,* 78, 1425, 1988.
21. Kanno, Y., Yoshikawa, N., and Takahashi, T., Detection of apple chlorotic leaf spot virus by dot-blot hybridization, *Ann. Phytopathol. Soc. Jpn.,* 57, 278, 1991.
22. Borja, M. J. and Ponz, F., An appraisal of different methods for the detection of the walnut strain of cherry leafroll virus, *J. Virol. Meth.,* 36, 73, 1992.
23. Flores, R., Detection of citrus exocortis viroid in crude extracts by dot-blot hybridization: conditions for reducing spurious hybridization results and for enhancing the sensitivity of the technique, *J. Virol. Meth.,* 13, 161, 1986.

24. Thomas, P., Hybridization of denatured DNA and small RNA fragments transferred to nitrocellulose, *Proc. Natl. Acad. Sci. U.S.A.*, 77, 5201, 1980.
25. Habili, N., McInnes, J. L., and Symons, R. H., Nonradioactive, photobiotin-labeled DNA probes for the routine diagnosis of barley yellow dwarf virus, *J. Virol. Meth.*, 16, 225, 1987.
26. Welnicki, M., Skrzeczkowski, J., Soltynska, A., Jonczyk, P., Markiewicz, W., Kierzek, R., Imiolczyk, B., and Zagorski, W., Characterization of synthetic DNA probe detecting potato spindle tuber viroid, *J. Virol. Meth.*, 24, 141, 1989.
27. Zekanowski, C., Welnicki, M., Skrzeczkowski, J., and Zagorski, W., Detection of potato spindle tuber viroid (PSTV) in dormant potato tubers by concatameric cDNA probe, *J. Virol. Meth.*, 30, 127, 1990.
28. Candresse, T., Macquaire, G., Monsion, M., and Dunez, J., Detection of Chrisanthemum stunt viroid (CSV) using nick translated probes in a dot-blot hybridization assay, *J. Virol. Meth.*, 20, 185, 1988.
29. MacQuaire, G., Monsion, M., Mouches, C., Candresse, T., and Dunez, J., Spot hybridization: application to viroid identification, *Ann. Virol. (Inst. Pasteur)*, 135E, 219, 1984.
30. Roy, B. P., AbouHaidar, M. G., and Alexander, A., Biotinylated RNA probes for the detection of potato spindle tuber viroid (PSTV) in plants, *J. Virol. Meth.*, 23, 149, 1989.
31. Boulton, R. E., Jellis, G. J., Baulcombe, D. C., and Squire, A. M., The application of complementary DNA probes to routine virus detection, with particular reference to potato viruses, in *Developments and Applications in Virus Testing*, Jones, R. A. C. and Torrance, L., Eds., Suffolk, Lavenham, 1986, 41.
32. Nikolaeva, O. V., Morozov, S. Yu., Zakhariev, V. M., and Skryabin, K. G., Improved dot-blot hybridization assay for large-scale detection of potato viruses in crude potato tuber extracts, *J. Phytopathol.*, 129, 283, 1990.
33. Wetmur, J. G., DNA probes: applications of the principles of nucleic acid hybridization, *Crit. Rev. Biochem. Mol. Biol.*, 26, 227, 1991.
34. Meinkoth, J. and Wahl, G., Hybridization of nucleic acids immobilized on solid supports, *Anal. Biochem.*, 138, 267, 1984.
35. Hames, B. D. and Higgins, S. J., Eds., *Nucleic Acid Hybridization: a Practical Approach*, IRL Press, Oxford, 1985.
36. Koenig, R., An, D., and Burgermeister, W., The use of filter hybridization techniques for the identification, differentiation and classification of plant viruses, *J. Virol. Meth.*, 19, 57, 1988.
37. Rosner, A., Spiegel, S., Alper, M., and Bar-Joseph, M., Detection of avocado sunblotch viroid (ASBV) by dot-spot self-hybridization with a ^{32}P-labeled ASBV-RNA, *Plant Mol. Biol.*, 2, 15, 1983.
38. Dodds, J. A., The potential for using double-stranded RNAs as diagnostic probes for plant viruses, in *Developments and Applications in Virus Testing*, Jones, R. A. C. and Torrance, L., Eds., Suffolk, Lavenham, 1986, 71.
39. Wetzel, T., Tavert, G., Teycheney, P. Y., Ravelonandro, M., Candresse, T., and Dunez, J., Dot hybridization detection of plum pox virus using ^{32}P-labeled RNA probes representing non-structural viral protein genes, *J. Virol. Meth.*, 30, 161, 1990.
40. Maniatis, T., Fritsch, E. F., and Sambrook, J., Eds., *Molecular Cloning: a Laboratory Manual*, Cold Spring Harbor Laboratory, Cold Spring Harbor, NY, 1982.
41. Barbara, D. J., Kawata, E. E., Ueng, P. P., Lister, R. M., and Larkins, B. A., Production of cDNA clones from the MAV isolate of barley yellow dwarf virus, *J. Gen. Virol.*, 68, 2419, 1987.
42. Skrzeczkowski, L. J., Hennig, J., Zagorski, W., Markiewicz, W., Kierek, R., and Welnicki, M., The application of synthetic DNA probe for the detection of potato spindle tuber viroid (PSTV), in *Proceedings of the International Seminar Viroids of Plants and their Detection*, August 12–20, 1986, Warsaw Agricultural University, Warsaw, 1986, 107.
43. Bar-Joseph, M., Segev, D., Blickle, W., Yesodi, V., Franck, A., and Rosner, A., Application of synthetic DNA probes for the detection of viroids and viruses, in *Developments and Applications in Virus Testing*, Jones, R. A. C. and Torrance, L., Eds., Suffolk, Lavenham, 1986, 13.
44. Sano, T., Kudo, H., Sugimoto, T., and Shikata, E., Synthetic oligonucleotide hybridization probes to diagnose hop stunt viroid strains and citrus exocortis viroid, *J. Virol. Meth.*, 19, 109, 1988.
45. Rao, A. L. N., Huntley, C. C., Marsh, L. E., and Hall, T. C., Analysis of RNA stability and (−)strand content in viral infections using biotinylated RNA probes, *J. Virol. Meth.*, 30, 239, 1990.
46. Varveri, C., Candresse, T., Cugusi, M., Ravelonandro, M., and Dunez, J., Use of a ^{32}P-labeled transcribed RNA probe for dot hybridization detection of plum pox virus, *Phytopathology*, 78, 1280, 1988.

47. Herrbach, E., Lemaire, O., Ziegler-Crafff, V., Lot, H., Rabenstein, F., and Bouchery, Y., Detection of BMYV and BWYV isolates using monoclonal antibodies and radioactive RNA probes, and relationships among luteoviruses, *Ann. Appl. Biol.,* 118, 127, 1991.
48. Diener, T. O., Smith, D. R., Hammond, R. H., Albanese, G., La Rosa, R., and Davino, M., Citrus B viroid identified as a hop stunt viroid, *Plant Dis.,* 72, 691, 1988.
49. Rigby, P. W. J., Dieckmann, M., Rhodes, C., and Berg, P., Labeling deoxyribonucleic acid to high specific activity in vitro by nick translation with DNA polymerase 1, *J. Mol. Biol.,* 113, 237, 1977.
50. Lund, V., Lindqvist, B. H., and Eggset, G., Detection of specific DNA sequences using antibodies recognizing UV-labeled DNA, *Nucleic Acids Res.,* 17, 539, 1989.
51. Leary, J. J., Brigati, D. J., and Ward, D. C., Rapid and sensitive colorimetric method for visualizing biotin-labeled DNA probes hybridized to DNA or RNA immobilized on nitrocellulose: bio-blots, *Proc. Natl. Acad. Sci. U.S.A.,* 80, 4045, 1983.
52. Chan, V. T.-W., Fleming, K. A., and McGee, J. O. D., Detection of subpicogram quantities of specific DNA sequences on blot hybridization with biotinylated probes, *Nucleic Acids Res.,* 13, 8083, 1985.
53. Eweida, M., Sit, T. L., Sira, S., and AbouHaidar, M. G., Highly sensitive and specific non-radioactive biotinylated probes for dot-blot, Southern and colony hybridizations, *J. Virol. Meth.,* 26, 35, 1989.
54. Eweida, M., Xu, H., Singh, R. P., and AbouHaidar, M. G., Comparison between ELISA and biotin-labeled probes from cloned cDNA of potato virus X for the detection of virus in crude tuber extracts, *Plant Pathol.,* 39, 623, 1990.
55. Brigati, D. J., Myerson, D., Leary, J. J., Spalholz, B., Travis, S. Z., Fong, C. K. Y., Hsiung, G. D., and Ward, D. C., Detection of viral genomes in cultured cells and paraffin-embedded tissue sections using biotin-labeled hybridization probes, *Virology,* 126, 32, 1983.
56. Langer, P. R., Waldrop, A. A., and Ward, D. C., Enzymaic synthesis of biotin-labeled polynucleotides: novel nucleic acid affinity probes, *Proc. Natl. Acad. Sci. U.S.A.,* 78, 6633, 1981.
57. Chollet, A. and Kawashima, E. H., Biotin-labeled synthetic oligodeoxyribonucleotides: chemical synthesis and use as hybridization probes, *Nucleic Acids Res.,* 13, 1529, 1985.
58. Richardson, R. W. and Gumport, R. I., Biotin and fluorescent labeling of RNA using T4 RNA ligase, *Nucleic Acids Res.,* 11, 6167, 1983.
59. Al-Hakim, A. H. and Hull, R., Studies towards the development of chemically synthesized non-radioactive biotinylated nucleic acid hybridization probes, *Nucleic Acids Res.,* 14, 9965, 1986.
60. Forster, A. C., McInnes, J. L., Skingl, D. C., and Symons, R. H., Nonradioactive hybridization probes prepared by the chemical labeling of RNA and DNA with a novel reagent, photobiotin, *Nucleic Acids Res.,* 13, 745, 1985.
61. Wang, W. Q., Natsuaki, T., Okuda, S., and Teranaka, M., Detection of cucumber mosaic virus RNA5 by hybridization with photobiotin-labeled double-stranded RNA5 probe, *Ann. Phytopathol. Soc. Jpn.,* 56, 33, 1990.
62. Kessler, C., The digoxigenin:anti-digoxigenin (DIG) technology—a survey on the concept and realization of a novel bioanalytical indicator system (Review), *Mol. Cell. Probes,* 5, 161, 1991.
63. Welnicki, M. and Hiruki, C., Highly sensitive digoxigenin-labeled DNA probe for the detection of potato spindle tuber viroid, *J. Virol. Meth.,* 39, 91, 1992.
64. Renz, M. and Kurz, C., A colorimetric method for DNA hybridization, *Nucleic Acids Res.,* 12, 3435, 1984.
65. Inoue, H., Gushi, K., Matsuura, S., and Sakata, Y., A sensitive colorimetric detection of virus DNA and oncogene, *Biochem. Biophys. Res. Commun.,* 143, 323, 1987.
66. Dahlen, P., Syvanen, A.-C., Hurskainen, P., Kwiatkowski, M., Sund, C., Ylikoski, J., Soderlund, H., and Lovgren, T., Sensitive detection of genes by sandwich hybridization and time-resolved fluorometry, *Mol. Cell. Probes,* 1, 159, 1987.
67. Jungell-Nortamo, A., Syvanen, A.-C., Luoma, P., and Soderlund, H., Nucleic acid sandwich hybridization: enhanced reaction rate with magnetic microparticles as carriers, *Mol. Cell. Probes,* 2, 281, 1988.
68. Urdea, M. S., Warner, B. D., Running, J. A., Stempien, M., Clyne, J., and Horn, T., A comparison of non-radioisotopic hybridization assay methods using fluorescent, chemiluminescent and enzyme labeled synthetic oligodeoxyribonucleotide probes, *Nucleic Acids Res.,* 16, 4937, 1988.
69. Syvanen, A.-C., Laaksonen, M., and Soderlund, H., Fast quantification of nucleic acid hybrids by affinity-based hybrid collection, *Nucleic Acids Res.,* 14, 5037, 1986.

70. Hull, R. and Al-Hakim, A., Nucleic acid hybridization in plant virus diagnosis and characterization, *Trends Biotechnol.,* 6, 213, 1988.
71. Yang, X., Hadidi, A., and Garnsey, S. M., Enzymatic cDNA amplification of citrus exocortis and cachexia viroids from infected citrus hosts, *Phytopathology,* 82, 279, 1992.
72. Hadidi, A. and Yang, X., Detection of pome fruit viroids by enzymatic cDNA amplification, *J. Virol. Meth.,* 30, 261, 1990.
73. Vunsh, R., Rosner, A., and Stein, A., Detection of bean yellow mosaic virus in gladioli corms by the polymerase chain reaction, *Ann. Appl. Biol.,* 119, 289, 1991.
74. Vunsh, R., Rosner, A., and Stein, A., The use of the polymerase chain reaction (PCR) for the detection of bean yellow mosaic virus in gladiolus, *Ann. Appl. Biol.,* 117, 561, 1990.

FURTHER READING

Borja, M. J. and Ponz, F., An appraisal of different methods for the detection of the walnut strain of cherry leafroll virus, *J. Virol. Meth.,* 36, 73, 1992.

Czosnek, H. and Navot, N., Virus detection in squash-blots of plants and insects: applications in diagnostics, epidemiology, and breeding, *Biotechnology in Agriculture,* 83, Alan R. Liss, New York, 1988.

Jones, R. A. C. and Torrance, L., Eds., *Developments and Applications in Virus Testing,* Suffolk, Lavenham, 1986.

Flores, R., Detection of citrus exocortis viroid in crude extracts by dot-blot hybridization: conditions for reducing spurious hybridization results and for enhancing the sensitivity of the technique, *J. Virol. Meth.,* 13, 161, 1986.

Gurr, S. J., McPherson, F. J., and Bowles, D. J., *Molecular Plant Pathology: a Practical Approach,* Oxford University Press, Oxford, 1992.

Hames, B. D. and Higgins, S. J., Eds., *Nucleic Acid Hybridization: a Practical Approach,* IRL Press, Oxford, 1985.

Hull, R. and Al-Hakim, A., Nucleic acid hybridization in plant virus diagnosis and characterization, *Trends Biotechnol.,* 6, 213, 1988.

Innis, M. A., Galfand, D. H., Sninsky, J. J., and White, *PCR Protocols: a Guide to Methods and Applications,* Academic Press, San Diego, 1989.

Singh, R. P., Molecular hybridization with complementary DNA for plant viruses and viroids detection, in *Perspectives in Phytopathology,* Agnihotri, V. P., Singh, N., Chaube, H. S., Singh, U. S., and Dwivedi, T. S., Eds., Today & Tomorrow's Printers & Publishers, New Dehli, 1989, 51.

Watmur, J. G., DNA probes: applications of the principles of nucleic acid hybridization, *Crit. Rev. Biochem. Mol. Biol.,* 26, 227, 1991.

Chapter 11

Tissue-Print Hybridization for the Detection and Localization of Plant Viruses

Tet-Fatt Chia, Yang-Sun Chan and Nam-Hai Chua

TABLE OF CONTENTS

I. INTRODUCTION

The ability to detect virus infection in plants is important for predicting and monitoring plant virus epidemics. Studies have shown that plants infected by different viruses can produce similar symptoms; conversely, symptoms caused by a virus can vary greatly, depending on the plant species. Thus, the causative agent which is the virus must be identified accurately and rapidly so that its mode of transmission can be understood and control measures effected.

To effectively detect and control the spread of viruses, it is necessary that the method of detection be sensitive, reliable, and easy to execute, as each one of these factors will affect the accuracy of the test. Traditionally, the mainstay in the diagnosis of viruses relies mainly on serological methods like the enzyme-linked immunosorbent assay (ELISA). However, in recent years, nucleic acid hybridization has been proposed and reviewed as being well suited for such purposes.[1,2] Its applicability has been extended by various developments in membrane materials, probes, and nonradioactive reporter groups. Most of the approaches to the use of nucleic acid hybridization in plant virus detection involve mixed-phase hybridization with the target nucleic acid immobilized onto a solid matrix.[3,4] The most common procedure is the dot-blot or slot-blot hybridization, and both methods have been used for the detection and discrimination of many different types and strains of viruses.[2,5,6]

However, in order to expedite the process of detection, tissue-print hybridization was utilized.[7,8] Printing plant tissue directly onto membranes (nylon or nitrocellulose) was first reported by Cassab and Varner,[9] and subsequently the method has been modified to suit different plant species. This method has the added advantage of being able to localize viruses within the plant.[8,10]

II. PRINCIPLES

The tissue-print hybridization technique can essentially be divided into two parts: (1) transfer of the viral nucleic acid from the plant tissue directly onto a membrane and (2) hybridization of the printed membrane with a nucleic acid probe reporter system.

A. TISSUE PRINT

Viruses are composed of either RNA or DNA, and these nucleic acids which are present in infected cells could be transferred directly onto a nitrocellulose or nylon membrane by printing. The principle behind printing is to transfer the sap from cells that have been mechanically ruptured directly from the tissue onto the membrane. Since the membrane has high binding affinity for nucleic acids, the viral RNA or DNA will be affixed onto the solid matrix and become immobilized. The presence of the virus capsid will not inhibit the binding of viral nucleic acid to the membrane.[11] However, for successful binding, the nucleic acid must not have an extensive secondary structure. Double-stranded RNA or DNA will not bind readily to the membrane, and the plant tissue would require an additional step of denaturation with denaturants like glyoxal or alkali before printing.

0-87371-877-1/95/$0.00+$.50

B. NUCLEIC ACID HYBRIDIZATION

To detect the viral nucleic acids, a nucleic acid probe complementary to that of the target must be used. This probe can either be DNA or RNA. As over 75% of known plant viruses have plus-strand RNA genomes, the most commonly used probe is a cDNA made from viral RNA by reverse transcription. Therefore the probe can be synthesized directly before each diagnosis or made readily in large quantity by insertion into a bacterial plasmid or a bacteriophage vector. Through the manipulation of the length and sequences of the probe nucleic acid, it is possible to either increase or decrease the specificity of detection. These probes can be labeled with radioactive ^{32}P-nucleotides or tagged/conjugated to non-radioactive groups that can act as a reporter for detection. Hybridization essentially involves a mixed-phase system with the target nucleic acid immobilized onto a solid matrix and the probe nucleic acid in the liquid phase.

III. METHODOLOGY

The technique of printing plant tissues onto a nylon or nitrocellulose membrane is a simple and straightforward procedure. The plant or its organs are sliced, and the cut surfaces are then pressed for a short duration (3 to 30 s) directly onto the membrane which is supported by two layers of Whatman® filter paper. This procedure is well suited only for firm or succulent tissues which can be handled with ease. For plants with thin and pliable organs, it is difficult to print directly by hand. For such cases, slight modification is needed to accommodate the difference in texture of the tissue. In the tomato leaf, which is slightly fleshy, a simple squash-printing by rolling a glass rod or pen over the tissue is sufficient to leave an imprint onto the membrane. For others, like the soybean leaf, the tough and dry texture of the leaf makes printing more difficult. Mansky et al.[10] uses a hydraulic press capable of generating 70 kg/cm^2 to press an imprint onto the membrane. From our experience, we found that we could circumvent the use of a hydraulic press if the tissues were first freeze-thawed using liquid nitrogen. This process will rupture the cells, thereby making it possible to print with the aid of a glass rod or block with a pressure of only 1 to 4 kg/cm^2. The beauty of tissue printing lies in its flexibility, and its limitation is our imagination in getting the imprint onto the membrane.

For the purpose of diagnosis, it would suffice as long as some sap is being transferred onto the membrane. But for virus localization studies, an even and complete print of the tissue is a must. A good print should consist of the distinct outline of the tissue that was printed. This is done by applying a constant pressure evenly during printing. Care must taken during printing to ensure that there are no smudges on the print. One often has to print a couple of times before obtaining a good print. To check for uniformity of the print, the imprint can be stained with india ink to show the protein distribution profile, which normally gives a good representation of the nucleic acid distribution.

After printing, the membrane is air dried, baked at 80°C for 1 h under vacuum, or cross linked with ultraviolet radiation if the membrane used is nylon. The membrane can then be kept at room temperature for at least 6 months without losing its hybridization capability.

The final part of tissue-print hybridization is the detection of the viruses using nucleic acid probes complementary to the target sequence of the virus. If the radioactive method of detection is used, a ^{32}P-labeled viral cDNA probe could be generated by random primed synthesis or nick translation. The sensitivity of the detection increases with the specific activity of the radioactive-labeled cDNA probe. Normally, a specific activity of 1 to 2 $\times$ 10^9 cpm/μg DNA should be sufficient to allow detection into the picogram range. The hybridization protocol can either be aqueous or formamide based.[12] The duration of hybridization is normally 12 to 20 h and the filters are then washed and autoradiographed.

For the nonradioactive method of detection, there are two systems in use; chromogenic and nonchromogenic. In the former, the result of the detection is visualized directly on the membrane through color reaction of peroxidase-, avidin-, or biotin-conjugated antibody. The latter is a light-base detection system in which a chemiluminescent substrate is being used as a substrate for peroxidase. The light signal generated is then enhanced and captured on a photosensitive film.

The chromogenic form, however, was not found to be generally useful because many plant species contain phenolic compounds in the sap which is easily oxidized to give a brown stain. The stain will persist through washing and interfere with the color reaction of the detection system. This interference also reduces the sensitivity of the method, particularly at low virus titre. As such, we would recommend the chemiluminescent method as a better alternative to the chromogenic method.

Table 1 **Comparison of the sensitivity of the enzyme-linked immunosorbent assay (ELISA) and nucleic acid hybridization assay for the detection of virus**

Detection method	Endpoint of detection (ng of virus)
ELISA	92[a]
Nucleic acid hybridization assay:	
12 h[b]	0.65[c]
24 h	0.22
72 h	0.02

[a] The endpoint was calculated from the mean A_{405} nm value of three samples as determined by an LKB spectrophotometer.

[b] Exposure of blot to film at −80°C with Du Pont Hi-Plus intensifying screen.

[c] Values were calculated from the mean absorption values of three samples as determined by an LKB densitometer scanner.

IV. APPLICATION IN PLANT PATHOLOGY

The development of tissue-print hybridization has opened the door for an alternative method of virus diagnosis compared to ELISA. Identification of viruses is required in many aspects of plant pathology, for example, in the prediction of plant diseases in crops, plant quarantine control, prevention of infection in planting stock, and disease monitoring. In all these areas, tissue-print hybridization is well suited for the job due to its rapidity and accuracy in processing the samples.

The detection method in this approach depends upon the probe nucleic acid which comprises sequences that are complementary to the target (viral) nucleic acid. The composition of the probe nucleic acid and the parameters of hybridization can be varied to allow flexibility in determining relations between viruses. By using stretches of sequences that are unique to a single virus strain, the probe can be made to detect only a specific strain of virus and not others. In this way, it can be used to differentiate different viral strains or isolates. In contrast, if the stretches of sequence used in the probe nucleic acid is common to a particular group of virus, then the specificity of detection will broaden.

One of the most important contributions of the tissue-print hybridization technique is its ability to show the localization of the virus in the plant. By skillful use of the tissue-printing technique, it is even possible to show the groups of cells that are infected.[13] In this respect, this method can be used to study local and systemic spread of the virus within the plant upon infection and also at the same time study its disease symptomatology. This approach should be useful for the study of dual infection by different viruses to see if systemic spread of two different viruses within a plant is synergistic, antagonistic, or independent of one another. In addition, tissue printing can be used to study the relationship between cross protection and the location of the two related strain of viruses in the plant. This will give us a better understanding of the phenomenon seen. On the whole, this technique has the potential for use in many aspects of plant pathological studies.

V. ADVANTAGES AND LIMITATIONS

The clear advantage of tissue-print hybridization lies in its simplicity and rapidity of sample processing. Unlike ELISA, minimal steps are involved and no expensive equipment is needed. Untrained personnel can be easily taught to print and handle a large volume of samples. Also, the sensitivity of this method is much higher than ELISA at approximately a thousandfold. Table 1 shows that the end-point for ELISA is in the nanogram (10^{-9}) range, while tissue-print hybridization is in the picogram (10^{-12}) level. Therefore, this sensitivity puts it at par with Southern blot hybridization.

Unlike serological methods, samples in this case do not need to be processed before detection; therefore, losses of hybridizable viral nucleic acid is minimized. As a very small quantity of tissue is needed for the analysis, a few thousand samples can be handled easily by a single person within a day. As the printed samples are very stable, they can be mailed from one country to another, thus facilitating sampling and analysis.

Another aspect of tissue printing is the ability to screen simply and quickly whole plants as well as different plant tissues for the location and distribution of the viruses at different times of infection.

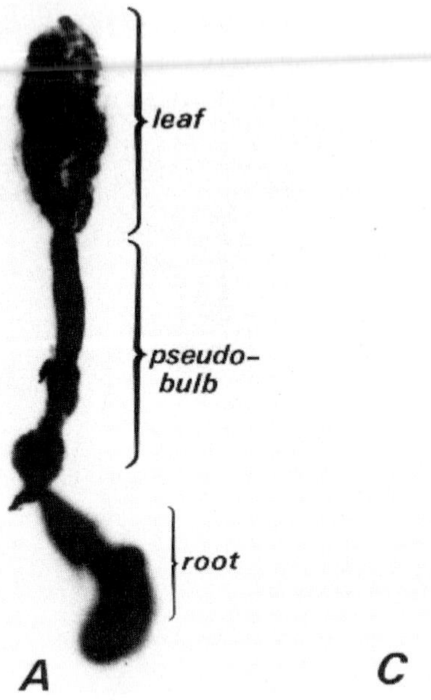

Figure 1 Whole-plant printing of unifoliate *Cattleya* plant onto a nitrocellulose membrane to localize distribution of cymbidium mosaic virus. Plants were sectioned longitudinally and block printed onto a nitrocellulose membrane with a uniform pressure of 1 kg/cm^2 and hybridized with ^{32}P-labeled cymbidium mosaic virus-specific probes. (A) shows the profile of cymbidium mosaic virus-infected plant, while an uninfected plant similarly printed (C) did not show any hybridization signal.

Printing the tissues at both the longitudinal as well as the transverse plane will allow the establishment of a three-dimensional representation of the virus distribution within the plant. This information will be useful for researchers interested in the pattern of localized and systemic movement of the viruses. An example of a whole-plant print, shown in Figure 1, demonstrates the ability to visualize the location of cymbidium mosaic virus in an infected orchid plant. In this particular instance, we could see that the virus is present throughout the plant and particularly in the roots, which is shown by the higher intensity in the hybridization signal.

Despite the many advantages encouraging the use of tissue-print hybridization, there is still some minor limitation to its widespread use. Radioactive-labeled probes that are currently in use are ideal in giving signals with a high resolution and a clear background. Besides, the cost of radioactivity and its labeling steps are comparatively less expensive. But to find widespread application in the future, it will be necessary to replace the ^{32}P reporter groups with nonradioactive labels. Currently, there are many methods and different reagents available for nonradioactive labeling of nucleic acids, but the high cost renders them unfavorable for general usage. However, we believe that the cost will decrease with time and the chemiluminescence method may be the choice of use in the future.

In conclusion, we are certain that the flexibility of this system and its convenience in usage will make this approach one of the important tools in plant pathological studies.

REFERENCES

1. Hull, R. and Al-Hakim, A., Nucleic acid hybridization in plant virus diagnosis and characterization, *Trends Biotechnol.,* 6, 213, 1988.
2. Nikolaeva, O. V., Nucleic acid hybridization methods in diagnosis of plant viruses and viroids, in *Advance Methods in Plant Pathology,* Singh, U. S. and Singh, R. P., Eds., Lewis Publishers, Chelsea, MI, 1994.
3. Owens, R. A. and Diener, T. O., Sensitive and rapid diagnosis of potato spindle tuber viroid disease by nucleic acid hybridization, *Science,* 213, 670, 1981.
4. Boulton, M. I. and Markham, P. G., Developments and applications in virus testing, in *Association of Applied Biologist,* Jones, R. A. C. and Torrince, L., Eds., Wellesbourne, 1986, 3.
5. Baulcombe, D. C. and Fernandez-North Cote, E. N., Detection of strains of potato virus X and a broad spectrum of potato virus Y isolates by nucleic acid spot hybridization, *Plant Dis.,* 72, 307, 1988.
6. Polston, J. E., Dodds, J. A., and Perring, T. M., Nucleic acid probes for detection and strain discrimination of cucurbit geminiviruses, *Phytopathology,* 79, 1123, 1989.

7. Navot, N. R. Ber and Czonek, H., Rapid detection of tomato yellow leaf curl virus in squashes of plant and insect vectors, *Phytopathology,* 79, 562, 1989.
8. Chia, T. F., Chan, Y. S., and Chua, N. H., Detection and localization of viruses in orchids by tissue-print hybridization, *Plant Pathol.,* 4, 355, 1992.
9. Cassab, G. I. and Varner, J. E., Immunocytolocalization of extensin in developing soybean seed coats by immunogold-silver staining and tissue printing on nitrocellulose paper, *J. Cell Biol.,* 105, 2581, 1987.
10. Mansky, L. M., Andrews, R. E., Jr., Durand, D. P., and Hill, J. H., Plant virus location in leaf tissue by press blotting, *Plant Mol. Biol. Rep.,* 8, 13, 1990.
11. Baulcombe, D. C., Boulton, R. E., Flavell, R. B., and Jelus, G. J., Recombinant DNA probes for detection of viruses in plants, in *Proc. Br. Crop Prot. Conf.,* 1984, 207.
12. Maniatis, T., Fritsch, E. F., and Sambrook, J., Eds., *Molecular Cloning: a Laboratory Manual,* Cold Spring Harbor Laboratory, Cold Spring Harbor, NY, 1982.
13. Wisniewski, L. A., Powell, P. A., Nelson, R. S., and Beachy, R. W., Local and systemic spread of tobacco mosaic virus in transgenic tobacco, *Plant Cell,* 2, 559, 1990.

FURTHER READING

Plant virus diagnosis and characterization through nucleic acid hybridization was reviewed by Hull and Al-Hakim (pp. 213–218), in *Trends Biotechnol.,* Vol. 6, 7, 1988.

The concept pertaining to the principle of DNA hybridization, constructing and nonradioactive labeling of viral nucleic acid probes has been clearly documented by Elias (pp. 404–434), in *Immunohistopathology: a Practical Approach to Diagnosis,* American Society of Clinical Pathologists, Chicago, 1990.

Concise and easy to follow protocols describing the molecular technique of DNA and RNA hybridization and detection, in *Molecular Cloning: a Laboratory Manual,* 2nd ed., by Sambrook et al. (Cold Spring Harbor Laboratory, Cold Spring Harbor, NY, 1989).

The concept of tissue printing onto nitrocellulose was first reported by Cassab and Varner (*J. Cell Biol.,* 105, 2581–2588, 1987). Although its use was on immunocytolocalization of protein, its potential was apparent.

Owens and Diener first reported the principle and use of nucleic acid hybridization for the detection of viroid disease in plants (*Science* 213, 670–672, 1981).

Chapter 12

Transposon and Marker Exchange Mutagenesis

George H. Lacy and Verlyn K. Stromberg

TABLE OF CONTENTS

I. INTRODUCTION

Transposon and marker exchange mutagenesis are potent tools for dissecting pathogen-plant interactions. The power of transposon mutagenesis evolves from its ability to mark the location of a genetic change. In comparison, the locations and nature of genetic changes caused by chemical or irradiational mutagenesis are difficult to determine or characterize. The power of marker exchange mutagenesis is its ability to select specific genes for mutation. In contrast, chemical or irradiational mutagenesis likely produces more than one mutation per genome, making it difficult to ascertain if one or more mutations is responsible for observed phenotypic changes.

0-87371-877-1/95/$0.00+$.50

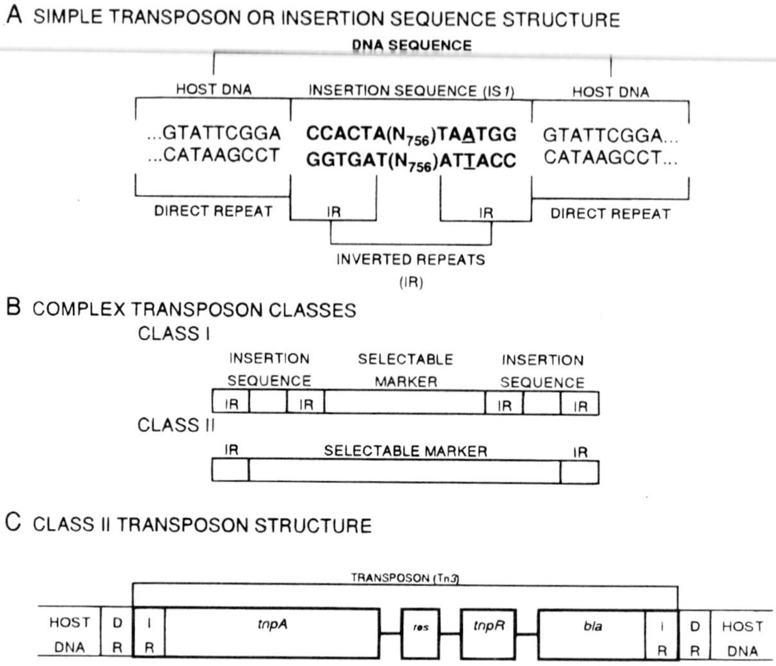

Figure 1 Relationship of insertion sequences and transposons. (A) Insertion sequences (IS) have inverted repeats (IR) located at their left and right borders. When integrated into a bacterial replicon, insertion sequences generate direct repeats (DR) of 8 to 9 base pairs (bp) derived from the host DNA sequences immediately adjacent to the insertion. The central region of an IS may vary greatly in size. IS*1* has 756 bp of DNA located between imperfect inverted repeats. The base pair where IR is imperfectly matched is underlined. (B) Transposon insertion results in a selectable bacterial phenotype; IS which lack phenotypic selection traits often are only detected upon sequencing. Of the two classes of transposons (Tn), Class I transposons are less stable, since the flanking IS may transpose independently or with the selectable marker. Class II transposons, which are essentially single IS with the selectable marker located between the IR, are most useful for genetic dissection of phytopathogenic bacteria. (C) The general structure of a typical Class II transposon (Tn*3*) is illustrated here (not drawn to scale). From left to right, the transposon is flanked by 38 bp IR; *tnpA* coding for a 110-kDa transposase; *res* or a 128-bp region called the internal resolution site; *tnpR* coding for a single 20-kDa protein which acts as resolvase and repressor to *tnpA* and *tnpR;* and *bla* coding for β-lactamase (penicillinase) which confers a selectable ampicillin-resistant phenotype. When inserted in a bacterial replicon, Tn*3* is flanked by 5-bp direct repeats.

II. PRINCIPLES

A. TRANSPOSON MUTAGENESIS

Transposons are discrete DNA sequences capable of *rec*A-independent transposition among replicons as diverse as bacterial chromosomes, bacteriophages, or plasmids (Figure 1). They may be thought of as parasitic biotic agents, even more simple in their construction than viruses or plasmids, which are not capable of autonomous replication, but depend upon the replication of the genome of their host. These mobile genetic elements, which insert apparently at random into the genomes of bacteria or other associated replicons, have been utilized by bacterial geneticists to cause polar mutations, to confer new genetic capabilities, or to delete sequences of DNA (Figure 1). The strong polar mutations are caused by distortion of the open reading frame of the mutated gene as well as one to several transcription terminators located on the transposon. Two types of transposons occur in bacteria: simple and complex. *Simple transposons,* or *insertion sequences,* carry only those genes required for transposition (Figure 1A). They are difficult to detect except by inactivation of the genes into which they have inserted or, in the case of simple transposons containing promoter sequences, by activation of adjacent genes in their host. *Complex transposons* are often bound by insertion sequences or their remnants and contain one or more genes besides those essential to transposition processes. These "extra" genes are usually easily detected, especially if they mediate production of factors involved in antibiotic resistance. Complex transposons may be formed by insertion sequences flanking the detectable or marker gene (Class I) or

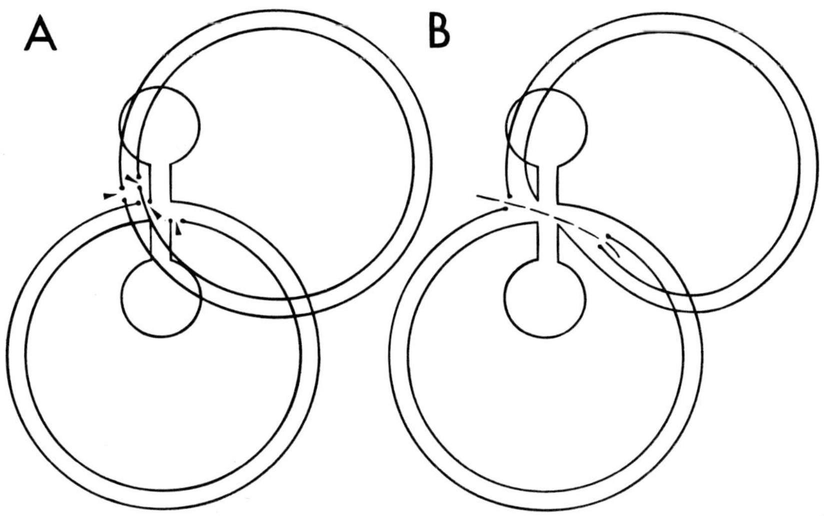

Figure 2 Transposition mechanism. Transposition requires the physical association of two replicons: transposon donor and recipient. (A) Transposase creates staggered single-strand nicks (arrows) in the recipient molecule which interact with single nicks closely flanking complementary copies of the transposon (shown for clarity in palindromic or "lollipop" form) in the donor replicon. (B) Two covalent bonds are created between strands of the recipient and donor replicons. Because the nicks in the recipient replicon were staggered, direct repeats of host DNA flank the transposon in its final location in the recipient replicon. The nonligated ends (dots) form two replication forks joined by a plane (dashed line) through the transposon. Replication along this plane results in the formation of two double-stranded copies of the transposon and creates a single donor-recipient replicon (figure eight). The proximity of the two transposon copies aids site-specific recombination and resolution of the fused replicon into separate donor and recipient replicons (not shown), each containing a copy of the transposon.

by inclusion of the "marker" gene within the insertion sequence (Class II) (Figure 1B). Transposons and insertion sequences express at least two proteins: a regulatory protein (TpnR) and a transposase (TpnA), which is responsible for creating the symmetry and strand nicks required for transposition (Figure 2).[1]

B. MARKER EXCHANGE MUTAGENESIS

Marker exchange mutagenesis depends upon exchange of genetic material by double crossover events and refers to the introduction of a marker gene. Common marker genes are for antibiotic resistance or catalytic abilities, such as β-galactosidase or alkaline phosphatase, which may be detected by colorimetric assays. In these mutational tactics, the marker gene is usually inserted into a DNA fragment from the target bacterium which has been cloned into a plasmid. If the cloning site interrupts the open reading frame of a gene whose contribution to pathogen-microbe interactions is being studied, introduction of the marker gene into the target bacterium sets up the conditions for double crossover recombination and insertion of the marker gene into the genome of the bacterium in "exchange" for the wildtype gene (Figure 3). The regions flanking the marker gene provide homology with the target genome for crossover formation. Two phenotypic changes follow successful marker exchange, the bacterium loses the function mediated by the wild-type target gene and gains the phenotype of the marker gene (Figure 3B).

C. OTHER USES FOR TRANSPOSONS

1. Gene Regulation

Gene or operon fusions provide a powerful instrument for observing gene regulation *in vivo.* Mu d1 (*lac*) mediates insertion and fusion of the β-galactosidase open reading frame from *Escherichia coli* into genes of several Mu-susceptible bacteria. For mutagenesis of Mu-nonsusceptible *Erwinia carotovora,* a double lysogen method was developed by Jayaswal et al.[2] which replaces the Mu host range genes with bacteriophage P1 host range genes. Since *E. carotovora* is susceptible to bacteriophage P1 attachment and penetration,[3] defective Mu d1 (*lac*) enters the plant pathogen. If it is inserted in frame behind bacterial gene promoters, readily detectable β-galactosidase activity or transcripts will be detected upon

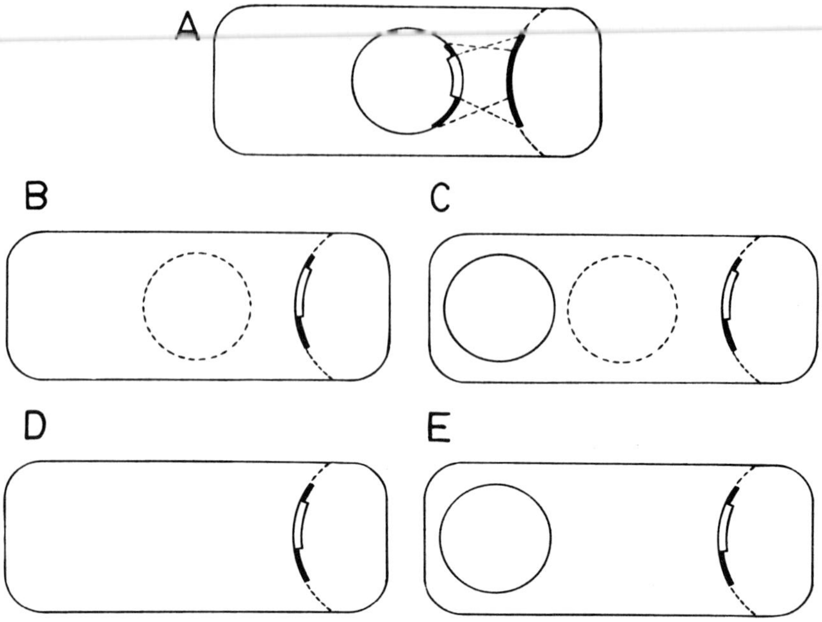

Figure 3 Marker exchange. In marker exchange, a modified gene cloned on a plasmid or bacteriophage is exchanged for the wild-type gene. Double crossover event: After introduction of the modified gene (open box = modification) on its replicon into the bacterium, flanking homologies between the gene targeted for mutagenesis (thick line) and the modified gene enhance the crossover event required for gene exchange (A). Double crossovers occur at a frequency of about $1/10^9$ CFU. Destablization of the vector replicon: The molecular vector for the modified gene is destabilized and lost through dilution from daughter cells of the mutagenized bacterium. Destablization (dashed circle) may be achieved by (B) introducing the modified gene on suicide vectors (plasmids or bacteriophage) incapable of replicating in the bacterium, phosphate starvation, thermoinhibition, or (C) by introducing a second plasmid (simple circle) belonging to the same incompatibility group as the vector plasmid. Genetic isolation of the mutagenized replicon: Because suicide vectors and phosphate starvation result in vector loss and its possible affects on the progeny phenotypes of the mutagenized bacterium, these methods are usually preferred (D). In multireplicon situations (e.g., rhizobia with plasmids involved in nodulation or nitrogen fixation or agrobacteria with pTi plasmids for gall formation) it may be expedient to use the incompatibility exclusion method, which results in retention of the destablizing plasmid (E).

gene activation. Mu-*lac* fusions have also been used to study the secretion and regulation of pectate lyases from *E. chrysanthemi.*[4,5]

2. Linkage Mapping

Schoonejans and Toussaint[6] used plasmid pULB113 (RP4::mini-Mu) to construct a linkage map of *E. chrysanthemi.* In this method, double mini-Mu insertions allow incorporation of smaller or larger pieces of the *E. chrysanthemi* chromosome into the plasmid. Because RP4 is a conjugal plasmid, these fragments from the chromosome of the plant-associated bacterium may be transferred into mutants of *E. chrysanthemi* or the related bacterium, *Escherichia coli.* Complementation of mutations associated with nutritional markers in *E. coli* allowed construction of a circular chromosomal map for *Erwinia chrysanthemi.*

III. METHODOLOGY AND APPLICATIONS IN PLANT PATHOLOGY

A. TRANSPOSON MUTAGENESIS

1. Transposon/Strain Characteristics

Ideally, transposons mobilize at a high frequency, insert in a random pattern, and remain stable at their insertion sites.[7] In fact, transposons vary from the ideal in these characteristics.[7,8] Class I transposons are the most unstable; therefore, we will restrict this discussion to Class II transposons (Figure 1B). Specific transposons or the interactions of transposons with specific genera, species, or strains of bacteria may have low transposition frequencies, insert at preferential sites, or be unstable. For instance, transposons Tn*5* (mediates resistance to kanamycin [Kan^r]) and Tn*7* (streptomycin [Str^r] spectinomycin [Spe^r], and trimethoprin [Tpm^r]), which are widely used because of their resemblance to the "ideal" transposon, were not well suited for genetic research with *Xanthomonas campestris* pv. *campestris.*[9] Shaw et al.[10] found that Tn*1721* (tetracycline [Tet^r]) was useful in strains of this bacterium. On the other hand, Daniels et al.[11] were able to use Tn*5* mutagenesis to recover mutants of *X. c.* pv. *campestris* altered in pathogenicity.

2. Plasmid Curing

Plasmids may be destablized for use in bacterial chromosome transposon mutagenesis. Allen et al.[12] used plasmid pDRT4 (ampicillin [Amp^r], Kan^r, Tet^r), which contained Tn*5,* as a mutagenic plasmid to obtain transpositional mutants of *E. carotovora.* Transconjugants were cured of pDRT4 using low phosphate levels (50 μM) by continuous culture in Torriani's A-P medium.[13] The ampicillin- and tetracycline-sensitive (Amp^s, Tet^s) Kan^r colonies recovered were Tn*5* mutants.

3. Suicide Plasmids

Separation of transposons on their vectors (plasmids or bacteriophage) and transposition mutations in the target replicon (plasmid or bacterial chromosome) may be difficult since they present the same phenotype—usually antibiotic resistance. To avoid this difficulty, transposon mutagenesis procedures routinely take advantage of "suicide" transposon vectors (Figure 3C). By this strategy, the vector or molecular platform carrying the transposon is not able to replicate within the target bacterium. Several methods may be used to develop suicide vectors. Boucher et al.[14] took advantage of the fact that the P-incompatibility group plasmid RP4 was not maintained by replication in a strain of *Pseudomonas solanacearum.* They used this plasmid as a platform for transposon mutagenesis with Tn*5,* Tn*7,* and Tn*10* (Tet^r).

Suicide plasmids may also be constructed. The suicide plasmid pJB4J1, constructed by integrating bacteriophage Mu *c*ts62 into a P-incompatibility group P plasmid containing Tn*5,* does not replicate, presumably because of the Mu *c*ts62 insertion, in *Agrobacterium rhizogenes,*[15] *A. tumefaciens,*[16,17] *E. carotovora,*[18] *Rhizobium leguminosarum,*[19] or *R. meliloti.*[20] Suicide plasmids have been used to move transposable elements, such as Tn*5,* into plant-pathogenic and plant-nodulating bacteria. Once the suicide vector is within the target bacterium, only those transposons which insert in the replicon(s) (chromosome, plasmids, and bacteriophages) of the target bacterium will survive dilution among daughter cells to express their antibiotic-resistant phenotype. Transfer-deficient plasmid pRL150 (Amp^r, Kan^r, Tet^r) containing Tn*904* (Kan^r) was constructed in a similar manner by Mu *c*ts62 insertion mutagenesis of the plasmid transfer function (*tra*) loci.[21] Transposition of Tn*904* from this plasmid to plasmid pTi in *A. tumefaciens* was detected by subsequent conjugal transfer and antibiotic selection for pTi::Tn*904.*

4. Suicide Bacteriophages

Salmond et al.[22] used λ, an *Escherichia coli* bacteriophage which is incapable of attachment to or replication in *Erwinia carotovora,* as a suicide vector for transposon Tn*5* and Tn*10* mutagenesis of that phytopathogenic bacterium. To accomplish attachment and subsequent penetration by λ, they introduced, on plasmid pHCP2, the *Escherichia coli* gene *lamB,* which mediates production of a constitutive protein in the maltose high-affinity uptake system as well as serving as an outer membrane receptor for λ.[23,24] In this manner, they were able to force λ to attach and inject its DNA into *Erwinia carotovora.* Because λ does not replicate in *E. carotovora,* only Tn*5* and Tn*10* which transposed into the *E. carotovora* genome survived to express antibiotic resistance.

5. Bioluminescence

Bioluminescence, the production of visible light through enzymatic processes, is an attractive selective trait for genetically altered bacteria, since it is easily detectable by photographic or luminimetric assays. Using photomultipliers, single marine bioluminescent bacteria have been detected.[25] A fragment of DNA containing the promoterless *lux* operon (*lux*C, *lux*D, *lux*I, *lux*R) responsible for the bioluminescent phenotype (Lux$^+$) of *Vibrio fisheri*[26] was cloned into a unique *Bam*HI site in Tn*1721*-Ba152 (Tet^r) on plasmid pRU675 to form Tn*4431* (Tet^r, Lux$^+$). Tn*4431,* in suicide plasmid pUCD623 (Amp^r, chloramphenicol [Cam^r], Lux$^+$, Tet^r), conferred Lux$^+$ when inserted into replicons under the control of indigenous promoters.[10,27] Tn*4431* was used to mutagenize *X. c.* pv. *campestris* and Lux$^+$ mutants were recovered, including one able to cause black rot symptoms on detached, but not on attached leaves. *In trans,* a genomic fragment of *X. c.* pv. *campestris* from a cosmid library complemented the mutation and restored the wild type ability to cause symptoms on attached leaves.[10]

Bioluminescence was also used to detect Tn*4431*-marked *X. c.* pv. *campestris* in soil and air samples.[28,29] In soil, as few as 0.5 colony-forming units (CFU)/g could be detected by coupling soil dilutions with enrichment culture and luminometric observations.[29] This compares favorably with enrichment-most probable number (MPN) procedures and polymerase chain reaction (PCR) DNA amplification.[30,31] Sedimenting from the air on open petri dishes, the marked pathogen could be detected by luminescence at the same frequency as with a selective medium.[29] *In planta,* as few as 10^4 CFU could be detected by luminescence, culturing on a selective medium, luminimetry, or using a nitrogen-cooled, charge-coupled device camera from inoculated plant tissue.[28,29,32] In root and seed colonization studies, 10^2 CFU/g could be detected.[33]

B. TRANSPOSON MUTAGENESIS OF PLASMIDS

Because of the greater target size of the bacterial genome ($>10^6$ basepairs [bp]) compared to plasmid genomes (10^3 to 10^5 bp), transpositions will occur more often in the bacterial genome; this is especially true in low copy number plasmids. In order to preferentially select mutations of plasmid pDR1, containing inserted DNA cloned from *E. carotovora,* over bacterial chromosome mutations, Roberts et al.[34] introduced transposon Tn*5* into the plasmid host, *Escherichia coli,* using bacteriophage λ NK467. Using the strategy of de Bruijn and Luski,[35] transposition will occur into the genomes of both the plasmid and the bacterium. Following transposon-mediated antibiotic resistance expression on selective media, plasmid DNA is harvested and the bacterial genomic DNA is discarded. After transformation or electroporation into nonmutated *E. coli,* almost all expression of antibiotic resistance will be due to transposon-mutated plasmids. In this manner, the selection of mutated plasmids may be enhanced without recourse to screening large numbers of genome-mutated bacteria. Roberts *et al.*[34] confirmed the nonrandom Tn*5* insertion patterns in pBR322-derived plasmids observed by others.[8]

Comai and Kosuge[36] used plasmid RP4 (Amp^r, Kan^r, Tet^r) as a vector for Tn*1* (Amp^r) mutagenesis of *P. syringae* pv. *savastanoi.* By conjugation, they introduced RP4 into *E. coli* containing the target plasmid, pLUC1 (Str^r, sulfanilamide [Sul^r]), for transposon mutagenesis. After growth of the double transconjugants, total plasmid DNA (pLUC1, RP4, and pLUC1::Tn*1*) was isolated, digested with restriction endonuclease *Xho*I, and transformed into *E. coli.* As RP4 has two *Xho*I digestion sites and pLUC1 and Tn*1* have none, most Amp^r transformants were pLUC1 derivatives with Tn*1* insertions (pLUC1::Tn*1*).

C. TRANSPOSITIONAL GENE FUSION

Promoterless genes conferring distinct phenotypes may be used as genetic markers for locating genes induced by the presence of plants or their components. Transposition into a replicon places a subset of the promoterless genes under the influence of indigenous promoters. This is advantageous for screening because under specific induction conditions only activated genes are easily located. Provided that the reading frame of the promoterless gene is coordinated with that of the indigenous gene, a fused gene product may be detected.

1. β-Galactosidase

Several researchers have used β-galactosidase gene (*lacZ*) insertions to monitor the fate of genetically altered bacteria released into microcosms or the environment.[37–41] One attractive system utilizes insertion of *lacZ* and *lacY* (mediating galactoside permease production) into disarmed Tn*7* to provide the target bacterium with the ability to enzymatically cleave lactose and take up the cleavage products. These traits are not commonly present in phytopathogenic bacteria and provide a selectable trait for genetically engineered bacteria that does not depend on antibiotic resistance.

Stachel et al.[42] constructed a Tn*3* (Amp^r) *lacZ* transposon to generate β-galactosidase gene fusions for use in *A. tumefaciens.* β-Galactosidase activity may be monitored visually and spectrophotometrically by enzymatic cleavage of *o*-nitrophenyl-β-D-galactopyranoside into galactose and yellow-colored *o*-nitrophenol. The Tn*3*::*lacZ* transposon is called Tn*3*-HoHo1 and is contained in plasmid pHOHO1.

2. Akaline Phosphatase

Some gene fusions may be used to selectively mark proteins in the membrane or periplasm of bacteria. Tn*pho*A is a Tn*5* derivative carrying a gene (*phoA*) mediating production of alkaline phosphatase which lacks a signal peptide.[43] In the reducing environment of the cytoplasm, monomeric alkaline phosphatase is inactive because the disulfide bond required for dimerization does not form. Catalytically active fusions are only formed with membrane or periplasmic proteins which supply the missing signal peptides. Alkaline phosphatase is monitored spectrophotometrically by the release of yellow *p*-nitrophenyl from *p*-nitrophenyl-phosphate. Reuber et al.[44] used Tn*phoA* to study the expression of genes for the production of *R. meliloti* extracellular polysaccharides in nodules.

3. Ice Nucleation

Ice nucleation activity (Ina$^+$) associated with a protein (InaZ) and a gene (*inaZ*) from *P. syringae* pv. *syringae* is cloned in plasmid pMWS10.[45] The *inaZ* gene without its promoter was removed from pMWS10 and inserted between the inverted repeats defining Tn*3* so that the resulting plasmid pTn*3*-Ice (Amp^r, Ina$^+$), would express InaZ only when transposed into a gene under the control of the promoter of that gene. A second plasmid, pTn3-Spice (Amp^r, Ina$^+$, Spe^r, Str^r), was created to provide additional antibiotic resistance markers for selection.[46] Ina$^+$ mediated by transposed *inaZ* was detected in *P. s.* pv. *phaseolicola, A. tumefaciens,* and *R. meliloti.*

4. Stabilizing Transpositions

Tn*3*, Tn*3*::Ice, and Tn*3*::Spice share an important feature: they form stable transpositions because their transposase (*tnpA*) gene is inactive. Transposition only occurs in the presence of plasmid pSShe which contains active *tnpA*.[42] Studies in *A. tumefaciens* and *P. s.* pv. *phaseolicola* indicate that Ina$^+$ is inducible by acetosyringone or in loci inducible by plant host contact, respectively.[46,47] To obtain very stable transposons with wider host ranges, additional selectable markers, and multiple insertions in the same replicon, further engineering is possible. Tn*5* and Tn*10* derivatives (mini-Tn*5* and mini-Tn*10*) were constructed using inverted repeats (19 bp from Tn*5*, 70 bp from Tn*10*) to flank selectable marker genes providing for Cam^r, Kan^r, Str^r/Spe^r, Tet^r, β-galactosidase activity (*lacZ*), alkaline phosphatase activity (*phoA*), bioluminescence (*luxA, luxB* from *V. harveyi*), and catechol 2,3-dioxygenase activity (*xylE* from *P. putida,* which releases yellow 2-hydroxymuconic acid from catechol),[48] arsenite resistance (Asn^r), mercury resistance (Mer^r), or bialaphos resistance (Ptt^r).[49,50] Because mini-Tn*5* and mini-Tn*10* lack *tpnA* (Figure 1C), similar genes (*tpn** for mini-Tn*5* and *lacI^q* for mini-Tn*10*) are provided on suicide vectors (pUT for mini-Tn*5*, pBOR8 for mini-Tn*10*). The vectors also encode an origin of replication (*ori R6K*), conjugal mobilization region (*mobRP4*), and ampicillin resistance (*bla*). These plasmids fail to replicate in bacteria not containing the π protein encoded by *pir.* Once inserted in a bacterial replicon, the mini-transposons are stable, since they lack *tpnA,* and they do not repress additional mutagenesis because they lack *tpnR*. These mini-transposons also contain at least one restriction endonuclease site (*Not*I) appropriate for chromosomal insertion of cloned DNA.

D. MARKER-EXCHANGE MUTAGENESIS

Exchanging an *in vitro* modified gene for a wild-type gene *in vivo* is called marker-exchange mutagenesis (Figure 3). Plasmid pBR322 was used as a suicide plasmid to introduce DNA fragments into *A. tumefaciens* to obtain marker-exchange mutagenesis of pTi plasmids.[51] In this case, replication of pBR322 does not occur in the host bacterium. Others have used phosphate starvation to displace pBR322-derived vector plasmids to disable pectate lyase genes in *E. c.* subsp. *carotovora* and *E. chrysanthemi* by interrupting the open reading frames with inserted kanamycin resistance genes (*npt*).[52,53] By this tactic, any plasmid may be used as a suicide plasmid since starvation for phosphate, required for the sugar-phosphate backbone of DNA, affects plasmid more than chromosomal replication simply because bacterial survival in broth medium depends on the chromosome. Plasmid pBR322 and its derived plasmids may be transformed, electroporated, or conjugally transferred into *E. c.* subsp. *carotovora* strain EC14[12,54–56] but, this is evidently not true for all strains of *E. carotovora* or the related soft rot pathogen *E. chrysanthemi.*[56a] These pBR322 plasmids are relatively stable in *E. c.* subsp. *carotovora;* however, they are maintained in lower copy numbers than in *E. coli* because recovery of pBR322-derived DNA from *E. c.* subsp. *carotovora* is disappointing (V. K. Stromberg, current research). These plasmids may be destablized by growing bacteria in a medium containing only 50 to 250 μ*M* phosphate.[13] After repetitive transfers, the plasmids are lost by dilution and the only remaining selectable markers are those engineered into the disabled pectate lyase gene which has been exchanged into the bacterial chromosome.

Ried and Collmer[57] used a marker exchange-eviction method to construct a strain of *E. chrysanthemi* deficient in pectate lyase isozymes. This system is attractive because it results in mutagenesis without using genes for antibiotic resistance and exploits the selective advantages of a *Bacillus subtilis* gene mediating production of periplasmic levansucrase (*sacB*) which results in the accumulation of toxic levan in the presence of sucrose.[58] Specifically, a 3.8-kb pair *Bam*HI DNA fragment from plasmid pUM24 containing *sacB,* its *cis*-acting regulatory sequence (*sacR*), and a gene conferring kanamycin resistance (*nptl*) is ligated into a *Sau*3A restriction site within the open reading frame of the target gene which is cloned on a plasmid. The hybrid plasmid containing the *sac* cartridge is moved into the bacterium to be mutagenized. Exchange of the mutagenized gene for the wild-type gene is detected by loss of phenotype and acquisition of Kan^r and sensitivity to sucrose. For eviction of the *sac* cartridge and elimination of *nptl,* the hybrid plasmid is subjected to *Pst*I digestion and religation to eliminate all but 28 bp of the original 3.8-kb cartridge. The second constructed plasmid, deleted for all except 28 bp of the cartridge, is introduced into the previously marker-exchanged bacterium and the transconjugant is placed under sucrose selection. Surviving bacteria most likely will have evicted the *sac* cartridge in exchange for the truncated cartridge. Through loss of the *sacB* gene and its attendant levansucrase expression, the survivors will no longer accumulate lethal levan. This method of mutagenesis may be extended to other Gram-negative bacteria, since *A. tumefaciens* and *R. meliloti* are also sensitive to levan.[58]

E. OLIGONUCLEOTIDE-DIRECTED MUTAGENESIS

Transposon and marker-exchange mutagenesis, as discussed above, suffer from a common defect: introduction of a single, but rather large genetic change which may result in concomitant introductions or deletions of genetic elements such as nonsense DNA, insertion sequences, promoters, termination factors, or open reading frames. The minimal mutation which alters gene expression is a single base change. Oligonucleotide-directed mutagenesis is an improvement upon marker-exchange mutagenesis and reduces the size of the genetic change to as small as a single base, alleviating any doubt about phenotypic effects of introduced or deleted DNA sequences. In this method, oligonucleotides are constructed which have intentional "mistakes" introduced into their DNA sequences. These oligonucleotides are annealed to single-stranded DNA from a recombinant plasmid containing the cloned wild-type allele or the "target" gene to be mutated. *In vitro* plasmid DNA replication in the presence of DNA polymerase and the nucleotide DNA precursors results in a nicked circular DNA mismatched for only one or a few bases in one copy of the gene to be mutated. Ligation and reintroduction of these imperfect copies of the gene on plasmids into *Escherichia coli* results in plasmid replication.

Conservative replication results in two populations of plasmids, one with the constructed changes (intentional mutation) in nucleotide sequence in the target gene and one with the wild-type nucleotide sequence of the gene (Figure 4A). Precise deletion mutations may be constructed using essentially the same technique by extension of mutagenic primers that span the area to be deleted.[59,60] Introduction of these plasmids by transformation, electroporation, or tri-parental mating into the organism to be mutated (wild-type phytopathogenic bacterium) results in double crossover events leading to marker-exchange mutagenesis. Several modifications of this general strategy are presented in the following paragraphs and are generally directed at making the procedure more effective by eliminating nonmutagenized plasmids derived by semiconservative DNA replication.

1. Restriction Endonuclease Bias

Restriction endonuclease sites present in the wild-type plasmid may be eliminated from the mutant being constructed by oligonucleotide mutagenesis. After replication, treatment with a restriction endonuclease which only digests the wild-type plasmid followed by transformation provides a replicating population composed almost exclusively of the mutant plasmid.[61]

2. Uracil Incorporation Bias

This method requires propagation of the target plasmid or bacteriophage in an *E. coli* strain (*dut, ung*) deficient in deoxyuridine triphosphatase and uracil glycosylase. Due to accumulation of deoxyuridine triphosphate (dUTP) in the intracellular pool, some of it is incorporated in DNA in place of deoxythymidine triphosphate (dTTP). Because the strain also lacks uracil glycosylase, which normally removes uracil from DNA, the molecule accumulates uracil, making it biologically inactive. The mutation is introduced by annealing a mutagenic primer to the single-stranded uracil-containing DNA, extending a nonuracil-containing DNA strand from the primer, and transforming or transfecting the double-stranded

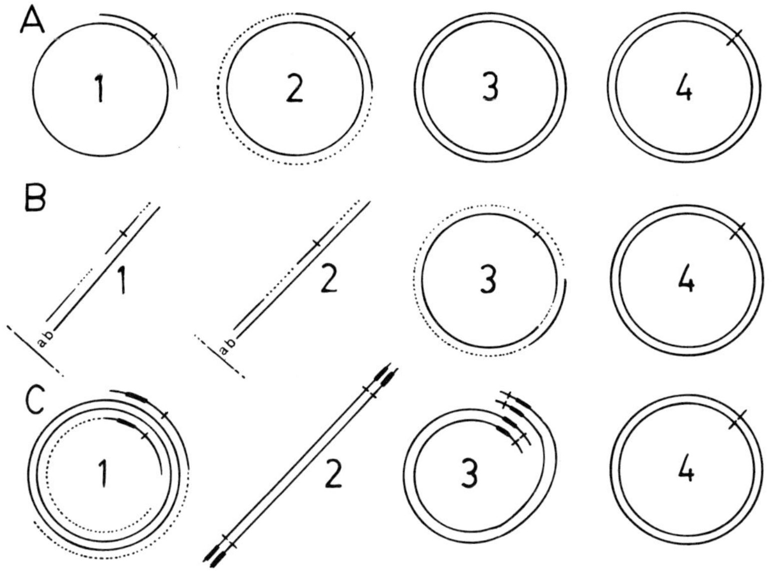

Figure 4 Oligonucleotide tactics for site-specific mutagenesis. Synthetic oligonucleotides may be used to introduce genetic changes into vectors for genetic analyses *in trans* on plasmids or *in cis* by marker-exchange mutagenesis. (A) Plasmid extension requires (1) annealing an oligonucleotide containing the lesion (hatch mark) to single-stranded plasmid DNA; (2) extending (dotted lines) and ligating the extended strand with DNA polymerase and DNA ligase to form a double-stranded plasmid, and moving the plasmid into a bacterium. Two plasmid populations are generated by semi-conservative DNA replication: (3) nonmutant and (4) mutant. Several strategies discussed in the text deal with selecting the mutant population for further analyses. (B) Solid-support site-directed mutagenesis: (1) single-strand, end-biotinylated (b) molecules linked to avidin-coated (a) solid substrate (beads) (2) may be extended from annealed oligonucleotides, (3) circularized with a third primer, (4) extended, and ligated to form double-stranded molecules. (C) Recombinant circle polymerase chain reaction (RCPCR) utilizes (1) two primers each containing the genetic lesion to (2) create through polymerase chain reaction (PCR) amplification double-stranded DNA with homologous regions (thickened lines) at their ends (3) which are complementary and anneal to provide circles adequate for bacterial transformation and (4) subsequent *in vitro* recombination to produce only mutant molecules.

replicon into a wild-type strain which discriminates against the uracil-containing strand and replicates only the mutagenized nonuracil-containing strand.[62,63]

3. Methylation Bias

In vitro extension of annealed mutagenic primer and its transformation into *E. coli* should result in 50% of the progeny being mutants. Actually, only about 18% of the progeny represent the mutant because the methyl-directed mismatch repair system favors repair of the nonmethylated *in vitro*-synthesized strand.[64,65] The recovery of mutants may be increased to greater than 90%.[65] In this method, DNA is hemi-methylated on the mutant strand by using 5-methyldeoxycytosine in the *in vitro* extension of the mutagenic primer. Certain methylation-sensitive restriction endonucleases (e.g., MspI, *Sau*3AI, and *Taq*I) are not totally blocked at methylated restriction sites, but nick the nonmethylated strand. Complete digestion of the restriction endonuclease-nicked nonmethylated strand is obtained with exonuclease III. Any remaining nonnicked, single-stranded, nonmethylated DNA may be linearized with *Hha*I. The remaining circular, single-stranded, methylated, mutant DNA is increased by transformation and replication in an *E. coli* strain (*mcrA, mcrB*) deficient for the ability to restrict methylated DNA.

4. Phosphorothioate Bias

Extension from an annealed mutagenic primer incorporating α-phosphorothioate-modified deoxycytidine triphosphate (dCTP) protects that strand from digestion by certain restriction endonucleases (e.g., NciI, *Pvu*I) while allowing nicking in the nonphosphorothioated strand. Exonuclease digestion of the nicked strand creates a population of single-stranded, modified DNA for direct transformation of *E. coli.* This procedure results in mutant recovery frequencies of 40 to 70%.[66,67]

5. Solid Support Oligonucleotide Site-Direct Mutagenesis

To reduce the inevitable DNA losses associated with molecular manipulations, DNA may be attached to a solid support (Figure 4B). Another benefit of solid support is that it may be used to separate mutagenized from nonmutagenized molecules. For instance, linearized double-stranded DNA may be end labeled with biotin, attached with avidin to beads, and melted to single strands. Mutagenic and extension oligodeoxynucleotides annealed to the anchored template strand are used to initiate extension of the complementary strand. The single-stranded regions of the extended strand are ligated, melted to single-stranded DNA, and annealed with a bridging primer. The templates may be discarded with the beads in the pellet after centrifugation. The double strand is extended, the single-stranded ends are ligated, and the constructed plasmid is available for transformation. Mutagenesis efficiencies of 70 to 92% have been reported.[68]

6. Duplex Crossover Linker Mutagenesis

This system simplifies deletion mutagenesis by removing the need to construct and clone deletions.[69,70] It takes advantage of three characters: (1) *E. coli* is capable of recombining *in vivo* short stretches of homologous ends, (2) DNA ligase attaches single-stranded oligonucleotides to complementary ends of double-stranded DNA, and (3) *E. coli* may be transformed at low frequencies with linearized DNA. Key to mutant construction is a synthetic "homology-searching" oligonucleotide and a restriction endonuclease site adjacent to the sequence to be deleted. The homology searching sequence contains one end complementary to the restriction endonuclease site. The other end is complementary to sequences located on the far side (opposite the restriction site) of the bases to be deleted. The homology-searching oligonucleotide is ligated to the linearized plasmid, creating a structure which circularizes *in vivo* at the point of homology. The bases to be deleted remain exposed to the intracellular environment and are removed by exonucleolytic digestion. The overlapping homologous ends are circularized by spontaneous intramolecular annealing. Subsequent endo- and/or exonuclease activity removes redundant bases at the junction of the paired regions, resulting in a nicked, circular molecule available for repair and replication. Selected mutant genotypes were recovered at frequencies from 0.2 to 2.8%.[69]

7. Polymerase Chain Reaction Methods

The methods discussed above require bacterial replication to amplify mutant molecules for use in marker-exchange mutagenesis. The PCR is an efficient method for DNA amplification.[71–73] These systems take advantage of *in vivo* duplex crossover linker mutagenesis as discussed above. The recombination circle PCR (RCPCR) and the recombination PCR (RPCR) systems create products that when combined, denatured, and reannealed form double-stranded DNA with discrete homologous ends complementary to each other which anneal to form DNA circles suitable for transformation of *E. coli* (Figure 4C).[74–78]

IV. ADVANTAGES AND LIMITATIONS

A. TRANSPOSON MUTAGENESIS

Transposon mutagenesis is a useful tool for marking and locating loci involved in bacterial-plant interaction and environmental survival. Loci may be marked randomly (Tn*1*, Tn*3*, Tn*5*, Tn*7*, Tn*10*) or specifically for export to the periplasm or outside the cell (Tn*pho*A) or for specific inducers (Tn*4431*, Tn*3*::Ice, Tn*3*::Spice). Disadvantages of transposon mutagenesis include nonrandom insertions, low frequency of transposition, instability, or lack of vector systems for specific genera, species, or strains of plant-associated bacteria. Many of these difficulties have been overcome by the development of transposase- and repressor-deficient mini-transposons.[49,50]

B. MARKER-EXCHANGE MUTAGENESIS

Exchange mutagenesis allows relatively precise exchange of *in vitro* modified genes for wild-type genes. This is an effective procedure for studying bacterial-plant interactions. Mutagenesis may be difficult if there is great nucleotide sequence identity between two genes expressing isozymes of the same enzyme. This may result in loss of gene specificity. Lacy and Stromberg[52] used such a strategy to construct a mutator plasmid using sequences from one pectate lyase to mutagenize a second gene. Oligonucleotide-directed mutagenesis strategies can provide more precision by selecting specific nucleotide sequences as targets.

C. OLIGONUCLEOTIDE-DIRECTED MUTAGENESIS

Oligonucleotide mutagenesis may be accomplished in bacteria or *in vitro*. Each system has its advantages. For instance, PCR methods amplify DNA quickly and to high levels, but *Taq* DNA polymerase lacks an exonucleolytic proofreading function.[79] Bacterial replication systems do not use expensive PCR equipment, but they may require extra manipulations (5-methyldeoxycytosine or α-phosphorothioate-modified dCTP) or specialized bacterial strains (*mcrA, mcrB* or *dut, ung*) in order to separate non-mutagenized from mutagenized vectors. Duplex crossover linker mutagenesis, RCPCR, and solid support oligonucleotide site-direct mutagenesis systems seem to combine the most advantageous techniques.

D. GENETIC SYSTEM

Plant pathologists and molecular biologists working with the less well-characterized genetic systems of plant-associated bacteria lack foreknowledge of the workability of the systems described here. For instance, transformation, electroporation, transfection, or conjugation may not work or only work poorly with a specific bacterium. Whether or not *in vivo* recombination will work in a specific plant-associated bacterium as it is described for *E. coli* is not known. In the final analysis, the researcher will have to choose or develop the techniques that work with the particular system under investigation. Fortunately, today there are more tools to choose among.

REFERENCES

1. McCommas, S. A. and Syvanen, M., Temporal control of transposition in Tn*5, J. Bacteriol.,* 170, 889, 1988.
2. Jayaswal, R. K., Bessan, R. A., and Handa, A. K., Mutagenesis of *Erwinia carotovora* subsp. *carotovora* with bacteriophage Mu d1 (Ap^r *lac* cts62): construction of *his-lac* gene fusions, *J. Bacteriol.,* 158, 764, 1984.
3. Goldberg, R. B., Bender, R. A., and Streicher, S. L., Direct selection for P1-sensitive mutants of enteric bacteria, *J. Bacteriol.,* 118, 810, 1974.
4. Hugouvieux-Cotte-Pattat, N. and Robert-Baudouy, J., Isolation of *kdg*K-*lac* and *kdg*A-*lac* gene fusions in the phytopathogenic bacterium *Erwinia chrysanthemi, J. Gen. Microbiol.,* 131, 1205, 1985.
5. Ji, J., Hugouvieux-Cotte-Pattat, N., and Robert-Baudouy, J., Use of Mu-*lac* insertions to study the secretion of pectate lyases by *Erwinia chrysanthemi, J. Gen. Microbiol.,* 133, 793, 1987.
6. Schoonejans, E. and Toussaint, A., Utilization of plasmid pULB113 (RP4::mini-Mu) to construct a linkage map of *Erwinia carotovora* subsp. *chrysanthemi, J. Bacteriol.,* 154, 1489, 1983.
7. Bennett, P., Bacterial transposons, in *Genetics of Bacteria,* Scaife, J., Leach, D., and Galizzi, A., Eds., Academic Press, London, 1985, 97.
8. Berg, D. E., Schmandt, M. A., and Lowe, J. B., Specificity of Tn*5* insertion, *Genetics,* 105, 813, 1983.
9. Turner, P., Barber, C., and Daniels, M., Behavior of the transposons Tn*5* and Tn*7* in *Xanthomonas campestris* pv. *campestris, Mol. Gen. Genet.,* 195, 101, 1984.
10. Shaw, J. J., Settles, L. G., and Kado, C. I., Transposon Tn*4431* mutagenesis of *Xanthomonas campestris* pv. *campestris:* characterization of a nonpathogenic mutant and cloning of a locus for pathogenicity, *Mol. Plant-Microbe Interact.,* 1, 39, 1988.
11. Daniels, M. J., Barber, C. E., Turner, P. C., Clearly, W. G., and Sawczyc, M. K., Isolation of mutants of *Xanthomonas campestris* pv. *campestris* showing altered pathogenicity, *J. Gen. Microbiol.,* 130, 2447, 1984.
12. Allen C., Stromberg, V. K., Smith, F. D., Lacy, G. H., and Mount, M. S., Complementation of an *Erwinia carotovora* subsp. *carotovora* protease mutant with a protease-encoding cosmid, *Mol. Gen. Genet.,* 202, 276, 1986.
13. Torriani, A., Influence of inorganic phosphate in the formation of phosphates by *Escherichia coli, Biochem. Biophys. Acta,* 38, 460, 1960.
14. Boucher, C., Message, B., Debieu, D., and Zischek, C., Use of the P-1 incompatibility group plasmids to introduce transposons into *Pseudomonas solanacearum, Phytopathology,* 71, 639, 1981.
15. White, F. F. and Nester, E. W., Hairy root: plasmid encodes virulence traits in *Agrobacterium rhizogenes, J. Bacteriol.,* 141, 1134, 1980.
16. Garfinkel, D. J. and Nester, E. W., *Agrobacterium tumefaciens* mutants affected in crown gall tumorigenesis and octopine catabolism, *J. Bacteriol.,* 144, 732, 1980.

17. Van Vliet, F., Silva, B., van Montagu, M., and Schell, J., Transfer of RP4 : : Mu plasmids to *Agrobacterium tumefaciens, Plasmid,* 1, 446, 1978.
18. Zink, R. T., Kemble, R. J., and Chatterjee, A. K., Transposon Tn*5* mutagenesis in *Erwinia carotovora* subsp. *carotovora* and *E. carotovora* subsp. *atroseptica, J. Bacteriol.,* 157, 809, 1984.
19. Beringer, J. E., Beynon, J. L., Buchana-Wollaston, A. V., and Johnston, A. W. B., Transfer of the drug-resistance transposon Tn*5* to *Rhizobium, Nature,* 276, 633, 1978.
20. Meade, H. M., Long, S. R., Ruvkun, G. B., Brown, S. E., and Ausubel, F. M., Physical and genetic characterization of symbiotic and auxotrophic mutants of *Rhizobium meliloti* induced by transposon Tn*5* mutagenesis., *J. Bacteriol.,* 149, 114, 1982.
21. Klapwijk, P. M., van Breukelen, J., Korevaar, K., Ooms, G., and Schilperoort, R. A., Transposition of Tn*904* encoding streptomycin resistance into the octopine Ti plasmid of *Agrobacterium tumefaciens, J. Bacteriol.,* 141, 129, 1980.
22. Salmond, G. P. C., Hinton, J. C. D., Gill, D. R., and Pérombelon, M. C. M., Transposon mutagenesis of *Erwinia* using phage λ vectors, *Mol. Gen. Genet.,* 203, 524, 1986.
23. de Vries, G. E., Raymond, C. K., and Ludwig, R. A., Extension of bacteriophage λ host range: selection, cloning and characterization of a constitutive λ receptor gene, *Proc. Natl. Acad. Sci. U.S.A.,* 81, 6080, 1984.
24. Harkki, A. and Palva, E. T., A *lamB* expression plasmid for extending the host range of phage λ to other bacteria. *FEMS Microbiol. Lett.,* 27, 183, 1985.
25. Booth, C. K. and Nealson, K. H., Light emission by luminous bacteria in the open ocean. *Biophys. J.,* 15 (Abstr.), 56a, 1975.
26. Engebrecht, J., Nealson, K., and Silverman, M., Bacterial bioluminescence: isolation and genetic analysis of functions from *Vibrio fisheri, Cell,* 32, 773, 1983.
27. Shaw, J. J. and Kado, C. I., Direct analyses of the invasiveness of *Xanthomonas campestris* mutants generated by Tn*4431,* a transposon containing a promoterless luciferase cassette for monitoring gene expression, in *Molecular Genetics of Plant Microbe Interactions,* Verma, D. P. S. and Brisson, N., Eds., Martinus Nijhoff, Boston, 1987, 57.
28. Shaw, J. J., Geiger, D., Dane, F., and Kloepper, J. W., Bioluminescence for the detection of genetically-engineered microbes released into the environment, in *Biological Monitoring of Genetically Engineered Plants and Microbes,* MacKenzie, D. R. and Henry, S. C., Eds., Agric. Res. Inst., Bethesda, MD, 1990, 229.
29. Shaw, J. J., Dane, F., Geiger, D., and Kloepper, J. W., Use of bioluminescence for detection of genetically engineered microorganisms released into the environment, *Appl. Environ. Microbiol.,* 58, 267, 1992.
30. Scanferlato, V. S., Orvos, D. R., Lacy, G. H., and Cairns, J., Jr., Enumerating low densities of genetically engineered Erwinia carotovora in soil, *Lett. Appl. Microbiol.,* 10, 55, 1990.
31. Steffan, R. J. and Atlas, R. M., DNA amplification to enhance detection of genetically-engineered bacteria in environmental samples, *Appl. Environ. Microbiol.,* 55, 2185, 1988.
32. Shaw, J. J. and Kado, C. J., Development of a *Vibrio* bioluminescence gene-set to monitor phytopathogenic bacteria during the ongoing disease process in a non-disruptive manner, *Bio/Technology,* 4, 560, 1986.
33. Mahaffee, W. F., Shaw, J. J., Backman, P. A., and Kloepper, J. W., Marking rhizobacteria with the luciferase operon for monitoring seed and root colonization, *Phytopathology,* 82 (Abstr.), 1177, 1992.
34. Roberts, D. P., Berman, P. M., Allen, C., Stromberg, V. K., Lacy, G. H., and Mount, M. S., Requirement for two or more *Erwinia carotovora* subsp. *carotovora* pectolytic gene products for maceration of potato tuber tissue by *Escherichia coli, J. Bacteriol.,* 167, 279, 1986.
35. de Bruijn, F. J. and Lupski, J. R., The use of transposon Tn*5* mutagenesis in the rapid generation of correlated physical and genetic maps of DNA segments cloned onto multicopy plasmids—a review, *Gene,* 27, 131, 1984.
36. Comai, L. and Kosuge, T., Cloning and characterization of *iaaM,* a virulence determinant of *Pseudomonas savastanoi, J. Bacteriol.,* 149, 40, 1982.
37. Barry, G. F., Construction of reporter strains, in *The Release of Genetically-Engineered Microorganisms,* Sussman, M., Collins, C. H., Skinner, F. A., and Stewart-Tull, D. E., Eds., Academic Press, San Diego, 1988, 211.

38. Cook, J. R., Weller, D. M., Kovacevich, P., Drahos, D., Hemming, B., Barnes, G., and Pierson, E. L., Establishment, monitoring, and termination of field tests with genetically altered bacteria applied to wheat for biological control of take-all, in *Biological Monitoring of Genetically Engineered Plants and Microbes,* MacKenzie, D. R. and Henry, S. C., Eds., Agric. Res. Inst., Bethesda, MD, 1990, 177.
39. Drahos, D. J., Hemming, B. C., and McPherson, S., Tracking recombinant organisms in the environment: β-galactosidase as a selectable non-antibiotic marker for fluorescent pseudomonads, *Biol Technology,* 4, 439, 1986.
40. Drahos, D. J., Barry, G., Hemming, B., Brandt, E., Kline, E., Skipper, H., Kluepfel, T., Huges, T., and Gooden, D., Field testing of *lacZY*-marked recombinant soil bacteria, *Phytopathology,* 78 (Abstr.), 1544, 1988.
41. Kluepfel, D. A., Kline, E. L., Huges, T., Skipper, H., Gooden, D., Drahos, D. J., Barry, G. F., Hemming, B. C., and Brandt, E. J., Field testing of a genetically engineered rhizosphere inhabiting pseudomonad: development of a model system in *Biological Monitoring of Genetically Engineered Plants and Microbes,* MacKenzie, D. R. and Henry, S. C., Eds., Agric. Res. Inst., Bethesda, MD, 1990, 189.
42. Stachel, S. E., An, G., Flores, C., and Nester, E. W., A Tn*3 lacZ* transposon for the random generation of β-galactosidase gene fusions: application to the analysis of gene expression in *Agrobacterium, EMBO J.,* 4, 891, 1985.
43. Manoil, C. and Beckwith, J., Tn*phoA:* a transposon probe for protein export signals. *Proc. Natl. Acad. Sci. U.S.A.,* 82, 8129, 1985.
44. Reuber, T. L., Long, S., and Walker, G. C., Regulation of *Rhizobium meliloti exo* genes in free-living cells and *in planta* examined by using Tn*phoA* fusions, *J. Bacteriol.,* 173, 426, 1991.
45. Wolber, P. K., Deininger, C. A., Southworth, M. W., Vandekerchhove, J., van Montagu, M., and Warren, G. J., Identification and purification of a bacterial ice-nucleation protein, *Proc. Natl. Acad. Sci. U.S.A.,* 83, 7256, 1986.
46. Lindgren, P. B., Frederick, R., Govindarajan, A. G., Panopoulos, N. J., Staskawicz, B. J., and Lindow, S. E., An ice nucleation reporter gene system: identification of inducible pathogenicity genes in *Pseudomonas syringae* pv. *phaseolicola, EMBO J.,* 8, 1291, 1989.
47. Rahme, L. G., Mindrinos, M. N., and Panopoulos, N. J., Genetic and transcriptional organization of the *hrp* cluster of *Pseudomonas syringae* pv. *phaseolicola, J. Bacteriol.,* 173, 575, 1991.
48. Zukowski, M. M., Gaffney, D. F., Speck, D., Kauffmann, M., Findeli, A., Wisecup, A., and Lecocq, J.-P., Chromogenic identification of genetic regulatory signals in *Bacillus subtilis* based on expression of a cloned *Pseudomonas* gene, *Proc. Natl. Acad. Sci. U.S.A.,* 80, 1101, 1983.
49. de Lorenzo, V., Herrero, M., Jakubzik, U., and Timmis, K. N., Mini-Tn*5* transposon derivatives for insertion mutagenesis, promoter probing, and chromosomal insertion of cloned DNA in Gram-negative eubacteria, *J. Bacteriol.,* 172, 6568, 1990.
50. Herrero, M., de Lorenzo, V., and Timmis, K. N., Transposon vectors containing non-antibiotic resistance selection markers for cloning and stable chromosomal insertion of foreign genes in Gram-negative bacteria, *J. Bacteriol.,* 172, 6557, 1990.
51. van Haute, E., Joos, H., Maes, M., Warren, G., van Montagu, M., and Schell, J., Intergeneric transfer and exchange recombination of restriction fragments cloned in pBR322. A novel strategy for the reversed genetics of the Ti plasmids of *Agrobacterium tumefaciens, EMBO J.,* 2, 411, 1983.
52. Lacy, G. H. and Stromberg, V. K., Pre-release microcosm tests with a genetically engineered plant pathogen, in *Biological Monitoring of Genetically Engineered Plants and Microbes,* MacKenzie, D. R. and Henry, S. C., Eds., Agric. Res. Inst., Bethesda, MD, 1990, 81.
53. Roeder, D. L. and Collmer, A., Marker-exchange mutagenesis of a pectate lyase isozyme gene in *Erwinia chrysanthemi, J. Bacteriol.,* 164, 51, 1985.
54. Allen, C., Evolution of a Gene for Pathogenicity: *Endo*-Pectate Lyase, Ph.D. dissertation, Virginia Polytechnic Institute and State University, Blacksburg, VA, 1987, 117 pp.
55. Berman, P., Mount M., and Lacy, G., Insertion of antibiotic-resistant plasmids into *Erwinia carotovora, Phytopathology,* 73 (Abstr.), 1342, 1983.
56. Berman, P., Mount, M., and Lacy, G., Electro-transformation of *Erwinia carotovora* subsp. *carotovora,* Abstr. Fallen Leaf Lake Conf. Genus *Erwinia.* Sept. 17–20, 1987, Fallen Leaf Lake, South Lake Tahoe, CA, p. 22.
56a. Keen, N. T., University of California, Riverside, personal communication.

57. Ried, J. L. and Collmer, A., An *nptI-sacB-sacR* cartridge for constructing directed, unmarked mutations in Gram-negative bacteria by marker exchange-eviction mutagenesis, *Gene,* 57, 239, 1987.
58. Gay, P., Le Coq, D., Steinmetz, M., Berkelman, T., and Kado, C. I., Positive selection procedure for entrapment of insertion sequence elements in Gram-negative bacteria, *J. Bacteriol.,* 164, 918, 1985.
59. Deng, W. P. and Nickoloff, J. A., Site-directed mutagenesis of virtually any plasmid by eliminating a unique site, *Anal. Biochem.,* 200, 81, 1992.
60. Zhu, L. and Chin, H., *In vitro* generation of multiple-site mutations and precise large deletions, *CLONTECHniques,* July, 9, 1992.
61. Wallrath, L. L. and Friedman, T. B., Constructing combinations of site-specific large deletions, *U.S. Biochem. Edit. Comments,* 19, 1, 1992.
62. Kunkel, T., Rapid and efficient site-specific mutagenesis without phenotypic selection, *Proc. Natl. Acad. Sci. U.S.A.,* 82, 488, 1985.
63. Slilaty, S. N., Fung, M., Shen, S.-H., and Lebel, S., Site-directed mutagenesis by complementary-strand synthesis using a closing oligonucleotide and double-stranded DNA template, *Anal. Biochem.,* 185, 194, 1990.
64. Kramer, B., Kramer, W., and Fritz, H. J., Different base/base mismatches are corrected with different efficiencies by the methyl-directed DNA mismatch repair system of *E. coli, Cell,* 38, 879, 1984.
65. Vandeyar, M. A., Weiner, M. P., Hutton, C. J., and Batt, C. A., A simple and rapid method for the selection of oligodeoxynucleotide-directed mutants, *Gene,* 65, 129, 1988.
66. Sugimoto, M., Esaki, N., Tanaka, H., and Soda, K., A simple and efficient method for the oligonucleotide-directed mutagenesis using plasmid DNA template and phosphorothioate-modified nucleotide, *Anal. Biochem.,* 179, 309, 1989.
67. Taylor, J. W., Ott, J., and Eckstein, F., The rapid generation of oligonucleotide-directed mutations at high frequency using phosphorothioate-modified DNA, *Nucleic Acids Res.,* 13, 8765, 1985.
68. Felts, K., Braman, J., and Weiner, M., Double Take™ double-stranded mutagenesis kit: site-directed mutagenesis in plasmids with no special vectors or cells, *Strategies Mol. Biol.,* 5, 26, 1992.
69. Sung, W. L. and Zahab, D. M., Site-specific recombination directed by single-stranded crossover linkers: specific deletion of the amino-terminal region of the β-galactosidase gene in pUC plasmids, *DNA,* 6, 373, 1987.
70. Sung, W. L., Zahab, D. M., MacDonald, C. A., and Tam, C. S., Synthesis of mutant parathyroid hormone genes via site-specific recombination directed by crossover linkers, *Gene,* 47, 261, 1986.
71. Mullis, K., Faloona, F., Scharf, S., Saiki, R., Horn, G., and Erlich, H., Specific enzymatic amplification of DNA *in vitro:* the polymerase chain reaction, *Cold Spring Harbor Symp. Quan. Biol.,* LI, 263, 1986.
72. Saiki, R. K., Scharf, S., Faloona, F., Mullis, K. B., Horn, G. T., Erlich, H. A., and Arnheim, N., Enzymatic amplification of β-globin genomic sequences and restriction site analysis for diagnosis of sickle cell anemia, *Science,* 230, 1350, 1985.
73. Saiki, R. K., Gelfand, D. H., Stoffel, S., Scharf, S. J., Higuchi, R., Horn, G. T., Mullis, K. B., and Erlich, H. A., Primer-directed enzymatic amplification of DNA with a thermostable DNA polymerase, *Science,* 239, 487, 1988.
74. Jones, D. H. and Howard, B. H., A rapid method for site-specific mutagenesis and directional subcloning by using the polymerase chain reaction to generate recombinant circles, *BioTechniques,* 8, 178, 1990.
75. Jones, D. H. and Howard, B. H., A rapid method for recombination and site-specific mutagenesis by placing homologous ends on DNA using polymerase chain reaction, *BioTechniques,* 10, 62, 1991.
76. Jones, D. H., Sakamoto, K., Vorce, R. L., and Howard, B. H., DNA mutagenesis and recombination, *Nature,* 344, 793, 1990.
77. Landt, O., Grunert, H.-P., and Halm, U., A general method for rapid site-directed mutagenesis using the polymerase chain reaction, *Gene,* 96, 125, 1990.
78. Perrin, S. and Gilliland, G., Site-specific mutagenesis using asymmetric polymerase chain reaction and a single mutant primer, *Nucleic Acids Res.,* 18, 7433, 1990.
79. Tindall, K. R. and Kunkle, T. A., Fidelity of DNA synthesis by the *Thermus aquaticus* DNA polymerase, *Biochemistry,* 27, 6008, 1988.

FURTHER READING

Berg, D. and Howe, M., *Mobile DNA,* American Society for Microbiology, Washington, D.C., 1989.

Deng, W. P. and Nickoloff, J. A., Site-directed mutagenesis of virtually any plasmid by eliminating a unique site, *Anal. Biochem.,* 200, 81, 1992.

Erlich, H. A., *PCR Technology: Principal for DNA Amplification,* Stockton Press, New York, 1989.

Gurr, S. J., McPherson, F. J., and Bowles, D. J., *Molecular Plant Pathology: a Practical Approach,* Oxford University Press, Oxford, 1992.

Jones, D. H. and Howard, B. H., A rapid method for recombination and site-specific mutagenesis by placing homologous ends on DNA polymerase chain reaction, *BioTechniques,* 10, 62, 1991.

Landt, O., Grunert, H.-P., and Halm, A., A general method for rapid site-directed mutagenesis using the polymerase chain reaction, *Gene,* 96, 125, 1990.

MacKenzie, D. R. and Henry, S. C., *Biological Monitoring of Genetically Engineered Plants and Microbes,* Agric. Res. Inst., Bethesda, MD, 1990.

Slilaty, S. N., Fung, M., Shen, S.-H., and Level, S., Site-directed mutagenesis by complimentary-strand synthesis using a closing oligonucleotide and double stranded DNA template, *Anal. Biochem.,* 185, 194, 1990.

Chapter 13

Polymerase Chain Reaction Technology in Plant Pathology

A. Hadidi, L. Levy, and E. V. Podleckis

TABLE OF CONTENTS

I. HISTORY AND PRINCIPLES

The polymerase chain reaction (PCR) is an *in vitro* method in which DNA sequences or transcripts are amplified rapidly with very high specificity and fidelity using oligonucleotide primers and Taq DNA polymerase in a simple automated reaction.[1–4] The seeds of PCR began as early as 1955 with Nobel Laureate Arthur Kornberg's discovery of a cellular enzyme called DNA polymerase. DNA polymerases serve several natural functions, including the repair and replication of DNA. It was not until the winter of 1983–84, however, that the PCR was developed by Kerry Mullis.[4] The first published account of PCR in 1985 was its application to the prenatal diagnosis of sickle cell anemia by Saiki and colleagues.[1]

0-87371-877-1/95/$0.00+$.50

Over the course of the next few years, the scientific literature centering on PCR increased from 3 papers in 1985, 20 papers in 1986, to over 1000 papers during 1987 to 1991, establishing PCR as one of the most substantial technical advances in molecular biology. Its current applications are in the areas of disease diagnosis, detection of pathogens, detection of DNA in small samples, DNA comparisons, high-efficiency cloning of genomic sequences, and gene sequencing.[5,6] PCR has impacted basic molecular, biological research, clinical research, forensics, evolutionary studies, the human genome project, and plant pathology.

The importance of PCR lies in its ability to amplify a specific DNA or cDNA transcript *in vitro* from trace amounts of a complex template. It is possible to amplify specific DNA or cDNA sequences, from as short as 50 bp to over 10,000 bp in length, more than a millionfold in a few hours, in a reaction that is carried out in an automated DNA thermal cycler. The reaction is based on the annealing and enzymatic extension of two oligonucleotide primers (each approximately 16 to 30 nucleotides in length) that flank the target region in a duplex DNA by means of DNA polymerase. The reaction mixture is first heated (DNA denaturation) and subsequently cooled (DNA annealing) in cycles of 30 s to a few minutes each. Heating the mixture separates the double-stranded DNA into two single strands. As the mixture cools, each primer hybridizes its complementary separated DNA strand. Each of the annealed primers is then enzymatically extended on the template strand into a new DNA strand with a DNA polymerase. These three steps (denaturation, primer annealing, and primer extension) which are carried out at discrete temperature ranges (for example, 94 to 98°C, 37 to 65°C, and 72°C, respectively) represent a single PCR cycle. The next cycle of heating separates the copies from the original strands and both sets become templates for a new round of DNA synthesis. As a result of repeated cycles, the target DNA multiplies exponentially in a chain reaction. In a few hours, 30 cycles of PCR can amplify a molecular signal, that was too small to detect, more than several millionfold. The length of the product generated during the PCR is equal to the sum of the lengths of the two primers plus the flanked target sequence. Double- or single-stranded DNA or RNA (after the reverse transcription [RT] into a cDNA copy [Figure 1]) can serve as templates.

Amplified DNA is detected by staining with ethidium bromide or silver nitrate after agarose or polyacrylamide gel electrophoresis, hybridization with labeled probes, or by colorimetric assay after affinity binding. Amplified DNA may also be digested with restriction endonucleases before electrophoresis to analyze the restriction fragment length polymorphism (RFLP) pattern of the amplified products. The electrophoretic analysis of amplified DNA labeled with fluorescent primers has also been reported.[6] PCR-amplified products labeled with biotin-11-dUTP or 14-dATP are detected by spotting on nitrocellulose or nylon membranes followed by colorimetric or chemiluminescent assay.[7,8] A combination of PCR and an ELISA-type assay has been described:[9] the target sequence is amplified using oligonucleotides that contain a functional group (a biotin moiety and a DNA sequence recognized by a DNA-binding protein). The amplified DNA-labeled oligonucleotide hybrid is then captured by the DNA-binding protein-coated microtiter plate, and then bound via the biotin moiety to a horseradish peroxidase-conjugated avidin complex.

II. APPLICATION OF POLYMERASE CHAIN REACTION IN PLANT PATHOLOGY

A. DETECTION AND DIAGNOSIS OF PLANT PATHOGENS

The availability of nucleotide sequences of many plant pathogens has made possible the development of PCR assays for the detection and diagnosis of several viroids, viruses, and other pathogens. Because of its great sensitivity, the PCR provides a good alternative to other diagnostic methods and can speed diagnosis, reduce the sample size required, and often eliminates the need for radioactive probes. Detection of viroids, viruses, bacteria, mycoplasma-like organisms (MLOs), fungi, and nematodes by PCR has impacted diagnostic practices, epidemiological studies, as well as studies of pathogen-host-vector interactions. In 1990, Hadidi and Yang[10] reported the detection of viroids in the apple scar skin group from total nucleic acid extracts of infected pome fruit trees by RT-PCR amplification (Figure 2). They also reported that RT-PCR amplification has been successfully utilized for the detection of RNA plant viruses and plant viral RNA satellites from infected tissue and predicted the application potential of PCR technology in the field of plant pathology. Subsequently, RT-PCR has shown its value in improving the detection and diagnosis of several viroids from their respective hosts. These viroids included citrus exocortis (CEV), mild citrus exocortis (CVIIa), citrus cachexia (CVIIb) (Figure 3), plum dapple, peach latent mosaic, grapevine speckle, and potato spindle tuber.[7,11–15] The RT-PCR assay transcribes and

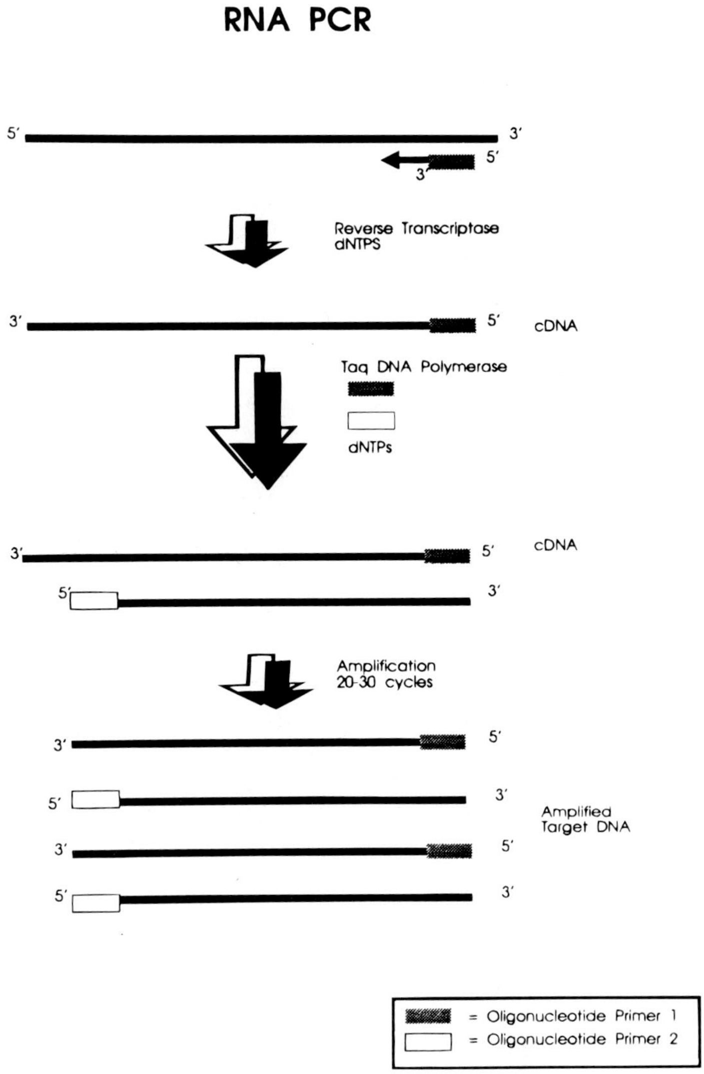

Figure 1 Reverse transcription of RNA to cDNA and subsequent amplification of the synthesized cDNA utilizing Taq DNA polymerase.

amplifies viroid RNA in total nucleic acid extracts of infected tissue with high specificity and fidelity.[16,17] The detection of viroids by RT-PCR requires 1 to 100 pg of total nucleic acids from infected tissue and it is 10 to 100-fold more sensitive than viroid detection by hybridization and 2500-fold more sensitive than return gel electrophoresis analysis.[10]

Detection of plum pox virus (PPV) by RT-PCR amplification of PPV cDNA from infected plant extracts has also been reported.[8,18–20] As few as 10 fg of purified viral RNA, corresponding to approximately 2000 viral particles, were detected in plant extracts.[19] Further, RT-PCR detection of PPV from infected tissue was more sensitive than molecular hybridization using ^{32}P-labeled cRNA probes.[19] The RT-PCR method has been also successfully utilized for the detection of grapevine virus A (GVA) from infected grapevine tissue[21] and viruliferous mealybugs,[22] and potato leafroll luteovirus (PLRV) from infected potato tissue and viruliferous aphids.[23] Figure 4 shows RT-PCR detection of PLRV from infected tissue and viruliferous insect vector. RT-PCR was more sensitive than molecular hybridization or ELISA for the detection of either GVA or PLRV.[21,23] Moreover, primers used to detect PLRV were successfully used to detect a luteovirus from mild yellow edge-diseased strawberry plants.[23] Other viruses reported to be detected by RT-PCR using viral-specific, nondegenerate primers include grapevine viruses B[24] and leafroll type III,[24,25] Fiji disease virus from diseased sugarcane,[26] pea seedborne mosaic virus,[27] watermelon mosaic virus-2, soybean mosaic virus-N and sugarcane mosaic virus,[28] beet western yellows and beet mild yellowing viruses,[29] tobacco rattle virus,[30] cherry leafroll virus,[31] 11 different badnaviruses,

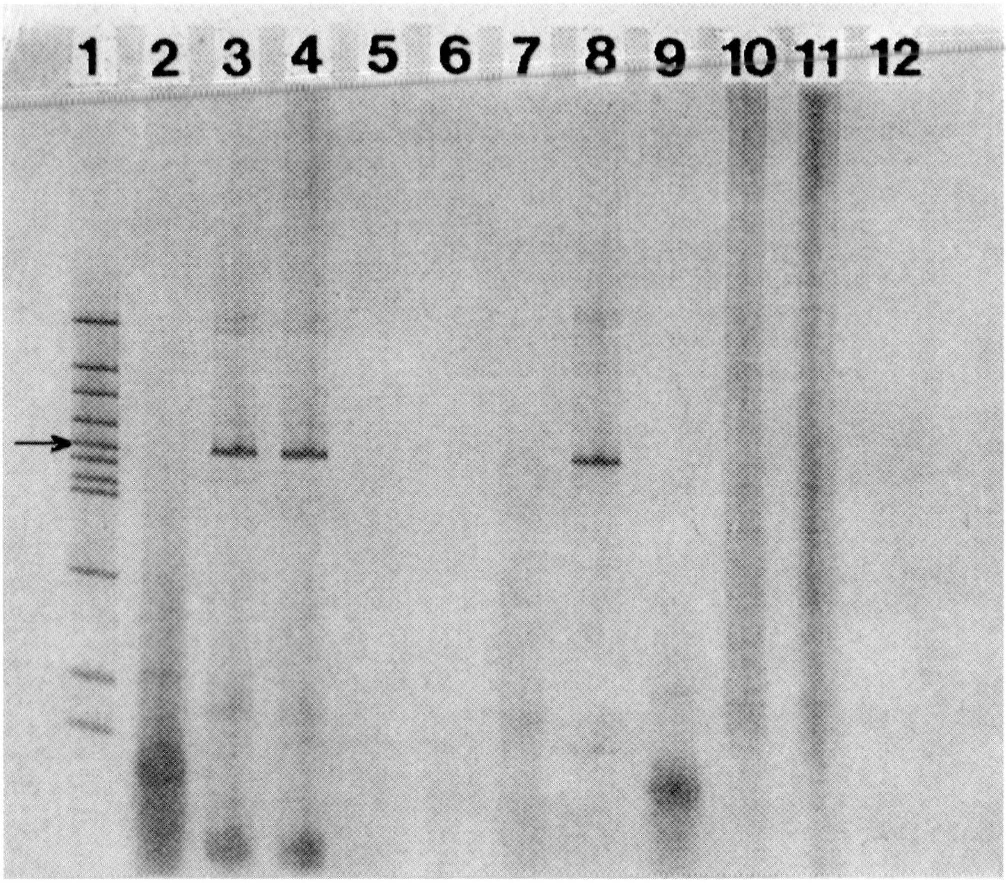

Figure 2 Polyacrylamide gel electrophoretic analysis of PCR-amplified apple scar skin viroid (ASSV) cDNA products from nucleic acid extracts of infected and uninfected apple tissue and from purified ASSV RNA. Molecular DNA marker of Hae III-digested PST B14 dimer (lane 1)—the arrow indicates 320 bp; transcript samples of TNA from infected tissue: unamplified (lane 2), amplified (lane 3), amplified and digested with RNase A (lane 4), amplified and digested with DNase (lane 5), TNA from infected tissue before transcription and amplification (lane 6); transcript samples of purified ASSV RNA: unamplified (lane 7), amplified (lane 8); transcript samples of TNA from uninfected tissue: unamplified (lane 9), amplified (lane 10), amplified and digested with RNase A (lane 11), amplified and digested with DNase (lane 12). Samples were visualized on polyacrylamide gel by silver staining. (From Hadidi, A. and Yang, X., *J. Virol. Methods,* 30, 261, 1990. With permission.)

including banana streak, rice tungro bacilliform and sugarcane bacilliform,[32] wheat soilborne mosaic virus,[33] and whitefly-transmitted geminiviruses.[34] Degenerate oligonucleotide primers complementary to conserved genomic sequences shared by members of subgroup I geminiviruses, which include maize streak virus and other geminiviruses of grasses and cereals, were successfully used for virus detection by PCR with a sensitivity approximately 10,000-fold greater than that of ELISA.[35] Similarly, degenerate oligonucleotide primers specific for members of the potyvirus group or luteovirus group have been used for virus detection and identification of several potyviruses infecting bulbous crops[36] and several luteoviruses infecting different host species.[37] The RT-PCR or PCR method has been successfully utilized to detect viroids or viruses from infected seeds[27,38] (Figure 5), fruits,[10] flower parts,[27,38,39] leaves,[7,8,11–21,27,35–37,40] bark,[8,10,11,13] roots,[38,39] potato tubers,[23] and viruliferous insects.[22,23,40]

PCR technology has also been extended to the detection, identification, and/or classification of MLOs, bacteria, fungi, and nematodes. The detection of clover proliferation and the closely related potato witch's broom MLOs from nucleic acid extracts of infected tissue by PCR DNA amplification, using primer pairs complementary to clover proliferation MLO, has been reported.[41] The PCR assay increased the sensitivity of MLO detection by at least 40-fold. PCR has also been used for the detection and identification of other MLOs, including the aster yellows, dwarf aster yellows, and periwinkle little leaf,[42] as well as tomato bigbud, clover phyllody, chrysanthemum yellow,[43] and grapevine yellow.[44] The

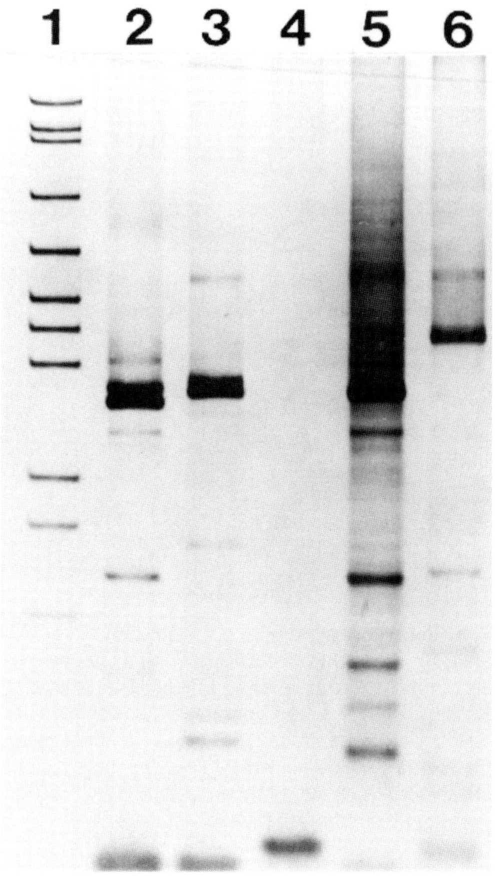

Figure 3 Polycrylamide gel electrophoretic analysis of RT-PCR-amplified cDNA products from citrus infected with citrus viroids IIa (CVIIa) and IIb (CVIIb = cachexia), members of the hop stunt viroid (HSV) group, and citrus exocortis viroid (CEV). pGEM DNA molecular marker with fragments sizes (bp) 2645, 1605, 1198, 676, 517, 460, 396, 350, 222, 179, 126, 75, 65, 51, 36, (lanes 1). RT-PCR-amplified products of CVIIa (302 bp): sweet orange, E32 (lane 3); sweet orange, IF-3 (lane 5). RT-PCR-amplified cDNA products of CVIIb (297 bp): citron, X7 (lane 2). RT-PCR-amplified products of CEV (371 bp); sweet orange, E1A (lane 6). RT-PCR products of healthy citron (lane 4) and E1A were amplified with CEV-specific DNA primers, all others (X7, E32, and IF-3) were amplified with HSV-specific DNA primers. Amplified products were analyzed on 6% polyacrylamide gel and visualized by silver staining. (From Levy, L., Hadidi, A., and Garnsey, S. M., *Proc. Int. Soc. Citriculture,* 2, 800, 1992. With permission.)

combination of PCR amplification of a sequence of the 16S rRNA gene with RFLP analysis of the amplified product has been reported to detect and differentiate 17 MLOs maintained in perwinkle plants and 9 MLOs from woody hosts.[45]

The amplification of fungal DNA sequences by PCR obviates the need to culture fungi in order to detect, identify, and classify species and isolates. Some of the fungi to which the PCR technology has been applied include *Gaeumannomyces,*[46] *Rhizoctonia,*[47–49] *Phytophthora,*[52–54] *Tilletia,*[53] *Heterobasidion,*[54] *Fusarium,*[55,56] *Aspergillus,*[57,58] *Verticillium,*[59] *Cochliobolus* (*Helminthosporium*),[60,61] *Glomerella* (*Colletotrichum*),[62] *Mycosphorella* (*Cercospora*),[63] *Ustilago,*[64] *Hypoxylon,*[65] *Acremonium,*[63] *Pythium,*[67] and *Laccaria.*[68]

At present only a few reports have been published on utilization of PCR technology for the detection and diagnostic identification of plant-pathogenic bacteria and nematodes. PCR appears to be a promising technique for detection and determination of potential pathogenicity of *Agrobacterium* strains from muscadine grapes where PCR was typically one or two orders of magnitude more sensitive than molecular hybridization.[69] PCR in combination with RFLP analysis was useful for the detection and strain differentiation of *Xanthomonas compestris.*[70,71] Amplification followed by cloning of the specific DNA sequence was successfully utilized to construct DNA probes useful in identification and differentiation *Pseudomonas,*[72] and *Corynebacterium sepedonicum,* the causal agent of potato ring rot.[73] The diagnosis of four species and several races of *Meloidogyne,* the root-knot nematode, has been reported.[74,75] The combination of PCR amplification of DNA and RFLP analysis of the amplified products was used to investigate the nematode *Radopholus.*[76]

B. SCREENING TRANSGENIC PLANTS

PCR provides a rapid, sensitive, and specific method for the detection of inserted recombinant viral or satellite viral DNA in the genomic DNA of transgenic plants.[77,78] It can also be used to screen for the challenge virus in transgenic plants.[79]

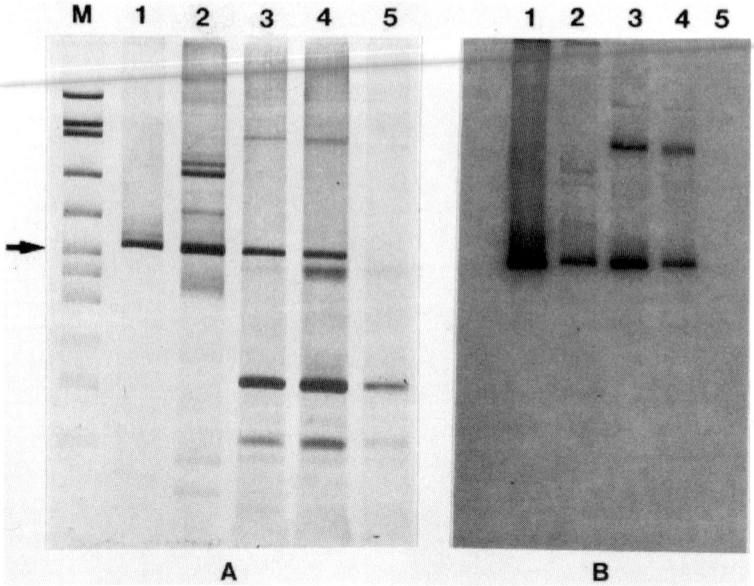

Figure 4 Polyacrylamide gel electrophoretic analysis (A) and autoradiograph of Southern blot hybridization analysis (B) of RT-PCR products from potato leafroll luteovirus (PLRV)-infected potato leaves and from viruliferous and nonviruliferous aphids. pGEM DNA marker (lane M)—the arrow indicates 460 bp; purified PLRV RNA (lane 1), PLRV-infected potato leaf (lane 2), viruliferous aphids maintained on PLRV-infected potato and *Datura strammonium* plants (lanes 3 and 4, respectively), nonviruliferous aphids maintained on tobacco plants (lane 5). Polyacrylamide gel was stained with silver, Southern blot was hybridized with ^{32}P-labeled PLRV cDNA probe specific for PLRV coat protein gene. RT-PCR products of nonviruliferous and viruliferous aphids were obtained from approximately one insect equivalent. (From Hadidi et al., *Plant Dis.,* 77, 595, 1993. With permission.)

C. MOLECULAR CLONING OF VIRAL AND VIROID GENOMES

Use of PCR has facilitated the production and cloning of full-length cDNAs of plant viruses and viroids. The procedure was used to obtain cDNA clones of cucumber mosaic virus (CMV),[80] apple scar skin viroid,[16,17] dapple apple viroid,[81] peach latent mosaic viroid,[14] and the closely related citrus viroid IIa and IIb (cachexia).[82]

CMV has a genome of three positive-strand RNA components: RNA 1 (3.4 kb), RNA 2 (3.0 kb), and RNA 3 (2.1 kb). Full-length first-strand cDNA to RNA 1, RNA 2, and RNA 3 were synthesized by RT using RNA isolated from purified CMV particles and a 3′ complementary oligonucleotide primer that contained a Bam HI site. After purification by alkaline agarose gel electrophoresis, phenol-chloroform extraction, and precipitation with ethanol, each cDNA product was used as a template for a PCR. The second-strand primers used for amplification contained a Bam HI site, a T_7 promoter, and sequences corresponding to the 5′ terminus of each RNA. Following digestion with Bam HI, the PCR products were cloned into the Bam HI site of the vector pEMBL9(+).[80]

Apple scar skin viroid (ASSV) contains 329 to 330 nucleotides[83,84] and DAV contains 331 nucleotides.[81] ASSV and DAV were reverse transcribed *in vitro* from total nucleic acid extracts of infected tissue and amplified in an RT-PCR assay. The sample was then extracted with phenol-chloroform and precipitated with sodium acetate and ethanol. The ds PCR product was treated with 15 units of S1 nuclease at 37°C for 30 min. The 5′ terminus of the DNA was phosphorylated with T_4 polynucleotide kinase followed by ligation to Eco RI adaptors. Excess adaptors were removed by centrifugation through Biospin 30 columns (Bio-Rad, Richmond, CA) and the DNA was ligated to Eco RI digested pUC9. The insert from positive clones were subcloned into pGEM vector for transcription. The cloned ds PCR product was sequenced using the Taq Track™ sequencing system (Promega, Madison, WI).

D. PCR-MEDIATED NUCLEOTIDE SEQUENCE ANALYSIS OF VIROIDS

Puchta and Sänger[85] were the first to sequence PCR-amplified cDNA of gel-purified hop stunt viroid from grapevine and established the accuracy of this method. Subsequently, we[14,17,81,82] and others[86] have shown that accurate nucleotide sequence of several viroids is obtained when viroid cDNA is synthesized directly from total nucleic acid fractions or low-molecular weight RNAs from viroid-infected tissue

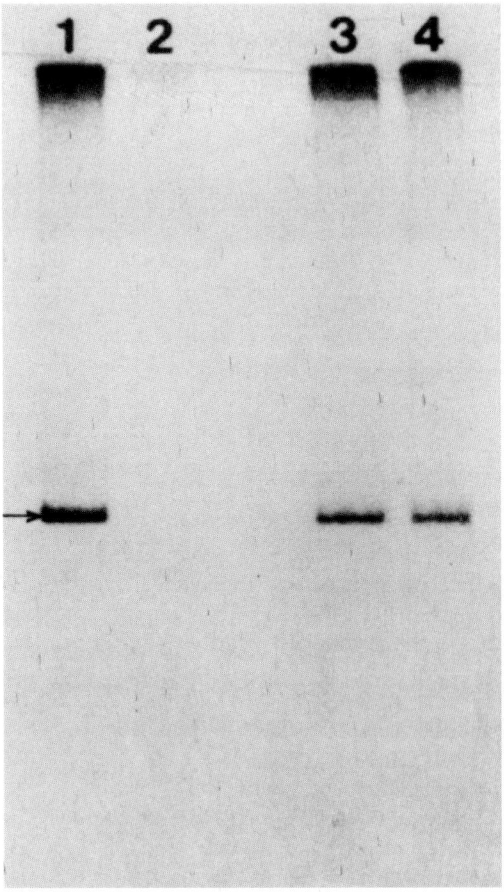

Figure 5 Autoradiograph of Southern blot hybridization analysis of ^{3}P-labeled ASSV cRNA probe with RT-PCR-amplified products from TNA of uninfected apple seeds (lane 2) and viroid-infected apple: seeds (lane 1), cotyledons (lane 3), and embryos (lane 4). Samples were electrophoresed in a 6% polyacrylamide gel under nondenaturing conditions. Arrow indicates the position of full-length (330-bp) viroid DNA-amplified product. RT-PCR-amplified products of 10 to 20 ng TNA per sample were analyzed. (From Hadidi et al., *Res. Viol.,* 142, 289, 1991. With permission.)

without further viroid purification by gel electrophoresis. The nucleotide sequences were determined using Taq Track™ or *fmol*™ sequencing systems (Promega, Madison, WI) using 5′-^{33}P-end-labeled primers. This method was used to determine the nucleotide sequence of variants of apple scar skin,[17] dapple apple,[81] citrus cachexia (CVIIb),[82] mild citrus exocortis (CVIIa),[82] peach latent mosiac,[14] and potato spindle tuber[86] viroids.

The following procedure is routinely used in our laboratory for isolation and sequencing of PCR-amplified viroid cDNA products. RT-PCR products are analyzed by electrophoresis on 6% native polyacrylamide gels (11 × 14 × 0.12 cm) at 120 V for 2.5 h in 1 × TBE (89 m*M* Tris, 89 m*M* borate, and 2.5 m*M* Na_2EDTA, pH 8.3). PCR products are visualized by staining with ethidium bromide and brief exposure to UV light. The major RT-PCR product for each viroid and primer combination is excised from the gel and isolated by the crush and soak method of Maniatis et al.[87] From 0.5 to 1 ng of gel-purified PCR products are then sequenced using 5′ end-labeled primers in the *fmol*™ sequencing system. Cycle sequencing is performed in a thermocycler following the manufacturer's recommended parameters: denaturation at 95°C for 30 s, annealing at 42°C for 30 s, and extension at 70°C for 1 min for a total of 30 cycles.

The PCR-mediated sequencing of plant viral genome lags behind viroid sequencing. Recently, however, a partial sequence of PCR-amplified DNA of tomato yellow leaf curl geminivirus from Egypt was determined.[88]

III. METHODS OF POLYMERASE CHAIN REACTION

A. PREPARATION OF PLANT NUCLEIC ACID EXTRACTS SUITABLE FOR PCR

Total nucleic acids (TNA) are extracted from plant tissues as described.[11,13] This procedure was used successfully for the extraction of numerous samples containing viruses or viroids for RT-PCR detection.[11–14,21,23] Briefly, 1 g of tissue is ground in liquid nitrogen and extracted with 5 ml of a buffer consisting of 0.1 *M* glycine-sodium hydroxide (pH 9.0), 50 m*M* sodium citrate, 1 m*M* disodium ethylenediamine tetraacetic acid (EDTA), 2% sodium lauryl sulfate, and 1% sodium lauryl sarcosine. Samples are further extracted with the addition of an equal volume of Tris-HCl-buffered phenol, pH 7.6 to 8.0 containing 0.1% 8-hydroxyquinoline and 0.2% 2-mercaptoethanol, and followed by the addition of an equal volume of chloroform. Nucleic acids are pelleted twice with 2.5 vol of absolute ethanol plus 1/10 vol of 3 *M* sodium acetate (pH 5.2) at −20°C. Nucleic acid preparations are purified using RNase-free ELUTIP™-R minicolumns (Schleicher and Schuell, Keene, NH) according to the manufacturer's directions with the following modification for tissue containing viroids: low-salt buffer contains 0.2 *M* sodium chloride, 20 m*M* Tris-HCl (pH 7.4), and 1.0 m*M* EDTA. Other methods for extraction include proteinase K digestion of plant tissue extracts, virus purification, or isolation of the pathogen-genomic DNA.

B. PCR PARAMETERS

1. Primer Design

Primer design is critical to the success of any PCR assay. The rules for primer selection are outlined in several PCR manuals.[89,90] Generally, primers are between 16 to 30 nucleotides in length and 50 to 60% G+C. The 3′ end should contain GG, CC, GC, or CG. Mispriming in G+C-rich regions may result from runs of Cs or Gs at the 3′ end of primers. Complementarity between the 3′ ends of primers should be avoided to prevent the annealing and extension of primer pairs forming primer-dimer artifacts. Although primers may be selected manually, several good computer programs are commercially available to aid in the selection of primers. These programs allow for the analysis of primer pairs to decrease the level of complementarity between primers. This can be important especially for multiplex PCR where multiple primer pairs are contained in single reactions. Some of these programs also determine annealing temperatures and unexpected priming sites in other regions of the sequence which can generate additional unpredicted PCR products.

2. Reaction Components

High concentration of deoxyribonucleotides, primers, $MgCl_2$ or Taq DNA polymerase can result in a high degree of mispriming of nontarget sites and production of PCR artifacts. Concentrations of each component must be determined empirically with each new target and primer pair. For RT-PCR, the level of reverse transcriptase used to synthesize cDNA can affect the concentration of PCR products. We have observed that some viroids required 400 units of MMLV reverse transcriptase, yet some viruses require only 100 to 200 units.[12,13,18] The use of 400 units for these particular viruses decreased the subsequent amount of PCR products by 40 to 50% based on gel analysis.

3. Thermocycling

Nonspecific PCR products can also be eliminated by adjustment of physical parameters in lieu of, or in addition to, chemical parameters. The increase of annealing temperature, decrease in the number of cycles (from 40 to 25), and the decrease in the length of each PCR temperature segment (incubation time at each temperature) can reduce mispriming.[90] Nonspecific PCR products can also be reduced by a technique called "hot-start" PCR.[92] To "hot start" a PCR reaction, all components of the PCR reaction are mixed together except either the target DNA or Taq DNA polymerase. The reaction is heated to 80°C for 5 min before the missing component is added and the PCR cycle is started.[92] This high-temperature incubation decreases the level of mispriming which may occur during assembly of PCR reaction at room temperature.

C. PCR METHODS USED WHEN TARGET NUCLEOTIDE SEQUENCE IS KNOWN

In these PCR methods, oligonucleotide primer pairs anneal to sites at each end of the DNA segment to be amplified. Each primer in the pair anneals to only one of the target DNA strands such that extension proceeds across the amplified region towards the other primer. The nucleotide sequence of these primers is determined by the target and requires a prior knowledge of the nucleotide sequence at the boundaries of the amplified segment.

1. Standard PCR

PCR reactions usually contain 0.05 to 1.0 μg of TNA containing target cDNA or DNA or 0.5 to 2.0 ng of known DNA or cDNA added to a 45 μl reaction mixture containing 1X PCR buffer (10 m*M*

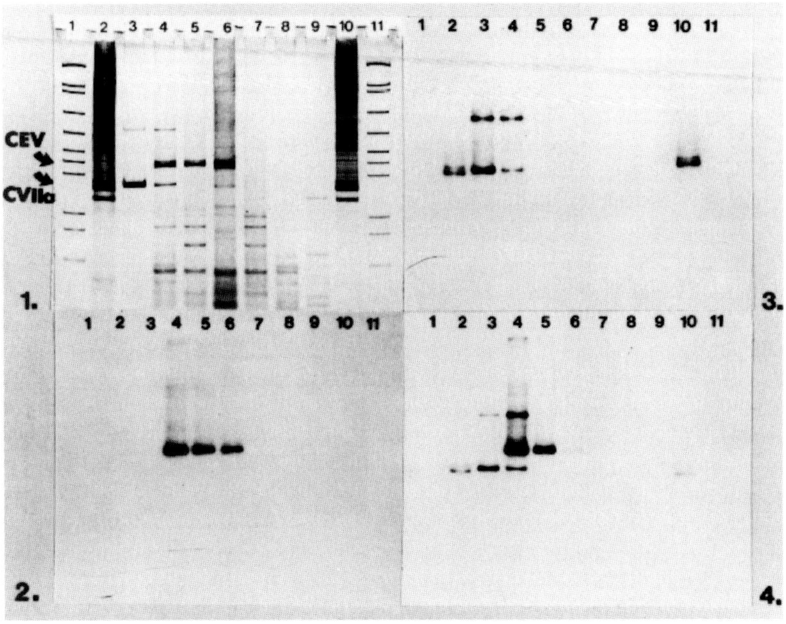

Figure 6 Specificity of multiplex RT-PCR amplification of CEV and citrus viroid IIa (CVIIa), a member of the HSV group, from infected citrus tissue. Polyacrylamide gel electrophoretic analysis of RT-PCR and multiplex RT-PCR cDNA (1). Autoradiograph of Southern blot hybridization analysis with: ^{33}P-labeled CEV cRNA probe (2); ^{33}P-labeled HSV cRNA probe (3); and ^{32}P-labeled CEV cRNA and HSV cRNA probes (4). pGEM DNA molecular marker (lanes 1 and 11). RT-PCR-amplified CVIIb: sweet orange, X7 (lanes 2 and 10). RT-PCR-amplified CEV: sweet orange, E9 (lane 6). Amplification of viroids from sweet orange, E22: RT-PCR amplification of CVIIa using HSV DNA primers (lane 3); RT-PCR amplification of CEV using CEV DNA primers (lane 5); multiplex RT-PCR amplification of CVIIa and CEV using CEV and HSV primers (lane 4). Amplification of uninfected sweet orange using HSV DNA primers, HSV and CEV DNA primers, and CEV DNA primers (lanes 7, 8, and 9, respectively). (From Levy, L., Hadidi, A., and Garnsey, S. M., *Proc. Int. Soc. Citriculture*, 2, 800, 1992. With permission.)

Tris-HCl (pH 8.3), 50 m*M* KCl, 0.5 to 1.5 m*M* Mg Cl_2, and 0.01% gelatin), 20.0 to 200 μ*M* dNTP (dGTP, dATP, dTTP, and dCTP), 20 to 50 pmol of a 16 to 30-mer upstream (homologous) and downstream (complementary) primer pair, 1 to 2.5 units of Taq DNA polymerase, and deionized water. The reaction is overlaid with 75 μl of mineral oil to prevent evaporation during amplification. Typical cycling parameters are denaturation for 30 s to 1 min at 94°C, annealing for 30 s to 2 min at 55°C, and extension for 45 s to 3 min at 72°C for 30 to 40 cycles with the final extension at 72°C for 7 min in a DNA thermal cycler (Perkin-Elmer Cetus Corp., Norwalk, CT).

2. RT-PCR

Approximately 1 μg of TNA[11–14] or 0.2 μg or less of specific RNA[23] is added to 100 pmol of complementary primer. The primer-TNA mixture is added to 6 μl of 5X first-strand cDNA buffer (250 m*M* Tris-HCl (pH 8.3), 375 m*M* KCl, 15 m*M* $MgCl_2$ 15 m*M* dithiothreitol) and deionized water to a final volume of 30 μl. Primers are denatured by heating at 100°C for 5 min, chilled on ice for 2 min, and annealed at room temperature for 0.25 to 1 h. The following reagents are then added to the reaction mixture: 4 μl 5X first-strand cDNA buffer, 5 μl 300 m*M* 2-mercaptoethanol, 2.5 μl 10 m*M* dNTP (2.5 m*M* each of dATP, dGTP, dTTP, dCTP), 1 μl RNasin (40 units/μl, Promega Corp., Madison, WI), 1 to 2 μl of cloned Moloney murine leukemia virus reverse transcriptase (200 units/μl, Gibco BRL Life Technologies, Inc., Gaithersburg, MD), and deionized water to a volume of 50 μl. The reaction is incubated at 37–42°C for 1 to 2.5 h. A 5-μl aliquot of this cDNA reaction is added to the PCR protocol described above.

3. Multiplex PCR

This procedure was originally used to coamplify gene products in a single PCR.[91] We have successfully used this technique to amplify multiple viroids or viruses from a single sample in single reactions[12,24] (Figure 6). The procedure is essentially the same as standard PCR and RT-PCR except that multiple primer sets at 100 pmol of complementary primer for cDNA and 20 pmol each of the primer sets included for PCR are utilized. No more than 250 ng of purified target nucleic acids or 1 μg of TNA are used for multiplex PCR. Care must be taken to avoid complementarity between primers in each reaction. PCR product formation in a multiplex reaction may favor one template or another, subsequently decreasing the amount of the unfavored template. The addition of dimethyl sulfoxide (DMSO) to multiplex PCR alleviates this problem. PCR reactions should be adjusted to 1 to 10% with DMSO, however, greater than 10% DMSO inhibits the activity of Taq DNA polymerase.[91]

4. Immunocapture PCR

Prior to PCR or RT-PCR, virus particles are immobilized and concentrated on a sterile solid surface previously coated with virus-specific antiserum.[93] This procedure has been modified for use with plant viruses from infected tissue and viruliferous insect vectors.[20,22,24] Sterile microfuge tubes or microplates are coated with 100 μl of virus-specific antiserum diluted to a concentration of 1 μg/ml in carbonate coating buffer (0.05 *M* carbonate, pH 9.6) and incubated overnight at 4°C. The tubes are washed three times for 5 min with PBST (0.02 *M* phosphate, 0.15 *M* saline, 0.05% Tween®-20, pH 7.5). The wash step is followed by blocking for 1 h at room temperature with PBST plus bovine serum albumin (10 mg/ml). The tubes are used directly or dried and stored at 4°C until use. Infected tissues are ground at a dilution of 1:4 in 50 m*M* trisodium citrate plus 2% polyvinylpyrrolidone and 20 m*M* diethyldithiocarbamic acid, pH 8.3 (or any suitable ELISA extraction buffer for the target virus). To the antibody-coated tubes or microplates, 100 μl of infected extract are added and the tubes or microplates are incubated overnight at 4°C. Samples are removed and the tubes are washed with PBST buffer. To release viral RNA, 25 μl of sterile 10 μm Tris-Hcl, pH 80, plus 1% Triton® X-100 is added to each tube and incubated at 65°C for 10 min with vortexing at 2-min intervals, followed by either PCR or RT-PCR.

5. Quantitative PCR

Quantitative PCR can be considered in two contexts: determination of the actual yield of the amplified product or the calculation of the amount of starting template, where that template is a specific gene, messenger RNA, virion, etc. The two procedures are related and any attempt to calculate the copy number of a PCR template invariably involves measurement of the yield of PCR products. Measurement of amplification yields first requires that products be separated from unincorporated nucleotides and the other constituents of the PCR reaction. This can be accomplished by selective precipitation, separation using any of a variety of commercially available spin columns and chromatography columns, or by affinity binding using, for example, a product-specific biotin-labeled probe and a streptavidin-coated capture matrix.[94] Once isolated, the concentration of the PCR product may be determined spectrophotometrically, colorimetrically, electrophoretically by titration against known amounts of a nucleic acid standard of similar size, or by autoradiography or scintillation counting in the case of radioisotopically labeled products.

Determination of the starting concentration of the target template is a more complex procedure than the calculation of the final yield of a given product. The exponential nature of PCR amplification creates the potential for even small variations in amplification efficiencies to produce widely different product yields. Some studies have quantitated PCR using external standards amplified in parallel reaction to create a standard curve.[95] Even under carefully controlled conditions, however, tube-to-tube variation in amplification efficiency occurs. Pipetting error or even variability within the thermocycler wells[96] can introduce variability. Gilliland et al.[97] reported as much as a six-fold variation in yield of identical PCR products even when using a master mix of reagents to avoid pipetting errors. Such variability introduces significant error into any attempt to calculate the initial amount of template. For these reasons, most published studies and the remainder of the this report describe quantitation using the coamplification of reporter templates in the same tube with the target sequence.

a. Reagents

i. Templates

Templates chosen for use as internal standards in quantitative PCR fall into two categories: heterologous or "exogenous" templates and homologous templates that differ only slightly from target sequences. Exogenous templates are unrelated "reporter" sequences that are coamplified with the target sequence. Reporter sequences used in previous studies have included, among others, unrelated sequences in the target genome[98] and synthetic RNA templates.[99] For the use of exogenous templates to be successful, it is critical that the reporter sequence be amplified with the same efficiency as the target. This is not always the case and must be determined empirically for each target-reporter combination. To address this pitfall, others have used homologous reporter systems. In these cases, a reporter sequence is created by modifying the target sequence slightly to allow the two to be separated after amplification. Gilliland et al.[97] created a reporter sequence by adding small intron to a mRNA target sequence. We have used the same approach in adding a five nucleotide insert to create a potato spindle tuber viroid mutant that can be coamplified with a wild type, then separated electrophoretically on the basis of molecular weight. Henco and Heibey[100] described a system in which a single nucleotide change in the reporter sequence allowed separation of the target and reporter by temperature-gradient gel electrophoresis. Several laboratories have used mutations in target sequences to create restriction sites in reporter sequences that allow the amplified products to be separated electrophoretically after digestion.[97,101]

ii. Primers

Primers are selected according to the same criteria, described earlier in this chapter. Ideally, the same primers are used to amplify both the target sequence and the corresponding reporter sequence.

b. Reverse Transcriptase

Most reports of quantitative PCR have been concerned with the determination of the copy number of specific mRNAs, but the procedure has also been used to detect viruses.[102] Since most plant viruses and all viroids have an RNA genome, RT is a prequisite to the amplification process. It is important to take into account variability in RT when quantifying RT-PCR reactions. RNA target sequences should be quantitated using RNA reporters and the efficiency with which the reporter sequences are reverse transcribed should be determined empirically.

c. Polymerase Chain Reaction

The PCR is prepared as described earlier in this chapter, except that serially diluted amounts of the reporter templates are added to a fixed amount of target template and coamplified in a competitive PCR assay. Reagents are prepared as a master mix to reduce variability. The dilution factors, number of cycles, and initial template concentrations are determined empirically so as to ensure that measurements are made during the exponential phase of amplification where the proportion of initial template concentration to the amount of product is a linear relationship.

d. Quantitation

After amplification, aliquots of the product are digested with the appropriate restriction enzyme (if the reporter requires digestion) and analyzed by gel electrophoresis. The ratio of unknown product to internal standard is estimated visually, by densitometry, or, in the case of radiolabeled products, by scintillation counting. It is assumed that the ratio of the amplified products reflects the ratio of unknown to internal standard present prior to amplification. Consequently, that concentration of the internal standard that produces product in a 1:1 ratio with the unknown should equal the concentration of the unknown initially present in the sample. One note regarding the use of reporters that require restriction digestion: several studies have demonstrated that the formation of heteroduplexes between products containing the mutant restriction site and wild-type products can occur.[97,101] These heteroduplexes are not sensitive to restriction digestion and their formation can therefore result in an overestimation of the concentration of wild-type product present. According to these studies, however, the formation of heteroduplexes is negligible as long as primer concentrations are not limiting and the number of cycles is less than 40.[103]

e. Summary

Several methods for quantitating the products of the PCR have been described. Because of the sensitivity of the PCR procedure, even small differences in reagents, especially primers and templates, that can effect changes in amplification efficiency can potentially introduce enormous errors in quantitation. The use of a master mix to prepare RT reactions (when necessary) and PCR experiments helps reduce such variability. Similarly, the coamplification of internal standards in a competitive PCR assay, rather than the use of a standard curve developed using an external standard, would seem less likely to introduce errors. As little as 1.0 pg of target in 1.0 ng of total mRNA has been quantitated using such a competitive PCR assay.[103]

D. PCR METHODS USED WHEN TARGET NUCLEOTIDE SEQUENCE IS LIMITED OR UNKNOWN

Several PCR techniques have been developed that allow amplification of targets when sequence information is limited or unknown. One technique, the random amplified polymorphic DNA (RAPD) PCR, has been used to characterize and identify several plant pathogens. Two other methods, inverse PCR and anchor PCR in its various forms, have not been widely applied to plant pathogens, but may be potentially useful when sequence information is limited.

1. Random Amplified Polymorphic DNA (RAPD)-PCR

RAPD-PCR can be used in conjunction with or in place of RFLP analysis currently used for genetic analysis of DNA polymorphisms.[104] RAPD technology has been extended to investigations of isolate designation, evolutionary biology, and gene mapping. Unlike previously described PCR assays which utilize two primers of a defined sequence, RAPD-PCR detects nucleotide sequence polymorphisms in a DNA-amplification-based assay using only a single primer of arbitrary nucleotide sequence.[105] The short-length single primer binds to the genome on opposite strands in an inverted orientation and produces a RAPD-PCR product that has partial or complete homology with the arbitrary primer sequence at each end. RAPD polymorphisms occur because of insertions, deletions, and base substitutions that effect the primer binding site and are reflected as the presence or absence of RAPD-PCR bands. RAPD-PCR has the advantage of being fast, inexpensive, conservative in the use of genomic DNA, and applicable to many isolates, cultivars, etc., because of the random nature of the primer sequence. These primers bind to homologous regions of the genome and produce only a portion of the potential RAPD markers. Additional bands can be produced, under identical conditions, by using different random primers. The conditions for RAPD analysis vary with organism, length of primer, and type of analysis. Generally, genomic DNA preparations can be fairly crude,[106,107] however, tissues which contain high levels of polysaccharides and phenolic compounds are unsuitable and require extraction methods for obtaining more purified genomic DNA.[108]

The following RAPD-PCR conditions are a compilation from several RAPD procedures.[75,105,109–116] Generally, RAPD-PCR parameters are 3 to 40 ng of input genomic DNA template; 1.5 to 4.0 m*M* Mg-Cl_2, 100 to 250 μ*M* each dTTP, dATP, dCTP, and dGTP; 5 p*M* to 1.6 μ*M* of 5 to 10 mer random primer[105,111,116] and 0.5 to 2.0 units of Taq DNA polymerase. Cycling parameters are denaturation for 5 s to 1 min at 94°C, annealing for 30 s to 1 min at 36°C, and primer extension for 30 s to 4 min at 72°C for 35 to 45 cycles.

a. RAPD-PCR Optimization

RAPD-PCR parameters (biochemical and physical) must be strictly standardized if this technology is to be transferred between laboratories with experimental reliability. The G + C content of random primers should be at least 50%, however, 60 to 70% G + C is more effective.[105,110,111] Changes in primer, $MgCl_2$ input DNA, and Taq DNA polymerase concentrations will change the RAPD marker pattern.[75,109,112,114] Even changes in the source of Taq polymerase can change RAPD marker patterns on identical samples.[75,105,110,114] Physical parameters can also affect the RAPD marker patterns. Increase in the annealing temperature can decrease the number of markers until all markers disappear, and longer extension times favor the formation of longer RAPD markers, whereas products smaller than 1.5 kb can be adequately amplified by extension times of 30 s.[109,115] No difference in the accumulation of PCR products was noticed for programs containing 35, 55, or 75 cycles, possibly because Taq DNA polymerase was a limiting factor.[115] Under identical RAPD-PCR conditions, DNA from different dated extractions can alter marker patterns,[75,114] When RAPD-PCR is used in evolutionary biology investigations, RAPD markers with faint or intermediate intensity can be generated if conditions are not carefully controlled, resulting in errors in defining evolutionary relationship. These errors can be avoided by screening large numbers of random primers to determine the reproducibility of a RAPD marker. Efforts to identify genes for resistance to plant pathogens using RAPD technology has experienced the same difficulties with RAPD marker reproducibility when scoring F_2 populations.[113] RAPD marker bands with faint or intermediate intensities could actually represent true polymorphism, but their unreliability could make them useless. Also, RAPD markers have been reported to fail to be amplified in related taxa.[113] These difficulties can be overcome by designing specific primers defined from the RAPD marker termini.[117]

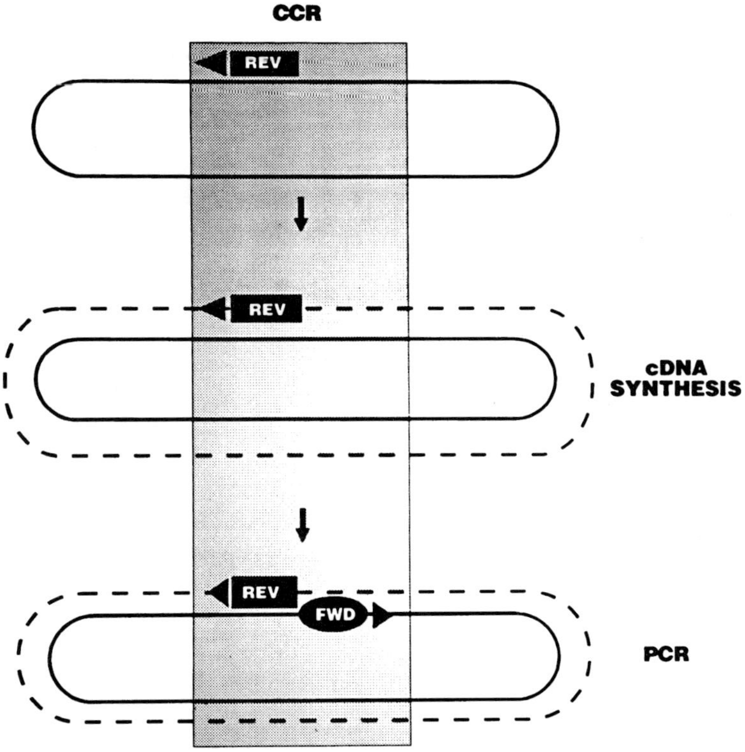

Figure 7 RT of full-length viroid RNA to cDNA and subsequent amplification of the cDNA by PCR. DNA primers used are designed from the nucleotide sequence of the upper central conserved region (CCR) of the viroid and its adjacent segment.

2. Inverse PCR

The standard PCR reaction allows for amplification of target DNA that lies *between* regions of known sequence. In some cases, for example, the insertion sites of transposable elements or *Agrobacterium* T-DNA, it may be useful to amplify regions of DNA that *flank* an area of known sequence. Three laboratories, nearly simultaneously, developed similar techniques for amplifying these flanking sequences by converting them to interior regions.[118,120] The procedure, referred to as inverse or inverted PCR, requires that only a short segment of sequence be known and that oligonucleotide primers are designed from this known sequence such that their 3′ ends face away from each other rather than toward each other as in a standard PCR. A generalized protocol for inverse PCR is as follows: DNA containing the sequence to be amplified is cut with a restriction endonuclease. Relatively small fragments (<1000 bp) that contain the known (core) sequence and the flanking sequence are selected by Southern hybridization. After inactivation of the restriction enzyme, the DNA is circularized using T4 DNA ligase and a dilute DNA concentration that favors circularization over concatamerization. The DNA is then linearized by a restriction cut between the 5′ ends of the oligonucleotide primers in the core sequence. Alternatively, the circularized DNA may simply be nicked by heating or left circularized rather than restriction digested. Silver and Keerikatte[119] reported a 100-fold increase in amplification when the circular DNA was first linearized by restriction digestion, while Triglia et al.[120] got similar results by using heat to produce random nicks in the circularized DNA. PCR amplification is then carried out under standard conditions.

We have used what might be considered a variation on the inverse PCR when amplifying viroids of known nucleotide sequences (Figure 7). Since by definition viroids are circular RNA molecules, no circularization is required, but using primers that anneal to the central conserved region of the viroid with 3′ ends facing away from each other, we have successfully amplified full-length DNA copies of a number of viroids.

3. Anchor PCR

Standard PCR or RT-PCR allows the amplification of sequences from minute quantities of DNA or RNA using sequence information from both sides of the region of interest. Several methods have been developed to allow amplification when information is available for only one side of the region or when one end of the sequence is variable. When the method involves the addition of a universal primer to the missing end of the target DNA, the procedure is termed anchor PCR. The basic procedure was described by Loh et al.,[121] who used the following protocol to amplify the variable region of human T cell receptor sites: total RNA is isolated and mRNA is reverse transcribed using a primer selected from the known sequence region (or, alternatively, a poly [dT] primer complementary to the polyadenylated 3′ end of the mRNA). The transcript is then tailed at the 3′ end with poly (dG) using terminal deoxynucleotidyl transferase. The tailed product is amplified with the specific primer used above for RT and an "anchor primer" consisting of a poly (dC) tail attached to the anchor sequence consisting of convenient restriction sites to facilitate cloning and provide a universal priming sequence. After an initial round of amplification using the anchor-poly (dC) primer, subsequent amplifications are carried out using the anchor sequence alone as the primer. Use of the anchor sequence alone reduced the artifactual products that sometimes resulted from the poly (dC) primer.

a. RACE PCR

When anchor PCR is applied to cDNAs, the procedure is referred to as rapid amplification of cDNA ends (RACE).[122] Again, using a small amount of specific sequence information from within the target region and universal anchor primers, the 3′ or 5′ ends of the cDNA molecule may be amplified. Amplification of the 3′ end (3′ RACE) is a straightforward procedure with less manipulations than anchor PCR. First-strand synthesis from mRNA is primed with a poly (dT) anchor primer complementary to the mRNA polyadenylated 3′ end. Subsequent amplification is then carried out using a 5′ primer determined from the known portion of the target sequence and the anchor primer from the 3′ end. The RACE procedure for amplification of 5′ ends is essentially the same as that outlined above for the general anchor PCR method except that a second sequence specific primer located, or nested, just to the 5′ side of the cDNA synthesis primer is used during amplification to increase the specificity of the PCR. Belyavsky et al.[123] described an amplification scheme that uses universal primers at both ends of the target sequence to amplify total cDNA in a population. First-strand cDNA is synthesized from total RNA using an oligo (dT) anchor primer. A homopolymeric (dG) tail is then added to the total cDNA which is then amplified using primers complementary to the oligo (dA) and oligo (dG) ends. Alternatively, first- and second-strand cDNA synthesis may be primed using random hexanucleotides.[124]

i. SLIC PCR

The use of homopolymeric tails in RACE may sometimes result in mispriming of the PCR and/or cDNA synthesis to generate a background population of nonspecific amplification products. To reduce mispriming, Dumas et al.[125] developed a modified RACE procedure called SLIC for single-strand ligation to ss-cDNA. The procedure is based on the ligation of a single-stranded oligonucleotide anchor sequence directly to the 3′ end of the synthesized cDNA, thereby avoiding the use of an oligomeric tail. The ligation of the anchor sequence is mediated by T4 DNA ligase which is able to join single-stranded DNA molecules in the presence of hexamine cobalt chloride.

E. ELEMENTS FOR RELIABLE PCR ASSAYS

1. Laboratory design
   a. Separate laboratories or clearly defined work areas for PCR construction and analysis
   b. Avoid air flows back from the post-PCR laboratory to the pre-PCR laboratory
   c. Dedicated equipment for each laboratory
2. Stock solution preparation
   a. Use fresh "PCR-dedicated" chemicals (including water) and aliquot all solutions
3. Operating procedures
   a. Set up a "master" RT and PCR mix, then aliquot into each tube
   b. Use positive-displacement pipettes or plugged (barrier) tips
   c. Use multiple coprocessed, negative controls
   d. Set up the known positive controls last
   e. Handle only one tube at a time
   f. Use gloves and change often

4. Assays and detection procedures
   a. Use multiple primer pairs, corresponding to different regions of the genome (one primer pair per assay)
   b. Ensure, if needed, positive confirmation of the PCR target product by a subsequent hybridization step

F. ADVANTAGES AND DISADVANTAGES OF PCR

PCR allows for: detection of low titer pathogens which elude conventional detection methods such as ELISA or dot-blot hybridization; identification of unknown pathogens;[23] detection of known pathogens that are currently detected by lengthy bioassays;[10–15,21,24] detection of multiple and unrelated pathogens in a single PCR reaction;[12,24] identification of the components of mixed infections or disease complexes;[23] rapid and sensitive evaluation of plants post-pathogen elimination therapy; evaluation of cross protection (classical or transgenic);[79] determination of specific sequence information with or without cloning from crude total nucleic acids;[14,16,17,81,82] and generation of pathogen-specific clones without pathogen purification.[14,16,17,80–82] This list does not begin to describe the complete list of potential pathogen, pathogen-vector, or pathogen-plant characteristics or interactions that can be examined by PCR.

PCR is primer directed. Thus primers can be designed to specifically amplify pathogen DNA or cDNA from heterogenous samples. This obviates the need to purify the pathogen from infected plant tissue. PCR can be performed on very small biological samples (i.e., single insects, spores, or pollen), herbarium-preserved fungi, and can be used to analyze unculturable obligate plant parasites.

The major disadvantages include: the initial expense in PCR laboratory setup; the requirement of trained personnel; cautious laboratory practices must be followed to prevent possible contamination from sample to sample; primer design requires some knowledge of the target sequence or a related sequence from published sequence data; and finally, for positive identification of specific target sequences it is useful to obtain a specific or related clone for hybridization (this is especially important during the development of a new PCR assay when many bands may be generated prior to changes in PCR parameters).

ACKNOWLEDGMENTS

We gratefully acknowledge Mrs. Irene Davis for her excellent technical assistance in numerous PCR experiments and for her tireless help and patience in the preparation of this manuscript. This work was supported in part by a USAID-NARP grant administered by OICD/USDA awarded to A. Hadidi.

REFERENCES

1. Saiki, R. K., Scharf, S., Faloona, F., Mullis, K. B., Horn, G. T., Erlich, H. A., and Arnheim, N., Enzymatic amplification of B-globin genomic sequences and restriction site analysis for diagnosis of sickle cell anemia, *Science,* 230, 1350, 1985.
2. Saiki, R. K., Gelfand, D. H., Stoffel, S., Scharf, S. J., Higuchi, R., Horn, G. T., Mullis, K. B., and Erlich, H. A., Primer-directed enzymatic amplification of DNA with a thermostable DNA polymerase, *Science,* 239, 487, 1988.
3. Mullis, K. B. and Faloona, F. A., Specific synthesis of DNA in vitro via a polymerase catalyzed chain reaction, *Methods Enzymol.,* 155, 335, 1987.
4. Mullis, K. B., The unusual origin of the polymerase chain reaction, *Sci. Am.,* April, 56, 1990.
5. Erlich, H. A., Polymerase chain reaction, *J. Clin. Immunol.,* 9, 437, 1989.
6. Erlich, H. A., Gelfand, D., and Sninsky, J. J., Recent advances in the polymerase chain reaction, *Science,* 252, 1643, 1991.
7. Kanematsu, S., Hibi, T., Hashimoto, J., and Tsruchizaki, T., Comparison of nonradioactive cDNA probes for detection of potato spindle tuber viroid by dot-blot hybridization assay, *J. Virol. Methods,* 35, 189, 1991.
8. Korschineck, I., Himmler, G., Sagl, R., Steinkellner, H., and Katinger, H. W. D., A PCR membrane spot assay for the detection of plum pox virus RNA in bark of infected trees, *J. Virol. Methods,* 31, 139, 1991.

9. Kemp, D. J., Smith, D. B., Foote, S. J., Samaras, N., and Peterson, M. G., Colorimetric detection of specific DNA segments amplified by polymerase chain reaction, *Proc. Natl. Acad. Sci. U.S.A.*, 86, 2423, 1989.
10. Hadidi, A. and Yang, X., Detection of pome fruit viroids by enzymatic cDNA amplification, *J. Virol. Methods,* 30, 261, 1990.
11. Hadidi, A., Terai, Y., Powell, C. A., Scott, S. E., Desvignes, J. C., Ibrahim, L. M., and Levy, L., Enzymatic cDNA amplification of hop stunt viroid variants from naturally infected fruit crops, *Acta Hortic.,* 309, 339, 1992.
12. Levy, L., Hadidi, A., and Garnsey, S. M., Reverse transcription-polymerase chain reaction assays for the rapid detection of citrus viroids using multiplex primer sets, *Proc. Int. Soc. Citriculture,* 2, 800, 1992.
13. Yang, X., Hadidi, A., and Garnsey, S. M., Enzymatic cDNA amplification of citrus exocortis and cachexia viroids from infected citrus hosts, *Phytopathology,* 82, 279, 1992.
14. Minafra, A., Shamloul, A. M., Hadidi, A., and Giunchedi, L., Detection and characterization of peach latent mosaic viroid by polymerase chain reaction (PCR) technology, *6th International Congress of Plant Pathology,* Montreal, Quebec, Canada, (Abstr.), 319, 1993.
15. Rezian, M. A., Krake, L. R., and Golino, D. A., Common identity of grapevine viroids from USA and Australia revealed by PCR analysis, *Int. Virol.,* 34, 38, 1992.
16. Hadidi, A., Yang, X., and Hammond, R. W., Polymerase chain reaction (PCR) amplification and nucleotide sequence of apple scar skin viroid, *Phytopathology,* 81 (Abstr.), 1246, 1991.
17. Yang, X., Hadidi, A., and Hammond, R. W., Nucleotide sequence of apple scar skin viroid reverse transcribed in host extracts and amplified by the polymerase chain reaction, *Acta Hortic.,* 309, 305, 1992.
18. Levy, L. and Hadidi, A., Development of a reverse transcription/polymerase chain reaction assay for the detection of plum pox potyvirus from microgram quantities of total nucleic acid, *Phytopathology,* 81 (Abstr.), 1154, 1991.
19. Wetzel, T., Candresse, T., Ravelonandro, M., and Dunez, J., A polymerase chain reaction assay adapted to plum pox potyvirus detection, *J. Virol. Methods,* 33, 355, 1991.
20. Wetzel, T., Candresse, T., Macquaire, G., Ravelonandro, M., and Dunez, J., A highly sensitive immunocapture polymerase chain reaction method for plum pox potyvirus detection, *J. Virol. Methods,* 39, 27, 1992.
21. Minafra, A., Hadidi, A., and Martelli, G. P., Detection of grapevine closterovirus A in infected grapevine tissue by reverse transcription-polymerase chain reaction, *Vitis,* 31, 221, 1992.
22. Minafra, A. and Hadidi, A., Detection of grapevine virus A in single mealybugs by immunocapture reverse transcription-polymerase chain reaction, *11th Meeting of the International Council for the Study of Viruses and Virus Diseases of the Grapevine (ICVG),* (Abstr.), 139, 1993.
23. Hadidi, A., Montasser, M. S., Levy, L., Goth, R. W., Converse, R. H., Madkour, M. A., and Skrzeckowski, L. J., Detection of potato leafroll and strawberry mild-yellow-edge luteoviruses by reverse transcription-polymerase chain reaction amplification, *Plant Dis.,* 77, 595, 1993.
24. Minafra, A., Hadidi, A., and Saldarelli, P., Sensitive immunocapture and multiplex reverse transcription-polymerase chain reaction for the detection of grapevine leafroll associated virus III and grapevine virus B, *11th Meeting of the International Council for the Study of Viruses and Virus Diseases of the Grapevine (ICVG),* (Abstr.), 137, 1993.
25. Ling, K., Hu, J., and Gonsalves, D., Molecular cloning and detection of grapevine leafroll virus by nucleic acid hybridization and polymerase chain reaction, *Phytopathology,* 83 (Abstr.), 245, 1993.
26. Smith, G. R., Vand de Velde, R., and Dale, J. L., PCR amplification of a specific double-stranded RNA region of Fiji disease virus from diseased sugarcane, *J. Virol Methods,* 39, 237, 1992.
27. Kohnen, P. D., Dougherty, W. G., and Hampton, R. O., Detection of pea seedborne mosaic potyvirus by sequence specific enzymatic amplification, *J. Virol. Methods,* 37, 253, 1992.
28. Frenkel, M. J., Jilka, J. M., Shukla, D. D., and Ward, C. W., Differentiation of potyviruses and their strains by hybridization with the 3′ non-coding region of the viral genome, *J. Virol. Methods,* 36, 51, 1993.
29. Jones, T. D., Buck, K. W., and Plumb, R. T., The detection of beet western yellows virus and beet mild yellowing virus in crop plants using the polymerase chain reaction, *J. Virol. Methods,* 35, 287, 1991.

30. Robinson, D. J., Detection of tobacco rattle virus by reverse transcription and polymerase chain reaction, *J. Virol. Methods,* 40, 57, 1992.
31. Borja, M. J. and Ponz, F., An appraisal of different methods for the detection of the walnut strain of cherry leafroll virus, *J. Virol. Methods,* 36, 78, 1992.
32. Olszewski, N. E. and Lockhart, B. E. L., Use of PCR technology for the detection and analysis of badnaviruses, *Phytopathology,* 82 (Abstr.), 1087, 1992.
33. Pennington, R. E., Sherwood, J. L., and Hunger, R. M., Use of a polymerase chain reaction (PCR)-based assay for wheat soilborne mosaic virus (WSBMV) to evaluate resistance in hard red winter wheat (*Triticum aestivum* L.), *Phytopathology,* 82 (Abstr.), 1147, 1992.
34. Roias, M. R., Gilbertson, R. L., Russell, D. R., and Maxwell, D. P., Detection and characterization of whitefly-transmitted geminiviruses by use of polymerase chain reaction (Abstr.), *Phytopathology,* 82, 1087, 1992.
35. Rybicki, E. P. and Hughes, F. L., Detection and typing of maize streak virus and other distantly related geminiviruses of grasses by polymerase chain reaction amplification of a conserved viral sequence, *J. Gen. Virol.,* 71, 2519, 1990.
36. Langeveld, S. A., Dore, J. M., Memelink, J., Derks, A. F. L. M., van der Vlugt, C. I. M., Asjes, C. J., and Bol, J. F., Indentification of potyvirus using the polymerase chain reaction with degenerate primers, *J. Gen. Virol.,* 72, 1531, 1991.
37. Robertson, N. L., French, R., and Gray, S. M., Use of group-specific primers and the polymerase chain reaction for the detection and identification of luteoviruses, *J. Gen. Virol.,* 72, 1473, 1991.
38. Hadidi, A., Hansen, A. J., Parish, C. L., and Yang, X., Scar skin and dapple apple viroids are seed-borne and persistent in infected apple tree, *Res. Virol.,* 142, 289, 1991.
39. Hadidi, A., Hansen, A. J., and Parish, C. L., Apple scar skin and dapple apple viroids are seed borne, *Acta Hortic.,* 309, 297, 1993.
40. Lopez-Moya, J. J., Gubero, J., Lopez-Abella, D., and Diaz-Ruiz, J. R., Detection of cauliflower mosaic virus (CaMV) in single aphids by the polymerase chain reaction (PCR), *J. Virol. Methods,* 37, 129, 1992.
41. Deng, S. and Hiruki, C., Genetic relatedness between two nonculturable mycoplasmalike organisms revealed by nucleic acid hybridization and polymerase chain reaction, *Phytopathology,* 81, 1475, 1991.
42. Schaff, D., Lee, I.-M., and Davis, R. E., Sensitive detection and identification of mycoplasma-like organisms in plants by polymerase chain reactions, *Biochem. Biophys. Res. Commun.,* 136, 1503, 1992.
43. Lee, I.-M., Davis, R. E., Chen, T.-A., Chiykowski, L. N., Fletcher, J., Hiruki, C., and Schaff, D. A., A genotype-based system for identification and classification of mycoplasma-like organisms (MLOs) in the aster yellows MLO strain cluster, *Phytopathology,* 82, 977, 1992.
44. Chen, K. H., Wu, S. Y., Chen, Y. O., Guo, Y. H., and Chen, T. A., Detection of mycoplasmalike organism associated with grapevine yellows by polymerase chain reactions, *Phytopathology,* 83 (Abstr.), 242, 1993.
45. Ahrens, U. and Seemüller, E., Detection of DNA of plant pathogenic mycoplasmalike organisms by a polymerase chain reaction that amplifies a sequence of the 16S rRNA gene, *Phytopathology,* 82, 828, 1992.
46. Henson, J. M., Goins, T., Grey, W., Mathre, D. E., and Elliott, M. L., Use of polymerase chain reaction to detect *Gaeumannomyces graminis* DNA in plants grown in artificially and naturally infected soil, *Phytopathology,* 83, 283, 1993.
47. Cubeta, M. A., Enchandi, E., Abernathy, T., and Vilgalys, R., Characterization of anastomosis groups of binucleate *Rhizoctonia* species using restriction analysis of an amplified ribosomal RNA gene, *Phytopathology,* 81, 1395, 1991.
48. Liu, Z. I. and Sinclair, J. B., Genetic diversity of *Rhizoctonia solani* anastomosis group 1, *Phytopathology,* 82 (Abstr.), 1102, 1992.
49. Liu, Z. I. and Sinclair, J. B., Genetic diversity of *Rhizoctonia solani* anastomosis group 2, *Phytopathology,* 82 (Abstr.), 1102, 1992.
50. Falkstein, K. F., Tooley, P. W., Goodwin, S. B., and Fry, W. E., Differentiation of group IV *Phytopthora* species by PCR amplification of nuclear ribosomal DNA internal transcribed spacer region 2, *Phytopathology,* 81 (Abstr.), 1157, 1991.
51. Schoelz, J., Ersek, T., and Erglish, J., Amplification of species-specific *Phytophthora* DNA sequences by PCR, *Phytopathology,* 82 (Abstr.), 1151, 1992.

52. Lee, S. B., White, T. J., and Taylor, J. W., Detection of *Phytophthora* species by oligonucleotide hybridization to amplified ribosomal DNA spacers, *Phytopathology,* 83, 177, 1993.
53. Shi, Y. and Leung, H., Amplification of ribosomal DNA from a small number of teliospores of the wheat bunt fungi, *Phytopathology,* 81 (Abstr.), 1190, 1991.
54. Chase, T. E., Otrosina, W. J., Spieth, P. T., and Cobb, F. W., Jr., Use of PCR to distinguish biological species within the *Heterobasidion annosum* complex, *Phytopathology,* 81 (Abstr.), 1190, 1991.
55. Wang, O., Lévesque, C. A., Punja, Z. K., and Rahe, J. E., Genetic variability among isolates of *Fusarium* spp. from tomato, *Phytopathology,* 82 (Abstr.), 1066, 1992.
56. Woodward, R. P. and Stewart, E. L., Variation in isozyme and PCR-amplified rDNA restriction fragments among *Fusarium graminearum* isolates, *Phytopathology,* 82 (Abstr.), 1164, 1992.
57. Egel, D. S. and Cotty, P. J., Restriction of DNA amplified from aflatoxin producing and nonproducing strains of *Aspergillus, Phytopathology,* 82 (Abstr.), 1080, 1992.
58. Keller, N. P., Owen, J., Bhatnagar, D., and Cleveland, T. E., Utility of PCR for diagnosis of *Aspergillus* species, *Phytopathology,* 82 (Abstr.), 1145, 1992.
59. Li, K.-N., German, T. L., and Rouse, D. I., A 30 base pair region in the mitochrondrial small subunit rRNA gene as a specific DNA probe for *Verticillium daliae, Phytopathology,* 82 (Abstr.), 1145, 1992.
60. Jones, M. J. and Dunkle, L. D., Races of *Cochliobolus carbonum* distinguished by PCR amplification, *Phytopathology,* 82 (Abstr.), 1086, 1992.
61. Delserone, L. M., Wilhelm, K. M., Merida, C. L., and Loria, R., Identification of DNA sequences specific to *Helminthoporium solani, Phytopathology,* 82 (Abstr.), 1143, 1992.
62. Vaillancourt, L. J. and Hanue, R. M., Isolates of *Glomerella (Colletotrichum)* from maize and sorghum are distinct species, *Phytopathology,* 82 (Abstr.), 1102, 1992.
63. Johanson, A. and Jeger, M. J., Use of PCR for detection of *Mycospharella fijiensis* and *M. musicola,* the fungi causing sigatoka loaf spots, in banana leaf tissue, *Phytopathology,* 82 (Abstr.), 1151, 1992.
64. Pope, D. D., Genetic similarity among races of *Ustilago hordei* and *Ustilago nigra, Phytopathology,* 82 (Abstr.), 1123, 1992.
65. Yoon, C. S. and Glawe, D. A., Correlation of randomly amplified DNA polymorphisms and stromatal type in *Hypoxylon truncatum, Phytopathology,* 82 (Abstr.), 1123, 1992.
66. Sevilla, M. O. and Pope, D. D., Genetic similarity within and among *Acremonium* species, *Phytopathology,* 82 (Abstr.), 1066, 1992.
67. Chen, W., Restriction fragment length polymorphisms in enzymatically amplified ribosomal DNAs of three heterothallic *Pythium* species, *Phytopathology,* 82, 1467, 1992.
68. Gardez, M., White, T. J., Fortin, J. A., Bruns, T. D., and Taylor, J. W., Identification of indigenous and introduced symbiotic fungi in ectomycorrhizae by amplification of nuclear and mitochondrial ribsomal DNA, *Can. J. Bot.,* 69, 180, 1990.
69. Dong, L-C., Sun, C-W., Thies, K. L., Luthe, D. S., and Gravez, C. H., Jr., Use of polymerase chain reaction to detect pathogenic strains of *Agrobacterium, Phytopathology,* 82, 434, 1992.
70. Leite, R. P., Jr., Stall, R. E., and Bonas, U., Identification of plant pathogenic *Xanthomonas* species pathovars based on amplified DNA fragments related to *HRP* genes of *Xanthomonas campestris* pv *vesicatoria, Phytopathology,* 82 (Abstr.), 1100, 1992.
71. Hartung, J. S., Daniel, J. F., and Pruvost, O. P., Amplification of the polymerase chain reaction (PCR) for the detection of *Xanthomonas campestris* pv *citri, Phytopathology,* 82 (Abstr.), 1106, 1992.
72. Gill, S., Mercier, J. P., Lemieux, C., Gagne, S., Brown, G. L., and Dion, P., Oligonucleotide probes to detect bacteria introduced into the environment, *Phytopathology,* 82 (Abstr.), 1179, 1992.
73. Williams, J. and Mills, D., Development of a DNA probe for the detection of *Corynebacterium sepedonicum, Phytopathology,* 82 (Abstr.), 1145, 1992.
74. Baum, T. J., Gresshoff, P. M., Lewis, S. A., and Dean, R. A., DNA amplification fingerprinting (DAF) of isolates of four common *Meloidogyne* species and their host races, *Phytopathology,* 82 (Abstr.), 1095, 1992.
75. Cenis, J. L., Identification of four major *Meliodogyne* spp. by random amplified polymorphic DNA (RADP-PCR), *Phytopathology,* 83, 76, 1993.
76. Meyer, R. J., Analysis of PCR-amplified DNA fragments from Radopholus species. *J. Nematol.,* 23(Abs.), 540, 1991.

77. Laimer da Cämara Machado, M., da Cämara Machado, A., Hanzer, V., Weiss, H., Regner, F., Steinkellner, H., Mattanovich, D., Plail, R., Knapp, E., Kalthoff, B., and Katinger, H., Regeneration of transgenic plants of *Prunus Armeniaca* containing the coat protein gene of plum pox virus, *Plant Cell Rep.*, 11, 25, 1992.
78. McGarvey, P. and Kaper, J. M., A simple and rapid method for screening transgenic plants using the PCR, *BioTechniques*, 11, 428, 1991.
79. Scorza, R., Levy, L., Damsteegt, V., Yepes, M., Cordts, J., Hadidi, A., Slighton, J., and Gonsalves, D., Plum pox virus protection in transgenic plum expressing the papaya rainspot virus coat protein gene, *6th International Congress of Plant Pathology*, Canada, (Abstr.), 192, 1993.
80. Hayes, R. J. and Buck, K. W., Infectious cucumber mosaic virus RNA transcribed *in vitro* from clones obtained from cDNA amplified using the polymerase chain reaction, *J. Gen. Virol.*, 71, 2503, 1990.
81. Zhu, S. F. and Hadidi, A., Direct nucleotide sequencing of full length PCR products of pome fruit viroids from nucleic acid extracts of infected tissue, *6th International Congress of Plant Pathology*, Montreal, Quebec, Canada (Abstr.), 319, 1993.
82. Levy, L. and Hadidi, A., Direct nucleotide sequencing of PCR-amplified DNAs of the closely related citrus viroids IIa and IIb (cachexia), in *Proc. of the 12th Int'l Org. of Citrus Virol.*, 180, 1994.
83. Puchta, H., Luckinger, R., Yang, X., Hadidi, A., and Sänger, H. L., Nucleotide sequences and secondary structure of apple scar skin viroid (ASSVd) from China, *Plant Mol. Biol.*, 14, 1065, 1990.
84. Hashimoto, J. and Koganezawa, H., Nucleotide sequence and secondary structure of apple scar skin viroid, *Nucleic Acids Res.*, 15, 7045, 1987.
85. Puchta, H. and Sänger, H. L., Sequence analysis of minute amounts of viroid RNA using the polymerase chain reaction (PCR), *Arch Virol.*, 106, 335, 1989.
86. Owens, R. A., Khurana, S. M. P., Smith, D. R., Singh, M. N., and Gary, I. D., A new mild strain of potato spindle tuber viroid isolated from wild *Solanum* spp. in India, *Plant Dis.*, 76, 527, 1992.
87. Maniatis, T., Fritsch, E. F., and Sambrook, J., *Molecular Cloning: a Laboratory Manual*, Cold Spring Harbor Laboratory, Cold Spring Harbor, NY, 1982.
88. Nakhla, M. K., Rojaz, M. R., McLaughlin, W. M., Wagman, J., and Maxwell, D. P., Molecular characterization of tomato yellow leaf curl from Egypt, *Plant Dis.*, 76, 538, 1992.
89. Erlich, H. A., *PCR Technology: Principal for DNA Amplification*, Stockton Press, New York, 1989.
90. Innis, M. A., Gelfand, D. H., Sninsky, J. J., and White, *PCR Protocols: a Guide to Methods and Applications*, Academic Press, San Diego, 1989.
91. Chamberlain, J. S., Gibbse, R. A., Rainer, J. E., Nguyen, P. N., and Caskey, C. T., Deletion screening of Duschenne muscular dystrophy locus via multiplex DNA amplification, *Nucleic Acids Res.*, 16, 11141, 1988.
92. Ruano, G., Brash, D. E., and Kid, K. K., PCR: the first few cycles, *Amplifications*, 7, 1, 1991.
93. Jansen, R. W., Gunter, S., and Lemon, S. M., Molecular epidemiology of human hepatitis. A virus defined by an antigen-capture polymerase chain reaction method, *Proc. Natl. Acad. Sci. U.S.A.*, 87, 2867, 1990.
94. Wahlberg, J., Lundberg, J., Hultman, T., and Uhlen M., General colorimetric method for DNA diagnostic allowing direct solid-phase genomic sequencing of the positive samples, *Proc. Natl. Acad. Sci. U.S.A.*, 87, 6569, 1990.
95. Rapolee, D. A., Mark, D., Banda, M. J., and Werb, Z., Wound macrophages express TGF-α and other growth factors *in vivo:* analysis by mRNA phenotyping, *Science*, 241, 708, 1988.
96. Linz, U., Thermocycler temperature variation invalidates PCR results, *BioTechniques*, 9, 286, 1990.
97. Gilliland, G., Perrin, S., Blanchard, K., and Bunn, H. F., Analysis of cytokine mRNA and DNA: detection and quantitation by competitive polymerase chain reaction. *Proc. Natl. Acad. Sci. U.S.A.*, 87, 2725, 1990.
98. Chelly, J., Montarras, D., Pinset, C., Berwald-Netter, Y., Kaplan, J.-C., and Kahn, A., Quantitative estimation of minor mRNAs by cDNA-polymerase chain reaction, *Eur. J. Biochem.*, 187, 691, 1990.
99. Wang, A., Doyle, M. V., and Mark, D., Quantitation of mRNA by the polymerase chain reaction, *Proc. Natl. Acad. Sci. U.S.A.*, 86, 9717, 1989.
100. Henco, K. and Heibey, M., Quantitative PCR: the determination of template copy numbers by temperature gradient gel electrophoresis (TGGE), *Nucleic Acids Res.*, 18, 6733, 1990.

101. Becker-Andre, M., and Hahlbrock, K., Absolute mRNA quantification using the polymerase chain reaction (PCR). A novel approach by PCR aided transcript titration assay (PATTY), *Nucleic Acids Res.*, 17, 9437, 1989.
102. Cheung, R. K., Hui, M. F., Dosch, H.-M., and Ewart, T. E., Detection of viral sequences by internally calibrated gene amplification, *BioTechniques,* 14, 785, 1993.
103. Gilliland, G. Perrin, S., and Bunn, H. F., Competitive PCR for quantitation of mRNA, in *PCR Protocols: a Guide to Methods and Applications,* Innis, M. A., Gelfand, D. H., Sninsky, J. J., and White, T. J., Eds., Academic Press, San Diego, 1989, 60.
104. Botstein, D., White, R. L., Skolnick, M. H., and Davis, R. W., Construction of a genetic map in man using restriction fragment length polymorphisms. *Am. J. Hum. Genet.,* 32, 314, 1980.
105. Williams, J. G. K., Kubelik, A. R., Livat, K. L., Rafalski, K. A., and Tingey, S. V., DNA polymorphisms amplified by arbitrary primers are useful as genetic markers, *Nucleic Acids Res.,* 18, 6531, 1990.
106. Murray, M. G. and Thompson, W. F., Rapid isolation of high molecular weight plant DNA, *Nucleic Acids Res.,* 8, 4321, 1980.
107. Gawel, N. J. and Jarret, R. L., A modified CTAB DNA extraction procedure for *Musa* and *Ipomoea, Plant Mol. Biol. Rep.,* 9, 262, 1991.
108. Lassner, M. W., Peterson, P., and Yoder, J. I., Simultaneous amplification of multiple DNA fragments by polymerase chain reaction in the analysis of transgenic plants and their progeny, *Plant Mol. Biol. Rep.,* 7, 116, 1989.
109. Munthali, M., Ford-Lloyd, B. V., and Newbury, H. J., The random amplification of polymorphic DNA for fingerprinting plants, *PCR Methods Appl.,* 1, 274, 1992.
110. Akopyanz, N., Bukanov, N. O., Westblom, T. U., Kresovich, S., and Berg, D. E., DNA diversity among clinical isolates of *Helibacter pylori* detected by PCR-based RAPD fingerprinting, *Nucleic Acids Res.,* 20, 5137, 1992.
111. Welsh, J. and McClelland, M., Fingerprinting genomes using PCR with arbitrary primers, *Nucleic Acids Res.,* 18, 7213, 1990.
112. Weeden, N. F., Timmerman, G. M., Hemmat, M., Kneen, B. E., and Lodhi, M. A., Inheritance and reliability of RAPD markers, in *Proceedings of the Symposium: Applications of RAPD Technology to Plant Breeding,* 1992, 12.
113. Kessell, R. V., Paran, I., and Mitchelmore, R. W., Efficient mapping of specifically targeted genomic regions and the tagging of these regions with reliable PCR-based genetic markers, in *Proceedings of the Symposium: Applications of RAPD Technology to Plant Breeding,* 1992, 31.
114. Cenis, J. L., Rapid extraction of fungal DNA for PCR amplification, *Nucleic Acids Res.,* 20, 2380, 1992.
115. Yu, K. and Pauls, K. P., Optimization of PCR program for RAPD analysis, *Nucleic Acids Res.,* 20, 2606, 1992.
116. Caetano-Anolles, G. B., Bassam, B. J., and Gresshoff, P. M., DNA amplification fingerprinting using very short arbitrary oligonucleotide primers, *Bio/Technology,* 9, 553, 1991.
117. Michelmore, R. W., Paran, I. R., and Kessell, R., Identification of markers linked to disease resistance genes by bulked segregant analysis: a rapid method to detect markers in specific genomic regions using segregating populations, *Proc. Natl. Acad. Sci. U.S.A.,* 88, 9828, 1991.
118. Ochman, H., Gerber, A. S., and Hartl, D. L., Genetic applications of an inverse polymerase chain reaction, *Genetics,* 120, 621, 1988.
119. Silver, J. and Keerikatte, V., Novel use of polymerase chain reaction to amplify cellular DNA adjacent to an integrated provirus. *J. Virol.,* 63, 1924, 1989.
120. Triglia, T., Peterson, M. G., and Kemp, D. J., A procedure for *in vitro* amplification of DNA segments that lie outside the boundaries of known sequences, *Nucleic Acids Res.,* 16, 8186, 1988.
121. Loh, E. Y., Elliot, J. F., Cwirla, S., Lanier, L., and Davis, M. M., Polymerase chain reaction with single-sided specificity: analysis of T cell receptor σ chain, *Science,* 243, 217, 1989.
122. Frohman, M., Dush, M. K., and Martin, G. M., Rapid production of full length cDNAs from rare transcripts: amplification using a single gene-specific oligonuceotide primer, *Proc. Natl. Acad. Sci. U.S.A.,* 85, 8998, 1988.
123. Belyavsky, A., Vinogradova, T., and Rajewsky, K., PCR-based cDNA library construction: general cDNA libraries at the level of a few cells, *Nucleic Acids Res.,* 17, 2919, 1989.
124. Harvey, R. J. and Darlison, M. G., Random-primed cDNA synthesis facilitates the solutions of multiple 5′-cDNA ends by RACE, *Nucleic Acids Res.,* 19, 4002, 1991.

125. Dumas, J. B., Edwards, M., Delort, J., and Mallet, J., Oligodeoxyribonucleotide ligation to single-stranded cDNAs: a new tool for cloning 5′ ends of mRNAs and for constructing cDNA libraries by *in vitro* amplification, *Nucleic Acids Res.,* 19, 5227, 1991.

FURTHER READING

Erlich, H. A., *PCR Technology: Principal for DNA Amplification,* Stockton Press, New York, 1989.

Gurr, S. J., McPherson, F. J., and Bowles, D. J., *Molecular Plant Pathology: a Practical Approach,* Oxford University Press, Oxford, 1992.

Innis, M. A., Gelfand, D. H., Sninsky, J. J., and White, *PCR Protocols: a Guide to Methods and Applications,* Academic Press, San Diego, 1989.

Mullis, K. B., The unusual origin of the polymerase chain reaction, *Sci. Am.,* April, 56, 1990.

Mullis, K. B. and Faloona, F. A., Specific synthesis of DNA in vitro via a polymerase catalyzed chain reaction, *Methods Enzymol.,* 155, 335, 1987.

Steffan, R. J. and Atlas, R. M., Polymerase chain reaction: application to environmental microbiology, *Annu. Rev. Microbiol.,* 45, 137, 1991.

Chapter 14

Gene Transfer to Higher Plants

Ralf R. Mendel and Robert M. Hänsch

TABLE OF CONTENTS

I. INTRODUCTION

Plant breeding, as well as the newer cell genetic techniques, are severely limited by species barriers, i.e., the range of organisms between which genes can be transferred. Here, plant genetic transformation provides a major new approach capable of solving this handicap without replacing the previous technologies, but instead complementing them. The strategy of this approach is (1) to identify, isolate, and characterize DNA sequences of diverse phylogenetic origins; (2) to modify these sequences according to the aims of the project; and (3) to stably introduce them in target plants where they should be correctly expressed to and inherited by the seed progeny.

In this chapter we will review the different experimental possibilities of delivering genes to recipient plants. It will become obvious that by far not all plants are presently amenable to genetic transformation. Also, one has to bear in mind that there is a great experimental difference between the principal possibility to transform a particular plant species and the existence of a routine transformation system for such a plant. In particular, several seed legumes and the small grain cereals are still difficult to transform nowadays, and much effort is being directed towards improving this situation worldwide. There are three principal ways to transfer a foreign gene into a plant:

- Agrobacterium-mediated transformation
- direct DNA transfer
- virus-mediated transformation

However, before describing these systems in detail it is necessary to discuss and define terms of general importance for this chapter.

In most of the cases **stable transformation** of the target plant is the aim of gene transfer, i.e., the introduction of one or more intact copies of one or more foreign genes into the nuclear genome of the recipient where the foreign gene is stably maintained through cell divisions and is being correctly expressed. Two forms of stable transformants are possible: (1) transgenic regenerated plants, and, in cases where plant regeneration from cell cultures is problematic, also (2) transgenic cell cultures. However, in basic research there are numerous scientific problems that can be studied already on the level of transgenic cell cultures without the urgent necessity of having regenerated plants available, e.g., the study of mechanisms for the recognition and integration of foreign genes into plant genomes, the study of physiological questions connected with housekeeping genes involved in primary metabolic steps, and questions of virus resistance.

Transient transformation is another approach of gene transfer that is frequently used among higher plants. In contrast to stable transformation, assays for **transient gene expression** allow the very rapid

0-87371-877-1/95/$0.00+$.50

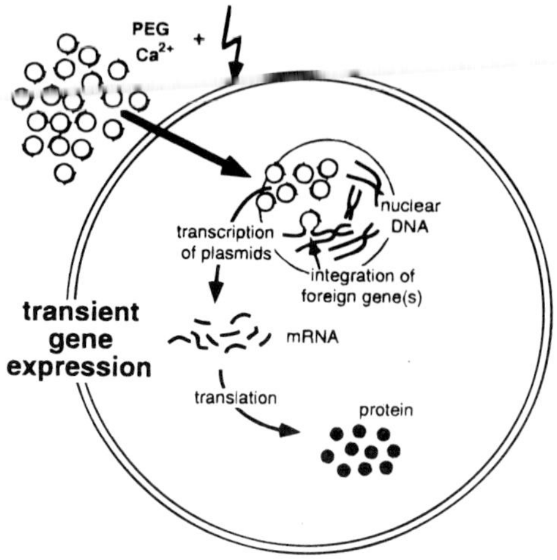

Figure 1 Transient gene expression in a plant cell.

evaluation of gene expression of a given construct within days. In particular, this approach is used in those cases where stable transformation is difficult or not yet possible. In these assays the DNA is introduced via direct DNA transfer into the cell in higher copy number and remains largely extrachromosomally located in the nucleus where it is lost over a period of 1 to 2 weeks. However, during its transient existence this introduced DNA remains transcriptionally active and gene expression can be monitored on the mRNA level as well as on the protein level (Figure 1). There are several advantages of this kind of assay: they are rapid—usually 2 d after DNA application expression tests are performed. They are independent of so-called position effects that occur with stable transformants where the host genomic region into which the foreign DNA had been introduced can positively or negatively interfere with the expression of the transferred gene. And finally, transient gene expression is always a mean value of a multitude of single expression events.

The process of stable transformation is a rare event with transformation frequencies ranging from as low as 10^{-5} up to 10^{-2}, depending on the transformation method used. Therefore, **selectable markers** have to be cotransferred together with the gene of interest in order to give a selectable advantage to the few transformed cells. Marker genes usually confer resistance to antibiotics or to herbicides (Table 1) and are driven by strong constitutive promoters. Table 1 shows also the group of so-called **reporter genes** that are used in all those cases where the experimentator is interested either in an easily scorable enzyme activity hitherto not occurring in the plant or plant cell, e.g., in order to monitor the activity of a particular promoter that was placed in front of the coding sequence of the reporter gene, or the experimentator wants to assay for organ- and tissue-specific gene expression in the transformed plant and utilizes a reporter gene the product of which can be histochemically detected.

Table 1 **List of selectable markers and screenable reporter genes commonly used for plant transformation**

Gene coding for	Origin	Confers resistance to
Markers		
Neomycin phosphotransferase II (NPT)	Tn5 (Tn903)	Kanamycin, neomycin, G418
Chloramphenicol acetyltransferase (CAT)	*E. coli*	Chloramphenicol
Hygromycin phosphotransferase (HPT)	*E. coli*	Hygromycin
Streptomycin phosphotransferase (SPT)	Tn5	Streptomycin
Dihydrofolate reductase (DHFR)	*E. coli,* mouse	Methotrexat
Phosphinothricin acetyltransferase (PAT)	*S. hygrocopicus*	Phosphinothricin
Reporters		
β-Glucuronidase (GUS)	*E. coli*	—
Luciferase (LUX)	*Vibrio fischeri, Photinus pyralis*	—

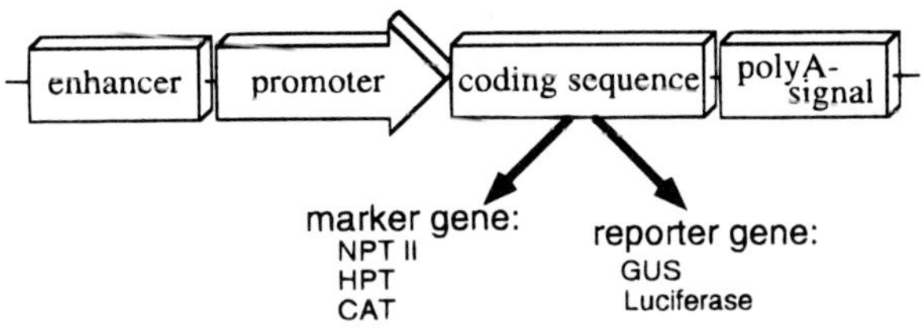

Figure 2 Sequence components making up a chimeric gene.

In the majority of the cases, the selectable marker genes, the reporter genes and the genes of interest were modified prior to transfer according to the needs of the project, e.g., they were placed under control of a different promoter or a transit sequence was introduced targeting the protein to a particular organelle in the recipient cell, etc. Those genes are called chimeric genes, since they consist of components derived from different origins (Figure 2).

II. PRINCIPLES AND METHODOLOGY

A. AGROBACTERIUM-MEDIATED TRANSFORMATION

The *Agrobacterium* system was historically the first successful plant transformation system, marking the breakthrough in plant genetic engineering in 1983. The Gram-negative soil bacterium *Agrobacterium tumefaciens* was known to cause crown gall disease in plants, i.e., the growth of tumorous grown galls at the site of infection. As was found by the ingenious studies of Schell and Van Montagu in the 1970s and beginning of the 1980s, this bacterium is able to transfer a portion of its megaplasmid (= Ti-plasmid) into the plant cell where the bacterial DNA stably integrates into nuclear DNA, thus transforming the recipient cell into a plant tumor cell. The transferred DNA (T-DNA; approximately 20 to 25 kb) consists of genes with promoters that can be read by the plant transcription/translation system. This DNA drives two processes: (1) high endogenous synthesis of auxins and cytokinins, thus causing a nondirected proliferation of the infected plant cells (= "crown gall"), and (2) synthesis of tumor-specific metabolites (opines) that serve as C- and N-source for the agrobacteria. Schell coined the term "genetic colonization" for this kind of parasitism where the parasite genetically transforms its host in order to achieve an exclusive access to the host's photosynthesis products.

How do we exploit this natural system of gene transfer into a plant cell? The following scenario can be laid out: (1) insertion of the gene of interest into the *Agrobacterium* T-DNA, (2) transformation of the plant cell with the modified T-DNA containing the new gene, (3) regeneration of a whole plant from the transformed cell. In order to achieve this goal, one has to eliminate the tumor-inducing genes on the T-DNA or one ends up with a crown gall callus that does not regenerate into a fertile plant. Also, the gene for opine synthesis should be removed, since this metabolite is of no interest in a plant that contains our gene of interest. Step by step, these experimental requirements were fulfilled. This was made possible by the circumstance that only the 25-bp sequences at both ends of the T-DNA are necessary for the *Agrobacterium* system being responsible for transferring this piece of DNA into the plant cell. This system consists of the interacting gene products of the *Agrobacterium* virulence genes (*vir*), comprising 8 operons with more than 20 open reading frames. One protein (VirD2) recognizes the border sequences, nicks one DNA strand, and directs a DNA polymerase to synthesize a single-stranded copy of the T-DNA. During this process VirD2 becomes covalently linked to the 5′ end of the single-stranded T-DNA copy and the remaining portion of this T-DNA copy is covered by many molecules of the VirE2-protein, thus forming a rod-like shaped protein-DNA complex ("T-complex") that is transferred into the wounded plant cell (Figure 3). VirD2 possesses strong nuclear-targeting motifs in its amino acid sequence and pilots the T-complex through the nuclear pores into the nucleus where the T-DNA is randomly integrated into the DNA of the recipient cell (for a recent review see Hooykaas and Schilperoort).[1]

Since only the 25-bp border sequences are necessary for this process it is possible to transfer virtually every gene that was placed between the borders into the plant cell without converting the recipient cell into a tumor cell. Another obstacle that had to be overcome was the size of the Ti-plasmid (about 200 kb) that made it difficult to manipulate this plasmid *in vitro*. This problem was solved by constructing small vectors into which the gene of interest has to be introduced. Nowadays the following system is

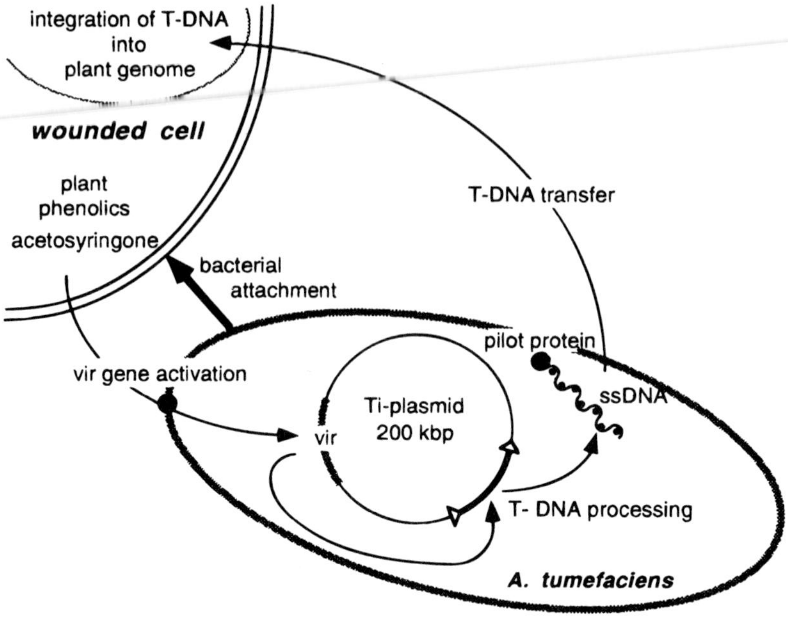

Figure 3 Schematic drawing of the natural process that is used by *Agrobacterium* for transferring T-DNA into a plant cell.

the most widely utilized one: a vector molecule of suitable size (5 to 10 kb) is made up of the following portions:

- the right and left border fragment in order to achieve the transfer into the plant cell
- between the borders the gene(s) of interest and the plant-selectable marker
- outside the borders an origin of replication for *Escherichia coli* and an origin of replication for *Agrobacterium*
- a marker for selection in *E. coli* and a marker for selection in *Agrobacterium*

The gene of interest has to be cloned into this vector between the left and right borders. This work is being performed in *E. coli* (therefore, we need the selectable marker for *E. coli*). Subsequently the vector is transferred into *Agrobacterium,* either by conjugational transfer or by direct transformation (for this we need the selectable marker for *Agrobacterium*). The *Agrobacterium* strains used for this purpose are specially designed ones: they lack the T-DNA, i.e., they possess only a Ti-plasmid where the T-DNA has been deleted. After introducing the vector from *E. coli,* the products of the *vir* genes of *Agrobacterium* are able to recognize *in trans* the 25-bp border sequences on the vector and initiate the transfer process into the plant cell. Since the border fragments and the *vir* genes are localized on separate plasmids, this system is being called a binary vector system (Figure 4).

Up to this step all experimental work was bacterial work. Now the engineered *Agrobacterium* has to be brought into contact with the plant cell to be transformed. There are basically two methods of transforming plant cells, both involving cell and tissue culture work. The aim of both methods is to transform a target cell that is able to regenerate into a whole plant after transformation. Thus the choice of the target to be infected depends on the availability of a plant regeneration system for the particular plant species in use. The one method uses protoplasts as the transformation target for *Agrobacterium.* It has to be admitted, however, that due to the intrinsic difficulties of using protoplasts and the high experimental skills needed, this system is less and less frequently used for *Agrobacterium*-mediated transformation. The second method of transformation uses organized explants with high regenerative potential as targets for the *Agrobacterium* infection: leaf pieces, stems, and root pieces, depending on the plant species used. As an example, the leaf-disk transformation will be discussed in greater detail.[2]

Leaves of the plant to be transformed are cut into pieces of about 5×5 mm, or disks of a diameter of about 5 mm are punched out. These explants are cocultured with the manipulated *Agrobacterium* harboring the binary vector. The cocultivation time ranges from minutes to 2 to 3 d. Thereafter, the

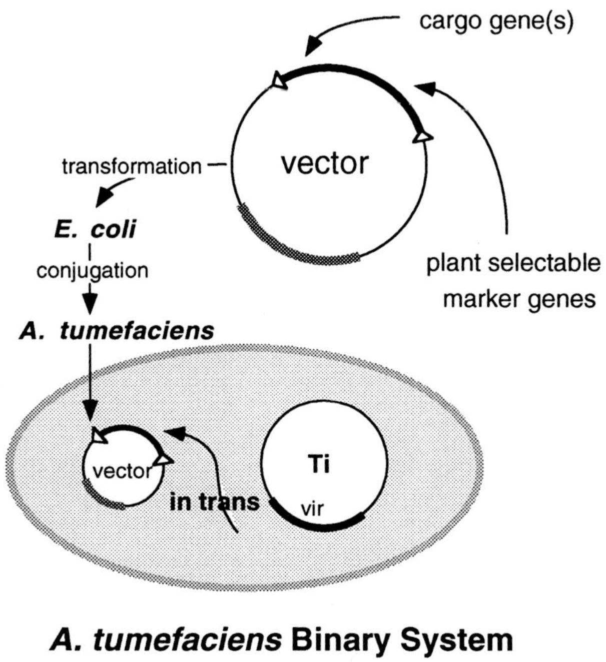

Figure 4 Binary vector system of *Agrobacterium*-mediated gene transfer.

explants are transferred onto a solid medium containing, in addition to the standard ingredients, the following three constituents:

- suitable antibiotics (preferably cefotaxime) for killing off the *Agrobacteria*
- hormones for stimulating the plant regeneration processes
- agents (mostly antibiotics) in order to select for transformed cells, i.e., only those cells should be able to survive this treatment that had been previously transformed with the T-DNA containing not only the gene of interest, but also a plant-selectable marker

The plant-selectable marker is of great importance, since the transformation frequency is usually not that high that one could screen for transformants. In very effective systems, e.g., with solanaceaen species, frequencies of 1 to 10% can be obtained; in most cases, however, the frequencies are lower than 1%. Depending on the plant species, within weeks until several months shoots with resistance to the selectable marker grow out of the explants. These shoots are removed and rooted on a separate (mostly hormone-free) medium. The rooting in the presence of the selectable marker is another—very potent—criterion for differentiating between real transformants and so called "escapes" that are not transformed.

Another member of the Agrobacteriaceae, *A. rhizogenes,* causing hairy root disease has also been exploited as a gene transfer system for higher plants using principally the same strategy as outlined for *A. tumefaciens.*[3]

In order to demonstrate that a given plant had been transformed, the experimentator has to present at least three evidences:

1. The physical presence of the new gene in the recipient plant has to be proven (usually shown by Southern DNA hybridization of genomic DNA with the new gene as probe). Further, the integration of the new gene into plant genomic DNA has to be demonstrated—this is being done by demonstrating hybridization of the probe with nondigested, high-molecular weight plant genomic DNA. Utilizing the *Agrobacterium* system, in most cases one or two copies of the new gene are being stably introduced into plant DNA.
2. The plant should obtain a new, measurable phenotype by the introduced gene. This can be achieved by measuring the enzymatic activity of the product of the new gene or by using immunodetection methods. In addition, morphological or developmental changes, connected with the expression of the new gene, can be monitored.

Table 2 **Choice of higher plant species transformed by *Agrobacterium***

Dicots	Monocots
Nicotiana spp.	*Asparagus officinalis*
Lycopersicum esulentum	*Gladiolus* spp.
Solanum tuberosum	*Dioscorea bulbifera*
Petunia hybrida	*Chlorophytum capense*
Daucus carota	*Narcissus*
Armoracia lapathifolia	
Kalanchoe daigremontianum	
Arabidopsis thaliana	
Brassica spp.	
Cucumis sativus	
Lactuca sativa	
Medicago sativa	
Trifolium spp.	
Pisum sativum	
Glycine max	
Vigna unguiculata	
Lotus corniculatus	
Gossypium hirsutum	
Helianthus annuus	
Linum usitatissimum	
Populus spp.	
Pseudotsuga menziesii	
Beta vulgaris	
Prunus spp.	

3. The new gene should be transferable to the seed progeny of the plant and its segregation in the next generation(s) has to be determined.

Nowadays, due to its ease and the good frequencies, the *Agrobacterium* system is a routine transformation system for many plant species. However, this system has also a great drawback, namely the *Agrobacterium* host range is limited to dicotyledonous plants and very few monocot plants (e.g., *Asparagus* and *Chlorophytum*). *Agrobacterium* does not measurably infect the economically very important cereals like maize, rice, and wheat. Thus for these plant species another gene-transfer system had to be developed. Also, among the dicots there are species that are difficult to transform, e.g., the grain legumes. Table 2 gives an overview about plant species being transformed by *Agrobacterium.*

B. DIRECT DNA TRANSFER

In parallel to the development of the *Agrobacterium* system, attempts were made to introduce DNA directly into protoplasts without the detour via *Agrobacterium* as a natural gene-transfer mediator (Figure 5). These attempts were fueled by the wish to establish a gene-transfer system applicable to all plant species whether dicot or monocot. Of course, such a system is strongly dependent on the existence of an already established protoplast regeneration system—and efficient tissue culture systems are not available for all species. Here, again, the cereals caused great problems that have to be solved.

The physical delivery of DNA into plant cells can be achieved in different ways. These systems will be reviewed and evaluated in the following sections.

1. DNA Transfer into Protoplasts

The addition of plasmid DNA to protoplasts does not lead to DNA uptake. Only the subsequent addition of polyethylene glycol (PEG)—a compound that was frequently used for protoplast fusion—initiates this process. In 1984, 1 year after the first *Agrobacterium*-mediated transformation, the first successful direct transformation of tobacco protoplasts was reported using neomycin phosphotransferase II (NPT) as a selectable marker.[4] The protoplasts were induced to regenerate cell walls and to divide in the presence of the selective agent, thus forming microcolonies resistant to kanamycin. Small colonies were

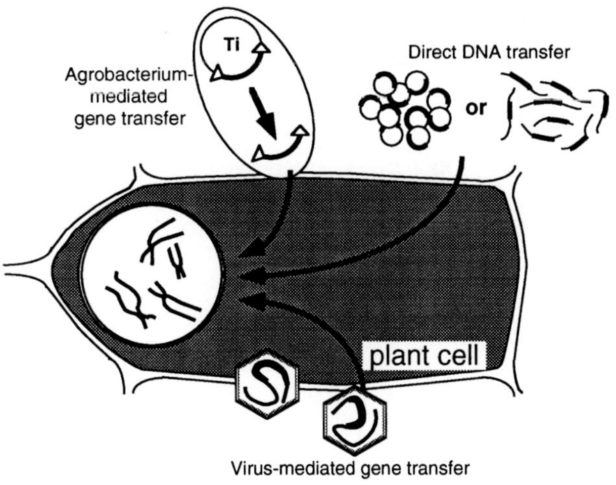

Figure 5 Schematic drawing of the three general systems of gene transfer in higher plants: *Agrobacterium*-mediated gene transfer, direct DNA transfer, and virus-mediated gene transfer.

picked and, after transfer to fresh medium, grown into larger callus colonies that represented sufficient cell material to assay for NPT activity and to isolate genomic DNA in order to perform a Southern hybridization. Characterized colonies were eventually grown up into a whole plant for further studies and for obtaining the seed progeny. The inheritance of the kanamycin-resistant phenotype could be shown and the new gene segregated in a Mendelian manner. Thus the first part of this approach—the physical introduction of DNA—differs from the *Agrobacterium* approach, whereas the second part, the selection, characterization, and regeneration of transformed plants, is similar. This is true for all those approaches that involve cell culture techniques for selecting and regenerating a single transformed cell.

Since this transformation system depends on the efficiency of the protoplast regeneration system also, the transformation frequencies vary strongly, ranging from as low as 10^{-6} up to 10^{-2}. In addition to PEG, electroporation had been introduced as another method to increase the transformation rate yet further. Here, after addition of DNA to the protoplasts the cells are subjected to a very short electrical impulse in the millisecond or microsecond range with a field strength of 500 to 3000 V/cm, thus creating transient pores in the plasma membrane through which the DNA can enter the cell.[5] Electroporation is being used instead of PEG or accompanying the PEG treatment. However, it was found that for several plant species PEG alone was more gentle to the fragile protoplasts and caused not that much damage as compared to electroporation.

The protoplast approach had been successfully applied to stably transform several dicot and also monocot species. Among the cereals, rice and maize have been stably transformed using the protoplast way, and green fertile plants have been obtained that transmitted the introduced new gene(s) to the progeny. However, with the small grain cereals wheat and barley, this approach yielded only stably transformed callus cultures, but no plants, since unforeseen difficulties occurred during the regeneration of the transformed callus lines, although protoplast regeneration of wheat and barley (in the absence of DNA) has been reported previously. Thus the bottleneck of direct DNA transfer into protoplast is not the transformation *per se,* but rather problems with or loss of the regenerative potential of the recipient cells.[6] This drawback is of no relevance if one can answer questions already on the level of transgenic callus lines without the necessity of having regenerated whole plants.

Protoplasts have also been the first system for demonstrating transient gene expression in higher plants. In order to test a given gene construct for transient expression, protoplasts have to be isolated in large amounts and should be uniform and vital. These requirements are met by leaves as well as by suspension cell cultures as protoplast sources. The cells are harvested and extracted 1 to 3 d after DNA transfer, and gene expression is measured either by determining the enzymatic activity of the introduced gene or by immunodetection of the protein, or by measuring the mRNA level.[7] It is noteworthy to mention that transient gene expression is not bound to cell division.

When using transient gene expression assays as a pretest for predicting the performance of a given gene after stable transformation, one should be very cautious with too far-reaching interpretations of the results, since protoplasts are highly stressed cells that do not always accurately reflect the function of the tissue from which they are derived. Thus the test for organ-specific gene expression in a transient

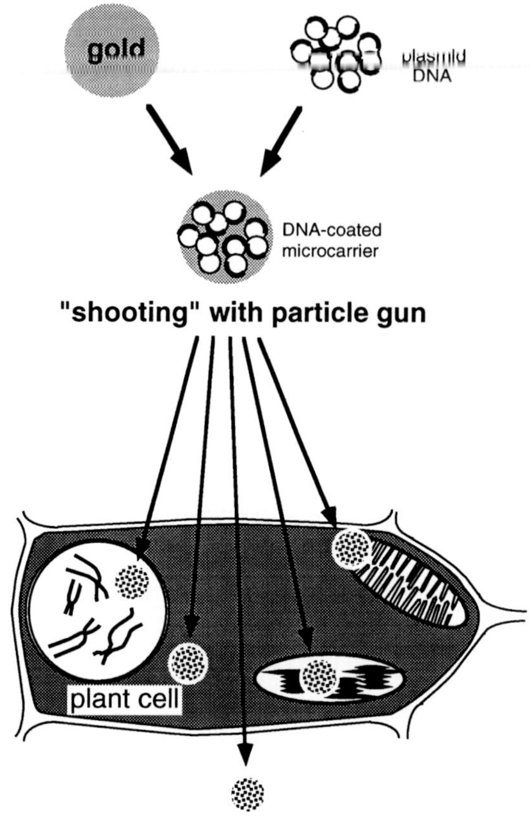

Figure 6 Direct DNA transfer mediated by high-velocity microprojectiles. DNA-coated metal microprojectiles are accelerated by the particle gun device and are shot into target cells where the projectiles can hit different cellular compartments.

protoplast assay might give—but will not necessarily give—accurate data, whereas reports about correct regulation of gene expression by hormones had been published.

Since 1984, transient expression assays have been widely used for rapidly checking the expressibility of new chimeric constructs. Moreover, they have also been successfully applied as a rapid means for testing for the infectibility of an engineered plant by a given virus or virus nucleic acid.

2. DNA Transfer by Microprojectiles

In order to overcome the obstacles of handling and regenerating protoplasts, in particular for those plant species that were previously recalcitrant to gene transfer attempts, another approach of direct DNA delivery had been introduced in 1987, the transfer of DNA by high-velocity microprojectiles (recently reviewed).[8,9] This method uses minute metal balls coated with DNA that are shot directly into the living target cells. Metal microprojectiles (typically high-density metals like tungsten or gold) with a diameter of 1 to 2 μm are coated with plasmid DNA and accelerated to high velocities so that they are able to penetrate intact plant cell walls. For this process the term biolistic (= biological ballistics) was coined. The biolistic gun apparatus is commercially available, but self-built versions are also functional.

Depending on the velocity, the microprojectiles can either get stuck in the first cell layer of the target explant or are able to penetrate several cell layers deep into the target tissue (Figure 6). After entering the cell, the microparticles can hit the plastids, the mitochondria, or the nucleus, or they may be found in the cytoplasm. Thereafter the DNA coated on the surface of the particles is released and the subsequent step is the same as observed with all the other gene-transfer methods: random integration into the cell genome.

Numerous factors are known to influence the biolistic DNA delivery process, e.g., the size, number, and velocity of the microprojectiles, the type of DNA coating, the size and type of target cells, as well as the physiological state of the recipients. Thus a thorough fine tuning is necessary for every new target type to be treated.

The most widespread application of this approach was, until recently, the test for transient gene expression using suspension culture cells as targets. Figure 7 shows an example of such an experiment using a maize cell culture as target and the gene for β-glucuronidase as reporter gene.

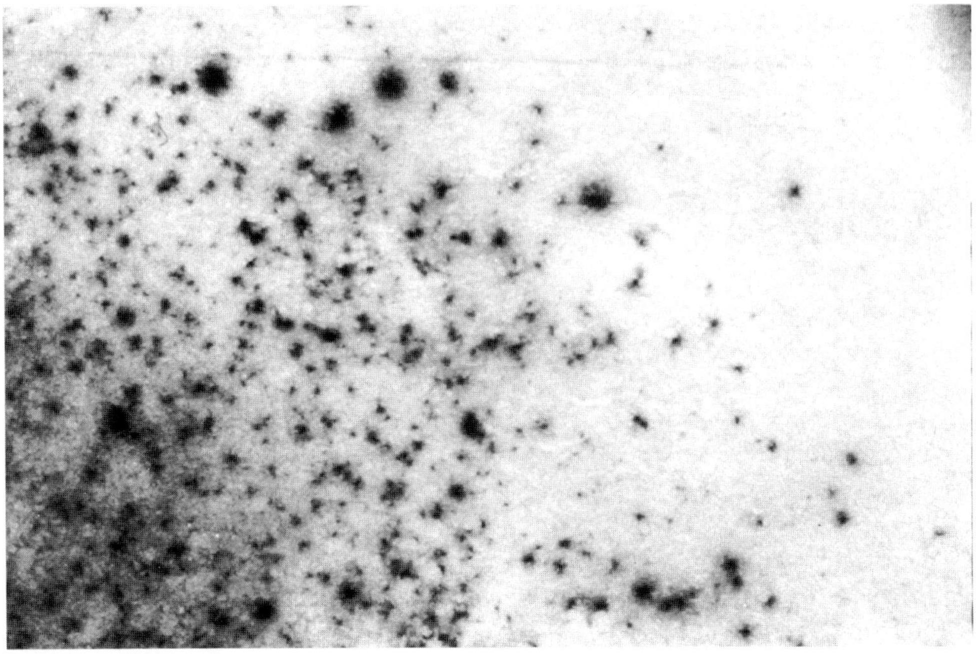

Figure 7 Transient expression of the bacterial β-glucuronidase gene in maize cells after microprojectile bombardment. Maize cells were spread on a filter paper and bombarded with gold particles (1 μm diameter) coated with a plasmid containing the bacterial β-glucuronidase gene under control of a strong promoter. The cells were stained for β-glucuronidase activity 2 d after transfer. Every spot represents a single transformation event. One can see all kinds of expression events ranging from small, pale blue spots to large, dark blue spots. In the latter case it cannot be excluded that aggregates of several cells are stained that are derived from one hit cell that underwent cell divisions in the time elapsed between DNA transfer and staining. The area shown in the picture has a diameter of 2 cm.

Later, more structured targets including leaves, seedlings, embryos, anthers, and pollen were successfully shown to enable transient expression. The most striking application of the biolistic approach (hitherto not achieved by any other gene-transfer method) is the test for tissue-specific transient expression. Researchers are interested to show not merely expression of a foreign gene in a plant as such, but rather to demonstrate its tissue- and organ-specific expression. To achieve this it was necessary at first to stably transform plants in order to test their tissues for expression of the introduced gene. This was time consuming, or in the case of the recalcitrant species, essentially not possible, since suitable methods for transformation were lacking. Now the biolistic approach opened up the possibility of shooting the gene constructs to be tested directly into the target tissue or organ and monitoring their expression without the necessity of a previous stable transformation of the particular plant. The first experiments of this kind showed the introduction and expression of structural genes and a regulatory gene of the anthocyanin biosynthesis pathway into the maize aleurone, thereby demonstrating the correct tissue-specific regulation of these genes.

For stable transformation it is essential to chose as targets those cells and tissues that possess the potential for regeneration. With the recalcitrant cereals and legumes, the meristems of an immature embryo and the embryogenic regions of an embryogenic cell culture are the most promising targets for microprojectile bombardment. Using multicellular targets, one has to be aware that chimeric plants with gene expression in one or more regions scattered over the organism are likely to occur, which is insufficient for obtaining a fertile transgenic plant where the reproductive organs also have to be transformed. Two cell genetic strategies were developed in order to circumvent this obvious dilemma: (1) multiple shoot induction on the primary regenerant and/or (2) selection for transgenic shoots by including a selectable marker in the DNA to be transferred. Using the biolistic approach the following species could be stably transformed: e.g., maize, rice, wheat, barley (transgenic callus), sugarcane, papaya, spruce, poplar, cotton, soybean, *Phaseolus,* etc.

Attempts to achieve the stable integration of new genes into the genome of chloroplasts and mitochondria have been without (reproducible) success over the past several years. Using the biolistic DNA delivery it was possible to stably introduce foreign genes into the chloroplasts of tobacco and of the alga *Chlamydomonas* and into the mitochondria of the yeast *Saccharomyces cerevisiae.* These experiments are not (yet) routine and involve sophisticated selection schemes.

3. Microinjection

In parallel to electroporation, microinjection techniques had been developed to deliver DNA directly into the plant cell (usually, protoplasts were taken). The disadvantages of this method (tedious work, low number of cells treated, very expensive equipment needed) are counterbalanced by obvious advantages: high transformation frequencies of up to 50% and the possibility to directly inject the DNA into the nucleus of the recipient cell. Also, multicellular structures like embryoids had been multiinjected. However, presently this approach is less frequently used due to its obvious disadvantages.

4. DNA Transfer via Pollen

All gene-transfer systems reviewed above require cell and tissue culture systems as an essential prerequisite. In order to circumvent these kinds of *in vitro* systems, other approaches were suggested. Most of them proposed the introduction of DNA into gametes followed by fertilization and zygotic embryogenesis. This kind of approach would be simpler, faster, and cheaper than the *in vitro* methods and would also avoid the problem of somaclonal variation. Thus it was a logical consequence to favor the use of pollen as the vector for DNA, since ovules are difficult to isolate and the injection into the embryo sac *in situ* seemed to be too tedious and unpredictable. It was hoped that pollen was easily accessible for DNA transfer and that the pollen tube would deliver the DNA to the egg cell (as reviewed by Hess).[10]

Thus the use of DNA-treated pollen as a DNA vector for pollinating fertile plants of maize was suggested,[11,12] and also Hess[10] described a similar transformation system for *Nicotiana glauca,* however, without molecular evidences. Ahokas[13] showed DNA uptake into pea pollen facilitated by liposomes, and Abdul-Baki et al.[14] demonstrated the introduction of labeled DNA into pollen grains of *N. gossei* by electroporation. Twell et al.[15] bombarded pollen with DNA using the particle gun approach and demonstrated transient gene expression of the marker gene GUS. However, although DNA could be taken up into pollen of diverse phylogenetic origin, there is no case reported yet that unequivocally demonstrates a successful gene transfer using this kind of approach.

Another approach was described for rice[16] where, after pollination, the stigmas were cut off and DNA solution was directly applied to the style using the pollen tube as a microcapillary. In DNA hybridizations of seed-derived marker-selected plants, positive signals were obtained. A third approach was proposed by Picard et al.[17] They applied DNA with the NPT marker gene onto just pollinated stigmas of wheat plants and found seedlings with kanamycin resistance and expression of the NPT gene among the progeny of the DNA-treated flowers. No molecular evidence was presented.

Recently, Heberle-Bors[18] critically reviewed all kinds of pollen-mediated gene transfer approaches and came to the conclusion that "the elegant idea of using mature pollen as a super vector for gene transfer fell short of experimental reality". We will not exclude that some kind of DNA transfer is being brought about by these approaches; however, these experiments are far from being reproducible and predictable. And, most of all, DNA introduced by these approaches is very rarely transmitted to the progeny, rather the meiotic divisions seem to eliminate the foreign genetic material. Thus, from the present point of view, it is highly unlikely that the pollen-mediated approach will develop into a reliable transformation method of general applicability for monocot and dicot plants.

5. Other Methods

There are numerous other methods published for directly delivering DNA into plant cells and organs. A choice of them is given below:

- For introducing DNA into protoplasts the use of PEG or electroporation could be replaced by a short pulse of mild sonication (tobacco).[19]
- Suspension culture cells of maize and tobacco showed transient gene expression after vortexing the cells in the presence of silicon carbide fibers.[20]
- A laser microbeam for cutting holes of defined dimensions into cell walls of various dicot cells and tissues followed by DNA uptake and transient gene expression as well as stable transformation.[21]

- Electrophoretic migration of DNA into barley caryopses, followed by transient expression of the reporter gene.[22]
- Macroinjection of DNA into floral tillers of small grain cereals.[23]
- Soaking of dry embryos in DNA.[24]

All these methods have one thing in common: their results are far from being reproducible and predictable. And most of all, DNA introduced by these kinds of approaches is very rarely transmitted to the progeny, rather the meiotic divisions seem to eliminate the foreign genetic material. Hence these methods are presently of purely academic value and it is rather unlikely that they will develop into reliable transformation methods of general applicability for monocot and dicot plants.

C. VIRUS-MEDIATED GENE TRANSFER

Plant viruses have always been very attractive for plant genetic engineering as potential transmitters of foreign genes. They have distinctive, potentially advantageous characteristics: their nucleic acid is in most cases directly infectious to plants, the infection process is much simpler to bring about than delivering genes via *Agrobacterium,* they show spread to every cell of the target plant within a short time, they replicate to very high copy numbers per plant cell, and their host range includes all major crop plants.

There are reports for the DNA viruses (caulimoviruses and geminiviruses like maize streak virus and wheat dwarf virus) where additional DNA had been introduced into dispensable virus reading frames, thus achieving the expression of reporter genes in infected plant cells. However, the virus-mediated gene-transfer systems have not gained wide distribution because there are also severe drawbacks: viruses are pathogenic agents that weaken plants to varying degrees, plant viruses usually do not integrate into the plant genome, so that there are no easy ways to transfer the new gene(s) to the seed progeny, most of the known viruses are RNA viruses, so that all *in vitro* manipulation has to be done on cDNAs of the viruses, packaging constraints limit the amount of additional DNA to be encapsidated, due to the high information density in the virus genomes, replacement of virus DNA by a gene to be transmitted to the plant can render the virus noninfectious. Also, a systemic spread of a virus could cause interferences with environmental protection regulations that require strict control over the introduced new gene (for a review see Gronenborn and Matzeit).[25]

A special version of gene transfer uses *Agrobacterium* to introduce a whole virus genome into a target plant. The virus genome is cloned between the two border fragments of the T-DNA and thus being introduced into the plant cell where it becomes active and starts its own replication. This way of transferring a virus genome via *Agrobacterium* is called agroinfection. Using maize streak virus, it was shown that with an extremely low frequency, *Agrobacterium* is able to infect maize plants which show after agroinfection a systemic spread of maize streak virus.[26]

In summary, virus-mediated gene transfer requires a lot of prerequisites (molecular and experimental knowledge about the particular virus, special laboratory facilities, and, in many countries, also permission according to legal regulations that are more strict than with the other two gene-transfer approaches), so its application is still limited.

III. ADVANTAGES AND LIMITATIONS OF THE MAIN GENE-TRANSFER SYSTEMS

The choice of the transformation system clearly depends on the plant species with which one is dealing.[27] Among the dicots, *Agrobacterium* is certainly the system of choice, although there are also dicot species that are difficult to transform in this way. The great advantage of the *Agrobacterium* system is the precision and high efficiency of its gene-transfer process. An exactly defined sequence of new genetic information (i.e., T-DNA with precise ends on both sides) is transferred to the plant cell and is stably integrated into the host genome, generally without rearrangements. The drawback is the limited host range of *Agrobacterium* being confined to dicot species and a very few monocots. The important cereals therefore require another transformation system. Direct DNA transfer into protoplasts or via the biolistic approach does overcome this obstacle, however, this way is more dependent on cell culture systems that essentially have to be established prior to the start of any gene-transfer experiments. Direct DNA transfer leads also to the integration of unwanted vector plasmid sequences in addition to the gene(s) of interest. Further, there are indications that the chance of the introduced DNA to undergo rearrangements is higher with this approach as compared to the *Agrobacterium* system. Nevertheless, direct DNA transfer has been very successfully used for stably transforming those plant species that are not transformable by other methods.

There are several approaches that directly deliver DNA into plant cells. Besides the protoplast approach and the particle bombardment method, there are numerous methods belonging to the group of the nonorthodox ones with results that were, however, far from being reproducible and predictable.[28] Thus only the transfer of DNA into protoplasts and the microprojectile-mediated transfer of genes through intact cell walls can be recommended for approaching the problem of stably transforming cereals and other problematic plants. But also, these two methods have bottlenecks that have to be taken into account. With the protoplast approach it is the loss of the regenerative potential of the selected transgenic colonies, and with the particle bombardment it is the question of choosing the appropriate target and of a tight enough selection. So if a protoplast-to-plant system is not available for a particular species and *Agrobacterium*-mediated transformation does not work (efficiently), and provided that an *in vitro* regeneration system is already available, then the biolistic approach is presently the method of choice for obtaining stable transformants, particularly if one has access to a marker system mediating efficient selection.

In summary, the *Agrobacterium* system, the direct DNA transfer into protoplasts, and the direct DNA transfer into structured targets via the biolistic way represent nowadays the three methods of choice with general applicability, in particular for the difficult monocots.

IV. APPLICATIONS AND PERSPECTIVES

The instrument of gene transfer has been very rapidly introduced into the arsenal of general methods of basic plant research where it opened up whole new avenues of research. We will only name some of them: functional characterization of genes, in particular of those that may become targets for genetic engineering, investigation of processes of plant development such as flowering, seed formation, ripening, and senescence, modification of pathways of the primary and secondary metabolism, signal transduction, molecular evolution, etc.

On the other hand, plant genetic transformation is also a synonym for spectacular economic applications made possible by this technology. Also here we will give some recent examples: resistance to herbicides by introducing genes for herbicide-detoxifying proteins or rendering the herbicide targets nonsensitive, resistance to insects, e.g., by introducing the *Bacillus thuringiensis* toxin protein, modifying fruit ripening of tomatoes, generation of cytoplasmic male sterility by introducing a RNAse gene under control of a tapetum-specific promoter so that pollen development is inhibited, modifying the oil composition in oil plants, etc. The vast possibilities of using molecular strategies in combination with plant transformation for introducing disease resistance into plants will be discussed in the following chapter.

REFERENCES

1. Hooykaas, P. J. J. and Schiiperoort, R. A., *Agrobacterium* and plant genetic engineering, *Plant Mol. Biol.,* 19, 15, 1992.
2. Horsch, R. B., Fry, J. E., Hoffmann, N. L., Elchholtz, D., Rogers, S. G., and Fraley, R. T., A simple and general method for transferring genes into plants, *Science,* 227, 1229, 1985.
3. Visser, R. G. F., Jacobsen, E., Witholt, B., and Feenstra, W. J., Efficient transformation of potato (*Solanum tuberosum*) with a binary vector in *Agrobacterium rhizogenes, Theor. Appl. Genet.,* 78, 594, 1989.
4. Paszkowski, J., Shillito, R. D., Saul, M., Mandak, V., Hohn, T., Hohn, B., and Potrykus, I., Direct gene transfer to plants, *EMBO J.,* 3, 2717, 1984.
5. Lindsey, K. and Jones, M. G. K., Electroporation of cells, *Physiol. Plant.,* 79, 168, 1990.
6. Mendel, R. R. and Teeri, T. H., Transformation of cereal crops: barley, wheat, rye, oats, in *Transformation of Plants and Soil Microorganisms,* Wang, K., Herrera-Estrella, A., and Van Montagu, M., Eds., Edward Arnold Series of Plant and Microbial Biotechnology, Cambridge University Press, Cambridge, in press.
7. Negrutiu, I., Dewulf, J., Pietrzak, M., Botterman, J., Rietveld, E., Wurzer-Figurelli, E. M., Ye, D., and Jacobs, M., Hybrid genes in the analysis of transformation conditions. II. Transient expression

vs. stable transformation—analysis of parameters influencing gene expression levels and transformation efficiency, *Physiol. Plant.*, 79, 197, 1990.

8. Mendel, R. R., Microprojectile-mediated gene transfer to plants, *AgBiotech News Inf.*, 2, 643, 1990.
9. Christou, P., Genetic transformation of crop plants using microprojectile bombardment, *Plant. J.*, 2, 275, 1992.
10. Hess, D., Pollen-based techniques in genetic manipulation, *Int. Rev. Cytol.*, 107, 367, 1987.
11. De Wet, J. M. J., Bergquist, R. R., Harlan, J. R., Brink, D. E., Cohen, C. E., Newell, C. A., and De Wet, A. E., Exogenous gene transfer in maize (*Zea mays*) using DNA-treated pollen, in *Experimental Manipulation of Ovule Tissue,* Chapman, G. P., Mantell, S. H., and Daniels, W., Eds., Longman, London, 197, 1985.
12. Ohta, U., High efficiency genetic transformation of maize by a mixture of pollen and exogenous DNA, *Proc. Natl. Acad. Sci. U.S.A.*, 83, 715, 1986.
13. Ahokas, H., Transfection by DNA-associated liposomes evidenced at pea pollination, *Hereditas,* 106, 129, 1987.
14. Abdul-Baki, A. A., Saunders, J. A., Matthews, B. F., and Pittarelli, G. W., DNA uptake during electroporation of germinating pollen grains, *Plant Sci.*, 70, 181, 1990.
15. Twell, D., Klein, T. M., Fromm, M. E., and McCormick, S., Transient expression of chemeric genes delivered into pollen by microprojectile bombardment, *Plant Physiol.*, 91, 1270, 1989.
16. Zhong-xun, L. and Wu, R., A simple method for the transformation of rice via the pollen-tube pathway, *Plant Mol. Biol. Rep.*, 6, 165, 1988.
17. Picard, E., Jacquemin, J. M., Cranler, F., Bobin, M., and Forgeois, P., Genetic transformation of wheat (*Triticum aestivum*) by plasmid DNA uptake during pollentube germination, *VIIth International Wheat Genetics Symposium,* University of Cambridge, Cambridge, 1988, 779.
18. Heberle-Bors, E., Germ line transformation in higher plants, *Newsl. Int. Assoc. Plant Cell Tissue Culture,* 64, 2, 1991.
19. Joersbo, M. and Brunstedt, J., Direct gene transfer to plant protoplasts by mild sonication, *Plant Cell Rep.*, 9, 207, 1990.
20. Kaeppler, H. F., Gu, W., Somers, D. A., Rines, H. W., and Cockburn, A. F., Silicon carbide fiber-mediated DNA delivery into plant cells, *Plant Cell Rep.*, 9, 415, 1990.
21. Weber, G., Monejambashi, S., Wolfrum, J., and Greulich, K.-O., Genetic changes induced in higher plant cells by a laser microbeam, *Physiol. Plant.*, 79, 190, 1990.
22. Ahokas, H., Transfection of germinating barley seed electrophoretically with exogenous DNA, *Theor. App. Genet.*, 77, 469, 1989.
23. De la Pena, A., Lörz, H., and Schell, J., Transgenic rye plants obtained by injecting DNA into young floral tillers, *Nature,* 325, 274, 1987.
24. Töpfer, R., Gronenborn, B., Schell, J., and Steinbiss, H. H., Uptake and transient expression of chimeric genes in seed-derived embryos, *Plant Cell,* 1, 133, 1989.
25. Gronenborn, B. and Matzelt, V., Plant gene vectors and genetic transformation: plant viruses as vectors, in *Cell Culture and Somatic Cell Genetics in Plants,* Vol. 6, *Molecular Biology of Plant Nuclear Genes,* Schell, J. and Vasil, I. K., Eds., Academic Press, San Diego, 1989, 69.
26. Grimsley, N., Hohn, T., Davies, J. W., and Hohn, B., Agrobacterium-mediated delivery of infectious maize streak virus into maize plants, *Nature,* 325, 177, 1987.
27. Potrykus, I., Gene transfer to plants: assessment and perspectives, *Physiol. Plant.*, 79, 125, 1990.
28. Potrykus, I., Gene transfer to cereals: an assessment, *Bio/Technology,* 8, 535, 1990.

FURTHER READING

Grierson, D. and Corey, S. N., *Plant Mol. Biol.*, Blackie, Glasgow, 1988.

Gronenborn, B. and Matzeit. V., Plant gene vectors and genetic transformation: plant viruses as vectors, in *Cell Culture and Somatic Cell Genetics in Plants,* Vol. 6, *Molecular Biology of Plant Nuclear Genes,* Schell, J. and Vasil, I. K., Eds., Academic Press, San Diego, 1989, 69.

Heberle-Bors, E., Germ line transformation in higher plants, *Newsl. Int. Assoc. Plant Cell Tissue Culture,* 64, 2, 1991.

Hooykaas, P. J. J. and Schilperoort, R. A., *Agrobacterium* and plant genetic engineering, *Plant Mol. Biol.*, 19, 15, 1992.

Mendel, R. R., Microprojectile-mediated gene transfer to plants, *AgBiotech News Inf.,* 2, 643, 1990.

Mendel, R. R. and Teeri, T. H., Transformation of cereal crops: barley, wheat, rye, oats, in *Transformation of Plants and Soil Microorganisms,* Wang, K., Herrera-Estrella, A., and Van Montagu, M., Eds., Edward Arnold Series of Plant and Microbial Biotechnology, Cambridge University Press, Cambridge, in press.

Paszkowski, J., Saul, M. W., and Potrykus, I., Plant gene vectors and genetic transformation: DNA-mediated direct gene transfer to plants, in *Cell Culture and Somatic Cell Genetics in Plants,* Vol. 6, *Molecular Biology of Plant Nuclear Genes,* Schell, J. and Vasil, I. K., Eds., Academic Press, San Diego, 1989, 52.

Potrykus, I., Gene transfer to plants: Assessment of published approaches and results, *Annu. Rev. Plant Physiol.,* Plant Molecular Biology.

Schell, J. and Vasil, I. K., Eds., *Cell Culture and Somatic Cell Genetics in Plants,* Vol. 6, *Molecular Biology of Plant Nuclear Genes,* Academic Press, San Diego, 1989.

Tempe, J. and Casse-Delbart, F., Plant gene vectors and genetic transformation: *Agrobacterium* Ri plasmids, in *Cell Culture and Somatic Cell Genetics in Plants,* Vol. 6 *Molecular Biology of Plant Nuclear Genes,* Schell, J. and Vasil, I. K., Eds., Academic Press, San Diego, 1989, 26.

Chapter 15

Arabidopsis as a Model Host in Molecular Plant Pathology

Roger Innes

TABLE OF CONTENTS

I. INTRODUCTION

Current studies in molecular plant-pathogen interactions have at their core three fundamental questions: How do plants recognize pathogens? How is recognition transduced into a physiological response in the plant? What physiological responses are causally related to disease resistance? Providing answers to these questions has proved very challenging. In this chapter I will describe new approaches to addressing these questions using the plant *Arabidopsis thaliana* (Arabidopsis) as a model host.

Much of our current understanding of plant-pathogen interactions can be incorporated into the model diagrammed in Figure 1. Pathogen recognition is mediated by products of disease resistance genes (*R*-genes), which are presumed to be receptors of some type. Dozens of *R*-genes in various plant species have been genetically characterized. *R*-Gene-encoded receptors are believed to specifically bind host-specific elicitor molecules produced by pathogens. Production of these elicitors is controlled by specific pathogen avirulence (*avr*) genes. Binding of elicitor to receptor constitutes the initial recognition event. This recognition event presumably activates a signal transduction cascade, but the components of this cascade are unknown. Ultimately, a series of physiological responses are activated in the plant cell. These responses include production of activated oxygen species (the "oxidative burst"), and induction of numerous "defense-response" genes. Defense-response genes include genes that encode cell wall components, hydrolytic enzymes, and genes involved in production of secondary plant metabolites with antimicrobial activity (phytoalexins). The oxidative burst and various defense-response genes may be involved in production of the "hypersensitive resistance reaction" (HR), which is almost always associated with *R*-gene-mediated resistance. The HR is characterized by a rapid collapse of host cells in the vicinity of pathogen ingress that appears concomitant with cessation of pathogen growth.

The model shown in Figure 1, however, is highly speculative. Although we know that at least some pathogen *avr* genes control production of host-specific elicitors, we do not have any direct evidence that *R*-genes encode receptors. In addition, our understanding of signal transduction mechanisms in plant cells is rudimentary. Furthermore, causal relationships between physiological responses and disease resistance have not been established.

A major obstacle to further progress has been a lack of defined mutations in plant genes that are involved in disease resistance. Without such mutations it is extremely difficult to establish a causal role

0-87371-877-1/95/$0.00+$.50

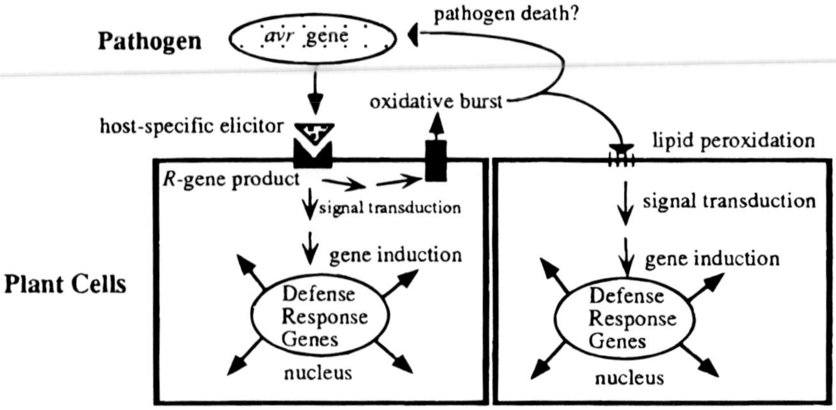

Figure 1 A possible mechanism for induction of defense responses in plants. Induction of defense responses occurs upon binding of a host-specific elicitor from the pathogen to a receptor encoded by a disease-resistance (*R*) gene in the plant. Binding of elicitor to the receptor activates a signal transduction pathway, which in turn activates two distinct responses: production of an oxidative burst (superoxide anion, hydrogen peroxide, and other active oxygen species), and induction of many different defense-response genes (see text). Indirect induction of some defense-response genes is also indicated. Production of the elicitor is controlled by a specific avirulence (*avr*) gene in the pathogen.

for any gene product. There is an abundance of data showing correlation of various plant responses with disease resistance, but we do not know what happens if these responses are absent. For example, the role of phytoalexins in disease resistance has remained controversial because, until recently, there were no plant mutants available that lacked specific phytoalexins. The recent isolation of such mutants in Arabidopsis (see below) will now allow us to address this question directly.

It should be possible to identify genes critical to plant disease resistance using a standard genetic approach: mutagenize a plant that is resistant to a pathogen of interest, screen for mutants that become susceptible, and then isolate and characterize the mutated genes. One deterrent to this approach has been the complex genomes of most crop plants. Many crops are polyploid, thus many genes may be functionally redundant; a mutation in just one member of a gene family might not have a phenotype. A second deterrent, related to the complexity of plant genomes, is the difficulty in isolating genes identified by mutation. Since the product of most such genes will be unknown, molecular isolation of these genes will usually require precise determination of genetic map position, followed by a laborious chromosome walk. The larger the genome, and the greater the complexity, the more difficult this "position-cloning" approach becomes. Isolation of genes identified by mutation, however, is essential if we are to understand the most fundamental aspects of plant-pathogen interactions. Fortunately, molecular tools and techniques have now been developed that greatly reduce the effort required to clone genes based on genetic map position.[1] In humans, which have similar problems of large and complex genomes, position-cloning approaches have recently met with celebrated success in the isolation of genes controlling heritable diseases.[1] The plant research community has been actively applying this technology to crops such as tomato, rice, and lettuce.[2–4] Position-cloning methods in plants have progressed the furthest, however, in Arabidopsis.[5]

A. WHY USE ARABIDOPSIS?

Arabidopsis has been embraced by molecular biologists as the plant of choice for studying most aspects of plant biology.[6] The primary reason for choosing Arabidopsis is that it is particularly well suited to position-cloning approaches. It is a true diploid and can be readily outcrossed or selfed. Its generation time is less than 6 weeks and individual plants can produce thousands of seed. The seed can be easily mutagenized using chemical mutagens (e.g., ethylmethane sulfonate, diepoxybutane, ethylnitrosourea), or ionizing radiation (e.g., X-rays, γ-rays, fast-neutron bombardment).[7] Its small size allows one to screen thousands of plants for mutations in a refrigerator-sized growth cabinet. The size of the Arabidopsis nuclear genome is smaller than any known flowering plant: approximately 80,000 kb pairs per haploid nucleus,[8] which is over 20 times smaller than soybean.[9] In addition, Arabidopsis has very little dispersed repetitive DNA,[8,10] and it is readily transformed by *Agrobacterium tumefaciens.*[11] Finally, gene families in Arabidopsis generally contain fewer members than equivalent families in other plant species.[12,13] This latter observation has important implications for isolation of mutants. As mentioned above in regard to polyploid plants, if one member of a gene family can substitute for another, it is necessary to mutate both members to observe a phenotype. In the legumes, most of the known defense-response genes are part of multigene families, including the genes encoding chalcone synthase (CHS) and lipoxygenase (LOX). In Arabidopsis, CHS is a single-copy gene.[12] Likewise, the lipoxygenase LOX gene family in Arabidopsis appears to be much smaller than in other characterized plant species.[13]

The basic features described above have greatly facilitated development of tools that further simplify isolation and analysis of Arabidopsis genes. A detailed genetic map based on morphological and restriction fragment length polymorphism (RFLP) markers has been developed.[14] Several yeast artificial chromosome (YAC) libraries of its genome are available,[15–17] as are phage and cosmid libraries. Development of a physical map of the Arabidopsis genome is nearing completion; greater than 90% of the genome has been assembled into cosmid contigs (contained on over 20,000 cosmid clones),[18] and YAC clones comprising over 30% of the genome have been placed on the genetic map.[19] Two international stock centers will distribute many of these germplasm and DNA resources, including RFLP markers.* In addition, two electronic databases, AAtDB and AIMS, provide information of all kinds relating to Arabidopsis, including what resources are available from the stock centers, frequently updated genetic and physical maps, and literature citations.** Thus, the Arabidopsis researcher has many resources upon which to draw.

One Arabidopsis resource that is especially useful for analysis of plant-pathogen interactions is the extensive collection of Arabidopsis varieties (ecotypes) collected from the Northern Hemisphere.[20] Since Arabidopsis is primarily self-fertilizing,[21] genetic variation between populations is preserved. Local populations of Arabidopsis have undoubtedly evolved under differing pathogen pressures, thus ecotypes would be expected to vary in their phenotypic responses to specific pathogens. Such variation has now been demonstrated by several groups who have looked at interactions with viral, bacterial, and fungal pathogens.[22–31] The genes controlling such variation can be identified genetically, and then isolated using position-cloning strategies as described above. Until recently, this germplasm resource was maintained by the Arabidopsis Information Service in Frankfurt, Germany. It is now available from the Ohio State and Nottingham Arabidopsis Stock Centers.*

B. HISTORY OF ARABIDOPSIS PATHOGEN STUDIES

Until the late 1980s, few pathogens of Arabidopsis had been reported. As a weed with no agronomic significance, there was little interest. There are two reports, however, of Arabidopsis serving as an alternative host for pathogens of crop plants. A 1971 report identified Arabidopsis as the primary weed host for a species of the fungus *Sclerotinia* that was infecting alfalfa in several fields in Maryland.[32] A 1981 abstract reported that the bacterium *Pseudomonas syringae* pv. tomato could be recovered from the leaves and roots of Arabidopsis growing in a field previously cropped to tomatoes.[33] This latter report is of interest because *P. syringae* pv. tomato is now used by several groups studying Arabidopsis-pathogen interactions.[30,34–36]

The development of Arabidopsis as a model system for plant molecular biology sparked a search for pathogens of Arabidopsis. Such pathogens were identified either by identifying naturally occurring infections on Arabidopsis, or by screening pathogens collected from other plants, especially pathogens of other crucifers. Table 1 provides a comprehensive list of the Arabidopsis-pathogen literature as of February 1993. Arabidopsis is now known to be a host of four different types of viruses, three bacterial species, eight fungal species, and five nematode species. Note that the great majority of these papers have been published in the last 2 years. Also note that several reviews of this field have been published, as well as an entire book that emphasizes methods for Arabidopsis-pathogen studies.[37] Because these reviews are both recent and comprehensive, I will not discuss in detail many of the papers listed in Table 1. Instead, I will focus on how researchers are using Arabidopsis to address the three fundamental

* For more information write Arabidopsis Biological Resource Center at Ohio State, 1735 Neil Avenue, Columbus, OH 43210, U.S.A. (E-mail: seeds@ genesys.cps.msu.edu), or The Nottingham *Arabidopsis* Stock Centre, School of Biological Sciences, University of Nottingham, University Park, Nottingham, NG7 2RD, U.K. (E-mail: PBZMHL@ vax.ccc.nottingham.ac.uk).

** For more information on AIMS contact: Inquire-aims@ genesys.cps.msu.edu. For information on AAtDB contact: curator@ weeds.mgh.harvard.edu.

Table 1 **A guide to the arabidopsis-pathogen literature**

Topic	Ref.
Reviews	105,106,107
Books	5,37
Interactions with Bacteria	
Pseudomonas syringae	23,26,30,33–36,43,44,69,70,73,76, 80,86,91,108,109
Xanthomonas campestris	28–30,34,110,111
Spiroplasma citri	112
Interactions with fungi	
Obligate biotrophs	
Peronospora parasitica	22,25,67,68,108
Albugo candida	22,67,108
Puccinia thlaspeos	67
Plasmodiophora brassicae	67
Erysiphe cruciferarum	67
Nonobligate parasites/broad host range (*Botrytis cinerea, Chromelosporium* spp., *Pythium* spp., *Rhizoctonia solani, Sclerotinia* spp.)	32,67,68
Interaction with viruses	27,31,96–102,113,114
Interaction with nematodes	118
Physiological responses to pathogens	
Induction of defense genes	13,34,35,44,70,71–73,76,83,115–117
Phytoalexin production	69,76
Systemic acquired resistance	91
Lesion mimic mutations	95

questions posed above. I will also stress significant new results that were not covered in these reviews. Finally, I will conclude with a discussion on the limitations of Arabidopsis and the use of purely genetic approaches, and will speculate on future directions in Arabidopsis-pathogen research.

II. GENE-FOR-GENE INTERACTIONS IN ARABIDOPSIS

Many plant-pathogen interactions are characterized genetically by a "gene-for-gene" relationship.[38–40] In such interactions, resistance of a plant to a given pathogen requires a specific (usually dominant) *R*-gene in the plant and a matching dominant avirulence (*avr*) gene in the pathogen. Loss of either member of this gene pair results in loss of resistance. *R*-Genes presumably mediate specific recognition of the pathogen by the plant, however, the mechanism of this recognition event is unknown. The pathogen *avr* genes are thought to control production of host-specific elicitor molecules.[40,41] The simplest model consistent with the genetic data is that host-specific elicitors bind to receptors on plant cells that are encoded by *R*-genes (Figure 1). The specificity of the interaction would thus be controlled by the ligand binding domains of these *R*-gene-encoded receptors. However, thus the validity of this model remains unknown. Isolating and characterizing a plant *R*-gene has become one of the most sought after goals in molecular plant pathology.

A. IDENTIFICATION OF *R*-GENE LOCI IN ARABIDOPSIS

In adopting Arabidopsis for host-pathogen interaction studies, one of the first questions asked was whether or not Arabidopsis would interact with its pathogens in a gene-for-gene fashion. It has been argued that gene-for-gene-type interactions are an artifact of breeding,[42] and that it would be difficult to uncover such interactions in wild plant populations. This proved not to be the case, however, with Arabidopsis. Several groups have now demonstrated the existence of classical dominant (or semidominant) *R*-genes that confer resistance to specific fungal and bacterial pathogens.[22,23,26] Nine of these *R*-genes have been placed on the Arabidopsis genetic map, including five that are specific to various races of *Peronospora parasitica* (causative agent of downy mildew of crucifers).[22] At least one

R-gene has been mapped to each of the five Arabidopsis chromosomes, and there appear to be clusters of *R*-genes on chromosome 1 and chromosome 4.[42a] Matching *avr* genes from the pathogen have been identified for three *R*-genes, *RPM1, RPS2,* and *RPS3,*[24,36,43] all of which confer resistance to specific strains of *Pseudomonas syringae*. Identification and mapping of *R*-gene loci in Arabidopsis represents the first step towards isolation of these genes via a position-cloning approach. Rapid progress is being made towards this goal,[22,26,44] and it appears likely that several Arabidopsis *R*-genes will be isolated in the near future.

B. ISOLATION OF *R*-GENES USING GENE-TAGGING APPROACHES

An alternative approach to position cloning for isolation of plant *R*-genes is to "tag" them with an insertional mutagen such as a transposable element, or the T-DNA from *Agrobacterium tumefaciens*. Transposon/T-DNA tagging can be much more rapid than position cloning to isolate genes with unknown products. Several developmentally important genes have been isolated from maize, snapdragon, and Arabidopsis via gene-tagging approaches,[45–47] and there is one promising report of an *R*-gene in snapdragon that may be tagged with a transposon.[48] In Arabidopsis, T-DNA tagging has been especially fruitful.[49] A large collection of Arabidopsis lines (>10,000) containing random T-DNA insertions has been developed by Feldman.[50] Over half of these are available from the Arabidopsis stock centers, and can be screened for mutant phenotypes of interest.* In addition, transposon tagging systems are being developed in Arabidopsis using elements introduced from either maize or snapdragon.[51–53] It should soon be possible to generate thousands of new insertion mutations in Arabidopsis using such heterologous transposons. These latter systems can be transferred among Arabidopsis ecotypes by conventional crosses, thus *R*-genes in diverse ecotypes can be targeted (the T-DNA-tagged lines are all in the Wassilewskija ecotype).

The primary difficulty in using gene tagging to isolate *R*-genes is obtaining an insertion in the targeted locus. Insertional mutagens generally produce only one or two mutations per plant, which is at least ten times fewer than what is typically generated using standard chemical mutagenesis.[7] Thus, the screen for susceptible plants must be sufficiently simple as to allow testing of many thousands of individuals. In addition, the spontaneous mutation rate of the target locus must be lower than the expected insertion rate of the transposon/T-DNA. Transposon tagging of the *Rp1* disease-resistance gene in maize has been frustrated by the high spontaneous mutation rate at this locus.[54] Furthermore, insertional mutagens are not random, and it is not possible to predict whether or not a targeted gene will be mutable by a given transposon/T-DNA.

C. EVOLUTION OF *R*-GENE SPECIFICITY

Genetic analyses have provided clues as to how *R*-genes might evolve new specificities during the evolutionary race with pathogens. Many *R*-gene loci are complex;[55] a single locus can have multiple alternative alleles, each with a different specificity. For example, the *Mla* locus of barley, which confers resistance to powdery mildew (*Erysiphe graminis*), has at least 20 different alleles.[56] In addition, multiple "alleles" can sometimes be present on a single chromosome, implying that the *Mla* locus actually consists of multiple closely linked loci.[56] These observations suggest that this locus may have evolved through repeated duplications and subsequent mutational modifications that alter specificity.[56] The presence of duplicated genes at a single *R*-gene locus could promote unequal crossing during meiosis, which may be the underlying cause of both instability and rapid evolution of some *R*-gene loci. For example, some alleles of *Rp1* in maize (confers resistance to *Puccinia sorghi*) are meiotically unstable.[54,57] Susceptible individuals arise at a rate of 1 in 300 in test crosses of the *Rp1-G* allele, and recombination events at or very near the *R*-locus are observed in all these individuals.[57] Unequal crossing over is also thought to have generated new alleles of *Rp1* with unique specificity.[55]

D. GENE-FOR-*GENES* INTERACTIONS IN ARABIDOPSIS

The large number of alleles reported at *R*-gene loci such as *Mla* and *Rp1* may be an overestimate. When resistance responses to two different fungal strains differ phenotypically, it has generally been assumed that these responses must be under control of different *R*-gene alleles.[56] It is equally plausible, however, that a single *R*-gene product is interacting with different *avr* gene determinants that trigger

* The remaining T-DNA lines are maintained by E.I. DuPont de Nemours & Company. Requests to screen these lines should be directed to Central Research and Development, Experimental Station, Wilmington, DE 19880-0402.

quantitatively different host responses. In the context of the receptor-ligand model, one can envision different ligands binding the same receptor with differing affinities. This would represent a "gene-for-*genes*" interaction. We have obtained data in support of this view in our laboratory.[24] Our data suggest that a single *R*-gene in Arabidopsis interacts with two different *avr* genes of the bacterial pathogen *Pseudomonas syringae.* We have isolated several Arabidopsis mutants that have lesions in the *RPS3* disease-resistance locus. *RPS3* confers resistance to *P. syringae* strains that carry the *avrB* avirulence gene, and our mutants are now susceptible to such strains. Genetic complementation analyses confirmed that the mutations are at *RPS3.* Interestingly, these mutants have also lost resistance to *P. syringae* strains that carry a different avirulence gene, *avrRpm1.* The resistance genes that match *avrRpm1* and *avrB* (*RPM1* and *RPS3*) map to the same location on chromosome 3,[23,24] thus it appears that *RPS3* and *RPM1* may be the same gene. If true, this gene is interacting with two dissimilar avirulence genes.[43,58] The resistant response to *P. syringae* strains expressing *avrB* is slower and weaker than the response to strains expressing *avrRpm1.* This difference might reflect differing affinities of the *RPS3/RPM1* resistance gene product for the specific elicitors produced by these two *avr* genes. It is possible, however, that *RPS3* and *RPM1* are just closely linked, and that our mutants have suffered a deletion that affects both *R*-genes. We will be able to resolve this question once the *RPS3/RPM1* locus is isolated.

Our work on the *RPS3* locus of Arabidopsis demonstrates the value of using a mutagenic approach. Were it not for the mutants that we obtained, we would have considered *RPS3* and *RPM1* to be simply closely linked loci. Such strict interpretations of the gene-for-gene paradigm unnecessarily restrict our thinking and model building. Clearly, the available genetic data are consistent with gene-for-*genes* interactions. Likewise, there is evidence that "*genes*-for-gene" interactions also occur. In common bean, there appear to be two different unlinked *R*-genes that condition resistance specific to a single *avr* gene from *P. syringae* pv. pisi.[59]

E. CONSERVATION OF *R*-GENE FUNCTION AMONG PLANT SPECIES

Isolation of *R*-genes from Arabidopsis may enable us to quickly isolate *R*-genes with similar specificities from crop plants. This hypothesis is based on the observation that *avr* genes known to interact with Arabidopsis (*avrRpt2, avrB,* and *avrRpm1*) also interact with other plant species.[30,36,43,60] For example, *P. syringae* strains that express *avrB* induce resistant responses specifically on Arabidopsis ecotypes that contain the resistance gene RPS3 and on soybean cultivars that contain the resistance gene *RPG1.*[24,60,61] Thus the *RPG1* gene of soybean is functionally homologous to *RPS3* gene of Arabidopsis. Parsimony would argue that *RPG1* and *RPS3* are true homologs, rather than products of convergent evolution. If so, isolation of *RPS3* from Arabidopsis should allow immediate isolation of *RPG1* from soybean via DNA:DNA hybridization. Such cross-species recognition of a single *avr* gene has been reported for several *avr* genes from both *Xanthomonas campestris* and *P. syringae,*[30,36,43,59,62,63] which further bolsters the argument that *R*-genes have been conserved during the evolution of plant species.

Isolation of *R*-genes from Arabidopsis may also facilitate isolation of *R*-genes with different specificities. The basic defense response mediated by *R*-genes, the HR, appears to be conserved among most, if not all, plant species.[64] Thus, *R*-genes appear to be linked to a similar defense-response pathway in most plant species, suggesting that there will be conserved structural elements among *R*-genes. Once several *R*-genes are isolated, it should be possible to identify conserved elements that can be used to design molecular probes (or polymerase chain reaction [PCR] primers) for isolation of many other *R*-genes. This presumed conservation of *R*-gene function also suggests that *R*-genes may be transferable among distantly related species. Arabidopsis may in fact provide a new source of *R*-genes for introduction into crop plants. This latter hypothesis can be quickly tested once an *R*-gene is isolated.

III. PHYSIOLOGICAL RESPONSES OF ARABIDOPSIS TO PATHOGENS

One concern regarding use of Arabidopsis in plant-pathogen interaction studies is whether or not Arabidopsis will be fundamentally similar to crop plants (especially noncrucifer crops) in its mechanisms of resisting disease. Will the knowledge gained from Arabidopsis be immediately transferable to crops, or have resistance mechanisms become too divergent between species? Clearly, many secondary products of plant metabolism are extremely different between species. The analyses completed this far, however, indicate that the response of Arabidopsis to pathogens is very similar to the responses observed in crop species.

R-Gene-mediated resistance in plants is almost always associated with an HR. As briefly mentioned above, the HR is manifested phenotypically as a rapid and localized collapse of host tissue at the site of pathogen ingress,[64] concomitant with a cessation of pathogen growth. This collapse is usually associated with other responses such as increased lignification of surrounding plant cell walls, peroxidation of membrane lipids, induction of various defense-response genes, and production of phytoalexins.[65,66] These responses are all observed in Arabidopsis when challenged with avirulent pathogens.[13,23,25,26,34,35,67–73]

A. ACTIVATION OF DEFENSE-RESPONSE GENES

Numerous genes are induced in plants during infection by both virulent and avirulent pathogens.[65] These genes, collectively known as defense-response genes, can be grouped by function into genes involved in phytoalexin and lignin biosynthesis (e.g., phenylalanine ammonia lyase [PAL], CSH), genes encoding hydrolytic enzymes (e.g., β-glucanases [BG], chitinases) and cell wall constituents (e.g., hydroxyproline-rich glycoproteins), and genes involved in oxidation processes (e.g., LOX, peroxidases, superoxide dismutase [SOD], glutathione-*S*-transferase [GST]), as well as genes of unknown function. In general, these genes are induced by both virulent and avirulent pathogens, but the response to avirulent pathogens is more rapid and more localized.[65] Many of these genes are also induced by abiotic stresses such as exposure to ultraviolet light, and during specific times in plant development (e.g., to make pigments in flower organs).

Several defense-response genes of Arabidopsis have been cloned and characterized. These include genes encoding PAL,[34,74] three different BGs,[35] two classes of chitinase,[70,75] a LOX,[13] GST and SOD,[76] and several genes of unknown function.[73,76] In addition, genes involved in two different steps of aromatic amino acid biosynthesis are inducible by pathogen infection.[71,72,77,78] This latter observation is of interest because the predominant phytoalexin made by Arabidopsis (see below) is likely derived from an aromatic amino acid precursor.

As mentioned in the introduction, it is not clear in any plant species which (if any) of the defense-response genes are primarily responsible for conferring resistance to specific pathogens. It should be possible to isolate mutations in at least some of these loci, or alternatively, to repress expression of these genes using antisense methods. A clear correlation between loss of a specific defense gene product and a reduction in resistance to a specific pathogen would provide strong evidence that a specific gene is important to active resistance.

One clue to the roles of various defense genes may be their relative time of expression during infection by avirulent pathogens as compared to virulent pathogens. Genes involved in arresting the initial infection should be induced very rapidly during infection by avirulent pathogens, while expression of genes involved in preventing subsequent infections ("wound dressing") may be delayed, and may not display a significant difference between avirulent and virulent infections. For example, at least one *PAL* gene of Arabidopsis is highly induced within 6 h after infection with an avirulent *P. syringae* strain, but is induced only slightly by an isogenic virulent *P. syringae* strain.[35] In contrast, induction of the Arabidopsis *LOX1* gene is not apparent until 12 h after infection by the same avirulent *P. syringae* strain, and is also highly induced by the virulent strain, although not until 48 h after infection.[13] From these results, induction of PAL appears to be a more immediate response to the avirulent pathogen than induction of LOX. Induction of the Arabidopsis BG genes is quite slow relative to PAL, gradually increasing after 24 h, and showing higher levels of induction during infection by virulent *P. syringae* than during infection by avirulent strains.[35] Similar time course measurements using the Arabidopsis-*P. syringae* system indicate that *GST1* is also rapidly induced by avirulent pathogens, but *SOD1* and *CHS* are not.[76] Thus, regulation of defense genes is complex; it is clear that a single switch does not control expression of all defense-response genes.

A major unanswered question is how *R*-genes control expression of defense-response genes. Is there a relatively direct signal transduction pathway between pathogen recognition (presumably by an *R*-gene product) and induction of at least some defense-response genes, or are these genes responding to some kind of generalized stress signal? It is very difficult to distinguish between these two possibilities, as some defense responses undoubtedly produce stress signals. For example, avirulent pathogens are known to rapidly induce production of activated oxygen species (the "oxidative burst") in resistant plants;[79] activated oxygen species are very reactive and will damage most organic molecules, including membrane lipids and DNA. The rapid induction of *GST1* in Arabidopsis could be a response to lipid oxidation, rather than a response to an *R*-gene-linked signal transduction pathway (Figure 1).

Several approaches are being employed to identify components of the signal transduction systems that regulate expression of defense-response genes. The most common approach is to screen for mutants that fail to induce a specific defense gene when challenged by an avirulent pathogen. To facilitate such screens, defense gene induction is usually monitored using reporter genes (e.g., β-glucuronidase [*GUS*]) fused to a defense gene promoter. For example, the promoters of the Arabidopsis *PAL1* and *GST1* genes have been fused to *GUS* and transformed into wild-type Arabidopsis plants.[79a] Seeds from these transgenic plants are then mutagenized. Plants from these mutagenized populations are screened for individuals that fail to induce *GUS* to normal levels when challenged by avirulent pathogens. This approach is still in its developmental stages, however, and no signal transduction mutants have been isolated.

A second approach to dissecting signal transduction pathways in Arabidopsis defense systems is to analyze existing mutants that are affected in signal transduction in other pathways. For example, Bent and co-workers[80] analyzed disease development in Arabidopsis mutants that do not respond correctly to the plant hormone ethylene. Because ethylene is known to induce many defense-response genes in plants, it was reasonable to postulate that mutants unable to respond to ethylene might be compromised in their ability to resist avirulent pathogens. This was not the case, however. Plants that were resistant to ethylene still responded to avirulent *P. syringae* strains in a wild-type manner. Interestingly, when virulent pathogens were tested, disease symptoms (chlorosis, tissue collapse) were significantly less severe in the ethylene-resistant mutant *ein2,* even though *in planta* pathogen population levels were unaltered. This result suggests that an ethylene response pathway may mediate symptom development during infections by virulent *P. syringae.*

As mentioned above, one of the first plant responses that is correlated with resistance is production of active oxygen species.[79] This "oxidative burst" is thus being subjected to increased scrutiny, both for its possible antimicrobial role and as possible component of signal transduction (Figure 1). The oxidative burst can be induced in cultured soybean cells using crude elicitors from fungal cell walls.[81] Of even more interest from the standpoint of *R*-gene function, the oxidative burst can also be induced in cultured soybean cells by *P. syringae* in an *avr*-gene-dependent manner,[82] suggesting that *R*-genes may mediate this response. Isolation of plant mutants that are unable to produce an oxidative burst is thus of prime interest. Production of the oxidative burst can be assayed using various indicator dyes, including nitro blue tetrazolium (NBT). NBT forms a purple-blue precipitate when exposed to superoxide anion. de Maagd and co-workers[83] reported that Arabidopsis seedlings grown in liquid culture stain blue when exposed to crude fungal cell wall elicitor plus NBT. This group also reported the isolation of Arabidopsis mutants that do not respond to the elicitor using this assay. Such mutants could presumably be defective in any of a number of steps, from detection of the elicitor to the production of superoxide. A block in uptake of the NBT indicator must also be considered. Genetic characterization of the mutants is currently underway to determine how many different genes have been identified.

Perhaps the most straightforward approach to identifying signal transduction mutants is to simply screen for plants that become susceptible to a pathogen to which they are normally resistant. Analogous to the NBT assay, such mutants could be impaired at several levels, from perception of the pathogen to production of a specific defense response. My laboratory and others have screened extensively for Arabidopsis mutants that have become susceptible to *P. syringae* strains expressing specific avirulence genes (e.g., *avrB* and *avrRpt2*). Several fully susceptible mutants have been identified,[24,26,76] but genetic analyses have shown that all these mutants map to the corresponding *R*-gene locus. Although the number of plants screened is still relatively low (less than 50,000), we can confidently say that screening for susceptible mutants primarily yields mutants at the specific *R*-gene locus being tested. A screen for barley mutants susceptible to powdery mildew produced a similar result; almost all the susceptible mutants were allelic, and they mapped to a specific *R*-gene locus.[84,85] These results indicate that signal transduction and defense-response genes are difficult to identify by mutation. This observation suggests that such genes are essential for viability (i.e., loss of function is lethal), or that the genes are redundant (i.e., function can be replaced by another gene). Mutations in functionally redundant genes might cause a weak phenotype. For example, Kunkel and co-workers[86] reported isolation of several Arabidopsis mutants with a partially susceptible phenotype; these mutants display mild disease symptoms when inoculated with an avirulent strain of *P. syringae* pv. tomato, a delayed HR, and intermediate levels of pathogen growth. It will be quite interesting to determine if these mutants contain lesions in genes other than the *R*-gene locus.

B. PHYTOALEXIN BIOSYNTHESIS

One of the more controversial components of the plant defense response is the production of phytoalexins. The fact that phytoalexin production is a nearly universal response of plants to pathogens suggests that they do play an important role in defense, but it has been very difficult to obtain unequivocal evidence to support this view. Since phytoalexins are most likely not required for viability of plants in the absence of pathogens, it should be possible to obtain plant mutants that fail to produce specific phytoalexins. Analysis of such mutants with a battery of plant pathogens would then indicate the relative importance of phytoalexins in plant defense.

The predominant phytoalexin produced by Arabidopsis is a 3-thiazol-2′-yl-indole (camalexin),[69] which is identical in structure to a phytoalexin isolated from the crucifer species *Camelina sativa.*[87] Camalexin is absent from healthy Arabidopsis tissue. It is induced by infiltration of leaves with an avirulent *P. syringae* pv. *syringae* strain that causes an HR, but not by a virulent *X. campestris* strain or by a nonpathogenic mutant of *P. syringae* pv. *syringae*. However, Ausubel and co-workers[76] have recently reported that camalexin is induced to equivalent levels by both virulent and avirulent strains of *P. syringae* pv. *maculicola* (*Psm*). Thus, *Psm* can cause a significant amount of disease in the presence of the phytoalexin. Ausubel and co-workers[76] also reported the isolation of three different mutants that produce little to no camalexin. The mutant that produced no camalexin was unaltered in its interaction with either avirulent or virulent *Psm* strains, suggesting that camalexin does not play an important role in Arabidopsis-*Psm* interactions. Inoculating these mutants with other pathogens should help clarify the role of camalexin, if any, in disease resistance.

C. SYSTEMIC ACQUIRED RESISTANCE

Many plants can develop a broad-spectrum resistance to bacteria, fungi, and viruses following an initial inoculation with a pathogen that induces an HR.[88] The basis of such systemic acquired resistance (SAR) is poorly understood, but it has recently received increased scrutiny. SAR can also be induced by application of salicylic acid or 2,6-dichloronicotinic acid (INA).[89] The SAR response is always correlated with production of several extracellular proteins ("Pathogenesis-Related" proteins), some of which have chitinase and BG activity,[90] but it is not clear whether or not these proteins contribute significantly to the immune state. Uknes and co-workers[91] have recently established that SAR can be induced in Arabidopsis using INA, and that such immunized plants become resistant to infection by *P. syringae* pv. tomato and *Peronospora parasitica.* The protection against *P. parasitica* appears to be mediated by an HR-like mechanism, suggesting that SAR and *R*-gene-type resistance may have common components. Several groups are now screening for Arabidopsis mutants that are altered in SAR. Two classes of mutants in particular are being sought: mutants that constitutively express PR-proteins and mutants that fail to express PR-proteins. Several mutants of the former class have been identified.[91a] As might be expected, these mutants behave as if SAR is constitutively turned on (e.g., they are resistant to normally virulent *P. parasitica* strains). Even more intriguing, however, is that some of these mutants also display necrotic HR-like patches on their leaves before inoculation,[91b] similar to "lesion mimic" mutants of maize (see below). Isolation of mutants that fail to express PR-proteins has not been reported.

IV. LESION MIMIC MUTATIONS IN ARABIDOPSIS

Lesion mimic mutants are characterized by development of necrotic patches on leaves, even in the absence of pathogens. Because the lesions often appear similar to lesions induced by specific pathogens, or similar to an HR, it has been postulated that lesion mimic mutants are affected in some aspect of the plant defense response. This connection to plant disease resistance, however, is tenuous, at best. Several different lesion mimic mutants (*Les* mutants) have been characterized in maize.[92–94] Some *Les* mutants develop lesions even when grown axenically, while others require nonsterile growth conditions. The second class of mutants have also been called "paranoid" mutants, as they appear to respond to minor infections with an exaggerated defense response. Interestingly, the majority of the maize *Les* mutants are dominant. It is tempting to speculate that such phenotypes may be caused by a defective signal transduction component that gets locked in the "on position" once activated.

Several lesion mimic-like mutants have recently been identified in Arabidopsis.[95] Mutations in at least two different genes give rise to this phenotype in Arabidopsis, and these have been designated *accelerated cell death* (*acd*) mutants. Like many of the *Les* mutants in maize, the Arabidopsis *acd* mutants develop spontaneous necrotic lesions, even when grown under axenic conditions. The *acd* mutants also develop disease symptoms faster than wild-type plants when inoculated with a virulent *PSM* strain, hence the *acd* designation. Interestingly, the spontaneous lesions are extensively colonized by resident soil bacteria and fungi when the plants are grown under nonaxenic conditions. In addition, the *acd* mutants allow increased growth of the pathogen *P. syringae* pv. phaseolicola, which does not grow in wild-type Arabidopsis. These observations suggest that the lesions are not the result of an overprimed defense system, but perhaps an unrelated cell death phenomenon. Unlike the majority of the maize *Les* mutants, the *acd* mutants are recessive. There have been preliminary reports of additional *Les*-like mutants in Arabidopsis, including dominant mutants.[95a] As mentioned in the previous section, one of these *Les*-like mutants expresses the SAR phenotype at the onset of lesion formation. Isolation of the mutated genes from such mutants should be especially informative, as they may provide insight into signal transduction systems involved in plant defense.

V. ARABIDOPSIS-VIRUS INTERACTIONS

Although plant viruses have been intensively studied for decades, very little is known about the host factors that are required for virus replication and spread. A better understanding of this aspect of virus biology is critical to design of effective control strategies. In addition, viruses should provide a tool for studying fundamental aspects of DNA and RNA replication in plant cells, as well as certain aspects of cell biology, such as the structure and function of plasmodesmata. It should be possible to identify host functions required for virus replication and spread using a genetic approach. However, many of these functions will likely be essential for cell viability (e.g., DNA replication), thus it will be necessary to develop screens for conditional mutants (e.g., temperature-sensitive mutants).

As was the case for bacterial and fungal plant pathogens, not much was known about viruses of Arabidopsis until recently. Predictably, many viruses that can replicate in other crucifers also can replicate in Arabidopsis. These include cauliflower mosaic virus (CMV; a double-stranded DNA virus),[31,96–98] beet curly top virus (BCTV; a single-stranded DNA geminivirus),[99] turnip crinkle virus (TCV; a positive-strand RNA virus),[27,100] turnip yellow mosaic virus (TYMV; a positive-strand RNA virus),[101] and a crucifer isolate of tobacco mosaic virus (TMV-Cg; a positive-strand RNA virus).[102] Thus several different classes of viruses are represented, including both DNA and RNA viruses.

Mutants (either naturally occurring or induced) that are altered in their interaction with specific viruses are now being sought. Several classes of mutants can be envisioned. These include "immune" mutants that do not support viral replication in any cell; "resistant" mutants that reduce replication and systemic spread of the virus; "tolerant" mutants that have reduced disease symptoms, but support normal levels of virus replication; and "sensitive" mutants that display increased symptoms. Each of these classes could arise by several different mechanisms. For example, a mutant that falls into the "resistant" class has been isolated by Ishikawa and co-workers.[102] This group identified an Arabidopsis mutant that accumulates significantly reduced levels of TMV-Cg particles (as assayed by coat protein accumulation) in noninoculated leaves. This phenotype could be caused by a reduction in viral replication rate, or by an inability of the virus to spread systematically in the plant. The stage at which viral multiplication/spread is inhibited in this mutant has not yet been established, however.

Identification of naturally occurring Arabidopsis ecotypes that are either resistant or tolerant to specific viral strains has also been reported.[27,31] This avenue may be especially fruitful for identification of genes appropriate for transfer to crop plants. Most Arabidopsis ecotypes are susceptible to infection by TCV, but Simon and co-workers[27] have identified an Arabidopsis ecotype (Dijon) that is resistant. Protoplasts from Dijon support high levels of virus replication, thus the resistance of this ecotype is likely caused by restricted virus spread, rather than a block in virus replication. Leisner and Howell[31] have reported similar observations for the interaction of CaMV and Arabidopsis. In several Arabidopsis ecotypes, movement of specific CaMV isolates is significantly reduced, as are disease symptoms. In both the TCV and CaMV systems, genetic analysis of the plant and viral isolates should provide insight into how plant and viral components interact to allow spread of the virus. These studies represent the first attempts to analyze virus movement using host genetics.

VI. FUTURE DIRECTIONS

Arabidopsis has now established itself as a powerful system for studying both basic and applied questions in plant pathology. The natural variation present among Arabidopsis ecotypes is proving especially useful for identification of genes that confer resistance to viral, bacterial, and fungal plant pathogens. Molecular isolation of several of these genes appears imminent, which should provide fundamental insights into their structure and function. Isolation of these genes will also allow us to test the efficacy of transferring disease-resistance genes between distantly related plant species. A positive outcome would have great practical and basic implications. Functional transfer would indicate that Arabidopsis can be used as a source of disease-resistance genes for use in crop plants. It would also indicate that signal transduction components that interact with disease-resistance genes are functionally conserved between plant species.

Although natural variation has proved useful for identification of disease-resistance genes, it is unlikely to be useful for identification of signal transduction genes; these genes are likely to be conserved between Arabidopsis varieties. Identification of these genes thus will require use of mutagenesis. Unfortunately, identification of mutations in these genes has proved more difficult than expected. As discussed previously, there are two probable explanations: these genes might be essential for plant viability; null mutations in essential genes cannot be recovered. Alternatively, these genes might be functionally redundant. The first possibility can be addressed by developing temperature-sensitive screens. This is relatively straightforward for Arabidopsis, since Arabidopsis is usually grown and assayed in temperature-controlled growth cabinets. However, if a gene is absolutely required for viability, it will be difficult to establish a specific role in disease resistance. Another possible scenario is that some genes will be required for growth under nonaxenic conditions, but not under axenic conditions. One could imagine that normally harmless saprophytes may become pathogenic on a mutant that lacks a fundamental component of the defense response. It will thus be useful to develop screens that can be carried out under axenic conditions. This also should be straightforward with Arabidopsis, as dozens of plants can be grown in a single petri dish under axenic conditions.

Functionally redundant genes present a larger problem. The first problem is to identify candidate genes. In the absence of other information, the best candidates are those genes that are induced during resistant reactions. If redundancy is due to gene duplication, it may be possible to reduce expression of all members of the gene family using antisense technology.[103,104] If redundancy is due to unrelated genes, however, this approach will not work. Since redundant genes often do not fully substitute for one another, a possible alternative is to develop sensitive screens that allow detection of mutants with intermediate phenotypes. If the intermediate phenotype can be reliably scored, the responsible gene can be isolated by a standard position-cloning approach. Alternatively, mutants with intermediate phenotypes could be remutagenized to look for plants with a stronger mutant phenotype ("enhancer" mutations).

Bringing the power of genetics to bear on the problems of plant-pathogen interactions is overdue. The development of Arabidopsis as a model genetic system has now made such an approach both feasible and attractive. In a very short period of time, employment of Arabidopsis genetics has provided new insights into phenomenon such as *R*-gene specificity, phytoalexin production, and SAR. Insight into even more fundamental aspects of plant-pathogen interactions should soon be forthcoming.

ACKNOWLEDGMENTS

In order to make this chapter as timely as possible I have included significant amounts of unpublished information. I thank the following people for generously sharing results with me prior to publication: Fred Ausubel, Massachusetts General Hospital; Andrew Bent, Karen Century, Barbara Kunkel, and Brian Staskawicz, University of California, Berkeley; Jeffrey Dangl and Robert Dietrich, Max-Delbrück Laboratory, Köln; Keith Davis, Ohio State University; Alan Slusarenko, University of Zürich; Ian Crute and Eric Holub, Horticultural Research International. I am also indebted to the members of my group (Sherryl Bisgrove, Karen Jepson-Innes, Nadine Smith, and Richard Wagner), who are responsible for the work cited from our laboratory and who provided critical reading of this manuscript. Research in our laboratory is funded by Public Health Service grants R29 GM 46451 and S07 RR 7031N.

REFERENCES

1. Collins, F., Positional cloning: let's not call it reverse anymore, *Nature Genet.*, 1, 3, 1992.
2. Martin, G. B., de Vicente, M. C., and Tanksley, S. D., High-resolution linkage analysis and physical characterization of the *Pto* bacterial resistance locus in tomato, *Mol. Plant-Microbe Interact.*, 6, 26, 1993.

3. Ronald, P. C., Albano, B., Tabien, R., Abenes, L., Wu, K., McCouch, S., and Tanksley, S. D., Genetic and physical analysis of the rice bacterial blight disease resistance locus, *Xa21, Mol. Gen. Genet.*, 236, 113, 1992.
4. Michelmore, R. W., Anderson, P. A., Witsenboer, H., Kesseli, R. V., Paran, I., Francis, D. M., and Ochoa, O., Molecular markers and genome analysis in the manipulation of lettuce downy mildew, in *Advances in Molecular Genetics of Plant-Microbe Interactions.* Nester, E. W. and Verma, D. P. S., Eds., Kluwer Academic Publishers, Dordrecht, 1993, 517.
5. Koncz, C., Chua, N.-H., and Schell, J., Eds., *Methods in Arabidopsis Research,* World Scientific, Singapore, 1992.
6. Meyerowitz, E. M., Arabidopsis, a useful weed, *Cell,* 56, 263, 1989.
7. Rédei, G. P. and Koncz, C., Classical mutagenesis, in *Methods in Arabidopsis Research,* Koncz, C., Chua, N.-H., and Schell, J., Eds., World Scientific, Singapore, 1992, 16.
8. Meyerowitz, E. M., Introduction to the *Arabidopsis* genome, in *Methods in Arabidopsis Research,* Koncz, C., Chua, N.-H., and Schell, J., Eds., World Scientific, Singapore, 1992, 100.
9. Goldberg, R. B., DNA sequence organization in the soybean plant, *Biochem. Genet.,* 16, 45, 1978.
10. Pruitt, R. E. and Meyerowitz, E. M., Characterization of the genome of *Arabidopsis thaliana, J. Mol. Biol.,* 187, 169, 1986.
11. Valvekens, D., van Montagu, M., and van Lijsebettens, M., *Agrobacterium tumefaciens*-mediated transformation of *Arabidopsis thaliana* root explants by using kanamycin selection, *Proc. Natl. Acad. Sci. U.S.A.,* 85, 5536, 1988.
12. Feinbaum, R. L. and Ausubel, F. M., Transcriptional regulation of the *Arabidopsis thaliana* chalcone synthase gene, *Mol. Cell. Biol.,* 8, 1985, 1988.
13. Melan, M. A., Dong, X., Endara, M. E., Davis, K. R., Ausubel, F. M., and Peterman, T. K., An *Arabidopsis thaliana* lipoxygenase gene can be induced by pathogens, abscisic acid, and methyl jasmonate, *Plant Physiol.,* 101, 441, 1993.
14. Hauge, B. M., Hanley, S. M., Cartinhour, S., Cherry, J. M., Goodman, H. M., Korneef, M., Stam, P., Chang, C., Kempin, S., Medrano, L., and Meyerowitz, E. M., An integrated genetic/RFLP map of the *Arabidopsis thaliana* genome, *Plant J.,* 3, 745, 1993.
15. Grill, E. and Somerville, C. R., Construction and characterization of a yeast artificial chromosome library of Arabidopsis which is suitable for chromosome walking, *Mol. Gen. Genet.,* 226, 484, 1991.
16. Ward, E. R. and Jen, G. C., Isolation of single-copy-sequence clones from a yeast artificial chromosome library of randomly-sheared *Arabidopsis thaliana* DNA, *Plant Mol. Biol.,* 14, 561, 1990.
17. Ecker, J. R., PFGE and YAC analysis of the *Arabidopsis* genome, *Methods,* 1, 186, 1990.
18. Hauge, B. M., Hanley, S., Giraudat, J., and Goodman, H. M., Mapping the *Arabidopsis* genome, in *Molecular Biology of Plant Development,* Jenkins, G. and Schurch, W., Eds., Cambridge University Press, Cambridge, in press.
19. Hwang, I., Kohchi, T., Hauge, B. M., Goodman, H. M., Schmidt, R., Cnops, G., Dean, C., Gibson, S., Iba, K., Danhoff, L., and Somerville, C., Identification and map position of YAC clones comprising one third of the *Arabidopsis* genome, *Plant J.,* 1, 367, 1991.
20. Kranz, A. R. and Kirchheim, B., Genetic resources in *Arabidopsis, Arabidopsis Inf. Serv.,* 24, 1, 1987.
21. Abbot, R. J., and Gomes, M. E. F., Population genetic structure and outcrossing rate of *Arabidopsis thaliana* (L.) Heynh, *Heredity,* 62, 411, 1989.
22. Crute, I. R., Holub, E. B., Tor, M., Brose, E., and Beynon, J. L., The identification and mapping of loci in *Arabidopsis thaliana* for recognition of the fungal pathogens: *Peronospora parasitica* (downy mildew) and *Albugo candida* (white blister), in *Advances in Molecular Genetics of Plant-Microbe Interactions,* Nester, E. W. and Verma, D. P. S., Eds., Kluwer Academic Publishers, Dordrecht, 1993, 437.
23. Debener, T., Lehnackers, H., Arnold, M., and Dangl, J. L., Identification and molecular mapping of a single *Arabidopsis thaliana* locus determining resistance to a phytopathogenic *Pseudomonas syringae* isolate, *Plant J.,* 1, 289, 1991.
24. Innes, R. W., Smith, N. M., Liu, Y.-C., Bent, A. F., Staskawicz, B. J., and Bisgrove, S. R., unpublished results, 1993.
25. Koch, E. and Stusarenko, A., *Arabidopsis* is susceptible to infection by a downy mildew fungus, *Plant Cell,* 2, 437, 1990.
26. Kunkel, B. N., Bent, A. F., Dahlbeck, D., Innes, R. W., and Staskawicz, B. J., *RPS2,* an Arabidopsis disease resistance locus specifying recognition of *Pseudomonas syringae* strains expressing the avirulence gene *aur Rpt 2. Plant Cell,* 5, 865, 1993.

27. Simon, A. E., Li, X. H., Lew, J. E., Stange, R., Zhang, C., Polacco, M., and Carpenter, C. D., Susceptibility and resistance of *Arabidopsis thaliana* to turnip crinkle virus, *Mol. Plant-Microbe Interact.,* 5, 496, 1992.
28. Simpson, R. B. and Johnson, L. J., *Arabidopsis thaliana* as a host for *Xanthomonas campestris* pv. *campestris, Mol. Plant-Microbe Interact.,* 3, 233, 1990.
29. Tsuji, J., Somerville, S. C., and Hammerschmidt, R., Identification of a gene in *Arabidopsis thaliana* that controls resistance to *Xanthomonas campestris* pathovar *campestris, Physiol. Mol. Plant Pathol.,* 38, 57, 1991.
30. Whalen, M., Innes, R., Bent, A., and Staskawicz, B., Identification of *Pseudomonas syringae* pathogens of *Arabidopsis* and a bacterial locus determining avirulence on both arabidopsis and soybean, *Plant Cell,* 3, 49, 1991.
31. Leisner, S. and Howell, S. H., Symptom variation in different *Arabidopsis thaliana* ecotypes produced by cauliflower mosaic virus, *Phytopathology,* 82, 1042, 1992.
32. Morgan, O. D., A study of 4 weed hosts of *Sclerotinia* species in alfalfa fields, *Plant Dis. Rep.,* 55, 1087, 1971.
33. Jones, J. B. and McCarter, S. M., Survival of *Pseudomonas tomato* in association with weeds, diseased tissue and soil in Georgia USA, *Phytopathology,* 71, 884, 1981.
34. Davis, K. R., Schott, E., and Ausubel, F. M., Virulence of selected phytopathogenic pseudomonads in *Arabidopsis thaliana, Mol. Plant-Microbe Interact,* 4, 477, 1991.
35. Dong, X., Mindrinos, M., Davis, K. R., and Ausubel, F. M., Induction of *Arabidopsis thaliana* defense genes by virulent and avirulent *Pseudomonas syringae* strains and by a cloned avirulence gene, *Plant Cell,* 3, 61, 1991.
36. Innes, R. W., Bent, A. F., Kunkel, B. N., Bisgrove, S. R., and Staskawicz, B. J., Molecular analysis of avirulence gene *avrRpt2* and identification of a putative regulatory sequence common to all known *Pseudomonas syringae* avirulence genes, *J. Bacteriol.,* 175, 4859, 1993.
37. Hammerschmidt, R. and Davis, K. R., *Arabidopsis as a Model System for Studying Plant-Pathogen Interactions,* APS Press, St. Paul, MN, 1992.
38. Flor, H. H., Current status of the gene-for-gene concept, *Annu. Rev. Phytopathol.,* 9, 275, 1971.
39. Flor, H. H., Host-parasite interactions in flax rust—its genetics and other implications, *Phytopathology,* 45, 680, 1955.
40. Keen, N. T., Gene-for-gene complementarity in plant-pathogen interactions, *Annu. Rev. Genet.,* 24, 447, 1990.
41. Keen, N. T., The molecular biology of disease resistance, *Plant Mol. Biol.,* 19, 109, 1992.
42. Barret, J., The gene-for-gene hypothesis: parable or paradigm, in *Ecology and Genetics of Host-Parasite Interactions,* Rollinson, D. and Anderson, R. M., Eds., Academic Press, London, 1985, 215.
42a. Crute, I., personal communication.
43. Dangl, J. L., Ritter, C., Gibbon, M. J., Muir, L. A., Wood, J. R., Goss, S., Mansfield, J., Taylor, J. D., and Vivian, A., Functional homologs of the Arabidopsis *Rpm1* disease resistance gene in bean and pea, *Plant Cell,* 4, 1359, 1992.
44. Dangl, J., Debener, T., Gerwin, M., Kiedrowski, S., Ritter, C., Bendahmane, A., Liedgens, H., and Lewald, J., Genetic approaches to an understanding of specific resistance responses of *Arabidopsis thaliana* against phytopathogenic Pseudomonads, in *Advances in Molecular Genetics of Plant-Microbe Interactions,* Nester, E. W. and Verma, D. P. S., Eds., Kluwer Academic Publishers, Dordrecht, 1993, 405.
45. Yanofsky, M. F., Ma, H., Bowman, J. L., Drews, G. N., Feldmann, K. E., and Meyerowitz, E. M., The protein encoded by the *Arabidopsis* homeotic gene *agamous* resembles transcription factors, *Nature,* 346, 35, 1990.
46. Martin, C., Carpenter, R., Sommer, H., Saedler, H., and Coen, E. S., Molecular analysis of instability in flower pigmentation of *Antirrhinum majus,* following isolation of the pallida locus by transposon tagging, *EMBO J.,* 4, 1625, 1985.
47. Fedoroff, N., Furtek, D., and Nelson, O., Cloning of the *bronze* locus in maize by a simple and generalizable procedure using the transposable controlling element *Ac, Proc. Natl. Acad. Sci. U.S.A.,* 81, 3825, 1984.
48. Newbury, H. J., Aitken, E. A. B., and Callow, J., Mutagenesis of a race-specific rust resistance gene in *Antirrhinum majus* using a transposon-tagging protocol, in *Advances in Molecular Genetics of Plant-Microbe Interactions,* Nester, E. W. and Verma, D. P. S., Eds., Kluwer Academic Publishers, Dordrecht, 1993, 511.

49. Koncz, C., Schell, J., and Rédei, G. P., T-DNA transformation and insertion mutagenesis, in *Methods in Arabidopsis Research,* Koncz, C., Chua, N.-H., and Schell, J., Eds., World Scientific, Singapore, 1992, 224.
50. Feldmann, K. A., T-DNA insertion mutagenesis in *Arabidopsis:* mutational spectrum, *Plant J.,* 1, 71, 1991.
51. Coupland, G., Transposon tagging in *Arabidopsis* in *Methods in Arabidopsis Research,* Koncz, C., Chua, N.-H., and Schell, J., Eds., World Scientific, Singapore, 1992, 290.
52. Grevelding, C., Becker, D., Kunze, R., Von Menges, A., Fantes, V., Schell, J., and Masterson, R., High rates of *Ac/Ds* germinal transposition in *Arabidopsis,* suitable for gene isolation by insertional mutagenesis, *Proc. Natl. Acad. Sci. U.S.A.,* 89, 6085, 1992.
53. Dean, C., Sjodin, C., Page, T., Jones, J., and Lister, C., Behavior of the maize transposable element *Ac* in *Arabidopsis thaliana, Plant J.,* 2, 69, 1992.
54. Hulbert, S. H. and Bennetzen, J. L., Recombination at the *Rp1* locus of maize, *Mol. Gen. Genet.,* 226, 1991.
55. Pryor, A., The origin and structure of fungal disease resistance genes in plants, *Trends Genet.,* 3, 157, 1987.
56. Jorgensen, J. H., Multigene families of powdery mildew resistance genes in locus *Mla* on barley chromosome 5, *Plant Breeding,* 108, 53, 1992.
57. Sudupak, M. A., Bennetzen, J. L., and Hulbert, S. H., Unequal exchange and meiotic instability of disease-resistance genes in the *Rp1* region of maize, *Genetics,* 133, 119, 1993.
58. Tamaki, S., Dahlbeck, D., Staskawicz, B., and Keen, N. T., Characterization and expression of two avirulence genes cloned from *Pseudomonas syringae* pv. *glycinea, J. Bacteriol.,* 170, 4846, 1988.
59. Fillingham, A. J., Wood, J., Bevan, J. R., Crute, I. R., Mansfield, J. W., Taylor, J. D., and Vivian, A., A virulence genes from *Pseudomonas syringae* pathovars *phaseolicola* and *pisi* confer specificity towards both host and non-host species, *Physiol. Mol. Plant Pathol.,* 40, 1, 1992.
60. Staskawicz, B. J., Dahlbeck, D., Keen, N., and Napoli, C., Molecular characterization of cloned avirulence genes from race 0 and race 1 of *Pseudomonas syringae* pv. *glycinea, J. Bacteriol.,* 169, 5789, 1987.
61. Keen, N. T. and Buzzel, R. I., New disease resistance genes in soybean against *Pseudomonas syringae* pv. *glycinea:* evidence that one of them interacts with a bacterial elicitor, *Theor. Appl. Genet.,* 81, 133, 1991.
62. Kobayashi, D. Y., Tamaki, S. J., and Keen, N. T., Cloned avirulence genes from the tomato pathogen *Pseudomonas syringae* pv. *tomato* confer cultivar specificity on soybean, *Proc. Natl. Acad. Sci. U.S.A.,* 86, 157, 1989.
63. Whalen, M. C., Stall, R. E., and Staskawicz, B. J., Characterization of a gene from a tomato pathogen determining hypersensitive resistance in a non-host species and genetic analysis of this resistance in bean, *Proc. Natl. Acad. Sci. U.S.A.,* 85, 6743, 1988.
64. Klement, Z., Hypersensitivity, in *Phytopathogenic Procaryotes,* Mount, M. S. and Lacy, G. H., Eds., Academic Press, New York, 1982, 149.
65. Dixon, R. A. and Harrison, M. J., Activation, structure and organization of genes involved in microbial defense in plants, *Adv. Genet.,* 28, 165, 1990.
66. Slusarenko, A., Croft, K., and Voisey, C., Biochemical and molecular events in the hypersensitive response of bean to *Pseudomonas syringae* pv. *phaseolicola,* in *Biochemistry and Molecular Biology of Plant-Pathogen Interactions,* Smith, C., Ed., Oxford University Press, Oxford, 1991, 126.
67. Koch, E. and Slusarenko, A. J., Fungal pathogens of *Arabidopsis thaliana* (L.) Heyhn, *Bot. Helv.,* 100, 257, 1990.
68. Mauch-Mani, B., Croft, K. P. C., and Slusarenko, A., The genetic basis of resistance of *Arabidopsis thaliana* L. HEYHN to *Peronospora parasitica,* in *Arabidopsis as a Model System for Studying Plant-Pathogen Interactions.* Hammerschmidt, R. and Davis, K. R., Eds., APS Press, St. Paul, MN, in press.
69. Tsuji, J., Jackson, E. P., Gage, D. A., Hammerschmidt, R., and Somerville, S. C., Phytoalexin accumulation in *Arabidopsis thaliana* during the hypersensitive reaction to *Pseudomonas syringae* pv *syringae, Plant Physiol.,* 98, 1304, 1992.
70. Samac, D. A. and Shah, D. M., Developmental and pathogen-induced activation of the Arabidopsis acidic chitinase promoter, *Plant Cell,* 3, 1063, 1991.
71. Niyogi, K. and Fink, G., Two anthranilate synthase genes in Arabidopsis: defense-related regulation of the tryptophan pathway, *Plant Cell,* 4, 721, 1992.

72. Keith, B., Dong, X., Ausubel, F. M., and Fink, G. R., Differential induction of 3-deoxy-D-arabino-heptulosonate 7-phosphate synthase genes in *Arabidopsis thaliana* by wounding and pathogenic attack, *Proc. Natl. Acad. Sci. U.S.A.*, 88, 8821, 1991.
73. Kiedrowski, S., Kawalleck, P., Hahlbrock, K., Somssich, I. E., and Dangl, J. L., Rapid activation of a novel plant defense gene is strictly dependent on the *Arabidopsis RPM1* resistance locus, *EMBO J.*, 11, 4677, 1992.
74. Ohl, S., Hedrick, S. A., Chory, J., and Lamb, C. J., Functional properties of a phenylalanine ammonia-lyase promoter from *Arabidopsis, Plant Cell,* 2, 837, 1990.
75. Samac, D. A., Hironaka, C. M., Yallaly, P. E., and Shah, D. M., Isolation and characterization of the genes encoding basic and acidic chitinase in *Arabidopsis thaliana, Plant Physiol.*, 93, 1990.
76. Ausubel, F. M., Glazebrook, J., Greenberg, J., Mindrinos, M., and Yu, G.-L., Analysis of the *Arabidopsis* defense response to *Pseudomonas* pathogens, in *Advances in Molecular Genetics of Plant-Microbe Interactions,* Nester, E. W. and Verma, D. P. S., Eds., Kluwer Academic Publishers, Dordrecht, 1993, 393.
77. Last, R. L., Bissinger, P. H., Mahoney, D. J., Radwanski, E. R., and Fink, G. R., Tryptophan mutants in *Arabidopsis:* the consequence of duplicated tryptophan synthase b genes, *Plant Cell,* 3, 345, 1991.
78. Last, R. L. and Fink, G. R., Tryptophan-requiring mutants of the plant *Arabidopsis thaliana, Science,* 240, 305, 1988.
79. Sutherland, M. W., The generation of oxygen radicals during host plant responses to infection, *Physiol. Mol. Plant Pathol.*, 39, 79, 1991.
79a. Ausubel, F. and Davis, K., personal communication.
80. Bent, A. F., Innes, R. W., Ecker, J. E., and Staskawicz, B. J., Disease development in ethylene-insensitive *Arabidopsis thaliana* infected with virulent and avirulent *Pseudomonas* and *Xanthomonas* pathogens, *Mol. Plant-Microbe Interact.*, 5, 372, 1992.
81. Apostol, I., Heinstein, P. F., and Low, P. S., Rapid stimulation of an oxidative burst during elicitation of cultured plant cells, *Plant Physiol.*, 90, 109, 1989.
82. Orlandi, E. W., Hutcheson, S. W., and Baker, C. J., Early physiological responses associated with race-specific recognition in soybean leaf tissue and cell suspensions treated with *Pseudomonas syringae* pv. *glycinea, Physiol. Mol. Plant Pathol.,* 40, 173, 1992.
83. de Maagd, R. A., Cameron, R. K., Dixon, R. A., and Lamb, C. J., Screening methods for Arabidopsis mutants affected in the signal transduction pathways to defense responses, in *Advances in Molecular Genetics of Plant-Microbe Interactions,* Nester, E. W. and Verma, D. P. S., Eds., Kluwer Academic Publishers, Dordrecht, 1993, 445.
84. Jorgensen, J. H., Genetic analysis of barley mutants with modifications of powdery mildew resistance gene Ml-a12, *Genome,* 30, 129, 1988.
85. Torp, J. and Jorgensen, J., Modification of barley powdery mildew resistance gene *Ml-a12* by induced mutation, *Can. J. Genet. Cytol.*, 28, 725, 1986.
86. Kunkel, B. N., Bent, A. F., Dahlbeck, D., Innes, R. W., and Staskawicz, B. J., Identification of an *Arabidopsis* locus that governs avirulence gene-specific disease resistance, in *Advances in Molecular Genetics of Plant-Microbe Interactions,* Nester, E. W. and Verma, D. P. S., Eds., Kluwer Academic Publishers, Dordrecht, 1993, 417.
87. Browne, L. M., Conn, K. L., Ayer, W. A., and Tewari, J. P., The camalexins: new phytoalexins produced in the leaves of *Camelina sativa* (Cruciferae), *Tetrahedron,* 47, 3909, 1991.
88. Chester, K. S., The problem of acquired physiological immunity in plants (continued), *Q. Rev. Biol.*, 8, 275, 1933.
89. Métraux, J.-P., Ahl-Goy, P., Staub, T., Speich, J., Steinemann, A., Ryals, J., and Ward, E., Induced systemic resistance in cucumber in response to 2,6-isonicotinic acid and pathogens, in *Advances in Molecular Genetics of Plant-Microbe Interactions,* Hennecke, H. and Verma, D. P. S., Eds., Kluwer Academic Publishers, Dordrecht, 1991, 432.
90. Ward, E. R., Uknes, S., Williams, S. C., Dincher, S. S., Wiederhold, D. L., Alexander, D. A., Métrauz, J.-P., and Ryals, J. A., Coordinate induction of gene expression in response to biological and chemical agents that induce systemic acquired resistance, *Plant Cell,* 3, 1085, 1991.
91. Uknes, S., Mauch-Mani, B., Moyer, M., Potter, S., Williams, S., Dincher, S., Chandler, D., Slusarenko, A., Ward, E., and Ryals, J., Acquired resistance in Arabidopsis, *Plant Cell,* 4, 645, 1992.
91a. Ward, E., personal communication.
91b. Dangl, J., personal communication.

92. Hoisington, D. A., Neuffer, M. G., and Walbot, V., Disease lesion mimics in maize, *Dev. Biol.*, 93, 381, 1982.
93. Walbot, V., Hoisington, D. A., and Neuffer, M. G., Disease lesion mimic mutations, in *Genetic Engineering of Plants*, Kosuge, T., Meredith, C. P., and Hollaender, A., Eds., Plenum Press, New York, 1983, 431.
94. Walbot, V., Maize mutants for the 21st century, *Plant Cell*, 3, 851, 1991.
95. Greenberg, J. R., and Ausubel, F. M., Arabidopsis mutants compromised for the control of cellular damage during pathogenesis and aging. *Plant J.*, 4, 327, 1993.
95a. Dangl, J. and Dietrich, R., personal communication.
96. Balazs, E. and Lebeurier, G., Arabidopsis is a host of cauliflower mosaic virus, *Arabidopsis Inf. Serv.*, 18, 130, 1981.
97. Melcher, U., Symptoms of cauliflower mosaic virus infection in *Arabidopsis thaliana* and turnip, *Bot. Gaz.*, 150, 139, 1989.
98. Susnova, V. and Polak, Z., Susceptibility of *Arabidopsis thaliana* to infection with some plant viruses, *Biol. Plant.*, 17, 156, 1975.
99. Stenger, D. C., Davis, K. R., and Bisaro, D. M., Limited replication of tomato golden mosaic virus DNA in explants of nonhost species, *Mol. Plant-Microbe Interact.*, 5, 525, 1992.
100. Li, X. H. and Simon, A. E., Symptom intensification on cruciferous hosts by the virulent satellite RNA of turnip crinkle virus, *Phytopathology*, 80, 238, 1990.
101. Bozarth, C. S., Weiland, J. J., and Dreher, T. W., Expression of ORF-69 of turnip yellow mosaic virus is necessary for viral spread in plants, *Virology*, 187, 124, 1992.
102. Ishikawa, M., Obata, F., Kumagai, T., and Ohno, T., Isolation of mutants of *Arabidopsis thaliana* in which accumulation of tobacco mosaic virus coat protein is reduced to low levels, *Mol. Gen. Genet.*, 230, 33, 1991.
103. Oeller, P. W., Min, W. L., Taylor, L. P., Pike, D. A., and Theologis, A., Reversible inhibition of tomato fruit senescence by antisense RNA, *Science*, 254, 437, 1991.
104. Ecker, J. R. and Davis, R. W., Inhibition of gene expression in plant cells by expression of antisense RNA, *Proc. Natl. Acad. Sci. U.S.A.*, 83, 5372, 1986.
105. Davis, K. R., *Arabidopsis thaliana* as a model host for studying plant-pathogen interactions in *Molecular Signals in Plant-Microbe Communications*, Verma, D. P. S., Ed., CRC Press, Boca Raton, FL, 1992, 393.
106. Dangl, J. L., Applications of *Arabidopsis thaliana* to outstanding issues in plant-pathogen interactions, *Int. Rev. Cytol.*, 144, 53, 1993.
107. Dangl, J. L., The emergence of *Arabidopsis thaliana* as a model for plant-pathogen interactions, *Adv. Plant Pathol.*, 10, 127, 1993.
108. Dangl, J. L., Holub, E. B., Debener, T., Lehnackers, H., Ritter, C., and Crute, I. R., Genetic definition of loci involved in *Arabidopsis*-pathogen interactions, in *Methods in Arabidopsis Research*, Koncz, C., Chua, N.-H., and Schell, J., Eds., World Scientific, Singapore, 1992, 393.
109. Schott, E. J., Davis, K. R., Dong, X., Mindrinos, M., Guevara, P., and Ausubel, F. M., *Pseudomonas syringae* infection of *Arabidopsis thaliana* as a model system for studying plant-bacterial interactions, in *Pseudomonas: Biotransformations, Pathogenesis, and Evolving Biotechnology*, Siver, S., Chakrabarty, A., and Iglewski, B., Eds., American Society for Microbiology, Washington, D.C., 1990, 82.
110. Daniels, M. J., Barber, C. E., Dow, J. M., Han, B., Liddle, S. A., Newman, M. A., Parker, J. E., Soby, S. D., and Wilson, T. G. J., Plant and bacterial genes involved in interactions between *Xanthomonas* and crucifers, in *Advances in Molecular Genetics of Plant-Microbe Interactions*, Nester, E. W. and Verma, D. P. S., Eds., Kluwer Academic Publishers, Dordrecht, 1993, 423.
111. Tsuji, J. and Somerville, S. C., *Xanthomonas campestris* pv. *campestris*-induced chlorosis in Arabidopsis, *Arabidopsis Inf. Serv.*, 26, 1, 1988.
112. Fletcher, J., *Arabidopsis thaliana* as an experimental host of the mollicute *Spiroplasma citri*, *Plant Dis.*, 76, 862, 1992.
113. Zijlstra, C. and Hohn, T., Cauliflower mosaic virus gene VI controls translation from dicistronic expression units in transgenic Arabidopsis plants, *Plant Cell*, 4, 1471, 1992.
114. Leisner, S. M., Turgeon, R., and Howell, S. H., Effects of host plant development and genetic determinants on the long-distance movement of cauliflower mosaic virus in Arabidopsis, *Plant Cell*, 5, 191, 1993.

115. Davis, K. R. and Ausubel, F. M., Characterization of elicitor induced defense responses in suspension-cultured cells of Arabidopsis, *Mol. Plant-Microbe Interact.,* 2, 363, 1989.
116. Verburg, J. G. and Huynh, Q. K., Purification and characterization of an antifungal chitinase from *Arabidopsis thaliana, Plant Physiol.,* 97, 450, 1991.
117. Trezzini, G. F., Horrichs, A., and Somssich, I. E., Isolation of putative defense-related genes from *Arabidopsis thaliana* and expression in fungal elicitor-treated cells, *Plant Mol. Biol.,* 21, 385, 1993.
118. Sijmons, P. C., Grundler, F. M. W., Von Mende, N., Burrows, P. R., and Wyss, U., *Arabidopsis thaliana* as a new model host for plant-parasitic nematodes, *Plant J.,* 1, 245, 1991.

FURTHER READING

Beckmann, J. S., *Plant Genomes: Methods for Genetic and Physical Mapping,* Kluwer Academic Publishers, Dordrecht, 1993.

Bennetzen, J. L. and Hulbert, S. H., Organization, instability and evolution of plant disease resistance genes, *Plant Mol. Biol.,* 20, 575, 1992.

Chapter 16

Isolation and Cloning of Plant Disease Resistance Genes

Nevin Dale Young

TABLE OF CONTENTS

I. INTRODUCTION

Understanding the molecular basis of disease-resistance genes has long been a goal of plant pathology. With the advent of gene cloning technology in the 1970s, this goal finally became realistic and in the intervening years several important host defense-related genes have been cloned and characterized.[1–4] Nevertheless, the goal of cloning actual resistance genes, as opposed to host response genes, has remained elusive. For the most part, this is because most, if not all, disease-resistance genes are defined genetically through segregation analysis and transmission genetics. By contrast, current gene cloning techniques depend upon some biochemical connection to the gene, generally through the protein or mRNA product to isolate the gene. This explains why so many host response genes—and so few plant disease-resistance genes—have been cloned to date.

Recently, new approaches to gene cloning have been developed and some of these strategies promise to make plant resistance gene cloning a reality. The two most widespread and general gene cloning techniques are *map-based cloning* (Figure 1) and *transposon tagging* (Figure 2). In map-based cloning, a DNA genetic marker, usually in the form of a restriction fragment length polymorphism (RFLP), is located near a gene of interest. The DNA sequence underlying the RFLP then acts as an entry point for chromosome walking,[5] which leads to the isolation and cloning of contiguous DNA segments including, potentially, the gene of interest. In transposon tagging, a cloned transposable element, either homologous or heterologous in origin, is induced to move in a genetic background containing the target resistance gene. Large numbers of progeny individuals are screened to find transposon-induced mutants in the resistance gene phenotype. A cloned copy of the transposon is then used to identify the genomic fragments into which the transposon inserted—and in this way, the underlying gene of interest.

In addition to these two primary methods for resistance gene cloning, other strategies, often unique to specific systems, have also been devised. One example is the use of a host-specific toxin to purify the corresponding toxin-binding protein.[6] While it is uncertain whether or not a toxin-binding protein necessarily corresponds to a susceptibility/resistance gene, this type of gene cloning strategy is important and, in appropriate systems, very powerful. However, the present review will focus only on map-based cloning and transposon tagging as methods for resistance gene cloning. This emphasis is based on the observation that these techniques, while still in their infancy, hold out the promise of someday cloning any resistance gene in any plant system simply on the basis of phenotype. In fact, a maize gene for resistance to *Cochliobolus carbonum* has already been tagged through the use of transposons[6a] and

0-87371-877-1/95/$0.00+$.50

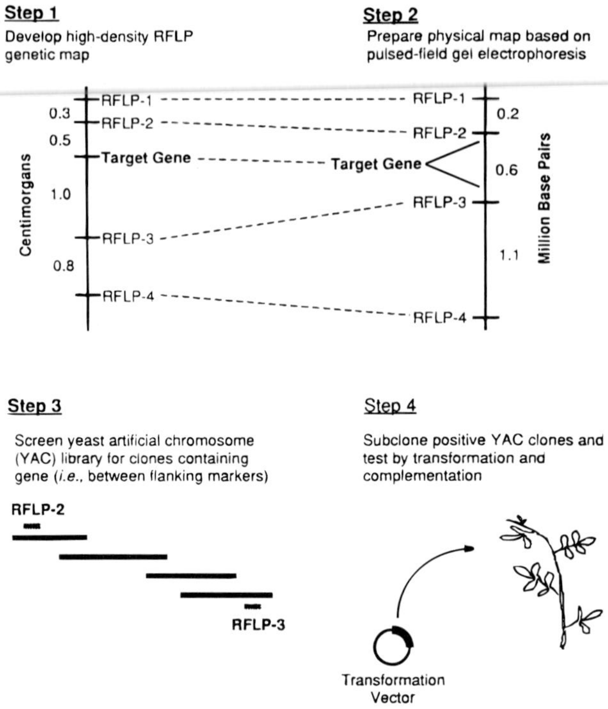

Figure 1 Major steps in map-based cloning. Details are given in the text. (From Young, 1990, with permission.)

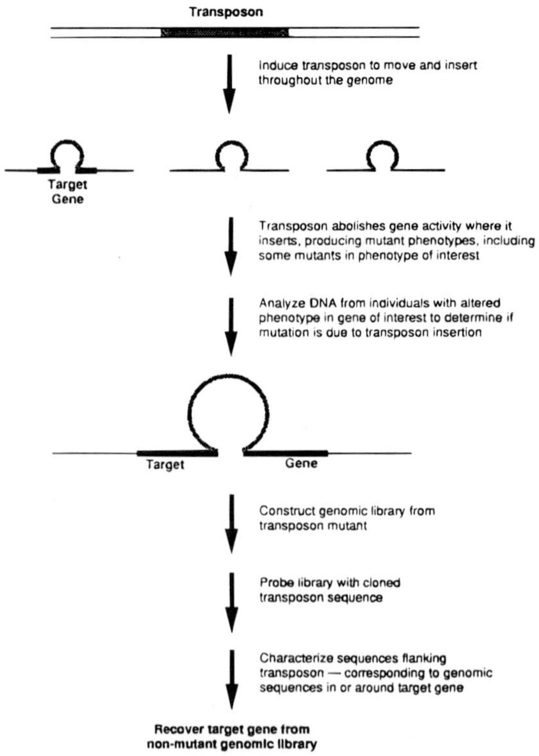

Figure 2 Major steps in transposon gene tagging. Details are given in the text.

Table 1 **Partial list of plant resistance genes that have been mapped with DNA genetic markers**

Gene	Plant genus	Pathogen	Ref.
RPM1	*Arabidopsis*	*Pseudomonas syringae*	45
Rps1, others	*Glycine*	*Phytophthora megasperma*	19
ml-o	*Hordeum*	*Erysiphe graminis*	17
Dm1, others	*Lactuca*	*Bremia lactucae*	18
Cf2, others	*Lycopersicon*	*Cladosporium fulvum*	78
Mi	*Lycopersicon*	*Meloidogyne incognita*	15
I1, I2	*Lycopersicon*	*Fusarium oxysporum*	32,79
Pto	*Lycopersicon*	*Pseudomonas syringae*	16
Sm	*Lycopersicon*	*Stemphylium* sp.	80
Tm-2	*Lycopersicon*	Tomato mosaic virus	14
Pi-2, Pi-4	*Oryza*	*Pyricularia oryzae*	20
Gro1	*Solanum*	*Globodera rostochiensis*	81
Rx1, Rx2	*Solanum*	Potato virus X	82
Ht1	*Zea*	*Helminthosporium turcicum*	83
Mdm1	*Zea*	Maize dwarf mosaic virus	84
Rp1, others	*Zea*	*Puccinia sorghi*	63

researchers are very close to the isolation of resistance genes for tobacco mosaic virus (TMV) and *Pseudomonas syringae* in tomato through map-based cloning methods.[7]

II. MAP-BASED CLONING

Cloning a gene solely on the basis of its chromosomal location is based on several molecular techniques that have been developed in the past few years. First, because the genomes of higher plants are very large (between 10^8 and 10^{10} base pairs or more in length), a DNA genetic marker must be extremely close to a resistance gene to be useful as a starting point for cloning. This requires special strategies for targeting the region around the gene of interest with many tightly linked DNA markers that can then act as starting points for chromosome walking.[8] Secondly, after a gene has been linked to nearby DNA markers, high-resolution genetic and physical maps need to be constructed as a basis for chromosome walking and gene cloning. This requires the specialized gel electrophoresis system, pulsed field gel electrophoresis (PFGE),[9] for analyzing DNA molecules a million base pairs or more in length. Thirdly, given a high-resolution physical map, there will still usually be a genomic region of several hundred kilobase pairs (kbp) to clone and characterize in order to identify the resistance gene itself. This requires specialized cloning vectors, specifically yeast artificial chromosomes (YACs),[10] that can maintain and propagate large genomic inserts up to several hundred kilobase pairs. Finally, finding the target resistance gene sequence within the cloned genomic region requires a strategy for identifying and characterizing candidate coding regions, as well as a highly efficient plant transformation technique.

A. IDENTIFYING DNA MARKERS LINKED TO TARGET RESISTANCE GENES

Map-based cloning is based on the ability to identify DNA genetic markers that are very tightly linked to the gene of interest. Several reviews have described the underlying methodology of linkage mapping with DNA genetic markers,[11,12] so it will not be described in detail here. Briefly, one begins with a population that segregates for numerous DNA genetic markers, including RFLPs or random amplified polymorphic DNAs (RAPDs). If the population also segregates for disease resistance, the underlying resistance gene can be placed on a linkage map composed of the DNA markers. The markers most tightly linked to the target gene can then act as entry points for chromosome walking and gene cloning. Table 1 is a partial list of major plant disease-resistance genes that have been mapped using DNA genetic markers.

In practice, complete analysis of a segregating population with hundreds of DNA markers is rarely used to uncover just one or a few markers tightly linked to a gene of interest. In fact, to find markers close enough to the gene to initiate chromosome walking and gene cloning, 1000 or more DNA markers may need to be tested. For these reasons, various strategies have been devised to home in on the region surrounding the target gene quickly and efficiently. The first of these strategies, and one that is still useful in appropriate situations, is the use of near isogenic lines (NILs).[13,14] In this approach, a large number of RFLP or RAPD markers are tested against a pair of NILs differing only in the presence or absence of a target resistance gene. Since conventional backcross breeding leads to a pair of lines in which the susceptible recurrent parent and the resistant derivative differ primarily in the genomic region surrounding the resistance gene, comparing DNA marker patterns between the NILs rapidly uncovers those markers tightly linked to the gene of interest. A few of the important disease-resistance genes that have been marked by this approach include *Tm-2a* (resistance to TMV),[14] *Mi (Meloidogyne incognita),*[15] and *Pto (P. syringae* pv. *tomato)*[16] in tomato, *ml-o (Erysiphe graminis)* in barley,[17] *Dm (Bremia lactucae)* in lettuce,[18] *Rps (Phytophthora megasperma)* in soybean,[19] and *Pi (Pyricularia oryzae)* in rice.[20]

A variation on this strategy is known as *bulked segregation analysis.*[8] This approach is similar in concept to the use of NILs in locating markers near a disease-resistance gene. Again, the goal is to find DNA markers tightly linked to a target gene by analyzing a minimum of DNA samples rather than a complete analysis of all individuals in a large segregating population. In bulked segregation analysis one need not start with NILs that were previously bred by recurrent selection. Instead, one only needs a population (such as an F_2) that segregates for the resistance gene of interest. DNA samples from individuals in the population that are phenotypically scored as homozygous for resistance are pooled together, as are a corresponding set of samples from homozygous susceptible plants. If the number of individuals in each pooled sample is large enough, the only genomic region that will be contrasting between the DNA samples will be the region surrounding the gene of interest. Testing RFLPs or RAPDs against these two pooled DNA samples can rapidly uncover markers that are linked to the target gene. In a variation on this technique, DNA markers flanking a region of interest can be used as a basis for selecting homozygous individuals for the contrasting pooled samples.[21]

B. CONSTRUCTING A HIGH-RESOLUTION GENETIC MAP

After identifying DNA markers that are tightly linked to a target resistance gene, the next step is construction of a high-resolution genetic map of those markers. Such a genetic map, which provides information primarily about orientation of the DNA markers relative to one another and the target gene, is essential before physical mapping, chromosome walking, and gene cloning can begin.

The most direct strategy for constructing a high-resolution genetic map is analyzing a very large segregating population and then identifying recombinant individuals near the gene of interest. These recombinants can then be used to orient tightly linked DNA markers. However, this approach may be impractical in many plant gene cloning situations. To obtain sufficient information for high-resolution physical mapping, chromosome walking, and gene cloning, as many as 1000 plants or more might need to be analyzed. Isolating DNA samples from this many plants, while feasible, can be extremely laborious and time consuming. Pooling small groups of plants together for DNA extraction could potentially speed the process. In this case, leaf samples from groups of several plants (depending upon the genome size of that organism) can be collected and extracted together, which cuts the number of DNA isolations down to a more manageable level. The plants from a bulked group that appears to contain a recombinant plant can then be examined singly to find the informative individual.[21a]

Another potential difficulty in finding informative crossovers very close to a target gene occurs when the resistance gene originates from a wild relative. Rates of recombination can be suppressed in introgressed DNA,[22] so even with very large populations recombinant individuals might not be found. In this case, it may be possible to develop a mating population from a cross between a resistant and a susceptible parent within the same or closely related accessions. This is a strategy that was used to generate recombinants near the *Tm-2a* gene of tomato. This gene for resistance to TMV was originally derived from *Lycopersicon peruvianum.*[23] In a cross between *L. esculentum* (cultivated tomato) and a line carrying the *L. peruvianum* introgression around *Tm-2a,* rates of recombination were very low. However, when a cross was made between two *L. peruvianum* lines of the same accession, one carrying *Tm-2a* and the other susceptible to TMV, recombination rates were nearly ten times higher.

Still another strategy for constructing a high-resolution genetic map near a resistance gene is the use of previously developed NILs. Different NILs for the same resistance gene often carry introgressed segments of various sizes, with different crossover locations.[24] Precise DNA marker analysis of the introgressed segments in the NILs enables the orientation of nearby DNA markers and potentially

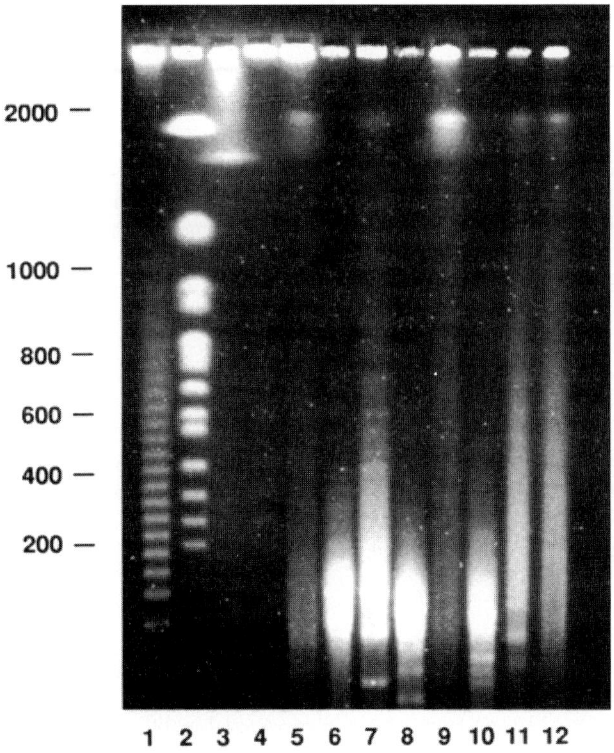

Figure 3 Contour-clamped homogenous electric field (CHEF)[26] electrophoresis analysis of mungbean (*Vigna radiata*) DNA. High-molecular weight DNA was isolated from mungbean protoplasts by a procedure adapted from Honeycutt et al.[30] and then digested with various rare-cutting restriction enzymes. Lane 1, molecular weight standards of Lambda DNA multimers; 2, *Saccharomyces cerviseae* chromosomes; 3, *S. pombe* chromosomes; 4, *S. cerviseae* chromosomes; 5, undigested mungbean DNA; 6, mungbean DNA cut with *Bgl*I; 7, *Mlu*I; 8, *Nhe*I; 9, *Not*I; 10, *Pvu*II; 11, *Sma*I; 12, *Sfi*I. Gel was run for 24 h at 200 V and 0.14 amps with a starting ramp time of 60 s and a final ramp time of 110 s. (Photograph courtesy of D. Menancio-Hautea.)

provides an excellent basis for physical mapping. This approach has been used with great success in the case of the *Mi* gene of tomato (conferring resistance to the nematode, *M. incognita)*.[15,25]

C. CONSTRUCTING A HIGH-RESOLUTION PHYSICAL MAP

High-resolution physical mapping determines the actual number of nucleotides between DNA markers that flank a resistance gene. The recent development of PFGE, which is capable of separating DNA molecules up to 10 million base pairs in length, has made long-range physical mapping practical. PFGE includes several types of related electrophoretic systems, such as clamped homogeneous electric field (CHEF)[26] and field-inversion gel electrophoresis,[27] which are all capable of separating DNA molecules greater than 100 kbp in size. In each of these systems, DNA molecules are separated not only on the basis of migration through a gel matrix, but also on how long it takes for DNA molecules to reorient themselves in an electric field whose orientation changes periodically. Larger DNA molecules tend to take longer to change directions and consequently travel more slowly. An example of CHEF gel analysis of mungbean (*Vigna radiata*) DNA is shown in Figure 3, in which DNA fragments ranging in size from 100 kbp all the way up to 1 million base pairs or more are observed. While DNA molecules up to 10 million base pairs in length can also be separated with PFGE, separations in this size range can take several days to complete.

The development of PFGE systems for physical mapping has been accompanied by two related techniques: methods for preparing very high-molecular weight DNA and the identification of new, rare-cutting restriction enzymes. In most cases, very high-molecular weight DNA from plants is generated through the use of protoplasts.[28–30] Once generated, the protoplasts are embedded in high-quality agarose and lysed by the addition of hydrolytic enzymes. As a result, long DNA molecules are immobilized in the gel and ready for restriction digestion and electrophoresis. Physical mapping also requires restriction enzymes with much rarer cutting frequencies than typical restriction enzymes. Most common restriction enzymes digest DNA molecules into fragments less than 10 kbp in length, which are too short to be useful in PFGE and physical mapping. For this reason, rare-cutting enzymes, resulting from recognition sites that are relatively long (8 base pairs or more) or contain a rare combination of nucleotides, have been essential in generating the longer DNA molecules suitable for long-range physical mapping.

Using the power of PFGE, there are now examples of physical maps in higher plants.[31,32] In most cases, physical mapping simply determined whether or not two DNA markers that appeared tightly linked by genetic mapping were also physically linked to one another, and if so, at what distance. Some of the best examples of physical mapping around resistance genes are in tomato and *Arabidopsis.* Tomato is a useful system because of its many known major resistance genes,[33] the availability of an RFLP map consisting of over 1000 markers,[34] and because its genome is modest in size (haploid genome estimated to be approximately 1×10^9 base pairs).[35] *Arabidopsis* has become a model system for cloning disease-resistance genes because of its very small genome size (haploid genome estimated to be 0.15×10^9 base pairs),[35] rapid generation time, small stature, and well-saturated RFLP map.[36–38] Both of these plant systems can also be transformed with foreign DNA.[39,40] Along with lettuce, maize, rice, and soybean (which also have active map-based gene cloning efforts), disease-resistance genes in tomato and *Arabidopsis* are the most likely to be cloned through map-based cloning in the near future.

An early example of physical mapping around a disease-resistance gene was the work of Ganal et al.[31] on the *Tm-2a* gene of tomato. In this case, several RFLP markers within 1.2 cM (centimorgan) of one another and only 0.3 cM from the *Tm-2a* gene had previously been identified by RFLP analysis.[14] Using PFGE analysis, the authors showed that only two out of five of the DNA markers were physically linked. Together, the five markers spanned at least 4 million base pairs, which indicated that recombination was suppressed by at least a factor of seven in this region of the tomato genome compared to the expected rate of recombination genome-wide (approximately 600 kbp per cM). As described earlier, this may have been due to the genetic distance between cultivated tomato and the source of the *Tm-2* gene, *L. peruvianum.*

By contrast, physical mapping around the *I2* gene of tomato (conferring resistance to *Fusarium oxysporum*) has demonstrated that rates of recombination in this region of the genome on chromosome 11 are much higher.[32] In this study, two RFLPs located near *I2* were analyzed by PFGE and found to reside on the same 175-kbp fragment, despite being separated by 4.1 cM genetically. This result indicated that recombinations occurred in this region of the tomato genome at a rate of nearly 43 kbp per cM, more than ten times the rate expected for the tomato genome at large.

In *Arabidopsis,* physical mapping near disease-resistance genes began more slowly than in tomato. Primarily, this was because *Arabidopsis* had few well-characterized resistance genes just a few years ago. Now there are several known resistance genes in *Arabidopsis,* and more are being discovered each year.[41–44] As a consequence, efforts at physical mapping in *Arabidopsis* are now underway.[38] A specific example is physical mapping near a major gene for resistance to *Pseudomonas syringae* known as RPM.[45] This gene has been located to a 6-cM segment on chromosome 3 of *Arabidopsis.* Physical mapping with PFGE has delimited the resistance gene to a region less than 430 kbp in length, and work with YAC clones (see below) is beginning to target this gene further.[45a]

D. CHROMOSOME WALKING WITH YEAST ARTIFICIAL CHROMOSOMES

Once a resistance gene is mapped at high resolution in terms of genetic and physical location, the next step is cloning the DNA sequence presumably containing the target gene. Currently, genetic and physical mapping can pinpoint a target gene to a region anywhere from one to several hundred kilobase pairs in length, so cloning the long stretches of DNA between flanking RFLP markers requires special cloning strategies. A pair of related techniques that are currently being used to address this challenge are *chromosome walking*[50] and YAC cloning vectors.[10] Chromosome walking is a technique that has been in use for more than a decade to identify genomic clones contiguous to a cloned DNA starting point (such as a genomic clone identified by an RFLP). The process involves a series of "steps" in which genomic clones contiguous to the starting point are identified by probing a library with the labeled ends of the first clone. Once contiguous clones on either end are identified, the ends of the new clones are labeled and used to find the next pair of contiguous genomic clones in the library, and so on.

Even still, distances of several hundred kilobase pairs would require large numbers of steps in a chromosome walk if traditional cloning vectors, such as cosmids (capable of genomic inserts up to 30 kbp), were used. For this reason, a special type of cloning vector, the YAC, is generally being used for map-based cloning. With YACs, genomic regions as large as 300 kbp or more can be inserted into a single clone. With this approach, the DNA between flanking markers can be bridged by one or only a few YAC clones. This simplifies the process of cloning the target genomic region significantly.

A YAC cloning vector consists of all the sequences necessary to maintain large segments of foreign DNA in yeast as an isolated "chromosome". One current YAC cloning system is based on a two-vector system, one vector with a yeast telomere, the second with a telomere plus a centromeric sequence.[46] The vectors both carry origins of replication and selectable markers for propagation in *Escherichia coli,* as well as yeast. The insert site is also flanked by markers suitable for rescuing the ends of the genomic insert in *E. coli,* an innovation that significantly aids in the process of chromosome walking. Once constructed, a genomic library of YAC clones provides a permanent resource for rapid long-range genome analysis and chromosome walking. For this reason, labs are constructing YAC libraries for many of the important plant species.[47–50] At the same time, YAC clones that are located near important plant disease-resistance genes are now being identified.[7]

E. PINPOINTING THE RESISTANCE GENE

The last step in map-based cloning is pinpointing the resistance gene within the genomic region isolated by YAC cloning and chromosome walking. At the present time, this can be the most challenging step of all, though this is likely to change in the years to come. The most convincing proof that the gene has been cloned is complementation of the recessive phenotype through DNA transformation with the putative gene sequence. In some plant systems, particularly Solanaceous and Cruciferous species, transformation is so efficient that this type of complementation experiment may indeed be feasible. Fortunately, this includes tomato and *Arabidopsis,* where so much progress in map-based cloning has already been made. Nevertheless, the expanse of genome that potentially contains a resistance gene can be great, even for efficient transformation techniques, so it is important to locate the gene as precisely as possible before further analysis. In plant systems where transformation is difficult or impractical, locating the gene precisely by nontransformation methods is essential.

The simplest way to narrow the search for a target resistance gene is by increasing the number of tightly linked markers and level of genetic resolution around the target gene. If crossovers can be used to delimit the resistance gene to a very small region of the cloned region, then the number of potential coding sequences that need to be considered can be reduced significantly. At the same time, it is important to determine which sequences within the target region actually code for proteins. Potentially, this can be done by mapping expressed genes on the YAC clone(s). However, because one usually does not know anything about the expression of the resistance gene or the abundance of its product in the mRNA pool, this step should be carried out with caution. Assuming open reading frames can be identified, rapid methods for DNA sequencing could then be used to sequence the candidate regions. Potentially, DNA from resistant and susceptible lines could be sequenced and compared, enabling the pinpointing of the likely resistance gene sequence. Indeed, this is the method that was used to locate precisely the cystic fibrosis gene in humans.[51] Alternatively, those sequences that have properties indicating they may be resistance genes (such as being a transmembrane glycoprotein or showing sequence homology to DNA regulatory proteins) could be examined further as candidates for the resistance gene.

III. TRANSPOSON GENE TAGGING

A. CHARACTERISTICS OF PLANT TRANSPOSONS

Transposons are DNA sequences capable of excising from one chromosomal location and inserting into another. First described in maize by McClintock[52] in the 1940s, transposons have become the basis for one of the most effective strategies for cloning genes of unknown biochemical function.[53] This strategy is briefly outlined in Figure 2. First, a transposon that has previously been cloned is introduced into a genetic background carrying a target resistance gene. In systems such as maize or *Antirrhinum majus* (snapdragon), this can be accomplished through genetic crossing alone. In other plant systems, transposons need to be introduced first by DNA transformation and then later by genetic crossing. Progeny individuals are then screened for mutants in the phenotype of the resistance gene, which are analyzed to confirm that the transposon sequence cosegregates with the gene of interest. If so, the flanking DNA sequence may be that of the target resistance gene.

The best known and most widely studied transposon is the *Activator* (*Ac*) element of maize. Other well-characterized transposons include *En*/*Spm* and *Mutator* (*Mu*) in maize and *Tam3* in snapdragon (for review on transposable elements, see Federoff).[54] *Ac* contains a gene for a transposase enzyme that catalyzes the movement process, plus distinctive border sequences at each end of the element that are essential for movement. Thus, *Ac* includes all sequences necessary to catalyze its own movement. For this reason, it is called an autonomous element (*En*/*Spm* and *Tam3* are also autonomous transposable elements). The related transposon element, *Dissociator* (*Ds*), resembles *Ac* in its border sequences, but lacks the transposase gene. For this reason, *Ds* requires the presence of an *Ac* element acting in *trans* to catalyze the movement of *Ds*. In an *Ac*/*Ds* system, the *Ac* element can also be a genetically altered element that still produces a transposase, but has lost appropriate border sequences for its own movement (also called a *disarmed Ac* element). Both types of transposon systems, *Ac* alone, as well as disarmed *Ac* in combination with *Ds*, are being studied as possible systems for cloning plant resistance genes.

To be an effective tool for gene cloning, it is important to understand the properties of transposons and their mechanisms for movement. Perhaps their most important characteristic is that transposons tend to move to locations on the genome linked to their previous site more frequently than to unlinked locations.[55] Clearly, this has important implications for gene tagging strategies (see below). Moreover, increasing numbers of *Ac* elements in a maize plant are also known to inhibit the rate of transposition,[56] a property that would be undesirable from the standpoint of gene cloning. The level of transposase gene expression affects the rate of transposon movement, though in ways that are not yet well understood. Depending upon whether transposition occurs at the germinal or somatic stage of development, plants may be sectored for the mutant phenotype. For disease-resistance genes, such sectoring could make it extremely difficult to identify mutant individuals. Finally, insertion of a transposon into a gene does not ensure complete inactivation of the phenotype—leaky mutants are possible. Considering the large number of plants that may need to be screened to identify transposon mutants, individuals showing a leaky phenotype could be a major problem. Still, with this type of basic information in mind, several groups have recently initiated transposon-based gene cloning projects, both in the homologous system of maize, as well as heterologous systems such as tobacco,[57] tomato,[58] and *Arabidopsis*.[59]

B. TRANSPOSON TAGGING IN HOMOLOGOUS PLANT SYSTEMS

The first demonstration that transposons could be used to tag and clone genes of unknown biochemical function involved seed mutants in maize. In 1984, Federoff et al.[53] reported the isolation of the *bronze* mutant of maize and described a "simple and generalizable method" for gene cloning. Since that time, other (nonresistance) genes have been tagged and cloned in maize[60] and snapdragon.[61] In each example, a strategy similar to the one shown in Figure 2 was employed.

The first attempt to apply gene cloning by transposon tagging to a disease-resistance gene involved the *Rp1* locus of maize, a gene that confers resistance to the rust fungus, *Puccinia sorghi*.[62] In this study, a line carrying the transposon element, *Mu,* was crossed with several other maize lines containing different race-specific resistance alleles at the *Rp1* locus. Over 100,000 total plants were screened and, in the process, the authors discovered that many *Rp1* alleles were intrinsically unstable. For this reason, mutants due to *Mu* were difficult to distinguish from spontaneous mutations. Subsequent work on the intrinsic instability at the *Rp1* locus has shown that the locus probably extends over a very long distance and that mechanisms such as unequal crossing over may be involved.[63] Nevertheless, the authors did identify one *Rp1* allele that showed much higher frequencies of mutation in the presence of *Mu* compared to spontaneous mutants. Unfortunately, this observation has not yet led to transposon tagging and cloning of the *Rp1* locus.

In a similar gene tagging strategy, the *Hm1* gene of maize, which confers resistance to *C. carbonum,* seems to have been successfully isolated from maize.[6a] Again, the *Mu* element was used for transposon tagging. One advantage of *Mu* for gene tagging is that so many copies of the element insert throughout the genome (up to 50 to 100 copies). However, this also requires that the sequence around several *Mu* insertion sites need to be analyzed to find the one associated with the target gene. The authors screened 40,000 plants and found three putative *Hm1* mutants. To confirm transposon tagging in these mutants, the authors first crossed the putative transposon-tagged *Hm1* line twice to a maize line not carrying *Mu* in order to dilute the number of *Mu* sequences in the genome. Then, they analyzed sequences hybridizing to *Mu* and used them as RFLP probes to demonstrate that the cloned sequence cosegregated with the *Hm1* locus. More recently, the gene for *Hm1* was cloned by a separate gene cloning strategy—purification of its gene product, the HC-toxin reductase enzyme, followed by conventional gene cloning.[64]

C. TRANSPOSON TAGGING IN HETEROLOGOUS PLANT SYSTEMS

In plant species that do not have well-characterized transposons, including most important crop species, it is necessary to introduce active transposons from plants such as maize and snapdragon. Transposons from these species have recently been transferred by DNA transformation into several species, including tobacco,[65] tomato,[66] and *Arabidopsis,*[67] as well as potato,[68] soybean,[69] and carrot.[67] In tomato and tobacco, where transposons have been analyzed in detail, heterologous transposons appear to excise and integrate at levels comparable to those observed in maize and snapdragon. However, the rate of transposon activity seems to be lower in *Arabidopsis*.[70]

The use of heterologous transposons can actually be preferable to homologous transposons. First, there are generally no other cross-hybridizing transposon sequences in the host genome to complicate the hybridization analysis of the transposon. The transposon sequence can also be inserted into an antibiotic or herbicide resistance gene, which can then be used as a reporter gene for excision.[71] Levels of transposase gene expression can also be modulated and analyzed by fusing different promoters in front of the transposase gene.[72] For these reasons, it may be desirable to introduce cloned transposons back into maize and snapdragon lines to study transposon activity and gene tagging in these species once transformation technologies improve. Another interesting observation has been that, in contrast to maize, increasing numbers of *Ac* sequences actually increase the frequency of transposition in tobacco and *Arabidopsis*.[73] This is in contrast to maize, where higher numbers of *Ac* elements lower transposition rate.

One attempt to use a heterologous transposon to tag a resistance gene was a study in which the *Ac* element of maize was transformed into a tobacco line carrying the *N*-gene for hypersensitive type resistance to TMV.[57] In screening approximately 40,000 seedling for mutants in the *N*-gene phenotype, 30 candidate plants were recovered. Unfortunately, none of these plants carried an *Ac* element linked to the *N*-gene locus, and were all presumably due to spontaneous mutations. This result, along with the transposon tagging experience with the *Rp1* locus of maize, emphasizes the importance of knowing about the level of spontaneous mutation before embarking on a transposon-based gene cloning strategy.

Because of the difficulties in the transposon tagging techniques, efforts are now underway to optimize the procedure experimentally before initiating gene cloning efforts. One strategy has been to place introduced *Ac* elements throughout the genome, including, potentially, one site close to the target resistance gene. As noted above, *Ac* tends to insert into sites near its original location, so placing *Ac* near a target gene should significantly increase the chances of success. Using the RFLP map of tomato, it has been possible to map the locations of *Ac* element insertion.[74] The results indicated that nearly half of the transposition events inserted into linked sites, while the other half involved sites randomly distributed throughout the remainder of the tomato genome. Similar results have also been observed in tobacco.[75]

Altered levels of transposase enzyme expression has also been examined. Because of the unusual relationship between *Ac* dosage and transposition activity (higher *Ac* dosage is related to less transposition in maize, but more transposition in tomato and tobacco), the effect of transposase gene expression was examined in detail in tobacco.[72] In this experiment, the effects of different promoter sequences fused to an *Ac* transposase gene were examined for their ability to induce the movement of a separate *Ds* element marked with the streptomycin phosphotransferase gene. The wild-type promoter induced *Ds* transposition throughout embryo development. By contrast, all chimeric transposase fusions produced sectored individuals, with higher levels of transposase activity (due to different promoters) associated with larger sectors. Nevertheless, increased levels of transposase transcription did not correlate in a simple way with higher levels of *Ds* transposition. These results contrast with those of Grevelding et al.,[59] in which overexpression of the *Ac* transposase gene in *Arabidopsis* resulted in germinal transposition of more than 27% of plants. Together, these attempts at optimizing heterologous transposon systems suggest that placing transposons near a target gene may be useful in gene tagging, but that increasing the overall rate of transposition by altering transposase activity is unpredictable.

IV. PERSPECTIVES ON RESISTANCE GENE CLONING

Over the past 50 years, hundreds of plant disease-resistance genes have been identified based on their genetic phenotype. However, because so little was known about their biochemical function, it has been impossible to clone these genes using conventional gene cloning techniques. By contrast, the two gene cloning strategies described in this chapter, map-based cloning and transposon tagging, promise to remedy this situation in the coming years. There is no doubt that both strategies are still technically difficult, which is reflected in the fact that so few plant resistance genes have actually been cloned to date. But the important point is that both techniques are generalizable so that any gene with a well-characterized phenotype should, eventually, be clonable. An important corollary is that the ground-breaking research that is being carried out today—better mapping strategies plus comprehensive plant YAC libraries for map-based cloning and custom-designed transposon constructs distributed throughout the genomes of important plant species for transposon tagging—will extend the techniques to resistance gene targets that seem beyond reach today. Other types of gene cloning strategies, particularly those based on genomic subtraction[76] and chromosome dissection,[77] plus an increased understanding of the cell biology underlying disease resistance will also probably lead to the isolation of resistance genes and the development of better forms of genetically engineered disease resistance. Nevertheless, map-based cloning and transposon tagging, along with techniques derived from them, will certainly form the foundation for resistance gene cloning in plants in the years to come.

ACKNOWLEDGMENTS

I thank Drs. D. Samac and D. Somers for their helpful comments on the manuscript and Dr. D. Menancio-Hautea for kindly providing unpublished results. This paper is published as contribution 20,245 in the series of the Minnesota Agricultural Experiment Station on research conducted under Project 015, supported by G.A.R. funds.

NOTES ADDED IN PROOF

Since the original writing of this chapter, efforts to use map-based cloning and transposon tagging to clone disease resistance genes have proven successful. Map-based cloning has been used to isolate the *Pto* gene of tomato (Martin, et al., *Science*, 262, 1432, 1993) and the *RPS2* gene of *Arabidopsis thaliana* (Mindrinos, et al., *Cell*, 78, 1089, 1994). Sequencing of the *Pto* gene suggests it may be a serine-threonine protein kinase, while the sequence of the *RPS2* gene indicates the presence of a nucleotide-binding site and leucine-rich repeats. The *N* gene of tobacco has been cloned through transposon tagging (Whitham, et al., *Cell*, 78, 1101, 1994). This gene sequence shows similarities to the *Toll* gene of *Drosophila* and the interleukin-1 receptor of mammals.

REFERENCES

1. Hedrick, S. A., Bell, J. N., Boller, T., and Lamb, C. J., Chitinase cDNA cloning and mRNA induction by fungal elicitor, wounding, and infection, *Plant Physiol.*, 86, 182, 1988.
2. Walter, M. H., Grima-Pettenati, J., Grand, C., Boudet, A. M., and Lamb, C. J., Cinnamyl-alcohol dehydrogenase, a molecular marker specific for lignin synthesis: cDNA cloning and mRNA induction by fungal elicitor, *Proc. Natl. Acad. Sci. U.S.A.*, 85, 5546, 1988.
3. Cramer, C. L., Edwards, K., Dron, M., Liang, X., Dildine, S. L., Bolwell, G. P., Dixon, R. A., Lamb, C. J., and Schuch, W., Phenylalanine ammonia-lyase gene organization and structure, *Plant Mol. Biol.*, 12, 367, 1989.
4. Linthorst, H. J. M., Melchers, L. S., Mayer, A., van Roekel, J. S. C., Cornelissen, B. J. C., and Bol, J. F., Analysis of gene families encoding acidic and basic β-1,3-glucanases of tobacco, *Proc. Natl. Acad. Sci. U.S.A.*, 87, 8756, 1990.
5. Steinmetz, M., Minard, K., Harvath, S., McNicholas, J., Srenlinger, J., Wake, C., Long, E., Mach, B., and Hood, L., A molecular map of the immune response region of the major histocompatibility complex of the mouse, *Nature*, 300, 35, 1981.
6. Wolpert, T. and Macko, V., Specific binding of victorin to a 100-kDa protein form oats, *Proc. Natl. Acad. Sci. U.S.A.*, 86, 4092, 1989.

6a. Johal, G. S. and Briggs, S. P., Reductase activity encoded by HM1 disease resistance gene in maize, *Science*, 258, 985, 1992.

7. Martin, G., Ganal, M., and Tanksley, S., Construction of a yeast artificial chromosome library of tomato and identification of cloned segments linked to two disease resistance loci, *Mol. Gen. Genet.*, 233, 25, 1992.
8. Michelmore, R. W., Paran, I., and Kesseli, R. V., Identification of markers linked to disease-resistance genes by bulked segregation analysis: a rapid method to detect markers in specific genome regions by using segregating populations, *Proc. Natl. Acad. Sci. U.S.A.*, 88, 9828, 1991.
9. Schwartz, D. C. and Cantor, C. R., Separation of yeast chromosome-sized DNAs by pulsed field gradient gel electrophoresis, *Cell*, 37, 67, 1984.
10. Burke, D. T., Carle, G. F., and Olson, M. V., Cloning of large segments of exogenous DNA into yeast by means of artificial chromosome vectors, *Science*, 236, 806, 1987.

11. Soller, M. and Beckmann, J. S., Genetic polymorphism in varietal identification and genetic improvement, *Theor. Appl. Genet.*, 67, 837, 1983.
12. Tanksley, S. D., Young, N. D., Paterson, A. H., and Bonierbale, M. W., RFLP mapping in plant breeding: new tools for an old science, *Bio/Technology*, 7, 257, 1989.
13. Muehlbauer, G. J., Specht, J. E., Thomas-Compton, M. A., Staswick, P. E., and Bernard, R. L., Near-isogenic lines—a potential resource in the integration of conventional and molecular marker maps, *Crop Sci.*, 28, 729, 1988.
14. Young, N. D., Zamir, D., Ganal, M. W., and Tanksley, S. D., Use of isogenic lines and simultaneous probing to identify DNA markers tightly linked to the *Tm-2a* gene in tomato, *Genetics*, 120, 579, 1988.
15. Messeguer, R., Ganal, M., de Vicente, M. C., Young, N. D., Bolkan, H., and Tanksley, S. D., High resolution RFLP map around the root knot nematode resistance gene (*Mi*) in tomato, *Theor. Appl. Genet.*, 82, 529, 1991.
16. Martin, G. B., Williams, J. G. K., and Tanksley, S. D., Rapid identification of markers linked to a *Pseudomonas* resistance gene in tomato using random primers and near-isogenic lines, *Proc. Natl. Acad. Sci. U.S.A.*, 88, 2336, 1991.
17. Hinze, K., Thompson, R. D., Ritter, E., Salamini, F., and Schulze-Lefert, P., Restriction fragment length polymorphism-mediated targeting of the *ml-o* resistance locus in barley (*Hordeum vulgare*), *Proc. Natl. Acad. Sci. U.S.A.*, 88, 3691, 1991.
18. Paran, I., Kesseli, R., and Michelmore, R., Identification of restriction fragment length polymorphism and random amplified polymorphic DNA markers linked to downy mildew resistance genes in tomato, using near-isogenic lines, *Genome*, 1021, 1991.
19. Diers, B. W., Mansur, L., Imsande, J., and Shoemaker, R. C., Mapping Phytophthora resistance loci in soybean with restriction fragment length polymorphism markers, *Crop Sci.*, 32, 377, 1991.
20. Yu, Z. H., Mackill, D. J., Bonman, J. M., and Tanksley, S. D., Tagging genes for blast resistance in rice via linkage to RFLP markers, *Theor. Appl. Genet.*, 81, 471, 1991.
21. Giovanonni, J., Wing, R., and Tanksley, S. D., Isolation of molecular markers from specific chromosomal intervals using DNA pools from existing mapping populations, *Nucleic Acids Res.*, 19, 6553, 1992.
21a. Tanksley, S. D., Ithaca, NY, personal communication.
22. Rick, C. M., Controlled introgression of chromosomes of *Solanum pennellii* into *Lycopersicon esculentum:* segregation and recombination, *Genetics*, 26, 753, 1969.
23. Alexander, L. J., Host-pathogen dynamics of tobacco mosaic virus on tomato, *Phytopathology*, 61, 611, 1971.
23a. Young, N., unpublished observations.
24. Young, N. D. and Tanksley, S. D., RFLP analysis of the size of chromosomal segments retained around the *Tm-2* locus of tomato during backcross breeding, *Theor. Appl. Genet.*, 77, 353, 1989.
25. Klein-Lankhorst, R., Rietveld, P., Machiels, B., Verkerk, R., Weide, R., Gebhardt, C., Koornneff, M., and Zabel, P., RFLP markers linked to the root knot nematode resistance gene *Mi* in tomato, *Theor. Appl. Genet.*, 81, 661, 1991.
26. Chu, G., Volrath, D., and Davis, R. W., Separation of large DNA molecules by contour-clamped homogenous electric fields, *Science*, 234, 1582, 1986.
27. Carle, G. F., Frank, M., and Olson, M. V., Electrophoretic separations of large DNA molecules by periodic inversion of the electric field, *Science*, 232, 65, 1986.
28. Ganal, M. W. and Tanksley, S. D., Analysis of tomato DNA by pulsed field gel electrophoresis, *Plant Mol. Biol. Rep.*, 7, 17, 1989.
29. Cheung, W. Y. and Gale, M. D., The isolation of high molecular weight DNA from wheat, barley and rye for analysis by pulse-field gel electrophoresis, *Plant Mol. Biol.*, 14, 881, 1990.
30. Honeycutt, R. J., Sobral, B. W. S., McClelland, M., and Atherly, A., Analysis of large DNA from soybean (*Glycine max* L. Merr.) by pulsed-field gel electrophoresis, *Plant J.*, 2, 133, 1992.
31. Ganal, M. W., Young, N. D., and Tanksley, S. D., Pulsed field gel electrophoresis and physical mapping of large DNA fragments in the *Tm-2a* region of chromosome 9 in tomato, *Mol. Gen. Genet.*, 215, 395, 1989.
32. Segal, G., Sarfatti, M., Schaffer, M. A., Ori, N., Zamir, D., and Fluhr, R., Correlation of genetic and physical structure in the region surrounding the I_2 *Fusarium oxysporum* resistance locus in tomato, *Mol. Gen. Genet.*, 231, 179, 1992.
33. Rick, C., The tomato, in *Handbook of Genetics*, King, R. C., Ed., Plenum Press, New York, 1975, 247.

34. Tanksley, S. D., Ganal, M. W., Prince, J. P., de Vincente, M. C., Bonierbale, M. W., Broun, P., Fulton, T. M., Giovanonni, J. J., Grandillo, S., Martin, G. B., Messeguer, R., Miller, J. C., Miller, L., Paterson, A. H., Pineda, O., Roder, M., Wing, R. A., Wu, W., and Young, N. D., High density molecular linkage maps of the tomato and potato genomes: biological inferences and practical applications, *Genetics,* 132, 1141, 1992.
35. Arumuganathan, K. and Earle, E., Nuclear DNA content of some important plant species, *Plant Mol. Biol. Rep.,* 9, 208, 1991.
36. Chang, C., Bowman, J. C., DeJohn, A. W., Lander, E. S., and Meyerowitz, E. S., Restriction fragment length polymorphism linkage map of *Arabidopsis thaliana, Proc. Natl. Acad. Sci. U.S.A.,* 85, 6856, 1988.
37. Nam, H.-G., Giraudat, J., den Boer, B., Moonan, F., Loos, W. D. B., Hauge, B. M., and Goodman, H. M., Restriction fragment length polymorphism linkage map of *Arabidopsis thaliana, Plant Cell,* 1, 699, 1989.
38. Reiter, R. S., Williams, J. G. K., Feldman, K. A., Rafalski, J. A., Tingey, S. V., and Scolnick, P. A., Global and local genome mapping in *Arabidopsis thaliana* by using recombinant inbred lines and random amplified polymorphic DNAs, *Proc. Natl. Acad. Sci. U.S.A.,* 89, 1477, 1992.
39. McCormick, S., Niedermey, J., Fry, J., Barnason, A., Horsch, R., and Fraley, R., Leaf disc transformation of cultivated tomato (*L. esculentum*) using *Agrobacterium tumefaciens, Plant Cell Rep.,* 5, 81, 1986.
40. Valvekens, D., Van Montagu, M., and van Lijsebettens, M., *Agrobacterium tumefaciens* mediated transformation of *Arabidopsis thaliana* root explants using kanamycin selection, *Proc. Natl. Acad. Sci. U.S.A.,* 85, 5536, 1988.
41. Simpson, R. B. and Johnson, L. J., *Arabidopsis thaliana* as a host for *Xanthomonas campestris* pv. *campestris, Mol. Plant-Microbe Interact.,* 3, 233,1990.
42. Davis, K. R., Schott, E., and Ausubel, F., Virulence of selected phytopathogenic Pseudomonads in *Arabidopsis thaliana, Mol. Plant-Microbe Interact.,* 4, 477, 1991.
43. Tsuji, J., Somerville, S. C., and Hammerschmidt, R., Identification of a gene in *Arabidopsis thaliana* that controls resistance to *Xanthomonas campestris* pv. *campestris, Physiol. Mol. Plant Pathol.,* 38, 57, 1991.
44. Whalen, M. C., Innes, R. W., Bent, A. F., and Staskawicz, B. J., Identification of *Pseudomonas syringae* pathogens of *Arabidopsis* and a bacterial locus determining avirulence of both *Arabidopsis* and soybean, *Plant Cell,* 3, 49, 1991.
45. Dabener, T., Lehnackers, H., Arnold, M., and Dangl, J. L., Identification and molecular mapping of a single *Arabidopsis thaliana* locus determining resistance to a phytopathogenic *Pseudomonas syringae* isolate, *Plant J.,* 1, 289, 1991.
45a. Dangl, J., Max Planck Institute, Cologne, personal communication.
46. Kotewicz, M. L., Plasmids pJS97 and pJS98: improved yeast artificial chromsome vectors for very large DNA fragment cloning and analysis, *Focus,* 13, 30, 1991.
47. Gúzman, P. and Eckert, J., Development of large DNA methods for plants: molecular cloning of large segments of Arabidopsis and carrot DNA into yeast, *Nucleic Acids Res.,* 16, 11091, 1988.
48. Ward, E. R. and Jen, G. C., Isolation of single-copy-sequence clones from a yeast artificial chromosome library of randomly-sheared *Arabidopsis thaliana* DNA, *Plant Mol. Biol.,* 14, 561, 1990.
49. Edwards, K. J., Thompson, H., Edwards, D., de Saizieu, A., Sparks, C., Thompson, J. A., Greenland, A. J., Eyers, M., and Schuch, W., Construction and characterization of a yeast artificial chromosome library containing three haploid maize genome equivalents, *Plant Mol. Biol.,* 19, 299, 1992.
50. Grill, E. and Sommerville, C., Construction and characterization of a yeast artificial chromosome library of *Arabidopsis* which is suitable for chromosome walking, *Mol. Gen. Genet.,* 226, 484, 1992.
51. Riordan, J. R., Rommens, J. M., Kerem, B.-S., Alon, N., Rozmahel, R., Grezelczak, Z., Zielenski, J., Lok, S., Plavsic, N., Chou, J.-L., Drumm, M. L., Lannuzzi, M. C., Collins, F. S., and Tsui, L.-C., Identification of the cystic fibrosis gene: cloning and characterization of complementary DNA, *Science,* 245, 1066, 1989.
52. McClintock, B., Mutable loci in maize, *Carnegie Inst. Washington, Year Book,* 49, 155, 1948.
53. Federoff, N. V., Furtek, D. B., and Nelson, O. E., Cloning of the *bronze* locus in maize by a simple and generalizable procedure using the transposable controlling element *Activator (Ac), Proc. Natl. Acad. Sci. U.S.A.,* 81, 3825, 1984.

54. Federoff, N., Maize transposable elements, in *Mobile DNA,* Berg, D. and Howe, M., Eds., American Society for Microbiology, Washington, D.C., 1989, 374.
55. Dooner, H. K. and Belachew, A., Tranposition pattern of the maize element *Ac* form the *bz-ms(Ac)* allele, *Genetics,* 122, 447, 1989.
56. McClintock, B., Chromosome organization and genic expression, *Cold Spring Harbor Symp. Quant. Biol.,* 16, 13, 1951.
57. Hehl, R. and Baker, B., Towards transposon tagging of the TMV resistance gene *N* using *Activator* in transgenic tobacco, *Maize Genet. Coop. Newsl.,* 64, 3, 1990.
58. Jones, J. D. G., Jones, D., Bishop, G., Thomas, C., Dickinson, M., Harrison, K., Balint-Kurti, P., Hammond-Kosack, K., and Ashfield, T., Tomato genes for resistance to *Cladosporium fulvum:* a useful model for understanding race specific plant pathogen interactions, in *Advances in Gene Technology: Feeding the World in the 21st Century,* Whelan, W., Ahmad, F., Bialy, H., Black, S., King, M. L., Rabin, M., Solomonson, L., and Vasil, I., Eds., IRL Press at Oxford University Press, 1992, 106.
59. Grevelding, C., Becker, D., Kunze, R., Von Menges, A., Fantes, V., Schell, J., and Masterson, R., High rates of *Ac/Ds* germinal transposition in *Arabidopsis* suitable for gene isolation by insertional mutagensis, *Proc. Natl. Acad. Sci. U.S.A.,* 89, 6085, 1992.
60. Dellaporta, S. L., Greenblatt, J., Kermicle, J., Hicks, B., and Wessler, S., Molecular cloning of the maize *R-nj* gene by transposon tagging with *Ac,* in *Chromosome Structure and Function,* Gustafson, J. P. and Apples, R., Eds., Plenum Press, New York, 1988, 263.
61. Martin, C., Carpenter, R., Sommer, H., Saedler, H., and Coen, E. S., Molecular analysis of instability in flower pigmentation of *Antirrhinum majus* following isolation of the pallida locus by transposon tagging, *EMBO J.,* 4, 1625, 1985.
62. Bennetzen, J. L., Qin, M.-M., Ingels, S., and Ellingboe, A. H., Allele-specific and *Mutator*-associated instability at the *Rp1* disease-resistance locus of maize, *Nature,* 332, 369, 1988.
63. Hulbert, S. H. and Bennetzen, J. L., Recombination at the *Rp1* locus of maize, *Mol. Gen. Genet.,* 226, 377, 1991.
64. Meeley, R. B., Johal, G. S., Briggs, S. P., and Walton, J. D., A biochemical phenotype for a disease resistance gene of maize, *Plant Cell,* 4, 71, 1992.
65. Baker, B., Schell, J., Lörz, H., and Federoff, N., Transposition of the maize controlling element "*Activator*" in tobacco, *Proc. Natl. Acad. Sci. U.S.A.,* 83, 4844, 1986.
66. Yoder, J. I., Palys, J., Alpert, K., and Lassner, M., *Ac* transposition in transgenic tomato plants, *Mol. Gen. Genet.,* 213, 291, 1988.
67. Van Sluys, M. A., Tempé, J., and Federoff, N., Studies on the introduction and mobility of the maize *Activator* element in *Arabidopsis thaliana* and *Daucus carota, EMBO J.,* 6, 3881, 1987.
68. Knapp, S., Coupland, G., Uhrig, H., Starlinger, P., and Salamini, F., Transposition of the maize transposable element *Ac* in *Solanum tuberosum, Mol. Gen. Genet.,* 213, 285, 1988.
69. Zhou, J. H. and Atherley, A. G., In situ detection of transposition of the maize controlling element (*Ac*) in transgenic soybean tissues, *Plant Cell Rep.,* 8, 542, 1990.
70. Schmidt, R. and Willmitzer, L., The maize autonomous element *Activator (Ac)* shows a minimal germinal excision frequency of 0.2%–0.5% in transgenic *Arabidopsis thaliana* plants, *Mol. Gen. Genet.,* 220, 17, 1989.
71. Baker, B., Coupland, G., Federoff, N., Starlinger, P., and Schell, J., Phenotypic assay for excision of the maize controlling element *Ac* in tobacco, *EMBO J.,* 6, 1547, 1987.
72. Scofield, S. R., Harrison, K., Nurrish, S. J., and Jones, J. D. G., Promoter fusions to the *Activator* transposase gene cause distinct patterns of *Dissociation* excision in tobacco cotyledons, *Plant Cell,* 4, 573, 1992.
73. Hehl, R. and Baker, B., Properties of the maize transposable element *Activator* in transgenic tobacco plants: a versatile interspecies genetic tool, *Plant Cell,* 2, 709, 1990.
74. Osborne, B. I., Corr, C. A., Prince, J. P., Hehl, R., Tanksley, S. D., McCormick, S., and Baker, B., *Ac* transposition from a T-DNA can generate linked and unlinked clusters of insertions in the tomato genome, *Genetics,* 129, 833, 1991.
75. Jones, J. D. G., Carland, E., Lim, E., Ralston, E., and Dooner, H. K., Preferential transposition of the maize element *Activator* to linked chromosomal locations in tobacco, *Plant Cell,* 2, 701, 1990.
76. Sun, T., Goodman, H. M., and Ausubel, F. M., Cloning the Arabidopsis *GA1* locus by genomic subtraction, *Plant Cell,* 4, 119, 1992.

77. Ludecke, H. J., Senger, G., Claussen, U., and Horsthemke, B., Cloning defined regions of the human genome by microdissection of banded chromosomes and enzymatic amplification, *Nature,* 338, 348, 1989.
78. van der Beek, J. G., Verkerk, R., Zabel, P., and Lindhout, P., Mapping strategy for resistance genes in tomato based on RFLPs between cultivars: *Cf9* (resistance to *Cladosproium fulvum*) on chromosome 1, *Theor. Appl. Genet.,* 84, 106, 1992.
79. Sarfatti, M., Abu-Abied, M., Katan, J., and Zamir, D., RFLP mapping of *I1,* a new locus in tomato conferring resistance against *Fusarium oxysporum f. sp. lycopersici* race 1, *Theor. Appl. Genet.,* 82, 22, 1991.
80. Behare, J., Latterot, H., Sarfatti, M., and Zamir, D., Restriction fragment length polymorphism mapping of the *Stemphylium* resistance gene in tomato, *Mol. Plant-Microbe Interact.,* 4, 489, 1991.
81. Barone, A., Ritter, E., Schachtschable, U., Debener, T., Salamini, F., and Gebhardt, C., Localization by restriction fragment length polymorphism mapping in potato of a major dominant gene conferring resistance to the potato cyst nematode *Globodera rostochiensis, Mol. Gen. Genet.,* 224, 177, 1990.
82. Ritter, E., Debener, T., Barone, A., Salamini, F., and Gebhardt, C., RFLP mapping on potato chromosomes of two genes controlling extreme resistance to potato virus X (PVX), *Mol. Gen. Genet.,* 227, 81, 1991.
83. Bentolila, S., Guitton, C., Bouvet, N., Sailland, A., Nyukaza, S., and Freyssinet, G., Identification of an RFLP marker tightly linked to the *Ht1* gene in maize, *Theor. Appl. Genet.,* 82, 393, 1991.
84. McMullen, M. D. and Louie, R., The linkage of molecular markers to a gene controlling the symptom response in maize to maize dwarf mosaic virus, *Mol. Plant-Microbe Interact.,* 2, 309, 1989.
85. Young, N. D., Potential applications of map-based cloning to plant pathology, *Physiol. Mol. Plant Pathol.,* 37, 81, 1990.

FURTHER READING

Dixon, R. A. and Harrison, M. J., Activation, structure and organization of genes involved in microbial defense of plants, *Adv. Genet.,* 28, 165, 1990.

Curr, S. J., McPherson, F. J., and Bowles, D. J., *Molecular Plant Pathology: a Practical Approach,* Oxford University Press, Oxford, 1992.

Hiatt, A., *Transgenic Plants: Fundamentals and Applications,* Marcell Dekker, New York, 1992.

Michelmore, R. W., Kesseli, R. V., Francis, D. M., Fortin, M. G., Paran, I., and Yang, C.-H., Strategies for cloning plant disease resistance genes, in *Molecular Plant Pathology, A Practical Approach,* Vol. 2., Gurr, S. J., McPherson, M. J., and Bowles, D. J., eds., IRL Press, Oxford, pp. 233–288.

Michelmore, R. W., Paran, I., and Kesseli, R. V., Identification of markers linked to disease-resistance genes by bulked segregation analysis: a rapid method to detect markers in specific genome regions by using segregating populations, *Proc. Natl. Acad. Sci. U.S.A.,* 88, 9828, 1991.

Whelan, W., Ahmad, F., Bialy, H., Black, S., King, M. L., Rabin, M., Solomonson, L., and Vasil, I., *Advances in Gene Technology: Feeding the World in the 21st Century,* IRL Press at Oxford University Press, Oxford, 1992.

Young, N. D., Potential applications of map-based cloning to plant pathology, *Physiol. Mol. Plant Pathol.,* 37, 81, 1990.

Chapter 17

Molecular Strategies to Develop Virus-Resistant Plants

James N. Culver

TABLE OF CONTENTS

I. INTRODUCTION

Virus diseases of cultivated plants cause substantial reductions in food, forage, and fiber throughout the world. Control of these diseases has been based primarily on cultural practices that include: removal of viral-infected debris, eradication of weed hosts, prevention of vector transmission, indexing for virus-free starting material (seed or vegetative propagules), and breeding for disease resistance.[1] No large-scale methods exist for curing plants once they have become virus infected. Thus, control of viral diseases is dependent upon methods to prevent or delay the establishment of infection.

Of the above disease-control measures, breeding for resistance is generally one of the most economical and practical methods, since it requires no additional labor or expense to the grower. Additionally, disease resistance is one of the most environmentally safe methods for controlling plant diseases, a concern of special importance given the public attention that accompanies this issue. Unfortunately, in many plant-virus systems resistance is not available and cannot be obtained using traditional plant breeding strategies. However, recent advances in molecular biology and gene manipulation now provide a way to integrate or create disease resistance in plant-virus systems where it is presently unavailable. This chapter focuses on several different molecular strategies that have recently been developed or proposed for the control of plant-virus diseases.

The development of molecular strategies for the control of virus diseases has been especially successful in comparison to other pathogen-caused diseases. This is due in part to the relatively small genomic sizes of plant viruses, making them particularly amenable to molecular techniques for cloning and characterizing their genetic information. To date, numerous plant viruses have been characterized at the nucleic acid level and their genomic organization and gene functions elucidated.[2] Thus, the fundamental information needed to employ molecular resistance strategies is presently available in many virus systems.

Plant viruses are dependent on their hosts for the cellular machinery needed to complete their life cycles. Precise interactions between plant and virus components are also necessary for infection and replication. These interactions comprise a number of different processes that allow the virus to uncoat

0-87371-877-1/95/$0.00+$.50

its genetic information, express its gene products, replicate, move cell to cell, and spread long distances throughout the plant. These processes are important for infection and the development of disease. Interruption of any of these precise interactions could inhibit the infection process and curtail disease development. Thus, information obtained from the molecular characterization of plant viruses can be used to develop and tailor strategies to disrupt plant-virus interactions that are important in the infection process. Modification of the genome of a plant so that the plant expresses the disruptive component could result in host resistance.

II. INTEGRATION OF RESISTANCE FACTORS

Strategies to utilize molecular information for controlling viral diseases have primarily focused on the production of transgenic plants that express specifically designed resistance factors. In general, potential resistance factors have been placed between a transcriptional promoter, such as the 35S promoter of cauliflower mosaic virus, and a polyadenylation sequence to allow the efficient transcription and translation of the factor.[3] Parameters effecting the optimal expression of these factors include promoter strength, consensus sequences for efficient translation, and gene position.[4,5] It should be noted that each different resistance strategy is likely to have a different optimal gene construct that will provide the highest level of resistance. However, the highest expressing plants may not always provide the greatest resistance. This may be due in part to the location of the inserted gene within the genome of the plant and/or its expression within different cells or tissues. Thus, a number of different transgenic lines may have to be screened before the level of resistance provided by a factor can be determined.

One commonly used method for transforming plants is *Agrobacterium*-mediated gene transfer. While this method is effective for transforming many plant species, it is not effective for the transformation of important monocots such as cereal crops.[6] Many plants not transformable by *Agrobacterium* have been successfully transformed using electroporation or projectile bombardment.[6,7] Often, the most difficult part of producing transgenic plants comes during the regeneration of the transformed cells, so that the newly transformed gene (transgene) is stable and can be passed on to the progeny of the plant. Because of this difficulty, easily regenerated plants, such as tobacco, have primarily been used as models to investigate the effectiveness of different strategies for molecular resistance. Reviews detailing the techniques used in plant transformation and regeneration have been published.[5,6,8,9]

III. RESISTANCE STRATEGIES

There are a number of different strategies for using molecular technology to integrate or create new resistance factors in plant-virus systems. While all of these strategies seek to provide resistance there are certain fundamental differences in their approaches. Primarily, these approaches can be placed into one of three general categories: (1) pathogen-derived resistance, relating to the integration of pathogen components that interfere with the normal life cycle of the virus; (2) pathogen-targeted resistance, involving the integration of components that specifically target viral genes and their products to render them nonfunctional; and (3) integration of existing host resistance genes into nonresistant hosts. The following sections outline specific examples of strategies that have been used or are being tested for the control of plant virus diseases.

A. PATHOGEN-DERIVED RESISTANCE

The concept of pathogen-derived resistance as described by Sanford and Johnston[10] is based upon the idea that during any interactions with the host the pathogen brings with it essential components and functions that are required for completion of its life cycle. These essential elements might then be disrupted by the presence of a corresponding pathogen gene that is dysfunctional, overexpressed, or appears during the wrong stage of the life cycle of the pathogen. This is especially true for viral pathogens considering their dependence upon the cellular machinery of the plant for replication. Thus, the objective of this approach is to identify those viral genes or gene products that when present at an improper time or in the wrong amount will interfere with normal functions of the infection process and prevent disease development.

1. Coat Protein-Mediated Cross Protection

The concept of cross protection, the ability of one virus to prevent or inhibit the effect of a second challenge virus, has been the subject of extensive studies on both its experimental and practical use.[11]

Powell-Abel et al.[12] first demonstrated that transgenic tobacco that expressed tobacco mosaic virus (TMV) coat protein showed resistance similar to what occurs in viral-mediated cross protection. Since then a number of coat protein genes from at least eight different virus groups have been found to provide resistance when expressed in transgenic plants, and resistance has been demonstrated in both greenhouse and field experiments. Nelson et al.[13] showed that coat protein-mediated resistance directed against TMV reduced yield losses by as much as 35% in field-grown tomatoes. Transgenic tobacco expressing the cucumber mosaic virus (CMV) coat protein have also been reported to be highly resistant under natural field conditions.[14] Additionally, a 3-year field study has shown that CMV coat protein-mediated resistance in cucumbers provides a level of resistance that is comparable to that obtained genetically.[15] Petitions for the commercial sale of similarly engineered resistant plants are presently being reviewed by the U.S. Animal and Plant Health Inspection Service.[16] Thus, it seems likely that coat protein-mediated resistance may become one of the first molecular strategies to be available for commercial use. Reviews detailing coat protein-mediated resistance have been published.[4,17]

a. Characteristics of Coat Protein-Mediated Resistance

In general, coat protein-mediated resistance is correlated with a delay in symptom development and is dependent upon the concentration of challenge inoculum. For example, Powell-Abel et al.[12] found disease development in transgenic tobacco expressing the coat protein of TMV to be delayed by as much as 10 d when using an inoculum concentration of 0.4 μg/ml of virus. However, an inoculum concentration of 2.0 μg/ml virus resulted in only a 2-d delay in symptom development. For alfalfa mosaic virus (AMV),[18] potato virus X (PVX),[19,20] potato virus Y (PVY),[20,21] and tobacco etch virus (TEV),[21] higher inoculum concentrations of up to 50 μg/ml were successful in completely preventing the establishment of the challenge virus. In each of these examples plants were challenged by mechanical inoculation. Yet in nature, mechanical inoculations may not properly reflect the level of resistance that will be obtained against viruses that are transmitted by a vector. Kawchuk et al.[22] pointed out that aphid-transmitted viruses are likely to be deposited within an individual cell at levels higher than obtainable by mechanical inoculation. However, transgenic potato and tobacco expressing either the PVX, CMV, or potato leafroll virus (PLRV) coat protein genes were found to be resistant to infection via viruliferous aphids, indicating that coat protein-mediated resistance is also effective against naturally vectored viruses.[14,20,22]

Coat protein-mediated resistance in many systems is generally limited to protection against the homologous virus or virus strains closely related to the integrated coat protein gene. This implies that the coat protein gene of each virus for which resistance is desired must be integrated into the genome of the host in order to obtain broad-spectrum resistance. However, heterologous resistance has been reported in several potyvirus systems. For example, transgenic tobacco expressing soybean mosaic virus coat protein showed resistance to TEV and PVY.[20] Also, tobacco expressing papaya ringspot virus coat protein was highly resistant to TEV, PVY, and pepper mottle virus.[23] In addition, Namba et al.[24] have reported that transgenic tobacco expressing coat protein of watermelon mosaic virus II (WMVII) or zucchini yellow mosaic virus are resistant to varying degrees to at least six different potyviruses. Heterologous resistance was generally less than the resistance conferred against the homologous virus.[24] Heterologous resistance has also been reported in transgenic tobacco expressing cucumovirus, CMV strain C, coat protein.[14] Thus, the potential exists for broad-spectrum resistance using limited numbers of viral coat protein genes.

In some systems, coat protein-mediated resistance is clearly dependent upon the expression of the viral coat protein. Powell et al.[25] and Loesch-Fries et al.[26] have demonstrated that transgenic plants that accumulate TMV or AMV coat protein are resistant to subsequent infection, while transgenic plants that do not accumulate coat protein remain susceptible. In addition, a transient protoplast system was used to demonstrate that TMV coat protein alone is effective in providing resistance similar to that observed in transgenic plants.[27] Thus, in these systems a mechanism involving the presence of the coat protein is responsible for the observed resistance.

A general correlation between the level of coat protein accumulation and the degree of resistance has been reported in transgenic tobacco expressing the TMV coat protein.[12] However, in other systems coat protein-mediated resistance does not always correlate with the level of coat protein accumulation. For example, one transgenic tobacco line accumulating the WMVII coat protein at 0.075% of the total soluble leaf protein was more resistant to challenge inoculation than a second tobacco line accumulating coat protein at 0.360%.[24] Interestingly, high levels of resistance have also been observed in transgenic plants that accumulate no detectable level of PLRV coat protein and in transgenic plants expressing mRNAs of a TEV coat protein that is defective in translation.[22,28] This suggests that another mechanism, perhaps mediated by the RNA, may be responsible for the resistance. Lindbo and Dougherty[28] have also demonstrated that TEV coat protein deletion mutants provide a higher level of coat protein-mediated resistance than that observed in plants expressing the full-length TEV coat protein. These truncated TEV coat proteins did not provide protection at the point of initial infection, as demonstrated by the ability of protoplast expressing these proteins to support TEV replication. Thus, Lindbo and Dougherty[28] speculate that these coat proteins may act dysfunctionally to inhibit viral movement. In contrast, Tumor et al.[29] demonstrated that the insertion of an additional glycine residue at position 2 in the AMV coat protein resulted in a mutant protein that could no longer confer protection against AMV infection, even though this mutant coat protein was capable of associating with and activating viral RNA. These differences in the abilities of mutant coat proteins to confer resistance suggest that different resistance mechanisms are active in different systems. Additional evidence that supports variations in cross-protection mechanisms can be seen in the ability of unencapsidated TMV RNA to largely overcome coat protein-mediated resistance,[30] while PVX resistance remains effective against unencapsidated PVX RNA.[19] Thus, a number of different mechanisms are apparently responsible for the resistance conferred to transgenic plants expressing different viral coat protein genes.

Coat protein-mediated resistance in many systems is correlated with the inhibition of virus replication at the initial point of infection. This resistance takes the form of reduced numbers of infection sites on inoculated leaves,[30] suggesting that an initial step in the virus life cycle has been disrupted. Sherwood and Fulton[31] have demonstrated that TMV cross protection may result from the coat protein of the protecting virus preventing uncoating of the challenge virus RNA. This would explain why unencapsidated TMV RNA can overcome cross protection. Prevention of uncoating may occur by direct interference with challenge virus disassembly or by blockage of a plant receptor necessary for uncoating.[30,32] Where unencapsidated viral RNA is unable to overcome resistance, then a mechanism involving interference in replication or translation may be involved. This could be directed by coat protein or through the inappropriate annealing of transgene RNA to viral RNA.

Coat protein-mediated resistance may also function at a systemic level. Establishment of a TMV infection, using RNA inoculum, in transgenic plants expressing the TMV coat protein results in lower levels of virus accumulation in both the inoculated and systemically infected leaves as well as a reduced rate of systemic virus spread.[32,33] Interestingly, Wisniewski et al.[33] found that a transgenic stem section, with an attached leaf expressing TMV coat protein, grafted into the center section of a nontransgenic tobacco plant acted essentially as barrier to the systemic movement of TMV. This retardation in systemic movement and virus accumulation may involve a similar or different mechanism than what is responsible for resistance at the initial point of infection. Thus, the mechanisms involved in coat protein-mediated resistance not only appear to differ from one virus system to another, but also different mechanisms of resistance may function differently within the same transgenic plant. The precise nature of these resistance mechanisms remains to be understood.

Most of the systems in which coat protein-mediated resistance has been reported have been directed against plus-sense RNA viruses with a single capsid protein. However, transgenic plants expressing the nucleo-capsid protein of the ambisense tomato spotted wilt virus (TSWV) have also been shown to be resistant to challenge inoculations.[34–36] TSWV particles are membrane enveloped, spherical, and composed of at least three structural proteins including the nucleoprotein and two membrane-associated glycoproteins. de Haan et al.[37] found TSWV nucleoprotein protection to be independent of protein expression and thus may be the result of an RNA-mediated mechanism. This resistance was also effective against inoculations using viruliferous thrips.[37] Additional studies have shown this transgene nucleoprotein resistance to be heterologous against a number of different isolates of the same TSWV serogroup as well as at least one isolate from a different serogroup.[36]

2. Nonstructural Protein-Mediated Resistance

Viruses encode nonstructural proteins that are necessary for replication. Recently, several of these nonstructural "replicase" proteins have been found to provide a high degree of resistance to virus infection when expressed in transgenic plants. In general, nonstructural protein-mediated resistance has been significantly greater than that conferred by other resistance strategies, including coat protein-mediated resistance. Golemboski et al.[38] first demonstrated this phenomena by expressing the 54-kDa open reading frame (ORF) of TMV in transgenic tobacco. This ORF is located in the read-through portion of the TMV 183-kDa replicase protein. Transgenic plants inoculated with TMV virions (500 μg/ml) or RNA (300 μg/ml) failed to become infected. Thus, 54-kDa transgenic plants were essentially immune to TMV infection. However, 54-kDa-mediated resistance was limited to homologous viral strains and was not effective against distantly related viruses.

Resistance conferred by expression of the TMV 54-kDa ORF is associated with suppression of viral replication within the area of initial infection.[38] Interestingly, production of the 54-kDa protein has not been detected in transgenic plants. Similarly, measurable levels of the 54-kDa protein have not been observed in viral infected tissue despite the presence and association of its mRNA with polyribosomes.[39] However, production of the 54-kDa protein is necessary for display of the resistance phenotype. Carr et al.[40] demonstrated that mutant 54-kDa constructs unable to translate protein due to an altered initiation codon or frameshift mutation did not confer resistance in a protoplast system. Two mechanisms to explain 54 kDa-mediated resistance have been proposed.[40] First, the 54-kDa protein may act as a true replicase component, perhaps functioning to regulate replication. Thus, its inappropriate expression could disrupt virus replication. Second, the 54-kDa protein may act as a truncated version of the TMV 183-kDa replicase protein, that inhibits replication by competition with normal replicase components. Experiments to determine what role the 54-kDa protein plays in viral replication should provide insight into the precise mechanism responsible for this resistance phenomena.

In a second system, the 54-kDa protein from pea early browning virus (PEBV), a tobravirus, has also been shown to provide resistance, at virion inoculum concentrations of 1 mg/ml, when expressed in transgenic tobacco.[41] Interestingly, mutations that cause the premature termination of the PEBV 54-kDa protein do not confer resistance. Thus, a mechanism requiring intact 54-kDa protein is apparently involved in this resistance phenomena. This evidence supports a functional role for the 54-kDa protein, perhaps as a true replicase component. Like the above TMV 54-kDa transgenic plants, this resistance apparently requires expression of the protein and is restricted to closely related viruses. In contrast, the transgene expression of a TMV 183-kDa replicase construct containing an inadvertent bacterial Tn-10 transposon insertion within the 183-kDa ORF conferred a high level of resistance to six different tobamoviruses including: TMV-U1, TMV-U5, TMV-U2, green tomato atypical mosaic virus, tomato mosaic virus, and ribgrass mosaic virus.[42,42a] Thus, heterologous resistance was obtained using this defective replicase gene. The transgenic expression of a truncated version of the CMV RNA-2 replicase ORF has also been found to provide a high level of resistance to CMV infection.[43] Although this truncated protein has not been detected in plants, it could potentially function as a defective replicase component that inhibits viral replication. It will be interesting to determine if these defective CMV replicase plants have a broad spectrum of resistance.

Braun and Hemenway[44] have recently demonstrated that transgene expression of the full-length PVX replicase ORF results in a high degree of resistance. Also, transgenic tobacco expressing the amino half of the PVX replicase ORF were similarly resistant, while plants expressing the carboxy half of the ORF were not resistant.[44] In contrast, transgene expression of the full-length 126-kDa replicase ORF of TMV did not confer resistance.[38] Thus, in different systems there appears to be a range of replicase constructs which can confer resistance.

It should be noted that not all nonstructural genes will confer this level of resistance. Angenent et al.[45] demonstrated that transgenic tobacco expressing either the 13, 16, or 29-kDa ORF of tobacco rattle virus were not resistant to subsequent challenge inoculations. In addition, transgenic plants expressing the 30-kDa cell-to-cell movement protein of TMV do not confer any resistance.[46] None of these genes are predicted to be involved in replicase complexes. Thus, nonstructural protein-mediated resistance may be limited to viral genes involved in replication.

3. Antisense- and Sense-Mediated Resistance

Another pathogen-derived strategy that has been investigated for the control of plant viruses is the transgene expression of antisense and more recently sense segments of viral RNAs. The logic behind this strategy is to bind up viral RNA with complementary RNA sequences expressed by the plant. Inappropriate RNA-RNA base paring would potentially prevent accessibility of the viral RNA for replication or gene expression. Thus, antisense and sense constructs could be used to block initial steps important in the establishment of a viral infection. Reviews outlining the uses and potential mechanisms of antisense and sense suppression of plant-expressed genes have been published.[32,47,48]

a. Antisense

Antisense protection has been demonstrated in tobacco expressing complementary RNA to the coat protein ORFs and 3′ untranslated regions of CMV, PVX, and TMV.[19,49,50] In these three cases protection was considerably lower than protection provided by plants accumulating the corresponding coat protein. For example, PVX transgenic plants accumulating coat protein showed protection at an inoculum concentration of 5 μg/ml virus, while transgenic plants expressing antisense to the coat protein showed significant protection at an inoculum concentration of only 0.05 μg/ml virus.[19] Differences also occur in the level at which antisense inhibition occurs. Tobacco expressing RNA complementary to the TMV coat protein ORF and 3′ untranslated region displayed resistance as an escape of infection.[50] This indicates that protection occurs at the initial point of infection, possibly via antisense RNA binding to newly uncoated viral RNA. In contrast, transgenic tobacco expressing coat protein antisense CMV RNA displayed resistance as a reduction of virus concentration in inoculated and systemically infected leaves,[49] indicating that antisense constructs in this system were capable of inhibiting virus replication after the establishment of infection. Also, transgene expression of an antisense construct to the 5′ end (nucleotides 70 to 355) of CMV RNA-1 has been shown to confer some resistance to challenge inoculations.[51] However, this resistance appeared in only one transgenic plant line and was not correlated with the transcription level of antisense RNA. Additionally, transgenic plants expressing antisense constructs directed to the 3′ end (nucleotides 2478 to 2900) of CMV RNA-2 and the 5′ end (nucleotides 31 to 236) of RNA-3 were not resistant to CMV infection.[51] Thus, not all antisense constructs confer resistance.

An interesting note is that TMV antisense tobacco expressing complementary RNA to the coat protein ORF and the 3′ untranslated region showed protection, while tobacco expressing only antisense RNA to the coat protein ORF showed no protection.[50] Thus, TMV antisense protection was dependent upon the presence of the 3′ untranslated region, possibly because of the importance of this region in binding replicase for the initiation of negative-strand RNA synthesis.[50] In contrast, Lindbo and Dougherty[28] have demonstrated that antisense RNA to the coat protein ORF of TEV, without the 3′ untranslated region, confers a high level of resistance to challenge inoculation. This resistance was equivalent or better than resistance displayed by transgenic plants accumulating the TEV coat protein. TEV replication was reduced in protoplasts from one TEV antisense transgenic line, suggesting that this TEV antisense construct inhibits virus replication. Thus, complementary binding within this potyvirus coat protein ORF may interfere with replicase processes necessary for RNA binding or negative-strand RNA syntheses. Future studies directed at determining the levels of negative-strand production in these systems will provide important insights into the mechanisms behind antisense-derived resistance as well as ideas for improving its effectiveness.

b. Sense

Resistance conferred by the transgene expression of the TMV and AMV coat protein genes has clearly been shown to be associated with the production of coat protein. In these systems, coat protein ORFs unable to translate protein did not provide resistance when expressed in transgenic plants, even though high levels of RNA transcripts accumulated.[25,50,52] In contrast, a TEV coat protein ORF sense construct, translationally unable to produce coat protein, provided a high degree of resistance that was greater than resistance conferred by similar plants accumulating coat protein.[28] Protoplasts from these TEV sense plants did not support viral replication, suggesting that resistance is due to an inhibition of an early event in viral replication such as base paring between transgene sense RNA and viral negative-sense RNA. Studies involving coat protein-mediated resistance against PVY and PLRV have also identified transgenic lines that expressed little or no detectable levels of coat protein, yet were still highly resistant.[20,22] Thus, resistance conferred in these coat protein systems may actually be due to an RNA sense-mediated mechanism.

The expression of other sense regions of viral genomes has also resulted in different degrees of resistance. For example, expression of full-length copies of AMV RNA-1 and RNA-2, both coding for replicase components, did not confer resistance.[53] However, transgene constructs containing the complete or partial ORF of the PVX 165-kDa replicase protein did confer resistance.[44] Protein from these PVX constructs was not detected, thus resistance due to the interfering properties of sense RNA may be a factor in this phenomena. However, further studies utilizing constructs deficient in the ability to produce protein will be required to confirm this possibility.

4. Satellite RNA Protection

Satellite RNAs are a class of small, single-stranded RNA molecules that are dependent upon a helper virus for replication and virion packaging. These RNA species have been associated with several different viruses.[54] Interestingly, a number of satellite RNAs have been shown to modulate the replication and symptomatology of their helper viruses.[54,55] Changes in symptom development range from severe necrosis to almost complete symptom attenuation, depending on the associated satellite RNA. Thus, satellite RNAs that attenuate symptoms can potentially be used to reduce the disease severity of the helper virus. Tien et al.[56,57] demonstrated that the deliberate inoculation of a mild strain of CMV with a symptom attenuating satellite RNA successfully protected tobacco, pepper, tomato, and cucumber plants from a virulent strain of CMV and reduced yield losses under greenhouse and field conditions. In China, the widespread use of this strategy has provided effective resistance against severe strains of CMV for more than 10 years with substantial reductions in yield losses to a variety of crop plants.[57]

Harrison et al.[58] and Gerlach et al.[59] have applied a molecular approach to this strategy by creating transgenic plants that express symptom-attenuating satellite RNAs of CMV and tobacco ringspot virus (TobRV). These transgenic plants expressed full-length copies of satellite molecules, which did not replicate until the subsequent establishment of the appropriate helper virus infection. Helper virus infection resulted in replication of the satellite molecule and attenuation of disease symptoms. Thus, transgenic expression of satellite RNAs results in symptom reductions similar to those observed in deliberately inoculated plants. In the case of TobRV, transgene expression of either plus and minus sense copies of satellite RNA resulted in satellite replication and symptom attenuation. On a larger scale, Tien and Gusui[57] have reported that 121 transgenic tomato plants expressing an attenuating CMV satellite RNA gave 50% yield increase over control plants when infected with a severe strain of CMV.

Transgenic expression of satellite RNAs provides several advantages over inoculation for protection.[58] First, it does not require the deliberate inoculation of the helper virus and satellite RNA. Thus, infection and associated yield reductions will occur only in those plants that are naturally infected by the helper virus. Also, risks associated with potential virulent changes in the helper virus can be avoided. However, there is the concern that satellite RNAs expressed by transgenic plants might mutate to become more virulent or in the presence of the helper virus be vectored into other more susceptible plant species. This strategy is also limited to those virus systems in which attenuating satellite RNAs are found. Reviews detailing this subject have been published and the reader is referred to these references.[54,57]

B. PATHOGEN-TARGETED RESISTANCE

Pathogen-targeted resistance refers to molecular strategies that target a viral function or component for inactivation. Elements used in these approaches may contain pathogen components, such as gene sequences, but the region active in pathogen disruption would not be of viral origin. Two examples are discussed below.

1. Ribozyme-Mediated Resistance

Ribozymes are essentially RNA-based RNA restriction enzymes capable of catalytically cleaving RNA molecules at specific sites. In nature, ribozymes have been found to be important in ribosomal RNA production, intron splicing, and satellite RNA replication.[60–62] However, genetically engineered ribozymes, directed against a number of additional RNA molecules, have also been created and shown to cleave *in trans*.[61,62] The ability to direct ribozyme cleavage provides a potentially useful strategy to control plant virus diseases, especially since the majority of agriculturally important plant viruses have RNA genomes. Thus, transgene expression of ribozymes designed to cleave viral RNAs could be used to disrupt viral replication and disease development.

Several different types of ribozymes with different sequences and structures have been identified.[61,62] Of these types, the hammerhead structure has been used in attempts to create viral resistant plants.[61,62] This ribozyme structure is composed of a highly conserved region that forms a catalytic domain and two flanking regions that base pair with the RNA substrate. Substrate cleavage occurs 3′ of a GUC triplet and results in a free 5′ hydroxyl and a 3′ terminal 2′, 3′-cyclic phosphate. The presence of an accessible GUC triplet in the target RNA is the only requirement for this ribozyme design. Edington and Nelson[62] have reviewed ribozyme types and their use in plant viral resistance.

In vitro studies have clearly shown that specifically designed ribozymes can cleave a variety of RNA molecules, including plant virus genomes. Ribozymes directed against the replicase or coat protein ORFs of PLRV and the replicase ORF of TMV were capable of cleaving their respective target viral

RNAs.[62,63] However, cleavage in these systems was accomplished at a temperature of 37 to 40°C with an excess ribozyme-to-substrate ratio, which indicates that ribozyme activity may not be effective under the conditions found *in planta*. The ability of a ribozyme to confer resistance to virus infection *in vivo* has been tested in a protoplast system.[62] Results from these experiments demonstrated that a ribozyme directed against the replicase ORF of TMV was effective at reducing viral accumulation in protoplasts by as much as 90% in the first 24 h postinfection.[62] However, the level and consistency of this resistance was not determined. In addition, control constructs using defective ribozymes are needed to determine if observed resistance is due to ribozyme cleavage or is a result of base paring by the antisense portion of the ribozyme.[63a] Experiments directed at improving substrate annealing and ribozyme turnover from one RNA molecule to another, perhaps by shortening annealing regions or inserting areas of mismatched bases, may lead to greater ribozyme activity and better virus control. Thus, the potential for the use of ribozymes in the control of plant virus diseases requires further investigation.

2. Plant Antibody Resistance

Animals, unlike plants, have evolved circulating immune systems which produce antibodies that can specifically recognize and bind pathogen molecules, targeting them for destruction. However, Hiatt et al.[64] have demonstrated that heavy- and light-chain immunoglobulin cDNAs expressed in the same transgenic plant were capable of combining to produce a functional antibody. Antibody accumulation in these plants was measured at 1.3% of the total leaf protein. Antibody expression in plants has also been obtained by the single expression of the immunoglobulin heavy-chain variable domain.[65] This domain alone is capable of binding antigen with good affinity, thus the need for cloning and integrating both the light and heavy chains of an antibody can be avoided.[65,66] Owen et al.[67] have recently used a synthetic single-chain E_V molecule consisting of the heavy- and light-chain variable domains of an antibody joined together by a flexible peptide linker and directed against phytochrome in transgenic tobacco. Seed from these plants displayed an aberrant phytochrome-dependent germination, indicating interference *in planta* by the expressed single-chain antibody. This discovery has obvious implications for the control of plant viruses. Virus-specific antibodies expressed in transgenic plants could potentially be used to interfere with the functions of key viral proteins, thereby inhibiting the virus life cycle. Probably, the most effective viral proteins to be targeted by antibodies will be those involved in replication. Antibodies that specifically bind replicase proteins might inactivate these proteins and prevent replication. Alternatively, antibodies that prevent virion disassembly or bind virus movement proteins may also be useful in conferring resistance. One positive feature of this strategy is that many potentially useful viral genes have been cloned, and hybridoma cell lines that express monoclonal antibodies to these proteins are available as a source for immunoglobulin genes. Also, concerns associated with the transgene expression of viral sequences, such as recombination or heterologous transcapsidation, would not be a factor in this strategy, since viral sequences are not used. Currently, plant antibody protection is being investigated in several systems.

C. HOST-DERIVED RESISTANCE

Plants have evolved numerous strategies to recognize and defend against pathogen invasion. Thus, a wealth of potentially useful resistance factors already exists in nature. Unfortunately, because of breeding incompatibilities between different species, a large number of these resistance genes cannot be transferred into agronomically important crops. However, the potential exists to use molecular techniques to isolate and transfer these genes into systems where resistance is presently unavailable.

Many plant resistance factors directed against viruses have been found to be under the control of single genes, making them particularly amenable to molecular techniques for gene transfer.[1,68] Studies into the viral components involved in several resistance interactions have demonstrated that different resistance genes target different virus components. For example, the tomato resistance gene, Tm-1, effectively prevents viral replication of certain tobamoviruses, but not specific mutants.[69] The virulence factor required to overcome Tm-1 resistance has been mapped to certain amino acid substitutions within the 12b-kDa replicase gene of TMV.[70] Thus, the mechanism behind Tm-1 resistance may involve an improper interaction between a host component and the TMV replicase protein, perhaps in the formation of a replicase complex. Plant resistance factors may also be directed at recognizing the presence of an infecting virus and initiating defense responses. Two resistance factors, Tm-2 and Tm-2^2, in tomato have been shown to be effective in restricting the movement of TMV.[71] Plants expressing these resistant genes develop a necrotic hypersensitive response when challenged with the appropriate TMV strain.

The ability of certain TMV mutants to overcome Tm-2 resistance has been mapped to the 30-kDa movement protein of the virus.[72] In contrast, induction of N′ gene hypersensitive resistance in tobacco has clearly been mapped to alterations within the TMV coat protein.[73] Specifically, N′ gene plants recognize certain structural features of the TMV coat protein. Additionally, the coat protein of PVX has also been shown to be responsible for the induction of the *Rx* and *Nx* hypersensitive resistance genes in potato.[74] These findings demonstrate that plant resistance strategies are dependent upon the presence or absence of precise interactions between host and virus components. Additionally, these interacting virus components could be useful as tags to specifically label the host components involved in disease resistance. To date, plant genes responsible for pathogen recognition and resistance have not been cloned or characterized at the molecular level. Thus, the use of these resistance genes, via molecular strategies, remains to be investigated.

IV. PRACTICAL APPLICATIONS OF RESISTANCE STRATEGIES

Molecular resistance strategies hold the potential for providing new methods for the large-scale control of viral diseases. Some strategies, such as satellite and coat protein protection, are already being tested and/or used at commercial levels. In addition to the assessment of the economic value conferred by these molecular strategies, investigations to assess the risks associated with the release of genetically engineered plants will also be required if these resistance strategies are to gain public acceptance and use at a commercial level.

A. RISK ASSESSMENT

There are concerns associated with developing technology that results in the introduction of virus gene sequences into crop plants. Questions and issues related to genetic engineering have been raised.[16,75,76] For example, will transcapsidation occur between transgene expressed coat protein and the genomic RNA of a non-protected virus, and could this lead to a wider host range, due to insect transmission, or increased rates of seed transmission for the nonprotected virus? Also, can recombination occur between a viral component expressed by the plant and a nonprotected virus? Clearly, we do not want to develop strategies for controlling viruses that result in new pathways for the evolution of more virulent virus strains. However, many of these concerns occur in nature. Typically, a field plant may be infected with several different viruses. This natural mixing can and has been shown to lead to transcapsidation and possibly recombination between the different infecting viruses.[77–80] Thus, under these circumstances we need to determine whether or not the risk associated with genetic engineering will be any different than what already occurs in nature. Future studies directed at addressing these concerns will be important if molecular resistance strategies are to gain public acceptance and widespread use.

Answers to the concerns associated with genetic engineering may also be found through a basic understanding of the mechanisms involved in the viral disease cycle. For example, potyvirus aphid transmission is dependent upon specific amino acids, Asp, Ala, and Gly within the N-terminus of the virus coat protein.[81] Removal or alteration of these amino acids results in a loss of insect transmissibility. Transgenic plants engineered with such nontransmissible coat proteins may prevent the inappropriate insect transmission of transcapsidated nonprotected viruses. Thus, it may be possible to engineer safeguards into molecular resistance strategies so as to avoid an associated risk.

B. DURABILITY AND EFFECTIVENESS OF RESISTANCE STRATEGIES

The practical use of any resistance factor is also dependent upon its durability in the field. Specifically, how long will a factor remain effective before resistance-breaking viral strains emerge? Presently, studies to specifically determine the durability of molecularly engineered resistance factors have not been done. However, resistance conferred by the deliberate inoculation of symptom-attenuating satellite RNAs has been found to be durable in the field over many years and without consequence to the environment.[57] This suggests that similar resistance, derived from the transgene expression of satellite RNAs, may also prove effective over time. In other systems, the potential for the evolution of resistance-breaking viral strains may be more prevalent. For example, MacFarlane and Davies[41] reported that the prolonged maintenance of inoculated plants expressing the 54-kDa gene of PEBV resulted in the appearance of two virus variants that were apparently capable of overcoming this resistance.

One method of increasing the durability of genetically engineered resistance factors may be to integrate two or more different factors into a single plant. Overcoming two factors, such as coat protein and nonstructural mediated resistance, which have different active mechanisms would presumably require two separate mutation events in the virus. Thus, the presence of both factors could greatly decrease the appearance of a resistance-breaking virus strain. Yie et al.[82] recently utilized this approach in the creation of transgenic tobacco that expressed both the coat protein and a symptom-attenuating satellite RNA of CMV. Interestingly, these plants had a level of resistance that was approximately twice as high as the resistance obtained from either the CMV coat protein or satellite RNA expressed alone. Thus, higher levels of resistance may be an additional bonus to combining different strategies for resistance. Broad-spectrum resistance to several different viral groups may also be obtained through a multiple integration process. Lawson et al.[20] used a double coat protein transformation vector to obtain multivirus resistance in transgenic potato expressing both PVX and PVY coat proteins.

Another problem faced by plant breeders is that the integration of many resistance genes from one plant species to another also results in the movement of undesirable plant traits, such as low or poor quality yields.[83] Clearly, molecular strategies have the potential to avoid this problem, since only factors specific for pathogen disruption are integrated into the genome of the plant. Thus, genes that are necessary for the optimal productivity of a given crop remain undisturbed. However, it should be noted that the integration of some viral genes into specific hosts can result in detrimental effects on the physiology of the plant. For example, transgenic tobacco expressing the gene VI protein of cauliflower mosaic virus display leaf chlorosis, and transgenic tobacco expressing TMV elicitor coat proteins have a stunted and necrotic phenotype.[84,85] Thus, an understanding of how different viral genes interact within different hosts is necessary before the widespread incorporation of a resistance factor.

V. CONCLUSION

The use of molecular strategies is clearly providing a number of new and potentially useful virus control measures. Several of these approaches have already been found to confer effective resistance against a variety of plant viruses and it seems likely that strategies, such as coat protein-mediated resistance, will be available for commercial use in the near future. Thus, the potential of molecular biology to create and integrate new virus resistance factors has become a reality. This is particularly exciting when we consider how rapidly many of these approaches have been developed and how little we understand about the mechanisms behind many of the strategies employed. The need now exists to understand the precise mechanisms behind these resistance strategies and to utilize this information to fine tune each approach to provide better resistance.

ACKNOWLEDGMENTS

I would like to thank Drs. W. O. Dawson, C. M. Kearney, L. Levy, R. S. Nelson, and J. S. Sherwood for suggestions and/or critical reading of this manuscript.

REFERENCES

1. Fraser, R. S. S., The genetics of resistance to plant viruses, *Annu. Rev. Phytopathol.,* 28, 179, 1990.
2. Matthews, R. E. F., *Plant Virology,* 3rd ed., Academic Press, San Diego, 1991.
3. Rogers, S. G., Klee, H. J., Horsch, R. B., and Fraley, R. T., Improved vectors for plant transformation: expression cassette vectors and new selectable markers. *Methods Enzymol.,* 153, 253, 1987.
4. Beachy, R. N., Loesch-Fries, S., and Tumer, N. E., Coat protein-mediated resistance against virus infection, *Annu. Rev. Phytopathol.,* 28, 451, 1990.
5. Protrykus, I., Gene transfer to plants: assessment of published approaches and results, *Annu. Rev. Plant Physiol. Plant Mol. Biol.* 42, 1991.
6. Potrykus, I., Shillito, R. D., Saul, M. W., and Paszkowski, J., Direct gene transfer state of the art and future potential, *Plant Mol. Biol. Rep.,* 3, 117, 1985.
7. Klein, T. M., Wolf, E. D., Wu, R., and Sanford, J. C., High-velocity microprojectiles for delivering nucleic acids into living cells, *Nature,* 327, 70, 1987.

8. Klee, H., Horsch, R., and Rogers, S., *Agrobacterium*-mediated plant transformation and its further applications to plant biology, *Annu. Rev. Plant Physiol.*, 88, 467, 1987.
9. Ream, W., *Agrobacterium tumefaciens* and interkingdom genetic exchange, *Annu. Rev. Phytopathol.*, 27, 583, 1989.
10. Sanford, J. C. and Johnston, S. A., The concept of parasite-derived resistance—deriving resistance genes from the parasite's own genome, *J. Theor. Biol.*, 113, 395, 1985.
11. Hamilton, R. I., Defenses triggered by previous invaders-viruses, in *Plant Disease, an Advanced Treatise,* Horsfall, J. G. and Cowling, E. R. Eds., Academic Press, New York, 1980, 279.
12. Powell-Abel, P. A., Nelson, R. S., De, B., Hoffman, N., Rogers, S. G., Fraley, R. T., and Beachy, R. N., Delay of disease development in transgenic plants that express the tobacco mosaic virus coat protein gene, *Science,* 232, 738, 1986.
13. Nelson, R. S., McCormick, S. M., Delannay, X., Dube, P., Layton, J., Anderson, E. J., Kaniewska, M., Proksch, R. K., Horsch, R. B., Rogers, S. G., Fraley, R. T., and Beachy R. N., Virus tolerance, plant growth, and field performance of transgenic tomato plants expressing coat protein from tobacco mosaic virus, *Bio/Technology,* 6, 403, 1988.
14. Quemada, H. D., Gonsalves, D., and Slightom, Expression of coat protein gene from cucumber mosaic virus strain C in tobacco: protection against infections by CMV strains transmitted mechanically or by aphids, *Phytopathology,* 81, 794, 1991.
15. Gonsalves, D., Chee, P., Provvidenti, R., Seem, R., and Slightom, Comparison of coat protein-mediated and genetically-derived resistance in cucumbers to infection by cucumber mosaic virus under field conditions with natural challenge inoculations by vectors, *Bio/Technology,* 10, 1562, 1992.
16. White, J. L., The introduction of genetically modified plants containing viral coat proteins: regulatory issues for commercialization, *Fifth International Symposium on Biotechnology and Plant Protection,* College Park, MD, 1992.
17. Hanley-Bowdoin, L. and Hemenway, C. L., Transgenic plants expressing viral genes, in *Genetic Engineering with Plant Viruses,* Wilson, T. M. A. and Davies, J. W., Eds., CRC Press, Boca Raton, FL, 1992.
18. Tumer, N. E., O'Connell, K. M., Nelson, R. S., Sanders, P. R., Beachy, R. N., Fraley, R. T., and Shah, D. M., Expression of alfalfa mosaic virus coat protein gene confers cross-protection in transgenic tobacco and tomato plants, *EMBO J.*, 6, 1181, 1987.
19. Hemenway, C., Fang, R-X., Kaniewski, W. K., Chua, N-H., and Tumer, N. E., Analysis of the mechanism of protection in transgenic plants expressing the potato virus X coat protein or its antisense RNA *EMBO J.*, 7, 1273, 1988.
20. Lawson, C., Kaniewski, W. K., Haley, L., Rozman R., Newell, C., Sanders, P., and Tumer, N. E., Engineering resistance to mixed virus infection in a commercial potato cultivar: resistance to potato virus X and potato virus Y in transgenic russet burbank, *Bio/Technology,* 8, 127, 1990.
21. Stark, D. M. and Beachy, R. N., Protection against potyvirus infection in transgenic plants: evidence for broad spectrum resistance, *Bio/Technology,* 7, 1257, 1989.
22. Kawchuk, L. M., Martin, R. R., and McPherson, J., Resistance in transgenic potato expressing the potato leafroll virus coat protein gene, *Mol. Plant-Microbe Interact.*, 3, 301, 1990.
23. Ling, K., Namba, S., Gonsolves, C., Slightom, J. L., and Gonsalves, D., Protection against detrimental effects of potyvirus infection in transgenic tobacco plants expressing the papaya ringspot virus coat protein gene, *Bio/Technology,* 9, 752, 1991.
24. Namba, S., Ling, K., Gonsalves, C., Slightom, J. L., and Gonsalves, D., Protection of transgenic plants expressing the coat protein gene of watermelon mosaic virus II or zucchini yellow mosaic virus against six potyviruses, *Phytopathology,* 82, 940, 1992.
25. Powell, P. A., Sanders, P. R., Tumer, N., Fraley, R. T., Beachy, R. N., Protection against tobacco mosaic virus infection in transgenic plants requires accumulation of coat protein rather than coat protein RNA sequences, *Virology,* 175, 124, 1990.
26. Loesch-Fries, L. S., Merlo, D., Zinnen, T., Burhop, L., Hill, K., Krahn, K., Jarvis, N., Nelson, S., and Halk, E., Expression of alfalfa mosaic virus RNA 4 in transgenic plants confers virus resistance, *EMBO J.*, 6, 1845, 1987.
27. Register, J. C. and Beachy, R. N., Effect of protein aggregation state on coat protein-mediated protection against tobacco mosaic virus using a transient protoplast assay, *Virology,* 173, 656, 1989.
28. Lindbo, J. A. and Dougherty, W. G., Pathogen-derived resistance to a potyvirus: immune and resistance phenotypes in transgenic tobacco expressing altered forms of a potyvirus coat protein nucleotide sequence, *Mol. Plant-Microbe Interact.*, 5, 144, 1992.

29. Tumer, N. G., Kaniewski, W., Haley, L., Gehrke, L., Lodge, J. K., and Sanders, P., The second amino acid of alfalfa mosaic virus coat protein is critical for coat protein mediated protection, *Proc. Natl. Acad. Sci. U.S.A.*, 88, 2331, 1991.
30. Nelson, R. S., Powell-Abel, P., and Beachy, R. N., Lesions and virus accumulation in inoculated transgenic tobacco plants expressing the coat protein gene of tobacco mosaic virus, *Virology,* 158, 126, 1987.
31. Sherwood, J. L. and Fulton, R. W., The specific involvement of coat protein in tobacco mosaic virus cross protection, *Virology,* 119, 150, 1982.
32. Register, J. C. and Nelson, R. S., Early events in plant virus infection: relationships with genetically engineered protection and host gene resistance, *Semin. Virol.,* in press.
33. Wisniewski, L. A., Powell, P. A., Nelson, R. S., and Beachy, R. N., Local and systemic spread of tobacco mosaic virus in transgenic tobacco, *Plant Cell,* 2, 559, 1990.
34. Gielen, J. J. L., de Haan, P., Kool, A. J., Peters, D., van Grinsven, M., and Goldbach, R. W., Engineered resistance to tomato spotted wilt virus, a negative-strand RNA virus, *Bio/Technology,* 9, 1363, 1991.
35. MacKenzie, D. J. and Ellis, P. J., Resistance to tomato spotted wilt virus infection in transgenic tobacco expressing the viral nucleocapsid gene, *Mol. Plant-Microbe Interact.,* 5, 34, 1992.
36. Pang, S. H., Nagpala, P., Wang, M., Slightom, J. L., and Gonsalves, D., Resistance to heterologous isolates of tomato spotted wilt virus in transgenic tobacco expressing its nucleocapsid protein gene, *Phyropathology,* 82, 1223, 1992.
37. de Haan, P., Gielen, J. J. L., Prins, M., Wijkamp, I. G., van Schepen, A., Peters, D., van Grinsven, Q. J. M., and Goldbach, R., Characterization of RNA-mediated resistance to tomato spotted wilt virus in transgenic tobacco plants, *Bio/Technology,* 10, 1133, 1992.
38. Golemboski, D. B., Lomonossoff, G. P., and Zaitlin, M., Plants transformed with a tobacco mosaic virus nonstructural gene sequence are resistant to the virus, *Proc. Natl. Acad. Sci. U.S.A.,* 87, 6311, 1990.
39. Sulzinski, M. A., Gabard, K. A., Palukaitis, P., and Zaitlin, M., Replication of tobacco mosaic virus. VIII. Characterization of a third subgenomic TMV RNA, *Virology,* 145, 132, 1985.
40. Carr, J. P., Marsh, L. E., Lomonossoff, G. P., Sekiya, M. E., and Zaitlin, M., Resistance to tobacco mosaic virus induced by the 54-kDa gene sequence requires expression of the 54-kDa protein, *Mol. Plant-Microbe Interact.,* 5, 397, 1992.
41. MacFarlane, S. A. and Davies, J. W., Plants transformed with a region of the 201-kilodalton replicase gene from pea early browning virus RNA1 are resistant to virus infection, *Proc. Natl. Acad. Sci. U.S.A.,* 89, 5829, 1992.
42. Donson, J., Kurath, G., Turpen, T. H., Khan, I. A., and Dawson, W. O., Tobacco plants transgenic for non-structural genes of tobacco mosaic virus are resistant to infection by TMV and other tobamoviruses, *Phytopathology* (Abstract), 82, 1071, 1992.
42a. Kearney, C. M., personal communication.
43. Anderson, J. M., Palukaitis, P., and Zaitlin, M., A defective replicase gene induces resistance to cucumber mosaic virus in transgenic plants, *Proc. Natl. Acad. Sci. U.S.A.* 89, 8756, 1992.
44. Braun, C. J. and Hemenway, C. L., Expression of amino-terminal portions or full-length viral replicase genes in transgenic plants confers resistance to potato virus X infection, *Plant Cell,* 4, 735, 1992.
45. Angenent, G. C., Van Den Ouwiland, J. M. W., and Bol, J. F., Susceptibility to virus infection of transgenic tobacco plants expressing structural and nonstructural genes of tobacco rattle virus, *Virology,* 175, 191, 1990.
46. Deom, C. M., Shaw, M. J., and Beachy, R. N., The 30-kilodalton gene product of tobacco mosaic virus potentiates virus movement, *Science,* 327, 389, 1987.
47. Jorgensen, R., Altered gene expression in plants due to *trans* interactions between homologous genes, *Trends Biotechnol.,* 8, 340, 1990.
48. Grierson D., Fray, R. G., Hamilton, A. J., Smith, C. J. S., and Watson, C. F., Does co-suppression of sense genes in transgenic plants involve antisense RNA, *Trends Biotechnol.,* 9, 122, 1991.
49. Cuozzo, M., O'Connell, K. M., Kaniewski, W., Fang, R-X., Chua, N-H., and Tumer, N. E., Viral protection in transgenic tobacco plants expressing the cucumber mosaic virus coat protein or its antisense RNA, *Bio/Technology,* 6, 549, 1988.
50. Powell, P. A., Stark, D. M., Sanders, P. R., and Beachy, R. N., Protection against tobacco mosaic virus in transgenic plants that express tobacco mosaic virus antisense RNA, *Proc. Natl. Acad. Sci. U.S.A.,* 86, 6949, 1989.

51. Rezaian, M. A., Skene, K. G. M., and Ellis, J. G., Anti-sense RNAs of cucumber mosaic virus in transgenic plants assessed for the control of the virus, *Plant Mol. Biol.*, 11, 463, 1988.
52. Cees, M. P., Van Dun, C. M., Overduin, B., Van Vloten-Doting, L., and Bol, J. F., Transgenic tobacco expressing tobacco streak virus or mutated alfalfa mosaic virus coat protein does not cross-protect against alfalfa mosaic virus infection, *Virology*, 164, 383, 1988.
53. Taschner, P. E. M., Van Der Kuyl, A. C., Neeleman, L., and Bol, J. F., Replication of an incomplete alfalfa mosaic virus genome in plants transformed with viral replicase genes, *Virology*, 181, 445, 1991.
54. Collmer, C. W. and Howell, S. H., Role of satellite RNA in the expression of symptoms caused by plant viruses, *Annu. Rev. Phytophathol.*, 30, 419, 1992.
55. Kaper, J. M. and Tousignant, M. E., Viral satellite: parasitic nucleic acids capable of modulating disease expression, *Endeavor (New Ser.)*, 8, 194, 1984.
56. Tien, P., Zhang, X., Qiu, B., Qin, B., and Wu, G., Satellite RNA for the control of plant diseases caused by cucumber mosaic virus, *Ann. Appl. Biol.*, 111, 143, 1987.
57. Tien, P. and Gusui, W., Satellite RNA for the biocontrol of plant disease, *Adv. Virus Res.*, 39, 321, 1991.
58. Harrison, B. D., Mayo, M. A., and Baulcombe, D. C., Virus resistance in transgenic plants that express cucumber mosaic virus satellite RNA, *Nature*, 328, 799, 1987.
59. Gerlach, W. L., Llewellyn, D., and Haseloff, J., Construction of a plant disease resistance gene from the satellite RNA of tobacco ringspot virus, *Nature*, 328, 802, 1987.
60. Cech, T. R., Conserved sequences and structures of group I introns: building an active site for RNA catalysis—a review, *Gene*, 73, 273, 1988.
61. Haseloff, J. and Gerlach, W. L., Simple RNA enzymes with new and highly specific endoribonuclease activities, *Nature*, 334, 585, 1988.
62. Edington, B. V. and Nelson, R. S., Utilization of ribozymes in plants: plant viral resistance, in *Gene Regulation: Biology of Antisense RNA and DNA*, Erickson, R. P. and Izant, J. G., Eds., Raven Press, New York, 1992.
63. Lamb, J. W. and Hay, R. T., Ribozymes that cleave potato leafroll virus RNA within the coat protein and polymerase genes, *J. Gen Virol.*, 71, 2257, 1990.
63a. Nelson, R. S., personal communication.
64. Hiatt, A., Cafferkey, R., and Bowdish, K., Production of antibodies in transgenic plants, *Nature*, 342, 76, 1989.
65. Benvenuto, E., Ordas, R. J., Tavazza, R., Ancora, G., Biocca, S., Cattaneo, A., and Galeffi, P., Phytoantibodies: a general vector for the expression of immunoglobulin domains in transgenic plants, *Plant Mol. Biol.*, 17, 865, 1991.
66. Ward, E. S., Gussow, D., Griffiths, A. D., Jones, P. T., and Winter, G., Binding activities of a repertoire of single immunoglobulin variable domains secreted from *Escherichia coli, Nature*, 341, 544, 1989.
67. Owen, M., Gandecha, A., Cockburn, B., and Whitelam, G., Synthesis of a functional anti-phytochrome single-chain Fv protein in transgenic tobacco, *Bio/Technology*, 10, 790, 1992.
68. Fraser, R. S. S., Virus recognition and pathogenicity: implications for resistance mechanisms and breeding, *Pestic. Sci.*, 23, 267, 1988.
69. Motoyoshi, F. and Oshima, N., Expression of genetically controlled resistance to tobacco mosaic virus infection in isolated tomato leaf mesophyll protoplasts, *J. Gen. Virol.*, 34, 499, 1977.
70. Meshi, T., Motoyoshi, F., Adachi, A., Watanabe, Y., Takamatsu, N., and Okada, Y., Two concomitant base substitutions in the putative replicase genes of tobacco mosaic virus confer the ability to overcome the effects of a tomato resistance gene, Tm-1 *EMBO J.*, 7, 1575, 1988.
71. Nishiguchi, M. and Motoyoshi, F., Resistance mechanisms of tobacco mosaic virus strains in tomato and tobacco, in *Plant Resistance to Viruses*, Evered, D. and Hamett, S., Eds., John Wiley & Sons, Chichester, 1987.
72. Meshi, T., Motoyoshi, F., Maeda, T., Yoshiwoka, S., Watanabe, H., and Okada, Y., Mutations in the tobacco mosaic virus 30-kd protein gene overcome Tm-2 resistance in tomato, *Plant Cell*, 1, 515, 1989.
73. Culver, J. N., Lindbeck, A. G. C., and Dawson, W. O., Virus-host interactions: induction of chlorotic and necrotic responses in plants by tobamoviruses, *Annu. Rev. Phytopathol.*, 29, 193, 1991.
74. Kavanagh, T., Goulden, M., Santa-Cruz, S., Chapman, S., Baker, I., and Baulcombe, D., Molecular analyses of a resistance-breaking strain of potato virus X, *Virology*, 189, 609, 1992.
75. Palukaitis, P., Virus-mediated genetic transfer in plants, in *Risk Assessment in Genetic Engineering: Environmental Release of Organisms*, Levin, M. and Strauss, H., Eds., McGraw-Hill, New York, 1990.
76. de Zoeten, G. A., Risk assessment: do we let history repeat itself?, *Phytopathology*, 81, 585, 1991.

77. Rochow, W. F., Dependent virus transmission from mixed infections, in *Aphids as Virus Vectors,* Harris, K. F. and Maramorosch, Eds., Academic Press, New York, 1977.
78. Creamer, R. and Falk, B. W., Direct detection of transcapsidated barley yellow dwarf luteoviruses in doubly infected plants, *J. Gen. Virol.,* 71, 211, 1990.
79. Bujarski, J. J. and Kaesberg, P., Genetic recombination between RNA components of a multipartite plant virus, *Nature,* 321, 528, 1986.
80. Allison, R., Thompson, C., and Ahlquist, P., Regeneration of a functional RNA virus genome by recombination between deletion mutants and requirements for cowpea chlorotic mottle virus 3a and coat genes for systemic infection, *Proc. Natl. Acad. Sci. U.S.A.,* 87, 1820, 1990.
81. Atreya, P. L., Atreya, C. D., and Pirone, T. P., Amino acid substitutions in the coat protein result in loss of insect transmissibility of a plant virus, *Proc. Natl. Acad. Sci. U.S.A.,* 88, 7887, 1991.
82. Yie, Y., Zhao, F., Zhao, S. Z., Liu, Y. Z., Liu, Y. L., and Tien, P., High resistance to cucumber mosaic virus conferred by satellite RNA and coat protein in transgenic commercial tobacco cultivar G-140, *Mol. Plant-Microbe Interact.,* 5, 460, 1992.
83. Van Der Plank, J. E., *Genetic and Molecular Basis of Plant Pathogenesis,* Springer-Verlag, New York, 1978.
84. Baughman, G. A., Jacobs, J. D., and Howell, S. H., Cauliflower mosaic virus gene VI produces a symptomatic phenotype in transgenic tobacco plants, *Proc. Natl. Acad. Sci. U.S.A.,* 85, 733, 1988.
85. Culver, J. N. and Dawson, W. O., Tobacco mosaic virus elicitor coat protein genes produce a hypersensitive phenotype in transgenic *Nicotiana sylvestris* plants, *Mol. Plant-Microbe Interact.,* 4, 458, 1991.

FURTHER READING

Gatehouse, A. M. R., Hilder, V. A., and Boulter, D., *Plant Genetic Manipulations for Crop Protection,* CAB International, Wallingford, 1992.

Hiatt, A., *Transgenic Plants: Fundamentals and Applications,* Marcel Dekker, New York, 1993.

Chapter 18

Engineering Disease Resistance in Plants: An Overview

Richard A. Dixon, Nancy L. Paiva, and Madan K. Bhattacharyya

TABLE OF CONTENTS

I. INTRODUCTION AND SCOPE

The goal of genetically manipulating disease resistance in plants has become a reality in the last 3 or 4 years. This is primarily due to advances in two areas: the technology of plant transformation and our better understanding of the molecular basis of plant-pathogen interactions. Transgenic plants expressing either novel proteins from foreign organisms or overexpressing a part of their own defensive arsenal have been engineered, tested in both laboratory and field situations, and evaluated for disease resistance. Engineered viral resistance through expression of viral coat protein genes,[1] replicase components,[2] or antisense RNA[3] has proved effective in several instances. To date, there are fewer reports on engineered resistance against fungi and bacteria. However, we believe that the next 5 years will witness a number of successes in this area, utilizing a range of different, and often ingenious, strategies. The purpose of this chapter is to outline the principles upon which strategies for fungal and bacterial resistance may be based, to evaluate the types of genetic manipulations which may lead to increased resistance, and to review those examples in which success has already been reported.

II. MOLECULAR FEATURES OF PLANT-PATHOGEN INTERACTIONS

The processes of plant-pathogen coevolution have led in many cases to interactions in which the outcome (compatibility or incompatibility) is determined by a single dominant gene for resistance in the host, the functional realization of which is determined by the presence of a corresponding, dominant avirulence gene in the fungal or bacterial pathogen.[4] In such gene-for-gene interactions, incompatibility is associated with the rapid activation of a battery of defense-response genes, whose products may include biosynthetic enzymes for the production of antimicrobial phytoalexins and wall-bound phenolics, hydrolytic enzymes and other so-called pathogenesis-related (PR) proteins, and hydroxyproline-rich glycoproteins.[5] These products usually accumulate locally around the site of attempted microbial ingress. In compatible

0-87371-877-1/95/$0.00+$.50

interactions, such defense-response genes are either not activated or are induced too late in the interaction to prevent disease symptoms.[6,7] These observations suggest that resistance could be engineered in plants by (1) altering the timing and extent of induced defenses by constitutive expression of a natural induced defense-response gene or by putting naturally occurring defense-response genes under the control of stronger inducible promoters, or (2) by genetic manipulation of the dominant resistance genes per se. The strategy of altering expression of defense-response genes could also include targeting expression of novel antimicrobial proteins from foreign organisms, either constitutively or to the plant-pathogen interface.

Some fungal and bacterial pathogens produce toxins which are responsible for the disease symptoms.[8] In such cases, virulence is dominant and resistance is expressed through the ability of the host either to not recognize the toxin (i.e., by lacking a toxin binding site) or to detoxify it. In such cases, incorporation of toxin-insensitive binding sites or enzymes for detoxification may provide means of engineering resistance.

Some fungal pathogens have acquired virulence by being able to detoxify the phytoalexins the host produces as a part of its defensive arsenal.[9] A basis of information now exists for engineering modified phytoalexin structures which may be resistant to detoxification, or for transferring a phytoalexin biosynthetic pathway from one plant to another which lacks that particular pathway. Such strategies will generally necessitate the transfer of several genes; although this may pose complications, attempts in this area should lead to further insights into the control of plant gene expression and the roles of secondary metabolites in plants.

The following sections review the prospects for engineering fungal and bacterial resistance in plants based on the above features of plant-pathogen interactions. For a more detailed background on the molecular basis of resistance in plant-microbe interactions, the reader is referred to the reviews by Lamb et al.,[10] Dixon and Lamb,[11] Dixon and Harrison,[5] and Keen.[12] More details of engineered resistance strategies can be found in the recent review by Lamb et al.[13]

III. CHOICE OF PROMOTERS

A large number of plant defense-response genes have now been cloned.[14] Most of these are transcriptionally activated in response to infection or exposure to microbial elicitor macromolecules.[5] The promoters of such genes could therefore be used to target expression of engineered transgenes encoding proteins to enhance resistance. Before selecting a defense-response gene promoter for such studies, several features of the promoter must be assessed. These include whether or not its expression is tissue or cell type specific, whether it is affected by developmental or environmental cues other than infection, its kinetics of activation in response to infection, and its extent of expression (i.e., promoter strength). If the protective factor being introduced is not toxic to the plant, it may be best to use a promoter which will deliver high-level constitutive expression; the cauliflower mosaic virus 35S promoter[15] or higher expression derivatives with double enhancer elements[16] have been used successfully in a number of cases. Indeed, the importance of the timing of defense gene activation in determining the outcome of many plant-pathogen interactions suggests that having the newly engineered defensive barrier in place prior to pathogen ingress should be beneficial. On the other hand, inducible promoters would be a necessity if constitutive expression of the transgene or its ultimate product (e.g., phytoalexins) were toxic to the plant or in any way compromised the ability of the plant to defend itself (e.g., by affecting amino acid or energy metabolism in the case of very highly expressed proteins).

The properties of several plant defense-response gene promoters are outlined in Table 1. Many of these show highly specific patterns of tissue and cell type expression. In some cases, it has proved possible to separate *cis*-elements conditioning infection or elicitor inducibility from those determining tissue-specific expression;[22] it may thus be possible to engineer a promoter which is only expressed in response to pathogen attack.

To be of general use, a promoter must retain its potential for correct activation in species other than that from which the gene was isolated. The examples in Table 1 indicate that most defense-response gene promoters studied to date are active in heterologous species. Whether or not this is likely to be universally true is not yet known, although it is interesting to note that the bean *chs*8 promoter, the activation of which is a component of the induction of isoflavonoid phytoalexins in the host species, is also induced by infection in tobacco,[20] which does not use the flavonoid pathway for defense and does not make isoflavonoids at all. Some monocot defense gene promoters are correctly expressed in

Table 1 **Properties of plant defense-response gene promoters**

Gene	Source	Protein encoded	Promoter expressed in:	Tissue specificity[a]	Induction[b] L	W	I	E	Ref.
pal1	*Arabidopsis thaliana*	L-Phenylalanine ammonia-lyase	*Arabidopsis*	V, Se, A, C	+	+	ND	+	17
pal2	*Phaseolus vulgaris*	L-Phenylalanine ammonia-lyase	Tobacco	P, A, St, R, St	+	+	ND	ND	18
chs8	*P. vulgaris*	Chalcone synthase	Tobacco						19, 20
ifr	*Medicago sativa*	Isoflavone reductase	Tobacco, *Medicago sativa*	RT, V, St	ND	–	ND	+	20a
5B	*P. vulgaris*	Basic chitinase	Tobacco	ND	ND	ND	+	+	21
RCH10	*Oryza sativa*	Basic chitinase	Tobacco	RT, R, V, St, O, A	ND	+	ND	+	22
—	*A. thaliana*	Acidic chitinase	*Arabidopsis*, tomato	R, V_L, H, G, A	ND	ND	+	ND	23
hrgp4.1	*P. vulgaris*	Hydroxyproline-rich glycoprotein	Tobacco	SN, RT, Sty, St	ND	+	+	ND	24

[a] A = anthers; C = carpels; G = guard cells; H = hydathodes; ND = not determined; O = ovaries; P = petals; R = roots, RT = root tips; Se = sepals; SN = stem nodes; St = stigmas; Sty = styles; V = vascular tissue; V_L = leaf vascular tissue.

[b] L = light; W = wounding; I = infection; E = elicitor.

Table 2 **Pathogenesis-related (PR) Proteins Induced During Plant Defense**

Class	Biological activity *in vitro*	Typical sources
Class I chitinase, basic	Antifungal	Bean,[27] tobacco,[28] maize,[29] rice,[30] *Brassica*[31]
Class I chitinase, acidic	Antifungal	Bean[32]
Class II chitinase, acidic	Antifungal	Tobacco[33]
Class III chitinase	Bifunctional lysozyme/chitinase	Cucumber,[34] tobacco[35]
Acidic extracellular glucanase	(Antifungal)	Bean,[36] tobacco[37]
Basic vacuolar glucanase	Synergist for chitinase, antifungal	Bean,[38,39] pea[40]
PR-la	Antifungal	Tobacco[41]
PR-lb	ND[a]	Tobacco[41]
PR-lc	ND	Tobacco[41]
Pv PR1, Pv PR2 (birch pollen allergen-like)	ND	Bean[42] (similar in parsley, pea, potato)
Pv PR3	ND	Bean[43]
Ao PR1	ND	Asparagus[44]
PR-4 (hevein-like, no lectin domain)	ND	Tobacco[45]
PR-5 (thaumatin-like)	Antifungal, synergist for PR-4	Tobacco,[46] barley[47]
PR-5 (osmotin-like)	Antifungal	Tobacco[48]

[a]Not detected or not determined.

dicot tissues[22] and the constitutive 35S promoter has been used successfully in some monocots.[25] Overall, signals for defense gene activation in plants would appear to be conserved even if the sets of genes upon which they act are different in different species.

To date, very few studies have attempted to engineer inducible defense responses using promoters such as are outlined in Table 1. Even in cases where potentially toxic secondary metabolites are being engineered, constitutive expression of a gene encoding a modifying enzyme may be acceptable if the earlier stages of the pathway which provide the substrate for that enzyme are only expressed locally in response to infection.

IV. MANIPULATION OF SINGLE GENE TRAITS TO DIRECTLY CONFER ANTIMICROBIAL ACTIVITY

A. PATHOGENESIS-RELATED PROTEINS

PR proteins are low-molecular weight proteins which accumulate to significant levels in infected plant tissues. They were initially defined and classified on the basis of their physical properties and induction characteristics in virus-infected tobacco.[26] The major classes of PR proteins are outlined in Table 2. Antimicrobial activity *in vitro* has not been demonstrated for all these proteins; it is possible that some are only active in combination with others.

On the basis of our present knowledge, the most attractive PR proteins for engineering resistance based on constitutive expression are the chitinases and 1,3,-β-D-glucanases, at least against those fungal pathogens which contain chitin in their cell walls. A basic, vacuolar chitinase of bean (*Phaseolus vulgaris*) has been expressed constitutively to high levels in transgenic plants of tobacco and *Brassica napus.* This expression resulted in significant protection of the plants from post-emergent damping off caused by the pathogen *Rhizoctonia solani.*[27] In the case of *B. napus,* although the protection was a delay rather than a complete inhibition of symptoms, it was concluded that the level of protection was sufficient to be of economic significance in field situations.[27] It would, however, be dangerous to assume that expression of a single chitinase gene will be of general efficacy in conferring resistance; indeed, constitutive expression of a tobacco basic chitinase gene in *Nicotiana sylvestris* proved ineffective against *Cercospora nicotianae.*[49] Such conflicting results should not be surprising, as we do not yet understand the basis of the protection conferred by hydrolytic enzymes such as chitinase. This may involve a direct lytic effect in which invading hyphae are killed, a perturbation of growth allowing other induced defenses to become effective, a release of fungal wall components which can elicit other defenses, or a combination of all three.

To date, there have been no reports of increased resistance from expression of a 1,3,-β-D-glucanase gene in transgenic plants. Likewise, down-regulation of glucanase expression by antisense RNA did not increase the susceptibility of transgenic *N. sylvestris* plants to infection by *C. nicotianae.*[50] *In vitro,* chitinases generally show greater antimicrobial activity than glucanases. However, glucanase has been shown to act as a powerful synergist for chitinase.[51] This is presumably because, by digesting the β-glucan portion of the fungal cell wall, this enzyme renders the chitin more digestible. These observations suggest the strategy of coexpression of chitinase and glucanase in transgenic plants. Transgenic tobacco plants expressing a basic rice chitinase exhibit slightly delayed symptoms on infection with *C. nicotianae.* If these plants are crossed with tobacco expressing an alfalfa acidic glucanase (which itself does not appear to confer significant protection), the delay of symptoms in progeny expressing both genes is greater than in plants expressing chitinase alone.[51a] There is considerable scope for optimizing and fine tuning such a protection mechanism, utilizing different combinations of chitinases and glucanases. These enzymes often exist in multiple forms, the basic forms generally being vacuolar and the acidic forms extracellular. As well as mixing and matching naturally occurring forms, it should be possible to modify targeting, for example, by removing the vacuolar targeting signals[52] from the carboxy termini of the basic forms in order to direct them to the extracellular space.

A number of groups have attempted overexpressing other PR proteins in plants with a view to understanding their function. The possibility that some of these proteins are only active in combination with others makes negative results difficult to interpret and necessitates the laborious testing of a matrix of different PR protein combinations.[53] It is also possible that individual PR proteins exhibit some degree of pathogen specificity. Thus, overexpression of the tobacco PR-1 gene does not protect tobacco against tobacco mosaic virus,[54] although it can delay the onset of infection by the blue mold pathogen *Peronospora tabacina.*[55] Some PR proteins other than chitinase and glucanase exhibit antimicrobial activity *in vitro* (e.g., the osmotin-like PR5 protein),[48] and are therefore candidates for further evaluation.

B. NOVEL ANTIMICROBIAL PROTEINS

Plants, and indeed other organisms, may contain antimicrobial proteins not necessarily associated with induced defense responses, which are potential subjects for engineered protection strategies. Floral organs and seeds often contain high levels of antimicrobial proteins, presumably to protect the vulnerable tissues of the reproductive phase of the plant. Table 3 lists a selection of antimicrobial proteins which could find uses in plant protection in the next several years.

The ribosome-inactivating proteins (RIPs) have *N*-glycosidase activity which cleaves a specific adenine residue from the large subunit ribosomal RNA. They exist as single-chain proteins (Type I, e.g., the pokeweed antiviral protein) or double chains possessing a galactose-specific lectin which targets them to cell surfaces (Type II, e.g., ricin). A recent review lists nearly 40 RIPs from a range of plant families.[66] RIPs do not inhibit ribosomes from the plant of origin; some are active against fungal ribosomes and the barley RIP exhibits antifungal activity *in vitro,*[68] an activity which is enhanced in the presence of enzymes which can degrade fungal cell wall polysaccharides. A major potential complication in using RIPs for plant protection concerns their potential cytotoxicity, and, if this is likely to be a problem, the need to express them from nonleaky, inducible promoters and/or to target them to the extracellular space. The barley RIP has been expressed under a wound-inducible promoter in transgenic tobacco and shown to afford protection (measured by overall plant growth parameters) against the soilborne pathogen *R. solani.*[69] The fact that the expression of this transgene, which occurs in floral tissues and pollen, did not effect the fertility of the primary transformants suggests that cytotoxicity may not be a serious problem, at least in this case.

C. ENGINEERED TOXIN INSENSITIVITY

The molecular targets of several fungal or bacterial toxins from plant pathogens are now known. One example will illustrate the strategy for engineering resistance by transfer of toxin-insensitive targets. The bacterial halo-blight pathogen of bean, *Pseudomonas phaseolicola,* produces a tripeptide toxin, phaseolotoxin, which causes the chlorotic halos symptomatic of the disease.[70] Phaseolotoxin inhibits the enzyme ornithine transcarbamylase, a key step in the biosynthesis of the amino acid arginine. Bacteria have been selected which contain a phaseolotoxin-insensitive ornithine transcarbamylase, and the gene encoding this enzyme has been cloned and transferred to tobacco, where its expression has been shown to prevent the symptoms caused by application of the toxin.[71] The lack of a transformation system for bean has so far precluded analysis of this manipulation with respect to resistance of the true host species.

D. ENGINEERED TOXIN DETOXIFICATION

An alternative to engineering plants with modified sites of action for microbial toxins would be to introduce genes encoding enzymes which can inactivate the toxin. In maize, resistance to the cyclic tetrapeptide toxin produced by *Cochliobolus carbonum* is associated with the presence of a pyridine nucleotide-dependent reductase which acts on an essential carbonyl group of the toxin. This enzyme activity is present only in resistant germplasm containing the dominant allele of the *hm1* locus.[72] This toxin-resistance gene has recently been cloned by transposon tagging.[73] This opens up the possibility of engineering resistance to *C. carbonum* in previously susceptible maize genotypes.

E. EXPRESSION OF ANTIBODIES IN PLANTS

With the development of the technology for cloning individual, specific antibody genes,[74] it has become possible to express monoclonal antibodies in plants. Although expression levels are very variable and the system is still far from optimized, especially with respect to targeting, it has been possible to produce functional antibodies in plants by either transforming separate plants with constructs containing the heavy or the light chain, followed by crossing to yield progeny in which both chains are expressed and the antibody assembles,[75] or by expressing a chimeric construct harboring both heavy- and light-chain genes.[76] Expression of an antibody targeted against an antigen essential for the *in planta* growth of a pathogen, or its ability to cause symptoms, could decrease the titer of the antigen to a level which could prevent disease and/or ameliorate the symptoms. Obvious targets for such antibodies would be viral replicase or systemic movement functions, or perhaps viral coat proteins. This strategy is currently being assessed for several antigens of tomato spotted wilt virus.[76a] Monoclonal antibodies have been produced against cell surface and extracellular components of fungal[77] and bacterial[78] plant pathogens, but it is not known if these would be effective in inhibiting pathogen growth *in planta*. Future development of this technology requires effective assembly of immunoglobulin chains in the cytoplasm of plant cells, a process which has not yet been achieved.[79] Alternatively, the design of single-chain antigen-binding constructs[80] could alleviate the need for assembly of the antibody complex.

Table 3 **Antimicrobial proteins with potential for engineering disease resistance in plants**

Class	Source	Activity	Ref.
Fungal, highly basic	*Aspergillus giganteus*	Antifungal	56
Polygalacturonase inhibitors	Bean (*Phaseolus vulgaris*), alfalfa	Inhibit fungal cell wall-degrading enzymes of pathogens	57,58
Chitin-binding proteins	Rubber tree (hevein)	Antifungal	59
	Stinging nettle (lectin)	Antifungal	60
	Barley	Antifungal, synergist with barley chitinase or PR-5	47
	Amaranthus caudatus	Antifungal (very potent); bactericidal	61
Thionins (cysteine-rich cell wall and vacuolar proteins)	Barley	Antifungal	62,63
Basic oligomeric proteins	Radish	Antifungal	64
2S storage albumins	Radish	Antifungal	64
Elicitins	*Phytophthora* spp.	Induce host defense responses	65
Ribosome-inactivating proteins	Range of plant species	Cytotoxic, antiviral, antifungal	66
Zeamatin	Maize	Antifungal, membrane permeabilizing	67

V. MANIPULATION OF PHYTOALEXINS TO INCREASE DISEASE RESISTANCE

Phytoalexins (antimicrobial secondary metabolites) are thought to contribute to the resistance of plants to disease. Phytoalexins have been identified in many different plant species and are structurally diverse, being synthesized from a wide range of precursors.[81] In many cases it has been shown that they quickly accumulate to very high levels around the site of pathogen attack, but not to a high degree in the surrounding uninfected tissue.[82–84] In some cases, before infection, significant amounts of the phytoalexin may be constitutively accumulated, but usually in special cells or organelles, or in a conjugated, inactive form.[85–87] During infection, the "stored" phytoalexins are mobilized, while genes for biosynthetic pathways are induced and the synthesis of more phytoalexin begins. Little is known about the turnover or degradation of phytoalexins by the whole plant following accumulation; studies with elicited cell cultures indicate that plant peroxidases may cause degradation of phytoalexins.[88,89] Much more significant can be the degradation of phytoalexins by plant pathogens.[9] Successful pathogens either have very effective detoxification machinery, are not sufficiently sensitive to the phytoalexins of the host plant, or infect without inducing phytoalexin synthesis.

Given an understanding of the interaction between a host plant and a particular pathogen, several strategies can be outlined for improving plant disease resistance by modifying phytoalexin production. These strategies fall into the three general categories of (1) introducing an entirely new class of phytoalexins, (2) modifying the structure(s) of the phytoalexin of the host, and (3) altering the level and/or timing of phytoalexin synthesis. Specific examples of published, on-going, or proposed/possible manipulations are described below, followed by a potential "checklist" of concerns that should be addressed before undertaking such projects.

A. INTRODUCING NEW PHYTOALEXINS

A successful pathogen may have evolved to detoxify or avoid the natural phytoalexins of its host plant, but might be sensitive to phytoalexins from other plants. Two groups have succeeded in transferring single enzyme genes into tobacco, resulting in the production of novel secondary metabolites. First, introduction of a stilbene synthase gene from peanut into tobacco resulted in the measurable production of the peanut stilbene resveratrol.[90] Various types of stilbenes are important phytoalexins in peanut (*Arachis hypogaea*), grape (*Vitis* sp.), and conifers such as pine (*Pinus* sp.) and spruce (*Picea sitchensis*),[91] but stilbenes are not normally made in tobacco. Stilbene synthase converts *p*-coumaroyl-CoA and malonyl-CoA (1:3 ratio) to a C_{14} molecule in much the same way that chalcone synthase converts the same precursors to flavonoids (Figure 1a).[92] Results of any pathogen challenges on these transgenic tobacco have not yet been reported.

Second, introduction of a fungal gene for a sesquiterpene cyclase, trichodiene synthase, resulted in the accumulation of low levels of trichodiene, the precursor of many fungal mycotoxins.[93,94] Solanaceous plants do accumulate sesquiterpenoid phytoalexins, but these contain carbon skeletons unlike trichodiene.[81] Sesquiterpene cyclases are found in many plants and fungi; all use farnesyl pyrophosphate (FPP) as their substrate, but fold and cyclize the molecule in a number of different ways. Further modification results in the hundreds of known sesquiterpenoids, including the phytoalexins of cotton[81,95] and sweet potato (Figure 1b).[96] Successful production of trichodiene demonstrates that a wide variety of sesquiterpenoid skeletons may be introduced into plants, but the initial cyclization products are not as antimicrobial as the final modified phytoalexins. Geranylgeranylpyrophosphate, produced by the addition of a five carbon unit to FPP, is an intermediate in diterpene biosynthesis found in many plants. Expression of the casbene synthase gene recently cloned from castor bean[97] may likely lead to the accumulation of the diterpene phytoalexin casbene, which is directly antifungal.

Initial metabolite and enzyme accumulation was very low in both of the above examples, but this was possibly due to lack of optimization of the expression vectors used. These two cases represent rare examples where introduction of one gene can produce a relatively new molecule. To introduce other new phytoalexins could require the cloning and introduction of several genes. For example, to generate tobacco plants that could make pisatin, the first characterized phytoalexin, would require the introduction of at least nine enzymatic steps, and most of these enzymes/genes have not yet been cloned.

B. MODIFYING EXISTING PHYTOALEXINS

There is much evidence that small modifications in the structure of an existing phytoalexin might greatly alter its toxicity to pathogens and/or its rate of degradation by detoxifying enzymes. Certain pathogens

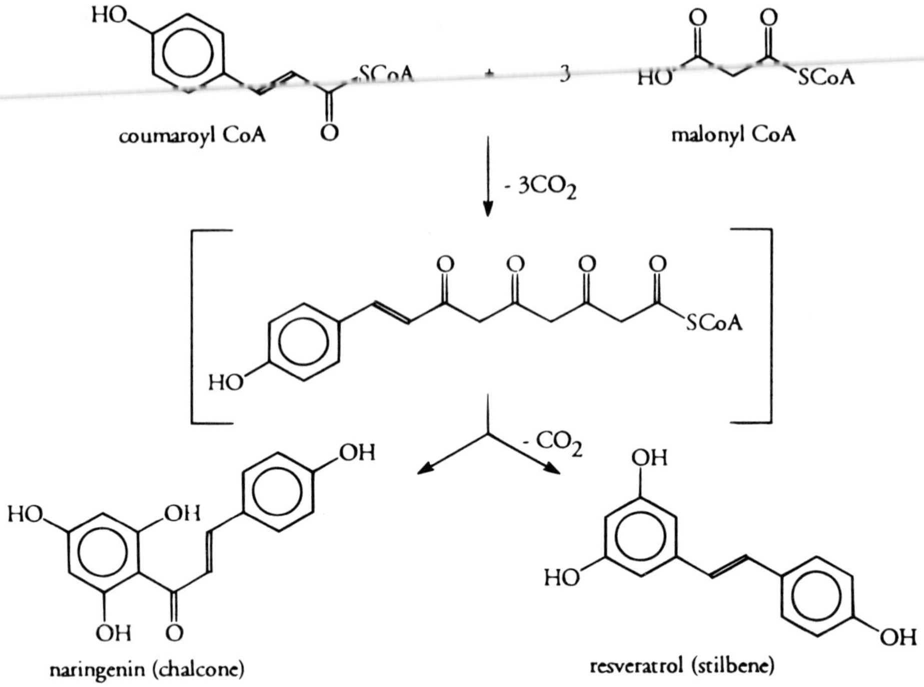

Figure 1a Conversion of coumaroyl-CoA and malonyl-CoA to a chalcone and a stilbene.

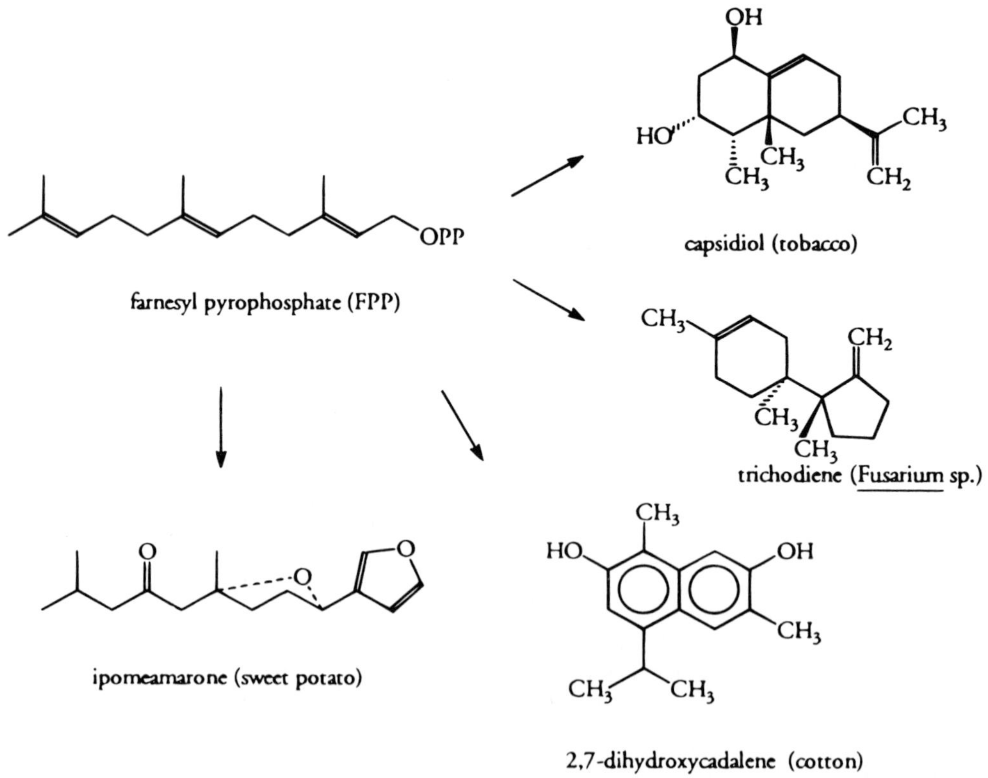

Figure 1b Conversion of farnesyl pyrophosphate to trichodiene and three sesquiterpenoid phytoalexins.

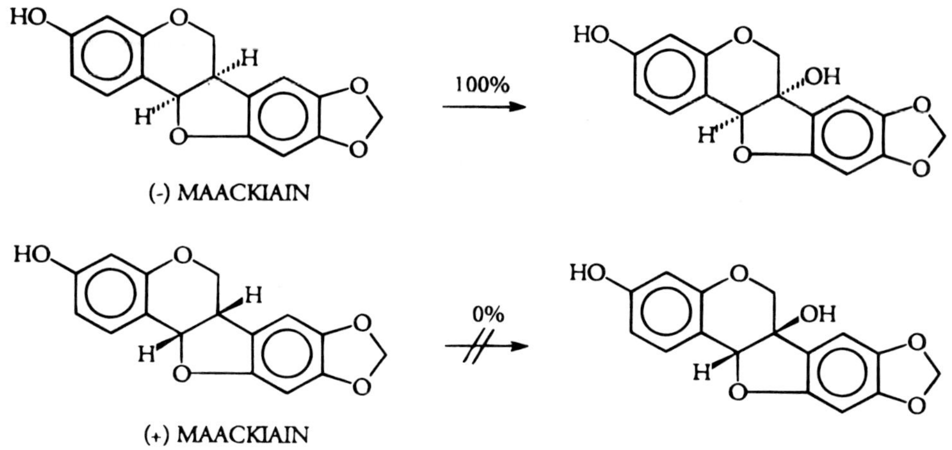

Figure 2 Differential metabolism of pterocarpan phytoalexin stereoisomers by *Nectria haematococca.*

have evolved extremely specific detoxification enzymes, and the ability to detoxify phytoalexins appears to play an important role in determining the virulence of pathogens.[9] In one example, isolates of the red clover pathogen *Nectria haematococca* could readily degrade (−) maackiain, the pterocarpan phytoalexin isomer found in red clover, but could not degrade (+) maackiain at all (Figure 2),[98] and were much more sensitive to (+) than (−) maackiain in bioassays.[99] Many legumes contain pterocarpan phytoalexins, either (−) (alfalfa, red clover, chickpea), (+) (certain cultivars of peanut), or both isomers (*Sophora japonica*).[100] Van Etten and others[9] have proposed that by moving the appropriate genes from a (+) pterocarpan-producing legume to a (−) pterocarpan-producing legume, the recipient plant could be engineered to make phytoalexins of the opposite stereochemistry.[9] In general, pathogens with stereospecific detoxification (such as the above mentioned *Nectria*) would be unable to degrade the "unnatural" isomer and may therefore be unable to infect. It is thought that only two enzymes control which isomer forms in the legume, isoflavone reductase and pterocarpan synthase; isoflavone reductase has been cloned from alfalfa[101] and chickpea[102] and pterocarpan synthase genes should be available in the near future.[103,104]

Several authors have proposed structure-activity relationships based on bioassays with various plant pathogens.[100,105] Striking increases in bioactivity were correlated with prenylation of isoflavonoids, presumably due to the increase in lipophilicity. For example, wighteone, kievitone, and phaseollidin are all much more antifungal than their unprenylated precursors (Figure 3).[106–108] Moving prenyltransferases into legumes which currently accumulate unprenylated isoflavonoids may result in the production of novel antimicrobial compounds. For example, a prenyltransferase (dimethylallyl pyrophosphate: 3,9-dihydroxypterocarpan 10-dimethylallyl transferase) involved in the biosynthesis of the bean phytoalexin phaseollin has been purified from bean cell cultures and found to act also on the alfalfa phytoalexin medicarpin.[109] The product has not yet been identified, but expression of this enzyme in alfalfa may greatly increase the antifungal activity of the resulting phytoalexins.

Methylation of free hydroxyls has also been shown to increase the antifungal activity of isoflavonoids, again presumably by increasing lipophilicity, and may also help protect hydroxyl groups from oxidative detoxification reactions. Methyltransferases for isoflavones and pterocarpans have been partially characterized from alfalfa[110] and pea[111] and, once cloned, may prove useful in modifying phytoalexins. The substrate specificity of such biosynthetic enzymes can be very high; the *O*-methyltransferase which carries out the final methylation to produce pisatin (pea pterocarpan phytoalexin) is totally inactive on the pterocarpan (−) medicarpin and therefore could not be used to directly methylate this alfalfa phytoalexin. In contrast, the purified alfalfa *O*-methyltransferase was active on a number of isoflavonoid substrates.

Another way in which phytoalexin modification may increase resistance is by the production of phytoalexin analogs which act as inhibitors of detoxification enzymes, even if they have no antimicrobial activity of their own. A well-characterized example of this comes from studies of β-lactam antibiotics. Clavulanic acid alone is not toxic to *Escherichia coli,* but is a powerful inhibitor of β-lactamases; the addition of a small amount of clavulanic acid to penicillin or other β-lactam antibiotics can prevent

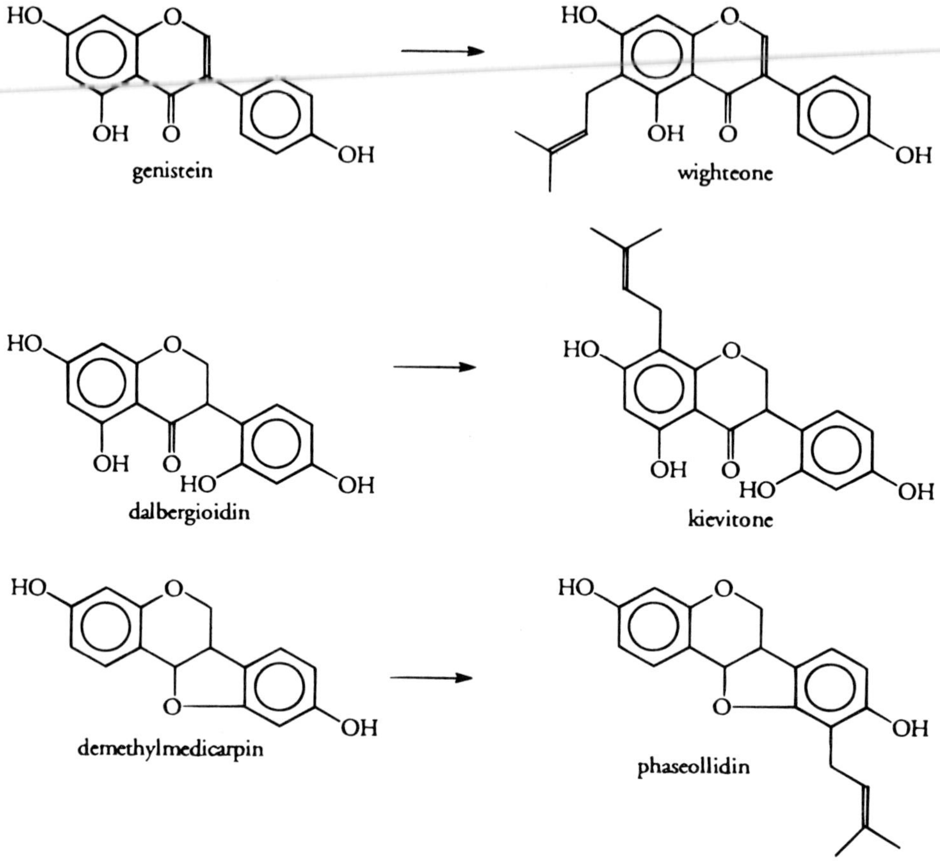

Figure 3 Prenylation to form the phytoalexins wighteone, kievitone, and phaseollidin.

resistant *E. coli* strains from degrading the antibiotic.[112] Initial screening for such inhibitors could be done by testing for synergistic effects between compounds in simple bioassays.

C. ALTERING THE AMOUNT OR TIMING OF PHYTOALEXIN PRODUCTION

There are correlations between disease resistance and production of phytoalexins. For example, cultivars of *Medicago sativa* (alfalfa) which produce higher levels of medicarpin were more resistant to the fungus *Verticillium.*[113,114] Similarly, susceptible cultivars of chickpea make less medicarpin and maackiain during infection by *Ascochyta rabei* and the differences are retained in elicited cell culture systems.[115] The lower production in the susceptible cultivar is apparently due to lower levels of isoflavone 2′-hydroxylase activity, an enzyme late in the biosynthetic pathway. Increased expression of this rate-limiting enzyme could increase the amount of phytoalexins produced.

In the soybean/*Phytophthora megasperma* f. sp. *glycinea* race-specific interaction (Section VI.A), the resistant cultivars quickly make high levels of glyceollins, while susceptible cultivars produce lower amounts much more slowly. However, when challenged with abiotic elicitors the two cultivars can produce comparable levels of phytoalexins.[116] Thus, low phytoalexin production is due to lack of early recognition of the pathogen. Increased and earlier phytoalexin production could be obtained by introducing cloned "resistance" genes (Section VI.A) or by linking the phytoalexin genes to different promoters. Candidate promoters would be ones which are activated early in both compatible and incompatible interactions, as well as ones which could be activated by spraying inducing chemicals. Many genes thought to be involved in systemic acquired resistance (SAR) are highly induced when plants are sprayed with salicylic acid or methyl-2,6-dichloroisonicotinic acid.[117] If the phytoalexin biosynthetic genes were put under the control of inducible SAR gene promoters, relatively nontoxic and cheap inducers could be sprayed to cause the plant to synthesize internally its own fungicides; the expression of endogenous SAR genes (mostly PR proteins; Section IV.A) would add to the defense. Alternatively, constitutive expression of phytoalexins could theoretically be achieved by altering the expression of promoter-binding factors or other proteins involved in the defense response signal transduction pathway.[13]

D. CONCERNS REGARDING PHYTOALEXIN MANIPULATIONS

The above examples are meant to illustrate ways in which simple phytoalexin manipulations may result in increased pathogen resistance. There is a potential for high economic payoff due to reduced use of pesticides and energy and/or increased crop production. However, there are several concerns that should be addressed before such an undertaking is proposed.

1. In modifying an existing phytoalexin pathway, does the new enzyme get targeted to the correct cell compartment? Can it interact with other enzymes in the pathway? The prenyltranferases involved in phytoalexin biosynthesis in bean and soybean are associated with plastid membranes,[118] and many of the hydroxylating enzymes have been shown to be P450s which require association with a membrane for activity.[119] It has been proposed that several of the enzymes involved in isoflavonoid biosynthesis are loosely associated in series on a membrane surface, and that intermediates are "channeled" from enzyme to enzyme.[120,121] Late stages in alkaloid and sorghum deoxyanthocyanidin phytoalexin biosynthesis have been shown to occur inside vesicles.[83,122] Chalcone reductase, a key enzyme early in pterocarpan biosynthesis, requires close protein-protein interaction with chalcone synthase in order to mediate a change in the product chalcone.[123] Any attempts at manipulating such pathways may require the inclusion of appropriate protein leader sequences, membrane-anchoring domains, or other protein sequences required for effective integration into the pathway.

2. Are the necessary precursors present in the plant cell, at the correct time? Is the new enzyme expressed at the correct time and in the correct cell type? Since tobacco already produces sesquiterpene phytoalexins, addition of a new sesquiterpene cyclase is likely to succeed because the FPP substrate must be available. In contrast, legumes in general produce phenylpropanoid phytoalexins; there is no evidence that FPP would be present during fungal attack in sufficient quantities to produce novel phytoalexins. Similarly, use of a "constitutive" promoter such as CaMV 35S may result in poor expression of a phytoalexin-modifying enzyme. The 35S promoter is much less active in older leaves than in younger leaves and roots; the native phytoalexin gene promoters can be highly activated in leaves or roots of any age.[124] The level of phytoalexin modification may be greatly increased by the use of promoters which normally drive the phytoalexin enzyme genes of the host plant by ensuring the proper tissue specificity and correct timing/inducibility of expression.

3. Is the new phytoalexin toxic to the host plant? The native phytoalexins are often toxic to the host as well as the pathogen[125] and are therefore "contained" or expressed only in the lesion tissue, which will die anyway. The concentration of phytoalexins in lesion tissue can be very high, over 100 times that of the surrounding uninfected tissue (pea,[84] potato,[82] sorghum[83]). In some members of the Asteraceae, thiarubrines are accumulated constitutively in specific cells or veins;[126] these thiophenes would be very toxic to the plant if applied externally, and it is not understood how they are contained in these cells without damage to the host. Medicarpin and maackiain are accumulated as malonylated glucosides in roots;[85,86] it is thought that the charged side chain facilitates movement to and storage in the vacuole, away from most of the metabolic activity of the cell. Sorghum produces high concentrations of deoxyanthocyanidin phytoalexins in vesicles which move toward the site of infection and later release their contents, coating the pathogen.[83] Accumulation of high levels of a toxic metabolite in cells of a plant species which does not have the ability to protect itself could cause more damage than the pathogen.

4. Is the new phytoalexin toxic to the symbionts of the host? Many plants have internal symbionts, including *Rhizobia,* endophytes, and mycorrhizal fungi, and there is evidence that rhizosphere bacteria such as Streptomycetes and Pseudomonads[127] may have beneficial effects on plant nutrition and defense. A new phytoalexin may decrease nitrogen fixation or otherwise be detrimental to symbionts as well as pathogens, and the effects may not be easy to assess with *in vitro* assays.

5. Does the change in phytoalexin content have an effect on food or forage quality? Many foods contain phytoalexins and other secondary metabolites which are not considered harmful. However, there are examples of plants which were bred for increased disease resistance, which was later correlated with an increase in a particular metabolite, and subsequently found to have detrimental effects on food quality. In celery, increases in the antifungal psoralens improved the harvest quality, but caused photoactivated blistering in field workers and grocery store personnel who handled the produce.[128] Subterranean clover and red clover with high coumestrol and isoflavone contents were found to perform

well as forage crops in Australia, but these compounds were later found to have estrogenic effects on sheep and cattle, causing infertility.[128]

6. Does the modification have an effect on allelopathy? Medicarpin has been shown to inhibit the germination of alfalfa seeds,[129] a plant which produces medicarpin, and may play a role in the gradual thinning of alfalfa stands; increased production of medicarpin may worsen the problem. Scopoletin (a methoxycoumarin derivative) is known as a phytoalexin in several species and has been implicated in the alleleopathic suppression of weeds in oats (*Avena* sp.);[130] increased production may improve both disease resistance and weed suppression.

7. Is the pathogen really more sensitive to the new phytoalexin? Does the pathogen use an "avoidance" mechanism of resistance, rather than detoxification or insensitivity?[9] If attack by the pathogen does not trigger production of the phytoalexin by the plant, phytoalexin modification will be useless; constitutive production of the phytoalexins might be useful if it is not detrimental in other ways. Ideally, bioassays against virulent pathogens of the host plant with the proposed new phytoalexins should be carried out long before any manipulation is begun.

Currently, the major limitation to the manipulation of phytoalexins is the lack of cloned genes for secondary metabolite biosynthesis, a problem which is, however, quickly being overcome.

VI. POTENTIAL FOR MANIPULATION OF DISEASE-RESISTANCE GENES

Breeding of resistant cultivars has long been considered to be one of the most important aspects of crop improvement.[131–134] Resistance of plant species can be classified broadly into (i) horizontal resistance, which is governed by polygenes or minor genes and (ii) vertical resistance, governed by major genes. We here only address the application of vertical resistance in engineering disease resistance of crop plants.

A. GENETICS OF HOST-PATHOGEN INTERACTIONS

In a host species, some of the cultivars are resistant while others are susceptible to a pathogen that causes a disease on that host. This resistance of the host is termed host resistance, also commonly known as cultivar or race-specific resistance. Resistance conferred by a nonhost towards a nonpathogen is known as nonhost resistance. There are obvious cytological differences between host and nonhost resistant responses of plants to fungi.[135] This distinction is less obvious in plant-bacterial interactions.[136] The genetics of host-pathogen interactions were first studied in detail by Flor[137] using flax (*Linum usitatissimum*) and flax rust (*Melampsora lini*) as a model system. From his study he concluded that "for each gene conditioning resistance in the host there is a specific gene for pathogenicity in the parasite". This hypothesis is known as the "gene-for-gene" hypothesis,[138] and was subsequently redefined as "for each gene for resistance in the host, there is a corresponding gene for avirulence in the parasite".[139] As the definition implies, the hypothesis deals with the cultivar or race-specific resistance of host plants. Mutational analysis indicated that the specificity of host-pathogen interactions, as dictated by this hypothesis, resides in the interaction between gene products of dominant resistance genes and avirulence genes.[140–142] Flor[141] reported from his mutational analysis of *M. lini* that deletion of an avirulence gene resulted in a compatible interaction with the host. In a gene-for-gene system a compatible interaction between the host and the pathogen occurs due to the absence of a correspondence between resistance gene in the host and avirulence gene in the pathogen. The compatible interaction, therefore, can occur due to either the absence of a resistance gene or the presence of a virulence gene in the absence of a correspondence between resistance gene in the host and avirulence genes in the pathogen. In recent years, a number of avirulence genes from bacterial[143–148] as well as fungal[149] pathogens have been cloned.

B. GENETICS OF NONHOST RESISTANCE AND EVOLUTION OF HOST-PATHOGEN SPECIFICITY

Cloning of avirulence (*avr*) genes from bacterial pathogens has led to a greater understanding of nonhost resistance and the probable mechanisms of coevolution of plant and pathogen, in addition to confirming the gene-for-gene hypothesis. Kobayashi et al.[150] cloned three *avr* genes from a nonpathogenic bacterium of soybean for which there are three corresponding resistance genes in the nonhost soybean plants.[150,151] Similarly, Whalen et al.[152] also found an *avr* gene from a tomato pathogen for which there is an incompletely dominant corresponding resistance gene in beans. The *avrRpt2* locus of *Pseudomonas syringae* pv. *tomato* encodes an avirulence gene that acts in a gene-for-gene manner with both *Arabidopsis* and the nonhost soybean.[153] This suggests that soybean may have a resistance gene functionally equivalent to that in *Arabidopsis.* Likewise, it has recently been shown that a bean pathogen carries an avirulence gene that has avirulence function not only on bean, but also on pea, and a pea pathogen carries an avirulence gene that confer specificity to interactions on both pea and beans.[154] Interestingly, it has recently been reported that a pathogenicity gene from *Xanthomonas citri,* when introduced into *X. phaseoli* and *X. campestris* pv. *malvacearum,* can function like an avirulence gene on both bean and cotton which are otherwise nonhost species of *X. citri.*[155] Dangl et al.[156] went one step further and showed that the avirulence gene *avrPpiA1* of the pea pathogen *P. syringae* pv. *pisi* confers avirulence against some genotypes of the nonhost *Arabidopsis* and has a very high level of identity to the avirulence gene *avrRpm1* of the *Arabidopsis* pathogen *P. syringae* pv. *maculicola.* The corresponding resistance genes in *Arabidopsis* for these two *avr* genes may be the same gene, *RPM1,* or, if not, they are very tightly linked. From these studies it appears that nonhost species carry disease-resistance genes corresponding to avirulence genes of pathogens. This opens up the possibility of utilizing nonhost disease-resistance genes in crop improvement for disease resistance.

Conventional breeding also suggests that genes available in nonhost species, or even in different genera, can be utilized for improving disease resistance in cultivated species.[133,134,157–159] For example, there are two major genes conditioning avirulence of the nonpathogen *Erysiphe graminis* f. sp. *agropyri,* Ak-1, to wheat (*Triticum aestivum,* cultivar 'Norin 4'). Through genetic analysis, it was shown that the two corresponding resistance genes are located on chromosomes 1D and 6B of wheat. Thus it was suggested that the gene-for-gene relationship should also fit the forma specialis-genus specificity.[158]

The rice nonpathogen *Eragrostis curvula* carries three independently segregating avirulence genes showing specificity towards different rice cultivars, presumably because of the gene-for-gene correspondence between the nonpathogen avirulence genes of *E. curvula* and nonhost disease-resistance genes in rice. In addition to major avirulence genes from the nonpathogen there are also minor genes that are responsible for determining disease symptoms on rice. Because of these minor genes, it was not possible to detect the segregation of avirulence genes from *E. curvula* until the fourth and fifth generations of backcrossing to the recurrent parent, the rice pathogen *Magnaporthe grisea.*[160] This is a good example of the role of minor genes from a nonpathogen in plant-pathogen interactions. A similar situation can be seen for the nonhost resistance genes of the plant. Thus, in addition to the existence of nonpathogen-specific resistance genes as suggested by Tosa,[158] there may also be a series of minor resistance genes in the nonhost conferring resistance to nonpathogens. This, however, requires experimental confirmation.

It has been suggested that there may be a gene-for-gene basis for the polygenes of the host and pathogen involved in horizontal resistance.[161] This assumes a much greater stability of horizontal resistance than that proposed in the addition model of Van der Plank.[162] Polygenes or minor genes encoding horizontal resistance could be a potentially powerful defense against nonpathogens, since in plant-pathogen coevolution the nonpathogens may not have lost most of the minor genes that encode avirulence functions on the nonhost. Therefore, major genes and minor genes could together create a potentially stable resistance to nonpathogens. Transfer of major disease-resistance genes from putative nonhost species by conventional breeding, and the presence of major disease-resistance genes in the nonhost for a nonpathogen, suggest that disease-resistance genes may be spread horizontally to related genera or species during the course of evolution. Similarly, avirulence genes may have been distributed to related forma speciales or pathovars of a pathogen species during their evolutionary process. Elimination or loss of the corresponding disease-resistance and avirulence genes of both major and minor gene categories through mutation during evolutionary processes may have resulted in different host and pathogen combinations.

In the coevolution of host and pathogen the selection pressure in the host is for gaining resistance genes, while in the pathogen for losing avirulence genes. There is evidence that a host carries few resistance genes while a pathogen carries many virulence genes.[163] Therefore, in engineering high-yielding cultivars with a durable resistance, one has to consider the frequent occurrences of new virulent races which presumably occur due to mutation of the existing avirulence genes. Use of multiline cultivars carrying different resistance genes is a possibility for obtaining durable resistance in such a situation.[164,165]

C. ENGINEERING DISEASE-RESISTANCE GENES FOR CROP IMPROVEMENT

Attempts to isolate disease-resistance genes have gained momentum in the past several years, primarily because of the development of map-based cloning and gene tagging strategies (see Chapter 16) and rapid progress in cloning these genes for different host-pathogen interactions has been observed during the last year. Isolation of these genes will enable us to learn more about the mechanisms of host-pathogen interactions and also to design strategies for improving crop resistance.

The *HM1* gene from maize which confers resistance to *C. carbonum* Nelson race 1 has recently been cloned.[73] In this plant-pathogen interaction, however, virulence is dominant over avirulence, a major difference from those host-pathogen interactions that follow Flor's gene-for-gene hypothesis. As described earlier, the *HM1* locus encodes an NADPH-dependent HC-toxin reductase that inactivates the HC-toxin of *C. carbonum* race 1. Apart from its role in improving resistance of corn to *C. carbonum,* the potential of this resistance gene could be explored for improving resistance of other crop species whose pathogens produce a structurally similar toxin. Cloning of the *HM1* locus also opens up the possibility of searching for putative toxin-detoxifying enzymes from other host species.

Dangl and co-workers[156] have isolated a yeast artificial chromosome (YAC) clone of *Arabidopsis* apparently carrying the *RPM1* gene that confers resistance to *P. syringae* pv. *maculicola* isolates.[166] Similarly, a YAC clone of tomato apparently carrying the *Pto* gene that confers resistance to *P. syringae* pv. *tomato* has been isolated.[166] These may be the first examples of dominant major resistance gene to be characterized at the molecular level.

The two-component sensor system proposed by de Wit[167] is an interesting strategy for engineering of disease resistance utilizing a cloned avirulence gene. The suggested strategy was to create a transgenic tomato line carrying both the *Cf9* resistance gene and the corresponding avirulence gene *avr9* of *Cladosporium fulvum* under a pathogen-inducible promoter. The rationale behind this strategy is that once the *avr9* gene is induced, the product of this gene should recognize the product of the *Cf9* gene and cause hypersensitive cell death, and thereby a resistant response to all virulent races of the pathogen. Timely expression and correct targeting of the gene product of *avr9* are two crucial requirements for this strategy. If it works, the resistance obtained may be durable because there will be a selection pressure against *avr9* mutation. The same strategy could in principle be applicable to other crop species in which the corresonding avirulence genes of the pathogen have been cloned.

As the functions of most avirulence genes and all dominant resistance genes that exhibit gene-for-gene correspondence are currently unknown, prediction of optimal strategies for improvement of crop plants by manipulation of disease-resistance genes is not easy. However, plant transformation protocols override the problem of interspecific or intergeneric incompatibility; thus, nonhost disease-resistance genes, once isolated, can be easily transferred to desired host species. In addition, isolation of a series of disease-resistance genes for a particular pathogen would facilitate the construction of multigenic isogenic lines rather than multigenic near isogenic lines in a short period of time. A series of multigenic isogenic lines each of which carries different combinations of resistance genes produced through plant transformation would be an invaluable genetic resource from which to construct desirable multiline cultivars (by mixing isolines) with durable disease resistance.

ACKNOWLEDGMENTS

We thank Allyson Wilkins and Scotty McGill for careful and patient preparation of the manuscript.

NOTES ADDED IN PROOF

Since the time of writing this article, five disease resistance genes have been cloned and characterized at the molecular level. They are from tomato (Martin et al., *Science*, 262, 1432, 1993; Jones et al., *Science*, 266, 789, 1994), *Arabidopsis* (Bent et al., *Science*, 265, 1856, 1994; Mindrinos et al., *Cell*, 78, 1089, 1994), tobacco (Whitham et al., *Cell*, 78, 1101, 1994), and flax (Moffat, *Science*, 265, 1804, 1994).

Also, the first example of increased disease resistance in transgenic plants due to the accumulation of an engineered phytoalexin was published. Tobacco plants producing the stilbene resveratrol due to the introduction of the stilbene synthase genes from grapevine were more resistant to a tobacco fungal pathogen (R. Hain et al., *Nature*, 361, 153, 1993).

REFERENCES

1. Sturtevant, A. P. and Beachy, R. N., Virus resistance in transgenic plants: coat protein-mediated resistance, in *Transgenic Plants, Fundamentals and Applications,* Hiatt, A., Ed., Marcel Dekker, New York, 1992, 93.

2. Golemboski, D. B., Lomonossoff, G. P., and Zaitlin, M., Plants transformed with a tobacco mosaic virus nonstructural gene are resistant to the virus, *Proc. Natl. Acad. Sci. U.S.A.,* 87, 6311, 1990.
3. Powell, P. A., Stark, D. M., Sanders, P. R., and Beachy, R. N., Protection against tobacco mosaic virus in transgenic plants that express the tobacco mosaic virus antisense RNA, *Proc. Natl. Acad. Sci. U.S.A.,* 86, 6949, 1990.
4. Thompson, J. N. and Burdon, J. J., Gene-for-gene coevolution between plants and parasites, *Nature,* 360, 121, 1992.
5. Dixon, R. A. and Harrison, M. J., Activation, structure, and organization of genes involved in microbial defense in plants, *Adv. Genet.,* 28, 165, 1990.
6. Bell, J. N., Dixon, R. A., Bailey, J. A., Rowell, P. M., and Lamb, C. J., Differential induction of chalcone synthase mRNA activity at the onset of phytoalexin accumulation in compatible and incompatible plant-pathogen interactions, *Proc. Natl. Acad. Sci. U.S.A.,* 81, 3384, 1984.
7. Bell, J. N., Ryder, T. B., Wingate, V. P. M., Bailey, J. A., and Lamb, C. J., Differential accumulation of plant defense gene transcripts in a compatible and incompatible plant-pathogen interaction, *Mol. Cell. Biol.,* 6, 1615, 1986.
8. Yoder, O. C., Toxins in pathogenesis, *Annu. Rev. Phytopathol.,* 18, 103, 1980.
9. Van Etten, H. D., Matthews, D. E., and Matthews, P. S., Phytoalexin detoxification: importance for pathogenicity and practical implications, *Annu. Rev. Phytopathol.,* 27, 143, 1989.
10. Lamb, C. J., Lawton, M. A., Dron, M., and Dixon, R. A., Signals and transduction mechanisms for activation of plant defenses against microbial attack, *Cell,* 56, 215, 1989.
11. Dixon, R. A. and Lamb, C. J., Molecular communication in interactions between plants and microbial pathogens, *Annu. Rev. Plant Physiol. Plant Mol. Biol.,* 41, 339, 1990.
12. Keen, N. T., The molecular biology of disease resistance, *Plant Mol. Biol.,* 19, 109, 1992.
13. Lamb, C. J., Ryals, J. A., Ward, E. R., and Dixon, R. A., Emerging strategies for enhancing crop resistance to microbial pathogens, *Bio/Technology,* 10, 1436, 1992.
14. Dixon, R. A., Characterized defense response genes, in *Molecular Plant Pathology: a Practical Approach,* Gurr, S. J., McPherson, H. J., and Bowles, D. J., Eds., Oxford University Press, Oxford, 1992, 191.
15. Odell, J. T., Nagy, F., and Chua, N.-H., Identification of DNA sequences required for activity of the cauliflower mosaic virus 35S promoter, *Nature,* 313, 810, 1985.
16. Comai, L., Moran, P., and Masylar, D., Novel and useful properties of a chimeric plant promoter combining CaMV 35S and MAS elements, *Plant Mol. Biol.,* 15, 373, 1990.
17. Ohl, S., Hedrick, S. A., Chory, J., and Lamb, C. J., Functional properties of a phenylalanine ammonia-lyase promoter from *Arabidopsis, Plant Cell,* 2, 837, 1990.
18. Liang, X., Dron, M., Schmid, J., Dixon, R. A., and Lamb, C. J., Developmental and environmental regulation of a phenylalanine ammonia-lyase-β-glucuronidase gene fusion in transgenic tobacco plants, *Proc. Natl. Acad. Sci. U.S.A.,* 86, 9284, 1989.
19. Schmid, J., Doerner, P. W., Clouse, S. D., Dixon, R. A., and Lamb, C. J., Developmental and environmental regulation of a bean chalcone synthase promoter in transgenic tobacco, *Plant Cell,* 2, 619, 1990.
20. Stermer, B. A., Schmid, J., Lamb, C. J., and Dixon, R. A., Infection and stress activation of bean chalcone synthase promoters in transgenic tobacco, *Mol. Plant-Microbe Interact.,* 3, 381, 1990.
20a. Oommen, A., Dixon, R. A., and Paiva, N. L., The elicitor-inducible alfalfa isoflavone reductase promotor confers different patterns of developmental expression in homologous and heterologous transgenic plants, *Plant Cell.,* in press.
21. Roby, D., Broglie, K., Cressman, R., Biddle, P., Chet, I., and Broglie, R., Activation of a bean chitinase promoter in transgenic tobacco plants by phytopathogenic fungi, *Plant Cell,* 2, 999, 1990.
22. Zhu, Q., Doerner, P. W., and Lamb, C. J., Stress induction and developmental regulation of a rice chitinase promoter in transgenic tobacco, *Plant J.,* 3, 203–212.
23. Samac, D. A. and Shah, D. M., Developmental and pathogen-induced activation of the *Arabidopsis* acidic chitinase promoter, *Plant Cell,* 3, 1063, 1991.
24. Powell, P. A., Wycoff, K. L., Corbin, D. R., Gonzales, R. A., Dixon, R. A., and Lamb, C. J., Wound and infection activation of the promoter of a plant cell wall hydroxyproline-rich glycoprotein gene superimposed on a novel pattern of tissue-specific developmental expression, *Plant Mol. Biol.,* submitted.
25. Battraw, M. J. and Hall, T. C., Histochemical analysis of CaMV 35S promoter-β-glucuronidase gene expression in transgenic rice plants, *Plant Mol. Biol.,* 15, 527, 1990.

26. Bol, J. F., Linthorst, H. J. M., and Cornelissen, B. J. C., Plant pathogenesis-related proteins induced by virus infection, *Annu. Rev. Phytopathol,* 28, 113, 1990.
27. Mauvais, C. J. and Broglie, R., Transgenic plants with enhanced resistance to the fungal pathogen *Rhizoctonia solani, Science,* 254, 1194, 1991.
28. Fukuda, Y., Ohme, M., and Shinshi, H., Gene structure and expression of a tobacco endochitinase gene in suspension-cultured tobacco cells, *Plant Mol. Biol.,* 16, 1, 1991.
29. Huynh, Q. K., Hironaka, C. M., Levine, E. B., Smith, C. B., Borgmeyer, J. R., and Shah, D. M., Antifungal proteins from plants: purification, molecular cloning, and antifungal properties of chitinases from maize seed, *J. Biol. Chem.,* 267, 6635, 1992.
30. Zhu, Q. and Lamb, C. J., Isolation and characterization of a rice gene encoding a basic chitinase, *Mol. Gen. Genet.,* 226, 289, 1991.
31. Rasmussen, U., Giese, H., and Mikkelsen, J. D., Induction and purification of chitinase in *Brassica napus* L. ssp. *oleifera* infected with *Phoma lingam, Planta,* 187, 328, 1992.
32. Margis-Pinheiro, M., Metz-Boutigue, M. H., Awade, A., de Tapia, M., le Ret, M., and Burkard, G., Isolation of a complementary DNA encoding the bean PR4 chitinase: an acidic enzyme with an amino terminus cysteine-rich domain, *Plant Mol. Biol.,* 17, 243, 1991.
33. Payne, G., Ahl, P., Moyer, M., Harper, A., Beck, J., Meins, F., Jr., and Ryals, J., Isolation of complementary DNA clones encoding pathogenesis-related proteins P and Q, two acidic chitinases from tobacco, *Proc. Natl. Acad. Sci. U.S.A.,* 87, 98, 1990.
34. Metraux, J. P., Burkhart, W., Moyer, M., Dincher, S., Middlesteadt, W., Williams, S., Payne, G., Carnes, M., and Ryals, J., Isolation of a complementary DNA encoding a chitinase with structural homology to a bifunctional lysozyme/chitinase, *Proc. Natl. Acad. Sci. U.S.A.,* 86, 896, 1989.
35. Lawton, K., Ward, E., Payne, G., Moyer, M., and Ryals, J., Acidic and basic class III chitinase mRNA accumulation in response to TMV infection of tobacco, *Plant Mol. Biol.,* 19, 735, 1992.
36. Awade, A., de Tapia, M., Didierjean, L., and Burkard, G., Biological function of bean pathogenesis-related (PR3 and PR4) proteins, *Plant Sci.,* 63, 121, 1989.
37. Ward, E. R., Payne, G. B., Moyer, M. B., Williams, S. C., Dincher, S. S., Sharkey, K. C., Beck, J. J., Taylor, H. T., Ahl-Goy, P., Meins, F., Jr., and Ryals, J. A., Differential regulation of β-1,3-glucanase messenger RNAs in response to pathogen infection, *Plant Physiol.,* 96, 390, 1991.
38. Daugrois, J. H., Lafitte, C., Barthe, J. P., Faucher, A. T., and Esquerre-Tugaye, M. T., Purification and characterization of two basic β-1,3-glucanases induced in *Colletotrichum lindemuthianum*-infected bean seedlings, *Arch. Biochem. Biophys.,* 292, 468, 1992.
39. Edington, B. V., Lamb, C. J., and Dixon, R. A., cDNA cloning and characterization of a putative 1,3-β-D-glucanase transcript induced by fungal elicitor in bean cell suspension cultures, *Plant Mol. Biol.,* 16, 81, 1992.
40. Mauch, F., Hadwiger, L. A., and Boller, T., Antifungal hydrolases in pea tissue. I. Purification and characterization of two chitinases and two β-1,3-glucanases differentially regulated during development and in response to fungal infection, *Plant Physiol.,* 87, 325, 1988.
41. Cornelissen, B. J. C., Hooft van Huijsduijnen, R. A. M., Van Loon, L. C., and Bol, J. F., Molecular characterization of messenger RNAs for 'pathogenesis-related' proteins 1a, 1b and 1c, induced by TMV infection of tobacco, *EMBO J.,* 5, 37, 1986.
42. Walter, M. H., Liu, J.-W., Grand, C., Lamb, C. J., and Hess, D., Bean pathogenesis-related (PR) proteins deduced from elicitor-induced transcripts are members of a ubiquitous new class of conserved PR proteins including pollen allergens, *Mol. Gen. Genet.,* 222, 353, 1990.
43. Sharma, Y. K., Hinojos, C. M., and Mehdy, M. C., cDNA cloning, structure, and expression of a novel pathogenesis-related protein in bean, *Mol. Plant-Microbe Interact.,* 5, 89, 1992.
44. Warner, S. A. J., Scott, R., and Draper, J., Characterisation of a wound-induced transcript from the monocot asparagus that shares similarity with a class of intracellular pathogenesis-related (PR) proteins, *Plant Mol. Biol.,* 19, 555, 1992.
45. Friedrich, L., Moyer, M., Ward, E., and Ryals, J., Pathogenesis-related protein 4 is structurally homologous to the carboxy-terminal domains of hevein, Win-1 and Win-2, *Mol. Gen. Genet.,* 230, 113, 1991.
46. Vigers, A. J., Wiedemann, S., Roberts, W. K., Legrand, M., Selitrennikoff, C. P., and Fritig, B., Thaumatin-like pathogenesis-related proteins are antifungal, *Plant Sci.,* 83, 155, 1992.
47. Hejgaard, J., Jacobsen, S., Bjorn, S. E., and Kragh, K. M., Antifungal activity of chitin-binding PR-4 type proteins from barley grain and stressed leaf, *FEBS Lett.,* 307, 389, 1992.

48. Woloshuk, C. P., Meulenhoff, J. S., Sela-Buurlage, M., van den Elzen, P. J. M., and Cornelissen, B. J. C., Pathogen-induced proteins with inhibitory activity toward *Phytophthora infestans*, *Plant Cell*, 3, 619, 1992.
49. Neuhaus, J.-M., Ahl-Goy, P., Hinz, U., Flores, Y., and Meins, F., High-level expression of a tobacco chitinase gene in *Nicotiana sylvestris*. Susceptibility of transgenic tobacco plants to *Cercospora nicotianae*, *Plant Mol. Biol.*, 16, 141, 1991.
50. Neuhaus, J. M., Flores, S., Keefe, S., Ahl-Goy, P., and Meins, F., Jr., The function of vacuolar β-1,3-glucanase investigated by antisense transformation. Susceptibility of transgenic *Nicotiana sylvestris* plants to *Cercospora nicotianae* infection, *Plant Mol. Biol.*, 19, 803, 1992.
51. Mauch, F., Mauch-Mani, B., and Boller, T., Antifungal hydrolases in pea tissue. II. Inhibition of fungal growth by combinations of chitinase and β-1,3-glucanase, *Plant Physiol.*, 88, 936, 1988.
51a. Zhu, Q., Maher, E. A., Masoud, S., Dixon, R. A., and Lamb, C. J., Enhanced protection against fungal attack by constitutive co-expression of chitinase and glucanase genes in transgenic tobacco, *Biol/Technology*, 12, 807, 1994.
52. Neuhaus, J. M., Sticher, L., Meins, F., Jr., and Boller, T., A short C-terminal sequence is necessary and sufficient for the targeting of chitinases to the plant vacuole, *Proc. Natl. Acad. Sci. U.S.A.*, 88, 10362, 1991.
53. Alexander, D., Glascock, C., Pear, J., Stinson, J., Ahl-Goy, P., Gut-Rella, M., Ward, E., Goodman, R., and Ryals, J., Systemic Acquired Resistance in Tobacco: Use of Transgenic Expression to Study the Functions of Pathogenesis-Related Proteins, Abstract S70, *Sixth International Symposium on Molecular Plant-Microbe Interactions*, Seattle, WA, 1992.
54. Linthorst, H. J. M., Meuwissen, R. L. J., Kauffmann, S., and Bol, J. F., Constitutive expression of pathogenesis-related proteins PR-1, GRP, and PR-S in tobacco has no effect on virus infection, *Plant Cell*, 1, 285, 1989.
55. Alexander, D., Goodman, R. M., Rella, M. G., Glascock, C., Weymann, K., Friedrich, L., Maddox, D., Ahl Goy, P., Luntz, T., Ward, E., and Ryals, J., Increased tolerance to two oomycete pathogens in transgenic tobacco expressing pathogenesis-related protein 1a, *Proc. Natl. Acad. Sci. U.S.A.*, 90, 7327, 1993.
56. Nakaya, K., Omata, K., Okahashi, I., Nakamura, Y., Kolkenbrock, H., and Ulbrich, N., Amino acid sequence and disulfide bridges of an antifungal protein isolated from *Aspergillus giganteus*, *Eur. J. Biochem.*, 193, 31, 1990.
57. Degra, L., Salvi, G., Mariotti, D., De Lorenzo, G., and Cervone, F., A polygalacturonase-inhibiting protein in alfalfa callus cultures, *J. Plant Physiol.*, 133, 364, 1988.
58. De Lorenzo, G., Ito, Y., D'Ovidio, R., Cervone, F., Albersheim, P., and Darvill, A. G., Host-pathogen interactions. XXXVII. Abilities of the polygalacturonase-inhibiting proteins from four cultivars of *Phaseolus vulgaris* to inhibit the endopolygalacturonases from three races of *Colletotrichum lindemuthianum*, *Physiol. Mol. Plant Pathol.*, 36, 421, 1990.
59. Van Parijs, J., Broekaert, W. F., Goldstein, I. J., and Peumans, W. J., Hevein: an antifungal protein from rubber-tree (*Hevea brasiliensis*) latex, *Planta*, 193, 258, 1991.
60. Broekaert, W. F., Van Parijs, J., Leyns, F., Joos, H., and Peumans, W. J., A chitin-binding lectin from stinging nettle rhizomes with antifungal properties, *Science*, 245, 1100, 1989.
61. Broekaert, W. F., Mariën, W., Terras, F. R. G., De Bolle, M. G. C., Proost, P., Van Damme, J., Dillen, L., Claeys, M., Rees, S. B., Vanderleyden, J., and Cammue, B. P. A., Antimicrobial peptides from *Amaranthus caudatus* seeds with sequence homology to the cysteine/glycine-rich domain of chitin-binding proteins, *Biochemistry*, 31, 4308, 1992.
62. Bohlmann, H., Clausen, S., Behnke, S., Giese, H., Hiller, C., Reimann-Phillip, U., Schrader, G., Barkholt, V., and Apel, K., Leaf-specific thionins of barley—a novel class of cell wall proteins toxic to plant-pathogenic fungi and possibly involved in the defence mechanism of plants, *EMBO J.*, 7, 1559, 1988.
63. Reimann-Philipp, U., Schrader, G., Martinoia, E., Barkholt, V., and Apel, K., Intracellular thionins of barley. A second group of leaf thionins closely related to but distinct from cell wall-bound thionins, *J. Biol. Chem.*, 264, 8978, 1989.
64. Terras, F. R. G., Schoofs, H. M. E., De Bolle, M. G. C., Van Leuvan, F., Rees, S. B., Vanderleyden, J., Cammue, B. P. A., and Broekaert, W. F., Analysis of two novel classes of plant antifungal proteins from radish (*Raphanus sativus* L.) seeds, *J. Biol. Chem.*, 267, 15301, 1992.

65. Nespoulous, C., Huet, J. C., and Pernollet, J. C., Structure-function relationships of α and β elicitins, signal proteins involved in the plant-*Phytophthora* interaction, *Planta,* 186, 551, 1992.
66. Stirpe, F., Barbieri, L., Battelli, M. G., Soria, M., and Lappi, D. A., Ribosome-inactivating proteins from plants: present status and future prospects, *Bio/Technology,* 10, 405, 1992.
67. Roberts, W. K. and Selitrennikoff, C. P., Zeamatin, an antifungal protein from maize with membrane-permeabilizing activity, *J. Gen. Microbiol.,* 136, 1771, 1990.
68. Leah, R., Tommerup, H., Svendsen, I., and Mundy, J., Biochemical and molecular characterization of three antifungal proteins from barley seed, *J. Biol. Chem.,* 266, 1564, 1991.
69. Logemann, J., Jach, G., Tommerup, H., Mundy, J., and Schell, J., Expression of a barley ribosome-inactivating protein leads to increased fungal protection in transgenic tobacco plants, *Bio/Technology,* 10, 305, 1992.
70. Mitchell, R. E., Isolation and structure of a chlorosis-inducing toxin of *Pseudomonas phaseolicola, Biochemistry,* 15, 1941, 1976.
71. De la Fuente, H. M., Mosqueda-Cano, G., Alvarez-Morales, A., and Herrera-Estrella, L., Expression of a bacterial phaseolotoxin-resistant ornithyl transcarbamylase in transgenic tobacco plants confers resistance to *Pseudomonas syringae* pv. *phaseolicola, Bio/Technology,* 10, 905, 1992.
72. Meeley, R. B., Johal, G. S., Briggs, S. P., and Walton, J. P., A biochemical phenotype for a disease resistance gene of maize, *Plant Cell,* 4, 71, 1992.
73. Johal, G. S. and Briggs, S. P., Reductase activity encoded by the *HM1* disease resistance gene in maize, *Science,* 258, 985, 1992.
74. Huse, W. D., Sastry, L., Iverson, S. A., Kang, A. S., Alting-Mees, M., Burton, D. R., Benkovic, S. J.; and Lerner, R. A., Generation of a large combinatorial library of the immunologlobulin repertoire in phage lambda, *Science,* 246, 1275, 1989.
75. Hiatt, A., Cafferkey, R., and Bowdish, K., Production of antibodies in transgenic plants, *Nature,* 342, 76, 1989.
76. Düring, K., Hippe, S., Kreuzaler, F., and Schell, J., Synthesis and self-assembly of a functional monoclonal antibody in transgenic *Nicotiana tabacum, Plant Mol. Biol.,* 15, 281, 1990.
76a. Nelson, R., Hein, M., and Sherwood, J., personal communication.
77. Wycoff, K. and Ayers, A. R., Monoclonal antibodies to surface and extracellular antigens of a fungal plant pathogen, *Phytophthora megasperma* f.sp. *glycinea,* recognize specific carbohydrate epitopes, *Physiol. Mol. Plant Pathol.,* 37, 55, 1990.
78. Wingate, V. P. M., Norman, P. M., and Lamb, C. J., Analysis of the cell surface of *Pseudomonas syringae* pv. *glycinea* with monoclonal antibodies, *Mol. Plant-Microbe Interact.,* 3, 408, 1990.
79. Hiatt, A., Antibodies produced in plants, *Nature,* 344, 469, 1990.
80. Davis, G. T., Bedzyk, W. D., Voss, E. W., and Jacobs, T. W., Single chain antibody (SCA) encoding genes: one-step construction and expression in eukaryotic cells, *Bio/Technology,* 9, 165, 1991.
81. Bailey, J. A. and Mansfield, J. W., Eds., *Phytoalexins,* John Wiley & Sons, New York, 1982.
82. Lyon, G. D., Occurrence of rishitin and phytuberin in potato tubers inoculated with *Erwinia carotovora* var. *atroseptica, Physiol. Plant Pathol.,* 2, 411, 1972.
83. Nicholson, R. L., Kollipara, S. S., Vincent, J. R., Lyons, P. C., and Cadena-Gomez, G., Phytoalexin synthesis by the sorghum mesocotyl in response to infection by pathogenic and nonpathogenic fungi, *Proc. Natl. Acad. Sci. U.S.A.,* 84, 5520, 1987.
84. Pueppke, S. G. and Van Etten, H. D., Pisatin accumulation and lesion development in peas infected with *Apanomyces euteiches, Fusarium solani* f.sp. *pisi,* or *Rhizoctonia solani, Phytopathology,* 64, 1433, 1974.
85. Barz, W., Bless, W., Daniel, S., Gunia, W., Hinderer, W., Jaques, U., Kessmann, H., Meier, D., Tiemann, K., and Wittkampf, U., Elicitation and suppression of isoflavones and pterocarpan phytoalexins in chickpea (*Cicer arientinum* L.) cell cultures, in *Primary and Secondary Metabolism of Plant Cell Cultures II,* Kurz, W. G. W., Ed., Springer-Verlag, Berlin, 1989, 208.
86. Kessmann, K., Edwards, R., Geno, P. W., and Dixon, R. A., Stress responses in alfalfa (*Medicago sativa* L.). V. Constitutive and elicitor-induced accumulation of isoflavonoid conjugates in cell suspension cultures, *Plant Physiol.,* 94, 227, 1990.
87. Dixon, R. A., Choudhary, A. D., Dalkin, K., Edwards, R., Fahrendorf, T., Gowri, G., Harrison, M. J., Lamb, C. J., Loake, G. J., Maxwell, C. A., Orr, J. D., and Paiva, N. L., in *Phenolic Metabolism in Plants,* Stafford, H. A. and Ibrahim, R. K., Eds., Plenum Press, New York, 1992, 91.

136. Heath, M. C., The role of gene-for-gene interactions in the determination of host species specificity, *Phytopathology,* 81, 127, 1991.
137. Flor, H. H., The complementary genic systems in flax and flax rust, *Adv. Genet.,* 8, 29, 1956.
138. Loegering, W. Q. and Ellingboe, A. H., H. H. Flor: Pioneer in phytopathology, *Annu. Rev. Phytopathol.,* 25, 59, 1987.
139. Kerr, A., The impact of molecular genetics on plant pathology, *Annu. Rev. Phytopathol.,* 25, 87, 1987.
140. Ellingboe, A. H., Genetical aspects of active defence, in *Active Defence Mechanisms in Plants,* Wood, R. K. S., Ed., Plenum Press, New York, 1982, 179.
141. Flor, H. H., The inheritance of X-ray-induced mutations to virulence in urediospore culture of race 1 of *Melampsora lini, Phytopathology,* 50, 603, 1960.
142. Loegering, W. Q. and Sears, E. R., Genetic control of disease expression in stem rust of wheat, *Phytopathology,* 71, 425, 1981.
143. Bonas, U., Stall, R. E., and Staskawicz, B., Genetic and structural characterization of the avirulence gene *avrBs3* from *Xanthomonas campestris* pv. *vesicatoria, Mol. Gen. Genet.,* 218, 127, 1989.
144. De Feyter, R. and Gabriel, D. W., At least six avirulence genes are clustered on a 90-kilobase plasmid in *Xanthomonas campestris* pv. *malvacearum, Mol. Plant-Microbe Interact.,* 4, 423, 1991.
145. Gabriel, D. W., Burges, A., and Lazo, G. R., Gene-for-gene interactions of five cloned avirulence genes from *Xanthomonas campestris* pv. *malvacearum* with specific resistance genes in cotton, *Proc. Natl. Acad. Sci. U.S.A.,* 83, 6415, 1986.
146. Hopkins, C. M., White, F. F., Choi, S.-H. Guo, A., and Leach, J. E., Identification of a family of avirulence genes from *Xanthomonas oryzae* pv. *oryzae, Mol. Plant-Microbe Interact.,* 5, 451, 1992.
147. Jenner, C., Hitchin, E., Mansfield, J., Walters, K., Betteridge, P., Teverson, D., and Taylor, J., Gene-for-gene interactions between *Pseudomonas syringae* pv. *phaseolicola* and *Phaseolus, Mol. Plant-Microbe Interact.,* 4, 553, 1991.
148. Staskawicz, B. J., Dahlbeck, D., and Keen, N. T., Cloned avirulence gene of *Pseudomonas syringae* pv. *glycinea* determines race-specific incompatibility on *Glycine max* (L.) Merr., *Proc. Natl. Acad. Sci. U.S.A.,* 81, 6024, 1984.
149. Van den Ackerveken, G. F. J. M., Van Kan, J. A. L., and De Wit, P. J. G. M., Molecular analysis of the avirulence gene *avr9* of the fungal tomato pathogen *Cladosporium fulvum* fully supports the gene-for-gene hypothesis, *Plant J.,* 2, 359, 1992.
150. Kobayashi, D. Y., Tamaki, S. J., and Keen, N. T., Cloned avirulence genes from the tomato pathogen *Pseudomonas syringae* pv. *tomato* confer cultivar specificity on soybean, *Proc. Natl. Acad. Sci. U.S.A.,* 86, 157, 1989.
151. Keen, N. T. and Buzzell, R. I., New disease resistance genes in soybean against *Pseudomonas syringae* pv. *glycinea:* evidence that one of them interacts with a bacterial elicitor, *Theor. Appl. Genet.,* 81, 133, 1991.
152. Whalen, M. C., Stall, R. E., and Staskawicz, B. J., Characterization of a gene from a tomato pathogen determining hypersensitive resistance in non-host species and genetic analysis of this resistance in bean, *Proc. Natl. Acad. Sci. U.S.A.,* 85, 6743, 1988.
153. Whalen, M. C., Innes, R. W., Bent, A. F., and Staskawicz, B. J., Identification of *Pseudomonas syringae* pathogens of *Arabidopsis* and a bacterial locus determining avirulence on both *Arabidopsis* and soybean, *Plant Cell,* 3, 49, 1991.
154. Fillingham, A. J., Wood, J., Bevan, J. R., Crute, I. R., Mansfield, J. W., Taylor, J. D., and Vivian, A., Avirulence genes from *Pseudomonas syringae* pathovars *phaseolicola* and *pisi* confer specificity towards both host and non-host species, *Physiol. Mol. Plant Pathol.,* 40, 1, 1992.
155. Swarup, S., Yang, Y., Kingsley, M. T., and Gabriel, D. W., An *Xanthomonas citri* pathogenicity gene, *pthA,* pleiotropically encodes gratuitous avirulence on nonhosts, *Mol. Plant-Microbe Interact.,* 5, 204, 1992.
156. Dangl, J. L., Ritter, C., Gibbon, M. J., Mur, L. A. J., Wood, J. R., Goss, S., Mansfield, J., Taylor, J., Taylor, J. D., and Vivian, A., Functional homologs of the *Arabidopsis RPM1* disease resistance gene in bean and pea, *Plant Cell,* 4, 1359, 1992.
157. Niks, R. E., Nonhost plant species as donors for resistance to pathogens with narrow host range. II. Concepts and evidence on the genetic basis of nonhost resistance, *Euphytica,* 37, 89, 1988.
158. Tosa, Y., Evidence on wheat for gene-for-gene relationship between formae speciales of *Erysiphe graminis* and genera of gramineous plants, *Genome,* 32, 918, 1989.
159. Xu, J. and Kasha, K. J., Transfer of a dominant gene for powdery mildew resistance and DNA from *Hordeum bulbosum* into cultivated barley (*H. vulgare*), *Theor. Appl. Genet.,* 84, 771, 1992.

160. Valent, B., Farrall, L., and Chumley, F. G., *Magnaporthe grisea* genes for pathogenicity and virulence identified through a series of backcrosses, *Genetics,* 127, 87, 1991.
161. Parlevliet, J. E. and Zadoks, J. C., The integrated concept of disease resistance; a new view including horizontal and vertical resistance in plants, *Euphytica,* 26, 5, 1977.
162. Van der Plank, J. E., *Principles of Plant Infection,* Academic Press, New York, 1975.
163. Frank, S. A., Models of plant-pathogen coevolution, *TIG,* 8, 213, 1992.
164. Browning, J. A. and Frey, K. J., Multiline cultivars as a means of disease control, *Annu. Rev. Phytopathol.,* 7, 355, 1969.
165. Day, P. R., *Genetics of Host Parasite Interaction,* W. H. Freeman, San Francisco, 1974.
166. Stacey, G., Gresshoff, P. M., and Keen, N. T., Friends and foes: new insights into plant-microbe interactions, *Plant Cell,* 4, 1173, 1992.
167. de Wit, P. J. G. M., Molecular characterization of gene-for-gene systems in plant-fungus interactions and the application of avirulence genes in control of plant pathogens, *Annu. Rev. Phytopathol.,* 30, 391, 1992.

FURTHER READING

Dixon, R. A., Harrison, M. J. and Lamb, C. J., Early events in the activation of plant defense responses, *Annu. Rev. Phytopathol.*, 32, 479, 1994.

Gurr, S. J., McPherson, F. J., and Bowles, D. J., *Molecular Plant Pathology: a Practical Approach,* Oxford University Press, Oxford, 1992.

Hiatt, A., *Transgenic Plants: Fundamentals and Applications,* Marcel Dekker, New York, 1992.

Keen, N. T., The molecular biology of disease resistance, *Plant Mol. Biol.,* 19, 109, 1992.

Singh, U. S., Singh, R. P., and Kohmoto, K., *Pathogenesis and Host Specificity in Plant Diseases,* Vols. I to III, Pergamon Press, Oxford, 1994.

Whelan, W., Ahmad, F., Bialy, H., Black, S., King, M. L., Rabin, M., Solomonson, L., and Vasil, I., *Advances in Gene Technology: Feeding the World in the 21st Century,* IRL Press at Oxford University Press, Oxford, 1992.

Section—IV
BIOTECHNOLOGICAL METHODS

Chapter 19

Present Problems in and Aspects of Breeding for Disease Resistance

J. E. Parlevliet

TABLE OF CONTENTS

I. INTRODUCTION

A. DEFENSES OF GREEN PLANTS TO PARASITES

In nature, organisms can be classified as *producers,* the green plants; *consumers,* all organisms that exploit producers or other organisms; and *decomposers,* organisms that use dead organisms. The green plants are used by a multitude of primary consumers of practically all classes of the living world, from various types of herbivores (mammals, snails, insects) to typical parasites (insects, mites, fungi, bacteria). In order to survive, green plants have developed a broad range of defense mechanisms to ward off most of these consumers. These defense mechanisms can be classified into three groups: *avoidance, resistance,* and *tolerance.*[1] Avoidance operates before parasitic contact between host and parasite is established and decreases the frequency of incidence. After parasitic contact has been established the host can resist the parasite by decreasing its growth or tolerate its presence by suffering relatively little damage.

Avoidance is mainly active against animal parasites and includes such diverse mechanisms as volatile repellents, mimicry, and morphological features like hairs, thorns, and resin ducts. Resistance is usually of a chemical nature. Of tolerance little is known. It is very difficult to measure, and is usually confounded with quantitative forms of resistance.[1]

Parasites classified as fungi, bacteria, mycoplasma, viruses, or viroids are collectively indicated as pathogens, disease-inciting parasites. Of the defense mechanisms employed by host plants against

0-87371-877-1/95/$0.00+$.50

pathogens, resistance mechanisms are by far the most important. Avoidance and tolerance play a minor role here. In crops, breeders have indeed predominantly used resistance to pathogens, and the topic of this chapter therefore is breeding for resistance to pathogens.

B. BREEDING BEFORE 1900

From the moment wild plant species were domesticated by man, selection played a role in the evolution of our crops. Until the 19th century the selection was a combination of natural and human selection. Natural selection occurred as the plant genotypes best adapted to the local conditions (climate, topography, soil, farming system) tended to contribute more to the next generation than those not so well adapted. Plant types favored by humans were selected by humans, the human selection component. Over the centuries, this led to local landraces adapted to the combined circumstances of cultivation, climate, soil, and human desires. Until the 19th century the importance of human selection was small compared with the natural selection component. This changed completely in that century. Halfway through the 19th century, conscious selection of improved plant types from landraces as well as crossing between selected plant types were carried out in most important crops. In the 1850s an English farmer selected a promising wheat plant from his field. Its progeny became the variety Browick, which was still highly resistant to yellow rust more than a century later. In potatoes the accumulation of viruses and the dramatic late blight epidemic in the middle of the 19th century were the major forces leading to the selection of new varieties derived from cross breeding.

At the end of this century, crossing varieties in order to create genetic variation from which new varieties were developed had become a regular procedure in many crops. Resistance to the important pathogens was right from the start a major goal of plant breeders.

C. BREEDING AFTER 1900

Very soon after the rediscovery and confirmation of the Mendelian inheritance of many traits in the early years of this century, it was realized that resistance to diseases did not behave differently. As early as 1920, resistant varieties were developed through cross breeding in various crops, as in wheat against stem rust, *Puccinia graminis* f. sp. *tritici,* yellow rust, *P. striiformis,* and stinking smut, *Tilettia caries;* and in flax, cotton, cabbage, and watermelon to the Fusarium wilts, *Fusarium oxysporum* f. sp. *lini, vasinfectum, conglutinans,* and *niveum,* respectively.

Almost as soon as resistant varieties were grown on a larger acreage, man was faced with the problem of genetic adaptation to the resistance by the pathogen population, through new races (flax wilt in flax, stem rust in wheat). Such setbacks did not discourage breeders. On the contrary, the search for resistance and the incorporation of it in commercial varieties grew with an ever increasing rate up till the present moment.

The massive use of resistance exposed the advantages as well as the disadvantages of resistance. Once introduced, resistance is a very cheap method to control the pathogen and very easy to use. The economic value of resistance totalled over all crops and pathogen is difficult to undervalue; it is at least a multibillion-dollar affair on a yearly basis. Two disadvantages, however, became apparent as well. Resistance genes are highly pathogen specific, i.e., they are effective to only one pathogen. Since in most situations crops are affected by at least several pathogens, resistance genes to each of them have to be introduced if pesticide use should be abandoned to a large extent. The specificity of the pathogens often go much further. Many resistance genes evoke adaptation in the pathogen population, resulting in a loss of effectiveness of those resistance genes. The breeder has to search again for new resistance genes.

This led to ideas of a more durable use of these nondurable genes using strategies such as multilines, variety mixtures, gene development, variety diversification, and multiple-resistance gene barriers. It also led to ideas concerning the nature and use of resistance that cannot be overcome by the pathogen: durable resistance.

At the same time, plant breeders and pathologists developed the resistance breeding into highly efficient procedures. Initially, the screening was done in the field in dependency of all kinds of disturbing factors. To expose the entries to be screened uniformly to the right pathogen population at the right time and desired plant development stage, technologically highly advanced screening methods were developed. The sources from which resistance genes are obtained have broadened steadily from the cultivated plant species in the beginning to a wide range of related wild species at present. And the first introduction of genetic information derived from totally unrelated organisms, such as bacteria (Btgenes) and viruses (protein coat genes) into crops to provide resistance to certain parasites has been realized in the late 1980s.

From those developments it was learned that efficient breeding for disease resistance requires:

1. The proper assessment of resistance
2. Identification of suitable sources of resistance
3. Incorporation of the resistance into commercial varieties

Both the identification as well as the incorporation of the resistance demand an efficient and representative screening method. Efficient in terms of time and costs, representative in the sense that the resistance identified in the screening procedure is fully effective in the growing situation as well. Efficient screening procedures in turn require knowledge of the genetics of the host-pathogen system (pathosystem) and of the ecology and epidemiology of the pathogen.

The above mentioned aspects of disease resistance breeding are the main topics of this chapter.

II. ASSESSMENT OF RESISTANCE[1–3]

The effect of resistance is a reduced growth and development of the pathogen. Ideally, one should therefore measure the amount of pathogen present at a given moment in comparison with the amount present on or in an extremely susceptible check variety. The greater the difference, the greater is the difference in susceptibility/resistance. To measure the amount of the pathogen is generally not possible, because the pathogen is either not or only partially visible. However, one can evaluate the direct or indirect effects of the pathogen on the host if the pathogen itself is not visible. In this respect, pathogens can be subdivided roughly into three types:

1. Pathogens that are *partially visible* such as the ectoparasitic powdery mildews (most of the pathogen visible), the rusts, bunts, and smuts.
2. Pathogens whose *direct effects* can be assessed; the pathogen itself is not visible, but its presence is recognizable by discolored tissue, such as *Septoria.*
3. The amount of pathogen has to be assessed through *the indirect effects* on the host, the true disease symptoms, such as wilting with vascular pathogens, leaf rolling, stunted growth, leaf mosaic, etc., caused by viruses.

Experience has taught that assessing the *amount of tissue affected* by pathogens of groups 1 and 2 gives a good estimate of the amount of pathogen present. Assessing resistance by assessing the amount of tissue affected relative to that of a highly susceptible control is therefore a good method. On the other hand, the indirect effects caused by disease-inciting pathogens, such as viruses, provide a much less reliable way of assessing the amount of pathogen present. The relationship between the severity of such disease symptoms and the amount of pathogen present may vary from reasonable to poor.

The amount of tissue affected is a good estimator of the amount of pathogen present. The amount of pathogen present, however, is not dependent only on the level of resistance of the host variety. Other factors may and do interfere with it, such as:

a. *Interplot interference.* Screening for resistance is generally done in small, adjacent plots. A fairly resistant entry may receive an abundance of inoculum if it has a highly susceptible neighbor. The amount of pathogen on the fairly resistant entry can then be increased considerably, especially with airborne pathogens, underestimating the level of resistance of that entry.
b. *Earliness.* If the entries differ considerably in earliness, the period of exposure to the pathogen varies greatly and the assessment is usually done at the same moment for all entries. Resistance to head blight caused by *Fusarium* in wheat is considerably overestimated in late cultivars due to this aspect. The same is valid for *Septoria* leaf and glume blotch; the later the entry, the lower the blotch scores.
c. *Inoculum density* may obscure small, but real differences in resistance if high inoculum densities are used. At too low a density, escapes can be confounded with resistance.
d. *Plant habit* may affect the assessment of resistance. Short plants may tend to increase and tall plants to decrease the amount of tissue affected. Short wheat cultivars are more affected than tall cultivars by *Septoria* leaf and glume blotch.

Table 1 **Resistance value of four winter wheat varieties for three foliar pathogens**

Variety	Yellow rust F L	Leaf rust F L	Powdery mildew F L	Years on the rec. lists
Clement	9 → 3[a]	9 → 4[a]	8 → 3[a]	7
Norda	8 → 3	7 → 4	7 → 4[a]	9
Tadorna	9 → 3[a]	5 5	8 → 6	10
Felix	9 9	8 → 5	—	15

Note: According to the Dutch recommended lists of cultivars in the first (F) and last (L) year on those lists. Resistance values are on a scale of 1 (extremely susceptible) to 10 (completely resistant).

[a]Resistance broke down within 3 years after introduction.

From Beschrijvende rassenlijst voor landbouwgewassen (Descriptive lists of varieties for arable crops) Nrs. 33–67 (1958–1992), Leiter-Nypels, Maastricht.

The amount of disease can be noted in two ways: as the proportion of plant units diseased, the *disease incidence,* i.e., the percentage plants, leaves, fruits with disease symptoms, and as the proportion of the plant tissue diseased, the *disease severity,* i.e., the percentage of tissue affected/diseased.

III. HOST-PATHOGEN INTERACTIONS

A. DURABILITY OF RESISTANCE

Resistance exposed on a commercial scale varies greatly in its durability. Table 1 shows the typical pattern of nondurable resistance. The introduced resistance loses its effectiveness after some years due to the appearance of new races in the pathogen population. Even the resistance of Felix to yellow rust broke down, but after this variety was withdrawn from the variety list. However, not all resistance breaks down. The resistance of pea varieties to pea wilt, *Fusarium oxysporum* f. sp. *pisi,* even after 30 years of use, is still effective in Western Europe. The resistance of potatoes to virus X, virus Y, virus A, and leafroll, irrespective of the type of resistance, is highly durable in Western Europe and North America. The monogenic resistance of cabbage to cabbage yellows, *F. oxysporum* f. sp. *conglutinans,* has been effective for nearly 70 years in almost all areas where it has been used.

Parlevliet[2] discussed the factors that affected the durability of resistance. These are (1) the farming system, (2) the type of pathogen, (3) the strategy used in the case of resistance genes, (4) the type of resistance genes used.

Farming systems that result in permanent low levels of the pathogen increase the effective period of nondurable resistance genes considerably. The pathogens that appear to adapt easily to introduced resistances appear to be confined predominantly to fungal pathogens with a biotrophic or hemibiotrophic way of life, having specialized to a narrow host range and being airborne or splashborne. Certain strategies in using the nondurable resistance genes may enhance the effective life of these genes. Not all resistance genes lead to adaptation in the pathogen population; they confer durability independent of the other factors mentioned.

Interesting is the consistent observation that the varieties after the breakdown of the resistance (Table 1) never become extremely susceptible. Table 2 shows this for four pathosystems. Apparently, some residual resistance remains and the level of this residual resistance is generally underestimated.

B. QUALITATIVE AND QUANTITATIVE RESISTANCE

The concept of resistance is often used in a qualitative sense. A variety is either resistant or susceptible. With resistance, a fairly high level of resistance is meant in such cases. In Table 1 the varieties Clement, Norda, Tadorna, and Felix were resistant to yellow rust at their introduction and, apart from Felix, became susceptible some years later. The high level of resistance was due to a single major resistance gene, in Felix to two (the reason why the resistance lasted so much longer). However, when one looks at what is meant with susceptible, anything from about a 6 to a 1 is indicated as such. So in the meaning "susceptible", a wide range of susceptibility is normally included. In the true sense one should consider all varieties with values of 2 or more to carry at least some resistance. This residual resistance (Table 2) or quantitative resistance, often overlooked or considered to be of no importance as its level is too low, can be highly useful to farmers if it was accumulated. And especially this quantitative type of

Table 2 **Mean resistance values of cultivars**

Crop	Pathogen	n	First year	Last year	Range[a]	Most susc. exotic[b]
Winter wheat	Yellow rust	9	8.22	4.28	3–6	1
Winter wheat	Leaf rust	6	7.50	4.33	4–5	1
Winter wheat	Powdery mildew	6	7.17	4.33	3–6	1
Spring barley	Powdery mildew	9	8.06	4.56	4–6	1

Note: Resistance value at entering (first year) the Dutch recommended cultivar lists and just before they were removed from that list (last year). Period 1960–1991. Resistance values on a scale of 1 (extremely susceptible) to 10 (completely resistant).

[a]Range of resistance values after resistance broke down.

[b]Resistance value of most susceptible genotypes, always unadapted exotics.

From Beschrijvende rassenlijst voor landbouwgewassen. (Descriptive lists of varieties for arable crops) Nrs. 33–67 (1958–1992), Leiter-Nypels, Maastricht.

resistance (different levels of susceptibility) seems to be present in practically all pathosystems[1,2] and is probably of a polygenic nature.

C. GENETICS OF RESISTANCE

If one relies solely on the published research, resistance is most often controlled by major genes. These major genes are most frequently inheriting dominantly, less frequently recessively. Polygenic inheritance of resistance has been reported as well, but its much lower frequency is most likely due to the more difficult nature of the research than to a truly lower frequency.

The major resistance genes often occur in a surprisingly high number. In maize (*P. sorghi*), oats (*P. coronata*), wheat (*P. graminis* f. sp. *tritici, P. recondita,* and *P. striiformis*), barley, (*Erysiphe graminis* f. sp. *hordei*), and flax (*Melampsora lini*), at least 20 major resistance genes are known. These resistance genes are often clustered together in certain chromosome arms, sometimes so tightly that they can be considered as complex loci, and true allelic series also occur. In the flax-flax rust pathosystem some 32 resistance genes have been described. They occur on five small regions in the flax genome, the K, L, M, N, and P regions with 2, 13, 7, 4, and 6 resistance genes. The L region seems to represent a multiple allelic series on one locus, while the regions M, N, and P each consist of some to several adjacent loci, each with one or more resistance alleles forming together a complex locus.[5] In barley, most of the resistance genes to powdery mildew are located on one arm of chromosome 5 and one arm of chromosome 4.[6] On the short arm of chromosome 10 in maize, at least 16 resistance genes to *P. sorghi* are found on the complex locus Rp1 and the loci Rp5 and Rp6 within 3 cM (centimorgans) of each other.[7] The three downy mildew resistance genes known in spinach are tightly linked.

Minor or polygenic resistance has been reported fairly often. Typical examples are the quantitative resistance in maize to the northern (*Setosphaeria turcica*) and southern (*Cochliobolus heterostrophus*) leaf blight,[8,9] in barley to *P. hordei,*[10] in wheat to *P. recondita* f. sp. *tritici,*[11] and in rice to bacterial blight.[12]

The expression of resistance genes can be modified by the action of other genes (epistasis), the development stage or tissue of the plant, or the environment. The major resistance gene Pa7 in barley to *P. hordei* gives complete resistance in “Cebada Capa”, but incomplete resistance in the varieties L94, Zephyr, and Vada. In cereals to the various rusts, adult plant resistance genes occur frequently; they give resistance only in the adult plant stages. In the seedling stage the resistance is not expressed. In potatoes, the quantitative resistance to late blight, *Phytophthora infestans,* in the foliage is poorly correlated with the resistance in the tubers, indicating that different genes are involved or expressed in the different tissues of the plant. Of the environmental factors, temperature plays a major role. Many scientists have reported that the expression of certain resistance genes depends on the temperature to which they are exposed, such as L1, L3, L7, L8, L10, and L11 in flax to flax rust,[5] Sr6 in wheat to stem rust, Lr16 and Lr17 in wheat to leaf rust, and many other genes.[13]

D. GENETICS OF VIRULENCE, THE GENE-FOR-GENE CONCEPT

It was Flor[14] who developed the gene-for-gene concept studying the flax-flax rust pathosystem. For each gene in the host there is a corresponding gene in the pathogen (Table 3). Resistance and avirulence

Table 3 **Interaction between two loci in a homozygous host, such as flax, and two loci in a dikaryotic pathogen, such as flax rust**

Host	Pathogen[a]			
	A.B.	aaB.	A.bb	aabb
rrss	+	+	+	+
RRss	−	+	−	+
rrSS	−	−	+	+
RRSS	−	−	−	+

Note: R and S are alleles for resistance, A and B for avirulence. A + indicates full compatibility, a − incompatibility.
[a]A. means either A or a, or B or b.

inherit in most cases as dominant, susceptibility and virulence as recessive factors. Incompatibility (resistance in the host, avirulence in the pathogen) is now known to be the result of the specific interaction at a cellular level of the product of the resistance gene and the product of the avirulence gene. If either of two are absent there is no incompatibility, the normal pathogenicity of the pathogen results in a compatible reaction (the host appears susceptible). What is normally meant with virulence is actually the normal pathogenicity shown in the absence of avirulence. Virulence is absence of avirulence; it is genetically seen an empty concept. The gene-for-gene system is apparently superimposed upon the pathogenicity system, the genes that, through interaction with the host, result in the disease.

In pathosystems where such a gene-for-gene system operates, pathogen genotypes can be easily differentiated by their behavior on host varieties carrying different resistance genes. Such pathogen genotypes are described as *races* (fungi), *pathotypes* (bacteria), or *strains* (viruses). The series of host genotypes used to differentiate the pathogen isolates into races is called a *differential series*.

Table 4 demonstrates this race-specific pattern. The authors[15] concluded on the basis of a much larger set of data than shown in Table 4 that in this pathosystem the gene-for-gene system too is operating. Race W2, they conclude, carries the virulence factors neutralizing the DM genes 2, 3, 4, 6, and 8, where US5 carries only a single virulence factor neutralizing the complementary genes (7-1, 7-2). Interesting to note is that the races NL4 from The Netherlands and IL4 from Israel carry the same virulence factors 2, 6, 7, and 8. Only Dm3 and Dm4 are still effective to those races. Although it is said that the races carry certain virulence factors, the right description would be that the avirulence factors are absent. So NL4 should be described as carrying avirulence to Dm3 and Dm4 only, where NL1 apparently carries avirulence to 3, 4, 6, 7, and 8.

The gene-for-gene system, where the specific recognition is between the resistance product of the host and the avirulence product of the pathogen, appears to be widespread. It most clearly and frequently operates in pathosystems where a biotrophic, highly specialized (in terms of host range) pathogen is involved such as the cereals with the various rusts, smuts, and bunts, and with the powdery mildew. Other typical examples are apple (*Venturia inaequalis*), coffee (*Hemileia vastatrix*), potato (*P. infestans*), and tomato (*Fulvia fulva*).[16] But it also appears to occur in other pathosystems such as rice (*Xanthomonas campestris* pv. *oryzae*), tomato-tobacco mosaic virus and potato-virus X and even in host-parasite

Table 4 **Reaction of five lettuce varieties with their assumed resistance genes (Dm) to eight races of the downy mildew, *Bremia lactucae***

Variety	Dm genes	Race							
		US5	NL1	W1	IL2	W2	W3	IL4	NL4
Meikoningin	2	−	+	+	+	+	+	+	+
Mildura	3, 4	−	−	+	−	+	−	−	−
Valverde	8	−	−	−	−	+	+	+	+
Great Lakes 659	(7-1, 7-2)[a]	+	−	−	+	−	+	+	+
Avondefiance	6, 8	−	−	−	−	+	−	+	+

Note: A + is a compatible, a − an incompatible reaction. (After Johnson et al., Reference 15.)
[a]These two genes act together in a complementary way.

Table 5 **Percentage of plants with leaf stripe (*Pyrenophora graminea*) of some barley varieties inoculated with several pathogen isolates.**

A	Isolate					
Variety	Ha-1 (Finland)	Ma-20 (Morocco)	Dk-1 (Denmark)	Tu-13 (Tunisia)	Ma-4 (Morocco)	Mean
Atlas 68	81	71	37	61	18	43.7
Orge 1700	84	35	23	19	4	32.8
Lami	59	24	27	17	0.3	22.8
Europa	48	25	17	14	2	19.5
Lofa	6	5	5	2	0.0	6.0
Mean	39.5	27.8	19.2	13.1	2.9	19.5

B	DK-3 (Denmark)	Ma-20 (Morocco)	.A-2 (Wales)	
Velvet	87	6	90	39.2
Warrior	25	60	29	21.2
Lofa	34	5	7	6.0
Zita	1	30	2	5.1
Betzes	0	0	25	2.4
Mean	36.7	27.8	24.6	19.5

Note: The means refer to 10 varieties and 12 isolates. (After Nørgaard Knudsen, Reference 17.)

systems such as wheat [*Mayetiola destructor* (Hessian fly)] and potato [*Globodera rostochiensis* (potato cyst nematode)].[16]

All these systems are characterized by major resistance genes, mostly of a dominant nature, that are race specific and not durable. The parasite can easily adapt to the introduced resistance, with the exception of pathosystems involving viruses.

E. GENETICS OF AGGRESSIVENESS

Where host genotypes may differ in the level of susceptibility, pathogen genotypes can vary in their aggressiveness and the rate at which they grow and parasitize their host. The data on barley leaf stripe, a seedborne fungal pathogen of barley, demonstrates this (Table 5A). The varieties differ considerably in susceptibility, from a mean percentage diseased plants of 43.7% to as low as 6.0%. The isolates too differ. They show large differences in their ability to infect all the varieties of barley. Ha-1 is on average the most aggressive isolate, Ma-4 the least aggressive of the ones shown in Table 5. However, the aggressiveness patterns are often difficult to discern as they tend to be covered up by race-specific effects of a fairly large size. In Table 5 the original data have been separated into a Table 5A, with predominantly quantitative differences in resistance and aggressiveness, and a Table 5B, where race-specific effects are the major effects. Velvet/Warrior with DK-3/Ma-20 is such a race-specific interaction; Lofa/Zita × DK3/Ma-20 is another one, as well as Velvet/Warrior × Ma-20/A-2 and Zita/Betzes × Ma-20/A-2. These large, race-specific effects almost certainly indicate the presence of major resistance genes interacting with avirulence genes in the pathogen isolates.

As for the aggressiveness, corresponding with quantitative resistance in the host, little is known because of the difficulties to investigate it. What is known, however, indicates that aggressiveness is polygenically inherited. Emara and Sidhu[18] reported polygenic inheritance of aggressiveness of *Ustilago hordei,* covered smut, in barley. Parlevliet[19] reported a small, race-specific effect for quantitative resistance in barley to *Puccinia hordei* and suggested that one of the polygenes in variety Julia was neutralized by a minor gene for aggressiveness in isolate 18. Kolmer and Leonard[20] tested the ability of *C. heterostrophus* to adapt on a quantitatively (polygenically) resistant maize genotype by exposing that genotype to a population of the pathogen. The isolates from the largest lesions were intercrossed and the host genotype again exposed to the pathogen population derived from that crossing. This recurrent selection for increased aggressiveness was done for three cycles, and the isolates were tested on the genotype on which the selection was carried out and on other quantitatively resistant genotypes. There was a significant increase in aggressiveness (lesion length) of 18% on the genotype on which the

Table 6 **Percentage leaf area affected of three barley cultivars with partial resistance to *P. hordei*, exposed to five isolates of that pathogen**

	Isolate				
Variety	**11-1**	**18**	**1–2**	**22**	**24**
Berac	8.1	6.7	3.1	5.0	0.9
Julia	4.5	12.1[a]	1.8	1.1	0.6
Vada	0.8	0.5	0.6	0.2	0.1

[a]In case of true race nonspecificity this value should be approximately 3%.
From Parlevliet, J. E., *Phytopathology,* 67, 776, 1977. With permission.

selection was done and of 7 to 10% on the other genotypes. This is best explained if it is assumed that aggressiveness is due to polygenes, which, as in the barley-leaf rust system,[19] are specific for the polygenes for resistance (a polygene-for-polygene system).

In maize-*S. turcica* it was also shown that selection for increased aggressiveness was possible.[21]

Although man apparently is able to select for increased aggressiveness and small race-specific effects for quantitative resistance do occur in most pathosystems,[2] there are no indications that quantitative resistance based on polygenes is subject to erosion.[2] The pathogen may have the potential to adapt to such resistances; in the field it does not seem able to exploit that possibility.

F. SPECIFICITY

Van der Plank,[22] and many scientists who followed him, considered resistance to parasites to consist of two categories of specificity: race-specific or vertical resistance and race-nonspecific resistance. The former he thought to be monogenic and nondurable, the latter polygenic and durable.[22] A nice and simple hypothesis, the reason why it became so popular. It does not, however, represent the much more complex reality in nature. Monogenic resistance is not always ephemeral; monogenic resistance with and without known corresponding races of the pathogen of great durability exist as in cucumber to *Cladosporium cucumerinum* and *Corynespora melonis,* in maize to *Periconia circinata,* in wheat to *Pseudocercosporella herpotrichoides,* in oats to *Helminthosporium victoriae,* in soybean to *X. campestris* pv. *glycines.*[2] The I gene in beans to bean common mosaic virus and the monogenic resistance to all the pea viruses do not show signs of becoming less effective,[23] which can be said also of the monogenic resistances in potato to several of its viruses.[2] In many, but not all of these pathosystems, the resistance is of a race-specific type.[2,23] Apparently, durability and race specificity can go together. Meiner[23] concluded that in edible legumes much of the resistance is of the race-specific type, but that most of the race-specific resistance held up for extended periods and it is with only a few pathogens, such as bean rust and lima bean downy mildew, that race-specific resistance was of a short duration.

Polygenic, quantitative resistance, supposed to be race-nonspecific by Van der Plank, does not appear to be so in most cases where detailed studies were carried out.[2,19–21] Parlevliet and Zadoks[24] described this in the following way: when resistance in the host and aggressiveness in the pathogen interact on a polygene-for-polygene basis and several host varieties are tested against a series of pathogen isolates, the general impression is of more or less nonrace specificity. Most variation is between varieties and between isolates. If the accuracy of the experiment is sufficiently high, small, but significant race-specific effects can be observed as well (Table 6). They even showed that resistance based on a polygene-for-polygene system would be considerably more durable than a polygenic resistance purely based on non-race-specific effects.

The resistance to pathogens in our crops employed by breeders must, therefore, be considered to be specific in nearly all cases. It is in the first place specific to the pathogen. Resistance genes, whether major genes or polygenes to one pathogen, do not operate to other pathogens, not even related ones.[1] Both the major resistance genes and the polygenes in barley to *Puccinia hordei* are completely ineffective to the yellow rust, *P. striiformis,* and major as well as polygenic resistance is of a race-specific type. Because the race-specific effects are exactly of the same size as the gene effects,[1] major resistance genes are associated with clear, identifiable races, while polygenes result in only very small race-specific effects, insufficient with which to unambiguously identify races. In this latter case one gets a general impression of race nonspecificity. Table 6 shows that: if a variety like L94, without any partial resistance, had been included its disease level would have been over 40% for all isolates. Vada is always more

Table 7 **Reaction of two oat varieties with and without the dominant alleles on locus Pc2 to isolates of *Helminthosporium victoriae* that do not or do produce the HV toxin**

Variety		Isolate: No toxin	Isolate: HV toxin
Victoria	Pc2Pc2	−	+
Bond	pc2pc2	−	−

Note: A − indicates incompatibility; a + indicates susceptibility.

resistant than Julia and Julia always more resistant than Berac, with the exception of its resistance to isolate 18. And so is the ranking order in aggressiveness of each isolate the same for the three varieties, again except for the Julia × 18 combination. So, by and large, there is a race-nonspecific pattern, with in this case one small, but significant race-specific exception.

G. POSSIBLE CAUSE OF DIFFERENCES IN DURABILITY OF RESISTANCE GENES

In the previous sections it was shown that most, if not all, resistance to specialized pathogens is of a race-specific nature, i.e., based on a gene-for-gene basis, whether these genes are major genes or polygenes. On the other hand, it is quite clear that in the barley-*P. hordei* system the resistance mechanism controlled by the race-specific, nondurable major genes is of a totally different nature than the resistance mechanism of the race-specific (small effects), durable polygenes.[31] The former operate after haustoria are formed, after which the host cells collapse, resulting in a hypersensitive type of reaction (post-haustorial resistance), while the latter operate before the haustoria are formed (pre-haustorial resistance). The host cell prevents penetration and does not collapse.

It is believed that the conspicuous difference in the durability of the two types of resistance genes is derived from the difference in the underlying mechanism, post-haustorial and prehaustorial, respectively. The former type is very easily overcome by the pathogen, which conforms to the pattern. This type of resistance is always elusive when specialized biotrophic or hemibiotrophic pathogens are involved. The latter type has been shown to be very durable.[1,2]

The resistance genes belonging to the first, nondurable type appear, when investigated, to operate on the gene-for-gene system as shown in Table 3, characterized by a recognition reaction for incompatibility between resistance and avirulence gene. The upper left quarter of the table shows the situation for one locus; one of the four possible combinations shows incompatibility when R meets A. All other combinations are compatible. The pathogen could, through a mutation, lose the ability to recognize the product of R, resulting in compatibility or virulence. Only a loss mutation is required, not difficult to produce. It would explain the ease by which pathogens could adapt to such resistance genes.

There is another gene-for-gene system possible, and the oat-*H. victoriae* system can be seen as representative of that other gene-for-gene system. In 1942 the crown rust-resistant variety Victoria was introduced in Iowa. In 1946 this and other varieties carrying this crown rust-resistance gene Pc were attacked by a seedling blight, *H. victoriae,* a minor pathogen of wild grasses. This pathogen produces a specific toxin, the HV toxin.[16] Isolates carrying the single gene for the HV toxin are compatible with (pathogenic on) Pc-carrying oat varieties (Table 7). Isolates not producing this toxin cannot attack oat. This is a gene-for-gene relationship, where the product of the susceptibility gene (Pc 2) reacts specifically with the gene product of the pathogenicity gene (HV toxin) to give compatibility. All other combinations result in incompatibility. This is the mirror image from the first gene-for-gene system. In this gene-for-gene system, loss mutations can only lead to loss of pathogenicity. To change from nonpathogenic to pathogenic requires one or more gain or positive mutations, much more difficult to acquire.

It is thought that the partial polygenic resistance of barley to barley leaf rust operates through this second gene-for-gene system, a system interacting directly with the pathogenicity system of the pathogen. This requires positive mutations in the pathogen to adapt to the resistance genes in the host, much more difficult to realize.

IV. SELECTION FOR RESISTANCE

A. SCREENING METHODS

Screening for resistance in the field is often not very efficient; irregular occurrence of the pathogen in time and space, absence of control of the racial composition, and interaction between diseases all may

INOCULATION

by	Dusting Spraying Dipping Injecting Rubbing Mixing	on/in	cells, tissue (in vitro) seed, germinating seeds, foliage of seedling or adult plants roots, stems, flowering structures, cut off plant parts	plant (parts)	with	right Pathogen race or population or toxin
			soil spreader plants/rows	plant's growing environment		

INCUBATION

Growth infected plant (parts)	in	(defined) Environmental conditions	in	Laboratory Greenhouse Field	till	Assessment

ASSESSMENT

Good assessment key to be used at the right moment with the right controls.

ADDITIONAL REQUIREMENTS

Increase inoculum	in/on	susceptible host suitable artificial substrate

Identity of inoculum to be used - race testing on differential host series

Figure 1 Screening for disease resistance.

interfere with a proper assessment. Breeders and plant pathologists, therefore, developed artificial screening methods. A good screening method must discriminate between resistant and susceptible genotypes (i.e., a high heritability), be easy and cheap to carry out, and be representative (the selected resistance is operating under field conditions). Screening methods are specific for the pathosystem. Figure 1 gives the various activities and steps required for a good screening method.

A few examples may illustrate this. In spinach, resistance to downy mildew is very important. Very young seedlings in shallow trays are sprayed with a spore suspension of the required race (there are four races known) of the pathogen. After inoculation the trays are placed in climatically controlled cabinets, as cool and very humid conditions are needed for a good development of the symptoms. Assessment can be done within a week. Sorghum genotypes can be tested for resistance to *Periconia circinata* by placing the seedling roots in a toxin-containing filtrate. Within a few days the susceptible seedlings, with necrotic roots, are visible. Tomato genotypes can be screened for resistance to the soilborne wilt, *Fusarium oxysporum* f. sp. *lycopersici* by dipping the roots of the seedlings after slight pruning of the roots (to create wounds) in a spore suspension of the desired race (there are three races known) of the pathogen. The inoculated seedlings are then planted in the greenhouse for assessment several weeks later. Selection for resistance to yellow rust in wheat in the adult plant stage is often done by sowing the entries to be tested in short rows perpendicular to a spreader row. The spreader row consists of a highly susceptible cultivar and is inoculated with yellow rust by, for instance, injecting spores of the required race into young stems or by placing sporulating plants in the spreader row at tillering. The sporulating plants were inoculated in the greenhouse with the required race.

The identity of races is assessed through differential series of host genotypes (see Section III.D and Table 4). When large numbers of races exist as in the case of the cereal rusts the differential series can consist of a considerable number of genotypes. The genotypes making up the differential series may consist of:

a. Genotypes of which the identity of the resistance genes is unknown. This is rare. In potato to the wart disease, *Synchytrium endobioticum,* the resistance genes in the host genotypes are only partly known.
b. Genotypes of which the resistance genes are known. This is the case with most differential series, like in Table 4.
c. Genotypes that carry each a different, known resistance gene, the ideal differential series. With wheat-stem rust and wheat-leaf rust, such differential series are being developed.

In many pathosystems very efficient screening methods have been developed. If possible, such methods employ germinating seeds, seedlings or very young plants, or even plant parts, as they use up very little space. The assessment is often on the basis of a clear distinction for resistance: plants become diseased or not (or hardly so). Such screening methods tend to identify major resistance genes. And it is indeed a fact that these efficient screening methods have enhanced the selection of those major genes that are race specific and nondurable and so enhanced the boom-and-bust cycle going together with resistance genes that are easily "broken" by the pathogen. On the other hand, these screening methods were at the same time a great help in identifying major genes that were not sensitive to "breakdown".

On the other hand, these highly efficient screening methods are generally only efficient in identifying large differences in resistance, those caused by major genes. Small differences in susceptibility/resistance often go unnoticed in such screening tests. Most of these tests are therefore unsuitable to screen for quantitative resistance.

If these effective screening methods (where a strong selection is carried out for resistance, often complete resistance) favor the selection of major resistance genes, the opposite, a *mild selection against susceptibility,* might favor quantitative resistance.[1,2] This was born out by Parlevliet and Van Ommeren,[26] who showed that removing the 30% most susceptible (most affected) plants or lines in three cycles of recurrent selection increased the partial resistance level to *Puccinia hordei* of two barley populations from quite susceptible to very resistant.

B. SOURCES OF RESISTANCE

When looking for resistance the breeder may search within and outside the primary gene pool of the crop to be improved. Preferably, the resistance is sought in closely related material, such as local or foreign commercial varieties and local landraces, because the less related the source is to the material to be improved, the more difficult it becomes to introduce the resistance without introducing at the same time undesirable genes from that source. If the resistance is not available in closely related material, one has to look for it in primitive varieties from the centers of diversity or in wild relatives.

1. Major Resistance Genes

Introduction from not closely related sources especially concerns major genes, as these are relatively easy to transfer from one species to another through repeated backcrossing. This has been done in many crop species. Table 8 gives an impression of that. All the major resistance genes, except the one to *Pseudocercosporella herpotrichoides,* appeared race specific. The idea that resistance derived from other species would be more durable than resistance from the crop species itself has been shown repeatedly to be unjustified. The resistance genes introduced into wheat from various species to give resistance to cereal rusts and in potato to *Phytophthora infestans,* for instance, "broke down" very easily.

Resistance can be obtained also through mutation as in the case of mint, *Mentha piperita,* where resistance to wilt, caused by *Verticillium albo atrum,* was obtained through mutagenic treatment, followed by extensive selection. This example proved that mutation breeding is as laborious and time consuming as cross breeding.

Table 8 **Examples of major resistance genes transferred from a wild species to the cultivated crop**

Crop	Donor species	Resistance to
Wheat, *Triticum aestivum*	*Aegilops umbellulata*	*Puccinia recondita*
	Agrogyron elongatum	*P. graminis*
	Triticum tauschii	*P. graminis*
	Aegilops comosa	*P. striiformis*
	Triticum timopheevi	*Erysiphe graminis*
	Aegilops ventricosa	*Pseudocercosporella herpotrichoides*
	Secale cereale	*Puccinia striiformis*
Potato, *Solanum tuberosum*	*Solanum demissum*	*Phytophthora infestans*
	S. stoloniferum	Potato virus Y
Tomato, *Lycopersicon esculentum*	*Lycopersicon pimpinellifolium*	*Fusarium oxysporum*
	L. penelli	*F. oxysporum*
	L. peruvianum	Tobacco mosaic virus
	L. hirsutum	*Fulvia fulva*
Lettuce, *Lactuca sativa*	*Lactuca serriola*	*Bremia lactucae*
Apple, *Malus pumila*	*Malus floribunda*	*Venturia inaequalis*
Tobacco, *Nicotiana tabacum*	*Nicotiana glutinosa*	Tobacco mosaic virus

Through biotechnological procedures it will be possible in future to transfer small pieces of genetic information from any type of organism to crop species. The first cases have already been realized. In a few crops (potato, tobacco, tomato) scientists have introduced the viral gene, coding for the protein coat, into the host with the result that the host, producing the viral protein, became considerably more resistant to that virus.

2. Minor Resistance Genes

Quantitative resistance based on the collective effect of several genes with small effects is very difficult to transfer from one species to another. Neither the repeated backcrossing procedure nor the recent transformation techniques are suitable for this purpose. Fortunately, there is in most crop-pathogen systems no need for these procedures, as quantitative resistance appears to be present sufficiently within the crop species whenever scientists look for it.[2,8–12,26,27]

C. SELECTION PROCEDURES

Breeders have to produce varieties that are superior in many traits, and resistance to disease is just one of the many aims. Most often, new varieties originate from more or less complex crosses, followed by selection over many successive generations where in each generation a certain proportion of the clones, lines, or families are removed, as they are not good enough.

If resistance to certain pathogens forms part of the selection criteria, the resistance to each pathogen must be present in at least one of the parents of the cross. In the successive generations of selection, only resistant entries are kept. In this way the breeder hopes to combine resistance with the other desired traits into a single new variety. In case major genes are involved, the selection is not difficult, the resistant entries are kept, the susceptible ones removed. With quantitative resistance another procedure is more effective in the long run. The breeder should only remove the most susceptible entries all the time, throughout his breeding program. In this way there will be a gradual and continuous accumulation of quantitative resistance, which can go much faster than often realized, as Parlevliet and Van Ommeren[26] showed.

If the aim is to introduce a major gene for resistance to a certain pathogen in an otherwise excellent cultivar, repeated backcrossing is the procedure used at present. This method, however, cannot be used for this purpose in vegetatively reproduced crops. Transformation may replace the repeated backcrossing procedure in future, and this method is also suitable for vegetatively reproduced crops.

V. STRATEGIES TO INCREASE DURABILITY OF NONDURABLE RESISTANCE GENES

Several strategies have been proposed which aim to lengthen the effective period of the nondurable major genes. None of these strategies is used extensively. The strategy most often applied is no strategy at all.

A. NO STRATEGY

For a strategy to be effective, all participants have to cooperate. These are, in the first instance, the farmers and the breeders. A second requirement is a thorough knowledge of the resistance genes to be used in the strategy employed. This latter requirement is rarely fulfilled. Normally, only part of the resistance genes are known and identified. And it is always very difficult to get breeders to agree to how to use the known resistance genes, and for the farmers it is equally difficult to act in concert. Because of this, breeders and extension services have not even tried to introduce strategies in most cases. Each breeder is using the resistances that seem good to him, and each farmer uses those varieties that fit best in his farming situation.

B. RESISTANCE GENES USED ONE AT A TIME

Only one resistance gene is used at one moment, and as soon as this becomes ineffective it is replaced by varieties carrying another, still effective resistance gene. This requires the cooperation of the breeders involved. The breeding of flax rust-resistant flax in North America is a good example of this approach. From the 1930s to date, flax has been protected to this rust by a succession of resistance genes. Up into the 1970s this was achieved by consecutively using the resistance genes L9, P, M, L, and N1. Each gene broke down to a new race, their effective periods ranging from 5 to 13 years. At present there are several resistance genes used in the recommended varieties.

C. MULTIPLE GENE USE

If two or three effective major resistance genes are present, a race able to attack a variety with such a resistance must acquire a multiple genetic adaptation. This appears not to be easy. In The Netherlands, some wheat varieties, such as Felix, Manella, and Arminda, remained resistant to yellow rust for periods of 15 years or longer.[4] This appeared due to the accidental presence of two or three effective major genes. This multiple protection could be very effective, provided the resistance genes used are not used singly. The resistance of Manella and Felix ultimately "broke" because the resistance genes involved were used individually in other varieties. This enabled the pathogen to adapt stepwise.[2] This strategy therefore, requires the cooperation of all breeders involved, which is not easy to realize. And the breeding procedure to keep two fully effective major genes together (the packet of two genes cannot be distinguished from the individual genes in the testing procedures) is too complicated for most breeders to be attractive. In future, when marking resistance genes becomes much easier through restriction fragment length polymorphism markers, the use of multiple-resistance gene barriers may become more popular.

D. MULTILINE AND VARIETAL MIXTURE APPROACH

Most populations that consist of components each with a different resistance gene have been shown to result in reduced epidemic buildup.[28] This approach, however, is only of value if it would retard the adaptation of the pathogen population considerably. This retardation or even inability to adapt to a host population harboring several resistance genes is based on the hypothesis that the more virulence factors are present in the pathogen race the less fit that race will be. Van der Plank,[22] a strong defender of this hypothesis, called this Stabilizing Selection; races would tend to have few unnecessary virulence factors. Parlevliet[29] refuted this hypothesis, as races with many (more than 20 sometimes) virulence factors are very common, even if these virulences are unnecessary. Apart from this unresolved question, the multiline approach, after some local, small-scale experiences, has been abandoned as the production and exploitation of multilines is too laborious. Mixtures of existing varieties is much easier, but one is restricted to the varieties available. At present this approach is used at a limited scale in England and Germany in barley to reduce the powdery mildew damage.

E. VARIETY DIVERSIFICATION

In farming systems with relatively large farms, farmers could grow two or three varieties on different plots. If these varieties carry different resistance genes and if many farmers would do this, the buildup of the pathogen population would be retarded as a whole. In the recommended variety lists of The Netherlands and England and Wales, information is provided enabling the farmers to choose the right combinations of varieties. It is not clear whether or not this approach is used at a scale large enough to have a clear impact.

F. REGIONAL GENE DEPLOYMENT

If the varieties recommended in different regions of an epidemiological area carry consistently different resistance genes, it would certainly tend to increase the life span of these resistance genes.[30] However, it is next to impossible to realize this idea, because all breeders and farmers have to support such a procedure by abiding by the rules, a highly unlikely situation.

REFERENCES

1. Parlevliet, J. E., Disease resistance in plants and its consequences for plant breeding, in *Plant Breeding II,* Frey, K. J., Ed., The Iowa State University Press, Ames, IA, 1981, 309.
2. Parlevliet, J. E., Breeding for durable resistance to pathogens, in *Proc. Sixth Regional Wheat Workshop for Eastern, Central and Southern Africa,* Tanner, D. G., Van Ginkel, M., and Mwangi, W., Eds., CIMMYT, Mexico, 1990, 14.
3. James, W. C., Assessment of plant disease and losses, *Annu. Rev. Phytopathol.,* 12, 27, 1974.
4. *Beschrijvende rassenlijst voor landbouwgewassen* (Descriptive lists of varieties for arable crops) Nrs 33–67 (1958–1992), Leiter-Nypels, Maastricht.
5. Islam, M. R. and Mayo, G. M. E., A compendium on host genes in flax conferring resistance to flax rust, *Plant Breeding,* 104, 89, 1990.
6. Jørgensen, J. H., Disease and pest resistance genes, Coordinator's report, *Barley Genet. Newsl,* 20, 89, 1990.
7. Saxena, K. M. S. and Hooker, A. L., On the structure of a gene for disease resistance in maize, *Proc. Nat. Acad. Sci. U.S.A.,* 61, 1300, 1968.
8. Jenkins, M. T. and Robert, A. L., Further genetic studies of resistance to *Helminthosporium turcicum,* Pass. in maize by means of chromosomal translocations, *Crop Sci.,* 1, 450, 1961.
9. Burnette, D. C. and White, D. G., Inheritance of resistance to *Bipolaris maydis* race 0 in crosses derived from nine inbred lines of maize, *Phytopathology,* 75, 1195, 1985.
10. Parlevliet, J. E., Further evidence of polygenic inheritance of partial resistance in barley to leaf rust, *Puccinia hordei, Euphytica,* 27, 369, 1978.
11. Broers, L. H. M. and Jacobs, Th., The inheritance of host plant effect on latency period of wheat leaf rust in spring wheat. II. Number of segregating factors and evidence for transgressive segregation in F3 and F5 generations, *Euphytica,* 44, 207, 1989.
12. Koch, M. F. and Parlevliet, J. E., Genetic analysis of, and selection for, factors affecting quantitative resistance to *Xanthomonas compestris* pv *oryzae* in rice, *Euphytica,* 53, 235, 1991.
13. Browder, L. E., Parasite: host: environment specificity in the cereal rusts, *Annu. Rev. Phytopathol.,* 23, 201, 1985.
14. Flor, H. H., Current status of the gene-for-gene concept, *Annu. Rev. Phytopathol.,* 9, 275, 1971.
15. Johnson, A. G., Crute, I. R., and Gordon, P. L., The genetics of race specific resistance in lettuce (*Lactuca sativa*) to downy mildew (*Bremia lactucae*), *Ann. Appl. Biol.,* 86, 87, 1977.
16. Day, P. M., *Genetics of Host-Parasite Interaction,* Freeman and Company, San Fransisco, 1974, chap. 4.
17. Nørgaard Knudsen, J. C., Resistance to barley leaf stripe, *Z. Pflanzenzuecht.,* 96, 161, 1986.
18. Emara, Y. A. and Sidhu, G., Polygenic inheritance of aggressiveness in *Ustilago hordei, Heridity,* 32, 219, 1974.
19. Parlevliet, J. E., Race-specific aspects of polygenic resistance of barley to leaf rust, *Puccinia hordei, Neth. J. Plant Pathol.,* 84, 121, 1978.
20. Kolmer, J. A. and Leonard, K. J., Genetic selection and adaptation of *Cochliobolus heterostrophus* to corn hosts with partial resistance, *Phytopathology,* 76, 774, 1986.
21. Nelson, R. R., Robert, A. L., and Sprague, G. F., Evaluating genetic potentials in *Helminthosporium turcicum, Phytopathology,* 55, 418, 1965.

22. Van der Plank, J. E., *Disease Resistance in Plants,* Academic Press, New York, 1968.
23. Meiners, J. P., Genetics of disease resistance in edible legumes, *Annu. Rev. Phytopathol.,* 21, 189, 1981.
24. Parlevliet, J. E. and Zadoks, J. C., The integrated concept of disease resistance; a new view including horizontal and vertical resistance in plants, *Euphytica,* 26, 5, 1977.
25. Parlevliet, J. E., Evidence of differential interaction in the polygenic *Hordeum vulgare-Puccinia hordei* relation during epidemic development, *Phytopathology,* 67, 776, 1977.
26. Parlevliet, J. E. and Van Ommeren, A., Accumulation of partial resistance in barley to barley leaf rust and powdery mildew through recurrent selection against susceptibility, *Euphytica,* 37, 261, 1988.
27. Roumen, E. C., Small differential interactions for partial resistance in rice cultivars to virulent isolates of the blast pathogen, *Euphytica,* 64, 173, 1992.
28. Browning, J. A. and Frey, K. J., Multiline cultivars as a means of disease control, *Annu. Rev. Phytopathol.,* 7, 355, 1969.
29. Parlevliet, J. E., Stabilizing selection in crop pathosystems; an empty concept or a reality?, *Euphytica,* 30, 259, 1981.
30. Frey, K. J., Browning, A. J., and Simons, M. D., Management of host resistance genes to control diseases, *Z. Pflanzenzuecht,* 80, 160, 1973.
31. Niks, R. E., Failure of haustorial development as a factor in slow growth and development of *Puccinia hordei* in partially resistant barley seedings, *Phys. Mol. Plant Pathol.,* 28, 309, 1988.

FURTHER READING

Leonard, K. J. and Fry, W. E., Eds., *Plant Disease Epidemiology,* Vol 2, *Genetics, Resistance and Management,* McGraw-Hill, New York, 1989.

Roelfs, A. P. and Bushnell, W. R., Eds., *The Cereal Rusts,* Vol. I, *Origins, Specificity, Structure and Physiology,* Academic Press, Orlando, FL, 1984.

Roelfs, A. P. and Bushnell, W. R., Eds., *The Cereal Rusts,* Vol. II, *Diseases, Distribution, Epidemiology and Control,* Academic Press, Orlando, FL, 1985.

Russell, G. E., *Plant Breeding for Pest and Disease Resistance,* Butterworths, London, 1978.

Chapter 20

Dual Culture: Fungi

O. K. Sinha

TABLE OF CONTENTS

I. INTRODUCTION

Tissue cultures of plant organs offer excellent substrate for culturing and propagating the obligate (biotrophic) fungi under defined conditions. For the first time, dual culture of downy mildew fungus (*Plasmopara viticola*) and its host vine (*Vitis vinifera*) was successfully accomplished by Morel.[1] Thereafter, numerous reports of dual and axenic cultures of biotrophs, especially rusts and downy mildew fungi, appeared in the literature (Table 1). Although a few reports indicate successful establishment of recalcitrant powdery mildew fungi on their host callus, balanced growth in dual cultures has not been obtained so far. Dual cultures or cocultures of biotrophic, semibiotrophic, and necrotrophic fungi have been recently realized to be extremely useful in the understanding of the infection process, the production and maintenance of axenic culture, the morphological and biochemical alterations in the cultured host tissue, the biosynthesis of phytoalexins, the expression of disease resistance or susceptibility in callus cultures, the role of plant growth regulators in determining disease resistance, the host-parasite relationship in infected tissue-cultured plants, the *in vitro* evaluation of fungicides for the control of biotrophic fungi, and determining the viability of internally seedborne mycelium for screening infected seed samples (Table 1). Ingram[62] considered dual cultures valuable for the maintenance of supplies of aseptic inoculum, the cloning of isolates, and their safe international transport.

Although the potential of tissue culture techniques in plant pathology has been highlighted by several authors,[63–67] the importance of dual culture in basic and applied studies has been recognized only recently. It has been unequivocally shown that a set of cultural conditions for a dual culture is critical to arrive at any conclusion. In this chapter, the factors that determine the successful establishment of dual culture of fungi have been discussed. Besides, the application of dual culture in basic and applied studies as well as the merits and demerits of dual culture study have been briefly discussed. This information would be useful for workers engaged in the dual culture of fungi.

0-87371-877-1/95/$0.00+$.50

Table 1 **Dual culture systems of fungi and their application**

Fungus	Host	Application	Ref.
I. BIOTROPHS			
Albugo ipomoeae-panduratae	*Ipomoea pentaphylla*	Growth pattern of the fungus	2
Bremia lactucae	Lettuce	Infection studies	3
Cronartium fusiforme	Slash pine; loblolly pine	Growth pattern of the fungus in axenic culture	4,5
C. ribicola	White pine	Axenic culture; infection studies; effect of growth regulators on colonization; cellular resistance	6–9
Gymnosporangium juniperi-virginianae	*Juniperus* sp.	Axenic culture	10
Puccinia antirrhini	Snapdragon	Infection studies	11
P. graminis f. sp. *tritici*	Wheat	Infection studies	12
P. helianthi	Sunflower	Growth pattern of the fungus	13
P. horiana	Chrysanthemum	Infection studies	14
Uromyces ari-triphylli	*Arisaema triphyllum*	Axenic culture	15
Peronospora farinosa f. sp. *betae*	Sugar beet	Growth pattern of the fungus	16
P. parasitica	*Brassica* spp.	Growth pattern of the fungus	17,18
P. tabacina	Tobacco	Infection studies	19
Plasmodiophora brassicae	*Brassica* spp.	Life cycle of the fungus; expression of disease resistance	20,21
Plasmopara halstedi	Sunflower	Expression of disease resistance	22
P. viticola	Grapevine	Growth pattern of the fungus; infection and symptomatology in tissue cultured plants; fungicide evaluation for disease control	1,23
Pseudoperonospora humuli	Hop	Infection studies	24
Sclerophthora macrospora	Finger millet	Growth pattern of the fungus	25
Sclerospora graminicola	Pearl millet	Axenic culture; screening of infected seed; production of disease-resistant plants	26–28
S. sacchari	Sugarcane	Growth pattern of the fungus on callus; histopathology of infected callus	29
S. sorghi	Sorghum	Growth pattern of the fungus in axenic culture	30
Synchytrium endobioticum	Potato	Infection studies	31
Physoderma citri	Citrus	Growth pattern of the fungus	32
Glomus mosseae	*Trifolium pratense*	Infection studies	33
II. SEMIBIOTROPHS			
Claviceps fusiformis	Pearl millet	Expression of disease resistance	34
Phytophthora infestans	Potato, tomato	Infection studies; expression of disease resistance; phytoalexin biosynthesis	35–37
P. megasperma f. sp. *medicaginis*	Alfalfa	Expression of disease resistance	38
P. megasperma f. sp. *sojae*	Soybean	Expression of disease resistance; phytoalexin production	39,40
P. parasitica f. sp. *nicotianae*	Tobacco	Expression of disease resistance; effect of growth regulator on infection process	41,42

Table 1 *Continued*

Fungus	Host	Application	Ref.
Ustilago maydis	Maize	Host-parasite interaction	43
U. scitaminea	Sugarcane	Growth pattern of the fungus; infection and symptomatology on tissue cultured plants; effect of growth regulators on the fungus; expression of disease resistance	44–46
Tilletia indica	Wheat	Infection studies; teliospore formation	47
Venturia inaequalis	Apple	Pathogenicity test; expression of disease resistance	48,49
III. NECROTROPHS			
Acremonium coenophialum	Tall fescue	Screening of infected seed samples	50
Alternaria alternata	*Brassica* sp.	Use of fungicide for balanced growth	51
A. brassicae	*Brassica* sp.	Use of fungicide for balanced growth; induction of sporulation	51,52
A. triticina	Wheat	Infection studies; expression of disease resistance	53
Botrytis cinerea	Chickpea	Expression of disease resistance	54
Fomes (*Heterobasidion*) *annosum*	Spruce	Growth pattern of the fungus	55
Helminthosporium oryzae	Rice	Factors affecting fungal colonization of callus	56
Phaeolus schweinitzii	Conifers	Expression of disease resistance	57
Phellinus abietis	Spruce	Growth pattern of the fungus	55
Phoma lingam	*Brassica napus*	Expression of disease resistance	58
Pithomyces chartarum	Jackbean	Phytoalexin production	59
Pyricularia grisea	Rice	Infection studies	60
Verticillium albo-atrum	Lucerne	Phytoalexin production	61

II. METHODOLOGY

Protocol for dual culture of some of the biotrophic and necrotrophic fungi has been described by Ingram,[68] Buczacki,[69] and Miller[70] and readers are referred to their informative reviews. Establishment of dual culture depends on the following cultural conditions that may vary with the objective of the study and the nature of the host plant and the fungus.

A. SELECTION OF EXPLANT

The initial step in dual culture study is to select an appropriate explant. Beforehand knowledge of the methodology for callus induction is necessary. Although callus can be induced from various plant parts, such as young pith tissues, root tips, shoots and leaves, a systemically infected explant is often found suitable for obtaining dual cultures of obligate fungi causing rust and downy mildew diseases. Cutter[10] established the dual culture of *Gymnosporangium juniperi-virginianae* on callus derived from telial galls formed on *Juniperus.* He also claimed saprophytic growth of the rust fungus on the medium when host cells were absent. Similarly, in several other host-pathogen systems, e.g., *Cronartium ribicola* on *Pinus monticola, Sclerospora sacchari* on sugarcane, *S. graminicola* on pearl millet, *Albugo ipomoeae-panduratae* on *Ipomoea pentaphylla,* and *Ustilago scitaminea* on sugarcane (Table 1), an infected explant was used for the callus development. While using infected explants, one must be careful in selecting the plant part for the viability of host tissue because the injured and infected tissue may fail to proliferate

Table 2 **Expression of disease resistance in dual cultures of fungi**

Fungus	Host	Response	Ref.
Alternaria triticina	Wheat	Expressed	53
Botrytis cinerea	Chickpea	Expressed	54
Claviceps fusiformis	Pearl millet	Expressed	34
Cronartium ribicola	White pine	Expressed	8
Peronospora farinosa f. sp. *betae*	Sugar beet	Not expressed	16
P. tabacina	*Nicotiana debneyi*	Expressed	71
Phaeolus schweinitzii	Conifers	Expressed	57
Phoma lingam	*Brassica napus*	Expressed	58
Phytophthora infestans	Potato	Expressed	35
P. megasperma f. sp. *medicaginis*	Alfalfa	Expressed	38
P. megasperma f. sp. *sojae*	Soybean	Expressed	39
P. parasitica f. sp. *nicotianae*	Tobacco	Expressed	41
Plasmopara halstedi	Sunflower	Not expressed	22
Pyricularia grisea	Rice	Not expressed	72
Sclerospora sacchari	Sugarcane	Not expressed	29
Ustilago scitaminea	Sugarcane	Expressed	46

on the culture medium. With most of the necrotrophs and some semibiotrophs, the fungus inflicts a lethal effect on the callus tissue. In case of sugarcane leaf tissue infected with *U. scitaminea,* the callus growth was found poor and the response to the kinetin/2,4-dichlorophenoxyacetic acid (2,4-D) ratio of the medium was different.[45] Under such situations, a healthy explant is the only option for callusing, and dual cultures can be accomplished by inoculating the fungus on the callus at a later stage.

When the fungus tends to outgrow the proliferating callus and produces inhibitory or lethal effect on the callus growth and development, the host resistance to the fungus can be considered as a criterion for selecting the explant. This could be utilized only in those systems in which resistance of the intact plant is expressed in callus cultures (Table 2).

The kind of explant tissue may even be important in determining colonization of callus by the fungus. Harvey and Grasham[7] noted that cultures from leaf mesophyll of white pine were colonized more rapidly by *C. ribicola* than those derived from the cortex tissue. Hrib[73] reported that callus cultures derived from the developed root, hypocotyl, cotyledons, and young epicotyl needles of *Picea* seedlings differed in the intensity of their defense reaction to the mycelium of *Phaeolus schweinitzii.*

B. COMPOSITION OF CULTURE MEDIUM

For culturing plant tissue, the most commonly used media are Murashige and Skoog (MS) and Schenk and Hildebrandt. Incorporation of auxin is required for desired callus growth. The concentration of auxin and the auxin-cytokinin ratio in the medium determine the characteristics of the callus. In the *Phytophthora parasitica* var. *nicotianae*-tobacco system, Helgeson et al.[41] recorded different rates of colonization of the callus by the fungus, depending on the hormonal regime in the culture medium. In compact callus, which often results from high auxin concentration, increased resistance response was noted. In *P. megasperma* var. *sojae*-soybean system, similar results were obtained.[39] The hormonal regime of each tissue and organ of the *Picea* seedling was suggested to determine the defense reaction against *Phaeolus schweinitzii.*[73] These reports indicate that hormonal concentration in the medium plays a critical role in expression of resistance against the fungus.

Although a fungicide is normally not a constituent of the tissue culture medium, it may be incorporated under certain situations. Joshi et al.[52] used captafol (*N*-[1,1,2,2-tetrachloro-ethyl-thio] cyclohex-4-ene-1,2-dicarboximide) to check the saprophytic growth of *Alternaria brassicae* on the tissue culture medium and also the fungal contaminants. Coppens et al.[74] successfully checked the growth of several moulds by imazalil sulfate (5 to 50 μg/l) without affecting the growth rate of barley callus. Use of dual culture for the bioassay of fungicides was initially attempted by Morel[75] and later by Nakamura.[76] Lee and Wicks[23] used a dual culture of *Plasmopara viticola* and grapevine as a tool for evaluating the systemic fungicide, metalaxyl (*N*-[2,6-dimethyl-phenyl]-*N*-[methoxyacetyl] alanine methylester). For fungicide application, they immersed the infected plants in the fungicide solution. It is necessary to ensure that the fungicide does not prove phytotoxic to callus growth and also does not affect the test fungus. Shields et al.[77] recorded the phytotoxic effect of several fungicides to tissue cultures of tobacco and *Nicotiana plumbaginifolia* and reported that carbendazim, fenbendazole, and imazalil are relatively safe. Tripathi and Ram[78] reported cytokinin-like activity of carbendazim, a systemic fungicide. They found that this fungicide was effective in inducing callus growth of secondary phloem tissues of carrot root.

C. NATURE OF THE PATHOGEN AND INOCULATION PROCEDURE

For callus development when a healthy explant is used, it becomes necessary to inoculate the fungus on the actively growing callus for establishing the dual culture. Most of the necrotrophs can grow well on tissue culture media and, therefore, the slow-growing one can be inoculated near the explant or sometimes later when the callus attains a little growth. In order to accomplish balanced growth of dual cultures, the time of inoculation is important and may depend on the rate of callus growth and its age. Fast-growing fungi may be grown separately on agar medium and a piece of agar bearing inoculum is cut from the culture and placed on the top of the callus piece. Since most of the facultative parasites, e.g., species of *Alternaria, Fusarium, Colletotrichum, Phytophthora,* and *Pyricularia* lose aggressiveness on repeated subculturing on synthetic media, it should be ensured before inoculation that the isolate of the fungus is aggressive enough to colonize the callus.

While working with obligate parasites, contaminant-free inoculum must be obtained from the host for inoculating the callus. Lee and Wicks[23] and Miller[70] have described the methods to obtain contaminant-free source of sporangial inoculum of *Plasmopara viticola* and *Peronospora tabacina,* respectively. In the case of *Plasmodiophora brassicae,* it is impossible to infect healthy callus of *Brassica* spp. with plasmodia or resting cyst. Galls from infected roots are often found contaminated with bacteria. The only source of getting contaminant-free plasmodia or cyst is the dual culture which is established from surface-sterilized creamy white galls formed on the roots. The protocol has been described by Buczacki.[69] So far, balanced growth of powdery mildew fungi on their respective host callus has not been successfully established. These specialized mildew fungi apparently require a host surface resembling the intact plant for successful infection.[79]

D. AMOUNT OF INOCULUM

The amount of inoculum is important for successful infection and also for desired extent of colonization of the callus. While using zoospore inoculum in fungi like *Phytophthora parasitica* var. *nicotianae,* it is necessary to adjust the zoospore concentration per application, and it should be sufficiently high to cause infection of the callus.[41] To differentiate the growth rate of *Helminthosporium oryzae* isolates on rice callus, Vidhyasekaran et al.[56] found that only two or four spores were required to infect the callus. A greater number of spores, i.e., six or more, resulted in mycelial growth on the entire callus, and virulence of different isolates of *H. oryzae* could not be distinguished. Conversely, in *P. megasperma* var. *sojae*-soybean system, the inoculum concentration was not critical for differentiating the resistance response.[39]

E. CALLUS SIZE AND GROWTH STAGE

The growth condition of a callus at the time of inoculation with a fungal inoculum is important and depends on the objective of the study. In many studies, young proliferating calli are preferred for inoculating the fungus. Holliday and Clarman[39] observed that callus size is important for expression of resistance against *P. megasperma* var. *sojae* in callus derived from a resistant cultivar of soybean. In small-sized callus (1.0 to 2.5 mm), the resistance to the pathogen did not express, perhaps due to reduction in the number of cells that could respond collectively to infection. The resistance response was distinguishable when 5- or 10-mm-thick calli were used irrespective of the inoculum concentration. While studying the disease reaction of different isolates of *H. oryzae* on rice callus, specificity of the isolates was distinguishable when 5-mm long and 5- or 10-mm thick calli were used. The inoculum concentration was, however, important in this case.[56] In *U. scitaminea*-sugarcane system, Sinha[46] found that 30- to 40-d-old callus was most appropriate for the expression of resistance against the fungus. By repeated subculturing of the callus, the resistance response was diminished. Uchiyama and Ogasawara[80] investigated the influence of subculturing of rice callus on responses to inoculation of fungi like *Pyricularia grisea, A. brassicicola, Botrytis cinerea,* and *F. oxysporum.* They observed that there was lesser colonization of callus subcultured one to five times than the callus subcultured many times. This was probably due to lesser production of antimicrobial substances in callus subcultured repeatedly.

F. INCUBATION CONDITIONS

Calli of most of the plant species and fungi grow well in the dark. *Plasmodiophora brassicae* requires darkness during the incubation period.[69] Although illumination may be important in some of the systems, temperature is a critical factor for attaining balanced growth of the two organisms in dual culture. Helgeson et al.[42] noted that the colonization rate of *Phytophthora parasitica* var. *nicotianae* on tobacco callus increased by raising the incubation temperature from 15 to 32°C. Holliday and Klarman[39] concluded that the temperature not only affects the growth rate of the fungus, but also determines the resistance of the callus to the fungal colonization. They observed that growth of *P. megasperma* var. *sojae* at 16 or 20°C was slow on the soybean callus derived from a resistant cultivar and relatively fast on those from a susceptible one. The differential colonization of calli from resistant and susceptible plants was not discernible at higher temperatures (>20°C).

The duration of incubation has been shown to be of significance. Vidhyasekaran et al.[56] concluded that incubation of rice callus inoculated with *H. oryzae* for 7 d or more caused colonization of the entire callus by the incompatible isolate and, therefore, disease reaction on the callus could not be differentiated.

G. MAINTENANCE OF DUAL CULTURE

Ideally, a balance must be attained in growth rates of the two members of a dual culture. It requires intensive standardization efforts. Ingram and Joachim[16] obtained balanced growth of *Peronospora farinosa* f. sp. *betae* and sugar beet callus. Whether the fungus is a biotroph, semibiotroph, or necrotroph, subculturing at frequent intervals is necessary to keep the two organisms in an active growth stage. In fungi like *Plasmodiophora brassicae,* subculturing helps maintain the plasmodial state as in ageing callus resting cysts begin to form.[69] Trigiano et al.[19] noted cessation of mycelial growth of *Peronospora tabacina* on 10-d-old tobacco callus, and after 20 d only a few hyphal cells were left. In ageing callus, certain morphological and biochemical changes occur that may adversely affect the growth of callus and the fungus as well. Accumulation of phenolics occurs in healthy sugarcane calli and adversely affects the callus growth. This problem can be managed by frequent subculturing.

Repeated cycles of subculturing, however, may pose some problems in the maintenance of dual cultures of obligate parasites. The slow growth of the mycelium of *S. sacchari* on fast-growing sugarcane callus resulted in fungus-free callus.[29] This situation was overcome by placing infected tissue in close contact with the growing callus. Cutter,[10] however, successfully maintained a dual culture of *G. juniperi-virginianae* on *Juniperus* callus for several years by periodic subculturing. There appears to be no work on the possibility of utilizing a cryopreservation method for long-term storage of dual cultures at ultra-low temperatures. Such a preservation technique would eliminate the undesirable effects of frequent sub-culturing.

III. APPLICATION OF DUAL CULTURES

To date, a number of basic and applied studies have been made with dual cultures of fungi (Table 1). The significant ones are summarized below.

A. PRODUCTION OF AXENIC CULTURES OF OBLIGATE PARASITES

Initial attempts on the dual culture of fungi were mainly confined to the culture of obligate fungi on their host callus. Later attempts were directed to use dual culture as a source for growing the biotrophs saprophytically on tissue culture medium. In this sequence, Cutter,[10,15] Tiwari and Arya,[26] and Harvey and Grasham[6] claimed axenic cultures of obligate fungi. More work in this area would be helpful for physiological studies on these fungi.

B. STUDY OF THE INFECTION PROCESS

Many biotrophic as well as necrotrophic fungi have been studied for pre- and postinfectional morphological and biochemical changes in the host callus. Helgeson[81] has reviewed *in vitro* studies related to host-pathogen interactions. The infection process in callus may differ from the one which occurs in intact plant for reasons like absence of cuticle, wax[80] and differentiated tissue in the callus, production of antimicrobial substances during callus proliferation,[11,72] and altered influence of growth regulators on infection. Scanning and transmission electron microscopy have provided ultrastructural details of the infection process in dual culture.[82]

C. INDUCTION OF SPORULATION IN FUNGI

It has been frequently observed that on ageing or repeated subculturing in axenic media, the fungal pathogen loses sporulation which may be associated with the loss of virulence. It becomes difficult to identify a nonsporulating fungus. Joshi et al.[52] observed such a phenomenon in *A. brassicae* and successfully induced sporulation by dual culture of the fungus on its host (*Brassica* sp.) callus. Loss of virulence by subculturing is commonly observed in *C. falcatum.* The dual culture technique can be useful in maintaining/restoring the virulence of the isolates.

D. RESISTANCE EXPRESSION IN CALLUS CULTURES

In several host-pathogen systems, resistance of the intact host is expressed in callus cultures (Table 2). Helgeson et al.[41] have developed and defined the conditions for studying the expression of quantitatively inherited resistance in *Phytophthora parasitica* var. *nicotiana*-tobacco system. This phenomenon has been applied by some workers to screen large numbers of breeding material for disease resistance under *in vitro* conditions. Haberlach et al.[42] modified the resistance reaction of callus cells by higher cytokinin levels (10 μM) and thereafter the callus was colonized by *P. parasitica* var. *nicotianae.* In *Sclerospora sacchari*-sugarcane system, the fungus grew luxuriantly on calli of both resistant and susceptible genotypes.[29] The nonexpression of disease resistance response of the plant in callus cultures in a number of instances (Table 2) may possibly be due to the lack of appropriate cultural conditions, like age and size of the callus, inoculum concentration, temperature, period of incubation, constituents of the medium, level of growth regulators, and the stage of subculturing of callus.

E. STUDIES WITH INFECTED TISSUE CULTURED PLANTS

Dual cultures of fungi were mainly limited to growth of the fungus on the callus. A few recent attempts have shown that the host-parasite relationship can be well studied on infected tissue cultured plants. Grapevine plants raised on culture medium were infected with sporangia of *Plasmopara viticola* and symptomatology was studied.[23] Sporulation of the fungus that normally occurs on the lower leaf surface of the intact plant was recorded on the upper leaf surface also. It was attributed to changes in the thickness of the epidermis and increased number of stomata. The dual cultures were maintained up to 12 weeks till subculturing became necessary. The infected plantlets served as the source of contaminant-free sporangial inoculum. Ferol[44] inoculated sugarcane plants (cultured *in vitro*) with a pure culture of *U. scitaminea* that causes smut. The inoculated plants were colonized by the fungus, and 37% of the plants produced a whip-like structure from the apical portion after 10 weeks of incubation at 26°C.

F. PHYTOPHARMACEUTICAL TESTS FOR DISEASE CONTROL

Dual culture, as a source of inoculum, was used for the bioassay of fungicides by Morel[75] and Nakamura.[76] Recently, Lee and Wicks[23] tested the systemicity and curative properties of metalaxyl in dual cultures of grapevine plantlets infected with *P. viticola.* They found the dual culture technique very useful for such studies. Of late, fungicides have been tested for their phytotoxicity to the host callus. Fungicides that are lesser or nontoxic to callus growth can be incorporated into the tissue culture medium for maintaining balance between the two organisms of the dual culture by checking saprophytic growth of the fungus on the surface of the tissue culture medium. They are also useful in checking the fungal contaminants that spoil the precious cultures.

G. PHYTOALEXIN PRODUCTION

Phytoalexins have been reported in callus and suspension cultures of several plants belonging mainly to Leguminosae and Solanaceae.[83] Although phytoalexins are known to produce postinfectionally, Bailey[84] reported pisatin production in pea callus in the presence of coconut milk, a common constituent of several tissue culture media. A few other reports indicated that phytoalexin production is influenced by the growth regulators that are present in the culture medium. Dixon and Fuller[85] reported 99 inhibitory effects of 2,4-D (2×10^{-6} *M*) and naphthaleneacetic acid (2×10^{-4} *M*) on *Botrytis* culture filtrate induced phaseollin production in suspension culture of *Phaseolus vulgaris.* Gibberellic acid and abscissic acid stimulated noninduced phaseollin production, while kinetin inhibited it. For studying the role of phytoalexins in resistance mechanism in callus cultures, hormonal regime seems to be important. Cell suspension cultures have also been used for studying the biosynthesis of phytoalexins.[81]

H. DETECTION OF INTERNALLY SEEDBORNE INOCULUM AND SCREENING OF SEED SAMPLES FOR FUNGAL INFECTION

Recently, the tissue culture technique has been employed in screening seed samples infected with biotrophic, semibiotrophic, as well as necrotrophic fungi. Conger and McDaniel[50] obtained calli from tall fescue seed embryos that were infected with *Acremonium coenophialum.* The slow growth of the fungus was suitable for its detection and identification and the technique was found useful for screening large numbers of seed samples. Parbhu et al.[27] demonstrated the viability and infectivity of *S. graminicola* in seeds of pearl millet by the dual culture method. Seeds from partially malformed ears were cultured on MS medium. The fungus growth was observed on callus tissue. These workers considered this technique useful for screening small size of seed samples. In another report, evidence was provided for seed transmission of *S. graminicola* by the dual culture method. The fungus produced oospores and sporangia in the callus tissue derived from infected embryos. In our studies, when meristematic tissue infected with *U. scitaminea* was excised from the nodal buds of sugarcane set (the vegetatively propagated seed of sugarcane) and cultured on MS medium, several hyphal fragments were observed in the squashed callus tissue, and these were similar to those present in the meristematic tissue of the intact plant.[86]

IV. MERITS AND DEMERITS OF THE DUAL CULTURE TECHNIQUE

A. MERITS

- i. The dual culture technique offers a precisely controlled environment for host-pathogen interactions to take place. Response of inoculum application can be closely observed at the cellular level.
- ii. The transit of biotrophic parasites becomes convenient and safe.
- iii. The inoculation process eliminates wounding of cells which becomes necessary when the intact plant surface is inoculated.
- iv. The dual culture technique permits the maintenance of axenic cultures of obligate parasites for extended periods. In addition, it serves as a constant source of contaminant-free inoculum (sporangia, zoospores, oospores, etc.) for inoculation purposes.
- v. *In vitro* bioassay of fungicides is possible for screening against biotrophic and necrotrophic pathogens.
- vi. The host-parasite relationship can be closely observed using infected callus or tissue cultured plants as an experimental material.
- vii. In host-parasite systems where resistance to a disease is expressed in callus tissue, the mechanism of resistance can be investigated in greater detail.
- viii. Environmental factors such as temperature, pH, illumination, growth regulators, and nutrients can be easily controlled for studying their effects on symptom expression, disease development, and resistance expression.
- ix. Biochemical pathways in phytoalexin biosynthesis can be studied by using cell suspension cultures.
- x. Viability of the pathogen present inside the host tissues can be ascertained. In testing of seeds for the presence of seedborne pathogens, especially the obligate ones, dual cultures can be gainfully employed.

B. DEMERITS

- i. Differentiated tissues present in the intact plant are absent in callus or cell suspension cultures. Therefore, the events that occur in different kinds of tissues during infection in the intact plant cannot be observed in callus cultures.
- ii. The presence of antimicrobial substances in the callus tissue and relatively higher concentration of growth regulators in the culture medium may interfere with the infection process and the fungal growth and thereby the disease reaction.
- iii. The genetic changes that usually occur during the course of callus proliferation may deviate the results of disease resistance expression from that of the intact plant.
- iv. In some instances, balanced growth of the host callus and the fungus is not attained and may pose difficulty in the maintenance of the dual culture for extended periods.
- v. Static defense barriers such as wax and cuticle or preexisting inhibitor are usually not present in tissue cultures. Under such a situation, studies on resistance mechanism against the pathogen become difficult.
- vi. The phenomenon of acquired resistance that is reported to occur in intact plants may be lost in the tissue culture system mainly due to disruption of intercellular communication.

vii. Biochemical events that follow in intact plants may not exactly reproduce in tissue cultures primarily because of nutrient composition of the medium and different auxin-cytokinin ratios.

REFERENCES

1. Morel, G., Le development du mildiou sur des tissus de vigne cultures *in vitro,* C. R. *Acad. Sci. Paris,* 218, 50, 1944.
2. Singh, H., The growth of *Albugo* in the callus cultures of *Ipomoea, Curr. Sci.,* 32, 472, 1959.
3. Mason, P. A., Studies on the Biology of *Bremia lactucae* Regel, Ph.D. thesis, University of Cambridge.
4. Waltkinshaw, C. H., Jewell, F. F., and Walker, N. M., Callus culture of Fusiform rust-infected slash pine, *Plant Dis. Rep.,* 49, 616, 1965.
5. Jacobi, W. R., Growth medium for dual cultures of loblolly pine callus and *Cronartium fusiforme, Phytopathology,* 72, 136, 1982.
6. Harvey, A. E. and Grasham, J. L., Growth of *Cronartium ribicola* in the absence of physical contact with its host, *Can. J. Bot.,* 48, 71, 1970.
7. Harvey, A. E. and Grasham, J. L., The relative susceptibility of needle- and stem-derived white pine tissue cultures to artificial inoculation with mycelium of *Cronartium ribicola, Can. J. Bot.,* 47, 1789, 1969.
8. Diner, A. M., Mott, R. L., and Amerson, H. V., Cultured cells of white pine show genetic resistance to axenic blister rust hyphae, *Science,* 224, 407, 1984.
9. Harvey, A. E., Grasham, J. L., and Waldron, C. C., The effects of growth regulating compounds on healthy and blister rust-infected tissue cultures of *Pinus monticola, Phytopathology,* 61, 507, 1971.
10. Cutter, V. M., Jr., Studies on the isolation and growth of plant rusts in host tissue cultures and upon synthetic media I. *Gymnosporangium, Mycologia,* 51, 248, 1959.
11. Maheshwari, R., Hildebrandt, A. C., and Allen, P. J., Factors affecting the growth of rust fungi on host tissue cultures, *Bot. Gaz. (Chicago),* 128, 153, 1967.
12. Milholland, R. D., Callus tissue from wheat for infection studies of *Puccinia graminis* var. *tritici,* (Abstr.), *Phytopathology,* 52, 21, 1962.
13. Nozolillo, C. and Cragie, J. H., Growth of the rust fungus *Puccinia helianthi* on tissue cultures of its host, *Can. J. Bot.,* 38, 228, 1960.
14. Ando, K., Katsuya, K., and Sato, S., Production of aerial mycelium and teliospores of *Puccinia horiana* in chrysanthemum callus cultures, *Can. J. Bot.,* 57, 2162, 1979.
15. Cutter, V. M., Jr., Studies on the isolation and growth of plant rusts in host tissue cultures and upon synthetic media. II. *Uromyces ari-triphylli, Mycologia,* 52, 726, 1960.
16. Ingram, D. S. and Joachim, I., The growth of *Peronospora farinosa* f. sp. *betae* and sugarbeet callus tissues in dual culture, *J. Gen. Microbiol.,* 69, 211, 1971.
17. Ingram, D. S., The susceptibility of *Brassica* callus to infection by *Peronospora parasitica, J. Gen. Microbiol.,* 58, 391, 1969.
18. Nakamura, H., The use of tissue cultures in the study of obligate parasites, *Proc. Int. Conf. Plant Tissue Culture,* McCutchan, Berkeley, 1965, 535.
19. Trigiano, R. N., Van Dyke, C. G., Spurr, H. W., Jr., and Gray, D. J., Infection and colonisation of tobacco callus by *Peronospora tabacina, Phytopathology,* 74, 280, 1984.
20. Strandberg, J. O., Williams, P. H., and Yukawa, Y., Monoaxenic culture of *Plasmodiophora brassicae* with cabbage tissue, *Phytopathology,* 56, 903, 1966.
21. Sachristan, M. D. and Hoffman, F., Direct infection of embryonic tissue cultures of haploid *Brassica napus* with resting spores of *Plasmodiophora brassicae, Theor. Appl. Genet.,* 54, 129, 1979.
22. Gray, A. B. and Sackston, W. E., An attempt to predict response of sunflowers to races of *Plasmopara halstedi* using cultured sunflower cells, *J. Phytopathol.,* 121, 81, 1988.
23. Lee, T. C. and Wicks, T., Dual culture of *Plasmopara viticola* and grapevine its application to systemic fungicide evaluation, *Plant Dis.,* 6, 308, 1982.
24. Griffin, M. J. and Coley-Smith, J. R., The establishment of hop tissue cultures and their infection by downy mildew *Pseudoperonospora humuli* (Miy. & Tak.) Wilson under aseptic conditions, *J. Gen. Microbiol.,* 53, 231, 1968.

25. Safeeulla, K. M., *Biology and Control of the Downy Mildews of Pearl Millet, Sorghum and Finger Millet,* Downy Mildew Research Laboratory, Mysore, India, 1976.
26. Tiwari, M. M. and Arya, H. C., *Sclerospora graminicola* in axenic culture, *Science,* 163, 291, 1969.
27. Parbhu, M. S., Safeeulla, K. M., and Shetty, H. S., Tissue culture technique to demonstrate the viability of downy mildew mycelium in pearl millet seeds, *Curr. Sci.,* 52, 1027, 1983.
28. Prasad, B., Parbhu, M. S., and Shanthamma, C., Regeneration of downy mildew resistant plants from infected tissues of pearl millet (*Pennisetum americanum*) cultured *in vitro, Curr. Sci.,* 53, 816, 1984.
29. Chen, W., Liu, M., and Chao, C., The growth of sugarcane downy mildew fungus in tissue cultures, *Can. J. Bot.,* 57, 528, 1979.
30. Kaveriappa, K. M., Safeeulla, K. M., and Shaw, C. G., Culturing *Sclerospora sorghi* in callus tissue of sorghum, in *Proc. Indian Acad. Sci. Plant Sci.,* 89, 131, 1980.
31. Ingram, D. S., An attempt to establish dual cultures of *Synchytrium endobioticum* and *Solanum tuberosum* callus, *Phytopathol. Z.,* 71, 21, 1971.
32. Eldon, B. G. and Oberbacher, M. F., *Physoderma citri* in citrus albedo and callus tissue, *Phytopathology,* 59, 241, 1969.
33. Mosse, B. and Heppar, C. M., Vesicular-arbuscular mycorrhizal infections in root organ cultures, *Physiol. Plant Pathol.,* 5, 215, 1975.
34. Kumar, A. and Arya, H. C., Nature of resistance and susceptibility *in vitro* in ergot of pearl millet caused by *Claviceps fusiformis* Lov., *Phytopathol. Mediterr.,* 20, 43, 1981.
35. Ingram, D. S., The expression of R-gene resistance to *Phytophthora infestans* and tissue cultures of *Solanum tuberosum, J. Gen. Microbiol.,* 49, 99, 1967.
36. Robertson, N. F., Friend, J., Aveyard, M., Brown, J., Huffee, M., and Homans, R. L., The accumulation of phenolic acids in tissue culture pathogen combinations of *Solanum tuberosum* and *Phytophthora infestans, J. Gen. Microbiol.,* 54, 261, 1968.
37. Warrens, R. S. and Routley, D. G., The use of tissue culture in the study of single gene resistance of tomato to *Phytophthora infestans, J. Am. Soc. Hortic. Sci.,* 95, 266, 1970.
38. Miller, S. A., Davidse, L. C., and Maxwell, D. P., Expression of genetic susceptibility, host resistance in alfalfa callus tissue inoculated with *Phytophthora megasperma, Phytopathology,* 74, 345, 1984.
39. Holliday, M. J. and Klarman, W. L., Expression of disease reaction types in soybean callus from resistant and susceptible plants, *Phytopathology,* 69, 576, 1979.
40. Keen, N. T. and Horsch, R., Hydroxyphaseollin production by various soybean tissues: a warning against use of 'unnatural' host-parasite systems, *Phytopathology,* 62, 439, 1972.
41. Helgeson, J. P., Kemp, J. D., Haberlach, G. T., and Maxwell, D. P., A tissue culture system for studying disease resistance: the black shank disease in tobacco callus culture, *Phytopathology,* 62, 1439, 1972.
42. Haberlach, G. T., Budde, A. D., Sequeira, L., and Helgeson, J. P., Modification of disease resistance of tobacco callus tissues by cytokinins, *Plant Physiol.,* 62, 522, 1978.
43. Hausen, G. and Beiderbeck, R., The development of an *in vitro* host-pathogen system consisting of maize seedling segments infected with *Ustilago maydis, J. Phytopathol.,* 118, 97, 1987.
44. Fereol, L., Inoculation of *in vitro* cultivated sugarcane with smut (*Ustilago scitaminea*), *Can. J. Bot.,* 62, 2043, 1984.
45. Peros, J. P. and Chagvardieff, P., Demonstration of *in vitro* toxic interactions between *Ustilago scitaminea* Syd. and sugarcane calluses: effect of two growth factors on the observed phenomenon, *Agronomie,* 3, 629, 1983.
46. Sinha, O. K., Studies on resistance expression in tissue cultures against red rot and smut diseases of sugarcane, Annual Report (1990–91), Indian Institute of Sugarcane Research, Lucknow, India, 53, 55, 1991.
47. Mishra, A. N., Personal communication, 1993.
48. Saad, A. T., The culture of apple callus tissue and its use in studies on pathogenicity of *Venturia inaequalis, Diss. Abstr.,* 25, 5487, 1965 (*Hortic. Abstr.* 36, 661).
49. Beech, I., Defago, G., and Gessler, C., Growth of *Venturia inaequalis* on apple tissue cultures, *Bot. Helvet.,* 95, 121, 1985 (*Rev. Plant Pathol.,* 65, 271).
50. Conger, B. V. and McDaniel, J. K., Use of callus culture to screen tall fescue seed samples for *Acremonium coenophialum, Crop Sci.,* 23, 172, 1983.
51. Joshi, M. K., Singh, U. S., and Garg, G. K., Use of fungicide to check saprophytic colonization of tissue culture medium by *Alternaria brassicae* and *A. alternata, Plant Cell, Tissue, Organ Culture,* in press.

52. Joshi, M. K., Singh, U. S., and Garg, G. K., Use of *Brassica* callus culture to induce sporulation in *Alternaria brassicae* (Berk.) Sacc., *Plant Cell, Tissue Organ Culture,* 14, 59, 1988.
53. Kumar, V. R., Kumar, A., and Arya, H. C., Nature of disease resistance and susceptibility in leaf blight of wheat caused by *Alternaria triticina in vivo* and *in vitro, Indian Phytopathol.,* 27, 455, 1974.
54. Pandey, B. K., Singh, U. S., and Chaube, H. S., Expression of resistance in chickpea against *Botrytis cinerea* in dual culture, *Curr. Sci.,* in press.
55. Hrib, J. and Rypacek, V., The growth response of wood-destroying fungi to the presence of spruce callus, *Ceska Mykol.,* 32, 55, 1978.
56. Vidhyasekaran, P., Borromeo, E. S., Ling, D. H., and Mew, T. W., Relationship between growth rate of *Helminthosporium oryzae* isolates on calluses of rice cultivars and their disease reaction on rice plants, *Plant Cell, Tissue Organ Culture,* 24, 237, 1991.
57. Hrib, J. and Rypacek, V., In vitro testing for the resistance of conifers to the fungus *Phaeolus schweinitzii* (Fr.) Pat. on callus cultures, *Eur. J. For. Pathol.,* 13, 86, 1983.
58. Sacristan, M. D., Resistance responses to *Phoma lingam* of plants regenerated from selected cell and embryogenic cultures of haploid *Brassica napus, Theor. Appl. Genet.,* 61, 193, 1982.
59. Gustine, D. L., Sherwood, R. T., and Vance, C. P., Regulation of phytoalexin synthesis in jackbean callus cultures. Stimulation of phenylalanine ammonia-lyase and *O*-methyltransferase, *Plant Physiol.,* 61, 226, 1978.
60. Ogasawara, N., Uchiyama, T., Oohashi, T., and Tanaka, H., Responses of callus tissues of rice to inoculation of conidia of *Pyricularia oryzae* Cavara and other fungi, *Ann. Phytopathol. Soc. Jpn.,* 43, 246, 1977.
61. Latunde-Dada, A. O. and Lucas, J. A., Involvement of phytoalexin medicarpin in the differential response of callus lines of lucerne (*Medicago sativa*) to infection by *Verticillium alboatrum, Physiol. Plant Pathol.,* 26, 31, 1985.
62. Ingram, D. S., Application in plant pathology, in *Plant Tissue and Cell Culture,* Street, H. E., Ed., Blackwell Scientific, Oxford, 1977, 463.
63. Braun, A. C. and Lipetz, J., The use of tissue culture in phytopathology, in *Cells and Tissues in Culture,* Vol. 3, Willmer, E. N., Ed., Academic Press, London, 1966, 691.
64. White, P. R., Tissue culture in the service of phytopathology, *Indian Phytopathol.,* 21, 14, 1968.
65. Maheshwari, R., Applications of plant tissue and cell culture in the study of physiology of parasitism, *Proc. Indian Acad. Sci. Sect. B,* 59, 152, 1969.
66. Ingram, D. S., Growth of plant parasites in tissue culture, in *Tissue Culture and Cell Culture,* Street, H. E., Ed., Blackwell Scientific, Oxford, 1973, 392.
67. Ingram, D. S., Tissue culture methods in plant pathology, in *Tissue Culture Methods for Plant Pathologists,* Ingram, D. S. and Helgeson, J. P., Eds., Blackwell Scientific, Oxford, 1980, 3.
68. Ingram, D. S., The establishment of dual cultures of downy mildew fungi and their hosts, in *Tissue Culture Methods for Plant Pathgologists,* Ingram, D. S. and Helgeson, J. P., Eds., Blackwell Scientific, Oxford, 1980, 139.
69. Buczacki, S. T., Culture of *Plasmodiophora brassicae* in host callus tissue, in *Tissue Culture Methods for Plant Pathologists,* Ingram, D. S. and Helgeson, J. P., Eds., Blackwell Scientific, Oxford, 1980, 145.
70. Miller, S. A., Tissue culture methods in phytopathology. II. Fungi, in *Plant Cell Culture, a Practical Approach,* Dixon, R. A., Ed., IRL Press, Washington D.C., 1985, 215.
71. Izard, C., Lacharpagne, J., and Schiltz, P., Behaviour of *Peronospora tabacina* in tissue cultures, and role of the leaf epidermis, *Ann. Dir. Etudes Equip. Sect. Deux,* 95, 1964.
72. Uchiyama, T., Kato, H., Ho, M., and Ogasawara, N., Changes in response to the inoculation of fungi by varying the number of subcultures of rice callus culture, *Ann. Phytopathol. Soc. Jpn.,* 50, 176, 1984.
73. Hrib, J., Digital time-lapse cinematography of a series of objects: recording of defence reaction of Norway spruce tissue cultures derived from single organs against the fungus *Phaeolus schweinitzii, Eur. J. For. Pathol.,* 14, 249, 1984.
74. Coppens, L., Vercruysse-Dewitte, D., and Gillis, E., Use of imazalil sulphate as a fungicide in barley tissue culture media, *Meded. Fac. Landbouwwey. Rijksuniv. Gent.,* 49, 351, 1984.
75. Morel, G., Recherches sur la culture associee de parasites obligatoires et de tissus vegetaux, *Ann. Ephiphyties,* 14, 1, 1948.
76. Nakamura, G., Bioassay of fungicides by plant tissue culture, *Bull. Agric. Chem. Inspect. Stn.,* 6, 67, 1963.

77. Shields, R., Robinson, S. J., and Anslow, P. A., Use of fungicides in plant tissue culture, *Plant Cell Rep.*, 3, 33, 1984.
78. Tripathi, R. K. and Ram, S., Induction of growth and differentiation of carrot callus cultures by carbendazim and benzimidazole, *Indian J. Exp. Biol.*, 20, 674, 1982.
79. Webb, K. J. and Gay, J. L., The recalcitrant powdery mildews—attempts to infect cultured tissues, in *Tissue Culture Methods for Plant Pathologists*, Ingram, D. S. and Helgeson, J. P., Eds., Blackwell Scientific, Oxford, 1980, 151.
80. Uchiyama, T. and Ogasawara, N., Disappearing of the cuticle and wax in outermost layer of callus cultures and decrease of protective ability against microorganisms, *Agric. Biol. Chem.*, 41, 1401, 1977.
81. Helgeson, J. P., Studies of host-pathogen interactions *in vitro*, in *Use of Tissue Culture and Protoplasts in Plant Pathology*, Helgeson, J. P. and Deverall, B. J., Eds., Academic Press, New York, 1983, 9.
82. Zoeten, G. A. de, Gaard, G., Haberlach, G. T., and Helgeson, J. P., Infection of tobacco callus by *Phytophthora parasitica* var. *nicotianae*, *Phytopathology*, 72, 743, 1982.
83. Dixon, R. A., Plant tissue culture methods in the study of phytoalexin induction, in *Tissue Culture Methods for Plant Pathologists*, Ingram, D. S. and Helgeson, J. P., Eds., Blackwell Scientific, Oxford, 1980, 185.
84. Bailey, J. A., Pisatin production by tissue culture of *Pisum sativum* L., *J. Gen. Microbiol.*, 61, 409, 1970.
85. Dixon, R. A. and Fuller, K. W., Effects of growth substances on non-induced and *Botrytis cinerea* culture filtrate-induced phaseollin production in *Phaseolus vulgaris* cell suspension cultures, *Physiol. Plant Pathol.*, 12, 279, 1978.
86. Sinha, O. K., Singh, K., and Misra, S. R., Stain technique for the detection of smut hyphae in nodal buds of sugarcane, *Plant Dis.*, 66, 932, 1982.

FURTHER READING

Dixon, R. A., Ed., *Plant Cell Culture, a Practical Approach*, IRL Press, Washington, D.C., 1985.

Garg, G. K., Singh, U. S., Khetrapal, R. K., and Kumar, J., Application of tissue culture in plant pathology, in *Experimental and Conceptual Plant Pathology*, Vol. *Techniques*, Singh, R. S., Singh, U. S., Hess, W. M., and Weber, D. J., Eds., Gordon and Breach Science Publishers, New York, 1988, 83.

Helgeson, J. P. and Deverall, B. J., Eds., *Use of Tissue Culture and Protoplasts in Plant Pathology*, Academic Press, New York, 1983.

Ingram, D. S. and Helgeson, J. P., Eds., *Tissue Culture Methods for Plant Pathologists*, Blackwell Scientific, Oxford, 1980.

Singh, U. S., Sharma, A. K., and Pandey, B. K., Growth of biotrophic plant parasites and symbionts in tissue culture, in *Perspectives in Phytopathology*, Agnihotri, V. P., Singh, N., Chaube, H. S., Singh, U. S., and Dwivedi, T. S., Eds., Today & Tomorrow's Printers and Publishers, New Delhi, 1989, 113.

Chapter 21

Dual Culture: Nematodes

S. Verdejo-Lucas

TABLE OF CONTENTS

I. INTRODUCTION

Axenic culture of free-living, insect-parasitic, and animal-parasitic nematodes in chemically defined or nondefined media has been achieved with some success.[1,2] But development of similar culture systems for plant-parasitic nematodes has been limited to a few fungivorous nematodes such as *Aphelenchoides* spp., *Aphelenchus avenae,* and *Bursaphelenchus xylophilus.*[2] The first axenic cultivation of a stylet-bearing nematode was accomplished by Myers[3] with the cultivation of *Aphelenchoides sacchari.* Axenic cultivation of strictly obligate parasitic nematodes of higher plants has not been achieved yet. At present, nematode members of the Tylenchoidea and Aphelenchoidea can be cultured only in association with plant tissues. Dual cultures of nematodes have proved to be a useful research tool in plant nematology. The system has been used as the environment to investigate aspects of the host-parasite relationship, and as a source of clean nematodes free of contaminants for a wide range of studies. This chapter will be concerned only with the dual culture of nematodes that feed on plant tissue as obligate parasites causing damage to the plant. The *in vitro* culturing of plant-parasitic nematodes and its applications has been reviewed by several authors including Zuckerman,[4,5] Krusberg and Babineau,[6] Platzer,[1] Jones,[7] Riedel et al.,[8] Koenning and Barker,[9] and Hooper.[10] Since some of these reviews[5,6,8] have dealt with callus tissue cultures, this topic is not covered comprehensively in this chapter.

A. TERMINOLOGY

In culturing nematodes, several terms, proposed by Dougherty,[11] for various types of cultures are in general use. Thus, *gnotobiotic cultures* are cultures with known associated organisms (known in number, but not necessarily to species), or nonassociated organisms. *Monoxenic cultures* are cultures with one other known associated organism. *Dixenic cultures* are cultures with two known associated organisms. *Axenic cultures* have no associated organisms. *Xenic* or *agnobiotic cultures* have an unknown number of associated organisms (usually a mixed microbial flora).[5,10]

II. PRINCIPLES

Dual cultures of nematodes and plant tissue are aimed to establish plant-parasitic nematodes in sterile cultures. Since phytophagous nematodes are obligate parasites, they need a plant tissue as a food source

0-87371-877-1/95/$0.00+$.50

from which they can obtain the nutrients necessary for their development and reproduction. Nematodes can feed as migratory ectoparasites and endoparasites, and as sedentary endoparasites, thus the nematode feeding habit determines primarily the type of plant tissue required for their culture. Nematodes that feed on vascular tissue, inducing a specific host response, require differentiated tissue for reproduction in dual cultures.[12] This is the case of the sedentary endoparasites *Meloidogyne, Heterodera,* and *Globodera.* In contrast, migratory nematodes do not require vascular elements and reproduce readily on undifferentiated callus tissue or carrot disks; examples are species of *Pratylenchus, Ditylenchus,* and *Radopholus.*

The monoxenic culture involves two living components, the nematode and the plant tissue. Both should develop on a chemically defined culture medium. The nematode feeds parasitically on the plant tissue and the medium provides the nutrients needed for tissue growth. Therefore, a culture medium should first be selected for establishing axenic cultures of the plant tissue. Plant seeds are commonly used to initiate axenic plant tissue cultures. Root explants and callus tissues have been the plant materials most frequently utilized. Accomplishment of axenic plant tissue leads to the establishment of the nematode in monoxenic culture. Before adding the nematodes to the cultures, they have to be freed of associated microorganisms by surface disinfection. Eggs, second-stage infective juveniles, cysts, egg masses, or mixed life cycle stages can be used as the nematode inoculum, the choice depending on the nematode species, the purpose of the culture, or study. Nematodes added to the cultures migrate toward the plant tissue and invade the tissue, where they feed, molt, and complete their life cycle. Nematodes can complete two to three generations in a single culture unit before cultures become exhausted of nutrients.

III. METHODOLOGY

The methods and procedures described here are those that are in general use in many laboratories and they are known to perform satisfactorily. Nonetheless, they should be taken as a guide. These methods have undergone minor modifications to suit the particular needs of each researcher. Aseptic techniques are essential, since all components of the system must be free of contamination. The general equipment needed includes autoclave, alcohol lamp or gas burner, laminar flow cabinet, and incubator. Glassware should be recently autoclaved before use. Test tubes and plastic petri dishes are the most common containers used because they are readily available, inexpensive, and easily manipulated and stored. The protocols described have been adapted from previously published ones as indicated.

A. CULTURE MEDIA

In general terms, the composition of the medium is a determining factor for growing successful plant tissue cultures. The medium should provide prolific root growth, which in turn will affect nematode development and reproduction. Most formulations are comprised of three groups of ingredients: the inorganic nutrients consist of mineral salts which supply the requirement for macro- and micronutrients, the carbon or energy source which is generally sucrose, and the vitamins that are required in trace quantities. Agar is usually added to solidify the medium. It is convenient to have a series of stock solutions for medium preparation.[9] Since some vitamins are heat labile, it is therefore recommended to filter sterilize the vitamin solution (22-μm pore filter units) which should be added after autoclaving and cooling of the medium (about 40°C). The medium is then poured, allowed to solidify, and stored in a cabinet until use. The use of plant hormones or growth regulators is not necessary for culturing parasitic nematodes on root explant, transformed root, or carrot disk cultures. However, callus tissue cultures require the addition of plant growth regulators for their establishment. In most cases, callus production is successfully achieved by using only 2,4-dichlorophenoxyacetic acid (2,4-D).[6] Formulations of media used for monoxenic culturing of nematodes are shown in Table 1. The medium of Gamborg's B-5 without cytokinins or auxins[13] is suggested for culturing most nematodes on root explants and *Agrobacterium rhizogenes*-transformed root cultures.[14,15]

B. SURFACE DISINFECTING SEEDS

The selection of high-quality seeds is important for success in subsequent steps in the establishment of nematode cultures.[16] Seeds with high germination rates and that are free of contaminating microorganisms are highly desirable. Discolored, cracked, or diseased seeds should be rejected. Seeds are treated with broad-spectrum biocidal chemicals (sodium hypochlorite, mercuric chloride, hydrogen peroxide).

Table 1 **Formulations of tissue culture media used more frequently for monoxenic culturing of nematodes***

Ingredients	MS/5[1]	B-5[2]	STW[3]	White[4]
Na_2SO_4				T200
$(NH_4)NO_3$	330		40	
$(NH_4)_2SO_4$		134		
KNO_3	380	2,500	80	80
$Ca(NO_3)_2 \cdot 4H_2O$			144	200
$CaCl_2 \cdot 2H_2O$	440	150		
$MgSO_4 \cdot 7H_2O$	370	250	72	360
$Na_2SO_4 \cdot 10H_2O$			38	
KH_2PO_4	170			
$NaH_2PO_4 \cdot H_2O$		150		16.5
KCl			65	65
$Fe_2(SO_4)_3$				25
$FeSO_4 \cdot 7H_2O$	27.8	27.85	23	
Na_2EDTA	37.3	37.25	37.25	
$MnSO_4 \cdot 4H_2O$	22.3		4.9	4.5
$MnSO_4 \cdot H_2O$		10		
$ZnSO_4 \cdot 7H_2O$	8.6	2	2.7	1.5
H_3BO_3	6.3	3	1.6	1.5
KI	0.83	0.75	0.75	
$Na_2MoO_4 \cdot 2H_2O$	0.25	0.25		
$CuSO_4 \cdot 5H_2O$	0.025	0.025		
$CoCl_2 \cdot 6H_2O$	0.025	0.025		
myo-inositol	100	100		
Nicotinic acid	0.5	1	0.5	0.5
Pyridoxine HCl	0.5	1	0.75	0.1
Thiamine HCl	0.1	10	0.1	0.1
Glycine	2		2	3.0
Sucrose	30,000	20,000	20,000	20,000
pH	5.7	5.5	5.8	5.5

[1]Modified Murashige and Skoog medium as used by Mugnier.[53]
[2]Gamborg's B-5 medium without cytokinins or auxins.[13]
[3]Skoog, Tsui, and White medium as used by Orion et al.[19]
[4]White medium as used by Mountain.[20]
*Milligram per liter of medium.

The use of sonication, ethanol soakings, or wetting agents (e.g., Tween® 20) increases the penetration of the chemical due to their humidifying effect on the seeds. Their use, although convenient, is not always necessary. Most seeds are disinfected satisfactorily with a 0.525% sodium hypochlorite solution (10% commercial bleach). Some seeds (cereal, soybean, sugar beet) are more difficult to disinfect; mercuric chloride at 100 to 250 ppm is then recommended. Alfalfa and clover seeds are, however, the most fastidious seeds to disinfect due to contaminating internal bacteria, and fungi. For persistant bacterial contamination in seed lots, see Viglierchio et al.[16] Treated seeds should be rinsed repeatedly in abundant sterile water to eliminate the chemical; germination otherwise would be affected. Seeds can be germinated directly on the nutrient medium selected, which reduces seedling manipulation and chances of contamination. Alternatively, seeds can be germinated in water agar or a richer medium, particularly when contaminants are suspected.

C. ESTABLISHING PLANT TISSUE CULTURES

Of all procedures described for dual nematode-tissue cultures, those that use root explants, callus tissues, and carrot disks have achieved wide acceptance among nematologists. *A. rhizogenes*-transformed root cultures have been used more recently for culturing sedentary endoparasitic nematodes.[15,17,18]

1. Root Explants

Root explants have been used for culturing migratory endoparasites, but mainly for sedentary endoparasites. Studies using root explants are, however, limited, maybe due to the small growth potential of root explants and to the culture media used. For instance, high concentrations of phytohormones or ammonium salts affect nematode behavior.[6,7,19]

Materials—Selected seeds, sodium hypochlorite (NaClO), mercuric chloride ($HgCl_2$), or hydrogen peroxide (H_2O_2), flasks filled with distilled or deionized water (H_2O), beakers, plastic petri dishes with Gamborg's B-5 medium, forceps, scalpel, Parafilm.

The protocol has been adapted from the work of Mountain,[20] Tiner,[21] Dropkin and Boone,[22] and Huettel.[23]

a. In a laminar flow cabinet, place seeds in a sterile beaker (enough to cover the bottom of the beaker) and add enough chemical solution to cover the seeds. Stir gently and allow to soak for 10 min ($HgCl_2$) or 15 to 20 min (NaClO, H_2O_2).
b. Drain the chemical solution and add abundant sterile H_2O.
c. Drain the H_2O and transfer the seeds to a new sterile beaker, and rinse four to six times with sterile H_2O.
d. Place one or two seeds in the center of each petri dish with Gamborg's B-5 medium. Alternatively, place 6 to 12 seeds (depending on plant species) on 1% water agar plates. Seal the edge of the plates with Parafilm.
e. Allow seeds to germinate for 4 to 7 d at 25°C in the dark.
f. Excise tip from the root when the root tip reaches the wall of the petri dish (about 4 cm in length). If seeds are germinated in water agar, excise the root tip when the tip has grown 2 to 3 cm, and transfer two to three tips to each petri dish with Gamborg's B-5 medium.
g. Add nematodes to cultures within 24 h after removing the top.[22,24] Seal plates with Parafilm to prevent cultures from desiccation and contamination. Incubate cultures at 25°C in the dark.

The life cycle of several nematode species has been described using axenic root explant cultures.[25–28] These cultures have been also used to study host reaction[22,26,29–33] as well as the effects of environmental factors on root penetration and nematode development.[32,34–37] Longevity of root explant cultures is limited because of desiccation of the medium.

2. Callus Tissue

Most of the cultured migratory ecto- and endoparasites have been propagated on callus tissue. The term *callus tissue* as used in plant nematology means a friable mass of plant cells arising from excised plant tissue or whole seedlings treated with 2,4-D.[6] A variety of callus tissues from different plants has been used as substrates for nematode propagation.[5,6,8,38]

Materials—Alfalfa (*Medicago sativa*) seeds, concentrated sulfuric acid (H_2SO_4), $HgCl_2$, ethanol, distilled H_2O, yeast extract medium, callusing medium, cheesecloth, filter paper, aluminum foil, beakers, flasks, plastic petri dishes, 25- × 15-mm test tubes with caps, racks.

The protocol for alfalfa callus tissue cultures is taken after Riedel.[39]

a. Soak alfalfa seeds in hot water at 61°C in a beaker for 10 min. Treat about 10 cm^3 of seeds, each in bags made of a double thickness of cheesecloth.
b. Plunge heated seeds into a beaker with cold water immediately after treatment.
c. Spread seeds on a thin layer of filter papers to dry for 2 weeks in aluminum foil trays lined with filter paper.
d. In a laminar flow cabinet, soak 5 to 10 cm^3 dried alfalfa seeds for 15 min in a beaker with enough concentrated H_2SO_4 to cover the seeds. Stir gently.
e. Rinse briefly in three separate baths of 150 ml sterile H_2O.
f. Soak washed seeds for 15 min in a beaker with enough $HgCl_2$ solution (1 ppm $HgCl_2$ in 30% ethanol) to cover the seeds. Stir gently.
g. Rinse in three separate baths of 150 ml sterile H_2O.
h. Spread sterilized seeds immediately on a yeast extract medium to test for contamination. Spread about 0.25 ml seeds per plate.
i. Allow seeds to germinate for 4 to 7 d at 22 to 24°C in the dark.
j. Transfer 8 to 12 sterile alfalfa seedlings only from uncontaminated plates to each tube with callusing medium.
k. Allow callus to grow for 10 to 14 d at 22 to 24°C in the dark before adding nematodes.

Yeast extract medium—Sucrose 10 g, Difco yeast extract 2 g, agar 10 g. Dissolve the ingredients in 1 l distilled H_2O in a flask, cover, and autoclave (15 to 20 min). Pour the medium in plastic petri dishes. The medium can be replaced by any rich medium such as potato dextrose agar.

Callusing medium—Sucrose 10 g, Difco yeast extract 2 g, 2,4-D 2 g, agar 10 g. Dissolve the ingredients in 1 l distilled H_2O and heat to melt the agar. Add 14 ml of medium per test tube, cover with caps, and autoclave (15 to 20 min). Allow tubes to cool on their sides to make slants.

Callus cultures inoculated with nematodes should be allowed to develop for 8 to 10 weeks before use. The protocol described here can be adapted for other plant species. Callus cultures have been used mainly as a source of quantities of germ-free nematodes for a variety of biological, physiological, taxonomic, and pathogenicity studies.[5,6,8] They have also been used for mass rearing of nematodes for breeding programs.[40,41] The time needed for obtaining callus cultures is the major drawback of the system.

3. Carrot Disk Cultures

Carrot disks have been used for culturing endoparasitic nematodes that do not induce a specific host response such as *Pratylenchus* spp., *Radopholus* spp., and *Zygotylenchus guevarai.*[23,42–47]

Materials—Fresh carrots (*Daucus carota*), ethanol, vegetable peeler or sharp knife, scalpel, forceps, sterile filter paper, sterile containers (glass tubes with caps, plastic petri dishes), Parafilm.

The protocol has been adapted from those described by O'Bannon and Taylor,[42] Moody et al.,[43] Lawn and Noel,[44] and Huettel.[23] The condition of the carrots is of great importance. Intact fresh carrots that have not been in cold storage will provide best results. Carrots that are split, cracked, broken, or bruised should be rejected.

a. Wash carrots in water with detergent to remove dirt adhered to the external tissues.
b. In a laminar flow cabinet, dip carrots in enough 95% ethanol in a tall beaker to cover three fourths of the carrot (use remainder of the carrot for a handle). Alternatively, hold carrots with a forceps and spray 95% ethanol all over the carrot.
c. Flame over an alcohol lamp until all ethanol is burned off.
d. Peel external tissues in thin strips with a vegetable peeler using sterile techniques. Flame peeler after removing every strip of epidermis. Peel only the flamed area of the carrot.
e. Cut off the tapered end of the carrot about 2 to 3 cm and slice disks 5 to 8 mm in thickness onto sterile filter paper or in a petri dish.
f. Transfer one or several disks to sterile containers. Carrot disks can be used immediately or stored for several months in the dark if properly sealed to prevent disks from desiccation.

Once nematodes are added to the carrot disks, it takes at least 8 weeks to obtain large quantities of nematodes. Nematodes from carrot disk cultures must be surface disinfected before inoculating new cultures.

This semisterile method is an easy way of culturing migratory endoparasitic nematodes. Carrot disk cultures provide large numbers of nematodes[42,44,45] even when inoculated with low numbers. For instance, single carrot disks inoculated with 10 mature females of *R. similis* or *P. vulnus* yielded 234,000 and 166,000 individuals, respectively, after 3 months at 26°C.[47] The intrinsic microflora associated intimately to the carrot tissue is the main limitation of these cultures, since it cannot be eliminated by surface disinfection. Unpredictable proliferation of intrinsic microflora causes the rapid breakdown of the carrot tissue which turns brown and develops a soft rot accompanied by leaking;[43] as a consequence, nematodes die. Tissue breakdown has been associated with high densities of nematodes which apparently trigger the proliferation of intrinsic microflora already existing within the carrot tissue.[43,47]

4. Roots Transformed by *Agrobacterium rhizogenes*

Root cultures genetically transformed by *A. rhizogenes* are a practical system for culturing sedentary[15,17,18] as well as migratory endoparasitic nematodes.[72] *A. rhizogenes* induces a genetic transformation of higher plants by inserting a fragment of its root (Ri) plasmid DNA (T-DNA) into the plant genome. In axenic cultures, transformed roots differ morphologically and physiologically from normal roots: they grow faster, are highly branched, are phytohormone independent, and are better adapted to grow axenically. Their high growth rate makes them effectively self-disinfecting because they outgrow the bacterium.[48]

The protocol for inducing and growing transformed root cultures has been adapted from the work of Paul et al.,[17] Mugnier,[49] and Kumar and Forrest.[18]

a. Seeds are surface disinfected as indicated in Section III.B.
b. Transfer disinfected seeds to petri dishes with solidified (1% agar) Murashige and Skoog medium[50] supplemented with 20 g sucrose/l (MS20).
c. Allow seeds to germinate at 25°C in the dark.
d. Excise tops, leaving about 1 cm of stem, and inoculate an overnight culture of *A. rhizogenes* in yeast extract mannitol (YMB) medium[51] onto the stem. Alternatively, small segments of cotyledons can be inoculated. Incubate at 25°C.
e. Remove developing roots emerging 3 to 4 weeks after inoculation and transfer them to fresh MS20 medium containing 250 mg/l cefotaxime (or 500 mg/l carbenicillin) to suppress bacterial growth and 100 mg/l kanamycin to select transformed roots.
f. Subculture young root tips free of *Agrobacterium* to fresh MS20 medium with antibiotics as in the previous step until rid of the bacterium.
g. Subculture pieces of transformed roots (three to four growing tips) from actively growing cultures to fresh culture medium.
h. Add nematodes to root cultures, seal petri dishes, and incubate at 25°C in the dark.

Plant materials other than seeds, e.g., storage roots of carrot and tubers of potato, are surface disinfected (see Section III.C.3), placed on 1% water agar plates, and inoculated with *A. rhizogenes* by pipetting the bacterium on the disks. Then, follow steps e through h. Root cultures should be analyzed for the presence of opines (agropine and mannopine) to confirm the transformation of the roots.[52] Opines are specifically synthesized by transformed roots.[48]

Yeast extract mannitol medium (YMB)—Manitol 10 g, Difco yeast extract 1 g, K_2HPO_4 0.5 g, NaCl 0.2 g, $MgSO_4 \cdot 7H_2O$ 0.2 g, $FeCl_3$ 0.004 g, agar 15 g, distilled water 1 l. Autoclave 15 to 20 min; pH 6.8.

The entire root surface of these cultures is very large, since many lateral roots are formed. Once established, transformed root cultures are easily maintained by transferring young root tips to fresh medium. They can be inoculated immediately after being transferred, and there is no waiting period between culture establishment and nematode addition. Longevity of this culture is greater than that of root explants. The genetic transformation induced by *A. rhizogenes* does not prevent nematodes from development and reproduction on such roots.[15,17,53] Nematodes reared on transformed roots maintained their pathogenicity to *in vivo*-grown whole plants.[15,17] These root cultures have been used to establish dixenic cultures of *M. javanica* and *Pasteuria penetrans* for the study of the interactions between the nematode and its bacterial antagonists.[54] Also, the mechanism of action of nematicides has been studied.[53] The rapid growth of transformed roots can be a disadvantage in two ways; firstly they need to be subcultured frequently (2 to 3 weeks) or cultures decline rapidly, and secondly, cultures would use up nutrients and grow old before nematodes could reproduce. However, root cultures that grow at a moderate rate can be selected.[15]

D. SURFACE DISINFECTING NEMATODES

The nematode species, and most important, the stage of the life cycle will determine the appropriate method because substances used to treat the eggs would kill vermiform stages.[9] Zukerman[4,5] provided a table with 34 procedures for surface disinfecting nematodes. Exposure duration to disinfectants is critical since chemicals affect nematode viability. Short exposure times to strong, broad-spectrum chemicals should provide best results. Procedures using organic or inorganic $HgCl_2$, singly or in combination with antibiotics, have given good results for disinfecting several nematode genera.[17,25,26,42–44,46,56] Techniques involving a single antibiotic have not generally been successful,[5] because antibiotic solutions tend to be bacteriostatic rather than bacteriocidal. Manipulations of nematodes should be reduced to a minimum for best survival and less contamination. Clean nematode suspensions free of soil particles and root debris are highly recommended. Samples with high nematode densities should be selected, but nematodes can be first multiplied in a susceptible host when found in low numbers. Nematodes extracted by standard procedures are hand picked (preferably in adult stages). Procedures involving nematode motility[21,43,44] or a settling step[57] select healthier individuals because unmotile or starved nematodes are not recovered. Small numbers of nematodes can be passed singly through the disinfecting solution using a hand pick. Generous rinsing in sterile water should follow the chemical treatment.

Materials—Graduated conical centrifuge tubes, BPI dishes, pasteur pipettes, test tubes filled with distilled H_2O, racks.

The protocol for surface disinfecting *Meloidogyne* eggs has been modified from Loewenberg et al.[57]

a. In a laminar flow cabinet, place 20 to 30 clean egg masses in a sterile centrifuge tube.
b. Add 0.5 ml of a 0.525% NaClO solution (10% commercial bleach), and incubate for 4 min. Agitate gently.
c. Dilute 20 times with sterile distilled H_2O.
d. Collect eggs settled within 30 min at the bottom of the tube with a pasteur pipette fitted to a rubber bulb.
e. Add nematodes to axenic plant tissue cultures by pipetting small droplets of the nematode suspension on the medium.

For cyst nematodes, place cysts in a BPI dish with zinc chloride (1 m*M*) to stimulate juvenile hatching. Collect hatched juveniles 24 to 48 h later, concentrate, and surface disinfect following the protocol described for *Meloidogyne* eggs, but using a 0.01% $HgCl_2$ and 1% streptomycin sulfate solution. Root-knot juveniles are obtained from egg masses placed in a vial or on a Baerman tray. Hatched juveniles are collected within 24 to 48 h and surface disinfected as described for cyst nematodes. Vermiform stages of other nematodes can be surface disinfected similarly. Counts of representative aliquots can be made to quantify numbers of nematodes added per culture unit.

E. INOCULATING WITH NEMATODES

The selection of inoculum depends on the nematode species and the experiment to be performed. For example, *Meloidogyne* has been commonly introduced as egg masses.[19,58,59] Egg masses are easily transferred, and several hundred reproductive units are transferred at once. They can be used when quantification is not required. One egg mass per culture unit is sufficient if taken from young cultures (6 weeks). More egg masses would be needed when cultures are older. The same principle is applicable to *Heterodera* species. For quantitative experiments dispersed eggs or juveniles can be used. Eggs hatch and invade the root gradually, but hatching rate is unpredictable. Viability of eggs treated with NaClO may be affected, and more so in developed eggs because of the increased permeability of the egg shell. Juveniles are easily quantified, and root invasion and infection are more synchronous than with egg masses or eggs. But juveniles could kill the root if there is not enough root mass available. Cyst nematodes such as *H. glycines* and *H. zeae* are introduced as females with egg masses.[27,60] Juveniles of sedentary nematodes can be obtained from cysts or egg masses produced monoxenically under sterile conditions.[15,26] Migratory nematodes are usually introduced as vermiform stages. The inoculum consists of mixed life cycle stages.[32,38,45] Eggs can be separated from vermiform stages by using Baerman trays or differential sieving.

F. CULTURE MAINTENANCE AND SUBCULTURING

Nematode cultures should be maintained in the dark, protected from the light. In general, most plant tissue and nematodes cultured dually can grow satisfactory at 25°C. Incubation temperature is an important factor that can be manipulated; for example, it can be lowered to extend the period of time between subcultures. Optimal temperature for reproduction has been determined for several nematodes on alfalfa callus and carrot disks.[41,61,62] As a rule, nematodes reproduce more rapidly when maintained at higher temperature. Cultures yield larger numbers of individuals in shorter time, but such cultures will decline more rapidly due to consumption of nutrients.[41,61] For routine maintenance of monoxenic cultures, subculturing is easily done by transfering aseptically small pieces of infected tissue (callus or roots) to fresh axenic plant tissue cultures. Egg masses or cysts of sedentary endoparasites are directly transferred. Nematodes could be subcultured every 2 to 3 months, although the length of time between subcultures is specific for each nematode species/plant tissue combination. Factors involved in such specificity are multiplication rate of the nematode species, pathogenicity, nutrients and mass of tissue available, incubation temperature, initial inoculum density, and culture age. Only experience and careful observation will show the optimal time for subculturing.

To extract nematodes from monoxenic cultures, several techniques and apparatus have been described.[8,45,63,64] Nematodes abandon the plant tissue in large quantities[21,47] when high populations have been reached in the cultures, and they can be easily recovered. A simple way of recovering nematodes is by adding water to the cultures and collecting the nematodes 24 to 48 h later on a 25-μm pore-diameter sieve. The majority (80 to 90%) of the nematodes are recovered in the water suspension, but those remaining within the tissue can be extracted by blending the tissues.[46] Also, cultures can be flooded with sterile distilled water and nematodes collected under sterile conditions.

IV. APPLICATIONS IN PLANT PATHOLOGY

Dual cultures of nematodes have many applications in plant pathology, since they provide an adequate means to analyze quantitatively the host-parasite relationship. Such cultures offer a controlled environment which has made possible studies that could not be done with plants growing in soil. Monoxenic cultures were first applied in the demonstration of pathogenesis caused by plant-parasitic nematodes.[20,25] Other applications, besides those already mentioned, include the observation of nematode behavior,[24,25,32,33,55,57] morphological and physiological changes associated with parasitism,[19,58,59] and nematicidal activity of chemicals.[53,63] But the greatest use of dual cultures of nematodes has been for propagating nematodes.[8,23,38,40,41] Monoxenic cultures are an effective means for obtaining large numbers of highly infective, germ-free nematodes for numerous purposes, such as inoculum for greenhouse or microplot experiments.[8] Histological and ultrastructural studies require debris-free nematodes. Studies at the molecular level also benefit from monoxenically grown nematodes because of the quantities demanded and their cleanliness.

Significant progress has been made in recent years. Thus, culture media suitable for nematode development were investigated.[14,19,53] Gelling agents more optically clear than agar were also studied.[65,66] Transformed root cultures were first used to culture sedentary endoparasites.[15,17,18] Several nematode genera, including economically important species that had not been cultured before, were propagated in dual cultures.[15,17,18,27,28,46,66–71] Nevertheless, such cultures have not been extensively used, probably due to technical difficulties found by many workers in their establishment and maintenance. Success in dual cultures results from the combination of several factors. Actively growing fresh plant tissue and fresh nematode inoculum will provide best results. Tolerant hosts will allow the buildup of higher populations, whereas susceptible hosts suffering great damage may impair nematode development (e.g., proliferation of callused tissue in *Meloidogyne* cultures). Initial inoculum density is related in some extent to mass of tissue available, inoculum age, and disinfecting procedure which determine surviving individuals. Frequent and meticulous observations of development of plant tissues and nematodes in the cultures will help beginners to establish successful cultures.

The achievements of the last decade will probably contribute to the increased use of dual cultures. These cultures will facilitate physiological and biochemical studies of the complex host-nematode relationship, particularly investigations on recognition, specificity, and resistance. The molecular biology of the host-parasite interaction can be investigated by using this system. *Arabidopsis thaliana* has been proposed as a model host plant for such studies.[67] Genetic engineering and tissue culture techniques offer a good opportunity for producing novel nematode-resistant plants.

V. ADVANTAGES AND LIMITATIONS

Plant-parasitic nematodes propagated in plant tissue culture maintain their infectiveness and virulence toward whole host plants.[6,15,17] Because nematode pathogenicity is not altered by the continuous culture, the system allows maintenance of a collection of nematodes in little space. Problems associated with nematode propagation on plants in greenhouses (accidental contaminations, demand of greenhouse space, etc.) are avoided. An important advantage of monoxenic cultures is the possibility of making nondestructive observations which can be used for research or educational purposes. The progress of nematodes attacking the roots of a susceptible host and the symptoms of nematode damage can be monitored at periodical intervals without disturbing the system.[22,24,26,29,30,55,68] Nematode cultures are easily destroyed by autoclaving, therefore, risks of contamination or accidental dispersal are minimized. Foreign pathogenic nematodes that are not, but can be potential pests in certain areas can be studied or compared by means of monoxenic cultivation. Studies in monoxenic cultures can be done without the interference of secondary organisms, and variables (e.g., temperature, pH, nutrients) are easily controlled. Cultures can be maintained at constant temperature without being influenced by seasonal fluctuations.

It is important to indicate that results obtained from tissue culture experiments may differ from those obtained with whole plants. In nature, biotic and abiotic factors interact with the nematode-plant host system. As a result of such interactions the response of the plant host or the damage caused by the nematode is modified. But the influence of many of those factors on the host-nematode relationship cannot be assessed in sterile dual cultures, mainly due to technical difficulties. The complexity of the system is considerably increased when a second associated organism or new factor is introduced into the system.

Fungal and bacterial contamination remains an important limitation in culturing nematodes monoxenically. The greatest importance is the observation of rigorous aseptic techniques at each step of the process. A primary disadvantage is that no allowance is made for continued nutrition of the plant tissue. Standardization of procedures is also needed to reduce variations in infectivity and improve reliability.

REFERENCES

1. Platzer, E. G., Culture media for nematodes, in *CRC Handbook Series in Nutrition and Food,* Section G: *Diets, Culture, Media, Food Supplements,* Vol. 2, *Food Habits of, Diets for Invertebrates and Vertebrates—Zoodiets,* CRC Press, Cleveland, OH, 1977, 29.
2. Bolla, R. I., Axenic culture of plant parasitic nematodes: problems and perspectives, in *Vistas on Nematology,* Veech, J. A. and Dickson, D. W., Eds., Society of Nematologists, Hyattsville, MD, 1987, 401.
3. Myers, R. F., Axenic cultivation of *Aphelenchoides sacchari* Hooper, *Proc. Helminthol. Soc. Wash.,* 34, 251, 1967.
4. Zuckerman, B. M., Gnotobiotic culture and its significance to plant nematology, *Helminthol. Abstr.,* 38, 1, 1969.
5. Zuckerman, B. M., Gnotobiology, in *Plant Parasitic Nematodes,* Vol. 2, Zuckerman, B. M., Mai, W. F., and Rohde, R. A., Eds., Academic Press, New York, 1971, 159.
6. Krusberg, L. R. and Babineau, Application of plant tissue culture to plant nematology, in *Plant Cell and Tissue Culture,* Sharp, W. R., Larsen, P. O., Paddock, E. F. and Raghavan, V., Eds., Ohio State University Press, Columbus, 1977, 401.
7. Jones, M. G. K., Interaction of plant parasitic nematodes with excised root and tissue cultures, in *Tissue Culture Methods for Plant Pathologists,* Ingram, D. S. and Helgeson, J. P., Eds., Blackwell Scientific, Oxford, 1980, 161.
8. Riedel, R. M., Alves de Lima, M., and Martin, M., Establishment of nematode germplasm banks, in *Handbook of Plant Cell Culture,* Vol. 1, *Techniques for Propagation and Breeding,* Evans, D. A., Sharp, W. R., Ammirato, P. V., and Yamada, Y., Eds., Macmillan, New York, 1983, 880.
9. Koenning, S. R. and Barker, K. R., Gnotobiotic techniques for plant-parasitic nematodes, in *An Advance Treatise on Meloidogyne,* Vol. 2, *Methodology,* Barker, K. R., Carter, C. C., and Sasser, J. M., Eds., North Carolina State University Graph., 1985, 49.
10. Hooper, D. J., Culturing nematodes and related experimental techniques, in *Laboratory Methods for Work with Plant and Soil Nematodes,* 6th ed., Southey J. F., Ed., MAFF, Her Majesty's Stationery Office, London, 1986, 133.
11. Dougherty, E. C., Introduction to axenic culture of invertebrate metazoa: a goal, *Ann. N.Y. Acad. Sci.,* 77, 27, 1959.
12. Sayre, R. M., Plant Tissue Culture as a Tool in the Study of the Physiology of the Root-Knot Nematode *Meloidogyne incognita* Chit, Ph.D. thesis, University of Nebraska, 1958.
13. Gamborg, O. L., Miller, R. A., and Ojima, K., Nutrient requirements of suspension cultures of soybean root cells, *Exp. Cell. Res.,* 50, 148, 1968.
14. Rebois, R. V., Huettel, R. N., and Lauritis, J. A., A comparison of media for the monoxenic culture of corn cyst, soybean cyst, reniform, root-knot, and lesion nematodes on root explants, *Proc. 1st Int. Cong. Nematol.,* Guelph, Ontario, Canada, 1984, 85.
15. Verdejo, S., Jaffee, B. A., and Mankau, R., Reproduction of *Meloidogyne javanica* on plant roots genetically transformed by *Agrobacterium rhizogenes, J. Nematol.,* 20, 599, 1988.
16. Viglierchio, D. R., Siddiqui, I. A., and Croll, N. A., Culturing and population studies of *Ditylenchus dipsaci* under monoxenic conditions, *Hilgardia,* 42, 177, 1973.
17. Paul, H., Zijlstra, C., Leeuwangh, J. E., Krens, F. A., and Huizing, H. J., Reproduction of the beet cyst nematode *Heterodera schachtii* Schm. on transformed root cultures of *Beta vulgaris* L., *Plant Cell Rep.,* 6, 379, 1987.
18. Kumar, A. and Forrest, J. M. S., Reproduction of *Globodera rostochiensis* on transformed roots of *Solanum tuberosum* cv. Desiree, *J. Nematol.,* 22, 395, 1990.

19. Orion, D., Wergin, W. P., and Endo, B. Y., Inhibition of syncytia formation and root-knot nematode development on cultures of excised tomato roots, *J. Nematol.,* 12, 196, 1980.
20. Mountain, W. B., A method of culturing plant parasitic nematodes under sterile conditions. *Proc. Helminthol. Soc. Wash.,* 22, 49, 1955.
21. Tiner, J. D., Cultures of the plant parasitic nematode genus *Pratylenchus* on sterile excised roots. I. Their establishment and maintenance, *Exp. Parasitol.,* 9, 121, 1960.
22. Dropkin, V. H. and Boone, W. R., Analysis of host-parasite relationships of root-knot nematodes by single-larva inoculations of excised tomato roots, *Nematologica,* 12, 225, 1966.
23. Huettel, R. N., Monoxenic culturing of plant parasitic nematodes using carrot discs, callus tissue and root-explants, in *Plant Nematology Laboratory Manual,* revised ed., Zuckerman, B. M., Mai, W. F., and Krusberg, L. R., Eds., Agricultural Experiment Station, University of Massachussetts, Amherst, 1990, 163.
24. Moriarty, F., The monoxenic culture of beet eelworm (*Heterodera schachtii* Schm) on excised roots of sugar beet (*Beta vulgaris* L.), *Parasitology,* 54, 289, 1964.
25. Byars, L. P., Preliminary notes on the cultivation of the plant parasitic nematode *Heterodera radicicola, Phytopathology,* 4, 323, 1914.
26. Müller, J., L'Élevage monoxénique d'*Heterodera schachtii* sur crucifères et son application pour la sélection des plantes résistentes, *Rev. Nematol.,* 1, 47, 1978.
27. Lauritis, J. A., Rebois, R. V., and Graney, L. S., Life cycle of *Heterodera zeae* Koshy, Swarup, and Sethi on *Zea mays* L. axenic root explants, *J. Nematol.,* 15, 115, 1983.
28. Lauritis, J. A., Rebois, R. V., and Graney, L. S., Development of *Heterodera glycines* Ichinohe on soybean, *Glycine max* (L.) Merr., under gnotobiotic conditions, *J. Nematol.,* 15, 272, 1983.
29. Dropkin, V. H. and Webb, R. E., Resistance of axenic tomato seedlings to *Meloidogyne incognita acrita* and to *M. hapla, Phytopathology,* 57, 584, 1967.
30. Dropkin, V. H., Davis, D. W., and Webb, R. E., Resistance of tomato to *Meloidogyne incognita acrita* and to *M. hapla* (root knot nematodes) as determined by a new technique, *Proc. Am. Soc. Hortic. Sci.,* 90, 316, 1967.
31. Lauritis, J. A., Rebois, R. V., and Graney, L. S., Screening soybean for resistance to *Heterodera glycines* Ichinohe using monoxenic cultures, *J. Nematol.,* 14, 593, 1982.
32. Rebois, R. V. and Huettel, R. N., Population dynamics, root penetration, and feeding behavior of *Pratylenchus agilis* in monoxenic root cultures of corn, tomato and soybean, *J. Nematol.,* 18, 392, 1986.
33. Webb, R. M., Effects of the nematode *Pratylenchus fallax* on roots of oilseed rape (*Brassica napus* var. *oleifera*), *Rev. Nematol.,* 13, 115, 1990.
34. McClure, M. A. and Viglierchio, D. R., Penetration of *Meloidogyne incognita* in relation to growth and nutrition of sterile, excised cucumber roots, *Nematologica,* 12, 237, 1966.
35. McClure, M. A. and Viglierchio, D. R., The influence of host nutrition and intensity of infection on the sex ratio and development of *Meloidogyne incognita* in sterile agar cultures of excised cucumber roots, *Nematologica,* 12, 248, 1966.
36. Johnson, R. N. and Viglierchio, D. R., Sugar beet nematode (*Heterodera schachtii*) reared on axenic *Beta vulgaris* root explants. I. Selected environmental factors affecting penetration, *Nematologica,* 15, 129, 1969.
37. Johnson, R. N. and Viglierchio, D. R., Sugar beet nematode (*Heterodera schachtii*) reared on axenic *Beta vulgaris* root explants. II. Selected environmental and nutritional factors affecting development and sex-ratio, *Nematologica,* 15, 144, 1969.
38. Reise, R. W., Huettel, R. N., and Sayre, R. M., Carrot callus tissue for culture of endoparasitic nematodes, *J. Nematol.,* 19, 387, 1987.
39. Riedel, R. M., Establishing *Pratylenchus* spp. in monoxenic culture on alfalfa (*Medicago sativa*) callus tissue, in *Plant Nematology Laboratory Manual,* Zukerman, B. M., Mai, W. F., and Harrison, M. B., Eds., Agricultural Experiment Station, University of Massachussetts, Amherst, 1985, 159.
40. Bingefors, S. and Bingefors, S., Rearing stem nematode inoculum for plant breeding purposes, *Swed. J. Agric. Res.,* 6, 13, 1976.
41. Faulkner, L. R., Bower, D. B., Evans, D. W., and Elgin, J. H., Jr., Mass culturing of *Ditylenchus dipsaci* to yield large quantities of inoculum, *J. Nematol.,* 6, 126, 1974.
42. O'Bannon, J. H. and Taylor, A. L., Migratory endoparasitic nematodes reared on carrot discs, *Phytopathology,* 58, 385, 1968.

43. Moody, E. H., Lownsbery, B. F., and Ahmed, J. M., Culture of the root-lesion nematode *Pratylenchus vulnus* on carrot disks, *J. Nematol.,* 5, 225, 1973.
44. Lawn, D. A. and Noel, G. R., Gnotobiotic culture of *Pratylenchus scribneri* on carrot discs, *Nematropica,* 16, 45, 1986.
45. Kaplan, D. T. and Davis, E. L., Improved nematode extraction from carrot disk culture, *J. Nematol.,* 22, 339, 1990.
46. Verdejo, S. and Pinochet, J., Reproduction of the nematode *Zygotylenchus guevarai* (Nemata: Pratylenchidae) in monoxenic cultures, *Rev. Nematol.,* 14, 317, 1991.
47. Verdejo-Lucas, S. and Pinochet, J., Population densities of five migratory endoparasitic nematodes in carrot disk cultures, *J. Nematol.,* 24, 96, 1992.
48. Tefer, D., The biology of genetic transformation of higher plants by *Agrobacterium rhizogenes,* in *Molecular Genetics of the Bacteria-Plant Interaction,* Pühler, A., Ed., Springer-Verlarg, Berlin, 1983, 248.
49. Mugnier, J., Establishment of new axenic hairy root lines by inoculation with *Agrobacterium rhizogenes, Plant Cell Rep.,* 7, 9, 1988.
50. Murashige, T. and Skoog, F., A revised medium for rapid growth and bioassay with tobacco tissue cultures, *Physiol. Plant.,* 15, 473, 1962.
51. Hooykaas, P. J. J., Klapwijk, P. M., Nuti, M. P., Schilperoort, R. A., and Rörsch. A., Transfer of the *Agrobacterium tumefaciens* TI plasmid to virulent *Agrobacteria* and to *Rhizobium* ex planta, *J. Gen. Microbiol.,* 98, 477, 1977.
52. Peerbolte, R., Krens, F. A., Mans, R. M. W., Floor, M., Hoge, J. H. C., Wullems, G. J., and Schilperoort, R. A., Transformation of plant protoplasts with DNA: co-transformation of non-selected calf thymus carrier DNA and meiotic segregation of transforming DNA sequences, *Plant Mol. Biol.,* 5, 235, 1985.
53. Mugnier, J., Transport of the nematicide oxamyl in root transformed with *Agrobacterium rhizogenes, Ann. App. Nematol.,* 2, 29, 1988.
54. Verdejo, S. and Jaffee, B. A., Reproduction of *Pasteuria penetrans* in a tissue-culture system containing *Meloidogyne javanica* and *Agrobacterium rhizogenes*-transformed roots, *Phytopathology,* 78, 1284, 1988.
55. Chen, T., Kilpatrick, R. A., and Rich, A. E., Sterile culture techniques as tools in plant nematology research, *Phytopathology,* 51, 799, 1961.
56. Verdejo, S. and Mankau, R., Culture of *Pasteuria penetrans* and *Meloidogyne incognita* on oligoxenic excised tomato root culture, *J. Nematol.,* 18, 635, 1986.
57. Loewenberg, J. R., Sullivan, T., and Schuster, H. L., Gall induction by *Meloidogyne incognita* by surface feeding and factors affecting the behavior pattern of the second stage larvae, *Phytopathology,* 50, 322, 1960.
58. Glazer, I. and Orion, D., Influence of urea, hydroxyurea, and thiourea on *Meloidogyne javanica* and infected excised tomato roots in culture, *J. Nematol.,* 16, 125, 1984.
59. Glazer, I., Orion, D., and Apelbaum, A. Interrelationships between ethylene production, gall formation, and root-knot nematode development in tomato plants infected with *Meloidogyne javanica, J. Nematol.,* 15, 539, 1983.
60. Lauritis, J. A., Rebois, R. V., and Graney, L. S., Technique for gnotobiotic cultivation of *Heterodera glycines* Ichinoche on *Glycine max* (L.) Merr., *J. Nematol.,* 14, 422, 1982.
61. Lownsbery, B. F., Huang, C. S., and Johnson, R. N., Tissue culture and maintenance of the root-lesion nematode *Pratylenchus vulnus, Nematologica,* 13, 390, 1967.
62. Towson, A. J. and Lear, B., Effect of temperature on reproduction and motility of *Pratylenchus vulnus, J. Nematol.,* 14, 602, 1982.
63. Tiner, J. D., Cultures of the plant-parasitic nematode genus *Pratylenchus* on sterile excised roots. II. A trap for collection of axenic nematodes and quantitative initiation of experiments, *Exp. Parasitol.,* 11, 231, 1961.
64. Webb, R. M., Extraction of nematodes from sterile culture, *Nematologica,* 17, 173, 1971.
65. Ko, M. P. and Van Gundy, S. D., An alternative gelling agent for culture and studies of nematodes, bacteria, fungi and plant tissues, *J. Nematol.,* 20, 478, 1988.
66. Eyre, M. J. and Caswell, E. P., Sterile culture of *Rotylenchulus reniformis* on tomato root with gellan gum as a supporting medium, *J. Nematol.,* 23, 229, 1991.
67. Sijmons, P. C., Grundler, F. M. W., von Mende, N., Burrows, P. R., and Wyss, U., *Arabidopsis thaliana* as a new model host for plant-parasitic nematodes, *Plant J.,* 1, 245, 1991.

68. Orion, D. and Bar Eyal, M., The etiology of *Helicotylenchus multicintus* on *Arabidopsis thaliana* root culture, *Nematropica,* 22, 132, 1992.
69. Plowright, R. A. and Akehurst, T. E., Monoxenic culture of the Ufra nematode *Ditylenchus angustus, Fundam. Appl. Nematol.,* 15, 327, 1992.
70. Ou, X. and Watson, A. K., In vitro culture of *Subanguina picridis* in *Acroptilon repens* callus, excised roots and shoot tissues, *J. Nematol.,* 24, 199, 1992.
71. Westcott, S. W., III and Hussey, R. S., Feeding behavior of *Criconemella xemoplax* in monoxenic cultures, *Phytopathology,* 82, 936, 1992.
72. Verdejo-Lucas, S., unpublished data.

FURTHER READING

Dropkin, V. H. and Boone, W. R., Analysis of host-parasite relationships of root-knot nematodes by single-larva inoculations of excised tomato roots, *Nematologica,* 12, 225, 1966.

Estey, R. H., Culturing bacteriophagous nematodes, in *Plant Nematology, Laboratory Manual,* Zuckerman, B. M., Mai, W. F., and Krusberg, L. R., Eds., revised ed., Agricultural Experiment Station, University of Massachusetts, Amherst, 1990, 155.

Estey, R. H., Culturing mycophagous nematodes, in *Plant Nematology, Laboratory Manual,* Zuckerman, B. M., Mai, W. F., and Krusberg, L. R., Eds., revised ed., Agricultural Experiment Station, University of Massachusetts, Amherst, 1990, 159.

Huettel, R. N., Monoxenic culturing of plant parasitic nematodes using carrot discs, callus tissue and root-explants, in *Plant Nematology, Laboratory Manual,* Zuckerman, B. M., Mai, W. F., and Krusberg, L. R., Eds., revised ed., Agricultural Experiment Station, University of Massachussetts, Amherst, 1990, 163.

Krusberg, L. R., Studies on the culturing and parasitism of plant-parasitic nematodes, in particular *Ditylenchus dipsaci* and *Aphelenchoides ritzemabosi* on alfalfa tissues, *Nematologica,* 6, 181, 1961.

Riedel, R. M., Alves de Lima, M., and Martin, M., Establishment of nematode germplasm banks, in *Handbook of Plant Cell Culture,* Vol. 1, *Techniques for Propagation and Breeding,* Evans, D. A., Sharp, W. R., Ammirato, P. V., and Yamada, Y., Eds., Macmillan, New York, 1983, 880.

Tefer, D., Transformation of several species of higher plants by *Agrobacterium rhizogenes:* sexual transmission of the transformed genotype and phenotype, *Cell,* 37, 959, 1984.

Chapter 22

Protoplast Culture: Viruses and Viroids*

Abed Gera

TABLE OF CONTENTS

I. INTRODUCTION

Isolated protoplasts have been described as "naked" plant cells because the cell wall has been experimentally removed by either a mechanical or an enzymatic digestion.[1] They provide a unique system for clarifying the interactions between a plant virus and its host at the cellular level. It was in 1960 that a cell wall-degrading enzyme was successfully used by Cocking[2] to isolate protoplasts from tomato. The crucial advance in the field came in 1968 when Takebe et al.[3] were able to solve the fundamental difficulties in preparing protoplasts from leaf mesophyll and gave strong evidence for substantial tobacco mosaic virus (TMV) replication resulting from infection with TMV[3–5] and TMV-RNA.[6] Subsequent to

* Contribution from the Agricultural Research Organization, The Volcani Center, Bet Dagan, Israel. No. 3587-E, 1992 Series.

0-87371-877-1/95/$0.00+$.50

infection by TMV, a rod-shaped virus, mesophyll protoplasts have been successfully infected with spherical viruses, such as cucumber mosaic virus (CMV),[7] filamentous viruses such as potato virus X (PVX),[8] multicomponent viruses such as brome mosaic virus (BMV),[9,10] and rhadoviruses such as sonchus yellow net virus (SYNV).[11]

Recently, mesophyll protoplasts have been successfully inoculated with a number of viruses with unique characteristics. Transient replication of several whitefly-transmitted geminiviruses has been demonstrated in protoplasts, including bean golden mosaic virus (BGMV),[12] African cassava mosaic virus (ACMV),[13] beet curly top virus,[14] tomato golden mosaic virus (TGMV),[15] and wheat dwarf virus (WDV).[16] Protoplast infection systems have been developed for aphid-borne, phloem-limited viruses such as tobacco necrotic dwarf virus[17] and barley yellow dwarf virus (BYDV).[18] Infection of protoplasts with viroids has also been achieved.[19]

Based on these findings, protoplast technology has become a standard technique in almost all plant virus laboratories for investigating the transcriptional, translational, and replicative events occurring during viral infection. Recently, routine and efficient protocols for gene transfer into protoplasts have been developed.[20] This enables the structure and function of isolated genes to be studied via modification and subsequent analysis of their function in transformed protoplasts.

This article reviews the current status of the use of protoplasts in plant virology. As there are several comprehensive reviews on this subject,[21–24] only information directly related to the methodology of protoplast isolation, infection with plant viruses, and culture is included.

II. PROTOPLAST ISOLATION

A. LEAF MESOPHYLL PROTOPLASTS

Protoplasts can be isolated from a variety of plant tissues and cultured cells. Leaf mesophyll cells have been used most frequently for protoplast isolation. Methods for protoplast isolation and culture including *Nicotiana* species have recently been reviewed.[25] Methods for protoplast isolation have varied enormously among plant species, mainly, it appears, because of the physiological conditions of the source tissue or plant. This phenomenon of variation among plants highlights the importance of optimizing growth conditions of the donor plant. The nutrition and age of the plants, as well as the temperature and illumination under which they are grown, directly affect the quality of isolated protoplasts and their suitability for further growth and regeneration.[26–31] Since all plants have different growth requirements, no optimal conditions can be stated; these parameters must be determined experimentally. For the isolation of mesophyll protoplasts from leaves, a general rule is to keep the plants under rapid growth with sufficient nutrients and to collect the upper, almost fully expanded leaves.

B. CELL CULTURE PROTOPLASTS

Protoplasts have been isolated from a variety of callus[32] and suspension cultures.[33] The growth conditions of these cells are also very important for protoplast isolation. In general, a rapidly growing culture in the exponential phase of growth is best for protoplast isolation. Cell culture has several advantages over leaves as a source of protoplasts. The cells and tissues are grown under aseptic and controlled physiological and environmental conditions, enabling uniform and contamination-free protoplasts to be obtained consistently and in large amounts. Furthermore, protoplasts derived from an *in vitro* system can be considered as better adapted to being cultured *in vitro* than leaf protoplasts.[33]

C. MECHANICAL METHODS

Early attempts to isolate plant protoplasts relied entirely on mechanical methods and were limited to tissue containing large and vacuolated cells, but the yields were usually insufficient.[34] These methods have neither improved nor become popular. However, for certain physiological studies, where side effects of degrading enzymes have to be avoided, these methods are still used.

D. ENZYMATIC METHODS

The routine isolation of protoplasts has become possible owing to the availability of a number of potentially powerful microbial hydrolases. The enzymatic isolation of protoplasts can be performed in two different ways: the two-step (or sequential) method and the one-step method.

1. Enzyme Used for Protoplast Isolation

The pioneering work of Cocking[2] stimulated rapid progress in the enzymatic isolation of plant protoplasts. With the availability of a number of commercial cell wall-degrading enzymes, the preparation of

Table 1 **Commercial hydrolases most commonly used for isolation of protoplasts**

Enzyme	Concentration	Supplier
Cellulases	0.2–2%	
Onozuka RS		Yakult Honsha Co. Ltd., Higashishinbashi, Tokyo, Japan
Onozuka R-10		Yakult Honsha Co. Ltd., Higashishinbashi, Tokyo, Japan
Cellulysin		Calbiochem, San Diego, CA
Cellulase		Sigma, St. Louis, MO
Pectinases	0.02–0.2%	
Macerozyme R-10		Yakult Honsha Co. Ltd., Higashishinbashi, Tokyo, Japan
Macerase		Calbiochem, San Diego, CA
Pectinase		Sigma, St. Louis, MO
Pectolyase Y-23		Seishin Pharm. Co. Ltd., Nihonbashi, Tokyo, Japan

protoplasts from various types of plant cells has been reported.[32,33] Commercially available enzymes and the common concentrations used for protoplast isolation are listed in Table 1.

The cellulases, which degrade the cell wall, have been used most frequently for protoplast isolation. The most popular is cellulase Onozuka, which derived from the fungus *Trichoderma viride.* Cellulase Onozuka RS is derived from a mutant strain of the fungus that has a stronger cellulase activity and shortens the time required for protoplast preparation. The pectinases, which dissolve the middle lamella, are also used in most protoplast isolation procedures. The most frequently used pectinases are macerozyme and pectolyase. More frequently, pectinases are combined with cellulases, a technique which apppears to be useful for many types of tissues. With the addition of Pectolyase Y-23, a more potent pectinase, higher yields of protoplasts from tobacco mesophyl[35] and tobacco suspension were obtained.[36]

The direct different combinations of cellulases and pectinases, as well as the osmolarity of the enzyme solution, are important factors for the preparation of protoplasts.

a. Sequential or Two-Step Procedure

The method is based mainly on the procedure of Takebe et al.[3] According to this procedure, the lower epidermis is removed from the surface-sterilized leaves. The exposed mesophyll is dissociated into single cells using pectinase and the cell wall is digested by cellulase. Addition of potassium dextran sulfate considerably enhances the separation of cells and the stability of the isolated protoplasts.[37,38]

b. Mixed-Enzyme Method or One-Step Procedure

Several methods have been reported for the direct isolation of protoplasts with a mixture of pectinase and cellulase. Usually, an empirical approach is required to develop an experimental protocol for protoplast isolation of specific plant species. Many laboratories have developed standard enzyme mixtures for routine protoplast isolation. When the appropriate enzyme mixture and concentrations have been determined, it does seem possible to establish a successful routine procedure. The procedure routinely used in our laboratory since 1978 for isolation of mesophyll protoplast of *Nicotiana* spp. is summarized in Table 2. The procedure involves the following four steps: (1) sterilization of leaves; (2) peeling off the lower epidermis; (3) enzymatic digestion; (4) isolation and cleaning of protoplasts.

Table 2 **Isolation of mesophyll protoplasts from *nicotiana* species**

1) Fully expanded leaves of tobacco are collected and surface sterilized in 1% sodium hypochlorite for 5 min, then rinsed with sterile distilled water.

2) The lower epidermis is peeled off and placed in a preplasmolyzing solution of 13.5% (w/v) mannitol in large sterile petri dishes for 1 to 2 h.

3) Leaf pieces are treated with 0.2 to 0.5% cellulase and 0.02 to 0.05% macerozyme in 13.5% mannitol, adjusted to pH 5.8, and incubated overnight at 25°C in darkness.

4) Enzyme solution containing the protoplasts is filtered through a nylon mesh (50 to 100 μm). Protoplasts are collected and purified by centrifuging (100 g for 5 min) and washing with 13.5% mannitol.

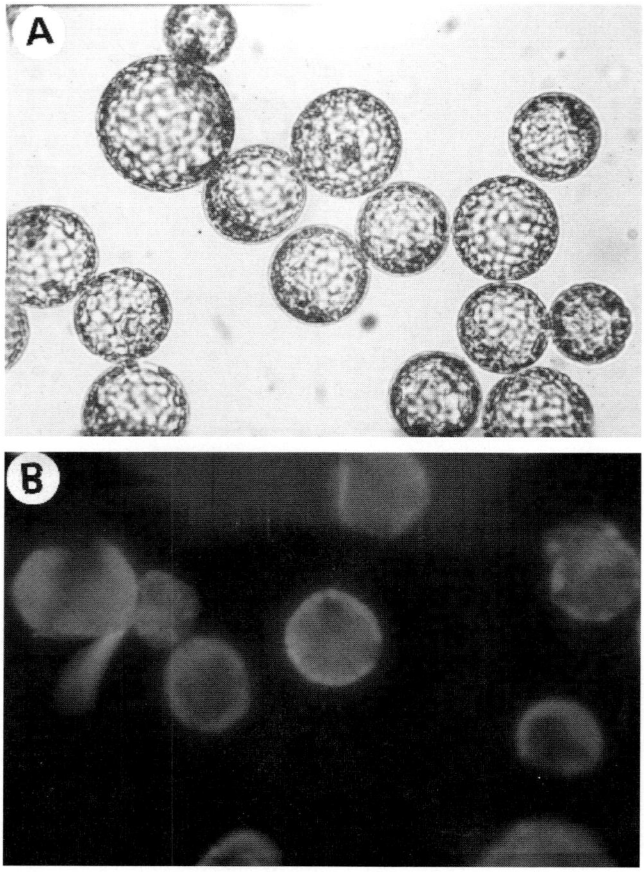

Figure 1 (A) Protoplasts of *Nicotiana tabaccum* isolated by a one-step procedure and suspended in 0.7 *M* mannitol. (B) Fluorescence of tobacco mesophyll protoplasts 72 h after inoculation with tobacco mosaic virus. The viral antigen combines with FITC-conjugated antibody.

The method is simpler than the two-step procedure and permits the isolation of 1 to 2 $\times$ 10^7 mesophyll protoplasts from 1 g of mature tobacco leaves (Figure 1A). For practical reasons we prefer the long enzyme treatment (overnight) in order to have the protoplasts available in the morning.[39,40]

Alternatively, we have successfully used the combination of 0.03% Pectolyase Y-23 and 0.3% Cellulase R-10 in order to release leaf mesophyll protoplasts in 3 to 4 h.[41]

E. OSMOTICUM

The cell wall that is enzymatically digested away during protoplast isolation normally provides support for the cell. Protoplasts released directly into standard cell culture medium will plasmolyze or burst. Hence the pressure that would be mechanically sustained by the plant cell wall must be balanced by the addition of an osmotic stabilizer.

The osmotic pressure of the protoplast medium is manipulated by the addition of sugars to the isolation and culture medium used for protoplasts. Mannitol and sorbitol are the most frequently used for leaf mesophyll protoplasts. Mannitol is not metabolized by the protoplast and infuses slowly into it. Glucose and sucrose are often used as an osmoticum for cultured cells.[42] The optimal osmotic potential varies with the source and growth conditions of donor plants. In general, optimum osmolarity ranges from 0.3 to 0.7 *M*. Increasing mannitol concentrations in the incubation medium of isolated protoplasts from two tobacco cultivars reduces virus replication,[41] probably owing to decreasing RNA and protein synthesis with increasing osmotic pressure.[43] The addition of low concentrations of potassium chloride is recommended to stabilize protoplast membranes.[44]

F. PROTOPLAST PURIFICATION

Protoplasts are usually purified following enzymatic digestion, by a combination of filtration, centrifuging, and washing. The enzyme solution containing the protoplasts is filtered through a stainless steel or nylon mesh (50 to 100 μm) to remove larger portions of undigested tissue and cell clumps. The

Table 3 **Protocol for the inoculation of protoplasts using the poly-L-ornithine procedure**

1. Preparation of inoculum: mix together the following in a 100-ml Erlenmeyer flask, all in 0.02 *M* potassium citrate pH 5.2 or 0.05 *M* phosphate pH 5.8 containing 13.5% mannitol: 2 μg/ml PLO (mol. wt. 120,000), 2 μg/ml TMV, purified by density-gradient centrifugation, in a total volume of 10 ml. Incubate at 25°C for 10 min with gentle agitation.
2. Inoculation: pour freshly sedimented protoplasts (5×10^6) in 10 ml citrate or phosphate (containing mannitol) into the inoculum solution. Incubate at 25°C for 10 min with gentle agitation.
3. Washing: collect protoplasts by centrifugation (100 g for 5 min) and carefully wash twice with 13.5% mannitol containing 10 m*M* $CaCl_2$.
4. Culture: resuspend inoculated protoplasts in 10 ml culture medium in Erlenmeyer flasks at a density of 10^5/ml. Incubate at 25°C under continuous illumination.

remaining mixture of protoplasts, small debris, and enzymes is then centrifuged (100 g for 5 min) to precipitate protoplasts while debris continues to float. The enzymes are removed and fresh medium is added before centrifuging. Protoplasts are washed in isolation solution and then floated on 20 to 23% sucrose for further purification. Centrifuging at about 60 g for 5 min gives a band of protoplasts floating on the top. The band is easily sucked off with a Pasteur pipette and is then placed in mannitol culture medium.[39,40]

Sorbitol has been used in place of sucrose for maize protoplasts.[45] Ficoll (polysucrose) and Percoll gradients have also been used to purify protoplasts from various sources.[46,47]

Protoplasts are examined for viability before subsequent manipulation.

G. PROTOPLAST INOCULATION

Infection by plant viruses is initiated by virus particles which enter host cells through wounds in the cell wall. Since most plant viruses are vector transmitted, it was hypothesized that the encapsulated particles are introduced directly into the cytoplasm and that no virus-specific receptor sites exist on the plasmalemma.

Infection of protoplasts by plant viruses was achieved by treating freshly isolated protoplasts with a virus or viral RNA inoculum in a buffered mannitol solution in the presence of a high-molecular weight polycation (Table 3). Poly-L-ornithine (PLO), a linear polymer of a basic amino acid, is most widely used for this purpose.

Chemical, physical, and electrical methods for inoculation of plant protoplasts with viruses are discussed below.

1. Poly-L-Ornithine (PLO)

The addition of PLO, a macromolecular polycation, has been successfully used for the inoculation of protoplasts from a wide range of plant species with a wide range of viruses, varying from rigid-rod types to flexous and multicomponent viruses.[48] The presence of PLO is an absolute requirement for high frequency of infection of protoplasts by TMV,[49] CMV,[7] PVX,[50] cowpea chlorotic mottle (CCMV),[51] and alfalfa mosaic (AMV)[52] viruses charged negatively at inoculation pHs. PLO is not essential for infection of protoplasts by pea enation mosaic[53] and BMV[9] viruses charged positively at inoculation pHs.

The actual mechanism of the stimulation of the virus infection by PLO has not yet been satisfactorily clarified. One possible function of PLO is to neutralize or even reverse the surface charge of virus particles. This would facilitate their adsorption onto the protoplast surfaces and may also stimulate virus entry into protoplasts.

The optimum concentration of PLO is 1 μg/ml in most of the protoplast infection systems. Concentrations of PLO higher than 2 μg/ml reduce the viability of protoplasts from a wide range of plant species. Preincubation of virus particles with PLO for 5 to 10 min before the subsequent incubation with protoplast improves virus infection of protoplasts.[21] This finding indicates that time must be provided for binding of the polycation to the virus particles.

Batches of PLO are not homogeneous in molecular size and electrophoretic behavior. There is evidence that the molecular weight is important for successful infection. PLO of mol. wt. 12,000 or 50,000 was much less effective in aiding infection by TMV than was PLO of mol. wt. 120,000.[54]

2. Polyethylene Glycol (PEG)

An alternative procedure for protoplast inoculation is to use PEG, an agent developed for protoplast fusion studies.[55] The method was first developed for the initiation of TMV infection in tomato protoplasts,[30,56] more recently it has been widely used for inoculation of protoplasts with different viruses and viral RNA.[22]

The mechanism of infection induced by PEG is not understood. It has been suggested that for successful infection, virus must first be adsorbed on the protoplast surface followed by aggregation of the protoplasts.

The optimal concentration of PEG (mol. wt. 6,000) was 13.3 m*M*; higher concentrations resulted in a decline of protoplast viability.[57]

Other polycations such as poly-L-lysine, poly-L-arginine,[58] and polyethyleneimine (Polyamin P) have been successfully used to stimulate virus uptake by plant protoplasts.[59]

3. Mild Sonication

Recently, Joersbo and Brunstedt[60] have developed a novel and efficient method for transfer of plasmid DNA into plant protoplast. Transient expression of chloramphenicol acetyltransferase (CAT) gene on sugar beet and tobacco protoplasts was obtained by brief exposure of the protoplasts to 20 kHz ultrasound in the presence of plasmid DNA. This method could be of considerable value in plant transformation.

4. Liposomes

Liposomes are artificial lipid membranes surrounding an aqueous phase.[61] The solution is mixed either with cochleate cylinders of phospholipid to give large unilamellar vesicles, or with a solution of lipid in ether, which is evaporated to give reverse-phase evaporation vesicles.[36] These liposomes can interact with protoplasts and deliver the encapsulated material into the cell. This ability makes them a promising tool for introducing nucleic acids into protoplasts. In almost all cases liposome delivery was found to be absolutely dependent on the presence of polyvinyl alcohol or PEG,[62] agents which are known to induce protoplast fusion.

5. Electroporation

Electroporation, also called electrotransfection, is a gene-transfer method in which the cell membrane is punctured with direct-pulse electric current and nucleic acids are taken up into the cell through the resulting pore (see Chapter by Mendel and Hänsch, this volume). Electrotransfection of plant protoplasts was first reported by Fromm et al.,[63] who showed that chimeric plasmids containing the CAT gene were introduced into plant protoplasts by means of a high-voltage electric pulse and were expressed transiently in protoplasts. Several reports on electrotransfection of plant protoplasts with viral nucleic acids have been published.[64] On the other hand, it has been reported that virus particles could also infect protoplasts under the influence of an electric field.[65–67] Electrotransfection with virus particles was effective with positively, but not with negatively, charged viruses.[65,66]

6. Inoculation Buffer and pH Value

The composition of the inoculation buffer and its pH value influence the infection of protoplasts. Citrate buffer, 0.2 *M,* pH 5.2, was the most commonly used.[50] Phosphate buffer, 0.05 *M,* pH 5.8, has been found to be more effective for many virus-protoplast systems.[68–70] Tris-chloride at pH 7 to 8.8 was found to be much more effective than citrate or phosphate for infecting barley protoplasts with BMV.[71]

The optimal pH for protoplast infection depends on the net charge of virus particles as well as of protoplasts.[71] Lower pH values at about pH 5 are the most commonly used for many virus-protoplast systems.[68]

The concentration of the protoplasts in the inoculation mixture is very important. Protoplast concentration should not exceed 2×10^5/ml. At higher population density the level of infection decreases markedly.[72]

For unknown reasons, it has been noticed that protoplasts remaining for some time in suspension medium are less susceptible to infection than freshly resuspended protoplasts.

H. CULTURE OF INFECTED PROTOPLASTS

For the incubation of the inoculated protoplasts to observe virus replication or other metabolic activities, most workers use the liquid incubation medium of Aoki and Takebe[6] with minor modification (Table

Table 4 **Medium commonly used for culturing virus-inoculated protoplasts**

KH_2PO_4	0.2 m*M*	Mannitol	0.7 *M*
KNO_3	1 m*M*	Carbenicillin	200 μg/ml
$MgSO_4$	1 m*M*	Mycostatin	10 μg/ml
$CaCl_2$	10 m*M*	pH	5.4
KI	1 μ*M*		
$CuSO_4$	0.01 μ*M*		

4). This contains some inorganic salts, but no metabolizable carbon source, so that protoplasts neither divide nor synthesize cell walls. Growth substances such as kinetin and 2,4-dichlorophenoxyacetic acid (2,4-D) were used in the original medium. TMV multiplication was markedly enhanced in protoplasts from two hypersensitive tobacco cultivars by postinoculation addition of 2 4-D to the incubation medium, whereas in protoplasts from a systemic tobacco cultivar TMV replication was decreased.[39]

Protoplasts are generally incubated in Erlenmeyer flasks or petri dishes under continuous illumination of about 3000 lux, at a temperature of 25 to 28°C. Protoplasts can be subjected to other temperature and light conditions. Virus replication proceeds satisfactorily when infected protoplasts are cultured in darkness.[69]

Infection of protoplasts by plant DNA viruses requires cell division in order to support virus replication.[13–16] This can be achieved by using richer media, containing auxins and cytokinins, such as those developed by Nagata and Takebe.[73]

I. DETERMINATION OF VIRUS REPLICATION

The replication of virus within protoplasts may be demonstrated and quantified by using any one of several methods: infectivity assays, electron microscopy, serology, and hybridization.

1. Bioassay

Bioassay procedures have been the traditional methods for detection and diagnosis of diseases caused by viruses and viroids. Inoculation of indicator plants with aliquots of disrupted inoculated protoplasts produces a symptom within a few days or weeks and provides proof of the presence of complete virions. As a routine procedure, protoplasts are collected by low-speed centrifugation, disrupted by homogenization, and tested for infectivity. Preferably, the homogenate is assayed on half-leaves of a local lesion host and compared with a standard concentration of a purified virus preparation on the opposite half of each leaf. The number of lesions produced by the protoplast extract is tested for a series of dilutions. By comparison with the lesions produced by a standard dilution of purified virus preparation, the concentration of virus per milliliter of extract can be estimated.[39,40] In the absence of a local lesion host, the dilution point giving a 50% infection of a systemic host can be determined.

2. Electron Microscopy

Each plant virus has a characteristic size and shape. This property is used to assist in identification of viruses and is also useful in routine detection of virus particles in a homogenate of infected protoplasts. Electron microscopy has also been used for studying ultrastructural changes induced by virus infection in protoplasts.[74]

3. Serology

Immunological diagnosis of viruses has been used for many years; it relies on the use of prepared antibodies specific to the viral coat protein.

4. Fluorescent Antibody Technique

The level of infection in terms of the percentage of the protoplasts which become infected can be conveniently determined by an immunofluorescence technique. The accumulated virus antigen in infected protoplasts is stained by antibodies conjugated with a fluorescent dye. For this purpose the protoplasts are usually fixed with 3% gluteraldehyde on glass slides, and then stained with viral antibody conjugated with fluorescein isothiocyanate (FITC). The virus-infected protoplasts show a specific fluorescence owing to the virus accumulated within them, and can be readily distinguished under a fluorescence microscope from noninfected ones, which do not contain fluorescent material (Figure 1B).

This method is the most commonly employed to quantify the efficiency of *in vitro* inoculation. In some cases where the nonspecific background of healthy protoplasts is too high, the use of an immunoperoxidase technique has been reported.[75]

5. Enzyme-Linked Immunosorbent Assay (ELISA)

ELISA has become the single most important procedure for plant virus detection and quantitative determination.[76] The method utilizes the wells of a microtiter plate made of polystyrene. The wells are filled with the appropriate antivirus gamma globulin (IgG) which is irreversibly adsorbed onto the surface. After the plate has been rinsed, aliquots of disrupted inoculated protoplasts are added to each well. Viral antigen is specifically trapped by these antibodies. The antigen-antibody complex is detected by the addition of an antibody-enzyme conjugate that gives a colored product in the presence of an appropriate substrate. The intensity of this color reaction is measurable photometrically. The method is simple, rapid and sensitive, and is finding increasing application.

6. Hybridization

Hybridization analysis is a new method that has been applied to the detection of plant viruses and viroids (see chapter by Nikolaeva, this volume). Although the method is more complicated and requires more time than the ELISA assay, it may be useful with viruses, in special circumstances, and offers a means for detection of viroids that cannot be detected by conventional serological methods.

The technique is based on incubation of single-stranded complementary DNA (ss cDNA) with ssRNA under conditions where the two molecules anneal to form a DNA-RNA hybrid. Hybridization analysis requires preparation of highly radioactive ^{32}P-labeled cDNA to the viroid or viral RNA.

Aliquots of disrupted virus-infected protoplasts are immobilized on a nitrocellulose membrane and the membrane is exposed to a prehybridization mixture followed by the ^{32}P-cDNA probe. After washing the membrane to remove the unhybridized probe, the membrane is autoradiographed using X-ray film.

7. *In Situ* Hybridization

Recently, the intercellular localization of potato spindle tuber viroid RNA in tomato protoplasts by *in situ* hybridization using a cDNA probe has been reported.[77] The technique also enabled viral RNA in tobacco protoplasts to be localized.

III. APPLICATION IN PLANT VIROLOGY

Protoplast systems have been extensively used for studying virus functions. Protoplast systems have been used in the kinetic analysis of nucleic acid, protein, and virion synthesis. Since these investigations have been reviewed recently in detail,[21–24] only a few examples of special uses of protoplasts in plant virology will be discussed.

A. PROTOPLASTS FROM COAT PROTEIN TRANSGENIC PLANTS

"Coat protein (CP)-mediated resistance" refers to the resistance caused by the expression of a virus coat protein gene in transgenic plants. Accumulation of the CP confers resistance to infection and/or disease development by the virus from which the CP gene was derived and by related viruses.[78] Protoplasts were isolated from CP transgenic plants and used for transfection studies. Use of protoplasts has yielded important information about early events leading to CP-mediated resistance. Loesch-Fries et al.[79] reported resistance against AMV in protoplasts isolated from transgenic tobacco plants that expressed AMV CP (+), but not against AMV-RNA. Likewise, Register and Beachy[80] reported less resistance against TMV-RNA and TMV that was briefly treated at pH 8.0 than against TMV in CP (+) protoplasts. Recently, transgenic plants containing one of six different forms of tobacco etch virus (TEV) CP nucleotide sequence have been generated.[81] Truncated forms of the TEV CP tended to confer greater protection than the full-length CP when transgenic plant lines were mechanically inoculated with TEV. In protoplast transfection studies, all lines that produced CP supported viral replication, whereas some lines that expressed high levels of sense or antisense RNA did not support replication at wild-type levels.

To address the mechanism of protection against TMV in CP (+) trangenic tobacco, a transient protection assay has been developed.[82] In this study it was shown that introduction of purified TMV

CP into tobacco protoplasts induced a transient protection to challenge virus introduced concomitantly or shortly thereafter.

B. INFECTION OF PROTOPLASTS USING RNA TRANSCRIPTS

The strategy of using infectious RNA transcripts from cDNA clones to investigate the molecular biology of RNA viruses has been used successfully for several plant viruses.[83] The development of protoplast systems has provided a reliable and versatile assay for infection using transcripts from cloned cDNA of different viruses.

Infectious RNA transcripts of cDNA clones not only provide an unlimited source of genetic material, but also provide for insights into viral gene expression and function using site-directed mutagenesis.

Recently it has been shown that transcripts derived from full-length cDNA copies of the genomic RNA of different viruses were infectious in isolated protoplasts, leading to the accumulation of CP, synthesis of viral RNAs, and the production of virus particles. This was first achieved with BMV,[84] followed by cowpea mosaic virus (CPMV),[85] BYDV,[86] CMV,[87] and beet western yellows virus.[88]

The availability of infectious RNA transcripts opens the way to studying the expression and genetic organization of plant viruses *in vivo* by expression of RNA transcripts obtained from copies engineered at specific sites.

C. INFECTION OF PROTOPLAST WITH DNA VIRUSES

The use of protoplasts derived from different plant species for studying geminiviruses has been investigated. Geminiviruses are unique infectious agents characterized by a twinned particle structure and a single-stranded DNA genome. They can be divided into two subgroups. The viruses of one subgroup are leafhopper transmitted and have monopartite genomes. Members of the second subgroup are restricted to dicotyledonous hosts, are transmitted by whiteflies, and have a bipartite genome.

The dicot infection geminiviruses, BGMV and ACMV, have been shown to replicate in *Phaseolus vulgaris* and *N. plumbaginafolia* protoplasts.[12,13] WDV has been shown to replicate in protoplasts derived from a *Triticum monococcum* suspension culture.[16] TGMV has been found to replicate efficiently on *N. tabaccum* protoplasts.[15] The protoplast system was further used to determine the effects of mutation in different viral genes on the accumulation of TGMV DNA and CP in infected cells. When the β-glucuronidase reporter gene was used to replace the CP gene (AR_1) of TGMV, it was shown that the viral AL_2 gene product transactivates expression of the CP AR_1 in tobacco protoplasts.[89] Recently, it has been shown that AL_2 also transactivates the BR_1 gene and that transactivation occurs at the level of transcription.[89a] The kinetics of viral replication and CP expression of TGMV were analyzed in *Nicotiana* protoplasts.[90] The system was further used to assess the effect of cytosine methylation on the replication of TGMV. Replacement of cytosine residues with 5-methylcytosine reduced the amount of viral DNA which accumulated in transfected protoplasts.[91]

T. monococcum protoplasts were used to study the replication and expression of WDV replicon in protoplasts demonstrate the potential of the WDV genome as a plant gene vector for monocotyledonous plants. This potential is emphasized further by the observation that such a WDV-based vector can be also used in cells derived from nonhost plants.[92]

Previous work on the replication of cassava latent virus component 1 in mesophyll protoplasts of *N. plumbaginifolia* led to the suggestions that viral DNA is linked to the cell division of the host cell.[12] Similar observations were made with WDV in protoplasts of *T. monococcum* suspension culture.[13]

Recently, the whitefly-transmitted geminivirus, tomato yellow leaf curl virus, has been shown to replicate in *N. tabaccum* protoplasts. The production of viral double-stranded DNA (dsDNA) forms consistent with the replication of the viral genome has been observed.[93]

D. RESISTANCE AND VIRUS INHIBITORS

Host specificity of viruses is an interesting question that was approached through the use of the protoplasts in the study of the biochemical basis of resistance to viruses. This approach has been used to determine whether resistance or apparent immunity to virus infection exists in isolated protoplasts. In general, it appears that if a plant is a host for a certain virus, then protoplasts isolated from this plant can be infected too. If the plant is a nonhost then protoplasts are usually nonhost. However, some plants of this category become infected only when the infection pressure is high. Furusawa and Okuno[94] found that mesophyll protoplasts isolated from Japanese radish, a nonhost for BMV, could be infected with a standard strain of BMV. Recently, Matsunaga et al.[95] showed that when protoplasts isolated from tomato, (a typical host for TMV), wild tomato (a conditional resistant host with the Tm-2 gene), barley (a subliminal infection host), and chrysanthemum (an apparent nonhost) were electrotransfected with TMV-RNA, viruses multiplied at the same rate. These results indicate the lack of host specificity in the initially infected cells of these plants.

In the case of potato, there are conflicting results concerning the expression of genes resistant to PVX at the single-cell level.[96–98] Potato protoplasts derived from susceptible and immune genotypes were infected *in vitro* with a purified PVX virion preparation for comparative analysis of replication. PVX multiplication rates in the immune genotype were about 15 times slower than in the susceptible genotype when virus concentrations were in the range of 0.01 to 0.1 ng per viable protoplast. However, when inoculum levels were raised to 1 ng per viable protoplast, PVX multiplication was about the same in all genotypes. This indicates that although the gene confers resistance at the cell level, it requires the tissue structure for full expression of immunity.[99]

Similar types of studies have been used to evaluate the nature of interactions between barley stripe mosaic virus (BSMV) and apparently resistant hosts. Protoplasts derived from susceptible barley or oat were susceptible to BSMV. More than 80% of protoplasts derived from an oat cultivar resistant to BSMV were readily infected by the virus, but protoplasts from ten barley lines resistant to BSMV remained resistant to the virus, although a limited number of protoplasts were infected. These results suggest that resistance in these barley lines may be the result of restriction of replication, whereas resistance in oat plants is more likely to be due to restriction of cell-to-cell movement.[100]

In a survey of over 1000 lines of cowpea, Breier et al.[101] found 65 lines which were immune to CPMV. However, isolated protoplasts from almost all these lines could support virus multiplication effectively. Protoplasts isolated from the immune line "Arlington" support very low levels of CPMV multiplication.[102,103] However, cowpea severe mosaic virus was able to overcome the resistance of "Arlington" cowpea.[104]

The Tm-1 gene, originally derived from wild species of tomatoes, provides resistance to infection by TMV. This gene strongly inhibits synthesis of both viral RNA and proteins in tomato protoplasts as well as in tomato plants.[105] Thus, this gene is manifested at the single-cell level. Watanabe et al.[106] showed that the synthesis of all known viral-coded proteins and RNAs was inhibited in protoplasts containing the Tm-1 gene, but that the resistance could be partly overcome in heterozygous plants if TMV RNA at high concentrations was used as inoculum.

The necrotic response of resistant tobacco plants cannot be assayed with protoplast systems, since this type of response is not expressed in protoplasts. Differences in TMV replication between protoplasts from necrotic and systemic hosts have been observed.[39] A substance, inhibitor of virus replication (IVR), is released into the medium from TMV-infected protoplasts of resistant tobacco cultivars. IVR was neither produced in protoplasts from susceptible plants nor from noninoculated protoplasts of the resistant cultivar.[40,107] Actinomycin D and chloramphenicol markedly increased TMV replication in protoplasts from resistant cultivars. Concomitantly with the increase in virus replication, production of IVR from these protoplasts was suppressed almost completely.[108] These results strengthen the suggestion that IVR is associated with the localization mechanism by suppression of virus replication. Recently, IVR has also been obtained from the intercellular fluid of resistant tobacco cultivars infected with TMV and from induced resistant tissue.[109] The biological activity of IVR was found to be associated with a specific 23-K protein.[110] Molecular studies on IVR can contribute to the understanding of the resistance mechanisms in plants.

IV. ADVANTAGES AND LIMITATIONS

Plant and protoplast cultures have become a major tool in the study of an increasing number of fundamental and applied problems in the plant sciences. These cultures have become an integral part of plant biotechnology research. They provide important tools to clarify many details of host-cell-pathogen interactions at the cell level.

The very few examples of the use of protoplasts as tools in plant virus research discussed above may demonstrate the usefulness of the system in studies of the molecular events in plant virus replication. The ability to infect protoplasts synchronously enables plant virologists to carry out one-step growth experiments. Besides improved synchrony of infection, protoplasts have several advantages: (i) The high proportion of protoplast that can be infected and the high efficiency achieved; (ii) close control of experimental conditions; (iii) the ease of labeling newly synthesized RNAs and proteins with radioisotopes; (iv) antimetabolites may act more effectively on protoplasts than on plant tissue; and (v) uniform sampling.

One should be aware, however, that isolated protoplasts constitute an artificial system, because they are removed from their natural environment by a traumatic enzymatic treatment and are kept under conditions which are completely different from those within the leaf. A number of actual or potential limitations must be kept in mind: (i) protoplasts are very fragile; (ii) the quantitative variation in behavior among batches of protoplasts; (iii) protoplasts survive for only 3 or 4 d and then decline and die; (iv) antibiotics added to the incubation medium may have unexpected effects on virus replication; and (v) experience is needed to obtain consistently good results.

ACKNOWLEDGMENTS

This work was supported in part by the United States-Israel (Binational) Agricultural Research and Development Fund (BARD).

My sincere thanks are due to Prof. G. Loebenstein for his support and encouragement. I am also very grateful to Prof. D. M. Bisaro, Ohio State University, and Dr. T. Hibi, National Institute of Agrobiological Resources, Japan, for providing me with information in press or prior to publication.

REFERENCES

1. Cocking, E. C., Plant cell protoplasts, isolation and development, *Annu. Rev. Plant Physiol.,* 23, 29, 1972.
2. Cocking, E. C., A method for the isolation of plant protoplasts and vacuoles, *Nature,* 187, 927, 1960.
3. Takebe, I., Otsuki, Y., and Aoki, S., Isolation of tobacco mesophyll cells in intact and active state, *Plant Cell Physiol.,* 9, 115, 1968.
4. Hibi, T., Yora, K., and Asuyama, H., TMV infection of Petunia protoplasts, *Ann. Phytopathol. Soc. Jpn.,* 34, 375, 1968.
5. Cocking, E. C. and Ponjar, E., An electron microscopic study of the infection of isolated tomato fruit protoplasts by tobacco mosaic virus, *J. Gen. Virol.,* 4, 305, 1969.
6. Aoki, S. and Takebe, I., Infection of tobacco mesophyll protoplasts by tobacco mosaic virus ribonucleic acid, *Virology,* 39, 439, 1969.
7. Otsuki, Y. and Takebe, I., Infection of tobacco mesophyll protoplasts by cucumber mosaic virus, *Virology,* 52, 433, 1973.
8. Shalla, T. A. and Petersen, L. J., Infection of isolated plant protoplasts with potato virus X, *Phytopathology,* 63, 1125, 1973.
9. Motoyoshi, F., Bancroft, J. B., and Watts, J. W., The infection of tobacco protoplasts with a variant of brome mosaic virus, *J. Gen. Virol.,* 25, 31, 1974.
10. Okuno, T., Furusawa, I., and Hiruki, C., Infection of barley protoplasts with brome mosaic virus, *Phytopathology,* 67, 610, 1977.
11. Jones, R. W. and Jackson, A. O., Replication of Sonchus yellow net virus in infected protoplasts, *Virology,* 179, 815, 1990.
12. Haber, S., Ikegami, M., Bajet, N. B., and Goodman, R. M., Evidence for a divided genome in bean golden mosaic virus, a geminivirus, *Nature,* 289, 324, 1981.
13. Townsend, R., Watts, J., and Stanley, J., Synthesis of viral DNA forms in *Nicotiana plumbaginafolia* protoplasts inoculated with cassava latent virus (CLV): evidence for independent replication of one component of the CLV genome, *Nucleic Acids Res.,* 14, 1253, 1986.
14. Briddon, R. W., Watts, J., Markham, P. G., and Stanley, J., The coat protein of beet curly top virus is essential for infectivity, *Virology,* 172, 628, 1989.
15. Bisaro, D. M., Sunter, G., Revington, G. N., Brough, C. L., Hormuzdi, S. G., and Hartitz, M., Molecular genetics of tomato golden mosaic virus replication: progress toward defining gene functions, transcription units and the origin of DNA replication, in *Viral Genes and Plant Pathogenesis,* Pirone, T. P. and Shaw, J. G., Eds., Springer-Verlag, New York, 1990, 89.
16. Woolston, C. J., Reynolds, H. V., Stacey, N. J., and Mullineaux, P. M., Replication of wheat dwarf virus DNA in protoplasts and analysis of coat protein mutants in protoplasts and plants, *Nucleic Acids Res.,* 17, 6029, 1989.

17. Kubo, S. and Takanami, Y., Infection of tobacco mesophyll protoplasts with tobacco necrotic dwarf virus, a phloem-limited virus, *J. Gen. Virol.,* 42, 387, 1979.
18. Young, M. J., Larkin, P. J., Miller, W. A., Waterhouse, P. M., and Gerlach, W. L., Infection of *Triticum monococcum* protoplasts with barley yellow dwarf virus, *J. Gen. Virol.,* 70, 2245, 1989.
19. Muhlbach, H.-P., Camacho-Henriquez, A., and Sanger, H. L., Infection of tomato protoplasts by ribonucleic acid of tobacco mosaic virus and by viroids, *Phytopathol. Z.,* 90, 289, 1977.
20. Potrykus, I., Petruska, J., Paszkowski, J., and Shillits, R. D., Direct gene transfer into protoplast of graminaceous monocot, *Mol. Gen. Genet.,* 199, 183, 1985.
21. Takebe, K., Protoplasts in plant virus research, *Int. Rev. Cytol., Suppl.,* 16, 89, 1973.
22. Sander, E. and Mertes, G., Use of protoplasts and separate cells in plant virus research, *Adv. Virus Res.,* 29, 215, 1984.
23. Harrison, B. C. and Mayo, M. A., The use of protoplasts in plant virus research, in *Use of Tissue Culture and Protoplasts in Plant Pathology,* Helgeson, J. P. and Deverall, B. J., Eds., Academic Press, New York, 1983, 69.
24. Motoyoshi, F., Protoplasts in virology, in *Plant Protoplasts,* Fowke, L. C. and Constable, F., Eds., CRC Press, Boca Raton, FL, 1985, 139.
25. Evans, D. A. and Bravo, J. E., Plant protoplast isolation and culture, *Int. Rev. Cytol. Suppl.,* 16, 33, 1983.
26. Arcioni, S., Davey, M. R., Santos, A. V. P. dos, and Cocking, E. C., Somatic embryogenesis in tissues from mesophyll and cell suspension protoplasts of *Medicago coerulea* and *N. glutinosa, Z. Pflanzenphysiol.,* 106, 105, 1982.
27. Saxena, P. K., Gill, R., Rashid, A., and Maheshwari, S. C., Plantlet formation from isolated protoplasts of *Solanum melongena* L., *Protoplasma,* 106, 355, 1981.
28. Bhatt, D. P. and Fassuliotis, D., Plant regeneration from mesophyll protoplasts of eggplant, *Z. Pflanzenphysiol.,* 104, 81, 1981.
29. Gunn, R. E. and Shepard, J. F., Regeneration of plants from mesophyll derived protoplasts of British potato (*Solanum tuberosum* L.), *Plant Sci. Lett.,* 22, 97, 1981.
30. Cassels, A. C. and Barlass, M., Method for the isolation of stable mesophyll protoplasts from tomato leaves throughout the year under standard conditions, *Physiol. Plant,* 42, 236, 1978.
31. Cassels, A. C. and Cocker, F. M., Seasonal and physiological aspects of the isolation of tobaccc protoplasts, *Physiol. Plant,* 56, 69, 1982.
32. Vasil, I. K., The progress, problems, and prospects of plant protoplast research, in *Advances in Agronomy,* Vol. 28, Academic Press, New York, 1976, 119.
33. Vasil, I. K. and Vasil, V., Isolation and culture of protoplasts, in *Perspectives in Plant Cell and Tissue Cultures,* Vasil, I. K., Ed. (Int. Rev. Cytol., 1, Suppl. 11B), Academic Press, New York, 1980.
34. Klercker, J., Eine Metode zur Isolterung lebender Protoplasten, *Svensk. Vet'Akad, Forh.,* 9, 463, 1892.
35. Nagata, T. and Ishill, S., A rapid method for the isolation of mesophyll protoplasts, *Can. J. Bot.,* 57, 1820, 1979.
36. Nagata, T., Okada, K., Takebe, I., and Matsui, C., Delivery of tobacco mosaic virus RNA into plant protoplasts mediated by reverse-phase evaporation vesicles (liposomes), *Mol. Gen. Genet.,* 184, 161, 1981.
37. Sink, K. C. and Niedz, R. P., Factors controlling plating efficiency in tomato, in *Proc. 5th Int. Cong. Plant Tissue and Cell Cultures,* Fujiwara, A., Ed., IAPTC, Maruzen Co., Tokyo, 1982, 583.
38. Mii, M. and Cheng, S.-M., Callus and root formation of mesophyll protoplasts of carnation, in *Proc. 5th Int. Cong. Plant Tissue and Cell Cultures,* Fujiwara, A., Ed., IAPTC, Maruzen Co., Tokyo, 1982, 585.
39. Loebenstein, G., Gera, A., Barnett, A., Shabtai S., and Cohen, J., Effect of 2,4-dichlorophenoxyacetic acid on multiplication of tobacco mosaic virus in protoplasts from local-lesion and systemic-responding tobaccos, *Virology,* 100, 110, 1980.
40. Loebenstein, G. and Gera, A., Inhibitor of virus replication released from tobacco mosaic virus-infected protoplasts of a local lesion-responding tobacco cultivar, *Virology,* 114, 132, 1981.
41. Shabtai, S., Gera, A., and Loebenstein, G., Partial suppression of tobacco mosaic virus multiplication in protoplasts at increasing mannitol concentrations, *Plant Sci. Lett.,* 24, 157, 1982.
42. Kao, K. B. and Michayluk, M. R., Plant regeneration from mesophyll protoplasts of alfalfa, *Z. Pflanzenphysiol.,* 96, 135, 1980.

43. Premecz, G., Ruzicska, P., Olah, T., and Farkas, G. L., Effect of "osmotic stress" on protein and nucleic acid synthesis in isolated tobacco protoplasts, *Planta,* 141, 33, 1978.
44. Binding, H., Nehls, R., Kock, R., Finger, J., and Mordhorst, G., Comparative studies on protoplast regeneration in herbaceous species of the Dicotyledoneae class, *Z. Pflanzenphysiol.,* 101, 119, 1981.
45. Kannai, R. and Edwards, G. E., Purification of enzymatically isolated mesophyll protoplasts from C_3, C_4 and crassulacean acid metabolism plants using an aqueous dextran polyethylene glycol two-phase system, *Plant Physiol.,* 52, 484, 1973.
46. Harms, C. T. and Potrykus, I., Enrichment for heterokaryocytes by the use of iso-osmotic density gradients after plant protoplast fusion. *Theor. Appl. Genet.,* 53, 49, 1978.
47. Scowcroft, W. R. and Larkin, P. J., Isolation culture and plant regeneration from protoplasts of *Nicotiana debneyi, Aust, J. Plant Physiol.,* 7, 635, 1980.
48. Cassels, A. C., Uptake of viruses by plant protoplasts and their use as transforming agents, in *Plant Protoplasts and Genetic Engineering,* Vol. 2, Bajaj, Y. P. S., Ed. (Biotechnology in Agriculture and Forestry, Vol. 9), Springer-Verlag, Berlin, 1989, 390.
49. Takebe, I. and Otsuki, Y., Infection of tobacco mesophyll protoplasts by tobacco mosaic virus, *Proc. Natl. Acad. Sci. U.S.A.,* 64, 843, 1969.
50. Otsuki, Y., Takebe, I., Honda, Y., Kajita, S., and Matsui, C., Infection of tobacco mesophyll protoplasts by potato virus X, *J. Gen. Virol.,* 22, 375, 1974.
51. Motoyoshi, F., Bancroft, J. B., Watts, J. W., and Burgess, J., The infection of tobacco protoplasts with cowpea chlorotic mottle virus and its RNA, *J. Gen. Virol.,* 20, 177, 1973.
52. Motoyoshi, F., Hull, R., and Flack, I. H., Infection of tobacco mesophyll protoplasts by alfalfa mosaic virus, *J. Gen. Virol.,* 27, 263, 1975.
53. Motoyoshi, F. and Hull, R., The infection of tobacco protoplasts with pea enation mosaic virus, *J. Gen. Virol.,* 24, 89, 1974.
54. Gera, A., unpublished data, 1980.
55. Kao, K. N. and Michayluk, M. R., A method for high-frequency intergeneric fusion of plant protoplasts, *Planta,* 115, 355, 1974.
56. Cassells, A. C. and Barlass, M., The initiation of TMV infection in isolated protoplasts by polyethylene glycol, *Virology,* 87, 459, 1978.
57. Muhlbach, H. P., The use of protoplasts in plant virus research, in *Protoplasts 1983—Lecture Proceedings,* Potrykus, K., Harms, C. T., Hinnen, A., Hutter, R., King, P. J., and Shillito, R. D., Eds., Birkhauser, Basel, 1983, 111.
58. Motoyoshi, F., Watts, J. W., and Bancroft, J. B., Factors influencing the infection of tobacco protoplasts by cowpea chlorotic mottle virus, *J. Gen. Virol.,* 25, 245, 1974.
59. Kikkawa, H., Nagata, T., Matsui, C., and Takebe, I., Infection of protoplasts from tobacco suspension cultures by tobacco mosaic virus, *J. Gen. Virol.,* 63, 451, 1982.
60. Joersbo, M. and Brunstedt, J., Direct gene transfer to plant protoplasts by mild sonication, *Plant Cell Rep.,* 9, 207, 1990.
61. Bangham, A. D., Standish, M. M., and Watkins, J. C., Diffusion of univalent ion across the lamellae of swollen phospholipids, *J. Mol. Biol.,* 13, 238, 1965.
62. Makins, J. F., The theory and practical applications of liposome-protoplast interactions, in *Protoplasts 1983—Lecture Proceedings,* Potrykus, K., Harms, C. T., Hinnen, A., Hutter, R., King, P. J., and Shillito, R. D., Eds., Birkhauser, Basel, 1983, 197.
63. Fromm, I., Taylor, P., and Walbot, V., Expression of genes transferred into monocot and dicot plant cells by electroporation, *Proc. Natl. Acad. Sci. U.S.A.,* 82, 5824, 1985.
64. Hibi, T., Electrotransfection of plant protoplasts with viral nucleic acids, *Adv. Virus Res.,* 37, 329, 1989.
65. Okada, K., Nagata, T., and Takebe, I., Introduction of functional RNA into plant protoplasts by electroporation, *Plant Cell Physiol.,* 27, 619, 1986.
66. Watts, J. M., King, J. M., and Stacey, N. J., Inoculation of protoplasts with viruses by electroporation, *Virology,* 157, 40, 1987.
67. Nishigushi, M., Sato, T., and Motoyoshi, F., An improved method for electroporation in plant protoplasts: infection of tobacco protoplasts by tobacco mosaic virus particles, *Plant Cell Rep.,* 6, 90, 1987.
68. Takebe, I., Protoplasts in the study of plant virus replication, in *Comprehensive Virology,* Vol. 11, Fraenkel-Conrat, H. and Wagner, R. R., Eds., Plenum Press, New York, 1977, 237.
69. Jarvis, N. P. and Murakishi, H. H., Infection of protoplasts from soybean cell culture with southern bean mosaic and cowpea mosaic virus, *J. Gen. Virol.,* 48, 365, 1980.

70. Barker, H. and Harrison, B. D., Infection of tobacco mesophyll protoplasts with raspberry ringspot virus alone and together with tobacco rattle virus, *J. Gen. Virol.,* 35, 125, 1977.
71. Okuno, T. and Furusawa, I., Modes of infection of barley protoplasts with brome mosaic virus, *J. Gen. Virol.,* 38, 409, 1978.
72. Lesney, M. S. and Murakishi, H. H., Infection of soybean protoplasts from cell suspension culture with bean pod mottle virus, *J. Gen. Virol.,* 57, 387, 1981.
73. Nagata, T. and Takebe, I., Cell wall regeneration and cell division in isolated tobacco mesophyll protoplasts, *Planta,* 92, 301, 1970.
74. Murphy, J. F., Jarlfors, U., and Shaw, J. G., Development of cylindrical inclusions in potyvirus-infected protoplasts, *Phytopathology,* 81, 371, 1991.
75. Ben-sin, C. and Po Tien, Infection of barley protoplasts with barley stripe mosaic virus detected and assayed by immunoperoxidase, *J. Gen. Virol.,* 58, 323, 1982.
76. Clark, M. F. and Adams, A. N., Characteristics of the microplate method of enzyme-linked immunosorbent assay for the detection of plant viruses, *J. Gen. Virol.,* 34, 475, 1977.
77. Yokoyama, M., Ogawa, M., Nozu, Y., and Hashimoto, J., Detection of specific RNAs by *in situ* hybridization in plant protoplasts, *Plant Cell Physiol.,* 31, 403, 1990.
78. Beachy, R. N., Loesch-Fries, L. S., and Tumer, N. E., Coat protein-mediated resistance against virus infection, *Annu. Rev. Phytopathol.,* 28, 451, 1990.
79. Loesch-Fries, L. S., Halk, E., Merlo, D., Jarvis, N., Nelson, S., et al., Expression of alfalfa mosaic virus coat protein gene and anti-sense cDNA in transformed tobacco tissue, in *Molecular Strategies for Crop Protection,* Arntzen, C. J. and Ryan, C. A., Eds., Alan R. Liss, New York, 1987, 221.
80. Register, J. C. and Beachy, R. N., Resistance to TMV in transgenic plants results from interference with an early event in infection, *Virology,* 166, 524, 1988.
81. Lindbo, J. A. and Dogherty, W. G., Pathogen-derived resistance to a potyvirus: immune and resistant phenotypes in transgenic tobacco expressing altered forms of a potyvirus coat protein nucleotide sequence, *Mol. Plant-Microbe Interact.,* 5, 144, 1992.
82. Register, J. C. and Beachy, R. N., Effect of protein aggregation state on coat protein-mediated protection against tobacco mosaic virus using a transient protoplast assay, *Virology,* 173, 656, 1989.
83. Dawson, W. O., Beck, D. L., Knorr, D. A., and Grantham, G. L., cDNA cloning of the complete genome of tobacco mosaic virus and production of infectious transcripts, *Proc. Natl. Acad. Sci. U.S.A.,* 83, 1832, 1986.
84. Ahlquist, P., French, R., and Bujarski, J. J., Molecular studies of brome mosaic virus using infectious transcripts from cloned cDNA, *Adv. Virus Res.,* 32, 215, 1987.
85. Vos, P., Jaegle, M., Wellink, J., Verver, J., Eggen, R., van Kammen, A., and Goldbach, R., Infectious RNA transcripts derived from full-length DNA copies of the genomic RNAs of cowpea mosaic virus, *Virology,* 165, 33, 1988.
86. Young, M. J., Kelly, L., Larkin, P. J., Waterhouse, P. M., and Gerlach, W. L., Infectious *in vitro* transcripts from a cloned cDNA of barley yellow dwarf virus, *Virology,* 180, 372, 1991.
87. Rizzo, I. M. and Palukaitis, P., Construction of full-length cDNA clones of cucumber mosaic virus RNAs 1, 2 and 3: generation of infectious RNA transcripts, *Mol. Gen. Genet.,* 222, 249, 1990.
88. Veidt, I., Bouzoubaa, S. E., Leiser, R.-M., Ziegler-Graff, V., Builley, H., Richards, K., and Jonard, G., Synthesis of full-length transcripts of beet western yellows virus RNA: messenger properties and biological activity in protoplasts, *Virology,* 186, 192, 1992.
89. Sunter, G. and Bisaro, D. M., Transactivation in a geminivirus: AL2 gene product is needed for coat protein expression, *Virology,* 180, 416, 1991.
89a. Bisaro, D. M., personal communication.
90. Brough, C. L., Sunter, G., Gardiner, W. E., and Bisaro, D. M., Kinetics of tomato golden mosaic virus DNA replica and coat protein promoter activity in *Nicotiana tabacum* protoplasts, *Virology,* 187, 1, 1992.
91. Brough, C. L., Gardiner, W. E., Inamdar, N. M., Xian-Yang Zhang, Ehrlich, M., and Bisaro, D. M., DNA methylation inhibits propagation of tomato golden mosaic virus DNA in transfected protoplasts, *Pl. Mol. Biol.,* 18, 703, 1992.
92. Matzeit, V., Schaefer, S., Kammann, M., Schal, H.-J., Schell, J., and Gronenborn, B., Wheat dwarf virus vectors replicate and express foreign genes in cells of monocotyledonous plants, *Plant Cell,* 3, 247, 1991.
93. Gera, A. and Antignus, Y., unpublished data, 1992.

94. Furusawa, I. and Okuno, T., Infection with BMV of mesophyll protoplasts isolated from five plant species, *J. Gen. Virol.,* 40, 489, 1978.
95. Matsunaga, R., Sawamura, K., De Kok, M., Makino, T., Miki, K., Kojima, M., Tsuchizaki, T., and Hibi, T., Electrotransfection of protoplasts from tomato, wild tomato, barley and chrysanthemum with tobacco mosaic virus RNA, *J. Gen. Virol.,* 73, 763, 1992.
96. Foxe, M. J. and Prakash, J., Expression of resistance to potato virus X in isolated potato protoplasts, *Arch. Virol.,* 88, 167, 1986.
97. Adams, S. E., Jones, R. A. C., and Coutts, R. H. A., Infection of protoplasts derived from potato shoot cultures with potato virus X, *J. Gen. Virol.,* 66, 1341, 1985.
98. Adams, S. E., Jones, R. A. C., and Coutts, R. H. A., Expression of potato virus X resistance gene Rx in potato leaf protoplasts, *J. Gen. Virol.,* 67, 2341, 1986.
99. Saladrigas, M. V., Ceriani, M. F., Tozzini, A. C., Arese, A. I., and Hopp, H. E., Potato gene X^i confers inoculum dependent resistance to potato virus X replication in protoplasts, *Plant Cell Physiol.,* 31, 749, 1990.
100. Zheng, Y. and Edwards, M. C., Expression of resistance to barley stripe mosaic virus in barley and oat protoplasts, *J. Gen. Virol.,* 71, 1865, 1990.
101. Beier, H., Siler, D. J., Russell, M. L., and Bruening, G., Survey of susceptibility to cowpea mosaic virus among protoplasts and intact plants from *Vigna sinensis* lines, *Phytopathology,* 67, 917, 1977.
102. Beier, H., Bruening, G., Russell, M. L., and Tucker, C. L., Replication of cowpea mosaic virus in protoplasts isolated from immune lines of cowpeas, *Virology,* 95, 165, 1979.
103. Kiefer, M. C., Bruening, G., and Russell, M. L., RNA and capsid accumulation in cowpea protoplasts that are resistant to cowpea strain SB, *Virology,* 137, 71, 1984.
104. Bruening, G., Ponz, F., Glascock, C., Russell, M. L., Rowhani, A., and Chay, C., Resistance of cowpeas to cowpea mosaic virus and to tobacco ringspot virus, in *Plant Resistance to Viruses, Ciba Foundation Symposium 133.* Evered, D. and Harnett, S., Eds., John Wiley & Sons, Chichester, England, 1987, 23.
105. Motoyoshi, F. and Oshima, N., Expression of genetically controlled resistance to tobacco mosaic virus in isolated tomato leaf mesophyll protoplasts, *J. Gen. Virol.,* 34, 499, 1977.
106. Watanabe, Y., Kishibayashi, N., Motoyoshi, F., and Okada, Y., Characterization of Tm-1 gene action on replication of common isolates and a resistance-breaking isolate of TMV, *Virology,* 161, 527, 1987.
107. Gera, A. and Loebenstein, G., Further studies of an inhibitor of virus replication from tobacco mosaic virus-infected protoplasts of a local lesion-responding tobacco cultivar, *Phytopathology,* 73, 111, 1983.
108. Gera, A., Loebenstein, G., and Shabtai, S., Enhanced tobacco mosaic virus production and suppressed synthesis of a virus inhibitor in protoplasts exposed to antibiotics, *Virology,* 127, 475, 1983.
109. Spiegel, S., Gera, A., Salomon, R., Ahl, P., Harlap, S., and Loebenstein, G., Recovery of an inhibitor of virus replication from the intercellular fluid of hypersensitive tobacco infected with tobacco mosaic virus and from uninfected induced-resistant tissue, *Phytopathology,* 79, 258, 1989.
110. Gera, A., Loebenstein, G., Salomon, R., and Franck, A., Inhibitor of virus replication from protoplasts of a hypersensitive tobacco cultivar infected with tobacco mosaic virus is associated with a 23-K protein species, *Phytopathology,* 80, 78, 1990.

FURTHER READING

Cassels, A. C., Uptake of viruses by plant protoplasts and their use as transforming agents, in *Plant Protoplasts and Genetic Engineering,* Vol. 2, Bajaj, Y. P. S., Springer-Verlag, Berlin, 1989, 390.

Garg, G. K., Singh, U. S., Khetrapal, R. K., and Kumar, J., Application of tissue culture in plant pathology, in *Experimental and Conceptual Plant Pathology,* Vol. *Techniques,* Singh, R. S., Singh, U. S., Hess, W. M., and Weber, D. J., Eds., Gordon and Breach Science Publishers, New York, 1988, 83.

Matsunaga, R., Sawamura, K., De Kok, M., Makino, T., Miki, K., Kojima, M., Tsuchizaka, T., and Hibi, T., Electrotransfection of protoplasts from tomato, wild tomato, barley and chrysanthemum with tobacco mosaic virus RNA, *J. Gen. Virol.,* 73, 763, 1992.

Motoyoshi, F., Protoplasts in virology, in *Plant Protoplasts,* Fowke, L. C. and Constable, F., Eds., CRC Press, Boca Raton, FL, 1985, 139.

Sander, E. and Mertes, G., Use of protoplasts and separate cells in plant virus research, *Adv. Virus Res.,* 29, 215, 1984.

Watts, J. M., King, J. M., and Stacey, N. J., Inoculation of protoplasts with viruses by electroporation, *Virology,* 157, 40, 1987.

Chapter 23

ELISA Methodology

R. K. Khetarpal and C. A. Kumar

TABLE OF CONTENTS

I. INTRODUCTION

Immunoassays can be characterized as quantitative analytical methods applied for measuring biologically important compounds/organisms using antibodies as specific analytical reagents. They are based on the unique recognition reaction between antibodies and antigens which elicit their production. Antibodies are specific binding proteins functioning in the natural defense mechanisms of animals against foreign intruders. An antigen is an immunogenic compound which can elicit a strong immune response in an immunized animal. An immunogenic antigen can be a peptide, protein, polysachharide, polynucleotide, or almost any polymeric compound containing functional groups on its surface recognized by antibody producing β-lymphocytes.

Antibodies are able to recognize and bind to a defined epitopic site on an antigen which forms the basis of their specificity. Accordingly, immunoassays are defined as being structurally specific assays in contrast to bioassays, which are functionally specific methods. The bioassays, in general, are too laborious and time-consuming to be widely adopted for routine analysis. The conventional serological methods of immunoprecipitation, immunodiffusion, and agglutination are either not very sensitive and/ or are practically inconvenient for large scale testing in routine.

The introduction of reporter groups attached to one of the components of immunologic reaction has made the monitoring of the binding reaction easier and more sensitive. During the evolution of immunoassays, the advent of radio-isotopic labels, with the potent sensitivity they achieve, has had an enormous impact on biomedical research and clinical practice. The evolution of radioimmunoassays (RIA) and most of the basic technologies and performance characteristics of today's immunoassays were originally defined with the use of radioisotopic tracers.

Even though RIAs have been the method of choice as sensitive and robust techniques, there has been a growing interest for nonisotopic alternatives since the late 1970s. Some of the drawbacks related to the use of radioisotopic labels and the desire to develop easy and rapid homogeneous assays have been the driving force towards finding non-radioisotopic challengers such as enzymes. This has lead to the development of enzyme-linked immunosorbent assay, popularly known as ELISA.

Enzyme-labeled antibodies have been used for some years in the detection of various antigens in tissue sections,[1,2] but their use in quantitative procedures is relatively recent.[3,4] Voller et al.[5] introduced the microplate method of ELISA which has subsequently been used for diagnosing a wide variety of antigens.[6–8]

0-87371-877-1/95/$0.00+$.50

II. PRINCIPLES

The ELISA procedure completely differs from the classical serological techniques of immunoprecipitation, immunodiffusion, and agglutination wherein a considerable degree of polymerization is needed before the threshold of visual detection is reached. In case of ELISA, the use of an enzyme marker linked to the antibody greatly increases the detection of a specific antigen.

The different types of enzyme immunoassays used in the field of diagnostics fall into two major groups: (1) *homogeneous assays* which generally are restricted to molecules of low molecular weight such as drugs, haptens, hormones, etc. and (2) *heterogeneous assays* which are suitable for detecting macromolecules and plants or animal pathogens.[9] In case of heterogeneous immunoassays, the reacting and non-reacting components are separated, the antigen is immobilized on a solid surface (in wells of microtiter plates or on nitrocellulose membranes), and the various reactants are present at the reaction site in a pre-defined sequence. Between every two steps of the sequence is the washing phase which removes the unwanted inhibitory substances from the reaction site and only the specifically immobilized reactants are retained. Of the various kinds of heterogeneous assays, the most commonly used for detecting plant pathogens is the double antibody sandwich (DAS) procedure first described in detail for plant viruses by Clark and Adams in 1977.[10] Since then, various modifications of this basic procedure have been described. However, the same principles of operation apply to all the immunosorbent assays.

Based on the enzyme-labeled antibody employed, the ELISA procedure can be termed as *direct* or *indirect.* In case of direct procedure, the antigen trapped on the solid phase is detected with an enzyme-labeled specific homologous antibody. The *indirect* ELISA involves the targeting of the trapped antigen by unconjugated specific (homologous) antibody which in turn is detected by an enzyme-labeled anti-immunoglobulin molecule which is commercially available. To explain further, if the specific antibody was produced in rabbit, then the antispecies antibody such as goat anti-rabbit immunogolobulin conjugated to an enzyme is used for detection purposes. Besides, protein-A conjugate is also used in indirect procedures.[11] Another indirect system using an initial protein-A coating (PAC) for binding and orientation of virus specific antibodies added in the subsequent step was used.[12] The protein-A binds to the immunoglobulins at the Fc region, leaving the F (ab′)2 region available for antigen binding. The bound antigen is then detected by another layer of antibodies, which in turn are detected by enzyme conjugated protein-A.

The choice of a particular ELISA procedure to be adopted is based on thorough understanding of merits and demerits of a procedure and the type of investigations to be carried out. The use of crude specific antiserum and the commercially available enzyme conjugate (universal enzyme conjugate) in *indirect* procedure and the application of *indirect* ELISA in studies of serological relationships makes it a versatile tool.[13] The *direct* procedure has a higher specificity for serotype detection and for large scale routine testing.[14–17]

III. METHODOLOGY

The study of methodology involved in ELISA technique includes an understanding of the steps involved in conducting various forms of ELISA, the requirement of various reactants, enzymes, and equipments, the analysis of ELISA results, and also so many technical considerations which play a key role in efficiently conducting the test. Here we will discuss basic information on generally adopted ELISA methodology in plant pathology.

A. BASIC STEPS

As mentioned in the principles, the ELISA procedure can be termed as *direct* or *indirect* based on the nature of the enzyme labeled antibody employed. The basic steps involved while conducting these two classical forms of ELISA are represented in List 1 and 2. A large variation in basic requirements of operations in both types of ELISA have been reported.[18–24]

List 1. Steps of a *direct*-ELISA test (The sequence outlined is of the classical double-antibody sandwich (DAS) form of *direct* ELISA)

1. Add specific IgG (gamma immunoglobulin) to the microtitre plate, incubate, and wash.
2. Add test sample, incubate, and wash.
3. Add specific IgG-enzyme conjugate, incubate, and wash.

4. Add enzyme substrate, incubate (and stop the reaction, if desired).
5. Record results by visual observation or by measuring absorbance spectrophotometrically.

List 2. Steps of an *indirect*-ELISA test (The sequence outlined is of the clasical direct antigen coating (DAC) form of *indirect* ELISA)

1. Add test sample, incubate, and wash.
2. Add specific antibody, incubate, and wash.
3. Add anti-Fc antibody enzyme conjugate, incubate, and wash.
4. Add enzyme substrate, incubate (and stop the reaction, if desired).
5. Record results by visual observation or by measuring absorbance spectrophotometrically.

B. VARIANTS OF ELISA

1. *Variants of direct ELISA:* These are given below in relation to List 1.
   a. *Double antibody sandwich (DAS) assay:* The sequence of steps are given in List 1.
   b. *Two step DAS assay:* The second and third steps are combined (the test sample and the antibody-enzyme conjugate preincubated is added).
   c. *F (ab′)$_2$ antibody sandwich assay:* In the first step instead of whole IgG only F (ab′)$_2$ fragment of IgG is added.
   d. *DAS-assay based on antigen specific antibodies from two animal species:* In the first step, the specific antibodies or IgGs produced in animal species I are coated and in the third step the specific IgGs produced in animal species II and conjugated with enzyme are used.
   e. *Clq based assay:* In the first step instead of antibodies or IgGs, Clq complement component is coated and in the second step the test sample and antibody enzyme conjugate complex pre-incubated is added.
2. *Variants of indirect-ELISA:* These are given below in relation to List 2.
   a. *Direct antigen coating (DAC) assay:* The sequence of steps are given in List 2.
   b. *Two step DAC assay:* The second and third steps are combined (the specific antibody and anti-IgG enzyme conjugate complex pre-incubated is added).
   c. *F (ab′)$_2$ and antibody sandwich assay:* Before the first step, the F (ab′)$_2$ fragment of specific IgG is added (it is an additional step).
   d. *DAS-assay based on antigen specific antibodies from two animal species:* Before the first step, the specific antibodies or IgGs produced in animal species I are coated (it is an additional step). In the second step the specific antibodies or IgGs produced in animal species II are added and in the third step only anti-Fc fragment of animal species II antibody conjugate is added.
   e. *Protein-A based DAS assay:* Before the first step, two extra steps are introduced, i.e., adding of protein-A in the microtiter plate, incubating, and washing followed by adding specific IgG, incubating, and washing. In the third step instead of anti-Fc antibody enzyme conjugate, protein-A enzyme conjugate is added.
   f. *Protein-A based DAC assay:* In the third step, instead of anti-Fc antibody enzyme conjugate, protein-A enzyme conjugate is added.
   g. *Clq based assay:* Before the first step, Clq complement component is coated (it is an additional step) and the first and second steps are combined (the pre-incubated test samples and specific antibody complex are added).

C. MATERIAL REQUIREMENTS

1. Solid Support

Generally microtiter plates or strips of polystyrene are used as solid supports for conducting ELISA test. However, other plastic solid phases have also been tried for enzyme immunosorbent assays, e.g., polyvinyl chloride microtiter plates, polystyrene tubes and beads, polystyrene cuvettes, nylon tubing stirring rods, etc. The microtiter plates from different manufacturing sources vary in quality and uniformity.[25] It is, therefore, often necessary to evaluate plates from different sources before their routine use. The use of plastic as solid phase is based on the fact that plastics have a finite capacity to adsorb proteins via an essentially irreversible hydrophobic interaction.[26]

The ELISA plates are supposed to be disposable and are not reused. However, keeping in view the expenses involved in large scale testing, attempts have been made to recondition the plates for repeated use. Several attempts to reuse ELISA plates, which have met with varied success, have primarily focused on cleaning procedures.[27–29] The cleaning with alkali, though found effective, changes the capability of polystyrenes to adsorb protein. Treatment of new plates with protein-A enhances antibody retention by the polystyrene[30] but it is not economical. Treatment of plates with 1% nitrocellulose, which has a high protein adsorption capacity improved plate efficacy by providing a thin uniform film on which ELISA could be performed.[31] It was also found that plates could be effectively reused up to six times when recoated with nitrocellulose after each cleaning. After this point, opacity developed in the plastic which was probably associated with degenerative breakdown of the polystyrene, resulting in poorer binding of the nitrocellulose. In any case, plates are generally not reused for sensitive tests and the reuse is not advisable unless there is an acute economic cause.

Besides ELISA plates, nitrocellulose membranes are also used as supports.[32–35] The inherent advantages of using these membranes are that they are indispensable for western-blot ELISA and also they can be easily stored for several days or weeks after spotting of test samples (antigen).

2. Test Samples

The crude extracts from plants or purified antigen can be tested in ELISA. If the test is to be carried out with the purified antigen, the task is simplified by diluting the preparation directly in a suitable buffer before use. However, for routine testing, test samples have to be prepared by extraction procedures. The extraction and preparation of test samples is the most laborious and time consuming task in an ELISA test. Therefore, one should be familiar with the various alternative methods of preparing test samples and with their relative efficacy in enhancing the sensitivity of antigen detection in ELISA.

The procedure to be adopted for preparing the test samples may vary depending on following main factors:

- pathogen to be tested, i.e., whether fungi, bacteria, virus, or MLOs
- concentration of the pathogen in the host tissue
- part of the host tissue to be tested
- nature of the host tissue, i.e., whether the host tissue to be extracted contains any inhibitors which might inhibit immunological or enzymatic reaction or which give rise to undesirable non-specific reaction
- number of the samples to be tested

It is very important to carry out a thorough investigation of all the factors associated with the infected material before standardizing the sample preparation procedure.

The efficient extraction of test samples technically depends on the buffer used for extraction and the method employed. The PBST-PVP (Phosphate buffered saline Tween-polyvinylpyrolidone) buffer has been found to be most satisfactory for extracting a large number of pathogens with certain exceptions. In general, ELISA test has been effectively carried out under moderate salt concentration and near neutral pH. The addition of PVP and sometimes bovine serum in phosphate saline buffer is made to reduce the background reaction as PVP to some extent nullifies the inhibitors like tannins present in certain host tissues. The effect of inhibitors can be more simply eliminated by diluting the plant extract. However, this has to be done cautiously as the extent of dilution should not exceed the detection threshold of the pathogen. Many other means of reducing the background reaction have been proposed such as use of different extraction buffers and additives,[36–38] storage of test extracts before testing,[39] centrifugation of test extract in extraction buffer,[40] and incubation of extracted test samples for 24 h at room temperature.[41]

The method to be employed for extraction of samples depends mainly upon the number of samples to be prepared and nature of the plant part (tissue) to be extracted, but generally individual preferences and prevailing facilities play a role. For extracting a few samples, mortar and pastles are generally preferred for viruses. Depending upon the nature of host-pathogen interaction, many workers have devised ingenious drilling devices. Maury et al.[40,42] have devised a battery of overhead homogenizers which act synchronously and substantially reduce the duration of extraction period. This has a very significant application in group testing of a large number of seeds in ELISA. Extracts of soft tissues such as potato sprouts may be obtained by using a roller press. For hard tissues, use of overhead dispersion homogenizers have been advocated which can be easily cleaned between samples in a wash solution. For detecting viruses in prunus seeds, Mink and Aichelle[43] used a specific device for extracting prunus seed. Intact single seeds were placed in steel grinding cups, crushed by a bolt and hammer, and triturated in 2.5 ml buffer by rotating each bolt for 10 s at 2000 rpm with a drill. The crushed endocarp acted as an abrasive and triturated all seed parts. In case of viruses that occur in high concentration in the host, diagnosis has been accomplished by immersing leaf discs in buffer without prior homogenization[44,45] or by gently crushing leaf pieces with a glass bar in wells of microtiter plate itself followed by addition of extraction buffer.[46]

For detecting the bacteria *Xylella fastidiosa* in landscape trees, leaf petioles and branch segments were extracted by using a high-pressure press made up of two stainless steel blocks, one mounted on one side of a giant C clamp, the other on a hydraulic cylinder fixed on the other side of the clamp. Typically, 280–350 kg/cm^2 of pressure was needed to extract the sap.[47] Specific information on extraction procedures are available on fungi,[48–52] bacteria,[53–56] virus,[57,58] spiroplasma,[59] and MLOs.[60–62]

3. Antiserum

Use of a good quality antiserum is the key to the success of any ELISA test. The production of a good quality antiserum is dependent upon the quality of purified antigen which is used for injecting the animal. The details of the methods of producing antiserum against different kind of antigens (pathogens) are given in excellent review and research articles published separately for fungi,[63–65] bacteria,[47,66,67] MLOs,[68–70] spiroplasma,[71,72] and virus.[66,73–75]

However, in brief it may be noted that of all the animals, young rabbits (4 to 12 months old) are generally used to produce polyclonal antisera to plant pathogens. However, immune serum obtained soon after injection is rich in low avidity IgG and may not be suitable for ELISA. Therefore, later bleeds which are richer in IgG are more useful. Animals other than rabbits may be preferred for specific purposes.[76,77]

If the crude antiserum used in ELISA (especially for trapping antigen and for enzyme conjugate) contains different proteins other than immunoglobulins, then the efficiency of ELISA may be reduced to a great extent. It is, therefore, preferred to use purified IgGs in ELISA. The commonly used procedures for purifying IgGs from crude antiserum are described below.

Steps for preparation of immunoglobulins by salt precipitation and DEAE-Cellulose filtration

1. To 2.0 ml of whole antiseum, add 8 ml of distilled water.
2. Add 8 ml of saturated ammonium sulfate solution and mix at room temperature for 30 to 60 min.
3. Centrifuge at 800 g for 10 min.
4. Collect the precipitate and dissolve it in 10 ml of water and repeat steps 2 and 3.
5. Dissolve precipitate in 2 ml half-strength PBS (1:1 PBS: H_2O).
6. Dialyze against at least three changes of 500 ml of half-strength PBS for 24 h.
7. Prepare a column of 5 to 10 ml bed volume of pre-equilibrated DEAE-cellulose.
8. Wash the DEAE-cellulose in the column with at least five bed volumes of half-strength PBS or until no UV-absorbing material can be detected in the washings.
9. Pipette 2 ml of immunoglobulin preparations on top of the cellulose.
10. Wash the immunoglobulin through the column with half-strength PBS, collecting the eluate in approximately 1 ml fractions.
11. Monitor the fraction at 280 nm and combine fractions containing the first protein peak to be eluted.
12. Measure the optical density at 280 nm of the combined fractions and adjust the concentration of the r-globulin with half-strength PBS to read approximately 1.4 (about 1 mg/ml). The ratio OD_{280}: OD_{252} should be about 2.5 to 2.6 and the preparation should be water-clear to transmitted light.

Steps for preparation of immunoglobulins by affinity adsorption with protein A-Sepharose

1. Rehydrate 0.5 g of protein A-Sepharose CL-4B (as per manufacturer's directions) and pack into a 5 ml disposable syringe.
2. Wash with several bed volumes of PBS.
3. Pipette 1 ml of rabbit antiserum into the column and wash it through the column with several bed volumes of PBS.
4. Elute adsorbed IgG with 0.1 M glycine-HCl buffer, pH 2.7; monitor eluate at 280 nm and collect 1 ml fraction into tubes containing 0.5 ml of 0.5 M Tris-HCl + 0.15 M NaCl buffer, pH 7.8.
5. Combine fractions containing the eluted IgG and adjust the protein concentration.

Storage of immunoglobulins: After obtaining the purified immunoglobulins, they may be stored for several months at 4°C with a suitable preservative (such as 0.02% sodium azide) or for longer period at 4°C or at −18°C in 50% glycerol. There is no significant effect of glycerol on the adsorption of IgG to solid support when it is used at less than 1% concentration. For long term storage of IgGs, they can also be freeze-dried in small aliquots and stored in glass vials under vacuum.

4. Enzyme-Antibody Conjugate

The enzyme to be labeled (conjugated) with antibody should be stable, easy to detect, and should retain its enzymatic activity in the conjugate. Generally two enzymes, namely, alkaline phosphatase (ALP) and horse peroxidase (HRP), have been used in immunosorbent assays of plant pathogens. ALP, despite its higher cost, has been widely used due to its stability, convenience of linking to protein by a glutareldehyde bridge, and its linear reaction kinetics with its substrate. ALP, however, cannot be efficiently labeled to protein-A. HRP cannot be efficiently linked to protein by the one-step glutareldehyde procedure but it is (in a way) a versatile enzyme. Owing to its steric conformation and having carbohydrate moieties in the molecule, it can be conjugated by other methods. The reaction kinetic of HRP with its substrate is non-linear, the enzyme apparently being progressively inhibited during the course of substrate hydrolysis.

The classical one-step procedure for conjugation of ALP or HRP by glutareldehyde to IgG give high molecular weight conjugates well-suited for detecting macromolecules such as plant pathogens.[78,79] The concentration of glutareldehyde used for labeling influences the degree of polymerisation.[80] Two-step procedure for conjugation by glutareldehyde of HRP to IgG is followed to produce homogeneous conjugates of low molecular weight but with a reduced efficiency of coupling.[86] Such a conjugate is suitable for assays for small antigens of higher sensitivity than that produced by one-step procedure.

Since HRP cannot be efficiently coupled to protein A by glutareldehyde, it is labeled by sodium periodate oxidation of its carbohydrate moieties which form aldehyde groups that reacts with amino acid residues of the antibody protein.[82,83] Such conjugates retain properties as those produced by one-step glutareldehyde procedure. It consists of a heterogenous collection of molecules of high molecular weight. In this case, the yield of conjugate is also higher and the product is useful for both *direct* and *indirect* ELISA procedures.

The antibody activity in conjugates made with glutareldehyde is reported to be lower than that of native molecule.[38,84] This adverse effect would mean that only a few conjugated antibody molecules would participate in the serological reaction.

Apart from ALP and HRP, recent conjugates made from enzymes such as penicillinase and inorganic pyrophosphatase have also been used for detecting certain plant viruses.[85,86] However, wide scale application of these enzyme-conjugates have not taken place thus far.

Procedures for conjugation of IgG with ALP or HRP by glutaraldehyde
One-step procedure:

1. Dissolve IgG in PBS.
2. Dissolve enzyme in IgG solution.
3. Dialyze at least three times against PBS.
4. Add freshly prepared glutaraldehyde solution and incubate.
5. Dialyze at least three times against PBS.

Two-step procedure (for HRP conjugate only):

1. Dissolve enzyme in PBS.
2. Add glutaraldehyde and incubate.
3. Dialyze at least three times against PBS.
4. Transfer to a glass tube containing IgG.
5. Adjust pH approximately 9.6 with carbonate buffer and incubate.
6. Add lysine solution and incubate.
7. Dialyze at least three times against PBS.

Procedure for conjugation of IgG or protein-A with HRP by periodate oxidation

1. Dissolve enzyme in distilled water.
2. Add sodium metaperiodate and shake.
3. Dialyze against distilled water adjusted to pH 4.4.
4. Add sodium carbonate buffer, pH 9.6.

5. Substrate

The addition of substrate is the final step in any ELISA protocol. The substrate to be used depends on the enzyme used for labeling the antibody. It must provide a sensitive and quantitative detection of the enzyme in the conjugate. Substrates, which when freshly prepared are colorless but give a colored hydrolysis product, are ideal for applications. The colored product can be observed visually or quantified by colorimetry.

For ALP conjugates, the preferred substrate is *p*-nitrophenyl phosphate, which is available both in powder and tablet form. It is generally prepared at 0.6 to 1.0 mg/ml in 10% diethanolamine, pH 9.8. It is stable in solution, gives a low background value, and has an almost linear rate of reaction with the enzyme. The reaction product *p*-nitrophenol is a yellow compound which can be seen visually and this end-product can be quantified spectrophotometrically at a wavelength of 405 nm. When sufficient color change has occurred, the activity of alkaline phosphatase can be stopped by adding excess alkali (NaOH, 3N at 50 μl/well). When plates cannot be read for a few hours, they may be stored at 4°C and read later on. This is often the case when the concentration of the antigen is very low in the sample.

Since flourescence assays are potentially more sensitive than colorimetric assays, fluorogenic substrates such as 4 methylumbelliferyl phosphate and 3-0 methylfluorescein phosphate, have been advocated as possible alternatives to PNP.[21] However, this requires a fluorimeter to measure the reaction and the end product cannot be visualized easily like colored *p*-nitrophenol. PNP is thus widely used as the substrate.

For HRP conjugates, the substrates that have been used are: (1) 5 aminosalicylic acid (5As), (2) O-dianisidine, (3) 2,2 azimodi-(3-ethylbenzothiazoline sulphone-6) diammonium salt (ABTS), (4) O-phenylene diamine (OPD), and (5) 3, 3′, 5, 5′-tetramethylbenzidine (TMB). Among these, TMB is mostly used nowadays as it is non-mutagenic, light stable, less sensitive to autodegradations, and gives cleaner background as compared to other substrates that lack one or the other such characteristics.[57] It is prepared fresh in sodium acetate buffer. The activity of HRP can be stopped by adding H_2SO_4 (3 M, at 50 μl/well) when sufficient color change has developed. The end product that can be seen as a bluish color can be quantified spectrophotometrically at a wavelength of 655 nm (when the reaction is not stopped) and 450 nm (if the reaction is stopped). It may be mentioned that HRP-substrate reactions are non-linear because of progressive inhibition of the substrate by the products of hydrolysis.

For penicillinase conjugates, penicillin is used as a substrate. In this case, penicillinase enzyme breaks down penicillin into penicilloic acid. If bromothymol blue (acid sensitive pH indicator) is used to measure the penicilloic acid, there is a change in color of the end product from blue to greenish yellow to deep yellow and the absorbance is measured at 620 nm.[85]

For inorganic pyrophosphatase conjugates, tetrasodium pyrophosphate is used as a substrate. This substrate is hydrolyzed to orthophosphate by the enzyme. Finally a color/stop reagent (containing malachite green) is added to this end product and the absorbance of the bright blue-green color which develops is measured at 630 nm.[86]

D. GENERAL CONSIDERATIONS IN PERFORMING ELISA

It is expected that the user is aware of the various forms of ELISA and is able to rightly select an appropriate procedure before starting the test. After procuring/preparing the materials required for the test, it is necessary to ascertain the optimum combination of dilutions of IgG, test samples, conjugate, and substrate. The aim is to select a combination which offers maximum discrimination between the negative control (healthy samples) and the weakest positive (infected) samples that can be encountered. For ELISA a "checkerboard" titration system has been recommended to determine suitable combinations of reactant concentrations.[10]

The length and condition of incubation affect ELISA results. Selection of a suitable combination of plate incubation for the antigen phase and of incubation period for the conjugate phase are often done on the basis of relative importance attached to reaction strength and the need for rapid results. It has been generally observed that incubating the test sample (antigen) overnight in a refrigerator (4 to 6°C) gives low background reactions. The plates should be covered with polythenes or plastic covers to prevent evaporation. Incubating plates at around 30°C is less likely to produce thermal gradients than at 37°C, and is suitable for most purposes. The plates should not be stacked but should be incubated singly to avoid thermal gradients.

Thorough washing of wells of ELISA plates between every two steps is essential. This prevents carryover of reactants that are not part of the reacting components. Usually the wells are washed at least three times with phosphate buffered saline containing 0.05% Tween-20 and it is often left for several minutes in the wash solution.

Although it may be noted that often guidelines are available in published works, a certain amount of experimentation and practice is necessary to establish optimum conditions for each new application of ELISA. Some flexibility in approach of the worker is desired.

E. ANALYSIS OF RESULTS

The ELISA results, i.e., the change in color of the end product are recorded by scanning the optical density of each well of the microtiter plate or strips with a multiscan spectrophotometer generally called the "ELISA Plate Reader". The wavelength chosen for measuring the optical density (absorbance values) depends upon the enzyme system adopted in ELISA. In certain cases (generally in the absence of a plate reader), visually scanning the development of color is taken as the criterion for classifying the samples as positive (infected) and absence of color as negative (uninfected). This qualitative observation, however, has limited importance such as in cases where a large number of field populations is to be indexed for the presence of a virus and where the specific anti-virus conjugate is used by field staff having limited laboratory facilities. The ALP conjugates are preferred in such cases due to stability and lack of autodegradation of its substrate.

The quantification of ELISA results is necessary in practice to make an objective analysis of results. It helps in detecting samples with low level of infection, i.e., samples detected below the visual detection threshold and samples with very high background reaction. In order to establish limits for negative and positive values, different norms as summarized below have been proposed:[87]

1. All test samples with an absorbance value above a specified threshold are considered positive.
2. The absorbance values are directly taken from the test when performed under defined conditions and in the presence of standard reference sample.
3. The absorbance values may be expressed in ratio of the absorbance value of the sample to the mean of the group of known negative samples. Usually the ratio value 2X or 3X negative is considered as positive.
4. The absorbance value may be expressed in units. The unit is expressed in absolute terms (e.g., ng protein/ml). In this case, a series of standard samples with known content are included in the test. The units to be measured are then calculated by applying the test values to a standard curve.[9]

There needs to be a clear differentiation between test positive and test negative values irrespective of the norms fixed for reporting the results. While setting the positive-negative thresholds of a test, one should carefully consider several test parameters. The significance of false-positive and false-negative results should be well understood. The negative control samples need to be tested for their negativity. An effective method involves the establishment of a frequency distribution of negative values.

The setting of the test threshold assumes a significant role in routine ELISA testing as in certification programs where the disease has a zero level of tolerance. If the error of a false positive is taken into account, there would be unnecessary financial loss to the grower.

The use of different samples for the negative standards is important to ensure that a range of values from healthy plants is included. The range of healthy background absorbance interval should be known. More convincing results are obtained with lower background absorbance values than with higher ones. Replication of samples is also important to cover up the minor variations in different wells of the plates, if any.

Gillett et al.[88] have given the following useful guidelines for adequately reporting ELISA data:

1. Clearly state the positive-negative threshold used.
2. Test enough plates to become familiar with the range of negative values involved.
3. Include enough known negative controls in each routine assay to ensure representation of the previously established range of negative background values.

4. Always include a positive control.
5. Match control samples and test samples with respect to host type, tissue type, age, and position.
6. Strongly consider replication of test samples.

As far as possible ELISA must be compared with other independent methods of defining pathogen presence. Then, using a population of positive and negative samples, the results from the reference and ELISA methods are compared and a threshold is chosen that yields the most correct results, i.e., the maximum possible false-negatives and/or false-positives. During routine use, any ELISA must be controlled by maintaining the threshold in a constant range of absorbance values. Also, satisfactory assay performance should be proved by contrasting and using test standards as controls.

The common use of thresholds such 2X, 3X, x + 3s, etc. should be taken as arbitrary. A threshold should be obtained on the basis of an acceptable reference.

IV. CONCLUSIONS

There has been a substantial impact of ELISA in the large scale diagnosis of diseases. ELISA has revolutionized the diagnosis for assessing disease for certification purposes and for control through quarantine or eradication procedures. Indexing for seed-borne infections has been greatly successful particularly for the viruses[58] and bacteria.[53] The application of ELISA in epidemiology is now increasing as it is being used more and more for studying vector relationships, investigating alternative hosts of pathogens, and studying the occurrence/distribution of various strains/races/biotypes of a pathogen. Though initially ELISA came in handy for detecting plant viruses, now with modifications and adaptations, it is also being increasingly used for detecting fungi, bacteria, spiroplasma, and MLOs.

The popularity of ELISA is largely due to the inherent advantages in this technique over the conventional serological methods used for detecting plant pathogens. Due to the repetitive nature of handling the reactants, in ELISA a large number of samples can be analysed simultaneously with ease and precision, and the technique can be subjected to automation. Also, the technique is much more sensitive than the generally known biological or serological methods of detection. The ELISA technique can be learned and applied even with limited experience in serology. The development of commercially prepared and standardized plates of immunoglobulins and conjugates have further extended the use of ELISA to those who are not equipped to prepare their own materials and also to those who need to make only limited tests for a particular pathogen. Above all the current boom in microcomputer technology and the use of such machines along with ELISA readers has made the handling and analysis of ELISA data extremely easy.

Besides, with the advent of monoclonal antibodies, ELISA has become an indispensable tool for screening the hybridomos.

ELISA, like any other serological technique, has its limitation as in serology only a few percent of the total information present in a nucleic acid is used. Nevertheless ELISA with all its versatility would remain a technique of choice for routine certification purposes. However, despite all the achievements, there still exists a large scope of modification and innovation in ELISA procedures.

REFERENCES

1. Nakane, P. K. and Pierce, G. E. Jr., Enzyme labelled antibodies: preparation and application for the localization of antigens, *J. Histochem. Cytochem.,* 14, 929, 1966.
2. Wicker, R. and Avrameas, S., Localization of virus antigens by enzyme-labelled antibodies, *J. Gen. Virol.,* 4, 465, 1969.
3. Engvall, E. and Perlmann, P., Enzyme-linked immunosorbent assay ELISA III. Quantification of specific antibodies by enzyme-labelled antimmunoglobulin in antigen coated tubes, *J. Immunol.,* 109, 129, 1972.
4. Van Weemen, B. K. and Shuurs, A. H. W. M., Immunoassays using hapten-enzyme conjugates, *FEBS Letters,* 24, 77, 1972.
5. Voller, A., Bidwell, D., Huldt, G., and Engvall, E., A microplate method of enzyme-linked immunosorbent assay and its application to malaria, *Bull. WHO,* 51, 209, 1974.
6. Voller, A., Bartlett, A., Bidwell, D. E., Clark, M. F., and Adams, A. N., The detection of viruses by enzyme-linked immunosorbent assay (ELISA), *J. Gen. Virol.,* 33, 165, 1976.

7. Voller, A., Bartlett, A., and Bidwell, D. E., The use of enzyme-linked immunosorbent assay in the serology of viral and parasitic diseases, *Scand. J. Immunol.,* 8, 125, 1978.
8. Voller, A., Bartlett, A., and Bidwell, D. E., Enzyme immunoassays with special reference to ELISA techniques, *J. Clin. Pathol.,* 31, 507, 1978.
9. Voller, A., Bidwell, D. E., and Bartlett, A., *The enzyme-linked immunosorbent assay (ELISA): A guide with abstracts of microplate application,* Dynatech Laboratories, Alexandria, Virginia, 125, 1979.
10. Clark, M. F. and Adams, A. N., Characteristics of the microplate method of enzyme linked immunosorbent assay for the detection of plant viruses, *J. Gen. Virol.,* 34, 475, 1977.
11. Mowat, W. P. and Dawson, S. Detection and identification of plant viruses by ELISA using crude sap extracts and unfractionated antisera, *J. Virol. Meth.,* 15, 233, 1987.
12. Edwards, M. L. and Cooper, J. L., Plant virus detection using a new form of indirect ELISA, *J. Virol. Meth.,* 11, 309, 1985.
13. Kumar, C. A., Mandal, B. B., Chandel, K. P. S., Jain, R. K., Varma, A., and Srivastava, M. Occurrence of sweet potato feathery mottle virus in germplasm of *Ipomoea batatas* (L.) in India, *Curr. Sci.,* 60, 321, 1991.
14. Khetarpal, R. K., Bossennec, J. M., Burghofer, A., Cousin, R., and Maury, Y., Effect du pea seedborne mosaic virus sur les rendement in graines du pois proteagineux, *Agronomie,* 8, 811, 1988.
15. Khetarpal, R. K. and Maury, Y., Seed transmission of pea seedborne mosaic virus in peas: early and late expression of the virus in the progeny, *J. Phytopathol.,* 129, 265, 1990.
16. Khetarpal, R. K., Maury, Y., Cousin, R., Burghofer, A., and Varma, A., Studies on resistance of pea to pea seed-borne mosaic virus and new pathotypes, *Ann. Appl. Biol.,* 116, 297, 1990.
17. Khetarpal, R. K., Parakh, D. B., Singh, S., and Nath, R., ELISA detection of soybean mosaic virus in testas, embryos and seedlings from mottled and unmottled seeds of imported soybean germplasm, *Ind. J. Virol.,* 8, 106, 1992.
18. Flegg, C. L. and Clark, M. F., The detection of apple chlorotic leaf spot virus by a modified procedure of enzyme linked immunosorbent assay., *Ann. Appl. Biol.,* 91, 61, 1979.
19. Lommel, S. A., McCain, A. H., and Morris, T. J., Evaluation of indirect enzyme-linked immunosorbent assay for the detection of plant viruses, *Phytopathology,* 72, 1018, 1982.
20. Barbara, D. J. and Clark, M. F., A simple indirect ELISA using F $(ab')_2$ fragments of immunoglobulin, *J. Gen. Virol.,* 58, 315, 1982.
21. Torrance, L. and Jones, R. A. C., Increased sensitivity of detection of plant viruses obtained by using a fluorogenic substrate in enzyme-linked immunosorbent assay. *Ann. Appl. Biol.,* 101, 501, 1982.
22. Kaniewski, W. K. and Thomas, P. E., A two-step ELISA for rapid, reliable detection of potato viruses, *Am. Potato J.,* 65, 561, 1988.
23. Khetarpal, R. K., Parakh, D. B., Singh, S., Nath, R., Jain, R. K., and Varma, A., Detection of bean common mosaic potyvirus by DAC-Indirect ELISA in imported accessions of *Phaseolus vulgaris* L., *Ind. J. Virol.,* 10, 13, 1994.
24. Zrien, M., Burckard, J., and Regenmortel, M. H. V., *J. Virol. Meth.,* 13, 121, 1986.
25. Khetarpal, R. K., Parakh, D. B., Singh, S., and Nath, R., Evaluation of efficacy of ELISA microtiter plates of different makes for detecting soybean mosaic virus in seeds of soybean (Abstract), Annual meeting of Indian Phytopath. Soc. (Delhi Zone), December 18, p. 22, 1993.
26. Engvall, E. and Ruoslahti, E., Principles of ELISA and recent applications to the study of molecular interactions, in *Laboratory and Research Methods in Virology and Medicine, Vol. 3, Immunoassays in the Clinical Laboratory,* Nakamara, R. M., Dito, W. R., and Tucker, E. S., Eds. Alan R. Liss, New York, 1979, 89.
27. Bar-Joseph, M., Muscovitz, M., and Sharafi, Y., Reuse of coated enzyme linked immunosorbent assay plates, *Phytopathology,* 69, 424, 1979.
28. Banttari, E. E. and Peterson, A. Jr., A cleaning technique for cuvettes used in enzyme-linked immunosorbent assay, *Plant Dis.,* 67, 18, 1983.
29. Goodwin, P. H. and Banttari, E. E., Increased sensitivity of ELISA for potato viruses S, X and Y by polystyrene pretreatments, additives, and a modified assay procedure, *Plant Dis.,* 68, 944, 1984.
30. Bachi, T., Darwal, G., and Wigzell, C., Staphylococcal protein A in immunosorbent techniques, *Scand. J. Immuno.,* 6, 241, 1977.
31. Stobbs, L. W. and Van Schagen, J. G., Reconditioning of ELISA plates, *Phytopathology,* 76, 483, 1986.
32. O'Donnell, I. J., Shukla, D. D., and Gough, K. H., *J. Virol. Meth.,* 4, 19, 1982.

33. Jain, R. K., Kumar, C. A., Bhat, A. I., and Varma, A., Serological relationship of sweet potato feathery mottle virus with other potyviruses, *Ind. Phytopathol.,* 46, 162, 1993.
34. Lange, L. and Heide, M., Dot Immuno Binding (DIB) for detection of virus in seed, *Can. J. Pl. Path.,* 8, 373, 1986.
35. Kumar, C. A. and Mandal, B. B., Screening of sweet potato [*Imomoea batalas* (L.)] germplasm against three potyviruses by dot-immunobinding assay, Abstract in Nat. Symp. Biotech. Diag. & Cont. Vir. Dis., February 16–18, Hyderabad, 1993, 36.
36. Lister, R. M. and Rochow, W. F., Detection of barley yellow dwarf by enzyme-linked immunosorbent assay, *Phytopathology,* 69, 649, 1979.
37. McLaughlin, M. R., Barnett, O. W., Burrows, P. M., and Baum, R. H., Improved ELISA conditions for detection of plant viruses, *J. Virol. Meth.,* 3, 13, 1981.
38. Ghabrial, S. A. and Shepherd, R. J., A sensitive radioimmunosorbent assay for detection of plant viruses, *J. Gen. Virol.,* 48, 311, 1980.
39. Tamada, T. and Harrison, B. D., Application of enzyme-linked immunosorbent assay to the detection of potato leafroll virus in potato tubers, *Ann. Appl. Biol.,* 96, 67, 1980.
40. Maury, Y., Duby, C., Bossennec, J. M., and Boudazin, G., Group analysis using ELISA: determination of the level of transmission of soybean mosaic virus in soybean seed, *Agronomie,* 5, 405, 1985.
41. Falk, B. W. and Purcifull, D. E., Development and application of an enzyme-linked immunosorbent assay (ELISA) test to index lettuce seeds for lettuce mosaic virus in Florida, *Plant Dis.,* 67, 413, 1983.
42. Maury, Y., Bossennec, J. M., Boudazin, G., Hampton, R., Pietersen, G., and Maguire, J., Factors influencing ELISA evaluation of transmission of pea seed-borne mosaic virus in infected pea seed: seed group size and seed decortication, *Agronomie,* 7, 225, 1987.
43. Mink, G. I. and Aichelle, P. D., Use of enzyme linked immunosorbent assay results in efforts to control orchard spread of cherry rugose mosaic disease in Washington, *Plant Dis.,* 68, 378, 1989.
44. Marco, S. and Cohen, S., Rapid detection and titer evaluation of viruses in pepper by enzyme-linked immunosorbent assay (ELISA), *Phytopathology,* 69, 1259, 1979.
45. Romaine, C. P., Newhart, S. R., and Anzola, D., Enzyme-linked immunosorbent assay for plant viruses in intact leaf tissue discs, *Phytopathology,* 1, 308, 1981.
46. Bar-Joseph, M. and Garnsey, S. M., Enzyme-Linked Immunosorbent Assay (ELISA): Principles and Applications for Diagnosis of Plant Viruses, In *Plant Disease and Vectors—Ecology and Epidemiology,* K. Maramorosch and K. F. Harris, Eds., Academic Press, New York, 35, 1980.
47. Sherald, J. L. and Lei, J. D., Evolution of rapid ELISA test kit for detection of *Xylella fastidiosa* in landscape trees, *Plant Dis.,* 75, 200, 1991.
48. Johnston, M. C., Pirone, T. P., Siegel, M. R., and Varney, D. R., Detection of *Epichloe typhina* in tall fescue by means of enzyme-linked immunosorbent assay, *Phytopathology,* 72, 647, 1980.
49. Liese, A. R., Gotlieb, A. R., and Sackston, W. E., Use of the enzyme-linked immunosorbent assay (ELISA) for the detection of downy mildew (*Plasmopara halstedii*) in sunflower. *Proc. 10th International Sunflower Conference,* Australia, 172, 1982.
50. Mac Donald, J. D., Stites, J., and Kabashima, J., Comparison of serological and culture plate methods for detecting species of *Phytophthora, Pythium* and *Rhizoctonia* in ornamental plants. *Plant Dis.,* 74, 655, 1990.
51. Sundaram, S., Plasencia, J., and Banttari, E. E., Enzyme-linked immunosorbent assay for detection of *Verticillium* spp. using antisera produced to *V. dahliae* from potato, *Phytopathology,* 81, 1485, 1991.
52. Kumar, V. U., Meera, M. S., Shishupala, S., and Shetty, H. S., Quantification of *Sclerospora graminicola* in tissues of *Pennisetum glaucum* using an enzyme-linked immunosorbent assay. *Plant Pathol.,* 42, 251, 1993.
53. Weaver, W. H. and Guthrie, J. W., Enzyme-linked immunosorbent assay application to the detection of seed-borne bacteria. *Phytopathol. News,* 12, 156, 1978.
54. Barzic, M. R. and Trigalet, A., Detection de *Pseudomonas phaseolicola* (Burkh.). Dowsan par la technique ELISA. *Agronomie,* 2, 389, 1982.
55. Takahashi, Y., Tsuchiya, K., Shohara, K., and Suzui, T., Production and utilisation of monoclonal antibodies against *Pseudomonas cepacia, Ann. Phytopath. Soc. Japan,* 56, 229, 1990.
56. Lamak, G. L., Hill, J. H., Mc Gee, D. C., and Braun, E. J., Development of an immunosorbent assay for seedborne *Erwinia stewartii* in corn seeds, *Phytopathology,* 81, 839, 1991.

57. Clark, M. F. and Bar-Joseph, M., Enzyme immunosorbent assays in plant virology. In: *Methods in Virology,* Vol. VII, K. Maramorosch and H. Koprowski, Eds., Academic Press, London, 1984, 51.
58. Maury, Y. and Khetarpal, R. K., Testing seed for viruses using ELISA, Perspectives in Phytopathology, V. P. Agnihotri, N. Singh, H. S. Chaube, U.S. Singh and T. S. Dwivedi, Eds., Today & Tomorrow Printers & Publishers, N. Delhi, 1989, 31.
59. Lin, C. P., Tseng, W. T., and Lin, J. T. H., Serological characterization of monoclonal and polyclonal antibodies against ribosomal proteins from *Spiroplasma citri, Pl. Prot. Bull.* (Taipei), 32, 70, 1990.
60. Clark, M. F., Morton, A., and Buss, S. L., Preparation of mycoplasma immunogens from plants and a comparison of polyclonal and monoclonal antibodies made against primula yellows MLO-associated antigens, *Ann. Appl. Biol.,* 114, 111, 1989.
61. Chen, T. A. and Jiang, Y. P., Progress in the detection of plant mycoplasma-like organisms by using monoclonal and polyclonal antibodies, in *Proc. Conf. Int. Organ Mycoplasmol.* 7th Zeatralb Bakteriol., Suppl. 20, 270, 1990.
62. Fos, A., Danet, J. L., Zreik, L., Garnier, M., and Bove, J. M., Use of monoclonal antibody to detect the stolbur mycoplasma like organism in plants and insects and to identify a vector in France, *Plant Dis.,* 76, 1092, 1992.
63. Mohan, S. B., Evaluation of antisera raised against *Phytophthora fragariae* for detecting the red core disease of strawberries by enzyme-linked immunosorbent assay (ELISA), *Plant Pathology,* 37, 206, 1988.
64. Mathew, D. S. and Brooker, J. D., The isolation and characterization of polyclonal and monoclonal antibodies to anastomosis group of *Rhizoctonia solani, Plant Pathol.,* 40, 67, 1991.
65. Velicheti, R. K., Lamison, C., Brill, L. M., and Sinclair, J. B., Immunodetection of *Phomopsis* species in asymptomatic plants, *Plant Dis.,* 77, 70, 1993.
66. Hampton, R., Ball, E., and DeBoer, S., Eds., *Serological Methods for Detection and Identification of Viral and Bacterial Plant Pathogens,* APS Press, St. Paul, Minnesota, 389, 1990.
67. Takahashi, Y., Tsuchiya, K., Shohara, K., and Suzni, T., Production and utilisation of monoclonal antibodies against *Pseudomonas cepacia, Ann. Phytopathol. Soc. Japan,* 56, 229, 1990.
68. Sarindu, N. and M. F. Clark, Antibody production and identity of MLOs associated with sugarcane whiteleaf disease from Thailand, *Plant Pathol.,* 42, 396, 1993.
69. Clark, M. F., Davies, D. L., Buss, S. L., and Marton, A., Serological discrimination among mycoplasma-like organisms using polyclonal and monoclonal antibodies, *Acta Hortic.,* 235, 107, 1988.
70. Clark, M. F., Morton, A., and Buss, S. L., Preparation of mycoplasma immunogens from plants and a comparison of polyclonal and monoclonal antibodies made against primula yellows MLO-associated antigens, *Ann. Appl. Biol.,* 114, 111, 1989.
71. Clark, M. F., Davies, D. L., and Barbara, D. J., Production and characterisation of antisera to *Spiroplasma citri* and clover phyllody-associated antigens derived from plants, *Ann. Appl. Biol.,* 103, 251, 1983.
72. Lin, C. P., Tseng, W. T., and Lin, J. T. H., Serological characterization of monoclonal and polyclonal antibodies against ribosomal proteins from *Spiroplasma citri, Plant Prot. Bull.* (Taipei), 32, 70, 1990.
73. Van Regenmortel, M. H. V., *Serology and Immunochemistry of Plant Viruses,* Academic Press, New York, 1982.
74. Lister, R. M., Hammond, J., and Clement, D. L., Comparison of intradermal and intramuscular injection for raising plant virus antisera for use in ELISA, *J. Virol. Meth.,* 6, 179, 1983.
75. Harlow, E. and Lane, D., *Antibodies: A Laboratory Manual,* Cold Spring Harbor Laboratory, Cold Spring Harbor, NY, 1988.
76. Chen, C. S. and Chen, T. A., Comparison of immunized mouse ascites fluids and rabbit sera in serological tests of two spiroplasmas, *Phytopathology,* 70, 279, 1980.
77. Van Regenmortel, M. H. V., Application of Plant Virus Serology, *Annu. Rev. Phytopathol.,* 16, 57, 1978.
78. Avrameas, S., Coupling of enzymes to proteins with glutaraldehyde. Use of the conjugates for the detection of antigens and antibodies, *Immunochemistry,* 6, 43, 1969.
79. Avrameas, S., Ternynck, T., and Guesdon, J. L., Coupling of enzymes to antibodies and antigens. In *Quantitative Enzyme Immunoassay,* E. Engvall and A. J. Pesce, Eds., *Scand J. Immunol.* Oxford: Blackwell, 1978.
80. Kopraditskul, P., Casper, R., and Lesseman, D. E., Evaluation of short reaction times and some characteristics of the enzyme-conjugation in enzyme-linked immunosorbent assay (ELISA). *Phytopathol. Z.,* 96, 281, 1979.

81. Avrameas, S. and Ternynck, T., Peroxidase labelled antibody and Fab conjugates with enhanced intracellular penetration, *Immunochemistry,* 8, 1175, 1971.
82. Nakane, P. K. and Kawaoi, A., Peroxidase labelled antibody. A new method of conjugation, *J. Histochem. Cytochem.,* 22, 1084, 1974.
83. Wilson, M. B. and Nakane, P. K., *Immunofluorescence and Related Staining Techniques,* Amsterdam, Elsevier, 1978, 215.
84. Koenig, R., ELISA in the study of homologous and heterologous reactions of plant viruses, *J. Gen. Virol.,* 40, 309, 1978.
85. Sudarshana, M. R. and Reddy, D. V. R., Penicillinase-based enzyme-linked immunosorbent assay for the detection of plant viruses, *J. Virol. Meth.,* 26, 45, 1989.
86. Mizenina, O. A., Borisova, O. V., Novikov, V. K., Evtushenko, O. A., Baykov, A. A., and Atabekov, J. G., Inorganic pyrophosphatase from *E. coli* as a label for the detection of plant viruses by ELISA. *J. Phytopathol.,* 133, 278, 1991.
87. De Savigny, D. H. and Voller, A., Communication of ELISA data from the laboratory to clinician, *J. Immunology,* 1, 105, 1980.
88. Gillet, J. M., Morrissey, S. M., and Ramsdell, D. C., Interpreting ELISA data and establishing the positive-negative threshold, *Plant Dis.,* 70, 722, 1986.

Chapter 24

Methods for the Production of Mouse Monoclonal Antibodies

R. Mernaugh and G. Mernaugh

TABLE OF CONTENTS

I. INTRODUCTION

The purpose of this article is to introduce the nonimmunologist to methods used to produce mouse monoclonal antibodies.

Before this topic can be discussed, it will be necessary to briefly describe an antibody and its features. A more in-depth discussion can be found elsewhere.[1-4]

An antibody is an immunoglobulin (Ig) molecule which can bind to an antigen. Antibodies are produced by an animals's B-cells or B-lymphocytes. The B-cells can be found in the spleen, lymph nodes, Peyer's patches of the digestive tract, and peripheral (circulating) blood. Each B-cell produces one antibody specific for one antigenic site (epitope) on an antigen. These epitopes may be shared by

0-87371-877-1/95/$0.00+$.50

more than one antigen, and an antibody which recognizes one of these shared antigenic sites recognizes a cross-reactive epitope.

Each antibody molecule generally consists of at least one pair of heavy and light chains which, together, comprise the paratope or antigen binding site of the antibody. Each heavy and light chain can be divided into two regions, based on the extent of amino acid sequence variability in the regions. These regions are designated as the variable and constant regions of the antibody molecule. The amino acid sequence of the variable region can vary extensively among antibodies with different antigenic specificities, while the amino acid sequences of the constant regions vary to a much lesser degree.

The variable regions of the heavy and light chains, which are located on the NH_2-terminus of the antibody molecule, constitute the Fv (fragment variable) region or idiotype of the antibody. The Fv region of the antibody is that region which is responsible for interacting with the antigenic site.

The constant region of the antibody molecule is located on the COOH-terminus of the antibody molecule and defines the class of the antibody (IgG, IgM, IgA, IgD, and IgE) or suclass (IgG1, IgG2a, IgG2b, IgG3, IgG4, IgM1, IgM2, IgA1, and IgA2). The class and suclass of an antibody define its isotype and can be determined using commercially available isotyping reagents.

IgG and IgM antibodies are the most abundant classes of antibodies produced by an animal in response to an antigenic stimulation. Therefore, most monoclonal antibodies generated will be of the IgG or IgM isotype.

II. PROTOCOLS FOR THE PRODUCTION OF MONOCLONAL ANTIBODIES

The mouse monoclonal antibody production protocols that follow are modifications of the procedures published by Van Deussen and Whetstone.[5] These modifications were made over a period of 10 years to reflect the needs, capabilities, and time constraints of the authors, their associates, or students. These protocols have been streamlined and designed to enable individuals who have little, if any, immunological or tissue culture experience to succeed in generating useful monoclonal antibodies.

The outline that follows will provide an overview of protocols that will be discussed with respect to the production of monoclonal antibodies.

- A. Developing a productive B-cell antibody response in mice
  1. Antigen preparation and animal selection
  2. B-cell antigenic stimulation
     a. *In vivo* antigen immunization guidelines
     b. *In vitro* B-cell antigenic stimulation protocol
- B. Developing and optimizing an enzyme-linked immunosorbent assay (ELISA) to detect antigen-specific antibodies
  1. Determining the optimum antigen and antibody concentrations used in ELISAs
- C. Monitoring the antibody response of a mouse to an antigen by ELISA
- D. Tissue culture media
  1. DMEM
  2. DMEM plus azaguanine
  3. SP2/O conditioned medium
  4. OPI medium
  5. OPI plus azaserine
  6. Cell freezing medium
- E. Cell freezing protocol
  1. Cell freezing protocol for a large number of cells
  2. Resurrecting frozen cell lines
  3. Microtiter plate cell freezing protocol
  4. Resurrecting cells previously frozen in microtiter plates
- F. Selection for hypoxanthine phosphoribosyl guanine transferase ($HPRGT^-$) SP2/O cells
- G. SP2/O myeloma cell growth prior to cell fusion
- H. Mouse hyperimmunization prior to cell fusion
- I. Mouse spleen removal
  1. Termination of a mouse

2. Mouse cardiac bleed
3. Surgical removal of a mouse spleen and spleen cell extraction

J. SP2/O cell count and viability stain
K. Spleen-SP2/O cell fusion protocol
L. Cloning antibody-producing hybridomas
M. Production of high levels of monoclonal antibodies in mouse ascities

A. DEVELOPING A PRODUCTIVE B-CELL ANTIBODY RESPONSE IN MICE

Antibody-producing B-cells are fused with myeloma cells (e.g., SP2/O-Ag14 myeloma cells) to generate monoclonal antibody-producing hybridoma cells. If the B-cells are not properly stimulated to produce antibodies, then the hybridomas generated from the spleen-myeloma cell fusion will not produce antibodies.

For the most part, IgG rather than IgM monoclonal antibodies are preferred for diagnostic or therapeutic use. The reasons for this are that IgG monoclonal antibodies are generally more specific for the target antigenic site, bind more tightly or strongly to the epitope (have a greater affinity for the antigenic site), and are easier to purify and label (i.e., with peroxidase, alkaline phophatase, biotin, etc.). However, due to the nature of an antigen or the genetic makeup of an animal, in some cases only IgM monoclonal antibodies may be obtained.

To develop a productive IgG B-cell antibody response, several points should be considered. These points include antigen preparation and animal selection, and B-cell antigenic stimulation.

1. Antigen Preparation and Animal Selection

The two things to consider with respect to antigen preparation are antigen purity and immunogenicity. The antigen (e.g., a plant virus) used to stimulate an immune response in an animal should be as pure as possible. Contaminants (e.g., plant host material) can also trigger a B-cell antibody response or, in some cases, suppress a productive antigen-specific B-cell antibody response. If a purified antigen is not used, subsequent immunoassays used to detect hybridomas producing antigen-specific antibodies (e.g., plant virus-specific antibodies) will have to be expanded to include negative controls. The negative controls (e.g., plant host material) must be used to exclude those hybridomas which produce non-specific antibodies.

Large antigens (>10,000 Da) with highly repetitive epitopes exposed on the surface of the molecule are usually more immunogenic than smaller antigens with fewer repetitive and buried epitopes. Small molecules (e.g., synthetic peptides) can be made more immunogenic by coupling the molecule to a large immunogenic carrier molecule (e.g., keyhole limpet hemocyanin).[3] In some cases, molecules are so labile that they may be broken down in the animal's body before an appropriate immune response can be formed. In situations such as these, the molecules can be chemically stabilized with a fixative (e.g., formaldehyde or glutaraldehyde).[3]

Monoclonal antibody-producing hybridomas can be generated using B-cells obtained from different mouse strains immunized with a variety of antigens. Balb/c mice work quite well and have a gentle disposition (seldom bite). However, due to genetic constraints, Balb/c mice may not always be able to mount a suitable IgG antibody response to an antigen.[3] Therefore, it may be necessary to use other strains of mice such as DBA/2, B10.G, CBA/J, SJL or C57B1/6.[6] All of these mice vary genetically at the immunological level. Consequently, they can process and present antigens differently and can generate a variety of IgG antibody responses to different antigens.

2. B-Cell Antigenic Stimulation

Two different schemes can be used to stimulate B-cell antibody production. These schemes are referred to here as *in vivo* immunization and *in vitro* B-cell antigenic stimulation. Since there are numerous methods that can be used to inoculate mice with an antigen to stimulate *in vivo* B-cell antibody production, the following discussion will describe some guidelines that can be followed to produce a successful B-cell antibody response. *In vitro* B-cell antigenic stimulation is not used as extensively; therefore, a protocol will be described which can be used to stimulate B-cells *in vitro* (in tissue culture).

a. In vivo Antigen Immunization Guidelines

To stimulate B-cells *in vivo,* mice are inoculated with an antigen. The *in vivo* immunization protocol used to produce an appropriate IgG response in an animal depends upon the nature of the antigen, the route of inoculation, and the genetic makeup of the animal being inoculated with the antigen. Before the animal is inoculated with the antigen, preimmune serum should always be collected to determine whether or not normal mouse antibody reacts nonspecifically with the antigen (for a serum collection protocol, see Section II. C). Generally, 100 to 200 μl of antigen (10 to 100 μg/ml in phosphate-buffered saline [PBS]) is emulsified in or mixed 1:1 with an adjuvant (e.g., Freund's complete adjuvant or alum).[3] The antigen is injected intraperitoneally (i.p.) into the abdomen of a 3- to 4-week-old mouse, and, within 3 to 4 weeks, an antigen-specific IgG response is usually elicited. In some cases (e.g., with synthetic peptides) the number of inoculations may have to be repeated every 3 to 4 weeks (using Freund's incomplete as an adjuvant and/or PBS as a diluent) until a suitable antibody response is obtained. In other cases, a route of inoculation (e.g., intradermal or i.d., intramuscular or i.m., intravenous or i.v., or intrasplenic) other than i.p. may have to be used. Areas of the body accessed by other routes of inoculation contain a variety of different cells which may suitably process and present an antigen such that an appropriate antibody response is obtained. If an appropriate immune response cannot be obtained using one strain of mouse, other strains of mice can also be used.[3,6]

b. In vitro B-Cell Antigenic Stimulation Protocol

If an antibody response to an antigen cannot be obtained in animals under any circumstances, it is possible that the antigen triggers an immune response in which other circulating cells (e.g., T suppressor cells) in the animal suppress B-cell antibody production. To circumvent this situation, B-cells can be stimulated with an antigen *in vitro*. However, IgM antibodies are generally obtained using this protocol.

The following protocol can be used to stimulate B-cell antibody production *in vitro* and to prepare stimulated B-cells for cell fusion.

1. Prepare 50 to 75 ml of 0.2-μm, filter-sterilized OPI medium which contains 5 to 25 μg of antigen per milliliter of medium.
   Note: The recipe for preparing OPI medium can be found in the tissue culture media section.
2. Aseptically remove the spleen from a naive (unimmunized) mouse and place the spleen into a sterile petri dish containing 5 to 10 ml of the filter-sterilized OPI medium and the antigen.
3. Prepare a spleen cell suspension.
   Note: The procedure used for removing the spleen from the mouse and for preparing a spleen cell suspension can be found in the "Mouse Spleen Removal" section.
4. Aseptically transfer the spleen cell suspension to a tissue culture flask (e.g., T-75) and place the flask on its side in humidified CO_2 incubator at 37°C. Incubate for 5 d.
5. Carefully remove the flask from the incubator and place the flask on an inverted-phase microscope.
6. View the cells at 400× magnification. If clumps of 2, 4, 8, 16, etc., cells are present, then the spleen cells have been antigenically stimulated and can be used as a source of B-cells to produce hybridomas.
7. Gently mix the contents of the tissue culture flask to resuspend the nonadherent B-cells.
   Note: Do not try to resuspend the cells that are strongly adhering to the tissue culture flask. These adherent cells can overgrow the antibody-producing hybridomas generated by cell fusion.
8. Aseptically transfer the cells to one or two 50-ml sterile screw-capped centrifuge tubes and centrifuge the cell suspension at 1500 rpm (500 × g) in a clinical centrifuge for 5 min at room temperature.
9. Aseptically remove the supernatant from the cell pellet.
10. Resuspend the cell pellet(s) in 1 to 2 ml of DMEM (without additives; see tissue culture media section).
11. Aseptically remove 1 μl of the cell suspension and transfer to 99 μl of trypan blue (0.4% trypan blue in PBS).
12. Perform a viability cell count on the lymphocytes using the "SP2/O cell count and viability staining protocol".
    Note: If the lymphocytes are not easily recognized, assume that there are 2 to 3 × 10^7 total lymphocytes in the cell suspension.
13. Fuse the spleen cells to myeloma cells (e.g., SP2/O cells) using the "Spleen-myeloma cell fusion protocol".

B. DEVELOPING AND OPTIMIZING AN ENZYME-LINKED IMMUNOSORBENT ASSAY (ELISA) TO DETECT ANTIGEN-SPECIFIC ANTIBODIES

An ELISA is a very convenient and useful assay to detect antigen-specific antibodies. However, the ELISA protocol should be optimized before the assay is used to screen for antibodies. If the assay is not optimized, the assay may not be sensitive or may yield false positive results.

Variables that may influence an assay include the plastics used to manufacture the microtiter plates, antigen-coating concentrations, primary and secondary antibody concentration, buffers, incubation times and temperatures, and blocking reagents. For the most part, these variables must be determined empirically.

Note: It will only be possible to optimize the antigen-coating concentration when screening hybridomas for antibody activity, since the concentration of antibodies produced by hybridomas will vary from hybridoma to hybridoma.

The following protocol can be used to optimize an ELISA. The optimum antigen and antibody concentrations used are determined by varying the antigen and antibody concentration in the assay.

Note: There are 12 columns (designated numerically from 1 to 12) and 8 rows (designated alphabetically from A to H) in a 96-well microtiter plate.

1. Dilute the antigen to 100 μg/ml in buffer (e.g., PBS or 0.05 *M* carbonate buffer pH 9.6).
2. Add 100 μl of buffer to each well of a 96-well microtiter plate using a micropipettor.
3. Add 100 μl of the diluted antigen to each well in column 1 and mix the contents of each well. Change the tips of the micropipet after each pipetting step so that undiluted reagents will not be carried over from well to well.
4. Remove 100 μl of the diluted antigen from each well in column 1 and add to corresponding wells in column 2, then mix.
5. Continue this twofold dilution process through column 11. Wells in column 11 will contain 200 μl of diluted antigen.
6. Remove 100 μl of the diluted antigen from each well in column 11 and discard. Do not add diluted antigen to the wells in column 12. Wells in column 12 will serve as negative (N) controls to determine if nonspecific interactions are occurring between the primary antibody (defined below) and the blocking buffer or plastic.
7. Incubate the microtiter plate for 1 h at room temperature or 37°C or overnight at 4°C in a humidified container (e.g., a plastic box containing wet paper towels).

   Note: Spurious results can occur when plates are subjected to temperature changes during the course of the assay. Uniform results are more easily obtained if all steps in the ELISA are performed at room temperature.
8. Invert the microtiter plate over a sink or suitable receptacle and flick out the contents of the wells. Tap the inverted plate on a paper towel to remove any residual drops.
9. Fill all of the wells of the plate with blocking buffer (e.g., 10% normal goat serum, 1% BSA, 3% nonfat dry milk, 1% casein, or 1% gelatin diluted in PBS) to coat any site in the wells not coated by the antigen. Proteins in the blocking buffer will prevent antibodies from sticking to sites on the microtiter wells not occupied by the antigen.

   Note: If the enzyme-conjugated secondary antibody was developed in goats (e.g., goat anti-mouse IgG antibody conjugated with alkaline phosphatase), then a good blocking buffer would be 10% normal goat serum diluted in PBS. Any sites on the antigen that react nonspecifically with goat antibodies would interact with the unconjugated antibodies present in the normal goat serum. The antigen would then be unavailable to interact nonspecifically with the enzyme-conjugated goat antibody to produce a false positive signal in the ELISA. Antibodies can react nonspecifically with blocking reagents which contain BSA, casein, etc.
10. Incubate the microtiter plate for 1 h at room temperature.
11. Invert the microtiter plate over a sink or suitable receptacle and flick out the contents of the wells. Tap the inverted plate on a paper towel to remove any residual drops.

    Note: To store the ELISA plates, rinse the plates by completely filling and emptying the wells three times with wash buffer (PBS containing 0.5% Tween® 20). Fill all wells with sterile filtered 3% lactose or sucrose in distilled water and incubate for 15 min at room temperature in a laminar flow hood. Empty the wells and tap the inverted plates on a paper towel to remove any residual drops in the wells. Dry the plates in the laminar flow hood. Wrap the plates in plastic wrap and store at 4°C in a desiccator until needed. The plates should be stable for several months; however, this will depend upon the antigen.
12. Dilute the serum antibody 1:100 or the purified antibody to 50 μg/ml in blocking buffer.
13. Add 100 μl of blocking buffer to each well of the microtiter plate.

Note: If the ELISA plate was stored dry, wash the plate several times in wash buffer (PBS containing 0.5% Tween® 20) before using the plate.

14. Add 100 μl of the diluted primary antibody to each well in row A and mix.
15. Transfer 100 l of the diluted primary antibody from each well in row A to a corresponding well in row B and mix.
16. Repeat this twofold dilution series through wells in row G. Wells in row G will contain 200 μl of the diluted primary antibody.
17. Remove 100 μl of diluted antibody from each well in row G and discard. Do not add primary antibody to the control wells in row H. Wells in row H will serve as negative controls to determine if nonspecific interactions are occurring between the secondary antibody and various concentrations of the antigen.
18. Incubate the microtiter plate at room temperature or 37°C for 1 h in a humidified container.
19. Invert the microtiter plate over a sink or suitable receptacle and flick out the contents of the wells. Tap the plate on a paper towel to remove any residual drops.
20. Rinse the plate by completely filling and emptying the wells three times with wash buffer (PBS containing 0.5% Tween® 20).
21. Invert the microtiter plate over a sink or suitable recepticle and flick out the contents of the wells. Tap the inverted plate on a paper towel to remove any residual drops.
22. Dilute the enzyme-conjugated secondary antibody in blocking buffer and transfer 100 μl of the diluted conjugate to each well of the microtiter plate.

    Note: The secondary antibody is an antibody that is used to detect the primary antibody bound to an antigen. The secondary antibody can be conjugated to an enzyme (e.g., alkaline phosphatase, peroxidase, urease, etc.), biotin, fluorescent dyes (e.g., fluorescein, rhodamine, etc.), or radioactive labels (e.g., ^{125}I or ^{3}H). These labeled antibodies are commercially available. If the primary antibody bound to the antigen is mouse IgG, then a suitable secondary antibody would be goat anti-mouse IgG conjugated with alkaline phosphatase or peroxidase.
23. Incubate the microtiter plate for 1 h at room temperature or 37°C in a humidified container.
24. Invert the microtiter plate over a sink or suitable recepticle and flick out the contents of the wells. Tap the inverted plate on a paper towel to remove any residual drops.
25. Rinse the plate by completely filling and emptying the wells three times with wash buffer (PBS containing 0.5% Tween® 20).
26. Add 100 μl of substrate to all wells.

    Note: For alkaline phosphatase, use para-nitrophenyl phosphate (PNPP) as a substrate. For peroxidase, use 2,2′-azino-bis (3-ethylbenzthiazoline-6-sulfonic acid) (ABTS) as a substrate.

    To make PNPP substrate, dissolve 1 g of PNPP in a 250-ml solution containing 40 ml of diethanolamine and 210 ml of water at a pH of 9.8. The PNPP substrate can be stored at −70°C until needed.

    To make ABTS substrate, dissolve 2.2 mg of ABTS in 10 ml of 0.05 *M* citrate buffer, pH 4.0, then add 17.5 μl of 30% H_2O_2 to the 10 ml of ABTS substrate. Once H_2O_2 is added, the substrate should be used immediately.
27. Incubate the microtiter plate for 20 to 30 min at room temperature.
28. Determine the absorbancy values (optical density readings) for each well using a spectrophotometer or a microtiter plate reader set at 405 nm for PNPP and 410 nm for ABTS.

1. Determining the Optimum Antigen and Antibody Concentrations Used in ELISAs

Once the absorbancy readings are obtained, the optimum antigen coating and primary antibody concentrations can be determined by calculating the P:N (+:−) ratio for the ELISA. To determine the optimum antigen coating concentration, use the following protocol.

1. Divide the absorbancy values of the test (P) wells in each column by the absorbancy value of the respective negative (N) control well in row H.
2. Using graph paper, plot antigen concentration on the X-axis and P:N ratio on the Y-axis. If the assay is working properly, a curve should be obtained when all of the values are plotted. The antigen coating concentration located directly beneath the peak of the curve will be the optimum antigen coating concentration that should be used in an ELISA.

The optimum antibody concentration can be determined using a similar protocol.

3. Divide the absorbancy values for the test (P) wells in each row by the absorbancy value of the respective negative (N) control well in column 12.

4. Using graph paper, plot the primary antibody concentration on the X-axis and P:N ratio on the Y-axis. If the assay is working properly, a curve should be obtained when all of the values are plotted. The primary antibody concentration located directly beneath the peak of the curve will be the optimum primary antibody concentration that should be used in an ELISA.

In some cases, nonspecific interactions between the primary antibody and the blocking buffer or negative control antigens are so extensive that the optimum antigen coating concentration and primary antibody concentration cannot be determined. In these cases, the blocking buffer, the type of plastic microtiter plate, or the concentration of the secondary antibody used can be changed to alleviate the problem. If these changes do not solve the problem, the primary antibody can be diluted in an equal volume of blocking buffer containing 0.2% Triton® X-100 and/or 1 to 2 m*M* EDTA.

Note: The negative control well designated 12H in the above assay only contains blocking buffer and the conjugated secondary antibody. If a positive signal is obtained in well 12H, then the wells of the microtiter plate were not blocked or washed well enough, or the concentration of the conjugated secondary antibody was not optimal. If the concentration of the conjugate is not optimal, then determine the optimal concentration by performing an ELISA. Vary the concentration of the conjugated antibody applied to test (P) wells which contain constant concentrations of the coating antigen and primary antibody. Also apply the conjugate to negative (N) control wells which contain blocking buffer only. Ascertain the optimal conjugate concentration utilizing the protocol used to determine the optimal primary antibody concentration (see steps 3 and 4 above).

C. MONITORING A MOUSE'S ANTIBODY RESPONSE TO AN ANTIGEN BY ELISA

Once the animal has been inoculated with the antigen, it will be necessary to determine if the animal has successfully mounted a humoral antibody immune response. If the animal successfully produces antibodies to the antigen, then the antibody-producing spleen cells can be used to generate antibody-producing hybridomas. The animal's serum, which contains the antibodies, is assayed by an ELISA or some other immunoassay for the presence of antigen-specific antibodies. The most humane way of obtaining mouse serum is via a tail bleed.

A tail bleed can be performed using the following protocol:

1. Place a mouse in a 15 cm × 15 cm × 15 cm plexiglass (or similar) container. The container should be equipped with several air holes, clean paper towels, and a 75-W light bulb mounted in the removable lid of the container.
2. Turn on the light bulb and heat the container until the mouse begins to perspire and the tail vein swells (approximately 3 to 5 min).
3. Remove the mouse from the container and place it on a surface (e.g., a metal screen or an inverted metal test tube rack) that allows the mouse to gain a foothold.
4. Grasp the tip of the mouse's tail and extend it straight back.
5. Nick the large vein on the underside of the tail with a sterile razor blade or scalpel.

   Note: Nick the tail vein as close to the tip of the tail as possible so that subsequent tail bleeds can be performed in the future. If nicks are made near the base of the tail vein, scar tissue will form which will prevent future tail bleeds.
6. Collect the blood in a 1.5-ml microfuge tube. Approximately 100 to 750 μl of blood can be obtained.
7. Apply 70% ethanol to a sterile or clean cotton ball and swab the nick on the mouse's tail. Return the mouse to the cage when the wound has stopped bleeding.
8. Centrifuge the blood sample at full speed (13,000 rpm) for 5 min in a microcentrifuge. The serum which contains the antibody will remain in the supernatant. The red blood cells (RB) will form a pellet on the bottom of the tube.
9. Transfer the supernatant to a sterile 1.5-ml microcentrifuge tube and either use the sample immediately or store the sample at −20 to −70°C until needed. The serum can be stored under these conditions for several years.
10. Assay the serum via an ELISA or other immunoassay. If the antigen-specific antibody titer is less than 1:500, inoculate the mouse again with the antigen to boost its immune response. If the titer of the serum antibody is greater than 1:500, then the spleen cells can be used to generate antibody-producing hybridomas.

If the mouse's IgG, antigen-specific antibody titer is less than 1:100 even after repeated antigen inoculations, assay the mouse for the presence of IgM antibodies. To detect IgM antibodies, use a secondary antibody preparation, e.g., alkaline phosphatase or peroxidase-conjugated goat anti-mouse IgMμ chain specific antibodies. If the IgM titer is higher than the IgG antibody titer, then the antigen may be incapable of eliciting an IgG response in the mouse. Either try another strain of mouse (see "*In vivo* antigen immunization guidelines") or couple the antigen to a carrier (e.g., keyhole limpet hemocyanin) and repeat the immunization process.

D. TISSUE CULTURE MEDIA

1. DMEM

A good medium for growing myeloma or antibody-producing hybridoma cells is Dulbecco's modified Eagle's medium (DMEM). The following formulation can be used to prepare DMEM.

1.	DMEM powder (e.g., Gibco cat. # 430-3700) which contains 4500 mg of glucose, 25 m*M* Hepes, L-glutamine	1 pkg
2.	Gentamicin @ 50 mg/ml (e.g., Sigma cat. # G1397)	1 ml
3.	100 × antibiotic, antimycotic (e.g., Sigma A9989)	10 ml
4.	Serum (heat inactivated and virus and mycoplasma free)	100 ml
5.	Tissue culture water up to	1000 ml

Filter sterilize the medium using a 0.2-μm sterile filter unit.

Note: DMEM contains phenol red, a pH indicator. Phenol red becomes orange (pH 6.5) or yellow (pH 6 or lower) under acidic pH conditions and red (pH 7) to purple (pH 8 or higher) under basic pH conditions. Sterile liquid DMEM can be obtained commercially (e.g., from Gibco cat. # 380-2430). Fresh L-glutamine should be added every 2 weeks to DMEM stored at 4°C or higher. A 200-m*M* (2.92 g of L-glutamine per 100 ml of water) is a 100× solution. The 100× L-glutamine solution should be stored at − 20°C. The antibiotics (gentamicin @ 50 μg/ml; penicillin @ 100 units/ml; and streptomycin @ 100 μg/ml of medium) and antimycotic (amphotericin B @ 0.25 μg/ml of medium) should be used if tissue culture experience and/or facilities are limited. The growth of hybridoma cells which were generated using SP2/O myeloma cells as fusion partners is not adversely affected by the simultaneous inclusion of all of these antimicrobials. If at all possible, the SP2/O cells should be grown in the absence of antimicrobials until just prior to cell fusion. If the antimicrobials are used to grow myeloma cells, they may mask the growth of a contaminant which may become resistant to the antimicrobials after the hybridomas have been laboriously generated.

The type of serum used in the tissue culture medium depends upon the origin of the myeloma cells used as the fusion partner. If the myeloma cells used as fusion partners were previously grown in medium containing fetal bovine serum, then fetal bovine serum should be used to continue growing the myeloma cells. If another serum is used, then fusions using these hybridomas may be unsuccessful.

2. DMEM Plus Azaguanine

DMEM plus azaguanine is DMEM containing 0.13 m*M* 8-azaguanine. A 50× azaguanine solution (Sigma cat. # A5284) is 6.6 m*M*.

3. SP2/O Conditioned Medium

SP2/O conditioned medium is medium in which SP2/O cells have been previously grown. The conditioned medium contains soluble factors liberated by the SP2/O cells. These soluble factors enhance the growth of B-cells and hybridomas.

To prepare SP2/O conditioned medium, use the following protocol.

1. Add 5 ml of rapidly growing SP2/O cells to 40 to 50 ml of DMEM in a T-75 tissue culture flask. Volumes and container size can be scaled up or down according to need.
2. Place the cap loosely on the T-flask and incubate the flask horizontally at 37°C in a humidified CO_2 incubator until the medium becomes slightly acidic (orange to yellow in appearance). This will generally take 2 to 3 d.
3. Gently remove the flask from the incubator. The medium should be clear until the SP2/O cells are resuspended. If the medium is cloudy, it is contaminated by microbial growth and should not be used.

4. Observe the SP2/O cells under an inverted-phase microscope. Viable SP2/O cells will appear intact and refractile. If the cells are not intact or refractile, the medium should not be used.
5. Resuspend the cells by gently shaking the flask and aseptically transfer the contents of the flask to several sterile 50-ml screw-capped centrifuge tubes.
6. Centrifuge the cells at 1500 rpm (500 × g) in a clinical centrifuge for 5 min at room temperature.
7. Aseptically transfer the SP2/O conditioned medium supernatant to a sterile container and freeze at −20°C until needed. Conditioned medium stored at −20°C is stable for 6 to 12 months.

The SP2/O cell pellets in the centrifuge tubes can be aseptically resuspended in fresh DMEM and used to make more conditioned medium. If 50 ml of cell suspension was used to produce the SP2/O cell pellet, then the cells should be resuspended in 500 ml of DMEM and incubated as before in large T-flasks (e.g., T-150) or other suitable containers.

4. OPI Medium

OPI medium is a very rich medium which contains components that enhance or promote the growth of cells previously exposed to some form of stress (e.g., cell fusion). The following recipe can be used to make 100 ml of OPI medium.

1.	SP2/O conditioned medium	50 ml
2.	100× OPI (Sigma cat. # O5003)	1.0 ml
	100× OPI contains 15 mg of oxaloacetate, 5 mg of sodium pyruvate, and 20 units of insulin per milliliter of water.	
3.	Gentamicin @ 50 mg/ml (Sigma cat. # G1397)	0.1 ml
4.	100× antibiotic/antimycotic (Sigma cat. # A9909)	1.0 ml
5.	50× HT	2.0 ml
	50× HT contains 5 m*M* sodium hypoxanthine and 0.8 m*M* thymidine	
6.	Mercaptoethanol	0.39 μl
7.	100× L-glutamine	1.0 ml
	100× L-glutamine contains 29.2 mg of L-glutamine per milliliter of water	
8.	Serum (heat inactivated and virus and mycoplasma free)	20 ml
9.	DMEM (without serum)	to 100 ml

Mix and filter sterilize through a 0.2-μm sterile filter. Store at −20°C until needed. OPI medium stored at −20°C is stable for 6 months.

5. OPI Plus Azaserine

Several different types of cell products are generated from the spleen-myeloma cell fusion. The cell products are individual unfused and fused pairs of SP2/O myeloma, spleen, or hybridoma cells. Most spleen or B-cells will only survive for about 14 d in tissue culture. SP2/O cells ($HPRGT^-$) are immortal and can outgrow antibody-producing hybridoma cells ($HPRGT^+$). Therefore, a selective agent (e.g., azaserine or aminopterin) which kills $HPRGT^-$ cells must be added to control the SP2/O cell growth.

To prepare OPI plus azaserine, asepticlly add 1 ml of 100× sterile filtered azaserine (100× azaserine is 0.58 m*M* azaserine in water) to 99 ml of OPI medium. Store frozen at −20°C until needed. The frozen medium is stable for 1 month.

6. Cell Freezing Medium

Myeloma or hybridoma cells can be stored in cell freezing medium at − 70°C or less for several years. To make cell freezing medium, add 10 ml of dimethyl sulforide to 90 ml of viral- and microbial-free serum and filter sterilize through a 0.2 m sterile filter. The cell freezing medium can be stored at −20°C until needed for up to 1 year.

E. CELL FREEZING PROTOCOL

1. Cell Freezing Protocol for a Large Number of Cells

The following protocol can be used to freeze hybridoma or myeloma cells. Not all of the cells frozen will survive long-term freezer storage. Therefore, a large number of viable cells should be frozen to ensure that the cell line can be recovered after long-term freezer storage.

1. If freezing hybridomas, perform an ELISA or other immunoassay to ensure that the hybridoma cells are producing antibodies.
   Note: If the cells are not producing antibodies or if the level of antibody production has fallen off, then antibody-producing hybridomas may be overgrown by nonantibody-producing cells when the cells are resurrected after freezer storage. If the antibody-producing cells are overgrown by other cells, then the antibody-producing cells may not be recovered. Clone cells (see cell cloning protocol) prior to freezer storage if antibody production has decreased.
2. Perform a cell count and viability stain. To ensure that the cell line can be resurrected after freezing, 10^6 to 10^7 total cells which are 70% or more viable should be frozen in a single container.
3. Aseptically transfer the cells from the tissue culture flask to a sterile screw-capped centrifuge tube.
4. Centrifuge the cells at 1500 rpm (500 $\times$ g) in a clinical centrifuge for 5 min.
5. Aseptically remove the supernatant from the cell pellet.
6. Gently resuspend the cell pellet in freezing medium. Use 1 ml of freezing medium for every 5 to 10 ml of cell suspension that was centrifuged.
7. Transfer 1 ml of the cell suspension to a 1.5-ml sterile microfuge or cryocentrifuge tube.
8. Place the 1.5-ml centrifuge tubes into a styrofoam container which has a thickness on all sides of at least 1.25 cm.
9. Close the styrofoam container and seal the lid with tape.
10. Place the styrofoam container in a -70°C freezer to allow the cells to slowly freeze overnight. If the cells are rapidly frozen, they will not survive the freezing process.
11. Rapidly transfer the cells to a suitable -70°C or liquid nitrogen storage container and store until needed. Cell lines stored in this fashion can be resurrected several years later; however, the ability to resurrect frozen cell lines decreases with increased storage.

2. Resurrecting Frozen Cell Lines

Frozen cells can be resurrected using the following protocol:

1. Remove the vial containing the frozen cells from the freezer or liquid nitrogen storage container.
2. Swab the vial with 70% ethanol.
3. Place the vial in a laminar flow hood and allow the cells to thaw at room temperature.
4. Add room temperature-warmed OPI medium to a sterile T-25 tissue culture flask. Add 10 to 15 ml of OPI medium to the flask for every 1 ml of frozen cell suspension that will be resurrected.
5. As soon as the cell suspension thaws, aseptically transfer the cell suspension to the T-flask.
6. Place the T-flask on its side in a CO_2 humidified incubator at 37°C. Incubate the cells for 4 to 7 d.
7. Remove the T-flask from the incubator and observe the cells under an inverted-phase microscope for cell growth. If the cells are growing and dividing, proceed on to the next step.
8. Dilute growing cells 1:10 in OPI medium and incubate for 2 to 3 d as described above. If the cells are growing rapidly, they can be transferred to DMEM and grown and handled as described before freezing.

3. Microtiter Plate Cell Freezing Protocol

In some cases, it may be necessary to freeze the entire contents of a microtiter plate. If the cells contained in the microtiter wells have grown to such a high density that the cells have to be fed every 1 to 2 d, then the cells can be frozen.

The following protocol can be used to freeze cells in microtiter plates.

1. Grow the cells until the medium in most of the wells has become slightly acidic (slightly yellow).
2. Aseptically remove the medium with a micropipettor. Most of the hybridoma cells will have settled out during incubation and will remain in the wells.
3. Add enough freezing medium to completely cover the cells in a well. For a 96-well microtiter plate, 50 μl or 1 drop of medium will adequately cover the cells. For cells in a 24-well plate, 500 μl will be needed.
4. Wrap the tissue culture plate in plastic wrap or Parafilm.
5. Transfer the plate to a styrofoam container which has a thickness of 1.25 cm on all sides.
6. Seal the styrofoam container with tape and place the container into a -70°C freezer to allow the cells to slowly freeze overnight.

7. Quickly transfer the frozen cells to another suitable −70°C container.

Cells frozen in this manner can be resurrected over a period of 6 months. However, since all cells in a tissue culture plate were growing at a different rate prior to freezing, not all cells will survive the freezing process.

4. Resurrecting Cells Previously Frozen in Microtiter Plates

The following protocol can be used to resurrect cells previously frozen in microtiter plates.

1. Remove the tissue culture plate from the −70°C freezer.
2. Remove the plastic or Parafilm wrapped around the plate.
3. Swab the plate with 70% ethanol.
4. Transfer the plate to a laminar flow hood and allow the cells to thaw at room temperature.
5. As soon as the cells have thawed, remove the freezing medium from the microtiter plates using a micropipettor.
6. Add 100 to 200 μl of OPI medium to each well of a 96-well microtiter plate or 1 ml of OPI medium to each well of a 24-well plate. The medium should contain antibiotics and antimycotics to prevent microbial contamination which can readily occur after cells frozen in microtiter plates are resurrected.
7. Transfer the microtiter plates to a humidified CO_2 incubator at 37°C. Incubate for 3 to 5 d or until the medium becomes slightly acidic.
8. Remove the acidic medium with a micropipettor; then add an equal volume of fresh OPI medium and incubate as above.
9. Repeat step 8, replacing spent acidic medium with fresh DMEM. Resurrected cells can be handled as described before freezing.

F. SELECTION FOR HPRGT$^-$ SP2/O CELLS

To select for spleen-myeloma cell hybridomas after cell fusion, selective agents, either azaserine or aminopterin, are incorporated into the tissue culture medium. These selective agents are toxic to cells (e.g., SP2/O cells) which lack the enzyme HPRGT and cannot synthesize purine nucleotides. If the selective agents are not incorporated into the medium, the myeloma cells may overgrow the antibody-producing hybridoma cells. SP2/O cells can become resistant to these selective agents. Prior to cell fusion, the SP2/O cells should be grown over a period of at least 1 week in medium which contains 8-azaguanine at a concentration of 0.13 m*M*. Azaguanine will select for the correct SP2/O cells which are HPRGT$^-$.

The following protocol can be used to select for HPRGT$^-$ SP2/O cells.

1. Dilute the SP2/O cells 1:5 in fresh DMEM containing 0.13 m*M* azaguanine.
2. Incubate the cells at 37°C in a humidified CO_2 incubator until the medium becomes slightly acidic. If the SP2/O cells are growing well, the medium should become acidic in 3 to 4 d.
3. Dilute the SP2/O cells 1:5 once again in medium containing azaguanine and incubate until the medium becomes slightly acidic.
4. Dilute the SP2/O cells 1:10 to 1:20 in DMEM without azaguanine. Incubate as above until the medium becomes acidic and transfer the cells to fresh medium as needed.

G. SP2/O MYELOMA CELL GROWTH PRIOR TO CELL FUSION

SP2/O cells used in the cell fusion protocol to generate hybridomas must be viable, actively growing cells. Hybridomas generated using poorly growing SP2/O cells will grow poorly or not at all.

To obtain viable, actively growing SP2/O cells, add 1 ml of actively growing SP2/O cells to 50 ml of OPI medium in a T-75 tissue culture flask 3 d prior to cell fusion. Incubate the flask on its side in a humidified CO_2 incubator at 37°C. If the SP2/O cells used to inoculate the OPI medium are growing slowly, prepare two tissue culture flasks of cells to ensure that there will be enough SP2/O cells for cell fusion.

H. MOUSE HYPERIMMUNIZATION PRIOR TO CELL FUSION

Spleen B-cells used to generate hybridomas must be producing antibodies. If the B-cells are not producing antibodies, then the hybridomas generated from the spleen and myeloma cell fusion will not produce antibodies.

To obtain a population of antibody-producing B-cells from mouse spleens, mice must be antigenically stimulated. Mice which demonstrate an antigen-positive serum antibody response in an immunoassay are antigenically stimulated. To obtain a large population of antibody-producing B-cells, the mice should be hyperimmunized with the antigen 3 to 4 d prior to cell fusion. To hyperimmunize the mouse, inject 10 to 100 μg of antigen in PBS into either the peritoneum (abdomen) or tail vein of a mouse. If mice are injected in the tail vein, care should be taken to avoid introducing air bubbles into the animal's veinous system. Air bubbles may cause blood clots which can prematurely terminate the mouse.

I. MOUSE SPLEEN REMOVAL

To generate antibody-producing hybridomas, antibody-producing B-cells from the spleens of mice are fused with SP2/O myeloma cells. To obtain a spleen from a mouse, the mouse must be terminated and the spleen must be surgically removed from the mouse. In some cases, a cardiac bleed is performed just after the mouse is terminated. The cardiac bleed clears RBCs from the spleen which may interfere with the myeloma:B-cell fusion process. The cardiac bleed also provides the researcher with a source of serum antibodies which can be useful controls in immunoassays.

1. Termination of a Mouse

Mice can be humanely terminated by CO_2 asphyxiation using the following protocol. Mice can also be terminated via cervical dislocation; however, cervical dislocation is a much more drastic method.

1. Set up a CO_2 tank with a gas regulator.
   CAUTION: The gas pressure in a CO_2 tank can be dangerously high. If you are not familar with the operation of a CO_2 tank and regulator, elicit the help of someone who is. Serious injury could result if the CO_2 tank and regulator are not handled properly.
2. Turn the control knob on the CO_2 regulator to shut off the gas flow.
3. Connect a long (20 to 30 cm) piece of rubber tubing to the CO_2 outlet on the regulator.
4. Place the free end of the rubber tubing into a container (e.g., plastic 1-l beaker, avoid glass containers which may shatter).
5. Place the mouse into the container, and cover the container and rubber tubing with a piece of tin foil.
6. Checking to make sure that the CO_2 regulator is shut off, slowly turn on the gas flow from the main CO_2 tank valve to the regulator.
7. Gradually add CO_2 gas to the plastic container housing the mouse by slowly turning on the gas flow using the CO_2 regulator. The gas flow to the container should not exceed 5 lb/in^2.
8. Introduce CO_2 to the plastic container for 3 to 5 min until the mouse is terminated.
9. Remove the mouse from the container and surgically extract the spleen as outlined below.

2. Mouse Cardiac Bleed

Once the mouse has been terminated it is still possible to obtain serum (up to 1 ml) from the mouse via a cardiac bleed. The following protocol can be used to perform a cardiac bleed.

1. Attach a 0.5-in, 26- to 30-gauge syringe needle to the end of a 1-ml syringe.
2. Grasp the mouse by the nape of the neck with the thumb and forefinger.
3. Using the same hand, trap the mouse's tail against the palm of the hand using the little finger.
4. Stretch the mouse flat by flexing the hand.
5. Place the syringe barrel flat against the abdomen of the mouse.
6. Insert the tip of the syringe needle directly under the sternum of the mouse and into the heart. Insert only one half of the syringe needle into the mouse. If the entire needle is inserted into the mouse, the needle may pass through the heart.
7. Gently pull back on the syringe plunger and withdraw as much serum as possible. If no serum is obtained, the heart may have been missed. Repeat steps 5 and 6.

It should be noted that a successful myeloma:B-cell fusion can still be performed without removing the RBCs from the spleen.

3. Surgical Removal of a Mouse Spleen and Spleen Cell Extraction

The following protocol can be used to aseptically remove the spleen from the mouse.

1. Place the mouse, on its right side, on the floor of a laminar flow hood. The mouse's spleen is located on the left side of the mouse just below the rib cage.

2. Pour 15 to 30 ml of 70% ethanol over the entire left side of the mouse to "sterilize" the left flank of the mouse. Allow the ethanol to evaporate. Note: Avoid open flames which may ignite the ethanol.
3. Grasp the fur on the left flank of the mouse just below the rib cage with a pair of forceps. Gently lift the fur up.
4. Make a 1- to 2-cm incision in the fur and the skin near the forceps using a sterile scissors. Do not cut through the clear, outer peritoneal membrane below the skin. The intact peritoneal membrane will prevent contamination of the spleen by agents present on the fur or skin.
5. Grasp the fur and skin on both sides of the incision with the thumbs and forefingers of both hands. Peel the fur and skin sufficiently away to reveal the spleen. The spleen lies beneath the rib cage and underneath the peritoneal membrane and looks very much like a miniature, purple, flattened hot dog.
6. Grasp the thin outer peritoneal membrane with a sterile forceps and lift.
7. Using a sterile dissecting scissors, cut and trim away the peritoneal membrane sufficiently to allow the spleen to be removed.
8. Using a sterile forceps, gently grasp the tip of the spleen by its lower end and lift.
9. Using a sterile dissecting scissors, cut the spleen free from any connective tissue. Do not cut the intestine, which lies close to the spleen. If the intestine is cut the spleen will be heavily contaminated by intestinal microbes.
10. Place the spleen in a sterile petri dish. If the spleen is dropped or accidentally contaminated, pick the spleen up with a sterile forceps and rinse it with 20 to 50 ml of sterile DMEM, then place the spleen in another sterile petri dish and proceed.
11. Cover the spleen with 5 to 10 ml DMEM (without serum or additives).
12. Place sterile 1- to 1.5-in, 18-gauge needles onto each of two syringes.
13. Tease the spleen apart into small pieces using the sterile needles.
14. Pipet the spleen and DMEM up and down using a large-bore, 10 to 25-ml pipet. This process will release the spleen cells from the spleen.
15. Transfer the spleen cell suspension (minus the spleen membrane) from the petri dish to a sterile, 50-ml screw-capped centrifuge tube. The spleen cells can be used for cell fusion or can be stimulated *in vitro* with an antigen.

    Note: The spleen cells should be used within 15 to 30 min after they are removed. If the spleen cells are allowed to stand for any lenth of time they may not be suitable for cell fusion or *in vitro* antigenic stimulation.

J. SP2/O CELL COUNT AND VIABILITY STAIN

The ratio of spleen cells to SP2/O myeloma cells used in cell fusion to generate hybridomas is approximately five spleen cells to one myeloma cell. For every spleen, 1 to 3 $\times$ 10^8 viable spleen cells can be obtained. If the spleen cells were stimulated with an antigen in tissue culture over a 5-d period, then only 2 to 4 $\times$ 10^7 total spleen cells will probably be viable. Therefore, to achieve a 5:1 spleen to SP2/O ratio, 3 to 5 $\times$ 10^7 SP2/O cells will be needed for spleen cells obtained from animals antigenically stimulated *in vivo.* For spleen cells stimulated with an antigen *in vitro,* 4 to 6 $\times$ 10^6 SP2/O cells will be needed.

The SP2/O cells should be at least 90% viable. If the viability of the SP2/O cells is low (70% or less), then efficiency of the cell fusion will be low.

The following protocol can be used to determine the viability and the number of SP2/O cells per milliliter of medium.

1. Prepare a 0.4% trypan blue stain in PBS. Trypan blue is a viability stain which stains dead cells blue, but does not stain viable cells. To prepare trypan blue, dissolve 0.4 g of trypan blue in 100 ml of PBS. Filter sterilize through a 0.2-m filter. Store at room temperature until needed.
2. Add 90 μl of trypan blue viability stain to a microfuge tube.
3. Gently mix the 3-d old SP2/O cell culture to resuspend the cells.
4. Aseptically transfer 10 μl of the resuspended SP2/O cells to 90 μl of trypan blue viability stain. Gently mix the cells and the stain.
5. Transfer 10 to 20 μl of the stained cell suspension to a hemocytometer or Petrof-Hauser counting chamber.
6. Perform a cell count and determine the number of cells per milliliter according to the manufacturer's instructions. Count the viable (unstained) and dead (blue-stained) cells.

7. Multiply the cell counts by 10 to correct for the 1:10 dilution factor (10 μl of cells in 90 μl of stain).
8. Determine percent cell viability using the following formula:

$$\% \text{ viability} = 100 \times \frac{(\# \text{ of viable cells})}{(\# \text{ of viable cells}) + (\# \text{ of dead cells})}$$

9. Determine the total number of viable cells in the entire cell suspension.
   Note: If there are 5×10^5 cells per milliliter of medium and 100 ml of cell suspension, then there are 5×10^7 total cells present in the 100-ml cell suspension.

K. SPLEEN-MYELOMA CELL FUSION PROTOCOL

Before the cell fusion protocol can be performed, the spleen should have been surgically removed from the mouse, and the spleen cells should have been extracted from the spleen (see "Mouse spleen removal"). In addition, the number of viable SP2/O cells per milliliter of OPI medium should have been determined (see "SP2/O cell count and viability stain"). The SP2/O cells should be 90% viable. Finally, the cell fusion protocol should be reviewed and all materials and reagents should be organized so that the fusion protocol can be performed without interruption. If spleen and/or SP2/O cells are allowed to stand between steps during the cell fusion, the efficiency of the cell fusion process will be diminished.

The cells from each spleen obtained from an antigenically stimulated mouse should be fused with 3 to 5×10^7 total viable SP2/O cells. In some cases, it may be necessary to centrifuge the SP2/O cell contents of two T-flasks and combine the resuspended SP2/O cell pellets to obtain sufficient SP2/O cells for cell fusion. If the mouse spleen cells were antigenically stimulated *in vitro* (in tissue culture), 4 to 6×10^6 total, viable SP2/O cells should be used for cell fusion.

The following cell fusion protocol can be used to generate antibody-producing hybridoma antibodies.

1. Add 3 to 5×10^7 SP2/O cells to the 50-ml screw-capped centrifuge tube containing spleen cells derived from an antigenically stimulated mouse. Or, add 4 to 6×10^6 SP2/O cells to the tube which contains spleen cells antigenically stimulated *in vitro.*
2. Briefly mix the cells, then centrifuge the cells at 1500 rpm ($500 \times g$) in a clinical centrifuge at room temperature for 5 min.
3. Aseptically remove the supernatant without disturbing the cell pellet.
4. Resuspend the cell pellet with the small amount of medium (about 100 μl) that will invariably remain in the tube.
5. Fill a pipet with 1 ml of sterile, hybridoma-tested polyethylene glycol (e.g., PEG 1500; Boehringer Manheim cat. # 783641).
   Note: PEG 1500 is made up as a 50% w/v solution in 75 m*M* Hepes, pH 8.0.
6. Swirl the cell pellet in the 50-ml centrifuge tube until the cells form a thin film over the lower 1-in (approximately) portion of the tube. Immediately begin the next step.
7. Slowly add 1 ml of PEG to the 50-ml centrifuge tube over a period of 10 s.
8. Gently swirl the PEG:cell suspension for 15 to 30 s. Immediately begin the next step.
9. Fill a 10 to 25-ml pipet with 11 ml of DMEM (which contains no additives, e.g., serum or antibiotics). The 11 ml of DMEM will be used for steps 10 and 12.
10. Slowly add 1 ml of DMEM over a period of 10 s.
11. Swirl the PEG:cell suspension for 5 to 10 s. Immediately begin the next step.
12. Slowly add 10 ml of DMEM (which contains no additives) over a period of 30 s.
13. Cap the centrifuge tube and gently invert several times to mix the cell suspension.
14. Incubate the cell suspension for 5 min at room temperature.
15. Centrifuge the cell suspension in a clinical centrifuge at 800 rpm for 8 min at room temperature.
16. Aseptically remove the supernatant.
17. Gently resuspend the cell pellet in several milliliters of OPI medium containing azaserine by using a large-bore, 10 to 25-ml pipet equipped with a propipet bulb or similar device. It may be necessary to gently pipet the cell suspension up and down the pipet to completely resuspend most of the cells. However, it is not necessary to completely resuspend the cells.
18. Transfer the resuspended cells to 80 to 100 ml of OPI medium containing azaserine, then mix the cell suspension.
19. Use a large-bore, 10 to 25-ml pipet to dispense 100 μl (approximately two drops) of the cell suspension to individual wells of 96-well tissue culture plates.

20. Incubate the cells at 37°C in a humidified CO_2 incubator for 2 to 3 d.
21. After 2 to 3 d, feed the cells with 100 μl of OPI medium (without azaserine). Incubate as in step 20.
22. Assay the hybridoma cells for antibody production when the medium becomes slightly yellow or acidic (9 to 14 d post-fusion).
23. Cells can be fed by removing acidic or yellow medium and adding fresh OPI or DMEM.
24. Clone antibody-producing hybridoma cells by limiting-dilution as described in Section II.L below.

L. CLONING ANTIBODY-PRODUCING HYBRIDOMAS

An antibody-producing hybridoma cell line which arises from a single cell can be obtained from a population of cells either by limiting-dilution or by single-cell cloning in soft agar. Either process has its advantages and disadvantages. However, cloning by limiting-dilution is easier to set up and is less time consuming than cloning in soft agar. Additionally, the agar used in the soft agar cloning protocol can be toxic to some hybridomas. Therefore, only cloning by limiting-dilution will be described here.

The following limiting-dilution protocol can be used to obtain a single cell clone (monoclonal) from a population of hybridomas.

1. Perform a viable cell count on the hybridomas to be cloned using the SP2/O cell count and viability staining protocol.
2. Dilute the hybridoma cells to 50 viable cells per milliliter in 5 to 50 ml of OPI medium.
   Note: If the hybridoma cells grow readily, dilute the cells to five to ten viable cells per milliliter of medium.
3. Mix the diluted hybridoma cells. Transfer 100 μl (two drops) of the diluted cell suspension to each well of one or more 96-well tissue culture plates using either a micropipettor or a 5 to 25-ml pipet.
4. Incubate the hybridoma cells at 37°C in a humidified CO_2 incubator for 4 to 5 d.
5. Remove the tissue culture plates from the incubator and place under an inverted-phase microscope.
6. Identify and mark those wells which contain a single colony derived from a single cell. Return the microtiter plate to the incubator and incubate for an additional 5 to 7 d.
7. Perform an ELISA on single-colony wells and assay for the presence of antigen-specific antibodies. If all single-colony wells fail to produce a positive signal in ELISA, then assay all of the wells which contain colonies arising from more than one cell.
8. Reclone single-colony, antigen-positive wells two to three more times. Assay for antigen-specific antibody activity after each cloning. If antigen-positive wells assayed in step 7 contained more than one colony, then reclone the well to obtain an antigen-positive single-colony clone; assay by ELISA and reclone two to three more times to ensure that the cell line is a monoclonal antibody-producing hybridoma.
9. Prepare frozen stocks of cloned cells according to the cell freezing protocol.

M. PRODUCTION OF HIGH LEVELS OF MONOCLONAL ANTIBODIES IN MOUSE ASCITES

High levels of monoclonal antibodies can be obtained from mouse ascites. Ascites, in this case, is a fluid that accumulates in the peritoneum (abdomen) of an immunosuppressed mouse that has been injected with hybridoma or myeloma cells. The concentration of monoclonal antibodies can vary from 0.5 to 20 mg/ml of ascites. Variations will occur from hybridoma to hybridoma. In some cases, antibodies generated in ascites fluid may react differently with the antigen than the same antibodies produced in tissue culture.

Mice (6 to 8 weeks of age) are initially inoculated with pristane to suppress the mouse immune response to the hybridoma cells. If the mouse's immune system was not suppressed, then it would prevent hybridoma cell growth and antibody production. Several weeks after pristane inoculation, mice are inoculated with the hybridoma cells. The ascites fluid, which contains the antibody, is collected 2 to 4 weeks later.

The following protocol can be used to generate monoclonal antibody in ascites.

1. Inoculate each mouse intraperitoneally (i.p.) with 0.5 ml of pristane. Inoculate five mice for each hybridoma. Wait at least 1 week or more (4 weeks or more is optimal) before inoculating mice with hybridoma cells.
   Note: Mice can be used up to 6 months after pristane injection.

2. Perform a hybridoma viable cell count using the SP2/O cell count and viability staining protocol.
3. Add 5 $\times$ 10^6 total viable hybridoma cells to a sterile, screw-capped centrifuge tube.
4. Centrifuge the cells at 1500 rpm (500 $\times$ g) in a clinical centrifuge for 5 min at room temperature.
5. Aseptically remove the supernatant and gently resuspend the cells in 15 ml or more of sterile PBS.
6. Repeat step 4.
7. Aseptically remove the supernatant and gently resuspend the cells in approximately 2.5 ml of sterile PBS.
8. Inoculate each pristane-primed mouse i.p. with 0.5 ml of the resuspended hybridoma cells or 1 $\times$ 10^6 total cells.
9. Observe the hybridoma-inoculated mice over a 2- to 4-week period. The abdomen of mice actively producing antibody-bearing ascites will be swollen two to three times larger than normal.
10. Insert a 1- to 1.5-in, 18-gauge syringe needle approximately 0.5 in into the abdomen of the ascites-producing mouse.
11. Collect the ascites fluid draining through the 18-gauge needle into a 50-ml, sterile, screw-capped centrifuge tube.

    Note: If the syringe needle becomes plugged, remove the needle from the mouse and collect the ascites fluid draining directly from the needle puncture in the abdomen. The mouse's abdomen can be gently massaged to stimulate the flow of ascites.
12. Return the mouse to its cage once ascities fluid has been collected.

    Note: Mice which appear to be suffering should be terminated via CO_2 asphyxiation.
13. Centrifuge the ascites fluid at 1500 rpm (500 $\times$ g) in a clinical centrifuge for 10 min at room temperature.
14. Transfer the antibody-containing supernatant to a sterile, screw-capped centrifuge tube. The ascites fluid can be stored at -20 to -70°C for 6 to 12 months (or more, in some cases) until needed.

    Note: Antibody in ascites which is stored unfrozen can readily degrade.

Antibodies can be purified from ascites using traditional antibody purification protocols. IgG antibodies can be purified using protein A or G affinity matrices, Affigel-blue, caprylic acid, and/or ammonium sulfate precipitation and size-exclusion, hydrophobic, or ion-exchange chromatography. IgM antibodies can be purified by boric acid and/or ammonium sulfate precipitation and size-exclusion chromatography. Purified monoclonal antibodies can be radiolabeled or labeled with biotin or a variety of enzymes, e.g., alkaline phosphatase, peroxidase, urease, or β-galactosidase.[3,4] The purified and labeled antibodies can then be used in a variety of diagnostic immunoassays.

REFERENCES

1. Mernaugh, R. L., Mernaugh, G. R., and Kovacs, G. R., The immune response: antigens, antibodies, antigen-antibody interactions, in *Serological Methods for Detection and Identification of Viral and Bacterial Plant Pathogens,* Hampton, R, Ball, E, and De Boer, S., Eds., APS Press, St. Paul, MN, 1990.
2. Roitt, I., *Essential Immunology,* 7th ed., Blackwell Scientific, Boston, 1991.
3. Harlow, E. and Lane, D., *Antibodies: a Laboratory Manual,* Cold Spring Harbor Laboratory, Cold Spring Harbor, NY, 1988.
4. Coligan, J. E., Kruisbeek, A. M., Margulies, D. H., Shevach, E. M., and Strober, W., *Current Protocols in Immunology,* Green Publishing Associates and Wiley-Interscience, New York, 1992.
5. Van Duessen, R. A. and Whetstone, C. A., Practical aspects of producing and using anti-viral monoclonal antibodies as diagnostic reagents, *Annu. Proc. Am. Assoc. Vet. Lab. Diag.,* 24, 211, 1981.
6. MacDonald, H. R., Shneider, R., Lees, R. K., Howe, R. C., Acha-Orbea, H., Fertenstein, H., Zinkernagel, R. M., and Hengartner, H., T cell receptor V use predicts reactivity and tolerance to Mls-encoded antigens, *Nature,* 322, 40, 1988.

Chapter 25

An Overview of Phage-Displayed Recombinant Antibodies

R. Mernaugh and G. Mernaugh

TABLE OF CONTENTS

I. INTRODUCTION

Phage-displayed recombinant antibodies were first described by McCafferty et al.[1] They are antibodies that have been genetically cloned and expressed on the tip of the M13 bacteriophage. The following overview will discuss the properties of M13 and its life cycle, the production of phage-displayed antibodies, and the advantages of using such antibodies.

M13 phage are flexuous, 6- × 870-nm, single-stranded (ss) DNA viruses which infect srains of *Escherichia coli* that carry the f′ episome (plasmid). Although M13 is a bacterial virus, it behaves like a bacterial parasite in that the cells it infects constantly produce and secrete intact M13 virus particles without lysing the host cell. The M13-infected *E. coli* host cell acts like a factory to constantly produce M13 phage components. These components include phage DNA, gene 8 coat proteins, gene 3 attachment proteins, and other proteins which may be linked or fused to these phage proteins. There are approximately 2700 copies of the gene 8 coat protein for every phage particle. The phage coat protein surrounds the phage DNA and protects the DNA from degradation. There are three to five copies of the gene 3 attachment protein per phage particle. These proteins, located on the tip of the phage, are responsible for phage attachment to receptors located on the *E. coli* cell. The receptor sites are hair-like projections (pili) expressed on the surface on *E. coli* which carry the f′ episome.[2]

The production of M13 phage-displayed recombinant antibodies involves genetically linking DNA from antibody-producing B-lymphocytes or hybridomas to the phage gene 3 DNA. The proteins encoded by the antibody and gene 3 DNA are coexpressed or fused to one another to produce an antibody-gene 3 fusion protein. Since the gene 3 and antibody DNA and proteins are linked, a bacteriophage carrying the gene fusion will simultaneously contain the antibody DNA and express an antibody molecule on its tip.

II. PHAGE LIFE CYCLE

To understand how phage-displayed antibodies are produced and genetically manipulated, it will be necessary to briefly discuss the M13 phage life cycle.

M13 attaches to the pili expressed on the surface of *E. coli.* Once attached, the DNA and some phage proteins are taken up by the cell, after which phage proteins are removed from the phage ssDNA. *E. coli* enzymes (e.g., DNA polymerases) interact with the ssDNA template to produce double-stranded (ds) DNA. Several phage proteins are also produced from transcription of the phage DNA and translation of the resulting mRNA template. M13 phage and *E. coli* encoded proteins then interact with the dsDNA to generate new ssDNA, which is packaged into new intact progeny phage particles. Phage-encoded proteins and ssDNA produced during the phage infection cycle are eventually transported to the inner cell membrane where an intact phage particle is assembled and eventually secreted through the cell wall into the surrounding medium. Any antibody DNA linked to phage DNA and any antibody proteins fused to phage proteins are also assembled and secreted in a similar fashion.[2,3]

0-87371-877-1/95/$0.00+$.50

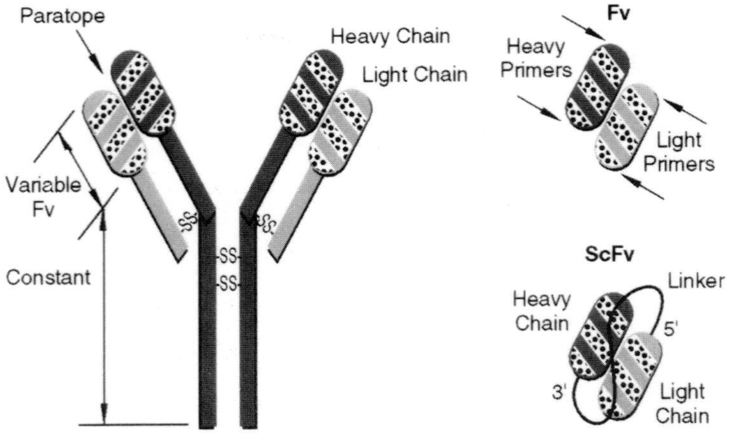

Figure 1 An intact antibody (Y-shaped) and the Fv and ScFv fragments that are cloned and displayed on the tip of M13 phage.

III. PROCEDURE FOR THE RECOMBINANT ANTIBODY PRODUCTION

The following discussion will describe the general procedure used to produce phage-displayed recombinant antibodies. The phage-displayed recombinant antibodies which are described are genetically cloned single-chain Fv (ScFv) antibodies. The antigen-binding domain (paratope) of the antibody is the only portion of the antibody that is expressed in this system. The expressed ScFv antibodies consist of the antibody heavy and light chain variable fragments (Fv) linked together to form a single chain (Figure 1).

The steps used to generate ScFv phage-displayed antibodies are outlined in Figure 2.

Messenger RNA from antibody-producing B-lymphocytes or hybridomas provides the template for production of recombinant antibodies (Figure 2A). Antibody mRNA is used because it contains complete heavy- and light-chain sequences needed to produce functional antibody proteins, and it can be purified away from other cellular components (e.g., rRNA, tRNA, proteins, etc.) that may interfere with subsequent reactions. The antibody mRNA, which is polyadenylated on the 3′ end (contains long stretches of adenine residues), can be readily obtained by affinity purification on oligo(dT) bound to a solid support or matrix. The oligo(dT) contains long stretches of thymine (T) residues which will, in the presence of high concentrations of salt, anneal to the polyadenylated tract on the 3′ end of the antibody mRNA. The oligo(dT) matrix is generally contained in a column. Cellular components applied to the column which do not anneal to oligo(dT) in the presence of salt are removed by salt buffer washes. The affinity purified mRNA is subsequently eluted off with water or a buffer that contains no salt.[3]

The antibody mRNA serves as a template for reverse transcriptase to generate a complimentary strand of DNA (cDNA) in a first-strand cDNA reaction (Figure 2B).[3] Random hexamers are also used in the first-strand reaction. The random hexamers are six base synthetic oligonucleotides containing any one of the four bases (deoxyadenosine, thymine, cytosine, or guanine) randomized at each of the six base positions. Random hexamers which anneal to or interact with the mRNA template act as primers ("stepping stones") by reverse transcriptase to catalyze the addition of nucleotides onto the newly forming DNA molecule.

The total amount of antibody DNA generated during the first-strand cDNA reaction is generally insufficient for antibody cloning purposes. Consequently, the antibody DNA must be amplified before it can be cloned.

Antibody cDNA is amplified using the polymerase chain reaction (PCR), *Taq* DNA polymerase, and synthetic DNA oligonucleotide primers (Figure 2C). These primers are complimentary to the 5′ and 3′ ends of the heavy- and light-chain antibody variable regions.[1] During PCR, the 3′ oligonucleotide primers anneal to the 3′ end of the antibody cDNA molecule generated in the first-strand cDNA reaction. *Taq* polymerase catalyzes the addition of nucleotides onto the synthetic oligonucleotide primers using the cDNA strand as a template. The resulting product is a dsDNA molecule comprised of the cDNA template and the newly formed DNA strand. The dsDNA is heated to separate (denature) the DNA into two separate strands. When the DNA is cooled, primers specific for the 5′ and 3′ ends of the antibody molecule anneal, respectively, to either the newly formed DNA molecule or the original cDNA

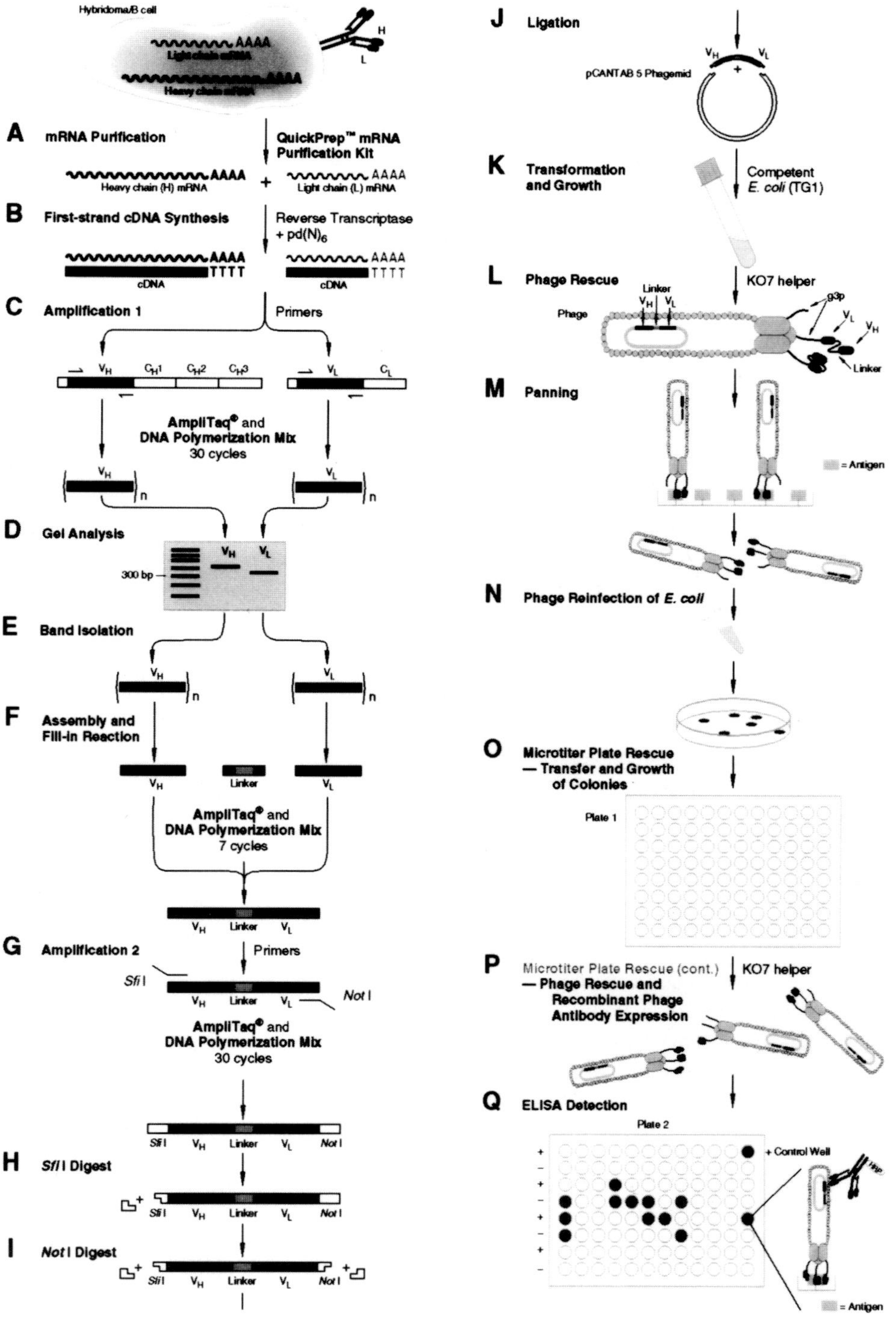

Figure 2 The steps followed to generate phage-displayed recombinant antibodies using mRNA obtained from antibody-producing B-cells or hybidomas as template for the production of recombinant antibodies. Note: The PCR process is covered by U.S. Patents 4,683,195 and 4,683,202 owned by Hoffman-LaRoche Inc. Use of the PCR process requires a license.

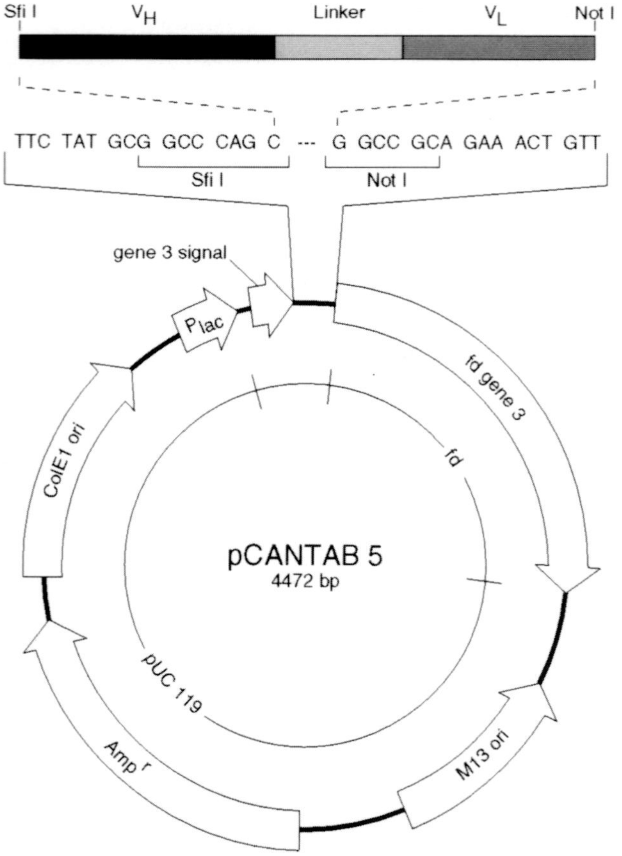

Figure 3 Genetic map of the pCANTAB 5 phagemid vector which is used to clone and express recombinant phage-displayed antibodies.

template. *Taq* polymerase, again, extends the primers using either the newly formed DNA or the cDNA strand as template. This process is repeated 20 to 30 times using a thermal cycler until the concentration of the antibody cDNA generated during the first-strand reaction is increased 2^{20} to 2^{30} or 1,048,576 to 1,073,741,824 times.[3]

The amplified antibody DNA heavy (approximately 340 bp) and light (approximately 325 bp) chains are separated from other components of the PCR by agarose gel electrophoresis (Figure 2D). The antibody DNA bands, which are stained with ethidium bromide, are visualized under ultraviolet light, sliced out of the gel, and extracted from the agarose by chemical or physical means (Figure 2E).[3]

The purified heavy and light chains are assembled into a single fragment using complimentary synthetic oligonucleotide linker primers (Figure 2F). The 5′ and 3′ ends of the linker primers are homologous with and anneal to the 3′ and 5′ ends of the heavy and light antibody chains.[4] *Taq* DNA polymerase catalyzes the addition of nucleotides to these linker primers using the antibody DNA as a template. The resulting products consist of heavy chain-linker or light chain-linker DNA. Since the linkers are complimentary they anneal to each other, *Taq* polymerase is then used to extend each linker-chain combination to produce a fully assembled ScFv product.[4]

The assembled ScFv fragment is again amplified using primers which respectively add *Sfi I* and *Not I* restriction sites to the 5′ and 3′ ends of the antibody DNA. (Figure 2G). The amplified product is then digested with *Sfi I* (Figure 2H) and *Not I* (Figure 2I) for ligation into an *Sfi I*/*Not* I digested recombinant antibody expression vector (Figure 2J) such as the pCANTAB 5 phagemid (Figure 3).

The pCANTAB 5 phagemid contains plasmid pUC119 (amplicillin resistance, Amp^r) and phage DNA sequences (phage + plasmid = phagemid). Phage sequences in pCANTAB 5 include gene 3, M13 origin of replication (ori), expression promoter (P_{lac}), and gene 3 signal (leader) DNA sequences. The glucose-suppressed P_{lac} promoter controls expression of the ScFv antibody gene product which can be toxic to the bacterial cells. Under high glucose concentrations ScFv antibody production is suppressed.[3] Once phage infection is initiated, glucose is removed from the medium. ScFv antibody is coexpressed as a fusion protein with the phage gene 3 protein.[1,4] The M13 ori of the phagemid associates with phage proteins during the phage infection process to initiate assembly of an intact phage particle. Any DNA associated with the M13 ori, including the phagemid and antibody DNA, will also be assembled into an intact phage particle. The gene 3 leader or signal sequence encodes for proteins which target antibody-gene 3 fusion proteins to the cell membrane where they will be assembled into a maturing phage particle and subsequently extruded from the cell.[3]

pCANTAB 5, containing the ScFv antibody gene, is introduced into *E. coli* cells by transformation (Figure 2K). Transformed cells are grown in the presence of high glucose concentrations to suppress the buildup of antibodies which may be toxic to the cells.[5] The cells are then infected with a helper phage such as M13KO7 (Figure 2L) in a process known as phage rescue. The helper phage contains all of the necessary M13 viral DNA components needed to initiate an active phage infection that will lead to the production of intact M13 phage particles. The helper phage contains the kanamycin (Kan^r) resistance marker and a defective M13 ori. Cells infected with helper phage can be selected for on media containing kanamycin. Since the helper phage contains a defective M13 ori, helper phage DNA will either be poorly packaged or not packaged at all into intact phage particles. Consequently, nearly all intact phage particles will contain the phagemid DNA with the ScFv antibody insert and will display an antibody molecule attached to the gene 3 protein at the phage tip. *E. coli* cells transformed with the phagemid (Amp^r) containing the antibody insert and infected with helper phage (Kan^r) are then grown in low-glucose liquid medium containing ampicillin and kanamycin to produce phage antibodies.[5]

After production, phage antibodies are transferred to an antigen-coated support, such as a plastic microtiter well or tube, and allowed to interact with the antigen (Figure 2M).[1] This method of selecting for antigen-positive phage antibodies is known as panning. Phage antibodies which fail to bind to the antigen are removed by washing; and the antigen-positive phage antibodies are eluted using either a drop in pH, excess antigen, triethylamine, or some other suitable eluant. Eluted phage antibodies, containing the phagemid and ScFv DNA insert, are then used to reinfect *E. coli* cells. Infected cells are plated onto medium containing ampicillin to select for phagemid-containing cells (Figure 2N). Individual colonies are transferred to 96-well microtiter plates (Figure 2O) and a small-scale helper phage rescue is performed on each well containing a colony and culture medium (Figure 2P). The microtiter plates are centrifuged to pellet out bacterial cells and the culture supernatant containing the phage antibodies are transferred to antigen-coated 96-well microtiter plates. Phage antibodies bound to antigen are then detected in an ELISA using an enzyme-conjugated antibody which is specific for M13 coat proteins (Figure 2Q). Positive colonies can then be grown in large cultures and rescued with M13 to produce an unlimited supply of phage-displayed antibodies.

IV. ADVANTAGES OF PHAGE-DISPLAYED RECOMBINANT ANTIBODIES

Phage-displayed recombinant antibodies have several advantages over polyclonal antibodies or hybridoma-derived monoclonal antibodies. The advantages are

1. Phage-displayed antibodies can be generated quickly. Commercially available recombinant antibody kits can be used to produce phage-displayed antibodies within 8 d.
2. An antigen-positive recombinant antibody clone can be easily selected by panning a population of phage-displayed antibodies against immobilized antigen.[1,5–8]
3. The antibody molecule and the antibody DNA are simultaneously contained on one phage particle. Consequently, the antibody DNA which represents the expressed antibody protein can be easily isolated and mutated using traditional molecular biology protocols[3] to produce antigen-specific antibodies with unique properties.[9]
4. PCR-amplified antibody heavy- and light-chain DNA obtained from different recombinant antibodies, hybridomas, or B-lymphocytes can be mixed in the assembly reaction (Figure 2F) to produce antibodies with unique or stronger antigen-binding characteristics.[10]
5. Once cloned, liters of phage-displayed recombinant antibodies can be produced inexpensively from bacterial culture supernatant, and the phage antibodies can be used directly in an immunoassay without purification.

V. CONCLUDING REMARKS

The recombinant antibodies described in this article are phage-displayed ScFv antibodies.[1,4,5,7,11–19] Phage-displayed Fab antibodies have also been generated.[6,14,20,21] Fab antibodies are FV antibodies which contain a single antibody consisting of heavy and light chain constant regions attached, respectively, to the 3′ ends of the variable heavy and light chains. Soluble ScFv and Fab recombinant antibodies, which are not phage-displayed and are not expressed as a fusion protein, have also been produced.[8–10,22–24]

Recombinant antibody technology represents a new approach which can be used to quickly develop inexpensive immunological reagents.[7] Since recombinant antibodies can be genetically engineered, they can be modified to produce unique reagents[9] which can be used in novel or traditional immunoassays.

REFERENCES

1. McCafferty, J., Griffiths, A. D., Winter, G., and Chiswell, D. J., Phage-antibodies: filamentous phage displaying antibody variable regions, *Nature,* 348, 552, 1990.
2. Kornberg, A., *DNA Replication,* W. H. Freeman, San Francisco, 1974, 724 pp.
3. Sambrook, J., Fritsch, E. F., and Maniatis, T., *Molecular Cloning, a Laboratory Manual,* 2nd ed., Cold Spring Harbor Laboratory, Cold Spring Harbor, NY, 1989.
4. Clackson, T., Hoogenboom, H. R., Griffiths, A. D., and Winter, G., Making antibody fragments using phage display, *Nature,* 352, 624, 1991.
5. Griffiths, A. D., Malmquist, M., Marks, J. D., Bye, J. M., Embelton, M. J., McCafferty, J., Baier, M., Holligan, K. P., Gorick, B. D., Hughes-Jones, N. C., Hoogenboom, H. R., and Winter, G., Human anti-self antibodies with high specificity from phage display libraries, *EMBO J.,* 12, 725, 1993.
6. Burton, D. R., Barbas, C. F., III, Persson, M. A. A., Koenig, S., Charock, R. M., and Lerner, R. A., A large array of human monoclonal antibodies to type 1 human immunodeficiency virus from combinatorial libraries of asymptomatic seropositive individuals, *PNAS,* 88, 10134, 1991.
7. Marks, J. D., Hoogenboom, H. R., Bonnert, T. P., McCafferty, J., Griffiths, A. D., and Winter, G., By-passing immunization. Human antibodies from V-gene libraries displayed on phage, *J. Mol. Biol.,* 222, 581, 1991.
8. Pack, P. and Pluckthum, A., Miniantibodies: use of amphipathic helices to produce functional, flexibly linked dimeric Fv fragments with high avidity in *Escherichia coli, Biochemistry,* 31, 1579, 1992.
9. Better, M., Chang, C. P., Robinson, R. R., and Horwitz, A. H., *Escherichia coli* secretion of an active chimeric antibody fragment, *Science,* 240, 1041, 1988.
10. Kang, S. A., Jones, T. M., and Burton, D. R., Antibody redesign by chain shuffling from random combinatorial immunoglobulin libraries, *PNAS,* 88, 11120, 1991.
11. Bird, R. E. and Walker, B. W., Single chain-antibody variable regions, *TIBTECH,* 9, 132, 1991.
12. Huston, J. S., Mudgett-Hunter, M., Tai, M-S., McCartney, J., Warren, F., Haber, E., and Oppermann, H., Protein engineering of single-chain Fv analogs and fusion proteins, *Methods Enzymol.,* 203, 46, 1991.
13. Johnson, S. and Bird, R. E., Construction of single-chain Fv derivatives of monoclonal antibodies and their production in *Escherichia coli, Methods Enzymol.,* 203, 88, 1991.
14. Marks, J. D., Hoogenboom, H. R., Griffiths, A. D., and Winter, G., Molecular evolution of proteins in filamentous phage, *J. Biol. Chem.,* 267, 16007, 1992.
15. Schmidt, T. G. M. and Skerra, A., The random peptide library-assisted engineering of a C-terminal affinity peptide, useful for the detection and affinity purification of a functional Ig Fv fragment, *Protein Eng.,* 6, 109, 1993.
16. Soderlind, E., Lagerkvist, A. C., Duenas, M., Malmborg, A-C., Ayala, M., Danielsson, L., and Borrebaeck, C. A. K., Chaperonin assisted phage display of antibody fragments in filamentous bacteriophage, *Bio/Technology,* 11, 503, 1993.
17. Whitlow, M. and Filpula, D., Single-chain Fv proteins and their fusion proteins, in *Methods: a Companion to Meth. in Enz.,* 2, 97, 1991.
18. Winter, G., Immunological Techniques, *Curr. Opin. Immunol.,* 5, 253, 1993.
19. Winter, G. and Milstein, C., Man-made antibodies, *Nature,* 349, 293, 1991.
20. Barbas, C. F., III, Kang, A. S., Lerner, R. A., and Benkovic, S. J., Assembly of combinatorial libraries on phage surfaces: the gene III site, *PNAS,* 88, 7978, 1991.

21. Hoogenboom, H. R., Griffiths, A. D., Johnson, K. S., Chiswell, D. J., Hudson, P., and Winter, G., Multi-subunit proteins on the surface of filamentous phage: methodologies for displaying antibody (Fab) heavy and light chains, *Nucleic Acids Res.,* 19, 4133, 1991.
22. Pluckthum, A., Antibody engineering: advances from the use of *Escherichia coli* expression systems, *Bio/Technology,* 9, 545, 1991.
23. Skerra, A., Pfitzinger, I., and Pluckthum, A., The functional expression of antibody Fv fragments in *Escherichia coli:* improved vectors and a generally applicable purification technique, *Bio/Technology,* 9, 273, 1991.
24. Wu, X-C., Ng, S-C., Near, R. I., and Wong, S-L., Efficient production of a functional single-chain antidigoxin antibody via an engineered *Bacillus subtilus* expression-secretion system, *Bio/Technology,* 11, 71, 1993.

Chapter 26

Direct Tissue Blot Immunoassay for Analysis of Plant Pathogens*

H. T. Hsu, R. H. Lawson, N. S. Lin, and Y. H. Hsu

TABLE OF CONTENTS

I. INTRODUCTION

Immunological assay is the single most important method for disease diagnosis and pathogen detection in use today. It offers great versatility in the type of test and format used in specific serological tests.[1] Many improvements have been made over the years in serological procedures.[2] The procedures have progressed from microprecipitin tests[3] and agar gel double diffusion tests[4] to more sophisticated assays such as enzyme-linked immunosorbent assays (ELISA)[5] and immunosorbent electron microscopy (ISEM).[6] Both ELISA and ISEM increase sensitivity of the assays of plant viruses by several orders of magnitude over the gel double diffusion test or liquid precipitin test. Although new methods are being developed and introduced because of greater ease of operation; better quantitation, greater sensitivity, or greater applicability, the underlying principles remain the same.

Recent development of an immunological technique that utilizes a direct blotting of plant tissue onto nitrocellulose membranes adds a new dimension in the studies of plant pathology.[7] Different terminologies of the blotting technique have been described in the literature.[8–11] The term *tissue blot immunoassay,* however, will be used throughout this chapter, since transfer of antigens from the specimens onto a nitrocellulose membrane support is by means of blotting a freshly cut tissue surface onto the supporting substrate.[7]

Tissue blot immunoassay has been successfully employed in investigations of many plant viruses[7,12–15] and to a lesser degree fungi[8,10] and mycoplasma-like organisms.[7] The greatest advantage of the tissue blotting method is its precise localization of the antigens of interest in the plant tissue image that is produced on the nitrocellulose membrane (Figures 1 and 2). Localization of plant virus antigens in certain specific tissues are also demonstrated (Figure 3). Furthermore, in addition to the advantages of specificity, sensitivity, and reliability that many commonly used serological methods offer, the direct tissue blotting technique also provides simplicity, rapidity, and convenience for the assay of a large number of samples. Although several modifications of tissue blotting methods have been described for plant viruses,[7,9,11] some of these have not been fully exploited in other areas of plant pathology. Localization of a plant-pathogenic mycoplasma-like organism in leaf phloem tissue of an infected plant was demonstrated (Figure 4).[7] The presence and translocation of a fungal protein that elicits plant defense

* Mention of a trade name, proprietary product, or specific equipment does not constitute a guarantee or warranty by the U.S. Department of Agriculture and does not imply approval to the exclusion of other products that may also be suitable.

0-87371-877-1/95/$0.00+$.50

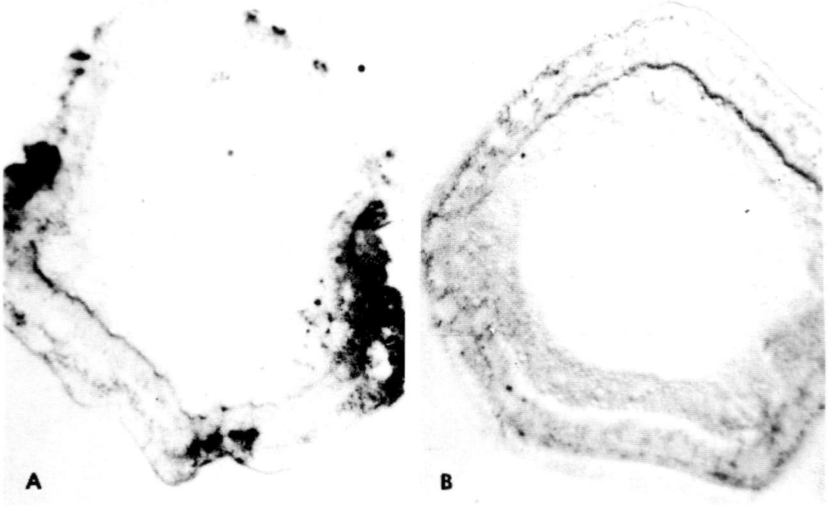

Figure 1 Tissue blots of cross sections of (A) passionfruit woodiness potyvirus (PWV)-infected and (B) healthy passionfruit vine on nitrocellulose membranes. The blots were reacted with PWV coat protein rabbit polyclonal antibodies and detected with alkaine phosphatase-labeled goat anti-rabbit immunogobulin antibodies. (From Lin, et al., *Phytopathology,* 80, 824, 1990. With permission.)

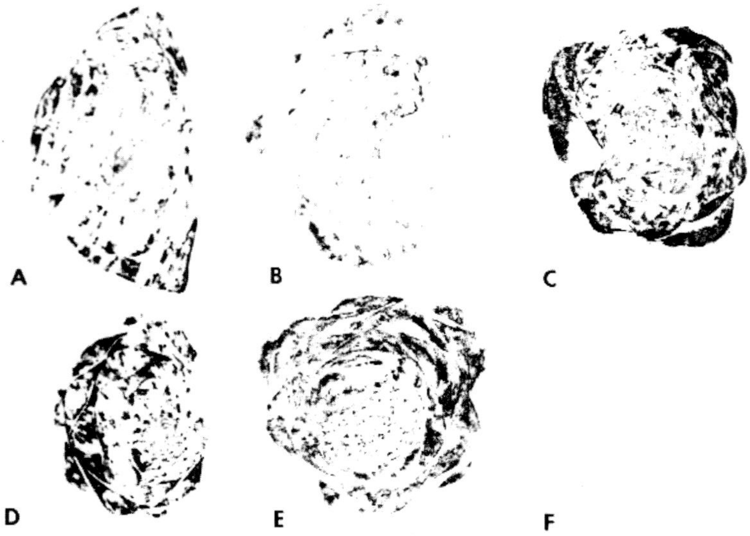

Figure 2 Tissue blots of cross sections of (A–E) lily symptomless carlavirus (LSV)-infected and (F) healthy lily bulbs. The blots were reacted with LSV rabbit polyclonal antibodies and detected with alkaline phosphatase-labeled goat anti-rabbit immunoglobulin antibodies.

response in *Nicotiana tobacum* was clearly illustrated by the technique.[10] The technique was also applied in the detection of *Acremonium coenophialum* infection in tall fescue.[8]

II. PRINCIPLES

Biological materials such as proteins and nucleic acids can be immobilized on nitrocellulose membranes. Although the exact nature of the binding has not been clearly understood, it is generally assumed that hydrophobic interactions promote the binding of macromolecules to a nitrocellulose matrix. The binding

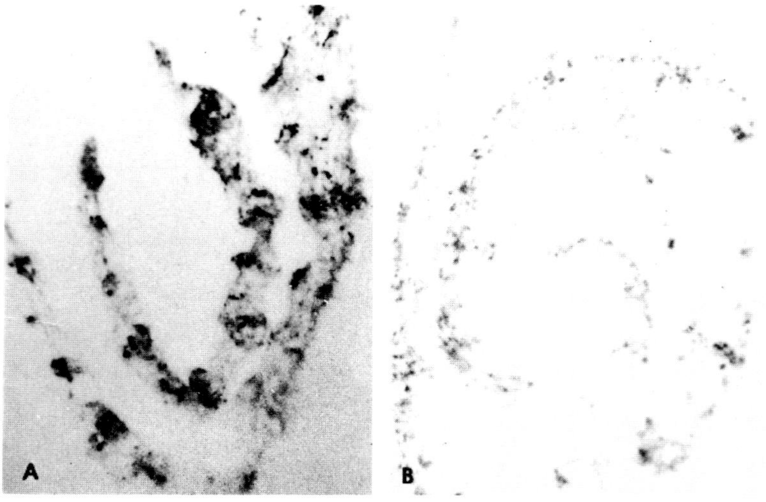

Figure 3 Tissue blots of cross sections of (A) barley yellow dwarf luteovirus (BYDV)-infected and (B) healthy barley leaves. The blots were reacted with BYDV mouse monoclonal antibodies and detected with alkaline phosphatase-labeled goat anti-mouse immunoglobulin antibodies. Viral antigen was localized primarily in the phloem of infected leaves. (From Lin et al., *Phytopathology,* 80, 824, 1990. With permission.)

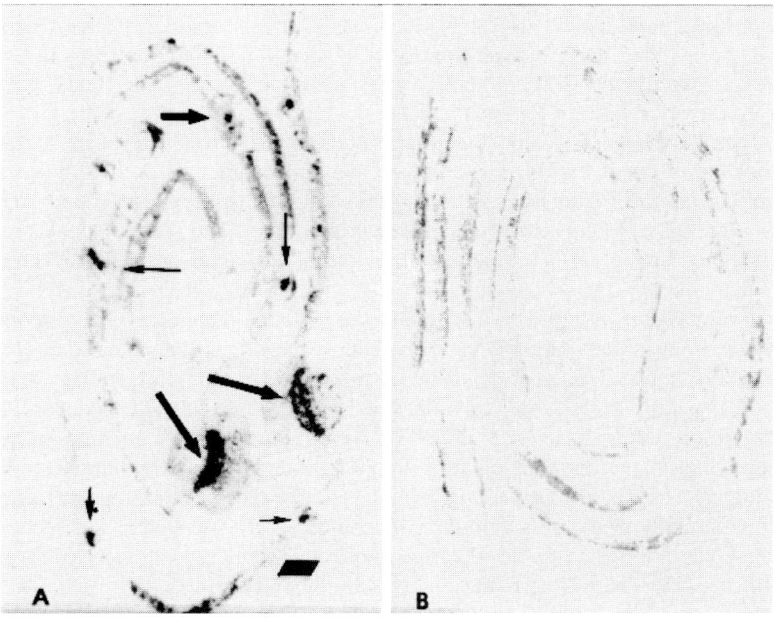

Figure 4 Tissue blots of cross section of (A) the tomato big bud mycoplasma-like organism (MLO)-infected and (B) healthy periwinkle leaves. The blots were reacted with mouse monoclonal antibodies (prepared to the MLO-inducing tomato big bud disease) and detected with alkaline phosphatase-labeled goat anti-mouse immunoglobulin antibodies. MLO antigens were restricted only to midrib phloem cells (large arrows) and secondary veins (small arrows) of the leaves of infected plants. (From Lin et al., *Phytopathology,* 80, 824, 1990. With permission.)

capacity of the nitrocellulose membrane is determined by the available surface area which is inversely proportional to the pore sizes of the matrix. The superior binding capacity of at least 80 μg of protein per square centimeter of nitrocellulose membrane[16] makes it an excellent solid substrate in dot immunoassay of viral antigens. We have routinely used 0.45-μm-diameter pore size membranes.

Tissue blotting is a process of transfer of viral antigens from a freshly cut plant tissue surface to nitrocellulose membranes. It is similar to, but not as complicated as that of protein transfer from gels

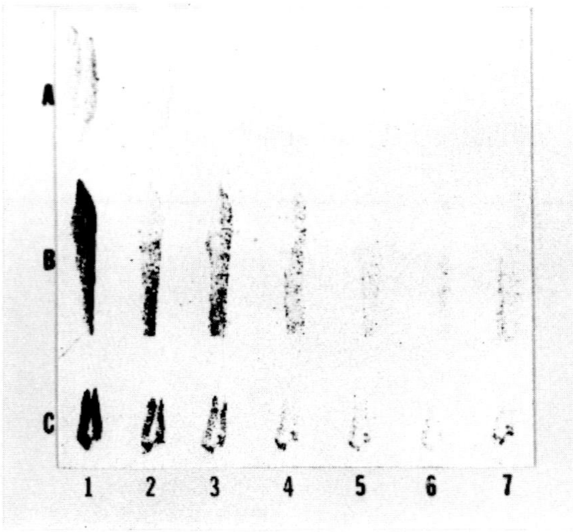

Figure 5 Detection of Cymbidium mosaic potexvirus (CyMV) antigens in consecutive blots (from 1 to 7) from a single cut surface of leaves of (A) healthy Cattleya; (B) infected Cattleya; and (C) infected Phalaenopsis orchids. CyMV antigens were detected by reacting tissue blots with virus-specific mouse monoclonal antibodies followed by alkaline phosphatase-labeled goat anti-mouse immunoglobulin antibodies. The dark images in healthy control A were attributable to the green of chlorophyll. (From Hsu et al., *Phytopathology,* 82, 491, 1992. With permission.)

to a nitrocellulose matrix.[17] Unlike protein transfer from a gel that involves capillary, vacuum, or electrical reactions,[17] tissue blotting is achieved simply by bringing a freshly cut tissue surface in direct contact with a dry nitrocellulose membrane.[7] The tissue imprint is made by application of a slight pressure of the cut tissue surface while it is in contact with the nitrocellulose matrix for no more than 1 s. A higher pressure that distorts the cut surface and squeezes liquid from the tissue is too much force and should be avoided. Details regarding exact localization of viral antigens in tissue blots may be lost due to excessive forces. Viral antigens in a series of consecutive blots made from the same cut tissue surface are clearly shown in Figure 5.

As with Western blot analysis, blocking is an important step when blots are to be probed with antibodies. Blocking, or quenching of nitrocellulose membranes after tissue blotting, is the process in which unoccupied protein binding sites on the nitrocellulose matrix are saturated so that detection antibodies which are also proteins do not bind nonspecifically to the matrix. Many protein solutions are effective blocking reagents.[18,19] Bovine serum albumin and nonfat dry milk are two examples, as they are easily available.

The presence of antigens on nitrocellulose membranes can be determined with the enzyme-labeled antibody technique. In the direct immunological procedure, the antigens are detected directly by enzyme-labeled virus-specific antibodies; whereas in the indirect method, the antigens are reacted first with primary antibodies specific to the virus, then detected with enzyme-labeled secondary antibodies that react with the primary antibodies.[7] The ability of protein A molecules to bind specifically to the carbohydrate moiety of many immunoglobulins[20] and the general availability of enzyme-labeled protein A make the conjugate a useful detection reagent.[21–24] In procedures where biotinylated primary antibodies are used,[25] a commercially available avidin-enzyme conjugate is employed as a detection reagent.

Detection enzymes for use in tissue blot immunoassay should be carefully evaluated before the system is established. Horseradish peroxidase is commonly employed in enzyme-linked immunoassay for detection of a number of plant pathogens. It may not be suitable for use in tissue blot immunoassay on nitrocellulose membranes, since endogenous peroxidase from plant tissue interferes with the assay.[26] Alkaline phosphatase-antibody (both primary virus-specific and secondary immunoglobulin-specific) conjugates are generally available in plant pathology laboratories and are convenient to investigators. Substrates that produce soluble colored products are used in the enzyme-linked immunosorbent assay. In tissue blot immunoassay, the substrates that yield insoluble colored products precipitating at the site of enzyme reaction should be the choice. When chemiluminescent substrates are utilized in tissue blot immunoassay, the presence of antigens can be recognized by the image registered on a light-sensitive X-ray film (Figure 6).[27] This is especially useful when colored pigments of plant tissue origin interfere with the results of analysis using chromogenic substrates.

III. PROPERTIES OF ANTIBODIES

Antibodies are protein molecules that are produced in sera of animals to combat invasions of foreign substances. Although production of polyclonal antisera looks very simple as the animals are doing most

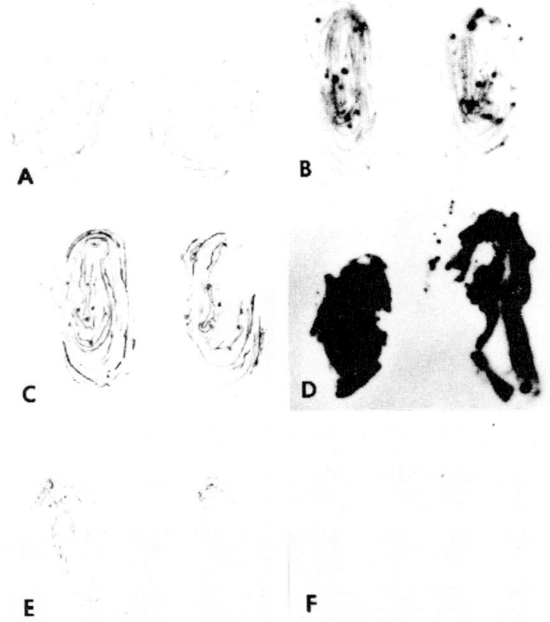

Figure 6 Tissue blot immunodetection of tospovirus antigens in infected *Nicotiana benthamiana* using (A, C, and E) chromogenic and (B, D, and F) chemiluminescent substrates. Leaf cross sections of (A and B) tomato spotted wilt virus-infected, (C and D) Impatiens necrotic spot virus-infected, and (E and F) healthy *N. benthamiana* were blotted on nitrocellulose membranes. The blots were reacted with virus-specific rabbit antibodies followed by alkaline phosphatase-labeled goat anti-rabbit immunoglobulin antibodies. Blots A, C, and E were incubated in a chromogenic substrate; blots B, D, and F were incubated in a chemiluminescent substrate and reaction results were registered on an X-ray film. (From Makkouk et al.,)

of the work, preparation of high-quality serological reagents is actually not an easy task. Plant virus antisera are generally produced by injecting rabbits with purified virus antigens.[1,2] Because biological systems often are much more sensitive than physical methods, immune responses can be induced by a minute quantity of antigens. Rabbits are better at detecting contaminants than we are at removing them. This means that virus preparation considered pure by investigators may otherwise contain contaminants that also elicit an immune response when injected into animals. This is often true for polyclonal antisera. Although such antisera may be adequate for many applications, they are not good enough for use in ISEM or tissue blot immunoassay. They react generously with plant tissue antigens, and require extensive clean-up before use.

Routinely, plant virus antisera are produced in rabbits, but monoclonal antibodies produced in mice or rats are becoming more important in serological testing of plant viruses.[28,29] Monoclonal antibodies provide a level of standardization that is not possible with polyclonal antibodies. Unlike the heterogenous mixtures of immune sera which differ not only from individual animals, but also from each bleed of the same animal, the monoclonal antibody produced by a selected hybridoma cell line is a well-defined chemical. Because monoclonal antibodies are so specific, they may be used to differentiate closely related strains of a virus.[30] In addition, a monoclonal antibody that reacts with a portion of the coat protein that is shared by all strains of a virus, or a group of viruses, should be useful for broad-spectrum testing.[31] Hybrid cell cultures that produce monoclonal antibodies can be stored in cryogenic freezers and can be revived and cultured to produce monoclonal antibodies as required.

Construction of hybridomas secreting antibodies to plant viruses is a well-established procedure, and monoclonal antibodies have been produced to numerous plant pathogens. Although the hybridoma technique does not require purified virus for production of virus-specific monoclonal antibodies, incubation of the immunogen with an antiserum prepared to normal plant constituents prior to immunization will reduce the contaminating plant antigens.[32] Tolerance to plant antigens can be induced immunologically in mice by injecting excess amounts of plant extracts into newborns. Using this method, it was possible to enhance the proportion of hyridomas secreting antibodies specific to tomato spotted wilt virus.[33] Similar procedures of immunological tolerance established in the neonatal mice were utilized to enhance the production of mouse hybridomas secreting antibodies to plant mycoplasma-like organisms that were difficult to obtain in pure form.[34,35]

Initial generation of specific antibody-secreting hybrid cells is time consuming, but production of antibodies by hybridoma cell lines is technically no more demanding than normal *in vitro* cell culture. High concentrations of monoclonal antibodies to plant viruses and pathogenic mycoplasmalike organisms have been produced in ascitic fluids[34,36,37] and contamination of plant antigen-reacting antibodies is eliminated.

IV. METHODOLOGY

Equipment—Nitrocellulose membranes (0.2 or 0.45 μm) are cut from bulk-size packages into strips about 6 × 9 cm. Razor blades, rectangular glass dishes slightly larger than the size of membranes or glass petri dishes 12 cm in diameter, and forceps to handle nitrocellulose membranes are also required.

Reagents—Phosphate-buffered saline solution (PBS), 0.01 *M* phosphate, 0.15 *M* NaCl, pH 7.2.

10 × stock solution	
NaH_2PO_4 H_2O	3.86 g
Na_2HPO_4	10.22 g
NaCl	8.50 g

Distilled water to 1 l

Dilute stock ten times with distilled water before use.

Blocking solution (0.5% bovine serum albumin, 2% nonfat dry milk in PBS). Make aliquots and store frozen.

Washing solution (0.05% Tween® 20, in PBS).

Tween® 20	0.50 ml
PBS	1000 ml

Substrate solution—Prepare 40 ml of substrate solution just before use. This solution contains 14 mg nitro blue tetrazolium (NBT) and 7 mg 5-bromo-4-chloro-3-indolyl phosphate (BCIP) in 40 ml of substrate buffer (0.1 *M* Tris, 0.1 *M* NaCl, 5 m*M* $MgCl_2$, pH 9.5). To prepare substrate, dissolve 14 mg NBT into 300 μl methanol (A) and dissolve 7 mg BCIP into 50 μl dimethyl sulfoxide (B). Add A into the substrate buffer, then add B.

Stopping solution (Tris-HCl 0.01 *M*, EDTA 0.05 *M*, pH 7.5).

Procedures—Avoid leaving fingerprints on nitrocellulose membranes. Do not handle the membranes with bare hands; always wear disposable plastic gloves or use a pair of forceps to pick up or transfer nitrocellulose membranes. The amount of reagent solution added to the dishes should be sufficient to completely cover the nitrocellulose membrane.

Preparation of tissue blots.

1. Excise tissues (leaves, petioles, stems, flower buds, emerging shoots, bulb, etc.).
2. For thin tissue such as leaves, roll them into a tight core.
3. Hold tissue in one hand and cut with a new razor blade in a steady motion with the other hand to obtain a single-plane cut surface.
4. Press for about a full second the newly cut surface onto a nitrocellulose membrane to obtain a tissue blot. Use a firm but gentle force. Do not squeeze juice out from the tissues.
5. Tissue blot membranes can be transported or stored for 4 weeks after they are prepared.
6. Block tissue blots by immersing nitrocellulose membranes in the blocking solution for 30 to 60 min with occasional shaking at room temperature.
7. Wash blots once with washing solution for about 1 min with gentle shaking (about one rotation per second on a mechanical shaker).

A. DIRECT IMMUNOASSAY METHOD

Reagent: Alkaline phosphatase-labeled virus-specific antibodies.

1. Incubate tissue blots in a glass dish with alkaline phosphatase-labeled virus-specific antibodies diluted in PBS for 60 min at room temperature. Be sure that the reagent solution covers the blot.
2. Wash blots four times in washing solution for 30 min with shaking.
3. Soak blots in substrate solution 2 to 5 min at room temperature to detect enzyme activity.
4. Rinse blots for a few seconds in distilled water.

5. Stop reaction by immersing blots in stopping solution, two to three changes, 10 min each.
6. Dry nitrocellulose membranes on two to three layers of tissue wipes in a dust-free area.

B. INDIRECT IMMUNOASSAY METHOD

Reagent: Primary virus-specific antibodies (rabbit polyclonal or mouse monoclonal antibodies), alkaline phosphatase-conjugated anti-rabbit immunoglobulin antibodies, or anti-mouse immunoglobulin antibodies.

1. Incubate tissue blots in a glass dish with primary virus-specific antibodies diluted in PBS for 60 min at room temperature.
2. Wash blots three times in washing solution for 10 to 15 min with shaking.
3. Cover blots with alkaline phosphatase-conjugated anti-rabbit immunoglobulin antibodies diluted in PBS for 60 min when rabbit anti-virus antibodies are used in Step 1. If mouse monoclonal antibodies are used in Step 1, alkaline phosphatase-labeled anti-mouse immunoglobulin antibodies should be used as a conjugate.
4. Wash blots four times in washing solution for 30 min with shaking.
5. Soak blots in substrate solution for 2 to 5 min at room temperature to detect enzyme activity.
6. Rinse blots for a few seconds with distilled water.
7. Stop reaction by immersing blots in stopping solution, two to three changes, 10 min each.
8. Dry nitrocellulose membranes on two to three layers of tissue wipes. Cover with an additional sheet of tissue wipes to prevent dust or dirt fall on the blots.

C. ALKALINE PHOSPHATASE-PROTEIN A CONJUGATE METHOD

Reagents: Virus-specific antiserum, alkaline phosphatase-labeled protein A conjugate.

1. Incubate tissue blots in a glass dish with virus-specific rabbit antiserum (unfractionated whole serum) diluted in PBS for 60 min at room temperature.
2. Wash blots three times in washing solution for 10 to 15 min with shaking.
3. Soak blots in alkaline phosphatase-protein A conjugate diluted in PBS for 60 min at room temperature.
4. Wash blots four times in washing solution for 30 min with agitation.
5. Soak blots in substrate solution at room temperature for color development.
6. Rinse blots with distilled water.
7. Stop enzyme reaction by immersing blots in stopping solution, two to three changes, 10 min each.
8. Dry blots on a filter paper in a dust-free place.

D. BIOTIN-AVIDIN METHOD

1. Biotinylated Primary Antibody Method

Reagents: Biotinylated virus-specific antibodies, alkaline phosphatase-conjugated avidin, or alkaline phosphatase-conjugated streptavidin.

1. Incubate blots in a glass dish with biotinylated virus-specific antibodies diluted in PBS for 60 min at room temperature.
2. Wash blots three times in washing solution for 30 min with shaking.
3. Transfer blots into alkaline phosphatase-labeled avidin solution diluted in PBS and incubate for 60 min at room temperature.
4. Wash blots four times in washing solution for 30 min at room temperature with shaking.
5. Develop color in substrate solution to detect enzyme activity.
6. Rinse blots in distilled water.
7. Stop reaction in stopping solution, two to three changes in 10 min.
8. Dry blots on a filter paper in a dust-free place.

2. Biotinylated Secondary Antibody Method

Reagents: Virus-specific primary rabbit or mouse antibodies (or unfractionated antiserum), biotinylated goat anti-rabbit or anti-mouse secondary antibodies, and alkaline phosphatase-conjugated avidin (or streptavidin).

1. Incubate blots in a glass dish with virus-specific primary (rabbit or mouse) antiserum diluted in PBS for 60 min at room temperature.
2. Wash blots three times in washing solution for 30 min with shaking.
3. Transfer blots into biotinylated goat anti-rabbit or anti-mouse antibodies and incubate for 60 min.
4. Wash blots three times in washing solution with gentle agitation for 30 min.
5. Incubate blots with alkaline phosphatase-labeled avidin conjugate diluted in PBS for 60 min at room temperature.
6. Wash blots four times in washing solution for 30 min at room temperature with shaking.
7. Develop color in substrate solution.
8. Rinse blot in distilled water and stop enzyme reaction in stopping solution.

V. COMMENTS

Membrane blots, after being prepared, can be stored for a period of time up to 4 weeks in a dust-free environment.[15] However, a blocking step should only proceed immediately before incubation with antibodies. Dilutions of all reagents must be determined by preliminary tests. In the cases where contaminating antibodies that react with normal plant constituents are suspected, the contaminants must be removed from the serum reagents. Reagents, both diluted primary antibody and conjugate solutions, can be used repeatedly over a period of time. Care must be taken to avoid microbial growth in the solution when stored in a refrigerator at 4°C. Addition of a preservative is, however, recommended, since reagent solutions are enriched with proteinaceous substances from both the antibody itself and those from the blocking solution.

In tissue that contains a high concentration of latex, the cut surface is first drained on tissue paper to remove excess exudate before blotting onto nitrocellulose membranes.

The procedures that employ virus-specific antibodies followed by enzyme-labeled secondary antibodies or biotinylated secondary antibodies and avidin-enzyme conjugate are convenient in laboratories where diagnosis and detection of virus diseases are a primary responsibility. The same reagents can be used for detection of many different viruses.

VI. ADVANTAGES AND LIMITATIONS

Tissue blot immunoassay retains advantages that many other assay methods offer, including specificity and reliability. The method also provides a precise tool for visualization and localization of antigens of specific interest. Compared with other immunoassays, preparation of sample materials for tissue blot immunoassay is relatively easy. A larger number of samples can be processed within the same time period that samples are prepared for ELISA.[38] The membrane blots can be prepared in one location and sent to another laboratory for detection. Some tissues may contain high concentrations of red-colored pigments, such as anthocyanins, that may interfere with the observation of a positive reaction. In this case, chemiluminescent substrates that register a reaction on an X-ray film may be a better choice.

REFERENCES

1. van Regenmortel, M. H. V., *Serology and Immunochemistry of Plant Viruses,* Academic Press, New York, 1982, 301 pp.
2. Hampton, R., Ball, E., and DeBere, S., Eds., *Serological Methods for Detection and Identification of Viral and Bacterial Plant Pathogens: a Laboratory Manual,* APS Press, St. Paul, MN, 1990, 389 pp.
3. Sampson, P. J. and Taylor, R. H., A comparison of the electron microscope, microprecipitation tests, and indicator plants for the detection of potato viruses S, X, and Y, *Phytopathology* 58, 489, 1968.
4. Ackers, G. K. and Steere, R. L., Restricted diffusion of macromolecules through agar-gel membranes, *Biochem. Biophys. Acta,* 59, 137, 1962.
5. Clark, M. F. and Adams, A. N., Characteristics of the microplate method of enzyme-linked immunosorbent assay for the detection of plant viruses, *J. Gen. Virol.,* 34, 475, 1977.
6. Derrick, K. S., Quantitative assay for plant viruses using serologically specific electron microscopy, *Virology,* 56, 652, 1973.

7. Lin, N. S., Hsu, Y. H., and Hsu, H. T., Immunological detection of plant viruses and a mycoplasmalike organism by direct tissue blotting on nitrocellulose membranes, *Phytopathology,* 80, 824, 1990.
8. Gwinn, K. D., Collins-Shepard, M. H., and Reddick, B. B., Detection of *Acremonium coenophialum* in tissue prints with western blot, *Phytopathology,* 80, 435, 1990.
9. Mansky, L. M., Andrews, R. E., Jr., Durand, D. P., and Hill, J. H., Plant virus location in leaf tissue by press blotting, *Plant Mol. Biol. Rep.,* 8, 13, 1990.
10. Bailey, B. A., Taylor, R., Dean, J. F. D., and Anderson, J. D., Ethylene biosynthesis-inducing endoxylanase is translocated through the zylem of *Nicotiana tobacum* cv. Xanthi plants, *Plant Physiol.,* 97, 1181, 1991.
11. Bravo-Almonacid, F., Haim, L., and Mentaberry, A., Rapid immunological detection of potato viruses in plant tissue squashes, *Plant Dis.,* 76, 574, 1992.
12. Hsu, H. T., Tissue blot immunoassay for detection of tomato spotted wilt virus, *Plant Diagn. Qtr.,* 12, 66, 1991.
13. Hsu, H. T. and Lawson, R. H., Direct tissue blotting for detection of tomato spotted wilt virus in *Impatiens, Plant Dis.,* 75, 292, 1991.
14. Hsu, H. T., Vongsasitorn, D., and Lawson, R. H., An improved method for serological detection of *Cymbidium* mosaic potexvirus infection in orchids, *Phytopathology,* 82, 491, 1992.
15. Srinivasan, I. and Tolin, S. A., Detection of three viruses of clovers by direct tissue immunoblotting, *Phytopathology,* 82, 721, 1992.
16. Gershoni, J. M. and Palade, G. E., Electrophoretic transfer of proteins from sodium dodecyl sulphate polyacrylamide gels to a positively charged membrane filter, *Anal. Biochem.,* 12, 396, 1982.
17. Towbin, H. and Gordon, J., Immunoblotting and dot immunobinding. Current status and outlook, *J. Immunol. Methods,* 72, 313, 1984.
18. Gershoni, J. M. and Palade, G. E., Protein blotting: principles and application, *Anal. Biochem.,* 131, 1, 1983.
19. Johnson, D. A., Gatsch, J. W., Sportsman, J. R., and Elder, J. H., Improved technique utilizing nonfat dry milk for analysis of proteins and nucleic acids transferred to nitrocellulose, *Gene Anal. Technol.,* 1, 3, 1984.
20. Goding, J. W., Use of *Staphylococcal* protein A as an immunological reagent, *J. Immunol. Methods,* 20, 241, 1978.
21. Barbara, D. J. and Clark, M. F., A simple indirect ELISA using F $(ab')_2$ fragments of immunoglobulin, *J. Gen. Virol.,* 58, 315, 1982.
22. Hsu, H. T. and Aebig, J., Enzyme-linked immunosorbent assay (ELISA) of plant viruses using protein A-enzyme conjugate and $F(ab')_2$ of immunoglobulin G (IgG), *Phytopathology,* 72, 708, 1982.
23. Hsu, H. T., Aebig, J., Civerolo, E. L., and Helkie, C., Enzyme conjugated protein A: a universal reagent for indirect enzyme-linked immunosorbent assays (ELISA), *Phytopathology,* 72, 938, 1982.
24. Edwards, M. L. and Cooper, J. L., Plant virus detection using a new form of indirect ELISA, *J. Viral Methods,* 11, 309, 1985.
25. Kendall, C., Ionescu-Matiu, I., and Dressman, R., Utilization of the biotin/avidin system to amplify the sensitivity of the enzyme-linked immunosorbent assay (ELISA), *J. Immunol. Methods,* 56, 329, 1983.
26. Hsu, H. T., Hibino, H., and Cabautan, P. Q., Development of serological procedures for rapid, sensitive, and reliable detection of rice grassy stunt virus, *Plant Dis.,* 74, 695, 1990.
27. Makkouk, K. M., Hsu, H. T., and Kumari, S., Detection of three plant viruses by dot-blot and tissue-blot immunoassays using chemiluminescent and chromogenic substrates, *J. Phytopathol.,* 139, 97, 1993.
28. Hsu, H. T. Some considerations in preparing monoclonal antibodies for plant viruses, in Proceedings of Regional Workshop on Biotechnology in Agriculture, SEAMEO Regional Center for Graduate Study and Research in Agriculture, Laguna, Philippines, 1986, 187–196.
29. Jordan, R. L. and Hsu, H. T., Immunological detection and epitope analysis of plant pathogens and their gene products: monoclonal antibodies in plant pathology, in *Biotechnology and Crop Improvement in Asia,* Moss, J. P., Ed., Patanchera, A.P. 502 34, India, ICRISAT, 1992, 283–295.
30. Halk, E. L., Hsu, H. T., Aebig, J., and Franke, J., Production of monoclonal antibodies against three ilarviruses and alfalfa mosaic virus and their use in serotyping, *Phytopathology,* 74, 367, 1984.
31. Jordan, R. and Hammond, J., Comparison and differentiation of potyvirus isolates and identification of strain-, virus-, subgroup-specific and potyvirus group-common epitopes using monoclonal antibodies, *J. Gen. Virol.,* 71, 25, 1991.

32. George, R. A. and Converse, R. H., Methods for the enrichment of desired B-cell populations before anti-cauliflower mosaic virus hybridoma formation, *Phytopathology,* 78, 1631, 1988.
33. Hsu, H. T., Wang, Y. C., Lawson, R. H., Wang, M., and Gonsalves, D., Splenocytes of mice with induced immunological tolerance to plant antigens for construction of hybridomas secreting tomato spotted wilt virus-specific antibodies, *Phytopathology,* 80, 158, 1990.
34. Hsu, H. T., Lee, I. M., Davis, R. E., and Wang, Y. C., Immunization for generation of hybridoma antibodies specifically reacting with plants infected with a mycoplasmalike organism (MLO) and their use in detection of MLO antigens, *Phytopathology,* 80, 946, 1990.
35. Shen, W. C. and Lin, C. P., Production of monoclonal antibodies against a mycoplasmalike organism associated with sweetpotato witches' broom, *Phytopathology,* 83, 671, 1993.
36. Hsu, H. T. and Lawson, R. H., Comparison of mouse monoclonal antibodies and polyclonal antibodies of chicken egg yolk and rabbit for assay of carnation etched ring virus, *Phytopathology,* 75, 778, 1985.
37. Hsu, H. T., Franssen, J. M., van der Hulst, C. T. V., Derks, A. F. L. M., and Lawson, R. H., Factors affecting selection of epitope specificity of monoclonal antibodies to tulip breaking potyvirus, *Phytopathology,* 78, 1337, 1988.
38. Makkouk, K. M., Comeau, A., and St. -Pierre, C. A., Screening for barley yellow dwarf luteovirus resistance in barley on the basis of virus movement, *J. Phytopathol.,* in press.

Chapter 27

Monoclonal Antibodies to Fungal Plant Pathogens: Basic and Applied Aspects

Keith L. Wycoff and Laurence M. Brill

TABLE OF CONTENTS

I. INTRODUCTION

Antibodies have been used in the study of plant-pathogen interactions for many years. Serological (antibody-based) techniques have been used for identification and taxonomic classification of plant pathogens, as well as for diagnosing infection levels. The potential of antibodies as tools for studying the antigens of plant pathogens was given an enormous boost by the hybridoma technology introduced by Köhler and Milstein.[1] As reagents for taxonomic, diagnostic, structural, and biochemical analysis of plant pathogens, monoclonal antibodies (mAbs) offer three primary advantages over conventional polyclonal antisera: (1) antibodies specific for a single antigenic determinant (molecule or portion of a molecule that the antibody binds to) can be obtained even when complex antigenic mixtures are used as immunogens; (2) the qualitative and quantitative variability in specific antibody content found in different batches of polyclonal sera are eliminated; (3) hybridomas secreting antibodies of interest can be preserved indefinitely in liquid nitrogen, thus assuring a continuous and unlimited supply over time. The main disadvantages of mAbs are that their production is technically more difficult, more expensive (due to specialized equipment and supplies), and more time consuming. While development of a good polyclonal sera may take 2 to 3 months, production and screening of a bank of mAbs may take over a year (see Table 1, of Chapter 28).

The use of monoclonal (and polyclonal) antibodies in plant disease research has been reviewed previously.[2–6] Other chapters in this volume deal specifically with the methodology of mAb production. In this chapter, we will review recent advances in the study of plant disease brought about by using mAbs raised against fungi.

0-87371-877-1/95/$0.00+$.50

Table 1 **Monoclonal antibodies raised against plant-invading fungi**

Fungus	Immunogen	Application	Specificity	Ref.
Armillaria mellea	Hyphal fragments	Disease diagnosis	NTSG	36
A. ostoyae	Specific proteins	Identification in wood	Species, genus	37*
Ascocalyx abietina	Specific proteins	Race differentiation	NTSG	4
Botrytis cinerea	Surface washings	Diagnosis in spoiling fruit	Species	38
Colletotrichum lindemuthianum	Germling extracts	Species differentiation	Species, genus	39
Erysiphe pisi	Haustorial complexes	Immunocytochemistry	Species	40,41
Fusarium graminearum	Soluble extract	Disease diagnosis	Species	42*
F. oxysporum				
f. sp. *cubense*	Hyphal wall	Race differentiation	Race?	43
f. sp. *lycopersici*	Spores	Disease diagnosis	NTG	29
Gliocladium roseum	Mycelial walls	Identification in wood	Species	44
Glomus occultum	Crushed spores	Species differentiation	Species, rare cross reaction	45
Humicola lanuginosa	Surface washings	Detection on rice grains	Genus, rare cross reaction	46
Leptosphaeria korrae	Homogenized mycelia	Detection in turfgrass	Species, rare cross reaction	47
Ophiostoma ulmi	Purified glycopeptide	Ultrastructure	NTSG	48,49
O. ulmi	Hyphal fragments	Disease diagnosis	Species	50
Penicillium islandicum	Surface washings	Detection on rice grains	Species	51
Phytophthora cinnamomi	Zoospores	Immunocytochemistry	Isolate, species, genus	52
P. fragariae	ng	Taxonomy	Race?, species	53*
P. sojae	Mycelial walls	Immunohistochemistry	NTSG	54–56
	Culture filtrates	Taxonomy	Species (in combination)	57
Phytophthora spp.	ng	Diagnosis in roots	Genus	58
Postia placenta	β-1,4-xylanase	Detection of wood decay	NTSG	59
Puccinia graminis	Purified protein	Immunocytochemistry	NTSG	60
Pyricularia grisea	Culture filtrates	Immunocytochemistry	Genus, possibly species	61
Pythium aphanidermatum	Zoospores	Immunocytochemistry	Species	62,63
P. aphanidermatum	ng	Diagnosis in turfgrass	Genus	33*,34*
P. ultimum	Mycelial walls	Diagnosis in roots	Species, genus	64*
Rhizoctonia solani	Culture filtrates	Taxonomy	Isolate, NTSG	65
Septoria nodorum	ng	Disease diagnosis	Species	35*
S. tritici	ng	Disease diagnosis	Species	35*
Sirococus strobilinus	Whole mycelium	Diagnosis in seeds	Genus, possibly species	30,31
Tilletia	Spores, surface extracts	Species differentiation	NTG	66

Note: ng = not given; NTG = not tested against other genera; NTSG = not tested against other species or genera; *abstract only.

II. DISEASE DIAGNOSIS

A. TRADITIONAL METHODS

The primary use of antibodies in plant pathology in recent years has been for the diagnosis of plant disease. Farmers and foresters need accurate, rapid, and economical diagnosis of disease or disease potential to aid them in making decisions regarding the timely application of fungicides, the choice of crop species and cultivars to plant, and the appropriateness of certain cultural practices such as no-till or crop rotation. The more traditional methods used for diagnosis of diseases caused by plant-pathogenic fungi are time consuming and require mycological expertise as well as culturing and microscopy facilities. Mycological identification is often difficult because the multiple colony and reproductive characteristics that are used can vary between individual isolates and are dependent on environmental conditions.[7-10] There are three primary methods currently used for diagnosis: (1) the agar plate bioassay, in which infected plant material is surface sterilized and placed on sterile agar plates; the pathogen grows out of the plant material and is then identified based on microscopic examination;[10,11] (2) the use of a desiccant herbicide (e.g., paraquat) to detect latent infection in plant tissues; in this method, treated tissue is placed in a moist chamber for several days to weeks, during which time the pathogen will sporulate, and the spores can be identified under the compound microscope;[12,13] (3) visual inspection of infected tissues, which is frequently unreliable; and (4) mycological examination of the pathogen in severely infected host tissues. The first two methods are time consuming, allowing for irreversible damage to diseased plants in the field, and the fourth uses largely destroyed plant tissue. None of these methods quantify the level of fungal infection, which may be critical in determining whether or not damage will occur, and when the expense of plant disease control is justified. Antibody-based diagnostic systems can overcome these difficulties.[5]

B. ANTIBODY-BASED METHODS

The existence of an antibody against a fungus does not guarantee it will be useful in detecting infection. For instance, an ELISA using polyclonal antibodies against *Diaporthe phaseolorum* f. sp. *caulivora* was unsuccessful in detecting that fungus in infected soybean seed.[14] When *Phomopsis longicolla*-infected seeds were assayed in ELISA using polyclonal antibodies, ELISA values were directly correlated with severity ratings of *Phomopsis* seed decay, but reactivity to soybean seed tissue was considered excessive.[15] To be useful in diagnosis of disease, an antibody-based assay should fulfill some minimal qualifications: (1) the assay should be accurate in identifying the fungus causing disease symptoms, with negligible cross reaction to normal host tissue, other fungi, or other pathogens; (2) it should be as sensitive as conventional diagnostic assays (e.g., culture plate assay); (3) it should be easy to perform on relatively crude samples without requiring extensive purification of fungal antigens.

1. Polyclonal Antibodies

Successful detection and quantitation of fungal plant pathogens in seed, plant tissue, and soil has been achieved using polyclonal antibodies in ELISA[15-21] and RIA.[22] Other assays that have been used include immunofluorescence,[23] protein A-gold immunocytochemistry,[24-26] and dot immunobinding.[27] Although many of them are quantitative and highly sensitive, not all of these assays are easy to perform. The simplest and most popular assays for disease diagnosis in the field are ELISA and dip stick-type assays. Tissue print immunoblot, a simple technique requiring no special equipment was as accurate as ELISA in diagnosing infection of tall fescue with an endophytic fungus.[28]

2. Monoclonal Antibodies

For some fungi, production of highly specific polyclonal antibodies for diagnostic purposes has not been possible, and this has generated greater interest in mAbs. Since 1983, when Ianelli et al.[29] developed mAbs to *Fusarium oxysporum,* there have been many reports of mAbs against plant-invading fungi (Table 1). Many of these mAbs are highly specific and thus have diagnostic potential, but only a few have been developed into diagnostic assays. Mitchell[30,32] and Mitchell and Southerland[31] developed mAbs specific for *Sirococcus strobilinus,* the causal agent of shoot blight in conifers, and successfully used them to detect the fungus in infected seeds, both by ELISA and dot immunoassay. A mAb for diagnosis of Pythium blight, which was directed against *Pythium aphanidermatum,* was shown to react with other *Pythium* spp. but not with the causal agents of the other turfgrass diseases.[33,34] Petersen et al.[35] have developed a test kit for *Septoria nodorum* and *S. tritici,* the causal agents of Septoria leaf spot of wheat, that uses two species-specific mAbs.

Independent attempts have been made to assess the usefulness of mAb-based commercial test kits in diagnosing disease. Test kits for *Phytophthora, Pythium,* and *Rhizoctonia* were tested on root tissue from commercial nurseries.[67] Although all the plants tested exhibited symptoms of root disease, these three genera were detected by either ELISA or culture plate methods in only half the plants, suggesting that symptoms in the rest were caused by other biotic or abiotic factors. In a controlled experiment, with plants deliberately infected with *Phytophthora cryptogea,* when disease symptoms were present, the pathogen could always be detected by one or both assays. ELISA detected 90% of the total, and culture plating detected 84%, a difference that was not significant. The authors attributed those situations where infection was detected by only one of the two assays to sampling error, since the same exact tissue piece could not be used for both. They also attempted to determine detection sensitivity by mixing extracts from infected and healthy roots. As little as 1% *P. cryptogea*-infected tissue could reliably be detected. Two *Rhizoctonia* kits were evaluated by Benson,[68] who found that they could detect the fungus in lesions as small as 8.6 mm,^{2} and were at least as reliable as culture plate methods. In some situations, for example, in control of Pythium blight in turfgrass, it would be advantageous to detect an outbreak before symptoms appeared. This would allow more successful and efficient application of fungicides. However, the commercial *Pythium* kit was not sensitive enough to detect pre-symptomatic infection.[69] A *Phytophthora*-specific ELISA kit was used to determine the amount of fungus in the soil of a large number of soybean fields over a-3-year period. The amount of *Phytophthora* in soil was not useful in predicting the magnitude of yield loss from root rot.[70]

A newer ELISA kit that uses two mAbs for diagnosing root rot caused by *Phytophthora* has been developed,[58] and has been tested by a plant disease clinic.[71] All *Phytophthora* isolates tested positive, although the reaction varied with different isolates of *P. cinnamomi.* The kit also cross reacted with some isolates of *Pythium.* As the authors note, however, "it may not be necessary to obtain an exact diagnosis, since the cultural and chemical controls for both fungi are presently the same". The kit also cross reacted to *Peronospora* isolates (a closely related genus), but since that is a foliar pathogen, rather than a root-rotting one, it would not normally be confused with *Phytophthora.* These studies illustrate an important point: no diagnostic ELISA is 100% accurate and, rather than replacing traditional diagnostic techniques, should be seen as a helpful complement that can increase speed and accuracy.

The presence of certain fungi and fungal products in plant tissue after harvest, while not affecting plant growth or yield, may still have an impact on the intended use of the tissue. For instance, certain fungi growing on stored grain produce mycotoxins that can adversely affect the health of people and livestock. An indirect competitive ELISA using mAbs is a highly sensitive method of detecting the presence of ochratoxin, aflatoxin B, and T-2 toxin in stored grain.[72] A mAb has been prepared against aflatoxin-B1, aflatoxin-B2, aflatoxin-G1 and aflatoxin-G2.[73] A commercial ELISA-based test kit has been used to detect the presence of the mycotoxin deoxynivalenol (colorfully known as vomitoxin), produced by *F. graminearum,* in harvested grain.[74] Polyclonal antibodies have also been produced that react with naphthazarin phytotoxins from *F. solani.*[75]

mAb-based assays have been developed for *Penicillium islandicum* and *Humicola lanuginosa.*[46,51] Both of these saprophytic fungi cause yellowing of harvested rice (and, consequently, economic damage), and the latter has been associated with toxicity. The mAb used in the *P. islandicum* assays was relatively specific, reacting with only 3 of 21 other fungal species tested.[46] Fortunately, none of those three were associated with yellowing of rice grains, so this did not present a problem in diagnosis. The mAb used for detecting *P. islandicum* was species specific, and cross reacted minimally with 15 other fungi commonly found in stored grain. This assay was successful in detecting the fungus in 90% of naturally infected rice grains.[51] mAbs have also been used to detect the presence of wood-rotting fungi[59] and a potential biological control fungus of sap-staining fungi in wood.[44] Polyclonal antibodies were used in an ELISA to determine the level of infection of grape juice by *Botrytis cinerea,* a fungus that adversely affects wine quality.[76]

An important consideration in designing a diagnostic assay for fungal infection is the type of assay to be used. Indirect ELISA is the assay most often used for initial screening of hybridoma supernatants for mAbs. In this type of ELISA, the antigen is bound directly to a microtiter plate, then mouse antibody is added, and finally, bound antibody is detected by an enzyme-linked anti-mouse secondary antibody. This assay is quite adequate if one wants to type pure cultures of fungus, but if the assay requires extraction of infected tissue, fungal antigens will likely be significantly diluted by plant material. For example, at a threshold value that did not give false positives with uninfected rice, only 26% of naturally infected rice grains were positive using the indirect ELISA for *H. lanuginosa*, even though fungus was clearly visible by scanning electron microscopy. Sensitivity probably could be improved by using a DAS-ELISA (*D*ouble *A*ntibody *S*andwich).[77] In this type of assay, the plate is first coated with a "capture antibody", either a polyclonal antibody or a second mAb recognizing a different epitope of the fungus. When the antigen mixture is then added, only the antigens of interest will bind to the antibodies on the plate. Following this, a second (enzyme-linked) antibody is added to detect bound antigen. DAS-ELISA is the assay format that has been adopted in the commercial test kits.[5]

The question of a positive/negative threshold in diagnostic ELISA is also an important, but often neglected consideration.[78] Many papers in which ELISA (or another quantitative immunoassay) is used for disease diagnosis do not even report the method used to set the threshold value, and of those that do it is often an arbitrary value such as the mean of all negative controls + 3 standard deviations. It is important that an independent method of determining pathogen presence (e.g., symptomatology) be directly compared to ELISA and a threshold chosen that gives the fewest false negatives and/or false positives. Sutula et al.[78] have suggested a number of guidelines to improve the reporting of data from ELISA and other quantitative immunoassays:

1. Clearly state the positive-negative threshold used.
2. Test enough plants to become familiar with the range of negative (healthy) values involved.
3. Include enough known negative controls in each routine assay to ensure representation of the previously established range of negative background values.
4. Always include a positive control.
5. Match control samples and test samples with respect to host type, tissue type, age, and position on the plant.
6. Replicate test samples.

Few direct comparisons have been made of polyclonal antibodies and mAbs as diagnostic probes for the same fungus. In the detection of *Sirococcus strobilinus,* mAbs were produced because polyclonal antibodies were too cross reactive to other fungi.[30] Polyclonal serum raised against *Ophiostoma ulmi* (the causal agent of Dutch elm disease) was highly cross reactive to other species and genera, as well as to the host plant. Of 33 stable hybridoma lines raised against mycelial homogenates of *O. ulmi,* 14 were genus specific, four were species specific, and two specifically recognized aggressive isolates over nonaggressive isolates. However, the specificity was antigen dependent: none of the mAbs could distinguish aggressive from nonaggressive isolates when surface washings were used as test antigen.[50] Conversely, in detection of the *Diaporthe/Phomopsis* complex of soybeans, polyclonal antibodies were more specific than any of the mAbs tested.[79]

III. TAXONOMY

As one might infer from the availability of several species- and genus-specific mAbs (Table 1), there is great potential for using them as taxonomic tools. Even mAbs that are not species specific may be used in combination to make taxonomic inferences.[29] Ferguson et al.[57] reacted a panel of 43 mAbs, all raised against *Phytophthora sojae* (formerly *P. megasperma* f. sp. *glycinea*) with homogenized mycelium of 14 phytopathogenic fungi in ELISA. None of these mAbs were specific at the species level,[80] but cluster analysis of ELISA (A_{405}) values separated the fungi into the same groups as traditional taxonomic criteria.

A. RACE SPECIFICITY

Frequently, two variants of a pathogen cause different reactions on the same host cultivar, one leading to disease and the other not. These variants, known as physiologic races, are identified by their interactions with a set of host cultivars, called "differentials", each of which carries a different race-specific resistance gene. Host resistance genes are usually inherited as single dominant alleles, while avirulence is dominant in the pathogen.[81]

Avirulence genes have been cloned from several bacterial plant pathogens, but thus far from only one fungal species, *Cladosporium fulvum.*[81a] One approach to identifying avirulence gene products would be to identify avirulence gene-associated antigenic differences using antibodies. Clearly, plants can distinguish between different pathogen races. It is therefore possible that antigenic differences associated with the presence or absence of avirulence genes exist and could be detected by using antibodies. Such avirulence gene-associated antigens, however, are not likely to be the major antigenic determinants on the pathogens. Further, identifying minor antigenic differences amid a background of antigenic similarity using conventional antisera would seem an impossible goal. Theoretically, mAbs have potential for identifying avirulence gene-associated antigens. However, reliably distinguishing races of fungal plant pathogens with mAbs remains elusive.

Wong et al.[43] raised mAbs against *F. oxysporum* f. sp. *cubense* using hyphal walls as the immunogen. Initial screening was against fungal culture filtrates in indirect ELISA. Subsequently, the mAbs were screened by indirect immunofluorescence against mycelia, where one of them apparently bound specifically to chlamydospores of all the race 4 isolates tested, but not to races 1 or 2. However, the mAbs showed considerable cross reactivity to hyphal walls of all the *F. oxysporum* isolates tested. In contrast, Ianelli et al.[29] found high cross reactivity of their mAbs to spores from different species of *Fusarium*. It is difficult to evaluate the significance of these results, especially since we are not aware of any studies on the genetics of virulence and avirulence in this pathogen. It would be useful to examine the binding of this mAb to different races of the fungus *in planta* to see where the epitope appears.

Wycoff et al.[54,55,82] raised a library of over 60 mAbs to extracellular culture filtrates and mycelial walls from race 1 of *P. sojae*, the causal agent of root rot in soybeans, in an attempt to identify race-specific antibodies. The library was screened against the same antigen preparations from 31 races and isolates of this fungus. The mAbs recognized 11 distinct epitopes as determined by competition analysis, but none of the epitopes could be correlated with particular avirulence genes. A later mass screening of over 1300 hybridomas with antigens from races 1 and 7 (which differ in at least five avirulence genes) revealed none that were race or even isolate specific.[82]

Pain et al.[39] immunized mice with homogenates of race γ of *Colletotrichum lindemuthianum* in an unsuccessful attempt to generate race-specific mAbs. Although all the mAbs cross reacted to more than one species, they did provide some interesting taxonomic data. Two mAbs whose reaction was limited to four species of *Colletotrichum* were found, based on periodate oxidation and Western blotting patterns, to recognize distinct carbohydrate epitopes. The presence of the epitope recognized by these mAbs on these four species, but not on other members of the genus, correlates with morphological criteria. Pain et al.[39] suggest that these species may better be considered as host-specific forms of a single species. The mAbs did not react with *C. lindemuthianum* isolates from cowpea, although they did with the isolate from bean. This is consistent with previous evidence that the cowpea isolate is a species distinct from *C. lindemuthianum.*[83]

The problem of cross reactivity of mAbs to species and genera other than the one used for immunization is commonly reported. Even those mAbs that have been carefully selected to distinguish fungi at the species or genus level do so in a quantitative rather than qualitative manner.[45,47,51] It has been suggested that immunodominance of nonspecific fungal carbohydrates is the problem.[46] Sometimes the fault can be fixed on the inject immunogen. Banowetz et al.[66] attempted to generate mAbs that would distinguish *Tilletia controversa* from *T. caries*. Immunization was with either intact teliospores or teliospore surface extracts that consisted primarily of polysaccharide—no protein was detected. All mAbs that reacted with *T. controversa* reacted equally well with *T. caries*. However, even when the immunogen is a purified protein, such as the β-1,4 xylanase used by Clausen et al.,[59] most of the mAbs that are produced are reactive to the carbohydrate portion.

B. IMPROVING SPECIFICITY

What might be done to increase the specificity of mAbs produced to plant-associated fungi? A number of schemes have been tried, some with more success than others, and some protocols that have worked with other types of immunogens are worth trying.

1. Selection of Specific Proteins

When a crude mixture of proteins is used as an immunogen, the dominant epitopes are likely to be those that are common to the majority of molecules—i.e., they will be carbohydrate. If a specific protein is used there will be more chance for other, possibly more specific, epitopes to predominate. Mohammed et al.[37] used specific *Armillaria ostoae* proteins excised from SDS-PAGE gels as inject antigen and isolated genus- and species-specific mAbs. When a purified phytotoxic glycopeptide produced by *O. ulmi* was used as immunogen, two of the five mAbs produced specifically bound that protein, as determined by western blot analysis.[49] Bossi and Dewey[38] immunized mice with a low-molecular weight (30-kDa) fraction from surface washings of *B. cinerea.* Three hybridoma lines secreted mAbs that specifically recognized *B. cinerea* and *B. fabae,* but not *B. allii* or other fungi normally involved in post-harvest spoilage of fruits and vegetables. Heat, protease, and periodate treatment of antigens suggest that all three recognize carbohydrate epitopes on a glycoprotein.[38] As we pointed out above, three of five mAbs reactive against β-1,4 xylanase recognized carbohydrate epitopes, but they all seemed to be specific to that enzyme. Polyclonal serum raised against purified ribosomes and used in indirect ELISA differentiated snow mold fungi at the genus level in wheat plants.[84] The use of fungal ribosomes as immunogen holds potential for producing more specific mAbs.

A procedure that chemically separates highly glycosylated proteins from others may be useful. When a 50% saturated ammonium sulfate precipitate of a mycelial extract from *Verticillium dahliae* was used as immunogen, the resulting polyclonal serum was species specific.[19] The reason may have been that proteins were enriched and carbohydrates were left in solution. However, ammonium sulfate precipitation of the antigen is not always successful in improving the specificity of the immunoreaction.[79]

2. Removal of Carbohydrates

Immunodominant carbohydrate epitopes might be removed from the immunogen, either chemically or enzymatically.[85] This may allow for mAbs with completely new epitope specificities. However, some of these epitopes may not be accessible on the untreated proteins.

3. Addition of Nonspecific Antibodies to Antigen Before Immunization

This method was successfully used to isolate mAbs specific to the aggregating cell stage of *Dictyostelium*. Mice were injected with aggregating cells along with a polyclonal serum against undifferentiated cells. Of the resulting mAbs, 20% had the desired specificity. This method was also tried by Banowetz et al.[66] without success; even so, it may be worth trying in other systems.

4. Immunosuppression

Cyclophosphamide is an antineoplastic drug that has been used for immunosuppression. If a mouse is immunized with an antigen mixture followed by cyclophosphamide, the spleen cells that proliferate will be killed. In a subsequent immunization with the same antigen mixture plus a new specific antigen, only cells that respond to the new antigen will survive and proliferate, thus increasing the percentage of specific mAbs. This has been used successfully in making mAbs specific to animal proteins[86] and to fungal animal pathogens,[87] but has met with less success in making antibodies to plant-invading fungi. Banowetz el al.[66] tried it in their attempts to make *T. controversa*-specific mAbs. However, their two teliospore surface extracts may simply have been too similar.

The immunization schedule may affect the success of immunosuppression. Wycoff[82] immunosuppressed Balb/c mice using culture filtrates from *P. sojae* race 12, and later immunized with race 1 culture filtrates. Sera from these mice gave as much as threefold higher reaction in ELISA to race 1 as to race 12, while serum from a control mouse (race 1 immunization alone) showed less than a 10% difference in ELISA values between the two races. Experimenting with different numbers of immunosuppression and immunization injections revealed that the most race-specific serum was produced after just one round of immunosuppression and immunization (although titer was low). Subsequent rounds destroyed race specificity. A methodology that might be worth trying would be to employ only one immunosuppression injection and then to immunize *in vitro*. *In vitro* immunization has been used successfully to produce large numbers of mAbs against epitopes that were not normally very immunogenic.[88]

5. Neonatal Tolerization

Tolerance is induced by injecting an antigen into mice within a few days after birth. When the mouse is exposed to the same antigen later in life, no immune response is mounted.[89] This technique was used successfully by Hsu et al.[90] to produce mAbs specific to tomato spotted wilt virus (TSWV). Neonatal Balb/c mice were injected with extract from a healthy plant on days 1, 3, 5, and 7 after birth. The same mice were injected once with partially purified TSWV at either 5, 7, or 9 weeks and then used for myeloma fusion 4 d later. The percentage of hybridomas making antibody specific to TSWV was 83, 50, and 40%, respectively. Control mice that were not tolerized, but simply immunized at 5, 7, or 9 weeks gave 0, 7, and 7% TSWV-specific hybridomas. Hardham et al.[91] used neonatal tolerization to reduce the percentage of hybridomas making antibodies against an immunodominant cyst coat material. Mice were immunized at 30 h and 4 d with *P. cinnamomi* cysts. At 8 weeks a mouse was immunized with zoospores (containing 5% cysts), and again 10 and 12 weeks later. Hybridoma supernatants were screened by immunofluorescence. Only 8% of the resulting mAbs were directed against the cyst coat material, and most of the remaining mAbs had epitope specificities not produced by the standard immunization protocol. This promising technique could be employed more widely by those wishing to produce more highly specific mAbs against phytopathogenic fungi, or those who might like to immunize with a very crude antigen, such as infected plant tissue.

IV. FUNGAL ULTRASTRUCTURE AND DISEASE PHYSIOLOGY

Because they can be highly specific probes, mAbs have been used to study the biology and ultrastructure of a few phytopathogenic fungi, both in culture and *in planta.* These studies have enhanced our understanding of the processes of fungal growth and differentiation, wall formation, and infection. Many immunocytochemical techniques are available for localizing antigens at both the light and electron microscope level, and these have been reviewed by Dewey et al.[6] Studies relating to the biology of zoosporic plant pathogens have been reviewed recently by Hardham et al.[91,92] We would like to touch on these as well as studies involving other classes of plant pathogens.

Many species of Oomycetes infect their host plants via motile biflagellate zoospores. When the zoospore reaches the surface of a potential host it encysts, a process involving detachment of the flagella, a change in cell shape from ovoid to spherical, and formation of a cell wall.[91] Several mAbs have been raised to zoospores and cysts of two Oomycetous species, *P. cinnamomi* and *Pythium aphanidermatum,* and have been used to study zoospore structure and the process of encystment. Several distinct patterns of labeling to the zoospore surface have been found.[52,62,93] The labeling patterns (and the mAb names) include: (1) the whole zoospore surface (Zp, PA1); (2) the anterior flagellum, specifically on mastigonemes (Zt, Zg, PA2); (3) both flagella (Zf); (4) the water expulsion vacuole (Zw); (5) the cyst coat material (Cpa, PA3–6); (6) vacuoles in the zoospore peripheral cytoplasm (Lpv); and (7) the surface of both zoospore and cyst (PA7, PA8). mAbs produced by the tolerization method[91] labeled several additional components, including: (1) the nucleus; (2) a reticular network within the cytoplasm; (3) the entire cytoplasm; (4) the periphery of the zoospores; and (5) spots of fluorescence on the zoospores.

Antibody Zf-1 bound to the surface of both flagella of *Phytophthora cinnamomi,* and triggered encystment and thus loss of mobility.[94] In *Pythium aphanidermatum,* encystment was induced by PA1, which bound to the entire zoospore surface and recognized a 75-kDa protein, but not by other mAbs that also bound to the zoospore surface.[95] This suggests a functional role for the 75-kDa protein in encystment.

Antibody Lpv-1 labeled large peripheral vesicles in *Phytophthora cinnamomi.* After encystment and wall formation, these vesicles move away from the cell surface and are distributed throughout the cytoplasm. After germination they are apparently degraded, and evidence suggests that the glycoproteins labeled by Lpv-1 serve as a store of protein for growth of the germ tube.[96] *Pythium aphanidermatum* also has large peripheral vesicles, containing high-molecular weight glycoproteins labeled by antibodies PA3 to 6, but after encystment these glycoproteins are secreted and are found on the exterior surface of the cyst wall. Some, however, remain within vesicles in the cyst cytoplasm.[62,63]

Double labeling with mAbs Cpa-2 and 4C7 revealed that there were two distinct types of small peripheral vesicles with no immunological cross reactivity.[91] During encystment the contents of the ventral vesicles are deposited mainly on one side of the cysts, while the contents of the dorsal vesicles are more evenly distributed. When zoospores were allowed to encyst on the surface of *Eucalyptus* roots, the material from the ventral vesicles was deposited between the cyst and the root surface, while the dorsal vesicle material coated the cysts on the distal side.[91]

mAbs that detected a phytotoxic glycopeptide from the pathogen *O. ulmi* were used to localize it in infected elm tissues.[48,49] Indirect immunoperoxidase staining strongly labeled conidia and weakly labeled hyphae grown in culture. The mAbs were used to probe ultrathin sections of *O. ulmi* grown in culture and were visualized by probing with protein A-colloidal gold. The toxic glycopeptide was found in all fungal cells, but primarily in the cell wall.[49] One mAb was used to localize the toxic glycopeptide ultrastructurally in infected elm tissue using a protein A-gold technique. These studies revealed that the toxic glycopeptide was localized on pit membranes between xylem vessels and paratracheal parenchyma cells, over the innermost wall layer of paratracheal parenchyma cells, and over intercellular spaces and the middle lamella. Increases in staining over the 4 d following inoculation correlated with an increase in disease symptoms.[48]

The biotrophic powdery mildew fungus *Erysiphe pisi* forms a specialized structure called a haustorium that is involved in the transfer of host nutrients to the fungus. Mackie et al.[40] and Callow et al.[41] raised mAbs against haustorial complexes (consisting of haustorium, extrahaustorial matrix, and the extrahaustorial membrane, which is an invagination of the plant plasma membrane) isolated from infected pea leaves. Four mAbs, selected by an indirect immunofluorescent (IMF) screen, specifically bound haustorial complexes. The antigens recognized by these mAbs were further characterized by immunogold labeling and western blot analysis. One mAb, UB7, recognizes both the fungal wall and plasma membrane as well as the haustorial wall and the haustorial plasma membrane. The other, UB8, binds only to the haustorial plasma membrane, suggesting a molecular differentiation between this membrane, across which nutrients from the plant flow, and normal hyphal membranes. Both UB7 and UB8 bind to distinct glycoproteins of ~62 kDa, and the UB8 epitope is protein, while the UB7 epitope is carbohydrate. Another mAb, UB10, recognizes a 45-kDa protein found in the haustorial plasma membrane. The antigen recognized by UB11 is a large (M_r 250 kDa) glycoprotein that is only found on the extrahaustorial membrane at early stages of haustorial development. It is not known if this protein is of host or pathogen origin.[41]

A set of five carbohydrate-specific mAbs were used to probe the ultrastructure of the walls of the soybean pathogen *Phytophthora sojae,* using a combination of immunofluorescence and immunogold labeling techniques.[56] Results with two β-1,3-glucan-specific antibodies suggest that β-1,3-glucans are present throughout the walls of both germ tubes and cysts, but are more prevalent in the outer portion. In addition, β-1,3-glucans on the surface of hyphal walls, but not cysts, are closely associated with other material (most likely protein) that sterically hinders antibody binding except to nonreducing terminal residues. An antibody whose epitope involved both β-1,4- and β-1,3-glucosyl linkages bound predominantly to the inner portion of the hyphal wall. However, fluorescent labeling with this antibody suggested that β-1,4 linkages are present on the exterior of *P. sojae* walls as well. Another mAb, of unknown carbohydrate specificity, stained predominantly older portions of growing hyphae, a pattern also seen with a mAb raised to *C. lindemuthianum.*[39] The usefulness of ultrastructural studies of fungi using mAbs against fungal antigens will be enhanced when the epitope specificity of more of these mAbs are determined.

A study by Cole et al.[97] illustrates the importance of using more than one immunocytochemical method to provide a more complete picture of the localization of an antigen. They used a mAb against a glycoprotein (M_r 92 kDa) found in surface washings of *Penicillium islandicum.* Indirect immunofluorescence suggested that the protein was located in walls of vegetative hyphae, but not in walls of conidiophores. Immunogold labeling showed that the protein was found in the outer wall layers in young hyphae and at the interface between inner translucent and outer pigmented walls in older hyphae and conidiophores. Binding decreased in a proximal to distal direction along the conidiophore and was absent in phialides and conidia. They suggested that the pigment on the exterior of the walls of conidiophores and conidia either (1) physically prevents the penetration of the antibodies and/or (2) masks the fluorescein isothiocyanate fluorescence in the light-microscope preparation of these structures.[97]

A glycoprotein that elicits a hypersensitive reaction on wheat has been isolated from germ tube walls of *Puccinia graminis* (the wheat stem rust fungus).[60] Two mAbs raised to this purified glycoprotein bound to fungal cells as determined by indirect immunofluorescence. Binding was only at the young growing tips of infection structures formed *in vitro* and only at haustoria in rust-infected wheat leaves. Unfortunately, because the mAbs were never shown (say, by western blot) to be specific to this protein, it is not known yet whether or not the distribution of antibody staining and distribution of the protein are identical.

The use of mAbs to study the process of fungal infection has not been limited to antibodies against fungal components. Antibodies directed against molecules produced by plants in response to infection have also been used. An ELISA that used mAbs has been developed for monitoring abscisic acid levels in soybeans infected by *Phytophthora sojae.*[98]

V. PRACTICAL CONSIDERATIONS

A. ANTIGEN SELECTION

Before embarking upon a project to raise mAbs against a plant-pathogenic fungus, one should be thoroughly familiar with the fungal disease of interest and with the general principles of humoral (antibody-based) immunity.[99] Original literature on the fungus should be consulted, and this can usually be found cited in Alexopoulos and Mims[7] or Farr et al.[100] It may be necessary to locate plant material that is infected. The fungus must then be isolated, usually on agar media, and positively identified. Alternatively, it may be possible to obtain rigorously identified pure cultures from colleagues or from the American Type Culture Collection. The purity of the culture should be verified by transferring growing hyphal tips to agar media containing antibiotics that allow the fungus to grow, but kill bacteria. The medium used, the culture conditions, and the length of the culture period will depend on the fungal species. Many published protocols for *in vitro* fungal culture are available (see original references cited in Farr et al.)[100] Other fungi associated with the host plant should be obtained and identified carefully also, as should additional isolates of the pathogen. Original isolates can be stored on agar slants at 4°C for periods of up to several months. Excessive transfers of cultures should be avoided, since fungi can change in culture. However, obligate biotrophs cannot be cultured; see Mackie et al.[40] and Callow et al.,[41] and below.

Since the goal is to produce mAbs against fungal antigens that will not cross react to plant antigens, it would be ideal to start with axenically cultured fungus. This may not be possible with certain fungi, in particular those causing downy mildews, powdery mildews, and rusts, which are obligate parasites.[101] Additionally, it may be desirable to produce mAbs against antigens that are not expressed in culture, but only in infected plants.[102] There are methods for isolating fungal material from infected plant tissue.[40]

There is no generally accepted "best" material to use as inject and screening antigen for mAb production. As Table 1 shows, various workers have used extracellular culture filtrates, surface washings, spores, hyphal wall fragments, mycelial extracts, and specific purified proteins. The type of material to use as immunogen and antigen deserves careful consideration, and will depend to a large extent on the purpose for using the mAbs. If the mAbs are to be used for disease diagnosis, the fungal material used as immunogen should also be present in the material to be diagnosed. Thus, if homogenized tissue extracts are to be used in the final assay, mycelial extracts, mycelial wall fragments, and extracellular culture filtrates are all likely to be good immunogens. But, if the intended assay will involve a gentle antigen release without homogenization of tissue, then surface washings or extracellular culture filtrates may be better immunogens than mycelial extracts.[46,51] If the mAbs are to be used for immunocytochemistry, then preparations enriched in the cell type/structure of interest should be used. It may be desirable to try different preparations on different individual animals and assess the titer and specificity of the polyclonal serum before proceeding to make mAbs. Due to the variability of the immune response between different animals, two or more animals should be used with each test preparation. Immunogenicity of an antigen preparation also may differ in different animal species. For example, extracellular culture filtrate from *Phomopsis longicolla* used as immunogen in rabbits gave rise to polyclonal antibodies with greater specificity, but lower titer than that from mycelial extracts. Conversely in mice, the extracellular culture filtrate did not elicit a detectable immune response, but the mycelial extract did.[79] We will outline a "generic" procedure for isolating the major fungal components that can be used as immunogen, but the original literature on the species of interest should be consulted for specific procedures. In particular, those interested in producing mAbs to zoospores should consult the excellent review of Hardham et al.[91]

For producing mycelial extract, mycelial wall, and culture filtrate immunogens it is probably best to start with fungus grown in liquid culture. Proteins of fungi can vary with culture age: young cultures contain more protein and less nonspecific carbohydrate, so are probably the most suitable starting material.[103,104] Proteins also may differ significantly with the type of growth medium used.[65] All fungal materials should be cooled and kept at 0 to 4°C during manipulation, to slow denaturation and/or enzymatic hydrolysis of antigenic molecules. It is not necessary to maintain sterility during preparation of antigens, but care should be taken to avoid the introduction of any debris or contaminating organisms that may themselves be immunogenic.

B. ANTIGEN PREPARATION

Mycelia are first separated from the culture fluid (containing secreted antigens) by filtration through nylon, filter paper, or Miracloth, followed by centrifugation to clear the filtrate. If the culture filtrate will be used, it may first need to be concentrated. A convenient and relatively nondenaturing method to concentrate the culture filtrate is direct lyophilization, but precipitation with two volumes of ethanol or acetone, followed by lyophilization, may also be used. The resulting powder should be resuspended in a small volume of water, then dialyzed against a saline buffer to rid it of low-molecular weight compounds, including toxins that could be excessively harmful to the animals that will be immunized. The protein (or carbohydrate) concentration should then be measured.[105–107] The amount of material needed for immunization of mice is relatively small, approximately 10 to 200 μg protein or carbohydrate per injection in 100 μl volume. Insoluble aggregates found in the resuspended preparation may enhance the immune response of the animals, but should be centrifuged out of the portion of the preparation that will be used in the screening assay, as a precaution to reduce nonspecific binding. The preparation should be divided into aliquots and stored frozen.

If a mycelial extract is to be used, the mycelium may be rinsed with sterile, deionized water and vigorously ground in a cold mortar and pestle. The resulting slurry should be centrifuged to remove insoluble materials, and the supernatant saved. A floating layer of lipid may be found after centrifugation, and this will not damage the extract if some is included. After determining protein concentration, the extract may then be aliquoted and stored frozen.

To isolate mycelial wall fragments, wash the insoluble pellet from the mycelial extract thoroughly with water, buffer containing EDTA (to remove insoluble divalent cations), and finally with acetone (to remove lipids).[108] The wall fragments should then be lyophilized and stored frozen. They can be resuspended in buffer just before use.

C. IMMUNIZATION/SCREENING

Before the animals are immunized, a control bleeding should be taken for the serum (normal serum) to serve as a negative control in antibody assays. Chapter 25 serves as a reference on the production of mAbs and the development of ELISA assays. For guidelines on immunization schedules also consult Goding.[99]

IMF screening assays using spores or hyphal fragments may be better than ELISA for identifying highly specific mAbs. IMF was effective for detection of mAbs binding to isolate-, species-, and genus-specific components on the surface of zoospores of *Phytophthora cinnamomi*.[52,93] Estrada-Garcia et al.[62] used indirect immunofluorescence, and six of eight hybridomas produced species-specific mAbs. Unfortunately, IMF techniques are not as practical for diagnosis as ELISA, because of the expense of fluorescent microscopes, more complex sample preparation, and a less quantitative nature. Additionally, mAbs that work well in IMF may not in ELISA, dot-blot, or other techniques.[99]

Hybridomas producing antibodies that test positive for reactivity to the fungal species of interest should be tested against other fungal species to examine specificity. If the mAb will be used for disease diagnosis or taxonomy it is especially important to assess the reactivity of the antibodies to other fungi that are found associated with, or produce similar symptoms on, the same plants as the pathogen of interest. Thus, these other fungi should be isolated and identified like the fungus of interest and antigen produced from them in the same way. Following the isolation of the desired hybridoma clone(s), the mAbs may be produced in larger quantities and used in diagnostic assays following verification of the ability of the mAbs to detect the fungal pathogen in plant tissue.

VI. FUTURE APPLICATIONS

A. COMMON ANTIGENS

Serological studies of plant pathogens have revealed that pathogens sometimes share antigenic determinants with their hosts. DeVay et al.[109] proposed the idea that compatibility may be engendered by the presence of common antigens in the host and pathogen. They hypothesized that pathogens that display host antigens on their surfaces might thereby evade recognition. Such a phenomenon is familiar in animal-parasite interactions,[110] and in plant-pathogen interactions many cases of common antigens have been documented.[111–113] However, none of the experiments that have been reported have been done in such a way as to shed light on the possible function of such antigenic similarity. Are common antigens related to host range or race specificity? If they are related to host range, one would expect to find them in all species that act as host for a particular fungus. If related to race specificity, one would expect to find them only in compatible host/pathogen combinations using isogenic plant lines. It seems likely, however, that they are simply due to the inevitable (and unremarkable) similarity in biochemistry that must exist between many different organisms.

Monoclonal antibodies that cross react to host antigens have been found. A mAb raised to *O. ulmi* that was species specific when tested against 13 fungal species from the same and different genera cross reacted strongly to healthy elm tissue.[50] At least 13 of 44 mAbs raised against extracellular culture filtrate or hyphal walls of *P. sojae* recognized soybean plasma membrane antigens.[55] They also reacted to plasma membranes of corn, pumpkin, and bean, none of which are hosts for this pathogen. Evidence suggested that these mAbs recognized carbohydrate epitopes on glycoproteins.[54] A study that deliberately sought to create mAbs against common antigens might be helpful in identifying the nature of such antigens, thus shedding some light on their potential functions.

B. PLANTIBODIES

mAbs that bind specifically to components of plant pathogens could be utilized in transgenic plants expressing the antibodies. Tobacco plants have been genetically engineered to express mouse mAbs, and these plant-produced antibodies were active and retained their specificity.[114,115] One possible application of this technology to plant protection is suggested by the work of Heiny and Gilchrist.[116] They found that a polyclonal antibody, raised against a phytotoxin produced by *Stemphylium botryosum,* could neutralize the phytotoxic activity of the protein in a leaf bioassay. It seems possible that expressing appropriate mAbs against certain fungal phytotoxins in plants might prevent the tissue necrosis and other symptoms of disease that these phytotoxins cause. Expressing mAbs against other fungal components, particularly those involved in infection and nutrient uptake, may confer a kind of general resistance.

In the last 10 years there has been an explosion of reports on the development and use of mAbs against phytopathogenic fungi. It is only in the last 5 years that some of these mAbs have been developed into diagnostic kits suitable for use in farmers' fields. The next 10 years should witness even greater application of mAbs in disease diagnosis, basic study of the infection process, and plant protection.

REFERENCES

1. Köhler, F. and Milstein, C., Continuous cultures of fused cells secreting antibody of predefined specificity, *Nature,* 256, 495, 1975.
2. Clark, M. F., Immunosorbent assays in plant pathology, *Annu. Rev. Phytopathol.,* 19, 83, 1981.
3. Halk, E. L. and De Boer, S. H., Monoclonal antibodies in plant-disease research, *Annu. Rev. Phytopathol.,* 23, 321, 1985.
4. Ouellette, G. B. and Benhamou, N., Use of monoclonal antibodies to detect molecules of fungal plant pathogens, *Can. J. Plant Pathol.,* 9, 167, 1987.
5. Miller, S. A., Molecular diagnosis of plant disease, *Annu. Rev. Phytopathol.,* 26, 409, 1988.
6. Dewey, M., Evans, D., Coleman, J., Priestley, R., Hull, R., Horsley, D., and Hawes, C., Antibodies in plant science, *Acta Bot. Neerl.,* 40, 1, 1991.
7. Alexopoulos, C. J. and Mims, C. W., *Introductory Mycology,* John Wiley & Sons, New York, 1979, 632.
8. Hobbs, T. W., Schmitthenner, A. F., and Kuter, G. A., A new *Phomopsis* species from soybeans, *Mycologia,* 77, 535, 1985.
9. Morgan-Jones, G., The *Diaporthe/Phomopsis* complex: taxonomic considerations, in *World Soybean Research Conference IV: Proceedings,* Pascale, A. J., Ed., Buenos Aires, Argentina, 1989, 1699.
10. Sinclair, J. B. and Backman, P. A., **Eds.,** *Compendium of Soybean Diseases,* 3rd ed., APS Press, St. Paul, MN, 1989, 106.
11. Dhingra, O. D. and Sinclair, J. B., *Basic Plant Pathology Methods,* CRC Press, Boca Raton, FL, 1985, 355.
12. Cerkauskas, R. F. and Sinclair, J. B., Use of paraquat to aid detection of fungi in soybean tissues, *Phytopathology,* 70, 1036, 1980.
13. Cerkauskas, R. F., Dhingra, O. D., and Sinclair, J. B., Effect of three desiccant type herbicides on fruiting structure of *Colletotrichum truncatum* and *Phomopsis* species on soybean stems, *Plant Dis.,* 67, 620, 1983.
14. Shelby, R. A., Use of enzyme-linked immunosorbent assay (ELISA) as an aid in the detection of *Diaphorthe phaseolorum* var. *caulivora, Proc. South. Soybean Dis. Workers,* 12, 70, 1985.
15. Gleason, M. L., Ghabrial, S. A., and Ferriss, R. S., Serological detection of *Phomopsis longicolla* in soybean seeds, *Phytopathology,* 77, 371, 1987.
16. Amouzou-Alladay, E., Dunez, J., and Clerjeau, M., Immunoenzymatic detection of *Phytophthora fragariae* in infected strawberry plants, *Phytopathology,* 78, 1022, 1988.

17. Duncan, J. M. and Mohan, S., Detection of *Phytophthora* spp. using ELISA, in *Techniques for Rapid* Detection and Diagnosis of Plant Disease, Programme Abstracts, British Society for Plant Pathology, University of East Anglia, 11–13 July, 1989, 1989, 6.
18. El-Nashaar, H. M., Moore, L. W., and George, R. A., Enzyme-linked immunosorbent assay quantification of initial infection of wheat by *Gaeumannomyces graminis* var. *tritici* as moderated by biocontrol agents, *Phytopathology,* 76, 1319, 1986.
19. Gerik, J. S., Lommel, S. A., and Huisman, O. C., A specific serological staining procedure for *Verticillium dahliae* in cotton root tissue, *Phytopathology,* 77, 261, 1987.
20. Musgrave, D. R., Grose, T. A., Latch, G. C. M., and Christensen, M. J., Purification and characterization of the antigens of endophytic fungi isolated from *Lolium perenne* and *Festuca arundinacea* in New Zealand, *N.Z. J. Agric. Res.,* 29, 121, 1986.
21. Velicheti, R. K., Lamison, C., Brill, L. M., and Sinclair, J. B., Immunodetection of *Phomopsis* species in asymptomatic soybean plants, *Plant Dis.,* 77, 70, 1993.
22. Savage, S. D. and Sall, M. A., Radioimmunosorbent assay for *Botrytis cinerea, Phytopathology,* 71, 411, 1981.
23. MacDonald, J. D. and Duniway, J. M., Use of fluorescent antibodies to study the survival of *Phytophthora megasperma* and *P. cinnamomi* zoospores in the soil, *Phytopathology,* 69, 436, 1979.
24. Mueller, W. C., Tessier, B. J., and Englander, L., Immunocytochemical detection of fungi in the roots of *Rhododendron, Can. J. Bot.,* 64, 718, 1986.
25. Svircev, A. M., Gardiner, R. B., McKeen, W. E., Day, W. W., and Smith, R. J., Detection by protein-A gold of antigens of *Botrytis cinerea* in cytoplasm of infected *Vicia faba, Phytopathology,* 76, 622, 1986.
26. Suske, J. and Acker, G., Identification of endophytic hyphae of *Lophodermium piceae* in tissues of green, symptomless Norway spruce needles by immunoelectron microscopy, *Can. J. Bot.,* 67, 1768, 1989.
27. Lange, L., Heide, M., Hobolth, L., and Olson, L. W., Serological detection of *Plasmodiophora brassicae* by dot immunobinding and visualization of the serological reaction by scanning electron microscopy, *Phytopathology,* 79, 1066, 1989.
28. Gwinn, K. D., Collins-Shepard, M. H., and Reddick, B. B., Tissue print-immunoblot, an accurate method for the detection of *Acremonium coenophialum* in tall fescue, *Phytopathology,* 81, 747, 1991.
29. Ianelli, D., Caparelli, R., Marziano, F., Scala, F., and Noviello, C., Production of hybridomas secreting monoclonal antibodies to the genus *Fusarium, Mycotaxon,* 17, 523, 1983.
30. Mitchell, L. A., Derivation of *Sirococcus strobilinus* specific monoclonal antibodies, *Can. J. For. Res.,* 16, 939, 1986.
31. Mitchell, L. A. and Southerland, J. R., Detection of seedborne *Sirococcus strobilinus* with monoclonal antibodies in an enzyme-linked immunosorbent assay (ELISA), *Can. J. For. Res.,* 16, 945, 1986.
32. Mitchell, L. A., A sensitive dot immunoassay employing monoclonal antibodies for detection of *Sirococcus strobilinus* in spruce seed, *Plant Dis.,* 72, 644, 1988.
33. Miller, S. A., Detection of Pythium blight in turfgrass using a monoclonal antibody-based diagnostic test (Abs.), *Phytopathology,* 76, 1057, 1986.
34. Miller, S. A., Rapid detection and monitoring of Pythium blight of turfgrass by means of a field usable enzyme-linked immunosorbent assay (ELISA) (Abs.), *Phytopathology,* 78, 1516, 1988.
35. Petersen, F. P., Rittenburg, J. H., Joaquim, T. R., Miller, S. A., and Grothaus, G. D., Development of field and laboratory immuno-diagnostic kits for detection and differentiation of *Septoria nodorum* and *Septoria tritici* in wheat, *Phytopathology,* 82, 1105, 1992.
36. Fox, R. T. V. and Hahne, K., Prospects for the rapid diagnosis of *Armillaria* by monoclonal antibody ELISA in the field, in *Proceedings of the 7th International Conference on Root and Butt Rots of Forest Trees,* Victoria, B.C., 1988, 458.
37. Mohammed, C., Priestley, R., and Dewey, F. M., Raising monoclonal antibodies to European species of *Armillaria* (Abs.), *Phytopathology,* 82, 1152, 1992.
38. Bossi, R. and Dewey, F. M., Development of a monoclonal antibody-based immunodetection assay for *Botrytis cinerea, Plant Pathol.,* 41, 472, 1992.
39. Pain, N. A., O'Connell, R. J., Bailey, J. A., and Green, J. R., Monoclonal antibodies which show restricted binding to four *Colletotrichum* species: *C. lindemuthianum, C. valvarum, C. orbiculare* and *C. trifolii, Physiol. Mol. Plant Pathol.,* 40, 111, 1992.

40. Mackie, A. J., Roberts, A. M., Callow, J. A., and Green, J. R., Molecular differentiation in pea powdery-mildew haustoria: identification of a 62-kDa N-linked glycoprotein unique to the haustorial plasma membrane, *Planta,* 183, 399, 1991.
41. Callow, J. A., Mackie, A., Roberts, A. M., and Green, J. R., Evidence for molecular differentiation in powdery mildew haustoria through the use of monoclonal antibodies, *Symbiosis,* 14, 237, 1992.
42. Gendloff, E. H. and Leighton-Sands, J., Production of monoclonal and polyclonal antibodies against *Fusarium* species (Abs.), *Phytopathology,* 78, 1580, 1988.
43. Wong, W. C., White, M., and Wright, I. G., Production of monoclonal antibodies to *Fusarium oxysporum* f. sp. *cubense* race 4, *Lett. Appl. Microbiol.,* 53, 2222, 1988.
44. Breuil, C., Luck, B. T., Rossignol, L., Little, J., Echeverri, C. J., Banerjee, S., and Brown, D. L., Monoclonal antibodies to *Gliocladium roseum,* a potential biological control fungus of sap-staining fungi in wood, *J. Gen. Microbiol.,* 138, 2311, 1992.
45. Wright, S. F., Mortorn, J. B., and Sworobuk, J. E., Identification of a vesicular-arbuscular mycorrhizal fungus by using monoclonal antibodies in immunosorbent assay, *Appl. Environ. Microbiol.,* 53, 2222, 1987.
46. Dewey, F. M., Macdonald, M. M., and Phillips, S. I., Development of monoclonal-antibody-ELISA, -DOT-BLOT and -DIP-STICK immunoassays for *Humicola lanuginosa* in rice, *J. Gen. Microbiol.,* 135, 361, 1989.
47. Nameth, S. T., Shane, W. W., and Stier, J. C., Development of a monoclonal antibody for detection of *Leptosphaeria korrae,* the causal agent of necrotic ringspot disease of turfgrass, *Phytopathology,* 80, 1208, 1990.
48. Benhamou, N., LaFontaine, J. G., Joly, J. R., and Oullette, G. B., Ultrastructural localization in host tissues of a toxic glycopeptide produced by *Ophiostoma ulmi,* the Dutch elm disease pathogen, *Can. J. Bot.,* 63, 1185, 1985.
49. Benhamou, N., Ouellette, G. B., LaFontaine, J. G., and Joly, J. R., Use of monoclonal antibodies to detect a phytotoxic glycopeptide produced by *Ophiostoma ulmi,* the Dutch elm disease pathogen, *Can. J. Bot.,* 63, 1177, 1985.
50. Dewey, F. M., Munday, C. J., and Brasier, C. M., Monoclonal antibodies to specific components of the Dutch Elm Disease pathogen *Ophiostoma ulmi, Plant Pathol.,* 38, 9, 1989.
51. Dewey, F. M., MacDonald, M. M., Philips, S. I., and Priestley, R. A., Development of monoclonal antibody-ELISA and -DIP-STICK immunoassays for *Penicillium islandicum* in rice grains, *J. Gen. Microbiol.,* 136, 753, 1990.
52. Hardham, A. R., Suzaki, E., and Perkin, J. L., Monoclonal antibodies to isolate- species-, and genus-specific components on the surface of zoospores and cysts of the fungus *Phytophthora cinnamomi, Can. J. Bot.,* 64, 311, 1986.
53. Pepin, H. S. and Prager, C. M., Application of polyclonal and monoclonal antibodies in the detection and identification of *Phytophthora* species and races in plant tissues (Abs.), *Phytopathology,* 78, 1516, 1988.
54. Wycoff, K. L., Jellison, J., and Ayers, A. R., Monoclonal antibodies to glycoprotein antigens of a fungal plant pathogen, *Phytophthora megasperma* f. sp. *glycinea, Plant Physiol.,* 85, 508, 1987.
55. Wycoff, K. L. and Ayers, A. R., Monoclonal antibodies to surface and extracellular antigens of a fungal plant pathogen, *Phytophthora megasperma* f.sp. *glycinea,* recognize specific carbohydrate epitopes, *Physiol. Mol. Plant Pathol.,* 36, 55, 1990.
56. Wycoff, K. L. and Ayers, A. R., Localization of carbohydrate antigens in the walls of *Phytophthora megasperma* f.sp. *glycinea* by monoclonal antibodies, *Physiol. Mol. Plant Pathol.,* 39, 95, 1991.
57. Ferguson, M. W., Wycoff, K. L., and Ayers, A. R., Use of cluster analysis with monoclonal antibodies for taxonomic differentiation of phytopathogenic fungi and for screening and clustering antibodies, *Curr. Microbiol.,* 17, 127, 1988.
58. Rittenburg, J. H., Peterson, E. P., and Miller, S. A., Monoclonal antibody-based immunoassays for detection of *Phytophthora* spp. in plants, (Abs.) *Phytopathology,* 80, 962, 1990.
59. Clausen, C. A., Green, F., and Highley, T. L., Characterization of monoclonal antibodies to wood-derived β-1,4-xylanase of *Postia placenta* and their application to detection of incipient decay, *Wood Sci. Technol.,* 27, 219, 1993.

60. Tiburzy, R., Rogner, U. C., Fischer, R., Beissmann, B., Kreuzaler, F. M., and Reisener, H.-J., Detection of an elicitor on infection structures of *Puccinia graminis* using monoclonal antibodies, *Eur. J. Cell Biol.,* 55, 174, 1991.
61. Xia, J. Q., Lee, F. N., and Kim, K. S., Monoclonal antibodies to an extracellular component of *Pyricularia grisea, Can. J. Bot.,* 70, 1790, 1992.
62. Estrada-Garcia, M. T., Green, J. R., Booth, J. M., White, G., and Callow, J. A., Monoclonal antibodies to cell surface components of zoospores and cysts of the fungus *Pythium aphanidermatum* reveal species-specific antigens, *Exp. Mycol.,* 13, 348, 1989.
63. Estrada-Garcia, M. T., Callow, J. A., and Green, J. R., Monoclonal antibodies to the adhesive cell coat secreted by *Pythium aphanidermatum* zoospores recognise 200×10^3 M_r glycoproteins stored within large peripheral vesicles, *J. Cell Sci.,* 95, 199, 1990.
64. Yuen, G. Y. and Craig, M. L., Monoclonal antibodies specific to *Pythium ultimum* and *Pythium* spp., (Abs.) *Phytopathology,* 80, 962, 1990.
65. Matthew, J. S. and Brooker, J. D., The isolation and characterization of polyclonal and monoclonal antibodies to anastomosis group 8 of *Rhizoctonia solani, Plant Pathol.,* 40, 67, 1991.
66. Banowetz, G. M., Trione, E. J., and Krygier, B. B., Immunological comparisons of teliospores of two wheat bunt fungi, *Tilletia* spp., using monoclonal antibodies and antisera, *Mycologia,* 76, 51, 1984.
67. MacDonald, J. D., Stites, J., and Kabashima, J., Comparison of serological and culture plate methods for detecting species of *Phytophthora, Pythium,* and *Rhizoctonia* in ornamental plants, *Plant Dis.,* 74, 655, 1990.
68. Benson, D. M., Detection by enzyme-linked immunosorbent assay of *Rhizoctonia* spp. on poinsettia stem cuttings, *Plant Dis.,* 76, 578, 1992.
69. Shane, W. W., Prospects for early detection of Pythium blight epidemics on turfgrass by antibody-aided monitoring, *Plant Dis.,* 75, 921, 1991.
70. Schmitthenner, A. F., Relationship between the amount of *Phytophthora* in soil and yield loss from Phytophthora rot of soybean in Ohio (Abs.), *Phytopathology,* 82, 1105, 1992.
71. Pscheidt, J. W., Burket, J. Z., Fischer, S. L., and Hamm, P. B., Sensitivity and clinical use of *Phytophthora*-specific immunoassay kits, *Plant Dis.,* 76, 928, 1992.
72. Lacey, J., Ramakrishna, N., Candlish, A. A. G., and Smith, J. E., Immunoassay of ochratoxin and other mycotoxins from a single extract of cereal grains utilizing monoclonal antibodies, *IARC (Int. Agency Res. Cancer) Sci. Publ.* 115, 97, 1991.
73. Tseng, T. C., Li, I., and Chao, Y. H., Preparation and characterization of a monoclonal antibody against aflatoxin-B1, aflatoxin-B2, aflatoxin-G1 and aflatoxin-G2, *Bot. Bull. Acad. Sin.,* 33, 369, 1992.
74. Putnam, M. L. and Binkerd, K. A., Comparison of a commercial ELISA kit and TLC for detection of deoxynivalenol in wheat, *Plant Dis.,* 76, 1078, 1992.
75. Phelps, D. C., Nemec, S., Baker, R., and Mansell, R., Immunoassay for naphthazarin phytotoxins produced by *Fusarium solani, Phytopathology,* 80, 298, 1990.
76. Ricker, R. W., Marois, J. J., Dlott, J. W., Bostock, R. M., and Morrison, J. C., Immunodetection and quantification of *Botrytis cinerea* on harvested wine grapes, *Phytopathology,* 81, 404, 1991.
77. Brill, L. M., McClary, R. D., and Sinclair, J. B., Analysis of two ELISA formats and antigen preparations using polyclonal antibodies against *Phomopsis logicolla, Phytopathology,* 84, 173, 1994.
78. Sutula, C. L., Gilett, J. M., Morrissey, S. M., and Ramsdell, D. C., Interpreting ELISA data and establishing the positive-negative threshold, *Plant Dis.,* 70, 722, 1986.
79. Brill, L. M., Serodiagnosis of Fungi of the Diaporthe/Phomopsis Complex of Soybeans, M. S. thesis, University of Illinois at Urbana/Champaign, 1990.
80. Wycoff, K. L., Unpublished data, 1988.
81. Day, P. R., *Genetics of Host-Pathogen Interactions,* W. H. Freeman, San Francisco, 1974.
81a. van Kan, J. A. L., van den Ackerveken, G. F. J. M., and deWitt, P. J. G. M., Cloning and characterization of cDNA of avirulence gene *AVR9* of the fungal pathogen *Cladosporum fulvum,* casual agent of tomato leaf mold, *Mol. Plant. Microbe Interact.,* 4, 52, 1991.
82. Wycoff, K. L., Antigens of the Plant Pathogenic Fungus, *Phytophthora megasperma* f.sp. *glycinea,* Ph.D. thesis, Harvard University, Cambridge, MA, 1988.
83. Bailey, J. A., Nash, C., and O'Connell, R. J., Infection process and host specificity of a *Colletotrichum* species causing anthracnose disease of cowpea, *Vigna unguiculata, Mycol. Res.,* 94, 810, 1990.

84. Takenaka, S., Use of immunological methods with antiribosome serums to detect snow mold fungi in wheat plants, *Phytopathology,* 82, 896, 1992.
85. Bradley, D. J., Wood, E. A., Larkins, A. P., Galfre, G., Butcher, G. W., and Brewin, N. J., Isolation of monoclonal antibodies reacting with peribacteroid membranes and other components of pea root nodules containing *Rhizobium leguminosarum, Planta,* 173, 149, 1988.
86. Matthew, W. D. and Patterson, P. H., The production of a monoclonal antibody that blocks the action of neurite outgrowth-promoting factor, *Cold Spring Harbor Symp. Quant. Biol.,* 48, 625, 1983.
87. Hamilton, A. J., Bartholomew, M. A., Fenelon, L. E., Figueroa, J., and Hay, R. J., A murine monoclonal antibody exhibiting high species specificity for *Histoplasma capsulatum* var. *capsulatum, J. Gen. Microbiol.,* 136, 331, 1990.
88. Boss, B., An improved in vitro immunization procedure for the production of monoclonal antibodies against neural and other antigens, *Brain Res.,* 291, 193, 1984.
89. Golumbeski, G. S. J. and Dimond, R. L., The use of tolerization in the production of monoclonal antibodies against minor antigenic determinants, *Anal. Biochem.,* 154, 373, 1986.
90. Hsu, H. T., Wang, Y. C., Lawson, R. H., Wang, M., and Gonsalves, D., Splenocytes of mice with induced immunological tolerance to plant antigens for construction of hybridomas secreting Tomato Spotted Wilt Virus-specific antibodies, *Phytopathology,* 80, 158, 1990.
91. Hardham, A. R., Gubler, F., Duniec, J., and Elliott, J., A review of methods for the production and use of monoclonal antibodies to study zoosporic plant pathogens, *J. Microsc.,* 162, 305, 1991.
92. Hardham, A. R., Lectin and antibody labelling of surface components of spores of *Phytophthora cinnamomi, Aust. J. Plant Physiol.,* 16, 19, 1989.
93. Hardham, A. R., Suzaki, E., and Perkin, J. L., The detection of monoclonal antibodies specific for surface components on zoospores and cysts of *Phytophthora cinnamomi, Exp. Mycol.,* 9, 264, 1985.
94. Hardham, A. R. and Suzaki, E., Encystment of zoospores of the fungus *Phytophthora cinnamomi,* is induced by specific lectin and monoclonal antibody binding to the cell surface, *Protoplasma,* 133, 165, 1986.
95. Estrada-Garcia, T., Ray, T. C., Green, J. R., Callow, J. A., and Kennedy, J. F., Encystment of *Pythium aphanidermatum* zoospores is induced by root mucilage polysaccharides, pectin and a monoclonal antibody to a surface antigen, *J. Exp. Bot.,* 41, 693, 1990.
96. Gubler, F. and Hardham, A. R., Protein storage in large peripheral vesicles in *Phytophthora* zoospores and its breakdown after cyst germination, *Exp. Mycol.,* 14, 393, 1990.
97. Cole, L., Hawes, C., and Dewey, F. M., Immunocytochemical localization of a specific antigen in cell walls of *Penicillium islandicum, Mycol. Res.,* 95, 1369, 1991.
98. Cahill, D. M. and Ward, E. W. B., An indirect enzyme-linked immunosorbent assay for measurement of abscisic acid in soybean inoculated with *Phytophthora megasperma* f. sp. *glycinea, Phytopathology,* 79, 1238, 1989.
99. Goding, J. W., *Monoclonal Antibodies: Principles and Practice,* 2nd ed., Academic Press, Orlando, FL, 1986, 315.
100. Farr, D. F., Bills, G. F., Chamuris, G. P., and Rossman, A. Y., *Fungi on Plants and Plant Products in the United States,* APS Press, St. Paul, MN, 1989, 1252.
101. Agrios, G. N., *Plant Pathology,* Academic Press, San Diego, 1988, 803.
102. De Wit, P. J. G. M., Hofman, A. E., Velthuis, G. C., and Kuc, J. A., Isolation and characterization of an elicitor of necrosis isolated from intercellular fluids of compatible interactions of *Cladosporium fulvum* (Syn. *Fulvia fulva*) and tomato, *Plant Physiol.,* 77, 642, 1985.
103. Abbott, L. K. and Holland, A. A., Electrophoretic patterns of soluble proteins and isoenzymes of *Gauemannomyces graminis, Aust. J. Bot.,* 23, 1, 1975.
104. Bent, K. J., Electrophoresis of proteins of three *Penicillium* species on acrylamide gels, *J. Gen. Microbiol.,* 49, 195, 1967.
105. Lowry, O. H., Rosebrough, N. J., Farr, A. L., and Randall, R. J., Protein measurement with the Folin phenol reagent, *J. Biol. Chem.,* 193, 265, 1951.
106. Bradford, M. M., A rapid and sensitive method for the quantitation of microgram quantities of protein utilizing the principle of protein-dye binding, *Anal. Biochem.,* 72, 248, 1976.
107. Dubois, M., Gilles, K. A., Hamilton, J. K., Rebers, P. A., and Smith, F., Colorimetric method for determination of sugars and related substances, *Anal. Chem.,* 28, 350, 1956.
108. Sharp, J. K., Valent, B., and Albersheim, P., Purification and partial characterization of a β-glucan fragment that elicits phytoalexin accumulation in soybean, *J. Biol. Chem.,* 259, 11312, 1984.

109. DeVay, J. E., Charudattan, R., and Wimalajeewa, D. L. S., Common antigenic determinants as a possible regulator of host-pathogen compatibility, *Am. Nat.*, 106, 185, 1972.
110. Greenblatt, C. L., Molecular mimicry and the carbohydrate language of parasitism, *ASM News*, 49, 488, 1983.
111. De Vay, J. E. and Adler, H. E., Antigens common to hosts and parasites, *Annu. Rev. Microbiol.*, 30, 147, 1976.
112. Alba, A. P. C. and DeVay, J. E., Detection of cross-reactive antigens between *Phytophthora infestans* (Mont.) de Bary and *Solanum* species by indirect enzyme-linked immunosorbent assay, *Phytopathol. Z.*, 112, 97, 1985.
113. Purkayastha, R. P. and Ghosal, A., Analysis of cross-reactive antigens of *Acrocylindrium oryzae* and rice in relation to sheath rot disease, *Physiol. Mol. Plant Pathol.*, 27, 245, 1985.
114. Hiatt, A., Cafferkey, R., and Bowdish, K., Production of antibodies in transgenic plants, *Nature*, 342, 76, 1989.
115. Hiatt, A. and Ma, J. K.-C., Monoclonal antibody engineering in plants, *FEBS Lett.*, 307, 71, 1992.
116. Heiny, D. K. and Gilchrist, D. G., Enzyme linked immunosorbent assay, immunoblot detection, and antibody neutralization of *Stemphylium botryosum* phytotoxin, *Physiol. Mol. Plant Pathol.*, 35, 439, 1989.

Chapter 28

Production and Application of Monoclonal Antibodies to Plant-Pathogenic Bacteria, Spiroplasmas, and Mycoplasma-Like Organisms*

Ramon Jordan

TABLE OF CONTENTS

I. INTRODUCTION

Unique functions of plant pathologists include diagnosis of plant diseases, detection and identification of disease-causing organisms, and research on biotic pathogens and host-pathogen interactions.[1] Plant

* Mention of trademark, proprietary product, or vendor does not constitute a quarantee or warranty of the product by the U.S. Department of Agriculture and does not imply its approval to the exclusion of other products or vendors that may be suitable.

0-87371-877-1/95/$0.00+$.50

pathologists have found immunochemical techniques to be extremely specific and sensitive methods useful for studying the taxonomic, functional, and structural relationships of antigens derived from plant pathogens as well as for the rapid and accurate routine detection of plant pathogens and ultimately the diagnosis of plant diseases.[2,3] These techniques have provided methods for the identification and quantitative assay of plant pathogens, for determining the degree of similarity between members of the various groups of plant pathogens, as well as for studying the functional and structural aspects of specific antigens associated with plant pathogens, i.e., structural and nonstructural gene products.

The introduction of hybridoma technology[4] of immortalizing specific antibody-forming cells by their fusion with myeloma cells has provided a novel approach for the production of homogeneous and biochemically defined immunological reagents (namely, monoclonal antibodies [mAbs]) of identical specificity, produced by a single cell line and directed against a unique epitope of the immunizing antigen. Hybridization of antibody-forming lymphocyte cells with malignant myeloma cells results in hybridomas which combine the parental traits of specific antibody secretion and continuous growth. Cloning and further selection of hybrids allows the development of homogeneous mAb preparations directed towards single antigenic determinants.

The ability to obtain practically unlimited quantities of the same antibody in a reproducible manner and the ability to immortalize the production of such monospecific reagents by cryopreservation of the hybridomas for unlimited periods are just two advantages of hybridoma-produced mAbs over conventional polyclonal antiserum. One of the more striking advantages of the hybridoma approach is the ability to produce and select mAbs to almost any antigenic determinant, even when impure antigen or antigen mixtures are used as immunogen.

mAb production has recently been adapted to the field of plant pathology.[5–8] The purpose of this report is to review for plant pathologists and nonplant pathologists alike the current status on the production and application of mAbs in the field of phytobacteriology. In this chapter, I have endeavored to provide the reader, especially those who have little or no hybridoma experience, with general protocols and techniques, but also with some of the rationale used to develop the protocols and approaches. Detailed procedures on the production of mAbs have recently been published.[9,10] Several other articles or books describing in more extensive detail the hybridoma technology and procedures for the preparation and characterization of mAbs have also been cited.[11–20] Attention here will also focus on the utilization of mAbs for the detection, identification, characterization, and analysis of variants and strains of plant-pathogenic bacteria, spiroplasmas, and mycoplasma-like organisms (MLOs).

II. METHODS OVERVIEW AND BASIC REQUIREMENTS

The methodology involved in the establishment of permanent cell lines producing mAbs is relatively simple, yet requires a large number of sometimes crucial steps, each of which may be carried out in many different ways. Standard procedures for producing mAbs involve immunization of the host animal (usually a mouse) with an antigen of interest, isolation of immune lymphocytes (spleen and/or thymus cells), and chemical fusion of these with easily cultured myeloma tumor cells. Hybrids are selected, tested for immunoreactive antibody secretion, cloned to homogeneity, and cultured to produce the desired mAb in virtually unlimited quantities. Table 1 illustrates the general stages (and time involved) of a stylized protocol used for the immunization, fusion, selection and screening, and growth, maintenance, cloning, and cryopreservation of mAb-secreting hybridomas. The development of mAbs depends not only on the generation of large numbers of hybridoma cell lines, but also on the successful immunization of the donor animal(s), knowledge and experience in basic tissue culture techniques (including media production, cell line maintenance and preservation, and sterile culture techniques), and the development of preplanned, sensitive, reliable, and relatively fast and simple serological assays for identifying and characterizing the appropriate target-specific antibodies. The following sections are derived from the author's experiences in the preparation and characterization of many mAbs to a wide spectrum of plant and plant pathogen antigens, including viruses,[21–24] spiroplasmas,[25] and natural plant products.[26,27] I have also drawn upon the experiences of those whom I have collaborated, as well as from the literature.

A. BASIC REQUIREMENTS

The production of mAbs requires knowledge and familiarity with the basic principles and techniques of cell culture, serology, and the handling (and possibly rearing) of large numbers of mice or rats under disease-free conditions.

Table 1 **Stages in the production of hybridomas and monoclonal antibodies**

Stage	Interval	Time in stage
Immunization		1–6 months
Primary	10–14 d	
Boost(s)	Every 10–14 d	
Test bleeds	Every 10–14 d	
Screen development		2 weeks
Hybridoma production		1–2 months
Final boost	10–21 d	
Fusion	3–5 d later	
Initial screen	7–10 d	
Propagate—expand "positives"	3–10 d	
Rescreen	3–7 d	
Freeze (and check viability)	3–10 d	
Secondary screens of other cell lines	Every 3–5 d	
Propagate—expand "positives"	3–10 d	
Rescreen	Every 3–7 d	
Freeze (and check viability)	3–10 d	
Preliminary characterization		1–4 weeks
Serological tests with all "positive" tissue culture supernatants		
Single-cell cloning		1–2 months
Limiting dilution plating		
Screen	7–10 d	
Propagate—expand "positives"	3–10 d	
Rescreen	Every 3–7 d	
Freeze (and check viability)	3–10 d	
Producing monoclonal antibodies		1 week to 2 months
Tissue culture	3–10 d	
Ascites fluids	1–2 months	
Cumulative time involvement:	3–12 months (usually 5–7 mo)	

Specifically, the researcher should have experience in sterile techniques, and while it is not necessary to have extensive cell culture experience or to be an immunologist to generate hybridomas, it surely does not hurt. Hybridomas are rather fastidious cells which need to be evaluated microscopically to determine viability and rate of growth. The chances of producing and maintaining hybridomas are certainly higher if the researcher has previous cell culture experience. The other very important prerequisite in mAb production relates to the antigen to be used and the serological assay(s) used to identify and characterize the hybridoma antibodies. The assay systems used will depend upon the nature of the antigen and the projected eventual use of the mAb.

1. Equipment and Plasticware

The basic requirements for hybridoma production include a variety of both major and minor equipment and tissue culture ware. The major cell culture equipment include: a laminar flow hood, a humidified, temperature-controlled (37°C) carbon dioxide incubator, an inverted microscope, and a bench-top clinical centrifuge. It is recommended that the sterile hood and CO_2 incubator be dedicated solely for hybridomas during a hybridoma production exercise. Outside the culture area, access to or possession of a 37 to 56°C water bath, −70°C freezer, autoclave, and a liquid nitrogen storage container are also needed. Other general laboratory items such as a hemocytometer, single and 8- to 12-channel variable (10- to 200-μl) automatic pipets and sterile tips, refrigerators, and freezers (−15°C) should be readily available, although a separate supply for the hybridoma work would be invaluable.

Depending upon the nature of the assay used, additional equipment (and support materials) will be needed: an ELISA reader, fluorescent microscope, fluorescent-activated cell sorter, scintillation counter, and spectrophotometer.

Animal holding facilities are needed to maintain the constant supply of mice or rats which are required for all stages of mAb production, including immunization, production of ascites fluid, and donation of thymuses and/or spleens for feeder cells. Cages, bedding, feed, and adequate temperature control and ventilation are obvious necessities.

The required tissue culture ware include: 6- to 96-well, flat-bottom sterile tissue culture plates with covers; 25 to 150 cm^2 sterile tissue culture flasks; sterile glass or plastic pipets; ampules for liquid nitrogen preservation; and sterile screw-cap centrifuge test tubes. Other necessary items include disposable plastic or glass Pasteur pipets, protective (vinyl) gloves, petri dishes (square), sterile syringe-type filters, 1- to 50-ml syringes, 16- to 26-gauge (1½- to ¾-in) needles, and surgical scissors.

2. Materials and Media

The establishment and growth of hybridomas requires not only special attention and quality reagents, but also a high degree of commitment because it involves the growth of vulnerable cells that are sensitive to extremes in cell densities and which continually need to be monitored and nursed. The use of prescreened media and additives greatly aids in the establishment, growth, and stability of hybridomas. The time and effort it saves makes the higher cost of these reagents worthwhile. All preparations are done with tissue culture-grade, deionized, glass double-distilled water.

a. Media

The most commonly used media for hybridoma production are Dulbecco's modified Eagle's medium (DMEM) and Rosewell Park Memorial Institute (RPMI-1640) medium. Both media are usually buffered with carbonate/bicarbonate. Media can be purchased in liquid or powder form. We generally prepare media from powder that can be kept longer periods (albeit less than 3 months) than the ready-to-use liquid media.

Glutamine is unstable and is added to 2 m*M* (from 200 m*M* stock; Gibco) to media older than 2 weeks. Some researchers supplement the media with nonessential amino acids (to 1%) and other growth mediums (i.e., NCTC 109 to 10%). Prepared media should be stored (4°C) in the dark to prevent the production of highly toxic photoproducts.[12,13]

b. Antibiotics

The most common tissue culture contaminants are bacteria, yeast, and fungi. The best control for these organisms is scrupulous use of good, sterile techniques. Most researchers add penicillin and streptomycin as a matter of routine to minimize bacterial contamination. When necessary, or as an alternative, gentamycin can be used as a broad-spectrum bacterial and mycoplasma antibiotic. Lincomycin and tylosin tartrate have been used, in conjunction with phagocytic macrophages, for removal of mycoplasma from important cell lines.[13,16] Fungizone (amphotericin B) is useful for fungal contamination.

c. Sera

Fetal bovine serum (FBS) is used in nearly all hybridoma work. The quality of sera for tissue culture supplied by most manufacturers has significantly improved in recent years so that extensive testing of many different batches is now not usually necessary. Several companies in fact provide "hybridoma-tested" sera or sera developed primarily for hybridoma work (e.g., HyClone, Sigma). Even so, the serum chosen should be able to support the growth of myeloma cells at one cell per well. The serum can be kept frozen at −20°C for at least 1 to 2 years. Most researchers heat the serum to inactivate complement. We normally thaw a bottle at room temperature, then transfer to 56°C for 30 to 45 min with occasional mixing prior to dispensing (aseptically!) into smaller one-time-use aliquots for −20°C storage.

d. Serum Substitutes, Supplements, and Serum-Free Media

Iron-supplemented bovine calf serum[28] and horse serum are sometimes substituted for FBS and are generally less expensive; however, these both contain high levels of contaminating immunoglobulins (FBS has very low levels) which can interfere with many immunoassays. Several companies offer defined or processed serum replacements which reportedly have consistent chemical and performance characteristics and can be used in place of FBS.

Serum-free media that support the growth of several parent myeloma lines (and their fusion-derived hybridomas) have been recently described.[29,30] These media are usually mixtures of RPMI and DMEM supplemented with hormones, transferrin, lipids, trace elements, and other factors. Unfortunately, at present none of the serum-free media developed for hybridoma cultivation are suitable for all hybridomas or myelomas. However, adaption to serum-free conditions would be advantageous for the production and purification of mAbs through large-scale culture of hybridomas.

Components and protocols for the preparation of media and working solutions for fusion and maintenance and cryopreservation of parental cell lines and hybridomas are described elsewhere.[3,9,10,13]

III. PRODUCTION OF MONOCLONAL ANTIBODY-SECRETING HYBRIDOMAS

Many steps are involved in the production of mAbs and there are many ways to carry them out. Some of the more important elements include the immune state of the spleen donor animal, the ability to prepare and fuse myeloma and spleen cells and obtain viable hybrids, the screening method, the ability to clone and cyropreserve cells, and the commitment to follow through on a project that will require hundreds of hours over a 5- to 7-month period.

A. ANTIGENS, IMMUNOGENS, AND IMMUNIZATIONS

Bacteria, spiroplasmas, and MLOs are complex entities that present a wide array of antigenic determinants to an immunized mouse and in an immunoassay. The prokaryotic antigens may be intracellular or extracellular, soluble or structural, and can be composed of protein, carbohydrate, or lipid.[6,31,32] Antigen preparations are used for both immunization (as immunogen) and as test antigen for selecting antibodies and determining antibody specificity.

1. Preparation of Antigens

Antigen preparations may consist of infected plant extracts, whole cells, crude cell extracts, membrane protein complexes, and purified intracellular or extracellular components or proteins. A variety of procedures for the preparation of antigens for use as immunogen and/or test antigen have been published for bacteria,[31–36] spiroplasmas,[25,37–39] and MLOs.[40–44] Proteins that are difficult to purify away from other proteins can be eluted from polyacrylamide gel slices, or the macerated gel slice can be injected with the protein intact.[16,27] Suspensions of antigen-bearing particles derived from antigen-spotted or -blotted nitrocellulose can also be used as immunogen.[45–47]

2. Immunogens

Some elements to consider in choosing and preparing an immunogen include the following. The purity of the antigen to be used as immunogen is not as crucial for mAb production as it is for polyclonal antibody production. If the target antigen is itself a poor immunogen or the screening assay method cannot distinguish between target-specific and impurity-specific antibodies, then the immunogen should be of greater purity. The immunogen should present the target antigenic sites in a conformation as similar to that with which the desired mAbs are expected to react. In other words, mAbs generated from antigen used in its denatured form as immunogen will not necessarily recognize native antigen, and vice versa. Low-molecular weight antigens (less than 1 kDa) and peptides are generally poor immunogens and should be coupled to larger immunogenic carriers such as albumin and keyhole limpet hemocyanin. Aggregated and/or particulate antigens, such as bacteria and cell walls or membranes, are stronger immunogens than soluble antigens. Repeating antigenic determinants within an antigen (such as on the surface of intact bacteria) elicit stronger responses (than, for instance, a purified membrane protein).

3. Immunization

The choice of animal species as immune spleen donor depends mainly on the myeloma cell line to be used for fusion. Mouse and rat myeloma lines are readily available, with mouse as the most common species used. The Balb/c strain of mice is preferred, as all the murine myelomas in use were derived from this strain.

The immune state of the animal from which the spleen is taken is very important in determining the success of a hybridoma project. The degree of immunization is dependent upon the choice of animal, the choice of immunogen, and the immunization schedule.[11–16] Immunization protocols and schedules vary considerably and cannot be considered separately from the immunogen. A protocol that works well for a membrane antigen will not necessarily be satisfactory with a soluble protein. However, generalizations can be made. Doses of 1 to 50 μg or 1×10^6 to 5×10^7 cells per rodent per injection is usually sufficient to elicit good responses if repeated two or three times at intervals of 10 to 21 d. Soluble antigens require the use of an adjuvant (i.e., Freund's or Hunter's TiterMax) in at least the first injections, usually with Freund's complete first and later with Freund's incomplete (with at least a 1:1, and preferably a 1:2, v/v, ratio of aqueous to adjuvant ratio), or with Hunter's TiterMax for all immunizations. When cells are used as immunogen, adjuvant is not usually required. The site of injection is probably not critical. Subcutaneous, intramuscular and intraperitoneal (i.p.) immunization routes are adequate to stimulate an immune response. Intradermal injections at multiple sites can cause painful ulcers, and the other routes are as efficient for priming.[11,12,16] This author uses the i.p. route for all injections. Volumes of 50 to 300 μl per mouse and 1 to 2 ml per rat are typical.

Several new techniques can be employed to improve the yield of antigen-specific hybridomas when low levels of antigen, insufficiently pure antigen, or "poorly" immunogenic antigen is the target antigen. These include: *in vitro* sensatation of cultured spleen cells with antigen for the primary, secondary, and/or tertiary immunization;[22,48,49] intrasplenic immunization;[47,50] selection and/or enrichment of antigen-specific B-lymphocytes;[51–53] and induction of tolerance to nontarget proteins.[54–57]

Many laboratories require the immunized animal to have a specified serum titer before its spleen will be used in a fusion.[13–16] Generally, one cannot go wrong by selecting animals with high titers of circulating antigen-specific antibodies. However, in contrast, the spleen cells that generate a high frequency of antigen-specific hybridomas are the rapidly dividing plasmablasts which do not themselves secrete large amounts of antibody.[11,12] Regardless of serum titer "requirements," a final boost 3 to 5 d prior to fusion generates a maximal number of plasmablasts.[11–20]

a. Immunization Protocol

A reasonable "standard" protocol and schedule would involve the following. At least two 1- to 4-month-old mice or rats are injected intraperitoneally with 10 to 50 μg purified protein or synthetic peptide, or 100 to 200 μg cell membranes or lysate, emulsified in 300 μl complete Freund's (or Hunter's TiterMax) adjuvant or 1 to 10 $\times$ 10^6 cells in buffered saline or Freund's incomplete adjuvant (or Hunter's TiterMax). The animals are boosted two to three times with the same dose in Freund's incomplete (or saline for cells) at 10- to 21-d intervals. Individual animals are bled 10 to 14 days later and the titer of the relevant antibodies is determined in the same assay(s) that will be used later for screening the hybridomas. Animals showing the highest antibody titers are given an additional 1- to 5-week rest period (more than 3 weeks from the last injection) prior to a final intraperitoneal boost of the same or higher dose in aqueous solution. The spleens are removed and used for fusion 3 to 5 d (if i.p. route used) later.

B. SCREENING ASSAYS

The immunoassay used for the screening of antibody activity in the hybridoma culture supernatants is one of the most important parts of hybridoma production and should be given considerable attention. In principal, any assay which is capable of detecting low concentrations of antibody of a desired specificity can be used. There are, in fact, many ways to screen hybridomas, including ELISA, dot-blot, Western-blot, immunofluorescence, and immune precipitation assays; details of such serological tests are discussed in other chapters of this book and elsewhere.[3,12,58] The practical aspects of screening assays are the need to be reliable and sensitive enough to detect all mAb-secreting hybridomas of interest, simple and quick enough to run up to hundreds of samples at a time, and initially designed with the end product in mind. Some of the more important criteria for selecting an assay include the nature of the antigen and the intended use of the selected antibody. Specific considerations of these topics will be discussed briefly below.

In the initial screens following a typical fusion, there can be 400 to 2000 cell lines that need to be assayed. If just 10% of these are antigen specific, then 40 to 200 cell lines will then need to be rescreened at least twice during expansion and freezing. Selected cell lines will be cloned, adding an additional 30 or so clones per "parent" cell line that will then also be screened. In other words, hundreds to thousands of assays will have to be performed during a single hybridoma project. If the assays cannot be done with large numbers, reliably and conveniently, then the project is likely to fail.

However, if the screening assay procedure is difficult, time consuming, or expensive, yet necessary to select a specific antibody activity, it may be worthwhile to consider performing the screening in two steps. The first step would involve a fast, but nonspecific assay that would reduce the number of hybrid cultures to be screened in the second more specific assay. A nonspecific first-phase screening assay could include simply testing for the presence of mouse or rat antibody in the culture supernatants with an antiglobulin that is capable of detecting all desirable subclasses.

Another important factor in selecting an assay is the nature of the antigen; i.e., its purity, abundance, and physical form (size, native conformation, fixed, denatured). Solid-phase assays are suitable for most antigen preparations encountered in phytobacteriology. High-molecular weight, soluble, highly purified antigens such as ribosomes, as well as insoluble, complex antigens such as whole bacteria, spiroplasmas, MLOs, bacterial cell walls, and cell membrane protein complexes, can be immobilized by binding to a solid-phase matrix. Pretreatment of the solid phase with glutaraldehyde or polylysine, for example, may be necessary with some antigens, whereas low-molecular weight antigens such as peptides, enzymes, or hormones need to be coupled to a suitable carrier for maximal binding.[3,12,16] Immunofluorescence and immunocytochemical techniques are suitable (albeit labor intensive) when screening for mAbs to structural or cellular antigens.[3,16] Approaches for selecting mAbs to functional antigens (e.g., enzymes) are to select antibodies that will precipitate, modify, inhibit, or neutralize the activity mediated by the specific antigen.[3,16]

Regardless of the choices of assays available, the screen should be appropriate for the intended use of the antibody. The condition of the antigen should also be as similar as possible in the screening assay as in the final assay. mAbs tend to be very assay specific in that they may perform well in one assay, but give negative results in another. This is due primarily to the strict specificity of mAbs for their respective epitopes and the "accessibility" of those epitopes in the assay(s).[21,58,59] The "status" of the antigen as presented in each assay is important for each tested antibody. For example, a conformation-dependent antibody reactive with an epitope present only on the native protein may not react with that same protein under denaturing conditions. In addition, the pH and ionic conditions of an assay, as well as the conditions for coating an antigen to solid phase can structurally alter the antigen and affect the ability of a mAb to react with it. The antibody itself can be affected by the assay conditions or become inactive when coupled to an enzyme.[3,12,16,21,58,59] Also, unless the antigen molecule has repeating antigenic determinants a mAb will not precipitate the antigen in double-diffusion or microprecipitin assays. It is therefore strongly recommended, especially in selecting mAbs for use in diagnostic tests, that the same assay conditions to be used in the diagnostic test are the same as in the mAb selection process. In other words, it is important to think about what you want and design how best to achieve it. Proper selection of an assay can reduce the workload and ensure the desired mAbs are identified.

C. PREPARATION OF FUSION PARTNERS

A critical component in the production of hybridomas is the preparation of cells for fusion. Both spleen B-lymphocyte cells and myelomas need to be in the optimum cell cycle stage and condition for successful hybridization and subsequent hybrid proliferation.

1. Myeloma Fusion Partner

Myeloma cell lines are mutants deficient in the production of hypoxanthine-guanine (and/or adenosine) phosphoribosyl-transferase or thymidine kinase. Such mutants die in the presence of aminopterin, which blocks the main DNA synthesis pathway. Unfused lymphocytes do not survive in culture and selection of antibody-secreting hybridomas is possible because myeloma-spleen cell hybrids can survive in aminopterin when hypoxanthine and thymidine (components of the hybridoma selection media) are present for use by the salvage pathway.

The choice of a myeloma is based on its overall performance, including growth rate, fusion efficiency, feeder cell dependence, and cloning efficiency. Several mouse cell lines currently in use yield high frequencies of hybridomas,[12,13,16] including P3/NS1/1.Ag.4.1 (NS1), P3X63/Ag8.653 (P3), Sp2/0.Ag14 (Sp2), and FOX-NY (an NS1 derivative).[60] All of these cell lines have been used with success in plant pathology.

Rat myeloma cell lines include YB2/0 and IR983/F.[11] Rats are advantageous for large production of ascites fluid, but their myeloma fusion efficiencies are lower than the murine myelomas.[11]

Most of the mice and rat myelomas are available from Flow, Gibco, ATCC, HyClone (FOX-NY), the Institute for Medical Research, or NIGMS Cell Repository (Camden, NJ 08103).

Regardless of the choice of myeloma to be used for fusion, the cells should be maintained at high viability and in logarithmic growth prior to fusion.[11–19] Good culture conditions of the myeloma cell line improves the fusion frequency tremendously. To avoid revertants, it is recommended not to keep the same culture going for long periods of time (months). Cells are usually thawed from frozen stock, scaled up (and fresh vials replaced as soon as possible), and propagated in early log growth for at least a week prior to fusion. If the cells have been passaged through medium containing, for example, 8-azaguanine (30 μg/ml) to ensure aminopterin sensitivity, they need to be passaged at least three times in medium without 8-azaguanine.

2. Donor Spleen Cells

A successful immunization schedule, including a 3- to 5-d prefusion antigen injection, should produce sufficient numbers of stimulated lymphoblasts in the donor spleen for use in the lymphocyte-myeloma cell fusion procedure. After the spleen is removed from the animal it is squeezed, teased, smashed, minced, strained, and/or injected with a solution to release the lymphocytes, which are then pelleted, washed, and counted before fusion with the myeloma cells. The procedure used to prepare the spleen cell suspension must be gentle to preserve cell viability and function. A mouse spleen will typically contain about 10^8 cells and a rat spleen about 5 to 10 $\times$ 10^8 cells.[12] Some researchers incubate the cell suspension in a lysis medium to remove red blood cells.[13] The lysis method can give variable results, and adequate numbers of hybrids are obtained without attempting to remove nonlymphatic cells.[9,11,12,17,19] Detailed protocols for the isolation of spleen and thymus lymphocytes have been described.[9–19]

3. Feeder Cells and Conditioned Media

Hybridoma cells are often intolerant of dilution. Feeder cells or conditioned medium of some sort increase the ability of cultured cells to grow at very low densities. If culture conditions are optimal, feeder cells may make little difference other than reducing variability between fusions.[14] Commonly used feeder cells (10^6/ml) include normal spleen cells,[15] thymocytes,[17,19] irradiated fibroblasts,[11] and peritoneal macrophages.[11] The presumed functions of feeder cells include: the addition of soluble growth factors to the medium (all feeder cells), removing waste products and dead cells (phagocytic macrophages only), or giving emerging hybridomas or clones cell contact at otherwise low cell density (all feeder cells). Procedures for preparing feeder cells can be found elsewhere.[9–19] Normal spleen cell feeders are prepared as per the parental spleen cells.

Conditioned medium consists of medium in which cells have been actively growing. The cells are removed and the medium is filtered and used to supplement (at 10 to 50%) the medium used to grow hybridomas. It is objectionable and often inconvenient to use cells from specially sacrificed animals either as feeders or for conditioned medium. Commercially available cell lines (i.e., mouse L929 and 3T3 fibroblast cell lines; ATCC) or conditioned medium (thymocyte, Hana Biologics, Berkeley, CA; and endothelial cell growth supernatant, Collaborative Research) are good alternatives.[61] We have had good results simply using the spent medium from the log growth phase myeloma fusion partner as our "conditioned medium" and do not use any feeder cells. Protocols for growing and preparing the myeloma fusion partner have been described.[9–12]

D. CELL FUSION

The next critical step in the production of hybridomas is the hybridization or fusion procedure. A successful fusion protocol should bring parent cells together and allow fusion to occur at a sufficiently rapid rate without causing more than minimal damage to the cells. The first hybridomas were made using Sendai virus as the fusing agent, whereas nearly all hybridomas are now generated with polyethylene glycol (PEG) as the fusogen.[12]

The conditions which favor successful fusions are not unique. Several modifications of the original PEG-induced fusion procedure[17,19] have been described, and all variations seem to work equally well. The more important variables include: the concentration and pH of the PEG mixture, the duration of exposure to PEG, and the physical handling and processing of the cells during and after fusion. The mechanism of fusion is complex, involving cell agglutination, cell swelling, and membrane fusion, and the optimal conditions for the three processes are often at odds.[11,12] Membrane fusion efficiency, but also toxicity, increases with increasing PEG concentration (30 to 50%), time of exposure (2 to 10 min), and temperature (20 to 40°C). All fusion protocols operate under these ranges, compromising conflicting needs to achieve sufficient fusion while keeping cell damage to an acceptable level. Fusion frequency is also dependent upon pH, and maximal numbers of hybrids are obtained at pH 7.8 to 8.2. The dilution and manipulation procedures are also very critical. Fast addition of medium and rough handling are harmful, as the aggregated cells which will eventually fuse can be dissociated by vigorous treatment.

Other factors which might influence the fusion efficiency under suboptimal conditions include the molecular weight of the PEG (range from 1000 to 6000), the source of PEG,[62] the ratio of spleen cells to myeloma cells (1:1, 2:1, 5:1, and 10:1 are common ratios), and the presence or absence of serum or red blood cells,[13] and the addition of DMSO with the fusogen (does not hurt and might be beneficial).[14]

Another area of variation between researchers concerns the size of the initial cultures (i.e., post-fusion plating densities). The two extremes include plating at high cell density (10^8/ml) in 1 to 5 24-well plates (1 ml per well)[17–19] and plating at low cell density (10^5/ml) in 10 to 20 96-well plates (0.2 ml per well).[9,11,12] Plating at relatively low numbers of cells per well has the advantage that "assay-positive" wells will probably contain one hybrid per well and thus mAbs from the beginning.

We have obtained moderate success using the following protocol, which is a compilation of several protocols.[18,62,63] A word of caution is added at this time. If a successfully working, reliable, fast, and sensitive screening assay has not been developed at this point do not proceed with a fusion. In approximately 2 weeks there will be 1 to 2000 wells ready to screen for antibody production. The cells will be growing at a fast rate and cannot wait for the development of a screening assay. A commitment of 3 to 8 weeks of continuous work must also be made at this time.

1. Fusion Protocol

Mix the resuspended spleen and myeloma cells in one tube (after keeping aside a sample of both for use as controls). Centrifuge at 400 g, 10 min and remove ALL the supernatant by careful aspiration. Loosen the pellet by gently tapping the tube. Using a 5-ml pipet, gently stir the cell pellet while adding, dropwise, 1.5 ml PEG (37°C) over a 45-s period. Incubate with swirling for 30 to 75 s (essentially the time it takes to prepare for the next step). Slowly add 2 ml serum-free medium with stirring over a 60-s period. Incubate and swirl 60 s. Slowly add another 2 ml over another 60-s period and incubate again for 60 s. Add 35 ml serum-free medium over the next 3.5 min. (Total elapsed time should be about 9 min.) Centrifuge cells 400 g, 10 min. Aspirate and discard supernatant. GENTLY, with much patience, resuspend cells in 20 ml complete medium + HAT. Do not pipet cells with anything smaller than a large-bored, 25-ml pipet (i.e., about 3 mm I.D.)—cells are very fragile clumps at this stage. Transfer cells to a 75-cm^2 flask containing 100 to 180 ml complete medium + HAT (resuspend in volume calculated to give a cell concentration of less than 1×10^5 myeloma cells per milliliter, based on original cell counts). Put flask in incubator (37°C, 5 to 10% CO_2) standing up with loosened cap. Let stand at least 1 h. (Fibroblast cells and large clumps/debris will generally stick to the bottom of the flask.) Carefully pour suspended fused cells into a new flask. Mix by gentle pipetting up and down with a 25-ml pipet. Pipet cells into a petri dish (square) and dispense 200 μl per well into 96-well microtiter plates, gently, with an 8-channel pipetman (usually 10 to 15 plates per spleen). If feeder cells are used, they are usually plated the day before (100 μl per well) and fused cells are then added at 100 μl per well (at a cell density of 2×10^5 myeloma cells per milliliter). Plate out saved spleen and myeloma control cells in complete medium + HAT separately at cell densities similar to the fused cells. Incubate at 37°C in 5 to 10% CO_2 in a humid incubator.

E. SELECTION, PROPAGATION, CRYOPRESERVATION, AND CLONING OF HYBRIDOMAS

After fusion the fused hybrids are placed into a medium that allows hybrids to grow and prevents the growth of unfused parent cells. The cultures are monitored daily (with the microscope) to assess the progress of the hybridomas, to decide upon the necessity of feeding, or testing supernatants for the presence of antibody, and to remove any contaminated cultures. This stage in the production of mAbs can be difficult and time consuming. Hundreds of colonies, at various times post-fusion, will be ready for screening. Once antigen-positive antibody-secreting hybridomas have been identified, they need to be transferred to larger vessels, rescreened, expanded again, rescreened and frozen, cloned, screened, recloned, rescreened, and bulk cultured for antibody production. This is no small feat with 20 to 100 different antigen-positive hybridomas, most at different stages of propagation.

1. Selection of Hybridomas

Laboratory protocols vary widely concerning the post-fusion care of hybridomas. There should be massive cell death 4 to 5 d after fusion and the fusion plates should be a scene of devastation. If there are living cells in the myeloma control wells, then the aminopterin stock may not be any good or the myeloma cell line has reverted to aminopterin resistance. All myelomas should be dead by day 10. The parent spleen cells should also be dead around days 5 to 10. Some protocols do not allow cells (hybrids) to be fed any sooner than 7 to 10 d post-fusion, whereas others suggest feeding as early as 3 d post-fusion. Every manipulation increases the risk of contamination; therefore, feeding should be as infrequent as possible. Feeding is accomplished by aspirating about half of the culture supernatant (aseptically!) and replacing it with fresh medium. Early feeding of cultures actually serves several purposes. One reason for feeding is to remove waste products from the predominantly dead cultures more than to replenish nutrients. A second reason for early feeding is to gradually dilute out any antibody made by unfused antibody-secreting spleen cells. (It is not unusual to have 100% of the fusion wells to be antigen positive, without any hybridoma growth, because of the presence of antibodies secreting by the immune lymphocytes surviving in the rich fusion medium.) Later feedings are determined according to hybrid cell growth. HAT medium is gradually replaced with HT medium 10 to 14 d after fusion. Because of the sensitivity of normal cells to aminopterin (without HT), cells should be *slowly* weaned out of HT to complete medium without HT. (Once weaned out of HAT, we routinely carry out all subsequent operations in medium + HT.)

2. Propagation and Screening

Supernatants are removed (aseptically!) for screening when the bottom of the well has reached 30 to 50% hybridoma confluence and the medium has changed to a yellow color. Wells should not be fed for at least 48 h before the removal of supernatants for screening to allow for adequate accumulation of antibody. Hybrids are usually screened and fed in the same step; i.e., supernatant aliquots ($\leq$100 µl) taken for the screening assay are replaced with fresh medium. Antigen-positive wells are transferred to the next size vessel (96-well plate to 24-well plate to 25-cm^2 flask). Cells remaining in the original well are diluted with fresh medium and serve as the backup source to the transferred cells. Our general routine is to take supernatants on Mondays and Thursdays for screening and then transfer the positives on Tuesdays and Fridays. Extremely fast-growing hybrids are simply moved up first and then screened in the next round of assays. Hybrids are screened at each expansion stage, and those secreting the desired antibodies should be frozen as soon as possible. We usually freeze four to six cryovials of each parent cell line from 25-cm^2 flasks. The contents of two 1-ml, 24-well cups are also sufficient (= two cryovials).

3. Cryopreservation

Cryopreservation of myeloma stock cells and stocks of each of the antigen-reactive hybridoma "parent" cell lines and clones is an essential safeguard and a very important part of hybridoma work. Parent myelomas and established hybridomas can be kept frozen with little difficulty for many years in liquid nitrogen with good recovery. More recently, mechanical freezers that can maintain temperatures less than -130°C have become available. Aliquots of cells are resuspended (10^7/ml) in medium containing 20 to 90% serum and 5 to 10% DMSO. Aliquots are frozen slowly (about 1°C/min) to less than -70°C before transfer to permanent storage. Retrieval and thawing of cells are, in contrast, done quickly at 37°C before removal from the DMSO and subsequent culture. During the propagation, expansion, and cryopreservation stages of hybridoma production, no back-up cultures should be discarded without first "test thawing" a cryovial to be sure that the frozen "stock" cells can be recovered in a viable (and sterile) state. A procedure has been described for freezing whole tissue culture plates containing hybrids.[64] This can be advantageous for those "highly successful" fusions. As the cell lines are prepared for frozen storage, the culture supernatants are retained and used in more extensive serological assays.

4. Cloning Hybridomas by Limiting Dilution

Once antigen-positive antibody-secreting hybridomas have been identified they need to be cloned as soon as practicable. This is to ensure that a given culture contains only one cell type and that it is producing only one immunoglobulin specificity; i.e., true monoclonality. Cloning is achieved by seeding single wells into culture wells (or on agar) and allowing them to grow into colonies. Hybridomas placed in culture at very high dilution have a tendency to die out, however. Cloning efficiency can be improved by using feeder cells, conditioned medium, or cell growth supplements (as discussed earlier for fusion).

The two most widely used cloning methods are cloning by limiting dilution and cloning in soft agar.[11–13,16] Soft agar cloning involves spreading cells in soft agar and transferring colonies that develop from single cells to liquid culture. Cloning by limiting dilution is the most widely used and relies on diluting cells to less than one cell per well (statistically) in a culture plate. Dispensing single cells into multiple wells is a random process and is best achieved by seeding large numbers of cells on several plates at 0.3 cells per well or lower. In practice, seeding 30 to 50 cells (diluted in 20 ml media) onto one 96-well plate is sufficient. According to Poisson distribution, if 63% (or less) of the wells show cell growth (and the distribution of growth is truly randomly distributed over the plate), they most likely contain a single clone per well.[65] To further ensure monoclonality, the cloning procedure should be repeated until all the subclones detected appear to secrete the same specific desired antibody.

IV. PRODUCTION OF MONOCLONAL ANTIBODIES

Large amounts of mAb can be produced either by culturing cells *in vitro* or growing them as solid or ascitic tumors *in vivo.*[11-16] The mAb is secreted and is accumulated in the spent medium of the cultured cells (at 10 to 100 μg/ml) and in the body fluids and serum (at 2 to 20 mg/ml) in tumor-bearing animals. Culture *in vitro* provides a more pure preparation of antibody. The protein impurities are the medium components (FBS, etc.) and the only mouse (or rat) immunoglobulins present is the mAb. In contrast, ascites and serum from hybridoma-injected mice contain 1 to 10% nonhybridoma mouse antibodies. For mAb production *in vivo,* mice must be histocompatible with the parent cells (see Section III.B). Mice should also be primed by injection with pristane[66] or Freund's incomplete adjuvant[67] 10 d before injection with hybridoma cells.

A. PRODUCTION OF MONOCLONAL ANTIBODIES IN TISSUE CULTURE

Expand desired cell line(s) from 25- to 150-cm^2 flasks (= about 50 ml medium). Hybridomas may be adapted to low serum growth in the transfer stages; i.e., 20% serum in 25-cm^2 flask to 10% serum in 75 cm^2 and 2 to 5% serum in 150-cm^2 flask. Allow cells to reach log phase growth in the maximum amount of medium per vessel and then to exhaustion (death), usually an additional 48 to 72 h. Harvest antibody-containing supernatant and discard the cells. For larger quantities of antibody, cells can be successively transferred up to 1- to 5-l roller or spinner bottles.

B. PRODUCTION OF MONOCLONAL ANTIBODIES IN MICE

Inject recipient mice 10 d before hybridomas intraperitoneally with 0.1 to 0.3 ml pristane or Freund's incomplete adjuvant. Harvest cells from a vigorously growing culture and resuspend in serum-free media at 1 to 10 $\times$ 10^6 cells per milliliter. For ascites fluid production, inject 0.3 to 0.5 ml cells intraperitoneally (20- to 22-gauge, 1½-in needle). After about 10 to 14 d (sometimes longer) mice will show abdominal swelling. Ascites fluid is removed by "tapping" the mouse by insertion of an 18- to 20-gauge, 1½-in needle intraperitoneally (in the lower right or left side of the abdomen) and collection of the milky solution in a 15-ml tube. Mice are observed daily and tapped one to three times, as necessary, usually every 1 to 3 d until death. A total of 1 to 30 ml can be collected from each mouse. Cells and the fibrin clot are sedimented by centrifugation and the clear ascitic fluid is then stored frozen (as small aliquots) at −20°C or at 4°C with added preservative (0.2% thimerosal).

C. PURIFICATION OF MONOCLONAL ANTIBODIES

For many purposes, purification of an antibody is not necessary, and culture or ascites fluid may be used directly. However, many methods used to characterize a mAb (or for using the mAb) involve labeling it with enzymes or isotopes, and a pure antibody is more suitable for this purpose. Procedures for purifying mAbs, including ion-exchange chromatography, affinity separation using protein A, G, or A/G, anti-immunoglobulin, and by antigen affinity chromatography, have been described.[9,12-16,68]

V. APPLICATIONS OF MONOCLONAL ANTIBODIES IN PROKARYOTIC PLANT PATHOLOGY

As mentioned in the introduction, the hybridoma technology for the production of mAbs has been applied to plant pathology. mAbs have been shown to be very useful for the detection, identification, and quantitative assay of plant-pathogenic bacteria, spiroplasmas, and MLOs, as well as for determining the degree of similarity between species, strains, and isolates of these different groups of pathogens. These mAbs have been used in a variety of serological assays, including ELISA, dot-blot, and tissue blot immunoassays, immunofluorescence microscopy, and immunosorbent electron microscopy. Since

1981 mAbs have been produced to over 13 different species or subspecies of phytopathogenic bacteria, 2 species of spiroplasmas, and at least 10 different MLOs.

A. PLANT-PATHOGENIC BACTERIA

Polyclonal antibodies have been widely used for the serological detection, serotyping, and taxonomic classification of phytopathogenic bacteria. Polyclonal antisera, however, usually contain a multiplicity of antibodies to various epitopes on (or within) bacterial cells and often cross react with more than one bacterial species.[6,69] The application of mAbs in plant bacteriology provided a solution to the antibody heterogeneity inherent in polyclonal antisera. mAbs have been made to at least six species of plant-pathogenic bacteria. These include *Agrobacterium tumefaciens,*[70] *Corynebacterium sepedonicum,*[32,33a,71,72] *Xanthomonas campestris,*[73–81] *Erwinia ananas,*[82] and *E. amylovora.*[83,84] In addition, mAbs have been made to lipopolysaccharide of *E. carotovora* subsp. *atroseptica*[33] or extracellular endopectable lyases of *E. carotovora* subsp. *carotovora.*[34,36,85]

Because mAbs are specific to single epitopes, they have been extremely useful in bacterial strain analysis. In fact, mAbs produced for *X. campestris* pv. *campestris* have been used for rapid identification of strains and for tracing strains in epidemiological studies of black rot of crucifers.[73,73a,80] Using a panel of mAbs, a unique strain of *X. campestris* pv. *citri* causing Mexican citrus bacteriosis was identified.[74] An important point emerging from studies using mAbs generated to over eight pathovars of *X. campestris* is an inverse relationship between the ease in generating taxon-specific mAbs and the heterogeneity of the host range of the xanthomonad.[75,81] Pathovar-specific mAbs were produced, and they reacted specifically with all strains of an *X. campestris* pathovar that infect relatively few genera of hosts (i.e., pathovars *oryzae, oryzicola, begonia, pelargonii,* and *phaseoli*).[75–77,81] On the other hand, for certain pathovars that infect several host genera (*campestris, dieffenbachiae, vesicatoria,* and *citri*), no mAbs have been found that reacted with all strains of the respective pathovar, whereas panels of mAbs formed pathovar serogroups.[75–77,79,81] In addition, broad-spectrum mAbs generated to *C. sepedonicum,* which react with all strains of the pathogen, have been shown to be useful probes for disease detection and in seed-certification programs.[32,33,71]

mAbs are potentially very powerful tools to detect and monitor the presence of bacterial metabolites. mAbs specific to extracellular endopectate lyases[34,36,85] should be very useful reagents in studies of host-parasite interaction. They can be high-precision tools for the investigation of the role of those molecules in pathogenesis. mAbs to *Erwinia* pectate lyase have been used as specific probes for the detection of pectate lyase in culture,[85] in potato tubers,[86] and to detect *Erwinia* by dot-blot.[36]

B. FASTIDIOUS, GRAM-NEGATIVE BACTERIA

Hung et al.[87] have reported the production of three hybridomas that secrete antibodies specific to strains of the plum leaf scorch bacterium that are responsible for the plant diseases designated as plum leaf scorch, phony peach, Pierce's disease of grapevine, elm leaf scorch, periwinkle wilt, sycamore leaf scorch, and mulberry leaf scorch. One mAb reacted only with the elm leaf scorch bacterium. mAbs have also been produced to the phloem-restricted prokaryote associated with the citrus greening disease.[88,89] mAbs generated to Indian and South African isolates also detected isolates from the Philippines and Reunion, but not to isolates from China, Thailand, or Malaysia.[89] All of these antibodies should be useful reagents to begin to understand the biology, epidemiology, and pathology of these organisms.

C. SPIROPLASMAS

Spiroplasmas are cell wall-less prokaryotes that are characteristically helical and motile, and have been shown to be important pathogens of plants, insects, and vertebrates.[90] mAbs have been generated that are highly specific to *Spiroplasma citri,*[25,37] or to corn stunt spiroplasma *S. kunkelii,*[25,38] or to epitopes present on both spiroplasmas and/or other group I spiroplasmas.[25]

All nine mAbs produced against *S. citri* (Maroc) by Lin and Chen[37] were highly specific for 10 of 14 isolates of *S. citri.* None of the mAbs reacted with any of the 27 other spiroplasmas tested. Based on the antigen specificity tests reported, no valid conclusions can be drawn concerning epitope specificity.

In a separate study, Lin and Chen[38] produced a panel of 7 mAbs to *S. kunkelii* (I747) that reacted only to 3 strains of *S. kunkelii* and not to any of the 29 other spiroplasmas tested. Again, no statements can be made concerning epitope specificity.

Jordan *et al.*[25] obtained 46 mAbs using a mixture of whole cells and cell lysates of *S. citri* (R8A2) and *S. kunkelii* (I747, F32, PU8-17) as immunogen and screening antigens. When tested against 36 strains of spiroplasmas representing group I and groups IV to XI spiroplasmas, 17 mAbs were selected that react specifically (only) with the strains of *S. citri* tested, and 17 mAbs that react only with *S. kunkelii* isolates. The remaining 12 mAbs react with antigenic sites common in 3 to 8 of the 8 group I spiroplasmas, including honeybee spiroplasmas (AS576), Maryland flower spiroplasma (M55), *Cocos* spiroplasmas (N525), and the periwinkle spiroplasma, *S. phoeniceum.* Based on the mAb reactivities to the various spiroplasma strains tested, the 46 mAbs define at least 17 different group I spiroplasma epitopes, many of which are located on the surface of the spiroplasma.[25,29,91] Three of these epitopes are located on nonmembrane-bound protein(s).[25]

All of these highly specific serological reagents should be very useful in providing new information on the antigenic relationships among the spiroplasmas, as probes for the detection and identification of spiroplasmas in plants and insects, and as molecular probes in identifying and locating specific spiroplasma proteins.

D. MYCOPLASMA-LIKE ORGANISMS

MLOs have been implicated in more than 300 yellows-type diseases in plants. Diagnosis of diseases caused by MLOs and detection of and identification of the causal agents are not only difficult, but also time consuming. Currently, the yellows diseases can be differentiated by host range, symptomatology, insect vector relations, and more recently by nucleic acid-based techniques;[90] none have as yet been cultivated *in vitro.* Although serology offers several reliable and rapid methods for pathogen detection and disease diagnosis, sufficient pure quantities of MLO have not been obtained for conventional polyclonal antibody production. Using partially purified MLO preparations from infected plant tissues (intact organisms or membrane fractions as immunogens), mAbs have been produced to at least ten different plant-pathogenic MLOs. These MLOs include aster yellows (AY),[41,99a] maize bushy stunt,[93] primula yellows,[44] peach eastern X-disease,[43] tomato big bud,[57] elms yellows,[94] peach yellow leafroll,[95] tomato stolbur,[88,96] clover phyllody,[88] and grapevine flavescence doree.[97,98]

Lin and Chen[41,99] reported the first successful production of mAbs to an MLO agent, the AY agent, which was achieved by using insect vector salivary glands as the antigen for immunization and for hybridoma screening. In indirect ELISA tests, a selected mAb reacted specifically with AY-MLO-infected plants and differentiated the AY agent from other MLOs. In *in situ* detection by immunofluorescent straining, the mAb bound specifically to the AY-MLO in sieve tubes in infected plants. Similar results have also been obtained for the peach eastern X-disease[43] and tomato stolbur[96] agents. Employing an immune tolerance-inducing procedure (using mice neonatally injected with nontarget antigens present in immunogen preparations) before immunization, 20 mAbs specific to the tomato big bud disease MLO agent were produced[57] and have been used in diagnostic tissue blot assays.[100] Also, dot-blot immunoassays using these mAbs were able to differentiate MLO strains in the AY strain cluster.[90,101] mAbs raised to primula yellows MLO cross reacted with a European strain of AY, but differentiated AY from clover phyllody MLO.[44] The highly discriminatory capacity of these MLO mAbs to differentiate MLOs will be the most important advantage over polyclonal antibodies for the classification of MLOs and for disease diagnosis, disease forecasting, and epidemiological studies.

VI. CONCLUDING REMARKS

The initial intended use of mAbs in plant pathology has been for diagnostic purposes. The great potential for mAbs in phytopathological diagnostics is essentially because homogeneous antibody preparations with defined activity and specificity can be produced in large quantities over long periods. Even though the hybridoma technology is a laborious and expensive enterprise compared to standard immunization procedures, the procedures have been simplified to the extent that their production is becoming routine in many laboratories. In the next few years mAbs will most probably be generated against many plant pathogens, including pathogenic prokaryotes, especially against those for which there is a demand for large-scale diagnosis and where the antibody heterogeneity of available polyclonal antisera is unacceptable. Diagnostic applications for plant-pathogenic bacteria, the cell wall-less spiroplasmas, and MLOs have already been demonstrated, albeit on a limited scale.

Compared to polyclonal antisera, mAbs often possess a superior discriminatory capacity for revealing small differences in the structure of epitopes. mAb use in plant pathology has demonstrated that certain fine details of the antigenic structure,[59] especially conformational aspects, could only be ascertained by using hybridoma technology. Current observations indicate that with mAb probes it should be possible to evaluate qualitatively and quantitatively: the antigenic nature of plant-pathogenic agents and their gene products and the interactions of these pathogen gene products and toxins with the plant host, especially in discerning the distribution, the site of action, and the cytopathogenic effect of these macromolecules. With the expanded use of mAbs, immunochemistry will continue to be an even more powerful tool for the detection, assay, differentiation, and topographical and structural analysis of the wide variety of molecules encountered in plant pathology.

REFERENCES

1. Grogan, R. G., The science and art of plant-disease diagnosis, *Annu. Rev. Phytopathol.,* 19, 333, 1981.
2. Clark, M. F., Immunosorbent assays in plant pathology, *Annu. Rev. Phytopathol.,* 19, 83, 1981.
3. Hampton, R., Ball, E., and DeBoer, S., Eds., *Serological Methods for Detection and Identification of Viral and Bacterial Plant Pathogens: a Laboratory Manual,* American Phytopathological Society, St. Paul, MN, 1990.
4. Kohler, G. and Milstein, C., Continuous cultures of fused cells secreting antibody of predefined specificity, *Nature,* 256, 495, 1975.
5. Hsu, H. T., Jordan, R. L., and Lawson, R. H., Monoclonal antibodies and plant viruses, *Am. Soc. Microbiol. News,* 50, 99, 1984.
6. DeBoer, S. H., Use of monoclonal antibodies to identify and detect plant pathogenic bacteria, *Can. J. Plant Pathol.,* 9, 182, 1987.
7. Halk, E. L. and DeBoer, S. H., Monoclonal antibodies in plant-disease research, *Annu. Rev. Phytopathol.,* 23, 321, 1985.
8. Jordan, R. L. and Hsu, H. T., Immunological detection and epitope analysis of plant pathogens and their gene products: monoclonal antibodies in plant pathology, in *Biotechnology and Crop Improvement in Asia,* Moss, J. P., Ed., ICRASAT, Patancheru, India, 1992.
9. Jordan, R. L., Strategy and techniques for the production of monoclonal antibodies, in *Serological Detection and Identification of Plant Viral and Bacterial Pathogens: a Laboratory Manual,* Hampton, R., Ball, E., and DeBoer, S. H., Eds., APS Press, St. Paul, MN, 1990, chap. II.B.2
10. Mernaugh, R. and Mernaugh, G., Chapter 25, this volume.
11. Campbell, A. M., *Monoclonal Antibody Technology: the Production and Characterization of Rodent and Human Hybridomas,* Elsevier, New York, 1984.
12. Goding, J. W., *Monoclonal antibodies: Principals and Practice. Production and Application of Monoclonal Antibodies in Cell Biology, Biochemistry and Immunology,* Academic Press, New York, 1983.
13. Zola, H., *Monoclonal Antibodies: a Manual of Techniques,* CRC Press, Boca Raton, Fl, 1987.
14. Fazekas de St. Groth, S. and Scheidegger, D., Production of monoclonal antibodies: strategy and tactics, *J. Immunol. Methods,* 35, 1, 1990.
15. Eshhar, Z., Monoclonal antibody strategy and techniques, in *Hybridoma Technology in the Biosciences and Medicine,* Springer, T. A., Ed., Plenum Press, New York, 1985.
16. Harlow, E. and Lane, D., *Antibodies, a Laboratory Manual,* Cold Spring Harbor Laboratory, Cold Spring Harbor, NY, 1988.
17. Kohler, G., The technique of hybridoma production, in *Immunological Methods,* Vol. 2, Lefkovits, I. and Pernis, B., Eds., *Academic Press,* New York, 1981, chap. 17.
18. Oi, V. T. and Herzenberg, L. A., Immunoglobulin-producing hybridoma cell lines, in *Selected Methods in Cellular Immunology,* Mishell, B. B. and Shiigii, S. M., Eds., W. H. Freeman, San Francisco, 1980, 351.
19. Galfre, G. and Milstein, C., Preparation of monoclonal antibodies: strategies and procedures, *Methods Enzymol.,* 73, 3, 1981.
20. Atassi, M. Z., Preparation of monoclonal antibodies to preselected protein regions, *Methods Enzymol.,* 121, 69, 1986.
21. Jordan, R. L. and Hammond, J., Comparison and differentiation of potyvirus isolates and identification of strain-, virus-, subgroup-specific and potyvirus group-common epitopes using monoclonal antibodies, *J. Gen Virol,* 72, 1531, 1991.
22. Jordan, R. L., Elliott, N., and Hsu, H. T., Production of monoclonal antibody secreting hybridoma cell lines to prune dwarf virus by *in vivo* and *in vitro* immunization, *Phytopathology,* 74, 862, 1984.

23. Aebig, J. A., Jordan, R. L., Hsu, H. T., and Lawson, R. H., Immunological and biological properties of a mouse monoclonal antibody reactive to Prunus necrotic ringspot ilarvirus, *Intervirology,* 28, 57, 1987.
24. Jordan, R. L., Nuss, D., and Hsu, H. T., Production of monoclonal antibodies to wound tumor virus using extracts from cultured virus-infected leafhopper cells, *Phytopathology,* 75, 1356, 1985.
25. Jordan, R. L., Konai, M. I., Lee, I. M., and Davis, R. F., Species-specific and cross-reactive monoclonal antibodies to the plant pathogenic spiroplasmas *Spiroplasma citri* and *S. kunkelii, Phytopathology,* 79, 880, 1989.
26. Kamo, K., Jordan, R. L., Hsu, H. T., and Hudson, D., Development of monoclonal antibodies to the conserved region of $p34^{cdc2}$ protein kinase, *J. Immunol. Methods,* 156, 163, 1992.
27. Mehta, A. M., Jordan, R. L., Anderson, J. D., and Mattoo, A. K., Identification of a unique isoform of 1-aminocyclopropane-1-carboxylic acid synthase by monoclonal antibody, *Proc. Natl. Acad. Sci. U.S.A.,* 85, 8810, 1988.
28. HyClone Laboratories, Inc., Iron-supplemented bovine calf serum: an excellent substitute for fetal bovine serum, *Art Sci.,* 7, 1, 1988.
29. Kawamoto, T., Sato, J. D., McClure, D. B., and Sato, G. H., Serum-free medium for the growth of NS-1 mouse myeloma cells and the isolation of NS-1 hybridomas, *Methods Enzymol.,* 121, 266, 1986.
30. Kovar, J. and Franck, F., Serum-free medium for hybridoma and parental myeloma cell cultivation, *Methods Enzymol.,* 121, 277, 1986.
31. DeBoer, S. H. and Schaad, N. W., Preparation of antigens, bacteria, in *Serological Detection and Identification of Plant Viral and Bacterial Pathogens: a Laboratory Manual,* Hampton, R., Ball, E., and DeBoer, S. H., Eds., APS Press, St. Paul, MN, 1990, chap. II.A.2.
32. DeBoer, S. H. and Wieczorek, A., Production of monoclonal antibodies to *Corynebacterium sepedonicum, Phytopathology,* 74, 1431, 1984.
33. DeBoer, S. H. and McNaughton, M. E., Monoclonal antibodies to the lipopolysaccharide of *Erwinia carotovora* subsp. *atroseptica* serogroup I, *Phytopathology,* 77, 828, 1987.
33a. DeBoer, S. H. and McNaughton, M. E., Evaluation of immunofluorescence with monoclonal antibodies for detecting latent bacterial ring rot infections, *Am. Potato J.,* 63, 533, 1986.
34. Ward, L. J. and DeBoer, S. H., Characterization of a monoclonal antibody against active pectate lyase from *Erwinia carotovora, Can. J. Microbiol.,* 35, 651, 1989.
35. Garnier, M., Martin-Gros, G., and Bove, J. M., Monoclonal antibodies against the bacterial-like organism associated with citrus greening disease, *Ann. Inst. Pasteur/Microbiol.,* 138, 639, 1987.
36. Klopmeyer, M. J. and Kelman, A., Use of monoclonal antibodies specific for pectate lyase as serological probes in the identification of soft rot *Erwinia* spp., *Phytopathology,* 78, 1430, 1988.
37. Lin, C. P. and Chen, T. A., Production of monoclonal antibodies against *Spiroplasma citri, Phytopathology,* 75, 848, 1985.
38. Lin, C. P. and Chen, T. A., Monoclonal antibodies against corn stunt spiroplasma, *Can. J. Microbiol.,* 31, 900, 1985.
39. Martin-Gros, G., Iskra, M. L., Garnier, M., Gandar, J., and Bove, J. M., Production of monoclonal antibodies against prokaryotes of plants: a general procedure using extracts from infected periwinkles as immunogen, *Ann. Inst. Pasteur/Microbiol.,* 138, 625, 1987.
40. Lee, I. M. and Davis, R. E., Phloem-limited prokaryotes in sieve elements isolated by enzyme treatment of disease plant tissues, *Phytopathology,* 73, 1540, 1983.
41. Lin, C. P. and Chen, T. A., Monoclonal antibodies against the aster yellows agent, *Science,* 227, 1233, 1985.
42. Jiang, Y. P., Lei, J. D., and Chen, T. A., Purification of aster yellows agent from diseased lettuce using affinity chromatography, *Phytopathology,* 78, 828, 1988.
43. Jiang, Y. P., Chen, T. A., Chiykowski, L. N., and Sinha, R. C., Production of monoclonal antibodies to peach eastern-X disease agent and their use in disease detection, *Can. J. Plant Pathol.,* 11, 325, 1989.
44. Clark, M. A., Morton, A., and Buss, S. L., Preparation of mycoplasma immunogens from plants and a comparison of polyclonal antibodies made against primula yellows MLO-associated antigens, *Ann. Appl. Biol.,* 114, 111, 1989.
45. Abou-Zeid, C., Filley, E., Steele, J., and Rook, G. A. W., A simple new method for using antigens separated by polyacrylamide gel electrophoresis to stimulate lymphocytes in vitro after converting bands cut from Western blots into antigen-bearing particles, *J. Immunol. Methods,* 98, 5, 1987.

46. Knudsen, K., Proteins transferred to nitrocellulose for use as immunogens, *Anal. Biochem.,* 147, 285, 1985.
47. Forrest, J. M. S. and Ross, H. A., Splenic implantation of nitrocellulose pellets as a means of raising antibodies to nanogram quantities of antigen, *J. Immunol. Methods,* 159, 275, 1993.
48. Reading, C. L., *In vitro* immunization for the production of antigen-specific lymphocyte hybridomas, *Methods Enzymol.,* 121, 18, 1986.
49. Siraganian, R. P., Fox, P. C., and Berenstein, E. H., Methods of enhancing the frequency of antigen-specific hybridomas, *Methods, Enzymol.,* 92, 17, 1983.
50. Spitz, M., "Single-shot" intrasplenic immunization for the production of monoclonal antibodies, *Methods Enzymol.,* 121, 33, 1986.
51. Van Mourik, P. and Zeijlemaker, W. P., Improved hybridoma technology: spleen cell separation and soluble growth factors, *Methods Enzymol.,* 121, 174, 1986.
52. George, R. A. and Converse, R. H., Methods of the enrichment of desired B-cell populations before anti-cauliflower mosaic virus hybridoma formation, *Phytopathology,* 78, 1631, 1988.
53. Maga, M. G., McHugh, L. L., and Rothstein, T. L., Mouse lymphocytes with and without surface immunoglobulin: preparative scale separation in polystyrene tissue culture plates coated with specifically purified anti-immunoglobulin, *J. Immunol. Methods,* 15, 47, 1977.
54. Golumbeski, G. S. and Dimond, R. L., The use of tolerization in the production of monoclonal antibodies against minor antigenic determinants, *Anal. Biochem.,* 154, 373, 1986.
55. Hockfield, S., A mAb to a unique cerebellar neuron generated by immunosuppression and rapid immunization, *Science,* 237, 67, 1987.
56. Hsu, H. T., Wang, Y. C., Lawson, R. H., Wang, M., and Gonsalves, D., Splenocytes of mice with induced immunological tolerance to plant antigens for construction of hybridomas secreting tomato spotted wilt virus-specific antibodies, *Phytopathology,* 80, 158, 1990.
57. Hsu, H. T., Lee, I. M., Davis, R. E., and Wang, Y. C., Immunization for generation of hybridoma antibodies specifically reacting with plants infected with a mycoplasma-like organism (MLO) and their use in detection of MLO antigens, *Phytopathology,* 80, 946, 1990.
58. Al Moudallal, Z., Altschuh, D., Briand, J. P., and van Regenmortel, M. H. V., Comparative sensitivity of different ELISA procedures for detecting monoclonal antibodies, *J. Immunol. Methods,* 68, 35, 1984.
59. Jordan, R. L., Potyviruses, monoclonal antibodies and antigenic sites, in *Potyvirus Taxonomy,* Barnett, O. W., Ed., Arch. Virol., 119 (Suppl. 5), 81, 1992.
60. Taggart, R. T. and Samloff, I. M., Stable antibody producing murine hybridomas, *Science,* 219, 1228, 1983.
61. Ransom, J. H., Endothelial cell growth supplements for promoting the growth of monoclonal antibody producing hybridoma cells, *Methods Enzymol.,* 121, 293, 1986.
62. Lane, R. D., Crissman, R. S., and Ginn, D., High efficiency fusion procedure for producing monoclonal antibodies against weak immunogens, *Methods Enzymol.,* 121, 183, 1986.
63. Westerwoudt, R. J., Naipal, A. M., and Harrisson, C. M. H., Improved fusion technique. II. Stability and purity of hybrid clones, *J. Immunol. Methods,* 68, 89, 1984.
64. Wells, D. E. and Bibb, W. F., A method for freezing hybridoma clones in 96-well microculture plates, *Methods Enzymol.,* 121, 417, 1986.
65. Coller, H. A. and Coller, B. S., Poisson statistical analysis of repetitive subcloning by the limiting dilution technique as a way of assessing hybridoma monoclonality, *Methods Enzymol.,* 121, 412, 1986.
66. Hoogenroad, N. J. and Wraight, C. J., The effect of pristane on ascites tumor formation and monoclonal antibody production, *Methods Enzymol.,* 121, 375, 1986.
67. Jones, S. L., Cox, J. C., and Pearson, J. E., Increased monoclonal antibody ascites production in mice primed with Freund's incomplete adjuvant, *J. Immunol. Methods,* 129, 227, 1990.
68. Bruck, C., Drebin, J. A., Glineur, C., and Portetelle, D., Purification of mouse monoclonal antibodies from ascitic fluid by DEAE Affi-gel blue chromatography, *Methods Enzymol.,* 121, 587, 1986.
69. Schaad, N. W., Serological identification of plant pathogenic bacteria, *Annu. Rev. Phytopathol.,* 68, 249, 1979.
70. Bishop, A. L., Burr, T. J., Mittak, V. L., and Katz, B. H., A monoclonal antibody specific to *Agrobacterium tumefaciens* biovar 3 and its utilization for indexing grapevine propagation material, *Phytopathology,* 79, 995, 1989.
71. DeBoer, S. H., Wieczorek, A., and Kummer, A., An ELISA test for bacterial ring rot of potato with a new monoclonal antibody, *Plant Dis.,* 72, 874, 1988.

72. Magee, W. E., Beck, L. F., and Ristow, S. S., Monoclonal antibodies specific for *Corynebacterium sepedonicum,* the causative agent for potato ring rot, in *Hybridoma Technology in Agricultural and Veterinary Research,* Stern, N. and Gamble, H. R., Eds., Rowman and Allenheld, New Jersey, 1984.
73. Alvarez, A. M., Benedict, A. A., and Mizumoto, C. Y., Identification of xanthomonads and grouping of strains of *Xanthomonas campestris* pv. *campestris* with monoclonal antibodies, *Phytopathology,* 75, 722, 1985.
73a. Alvarez, A. M. and Lou, K., Rapid identification of *Xanthomonas campestris* pv. *campestris* by ELISA, *Plant Dis,* 69, 1082, 1985.
74. Alvarez, A. M., Benedict, A. A., and Civerolo, E. L., Unique strains of *Xanthomonas campestris* pv. *citri* associated with Mexican citrus bacteriosis, *Phytopathology,* 76, 1078, 1986.
75. Alvarez, A. M., Benedict, A. A., Mizumoto, C. Y., Pollard, L. W., and Civerolo, E. L., Analysis of *Xanthomonas campestris* pv. *citri* and *Xanthomonas campestris* pv. *citrumelo* with monoclonal antibodies, *Phytopathology,* 81, 857, 1991.
76. Benedict, A. A., Alvarez, A. M., Berestecky, J., Imanaka, W., Mizumoto, C. Y., Pollard, L. W., Mew, T. W., and Gonzlez, C. F., Pathovar-specific monoclonal antibodies for *Xanthomonas campestris* pv. *oryzae* and *Xanthomonas campestris* pv. *oryzicola, Phytopathology,* 79, 322, 1989.
77. Benedict, A. A., Alvarez, A. M., and Pollard, L. W., Pathovar-specific antigens of *Xanthomonas campestris* pv. *begonia* and *Xanthomonas campestris* pv. *pelargonii, Appl. Environ. Microbiol.,* 56, 572, 1989.
78. Norman, D. and Alvarez, A., A rapid method for presumptive identification of *Xanthomonas campestris* pv. *diffenbachiae* and other xanthomonads, *Plant Dis.,* 73, 654, 1989.
79. Permar, T. A. and Gottwald, T. R., Specific recognition of a *Xanthomonas campestris* Florida nursery strain by a monoclonal antibody probe in a microfiltration enzyme immunoassay, *Phytopathology,* 79, 780, 1989.
80. Yuen, G. Y., Alvarez, A. M., Benedict, A. A., and Trotter, K. J., Use of monoclonal antibodies to monitor the dissemination of *Xanthomonas campestris* pv. *campestris, Phytopathology,* 77, 366, 1987.
81. Lipp, R. L., Alvarez, A. M., Benedict, A. A., and Berestecky, J., Use of monoclonal antibodies and pathogenicity tests to characterize strains of *Xanthomonas campestris* pv. *dieffenbachiae* from aroids, *Phytopathology,* 82, 677, 1992.
82. Sheng, W. S., Wells, J. M., and Chen, T. A., Serological differentiation of strains of *Erwinia ananas* with hybridoma antibodies, in *Plant Pathogenic Bacteria,* Civerolo, E. L., Collmer, A., Davis, R. E., and Gillaspie, A. G., Eds., Martinus Nijhoff, Boston, 1986.
83. Lin, C. P., Chen, T. A., Wells, J. M., and van der Zwet, T., Identification and detection of *Erwinia amylovora* with monoclonal antibodies, *Phytopathology,* 77, 376, 1987.
84. Hutschemackers, J., Snaken, F., and Verhoyen, M., Utilization of rat monoclonal antibodies to detect *Erwinia amylovora, Acta Hortic.,* 273, 43, 1990.
85. Livingston, R. S., Galuska, D. A., Klopmeyer, M. J., and Kelman, R., Production and characterization of monoclonal antibodies for pectic lyases, *Phytopathology,* 76, 1138, 1986.
86. Maher, E. A., Livingston, R. S., and Kelman, A., Expression of pectic lyase by *Erwinia carotovora* subsp. *carotovora* in potato tuber tissue, *Phytopathology,* 78, 1551, 1988.
87. Hung, H. Y., Wells, J. M., and Chen, T. A., Monoclonal antibodies against strains of the fastidious, gram-negative bacteria from plants, in *Plant Pathogenic Bacteria,* Civerolo, E. L., Collmer, A., Davis, R. E., and Gillaspie, A. G., Eds., Martinus Nijhoff, Boston, 1987.
88. Garnier, M., Martin-Gros, G., Iskra, M.-L., Zreik, L., Gandar, J., Fos, A., and Bove, J. M., Monoclonal antibodies against the MLOs associated with tomato stolbur and clover phyllody, *Zentralbl. Bakteriol. Suppl.,* 20, 263, 1990.
89. Garnier, M., Villechanoux, S., Gao, S., He, Y., and Bove, J. M., Study of greening organism with monoclonal antibodies: purification serological identification, morphology, and serotypes of greening organism, *Proc. Conf. Int. Org. Citrus Virol.,* 1991.
90. Lee, I. M. and Davis, R. E., Mycoplasmas which infect plants and insects, in *Mycoplasmas: Molecular Biology and Pathogenesis,* Maniloff, J., McElhaney, R. N., Finch, L. R., and Baseman, J. B., Eds., American Society for Microbiology, Washington, D.C., 1992, chap. 23.
91. Konai, M. I., Davis, R. E., and Jordan, R. L., Immunofluorescence microscopy using monoclonal antibodies to study *Spiroplasma* with preserved helical cell shape, *Phytopathology,* 76, 1083, 1986.
92. Davis, R. E., Lee, I. M., Jordan, R. L., and Konai, M., Detection of corn stunt spiroplasma in plants by ELISA employing monoclonal antibodies and by isolation of the pathogen and its phages in serum-

free medium, in *Plant Pathogenic Bacteria,* Civerolo, E. L., Collmer, A., Davis, R. E., and Gillaspie, A. G., Eds., Martinus Nijhoff, Boston, 1987.

93. Chen, T. A. and Jiang, X. F., Monoclonal antibodies against the maize bushy stunt agent, *Can. J. Microbiol.,* 34, 6, 1988.
94. Chaing, T. I. and Chen, T. A., Utilization of monoclonal antibodies against elms yellows mycoplasmalike organisms in detection of elms yellows disease, *Phytopathology,* 81, 121, 1991.
95. Davies, D. L. and Clark, M. F., Production and characterization of polyclonal and monoclonal antibodies against peach yellow leafroll MLO-associated antigens, *Acta Hortic.,* 3409, 275, 1992.
96. Fos, A., Danet, J. L., Zreik, L., Garnier, M., and Bove, J. M., Use of a monoclonal antibody to detect the stolbur mycoplasmalike organism in plants and insects and to identify a vector in France, *Plant Dis.,* 767, 1092, 1992.
97. Schwartz, Y. E., Boudon-Padieu, E., Grange, J., Meignoz, R., and Caudwell, A., Monoclonal antibodies to the mycoplasmalike organism (MLO) responsible for grapevine flavescence doree, *Res. Microbiol.,* 140, 311, 1989.
98. Guo, J. R., Chen, T. A., and Loi, N., Production of monoclonal antibodies against flavescence doree mycoplasmalike organism, *Phytopathology,* 81, 1210, 1991.
99. Lin, C. P. and Chen, T. A., Comparison of monoclonal antibodies and polyclonal antibodies in detection of the aster yellows mycoplasma-like organism, *Phytopathology,* 76, 45, 1986.

99a. Yang, K. D. and Chen, T. A., Detection and comparison of aster yellows strains by monoclonal antibodies, *Phytopathology,* 80, 958, 1990.

100. Lin, N. S., Hsu, Y. H., and Hsu, H. T., Immunological detection of plant viruses and a mycoplasmalike organism by direct tissue blotting on nitrocellulose membranes, *Phytopathology,* 80, 824, 1990.
101. Lee, I. M., Davis, R. E., and Hsu, E. T., Monoclonal antibodies against tomato big bud mycoplasmalike organism (MLO) distinguish a group of interrelated MLO strains within the "aster yellows" strain cluster, *Phytopathology,* 80, 958, 1990.

Chapter 29

Monoclonal Antibodies: Viruses

Kazusato Ohshima

TABLE OF CONTENTS

I. INTRODUCTION

The hybridoma technology introduced by Köhler and Milstein[1] has provided a revolutionary advance in the method of antibody production that eliminates many of the problems associated with polyclonal antibody (pAb). Hybridoma technology has the potential for producing an unlimited quantity of monospecific antibodies (monoclonal antibodies [mAbs]) that are ideal serological reagents for taxonomic, diagnostic, structural, and biochemical analyses of plant viruses.

mAbs offer several advantages over conventional pAbs:[2] (1) an unlimited quantity of antibody can be produced from a small quantity of antigen; (2) pure antibodies specific for a single antigenic determinant can be obtained, even when impure antigen or antigen mixtures are used as the immunogen; (3) hybridomas can be preserved by freezing in liquid nitrogen, thereby assuring a continuous supply of antibody over time; (4) highly specific mAbs may reveal the serological relationships between plant viruses that were previously unrecognized with pAbs; (5) the use of mAbs eliminates the qualitative and quantitative variability in specific antibody content in different batches of pAb; (6) mAbs may differentiate virus strains which can not be differentiated by pAbs; and (7) mAbs are useful for the epitope mapping of coat protein of plant viruses.

On the other hand, mAbs have a few disadvantages compared with pAbs: (1) the production and characterization of mAbs take more than one half year; (2) some mAbs are not useful for some immunological techniques; (3) producing mAbs costs much money and care than producing pAbs.

0-87371-877-1/95/$0.00+$.50

mAbs against plant viruses were first produced in the early 1980s,[3–5] and mAbs to all members of the major groups of plant viruses have been produced (Table 1). Techniques for the production of mAbs differs, depending on the independent laboratory. The methods described by Galfré and Milstein[73] have been adapted and modified through time and have given good results with many different viruses. In addition to the protocols of our laboratory given here, the reader is also recommended to refer to

Table 1 **Production of monoclonal antibodies to plant viruses**

Virus group	Virus	Ref.
Alfalfa mosaic	Alfalfa mosaic	6, 7
Carla	Potato virus M	8
Caulimo	Carnation etched ring	9
Clostero	Apple chlorotic leaf spot	10
	Citrus tristeza	11,12
	Grapevine leafroll associated	13
Como	Bean pod mottle	14
	Cowpea mosaic	15
	Cowpea severe mosaic	15
Cucumo	Cucumber mosaic	16,17
Diantho	Sweet clover necrotic mosaic	18
Gemini	African cassava mosaic	19
	Maize streak	20
Furo	Beet necrotic yellow vein	21–23
Ilar	Apple mosaic	7
Luteo	Barley yellow dwarf	24–29
	Beet western yellows	30
	Potato leafroll	31–34
	Soybean dwarf	28
	Tobacco necrotic dwarf	35
Nepo	Grapevine fanleaf	36
Plant reo	Rice dwarf	37,38
	Rice ragged stunt	39
Potex	Potato virus X	40–42
Poty	Bean yellow mosaic	43–45
	Lettuce mosaic	46
	Maize dwarf mosaic	47
	Pea mosaic	45
	Plum pox	48
	Potato virus Y	
	Ordinary strain	49
	Necrotic strain	49,50
	Tobacco necrotic strain	51
	Tobacco veinal necrosis strain	5
	Soybean mosaic	52,53
	Tobacco etch (nuclear inclusion bodies)	54
	Tulip breaking	55
	Turnip mosaic	56
	Watermelon mosaic 2	57
	Zucchini yellow mosaic	58
Rhabdo	Lettuce necrotic yellows	59
Sobemo	Southern bean mosaic	60,61
Tobamo	Tobacco mosaic	3,4,62–65
	Odontoglossum ringspot	66
	Wheat soilborne mosaic	67
Tenui	Rice grassy stunt	39
	Rice stripe	68, Ohshima et al. (unpublished result)
Tomato spotted wilt	Tomato spotted wilt	69–72

Campbell,[74] Halk and De Boer,[2] Sander and Dietzgen,[75] Torrance,[76] and Chapter 24 in this book for general information.

II. METHODOLOGY

Since basic protocols for the production of mAbs to plant viruses are duplicated to those described in Chapter 24, I would like to focus attention on the topics in plant virology and in our laboratory.

A. ANTIGEN PREPARATION

One of the most misleading concepts which have grown up around mAb technology is that there is no need to purify the antigen used for immunization. While the concept is undoubtedly correct, in practice this is frequently unsatisfactory. In the great majority of cases, the major antigens dominate a fusion in the same way as they dominate a serum titer, and some preliminary purification of minor antigens is highly desirable. That is, plant host components have to be excluded from crude extracts of plant virus. Sucrose density gradient centrifugation, for example, is useful for purification of the viruses from host components.

Another important fact for immunization is the condition of antigens. If the mAbs against denatured antigens are required, denatured antigens need to be prepared. On the other hand, if mAbs against native antigens are required, nondenatured antigens need to be prepared. Enzyme-linked immunosorbent assay (ELISA) is the most commonly used method for the detection of plant viruses. However, several types of ELISA procedures are used for the detection of plant viruses (Table 2). Generally, virus-infected plants are homogenized in phosphate-buffered saline (PBS), pH 7.4, containing Tween® 20, and those crude saps are used for the antigens. Therefore, we dissolve purified preparations in PBS for the production of mAbs which will be used for ELISA. This is much related with screening procedures of mAb-secreting hybridomas. Further discussion of screening mAb-secreting hybridomas will be presented later.

B. IMMUNIZATION

A typical immunization procedure that provides satisfactory results for most plant viruses (antigens) is to immunize mice (Balb/c) with two intramuscular injections of usually 10 to 100 μg of antigen in Freund's complete adjuvant 2 weeks apart. This method incorporates a final booster injection (usually intravenous) approximately 3 to 4 d before the fusion. This method requires more than 1 month from primary injection to the fusion of spleen cells and myeloma cells.

In our laboratory, mice are immunized by a single intrasplenic injection as described by Spitz et al.[77] with some modifications. A 4- to 6-week-old mouse is anesthetized by ether and the skin is incised about 1 to 1.5 cm to expose the spleen. Using needle (1/2 × 26 gauge) fitted to a 1-ml syringe, 10 to 20 μg of purified virus in 0.1 ml of the appropriate buffer is directly injected into the spleen. After injection is completed, the peritoneum is joined by a stapler. This method offers several advantages over the typical immunization method shown above: (1) very small amounts of antigen (a few micrograms) are needed, (2) the spleen cells can be fused with myeloma cells 3 to 4 d after injection; therefore, in the case of unsatisfactory fusion, we can immunize a new mouse again immediately.

Table 2 **ELISA procedures used for the detection of plant viruses in virus-infected plants**

ELISA procedure[a]	Step			
	1st	2nd	3rd	4th
Procedure 1	Monoclonal antibody	Crude sap	Monoclonal antibody EC[b]	
Procedure 2	Polyclonal antibody	Crude sap	Monoclonal antibody	Anti-mouse Ig EC
Procedure 3	Crude sap	Monoclonal antibody	Anti-mouse Ig EC	

[a]Procedure 1: direct double-antibody sandwich ELISA (DAS-ELISA)
Procedure 2: indirect double-antibody sandwich ELISA (IDAS-ELISA)
Procedure 3: antigen adsorption indirect ELISA (AAI-ELISA)
[b]EC, Enzyme conjugate

Table 3 **ELISA procedures for the screening of monoclonal antibody-secreting hybridomas**

AAI-ELISA	Step 1st	2nd	3rd
Procedure 1	Purified virus buffer: PBS, pH 7.4	Culture fluid	Anti-mouse Ig EC
Procedure 2	Purified virus buffer: SCB, pH 9.6	Culture fluid	Anti-mouse Ig EC

IDAS-ELISA	Step 1st	2nd	3rd	4th
Procedure 3	pAb	Purified virus buffer: PBS-T, pH 7.4	Culture fluid	Anti-mouse Ig, EC
Procedure 4	pAb	Crude sap buffer: PBS-T, pH 7.4	Culture fluid	Anti-mouse Ig, EC

AAI-ELISA, Antigen adsorption indirect ELISA; Buffer, buffers which are used for reacting antigens; PBS, phosphate-buffered saline; EC, enzyme conjugate; SCB, sodium carbonate-bicarbonate buffer; IDAS-ELISA, indirect double-antibody sandwich ELISA; PBS-T, phosphate-buffered saline containing Tween® 20.

Another immunization method is *in vitro* immunization reported by Campbell[74] and Reading.[78] In this technique, a spleen is cultured with antigen for 3 to 4 d, reported to be advantageous for antigens which give a poor immune response or where very small amounts (a few micrograms) of antigen are available. It is also a very quick technique compared to the method described by Spitz et al.,[77] but this method tends to produce IgM mAbs.

C. MYELOMA CELL LINES

Myeloma cell lines most commonly used in the production of mAbs to plant viruses are SP2/O-Ag14, P3-X63-Ag8.653, and P3-NS1/1-Ag4-1 (NS1/1). The cell line SP2/O-Ag14 used in our laboratory does not produce endogenous antibodies, whereas the cell line P3-NS1/1-Ag4-1 (NS1/1) produces kappa light chains intercellularly, which may appear in virus-specific antibodies secreted by some hybridomas.[73]

D. SCREENING OF MONOCLONAL ANTIBODY-SECRETING HYBRIDOMAS

The screening procedures used to identify antibody-secreting hybridomas are the important step for the production of mAbs. Screening procedures should be capable of identifying the maximum number of mAb-secreting hybridomas that react with immunogen and have desired properties for their end use. Procedures must be sensitive, rapid, reproducible, and capable of processing hundreds of samples over a few days. For these reasons, ELISA procedures are the most commonly used. If mAbs will be used in ELISA, mAb-secreting hybridomas need to be screened by ELISA. If mAbs will be used in Western blotting analyses or in immunoelectron microscopic analyses, those mAb-secreting hybridomas need to be screened by Western blotting analyses or by immunoelectron microscopic analyses, respectively.

As shown in Table 3, four different procedures of indirect ELISA are used for screening of mAb-secreting hybridomas in our laboratory. More than 235 mAb-secreting hybridomas against 3 luteoviruses, 2 plant reoviruses, and a potyvirus were produced and were screened by these four different procedures of indirect ELISA.[79] Table 4 shows groups of mAb-secreting hybridomas according to reactivities to homologous viruses in four different ELISA procedures. Using purified mAbs from ascitic fluids, direct double-antibody sandwich ELISA (DAS-ELISA) was examined for the detection of virus antigens in infected plants. As shown in Table 5, all the mAbs which reacted with virus in DAS-ELISA belong to two groups, that is, the group A in which the mAbs were reactive in each of the four screening procedures, or the group B in which mAbs were reactive in three of the screening procedures, 1, 3, and 4. These results indicate that if only one procedure of ELISA is used for screening, it is very difficult to produce mAbs reactive in DAS-ELISA. And these results also indicate that mAbs being reactive in DAS-ELISA can be readily selected from hybridomas in group A or B.

Table 4 **Groups of monoclonal antibody-secreting hybridomas according to ELISA reactions**

Group	Screening procedure[b] 1	2	3	4	Antigens used for immunization and screening[a]: Luteovirus PLRV	Luteovirus BWYV	Luteovirus TNDV	Plant reovirus RDV	Plant reovirus RRSV	Potyvirus PVY-O
A	+[c]	+	+	+	[d]6 (7.7%)[e]	1 (2.8%)	5 (23.8%)[f]	13 (56.5%)	15 (48.4%)	55 (56.7%)
B	+	−	+	+	2 (2.6%)	1 (2.8%)	1 (4.8%)	0 (0.0%)	0 (0.0%)	25 (25.7%)
C	+	+	−	−	70 (89.7%)	32 (88.9%)	15 (71.4%)	0 (0.0%)	0 (0.0%)	0 (0.0%)
D	−	−	+	+	0 (0.0%)	0 (0.0%)	0 (0.0%)	3 (13.0%)	5 (16.1%)	4 (4.1%)
E	−	−	−	+	0 (0.0%)	0 (0.0%)	0 (0.0%)	0 (0.0%)	5 (16.1%)	12 (12.4%)
F	+	−	−	−	0 (0.0%)	2 (5.6%)	0 (0.0%)	0 (0.0%)	0 (0.0%)	1 (1.0%)
G	+	+	+	−	0 (0.0%)	0 (0.0%)	0 (0.0%)	1 (4.3%)	1 (3.2%)	0 (0.0%)
H	−	+	+	+	0 (0.0%)	0 (0.0%)	0 (0.0%)	3 (13.0%)	3 (9.7%)	0 (0.0%)
I	−	+	+	−	0 (0.0%)	0 (0.0%)	0 (0.0%)	1 (4.3%)	2 (6.5%)	0 (0.0%)
J	−	−	+	−	0 (0.0%)	0 (0.0%)	0 (0.0%)	2 (8.7%)	0 (0.0%)	0 (0.0%)
Total					78 (100%)	36 (100%)	21 (100%)	23 (100%)	31 (100%)	97 (100%)

[a]PLRV, potato leafroll virus; BWYV, beet western yellows virus; TNDV, tobacco necrotic dwarf virus; RDV, rice dwarf virus; RRSV, rice ragged stunt virus; PVY-O, potato virus Y ordinary strain. Homologous antigens were used for screening antibody-secreting hybridomas.
[b]Screening procedures of ELISA are shown in Table 3.
[c]+, Antibodies of hybridoma reacted with virus antigens; −, Antibodies of hybridomas did not or slightly reacted with the virus antigens.
[d]Number of antibody-secreting hybridomas.
[e]Percentage (%) of [number of hybridomas belong to the group/total number of hybridomas].
[f]Underlines show that more than 20% of hybridomas belonged to the group.
Modified from Oshima et al.[79]

We also screened mAbs to potato virus Y (PVY)-secreting hybridomas. After cloning and purification of mAbs, several immunological assays using these mAbs were performed. As shown in Table 6, some mAbs were not reactive in microprecipitin tests or Western blotting analyses. This indicated that screening procedures of ELISA are not well adapted for the use of mAbs in these two methods. So the greatest care must be taken in dealing with the screening procedures for using mAbs.

Himmler et al.[48] screened mAb-secreting hybridomas by immunoelectron microscope. They succeeded to produce mAbs useful for immunogold labeling techniques.

III. APPLICATIONS IN PLANT VIROLOGY

A. CHARACTERIZATION OF MONOCLONAL ANTIBODIES BY IMMUNOLOGICAL ASSAY

Several immunological assays (four different ELISA procedures, precipitin tests, and Western blotting analyses) are used to determine overlapping epitopes of mAbs in our laboratory. As shown in Table 6, 13 mAbs to PVY were produced and characterized by several immunological methods. The results showed that mAbs were clearly divided into several groups, and these were directed to different epitopes.

1. Precipitin Tests

Many virus-specific mAbs did not precipitate viruses or coat proteins in microprecipitin or immunodiffusion tests.[6,7,20]

a. Microprecipitin Tests

As shown in Table 6, 8 out of the 13 mAbs to PVY ordinary and necrotic strains were capable of precipitating PVY particles in microprecipitin tests.

Table 5 **Absorbance values of four different ELISA procedures of purified monoclonal antibodies which are reactive and are not reactive in DAS-ELISA**

(I) Reactive in DAS-ELISA[a]						(II) Not reactive in DAS-ELISA					
	AAI-ELISA		IDAS-ELISA				AAI-ELISA		IDAS-ELISA		
Monoclonal antibody[b]	Proc. 1[c]	Proc. 2	Proc. 3	Proc. 4	Group[d]	Monoclonal antibody	Proc. 1	Proc. 2	Proc. 3	Proc. 4	Group
TND- 1E2	2<[e]	2<	2<	2<	A	PLR- 11A09	2<	0.18	1.17	1.46	B
RD- 11C10	2<	2<	2<	2<	A	33E07	1.84	1.73	0.35	0.09	C
RRS-183D07	2<	1.99	1.33	1.64	A	51C01	1.37	0.91	1.67	2<	A
PVYO- 12H01	2<	0.36	2<	2<	B	BWY- 21G06	0.73	0.62	0.06	0.04	C
21H05	2<	2<	1.47	2<	A	TND- 33D07	2<	2<	2<	2<	A
22D01	2<	0.42	2<	2<	B	33E07	1.91	0.29	2<	2<	B
42A08	2<	0.24	2<	2<	B	43B05	1.75	2<	0.39	0.00	C
51E04	2<	2<	0.97	2<	A	RD- 63C12	0.61	0.91	1.58	1.05	A
						RRS-191G05	1.26	1.98	2<	1.15	A
						PVYO-11C01	1.48	0.14	1.92	2<	B
						22A01	1.37	0.07	1.25	0.53	B
						31C12	0.16	0.03	0.94	2<	D
						41G07	2<	1.43	2<	2<	A
						62E04	0.22	0.01	0.33	1.37	E

[a]Monoclonal antibodies which reacted with virus antigen in crude saps by DAS-ELISA (I) and monoclonal antibodies which did not react with virus antigen in crude saps by DAS-ELISA (II).

[b]Monoclonal antibodies are marked as: from potato leafroll virus-hybridomas, PLR-; from beet western yellows virus-hybridomas, BWY-; from tobacco necrotic dwarf virus-hybridomas, TND-; from rice dwarf virus-hybridomas, RDV-; from rice ragged stunt virus-hybridomas, RRS-; from potato virus Y ordinary strain-hybridomas, PVYO-.

[c]Proc., Procedure. ELISA procedures are shown in Table 3.

[d]Groups are shown in Table 4.

[e]Absorbance values (A_{415}) of ELISA. Underlines show that A_{415} values are more than 0.5 (threshold of positive reaction). Homologous antigens were used to determine the reactivities of monoclonal antibodies.

Modified from Ohshima et al.[79]

Table 6 **Reaction of monoclonal antibodies to potato virus Y using immunological methods**

Monoclonal antibody	Specific to	ELISA procedure[a] AAI-ELISA Proc. 1[b]	AAI-ELISA Proc. 2	IDAS-ELISA Proc. 3	IDAS-ELISA Proc. 4	DAS-ELISA Proc. 1	Western blotting analysis[c]	Micro-precipitin test[d]
PVYO-11G03[e]	PVY-O[f]	+[g]	−	++	++	−	−	−
12F05	PVY-O	+++	+++	+++	+++	−	−	++
12H01	PVY-O	+++	+	+++	+++	+	+	+++
21H05	PVY-O	+++	+++	+++	+++	+++	++	++
22D01	PVY-O	+++	+	+++	+++	+	−	++
41G07	PVY-O	+++	++	+++	+++	−	+	−
42A08	PVY-O	+++	+	+++	+++	+++	−	−
51E04	PVY-O	+++	+++	+	+++	+	−	++
PVYT-4E7	PVY-T	+++	+++	+++	+++	+++	+++	−
6C11	PVY-T	+++	+++	+++	+++	+++	+++	−
PVYO-12H03	PVY-O,T	++	+	++	++	−	−	+
41G03	PVY-O,T	+++	+++	+++	+++	++	+	++
42C07	PVY-O,T	++	++	+++	+++	++	+	++

[a]ELISA procedures of AAI- and IDAS-ELISA are shown in Table 3 and that of DAS-ELISA are shown in Table 2.
[b]Proc., Procedure.
[c]After antigens (purified preparations of virus) were loaded under denaturing conditions of SDS-polyacrlylamide gel electrophoresis, immunostaining was performed.
[d]Microprecipitin tests were performed by mixing antigens with antibodies in phosphate-buffered saline (pH 7.4).
[e]Monoclonal antibodies are marked as: from PVY ordinary strain-hybridomas, PVYO-; from PVY necrotic strain-hybridomas, PVYT-. Monoclonal antibodies used in ELISA procedures were purified from ascitic fluids.
[f]PVY-O, Potato virus Y ordinary strain; PVY-T, potato virus Y necrotic strain.
[g]Plus and minus signs were ranked by relative ELISA A_{415} values (+++ = > 1.5; ++ = 1.5–1.0; + = 1.0–0.1; − = 0.1<), signal intensity of Western blotting analyses and dilution endpoints of monoclonal antibodies. Homologous antigens were used to determine the reactivities to PVY-O or PVY-T monoclonal antibodies.
Modified from Inoue[80] and Ohshima et al.[79]

b. Immunodiffusion Tests

Only 2 (mAbs 5 and 8) of the 15 mAbs to alfalfa mosaic virus (AMV) were capable of precipitating AMV particles in immunodiffusion tests.[6] Whereas mAb 5 produced clear precipitin lines in agarose gels irrespective of whether culture supernatants or affinity-purified antibodies were used, mAb 8 required the addition of polyethylene glycol to the agarose to induce precipitation with culture surpernatants. mAb 5 produced clear precipitin lines when tested against preparations of native particles of AMV, but failed to react with the viruses after fixation. On the other hand, mAb 8 produced precipitin lines when tested against the fixed particles, but not against the native particles.

2. Enzyme-Linked Immunosorbent Assays

In the application of mAbs in ELISA for viral detection in infected plants, indirect double—antibody sandwich ELISA (IDAS-ELISA) using pAbs as trapping antibody has been mostly employed (Table 2, procedure 2). The disadvantage of this system is that the pAbs must be prepared in addition to the mAbs. Considering the advantage of mAbs shown before, the ELISA procedures such as DAS-ELISA (Table 2, procedure 1) are recommended. However, some of virus-specific mAbs are not reactive to virus in DAS-ELISA (Table 5).

3. Dot Immunobinding Assays

An important variation of the ELISA procedures is known as the dot immunobinding assay (DIBA). In the tests, nitrocellulose membranes are substituted for microplates in ELISA. There are two DIBA procedures: after antigens are spotted onto nitrocellulose membranes, in procedure A the membranes are incubated with mAbs, followed by incubation with alkaline phosphatase—labeled anti-mouse immunoglobulin. In procedure B, the membranes are incubated with biotinylated mAbs, followed by incubation with streptavidin alkaline phosphatase, then tested for the ability of mAbs to detect tomato spotted wilt virus (TSWV) in plant extracts and in purified preparations (Huguenot et al.).[71] In procedure B, use of biotinylated mAbs gave high background staining on the membranes. The best result was obtained in procedure A with mAb 3.22.6, which was able to detect TSWV in plants when diluted up to 1/5000.

4. Competition Assays

One of the earliest questions which is asked of mAbs is whether they are directed at the same or at different epitopes. The most commonly used tests to determine overlapping epitopes are competition assays. There are several competition assays:[74] (1) direct competition between labeled and unlabeled mAbs and (2) competition between two unlabeled mAbs using an enzyme-linked detecting antibody.

5. Immunoelectron Microscopy

Immunoelectron microscopic studies of binding of mAbs to plant viruses have been performed by Dore et al.[65] and Himmler et al.[48]

The technique of gold-labeled immunosorbent electron microscopy was used for the screening of mAb-secreting hybridomas and compared to ELISA and Western blotting analysis.[48] The technique, compared to ELISA, has the advantage of the visualization of the antibody-antigen reaction and compared to Western blotting (immunoblotting) analysis has the additional advantage of high sensitivity.

mAbs to tobacco mosaic virus (TMV) that bind only to one end of the viral rods were shown to recognize the surface of the protein subunit designated as the bottom, which contains the right radial and left radial α-helices.[65] The specificity of the antibody binding was established by immunoelectron microscopy of complexes in which the 5′ end of the RNA had been exposed at the bottom of the helical virus particle. These mAbs were shown to bind to both ends of the stacked disk aggregate of TMV protein, which is, therefore, bipolar.

6. Western Blotting Analyses

Western blotting analysis is marked by detecting denatured antigen. Antigens are mixed with an equal volume of sample buffer containing 2-mercaptoethanol and sodium dadecyl sulfate (SDS), and then heated for a few minutes in boiling water followed by SDS-polyacrylamide gel electrophoresis (SDS-PAGE). After blotted onto nitrocellulose or PVDF transfer membranes from SDS-polyacrylamide gel, immunostaining is performed. Some mAbs to PVY screened by ELISA are reactive to the denatured antigen by Western blotting analysis, but some are not (Table 6).

B. DIAGNOSTIC APPLICATIONS

mAbs are now widely used for the diagnosis of plant viruses (Table 1). As shown in Table 7, mAbs specific to PVY ordinary and necrotic strains were produced and potato plants collected from a field were examined by DAS-ELISA using pAbs and mAbs. Specific mAbs could apparently differentiate the two strains of PVY in the plants infected with each of the two strains or both strains. However, the specific mAbs did not react with some potato plants infected with PVY, as determined by pAbs. This indicates that mAbs miss some PVY-infected potato plants, so we recommend that mAbs are used in conjunction with pAbs or that strain-common mAbs are used to diagnose plant viruses.

A potyvirus broad-spectrum (group-common) mAb has been successfully produced by Jordan and Hammond.[81] They injected mice with mixtures of native and denatured cytoplasmic inclusions and native (intact) virions and coat protein subunits (dissociated and/or denatured virus) from up to six different isolates of bean yellow mosaic virus and six other potyviruses (iris mild mosaic virus, iris severe mosaic virus, PVY, asparagus virus 1, pea seedborne mosaic virus, and tobacco etch virus [TEV]).

We have synthesized a polypeptide of the C-terminal region of PVY (H_2N-HTTEDVSPSMHTLLGVKNM -COOH) which was exposed on a virus particle[82] and is common to most of the strains. And we successfully produced the pAbs which were reactive to four of the Japanese strains in crude saps by ELISA procedures and Western blotting analyses.[83] Further studies on production of mAbs to the C-terminal region are now under way.

C. TAXONOMIC APPLICATIONS

Serological methods using mAbs play an important role in defining relationships at the group, virus, strain, and serotyping level in plant virus classification schemes.

Table 7 **Detection of potato virus Y (PVY) in potato plants by direct double-antibody sandwich ELISA (DAS-ELISA) using polyclonal and monoclonal antibodies of PVY ordinary (PVY-O) and necrotic (PVY-T) strains**

			Polyclonal antibody		Monoclonal antibody[a]		Virus[b]
	Sample No.	Cultivar	PVY-O	PVY-T	PVYO-42A08	PVYT-4E7	detected
Group 1	28	Danshaku	1.42[c]	0.79	2<	0.04	O
	40	Danshaku	0.75	0.25	0.87	0.01	O
	46	Danshaku	1.36	0.39	1.29	0.01	O
	54	Danshaku	2<	0.51	2<	0.02	O
	78	Toyoshiro	0.97	0.26	1.69	0.01	O
	200	Norin 1 go	2<	0.55	2<	0.04	O
Group 2	10	Danshaku	0.63	2<	0.03	0.91	T
	60	Toyoshiro	0.73	1.97	0.00	1.27	T
	68	Danshaku	0.89	2<	0.00	2<	T
	81	Danshaku	0.24	2<	0.02	2<	T
	177	Norin 1 go	1.17	2<	0.01	1.67	T
	207	Benimaru	0.57	2<	0.01	1.17	T
Group 3	39	Danshaku	1.86	2.00	1.82	0.76	O + T
	107	Danshaku	2<	2<	2<	2<	O + T
	187	Danshaku	1.21	0.94	2<	0.49	O + T
	203	Benimaru	1.89	2<	2<	1.49	O + T
	205	Benimaru	2<	2<	2<	1.99	O + T
Group 4	27	Danshaku	1.26	0.56	0.04	0.00	?
	36	Danshaku	0.92	0.33	0.02	0.02	?
	72	Danshaku	0.45	0.24	0.04	0.05	?
	PVY-O	*N. sylvestris*	2<	0.28	2<	0.00	
	PVY-T	*N. sylvestris*	0.64	2<	0.03	2<	
	H[d]	*N. sylvestris*	0.07	0.05	0.01	0.01	
	PBS[e]		0.00	0.00	0.00	0.00	

[a]PVYO-42A08, specific to PVY-O; PVYT-4E7, specific to PVY-T.

[b]O, PVY-O was only detected; T, PVY-T was only detected; O + T, both PVY-O and PVY-T were detected; ?, neither PVY-O nor PVY-T were detected.

[c]Absorbance values (A_{415}).

[d]H, Healthy *N. sylvestris* plants.

[e]PBS, Phosphate-buffered saline.

Modified from Ohshima et al.[49]

It was observed in the preceding paragraph that specific mAbs failed to detect some PVY-infected potato plants (Table 7: sample nos. 27, 36, 72). Biological properties of the ordinary, necrotic strains and PVY-36 strain (sample no. 36) which did not react with the mAbs specific to the ordinary and the necrotic strains were compared. As shown in Table 8, the ordinary strain did not produce necrotic symptoms on *Nicotiana tabacum* and *Physalis floridana,* but produced chlorotic and necrotic local lesions on *Chenopodium amaranticolor* and *C. quinoa.* The necrotic strain produced necrotic flecks and veinal necrosis symptoms on *N. tabacum,* but did not infect *C. amaranticolor* and *C. quinoa,* and did not produce necrotic symptoms on *P. floridana.* On the other hand, the PVY-36 strain did not produce necrotic symptoms on *N. tabacum,* but produced chlorotic and necrotic local lesions on *C. amaranticolor, C. quinoa,* and *P. floridana.*[80,84] These results indicated that these three strains are considerably different in biological properties.

Furthermore, the nucleotide sequences of the coat protein gene of these three strains (the ordinary and necrotic strains[84] and the PVY-36 strain)[90] were determined. The number of nucleotides encoding coat protein of the three strains was 801, and the number of predicted amino acids was 267. Amino acid sequence of the N-terminal regions of the coat protein varied in sequence in PVY strains (isolates), whereas the C-terminal regions were highly homologous. The difference of predicted amino acid sequences found between the ordinary and PVY-36 strain coat proteins was only the N-terminal region.

Table 8 **Symptoms on test plants induced by potato virus Y (PVY) strains**[a]

Test plant	Symptom[b]		
	PVY-O	PVY-T	PVY-36
Nicotiana glutinosa	M	M	M
N. sylvestris	VB	VB	VB
N. tabacum			
Xanthi nc	YS, VB, M	YS, N, VB, VN	YS, VB, M
White burley	YS, VB	YS, N, VB, VN	YS, VB
Samsun NN	YS, VB	YS, N, VB, VN	YS, VB
KY 57	M	VN, NS	M
Chenopodium amaranticolor	YS, CL, NL	—	YS, CL, NL
C. quinoa	YS, NL	—	YS, NL
C. murale	—	—	—
Datura metel	M, VB	M, VB	M, VB
D. stramonium	—	—	—
Physalis floridana	M	M	M, NL

[a]PVY-O, ordinary strain; PVY-T, necrotic strain; PVY-36, sample no. 36 which did not react with monoclonal antibodies (PVYO-42AO8 and PVYT-4E7) specific to ordinary and necrotic strains (see Table 7).
[b]M, mottling; VB, vein banding; YS, yellow spots; N, necrotic flecks; VN, veinal necrosis; NS, necrotic spot; CL, chlorotic local lesions; NL, necrotic local lesion; —, no infection.
From Inoue[80] and Ohshima et al.[84]

Shukla et al.[82] suggested that the N-terminal region of the coat protein of potyviruses contains the major virus-specific epitopes.

D. THE ANTIGENIC STRUCTURE OF VIRAL PROTEINS

The antigenic structure of the coat protein of TMV has been studied over the last 30 years by using pAbs specific for the dissociated viral subunit.[85,86] In previous studies of Van Regenmortel and co-workers,[63,64] seven continuous antigenic determinants have been identified in TMV coat protein in vicinity of residues 1–10, 34–39, 55–61, 62–68, 80–90, 108–112, and 153–158 by inhibition assays with cleavage and synthetic peptides corresponding to fragments of the coat protein. Furthermore, a number of discontinuous epitopes have also been identified at the surface of TMV particles by mAbs prepared against intact virions. About half of these 18 anti-TMV mAbs were so specific for quaternary structure of the capsid that they were unable to react with monomeric viral subunits. And only three of 18 mAbs were able to bind to some tryptic peptides of TMV coat protein. These findings are in line with the common observation that most mAbs to native proteins appear to be directed against conformation-dependent or discontinuous epitopes. They also determined the fine specificity of such mAbs by measuring their ability to bind to synthetic peptides representing virtually the entire length of the viral coat polypeptide chain and showed that a major part of the polypeptide chain of the TMV coat protein was antigenic. For this type of study, mAbs possess a superior discriminatory capacity compared to the pAbs.

E. BIOCHEMICAL APPLICATIONS

There are a few reports on using mAbs to plant viruses for biological application. Slade et al.[54] produced mAbs to tobacco etch potyvirus (TEV)-encoded nonstructural protein, the 49-kDa proteinase. TEV nonstructural protein crystallizes in the nuclei of virus-infected cells to form nuclear inclusion (NI) bodies which can be purified readily. Balb/c mice were immunized with purified NI bodies mixed with Freund's complete adjuvant and purified NI bodies were denatured by adding SDS and 2-mercaptoethanol (0.1% w/v each) followed by heating in boiling water for 3 min. The mAbs reactive with 49-kDa proteinase determined by Western blot analyses were characterized further with respect to the 49-kDa domain with which they reacted and with respect to their ability to inhibit the autocatalytic or self-processing activity of the 49-kDa proteinase. The 49-kDa antigens were synthesized from a TEV cDNA sequence using cell-free transcription and translation systems. Each anti-49-kDa mAb was used in immunoprecipitation studies with a series of 49-kDa antigens which represented a nested set of

49-kDa proteins with common amino termini, but varying in length. Immunoprecipitation results showed that a mAb reacted with 38-amino acid region B contained the proposed catalytic cysteine 339 residue. The mAb was the only anti-49-kDa mAb capable of inhibiting the self-processing reaction in which the 49-kDa proteinase is released from its 75-kDa polyprotein precursor. This study was undertaken to develop specific protein probes to identify important domains of the TEV 49-kDa protein involved in proteolysis.

F. EPITOPE MAPPING OF PLANT VIRUSES

The new approach of epitope mapping on the fragments of plant virus coat protein using molecular biological techniques is now being undertaken. The presence of five different epitopes on particles of beet necrotic yellow vein virus (BNYVV) was demonstrated by Koenig et al.[22] and Lesemann et al.[23] Using *Escherichia coli*-expressed free coat protein and a series of fusion proteins containing fragments of the coat protein derived from BNYVV cDNA, the presence and location of three out of five SDS-stable epitopes were confirmed with pAbs and mAbs on Western blot analyses by Commandeur et al.[21] Epitope 1, which was exposed on only one extremity of the virus particle, was located in the region between amino acids (aa) 1 and 7, i.e., on the N-terminal region of the coat protein. Epitope 3, which was exposed on the opposite extremity of the particle, was located in the region between aa 37 and 59. Epitope 4, which was exposed along the entire length of the particle, occurred on the C-terminus of the coat protein (aa 183 to 188). Also, two new SDS-stable epitopes were identified in the regions between aa 115 and 125 (epitope 6) and aa 125 and 140 (epitope 7). The former was located on the same extremity of the particle as epitope 3; the latter became accessible only after denaturation of particle.

We have analyzed which part of the N-terminal region of PVY was recognized by PVYT-4E7 mAb specific to PVY necrotic strain.[87] As shown in Figure 1, a series of fusion proteins containing the chimeric N-terminal region of the coat protein of PVY strains was constructed. PVYT-4E7 mAb, which is not reactive to tripsin-digested PVY coat proteins, but only reacted with chimera-1, indicated that the mAb recognized the B region (H_2N- GGSTKKDAKQE -COOH) of the PVY-T_H coat protein N-terminal region.

G. ANTI-IDIOTYPIC MONOCLONAL ANTIBODIES

Epitopes associated with antigen binding sites on antibodies are called idiotypes. Briefly, antibody is constructed by heavy and light chains and each chain consists of constant (highly conserved in amino

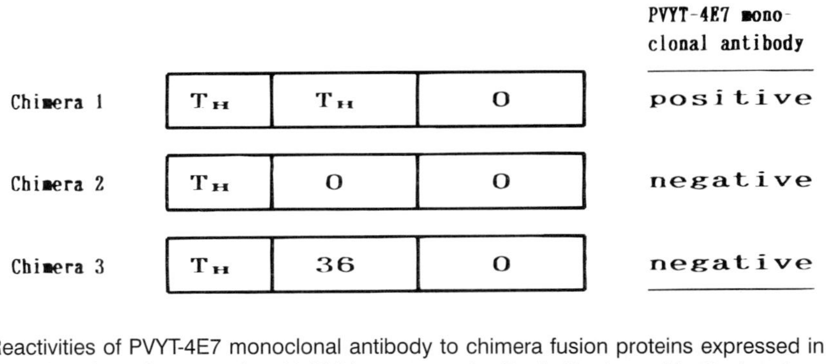

Figure 1 Reactivities of PVYT-4E7 monoclonal antibody to chimera fusion proteins expressed in *Escherichia coli* of N-terminal region of potato virus Y (PVY) coat protein. Dash (-) shows identical amino acid as PVY ordinary strain (PVY-O). PVY-T_H, PVY necrotic stain; PVY-36, sample no. 36 which did not react with monoclonal antibodies (PVYO-42AO8 and PVYT-4E7) specific to ordinary and necrotic strains (Table 7). (From Hataya et al., *Ann. Phytopathol. Soc. Jpn.*, 57, 459, 1991 [Abstr. in Japanese]. With permission.)

acid sequences) and variable regions (variable in amino acid sequence). The variable region contains three hypervariable regions (CDR1, CDR2, and CDR3) called idiotypes. Antibodies against the idiotypes are anti-idiotypic antibodies. Anti-idiotypic antibodies to an anti-barley yellow dwarf luteovirus mAb were produced for studying the receptors of virus transmission by aphids,[88] and anti-idotypic antibodies to an anti-soybean mosaic potyvirus mAb were produced for using as positive controls in immunological assays.[89]

H. AMINO ACID SEQUENCE OF MONOCLONAL ANTIBODY

Up to date, the nucleotide sequence of mAb gene to plant viruses is not known. A PVYT-42C07 mAb is IgG_3 and common to PVY ordinary and necrotic strains, and it binds PVY coat protein. We successfully cloned the mAb gene. The predicted amino acid sequences derived from the nucleotide sequences of both the heavy and light (kappa) chains are shown in Figure 2. We are now producing transgenic plants using the mAb gene which express the mAb. For further details of the strategy of cloning of a mAb gene from a mAb-secreting hybridoma, see Chapter 25.

Production of a mAb in transgenic plants was first reported by Hiatt et al.[90] The source of the mAb mRNA was a hybridoma cell line expressing a catalytic IgG_1 antibody which binds a low-molecular weight phosphonate ester and catalyzes the hydrolysis of certain carboxylic esters. Constructs used for

Figure 2 Predicted amino acid sequence of PVYT-42CO7 monoclonal antibody to potato virus Y coat protein. (From Ohshima et al., *Ann. Phytopathol. Soc. Jpn.*, 60, 600, 1994. With permission.)

the mAb expression in plants consisted of coding-length cDNA of the heavy (gamma) or light (kappa) chain with or without their leader sequences. These four cDNAs were ligated into the constitutive plant expression vector pMON530 to form four individual plasmids. They transformed tobacco plants using *Agrobacterium* containing each of these four plasmids and screened leaf extracts from regenerated transformants for the presence of mAb heavy or light chain by ELISA and Western blot analyses. Transformants expressing individual mAb chains were then sexually crossed to produce progeny expressing both chains. From the results of ELISA and Western blot analyses, heavy and light chains with leader sequences in these plants were successfully assembled into heavy-light complexes. This result leads to the approach of a high level of production of mAb in the field. Furthermore, this result gives us the dream of mAb-mediated resistance of transgenic plants. "Coat protein-mediated resistance" in transgenic plants was first reported by Beachy and co-workers.[92] There are now many reports on the coat protein-mediated resistance and useful information on the coat protein-mediated resistance is given by Beachy[91] and Beachy et al.[92]

IV. CONCLUSIONS

Over the past few years a considerable number of studies have been made on using mAbs for plant virus diagnosis. However, use of mAbs is limited to independent countries where mAbs were produced. That is to say, there is little international exchange over the world. mAb offers several advantages over conventional pAb, such as qualitative uniformity and quantative semi-permanent. The exchanges will lead us to the clearance of relationships and classification of plant viruses. Studies on the antigenic structure of plant viruses are limited to that of TMV by Van Regenmortel and co-workers[63,64] because of difficulties of plant virus by X-ray analyses. To date, techniques of molecular biology and synthetic polypeptide have progressed rapidly. And now epitopes of plant virus coat proteins (primary structure of coat proteins) can be analyzed using those techniques. Only epitopes of BNYVV[22,23] and PVY[87] coat proteins have been studied, but epitopes of most plant virus coat proteins will be ascertained in the near future. Analyses of nonstructure proteins of plant viruses using mAbs were performed by Slade et al.[54] These analyses have opened the gate for new biochemical approaches using mAbs. Until today, a virus-specific cellular protein is not known. If mAbs to total cellular proteins are produced, new virus-specific cellular proteins will be obvious. Hiatt et al.[90] produced a mAb in plants. The result is that virus-resistant plants will be produced in the near future. The point I wish to emphasize is that there are many directions of using mAbs.

ACKNOWLEDGMENTS

Our studies described here were done under the projects of Dr. Eishiro Shikata at the Department of Botany, Faculty of Agriculture, Hokkaido University, Sapporo 060, Japan. The author would like to thank Dr. Eishiro Shikata, Dr. Ikuo Kimura, Dr. Koukichi Kikuchi, Dr. Nobuo Hashimoto, Dr. Ichiro Uyeda, and Dr. Teruo Sano for much kind advice. The author also thanks Dr. Akihiro Matuura, Dr. Takashi Hagita, Dr. Jumanto Harjosudarmo, Mr. Atsushi Sato, Mr. Tatsuji Hataya, Mr. Naoki Yamada, Miss Alice Kazuko Inoue, Mr. Takaaki Nakaya, and many co-workers of the Department of Botany, Faculty of Agriculture, Hokkaido University.

REFERENCES

1. Köhler, G. and Milstein, C., Continuous cultures of fused cells secreting antibody of predefined specificity, *Nature,* 256, 495, 1975.
2. Halk, E. L. and De Boer, S. H., Monoclonal antibodies in plant-disease research, in *Annual Review of Phytopathology,* Vol. 23, Horn, P. and Kline, A., Eds., Annual Reviews, Palo Alto, CA, 1985, 321.
3. Dietzgen, R. G. and Sander, E., Monoclonal antibodies against a plant virus, *Arch. Virol.,* 74, 197, 1982.
4. Briand, J. P., Al Moudallal, Z., and Van Regenmortel, M. H. V., Serological differentiation of tobamoviruses by means of monoclonal antibodies, *J. Virol. Methods,* 5, 293, 1982.
5. Gugerli, P. and Fries, P., Characterization of monoclonal antibodies to potato virus Y and their use for virus detection, *J. Gen. Virol.,* 64, 2471, 1983.
6. Hajimorad, M. R., Dietzgen, R. G., and Francki, R. I. B., Differentiation and antigenic characterization of closely related alfalfa mosaic virus strains with monoclonal antibodies, *J. Gen. Virol.,* 71, 2809, 1990.

7. Halk, E. L., Hsu, H. T., Aebig, J., and Franke, J., Production of monoclonal antibodies against three ilarviruses and alfalfa mosaic virus and their use in serotyping, *Phytopathology,* 74, 367, 1984.
8. Järvekülg, L., Sôber, J., Sinijärv, R., Toots, I., and Saarma, M., Time-resolved fluoroimmunoassay of potato virus M with monoclonal antibodies, *Ann. Appl. Biol.,* 114, 279, 1989.
9. Hsu, H. T. and Lawson, R. H., Comparison of mouse monoclonal antibodies and polyclonal antibodies of chicken egg yolk and rabbit for assay of carnation etched ring virus, *Phytopathology,* 75, 778, 1985.
10. Poul, F. and Dunez, J., Use of monoclonal antibodies for the identification of different antigenic domains in apple chlorotic leaf spot virus, *Arch. Virol.,* 114, 191, 1990.
11. Permar, T. A., Garnsey, S. M., Gumpf, D. J., and Lee, R. F., A monoclonal antibody that discriminates strains of citrus tristeza virus, *Phytopathology,* 80, 224, 1990.
12. Vela, C., Cambra, M., Cortés, E., Moreno, P., Miguet, J. G., Pérez De San Román, C., and Sanz, A., Production and characterization of monoclonal antibodies specific for citrus tristeza virus and their use for diagnosis, *J. Gen. Virol.,* 67, 91, 1986.
13. Hu, J. S., Gonsalves, D., Boscia, D., and Namba, S., Use of monoclonal antibodies to characterize grapevine leafroll associated closteroviruses, *Phytopathology,* 80, 920, 1990.
14. Joisson, C. and Van Regenmortel, M. H. V., Influence of the C terminus of the small protein subunit of bean pod mottle virus on the antigenicity of the virus determined using monoclonal antibodies and anti-peptide antiserum, *J. Gen. Virol.,* 72, 2225, 1991.
15. Kalmar, G. B. and Eastwell, K. C., Reaction of coat proteins of two comoviruses in different aggregation states with monoclonal antibodies, *J. Gen. Virol.,* 70, 3451, 1989.
16. Maeda, T., Sako, N., and Inouye, N., Production of monoclonal antibodies to cucumber mosaic virus and their use in enzyme-linked immunosorbent assay, *Ann. Phytopathol. Soc. Jpn.,* 54, 600, 1988.
17. Porta, C., Devergne, J. C., Cardin, L., Briand, J. P., and Van Regenmortel, M. H. V., Serotype specificity of monoclonal antibodies to cucumber mosaic virus, *Arch. Virol.,* 104, 271, 1989.
18. Hiruki, C., Figueiredo, G., Inoue, M., and Furuya, Y., Production and characterization of monoclonal antibodies specific to sweet clover necrotic mosaic virus, *J. Virol. Methods,* 8, 301, 1984.
19. Thomas, J. E., Massalski, P. R., and Harrison, B. D., Production of monoclonal antibodies to African cassava mosaic virus and differences in their reactivities with other whitefly-transmitted geminivirus, *J. Gen. Virol.,* 67, 2739, 1986.
20. Dekker, E. L., Pinner, M. S., Markham, P. G., and Van Regenmortel, M. H. V., Characterization of maize streak virus isolates from different plant species by polyclonal and monoclonal antibodies, *J. Gen. Virol.,* 69, 983, 1988.
21. Commandeur, U., Koenig, R., Lesemann, D.-E., Torrance, L., Burgermeister, W., Liu, Y., Schots, A., Alric, M., and Grassi, G., Epitope mapping on fragments of beet necrotic yellow vein virus coat protein, *J. Gen. Virol.,* 73, 695, 1992.
22. Koenig, R., Commandeur, U., Lesemann, D.-E., Burgermeister, W., Torrance, L., Grassi, G., Alric, M., Kallerhoff, J., and Schots, A., Antigenic analysis of the coat protein of beet necrotic yellow vein virus by means of monoclonal antibodies, *J. Gen. Virol.,* 71, 2229, 1990.
23. Lesemann, D.-E., Koenig, R., Torrance, L., Buxton, G., Boonekamp, P. M., Peters, D., and Schots, A., Electron microscopical demonstration of different binding sites for monoclonal antibodies on particles of beet necrotic yellow vein virus, *J. Gen. Virol.,* 71, 731, 1990.
24. D'Arcy, C. J., Murphy, J. F., and Miklasz, S. D., Murine monoclonal antibodies produced against two Illinois strains of barley yellow dwarf virus: production and use for virus detection, *Phytopathology,* 80, 377, 1990.
25. Diaco, R., Lister, R. M., Hill, J. H., and Durand, D. P., Demonstration of serological relationships among isolates of barley yellow dwarf virus by using polyclonal and monoclonal antibodies, *J. Gen. Virol.,* 67, 353, 1986.
26. Diaco, R., Lister, R. M., Hill, J. H., and Durand, D. P., Detection of homologous and heterologous barley yellow dwarf virus isolates with monoclonal antibodies in serologically specific electron microscopy, *Phytopathology,* 76, 225, 1986.
27. Hsu, H. T., Aebig, J., and Rochow, W. F., Differences among monoclonal antibodies to barley yellow dwarf viruses, *Phytopathology,* 74, 600, 1984.
28. Ohshima, K., Uyeda, I., and Shikata, E., Serological relations of luteoviruses using monoclonal antibodies, *5th Int. Congr. Plant Pathol.,* Kyoto, Japan, 1988, 65.
29. Torrance, L., Pead, M. T., Larkins, A. P., and Butcher, G. W., Characterization of monoclonal antibodies to a U. K. isolate of barley yellow dwarf virus, *J. Gen. Virol.,* 67, 549, 1986.

30. Ohshima, K. and Shikata, E., On the screening procedures of ELISA for monoclonal antibodies against three luteoviruses, *Ann. Phytopathol. Soc. Jpn.*, 56, 219, 1990.
31. Martin, R. R. and Stace-Smith, R., Production and characterization of monoclonal antibodies specific to potato leaf roll virus, *Can. J. Plant Pathol.*, 6, 206, 1984.
32. Massalski, P. R. and Harrison, B. D., Properties of monoclonal antibodies to potato leafroll luteovirus and their use to distinguish virus isolates differing in aphid transmissibility, *J. Gen. Virol.*, 68, 1813, 1987.
33. Ohshima, K., Uyeda, I., and Shikata, E., Production and characteristics of monoclonal antibodies to potato leafroll virus, *J. Fac. Agric. Hokkaido Univ.*, 63, 373, 1988.
34. van den Heuvel, J. F. J. M., de Blank, C. M., Goldbach, R. W., and Peters, D., A characterization of epitopes on potato leafroll virus coat protein, *Arch. Virol.*, 115, 185, 1990.
35. Ohshima, K., Uyeda, I., and Shikata, E., Characterization of monoclonal antibodies against tobacco necrotic dwarf virus, *Ann. Phytopathol. Soc. Jpn.*, 55, 420, 1989.
36. Huss, B., Muller, S., Sommermeyer, G., Walter, B., and Van Regenmortel, M. H. V., Grapevine fanleaf virus monoclonal antibodies: their use to distinguish different isolates, *J. Phytopathol.*, 119, 358, 1987.
37. Harjosudarmo, J., Ohshima, K., Uyeda, I., and Shikata, E., Production of monoclonal antibodies against rice dwarf virus, *J. Fac. Agric. Hokkaido Univ.*, 64, 107, 1990.
38. Kitagawa, Y., Okuhara, E., and Shikata, E., Production of monoclonal antibodies against rice dwarf virus and characterization of antigenic proteins with affinity media, *J. Virol. Methods*, 16, 217, 1987.
39. Harjosudarmo, J., Studies on Monoclonal Antibodies of Rice Viruses, Doctoral thesis, Faculty of Agriculture, Hokkaido University Japan, 1991.
40. Koenig, R. and Torrance, L., Antigenic analysis of potato virus X by means of monoclonal antibodies, *J. Gen. Virol.*, 67, 2145, 1986.
41. Sõber, J., Järvekülg, L., Toots, I., Radavski, J., Villems, R., and Saarma, M., Antigenic characterization of potato virus X with monoclonal antibodies, *J. Gen. Virol.*, 69, 1799, 1988.
42. Torrance, L., Larkins, A. P., and Butcher, G. W., Characterization of monoclonal antibodies against potato virus X and comparison of serotypes with resistance groups, *J. Gen. Virol.*, 67, 57, 1986.
43. Hagita, T., Ohshima, K., Uyeda, I., and Shikata, E., Production of monoclonal antibodies to two strains of bean yellow mosaic virus and their reactions in ELISA, *Ann. Phytopathol. Soc. Jpn.*, 55, 111, 1989 (Abstr. in Japanease).
44. Sato, A., Ohshima, K., Sano, T., Uyeda, I., Nariuchi, H., and Shikata, E., Specificity of monoclonal antibodies to bean yellow mosaic virus strains, *Ann. Phytopathol. Soc. Jpn.*, 53, 75, 1987 (Abstr. in Japanese).
45. Scott, S. W., McLaughlin, M. R., and Ainsworth, A. J., Monoclonal antibodies produced to bean yellow mosaic virus, clover yellow vein virus, and pea mosaic virus which cross-react among the three viruses, *Arch. Virol.*, 108, 161, 1989.
46. Hill, E. K., Hill, J. H., and Durand, D. P., Production of monoclonal antibodies to viruses in the potyvirus group: use in radioimmunoassay, *J. Gen. Virol.*, 65, 525, 1984.
47. Jones, F. E., Hill, J. H., and Durand, D. P., Detection and differentiation of maize dwarf mosaic virus, strains A and B, by use of different class immunoglobulins in a double-antibody sandwich enzyme-linked immunosorbent assay, *Phytopathology*, 78, 1118, 1988.
48. Himmler, G., Brix, U., Steinkellner, H., Laimer, M., Mattanovich, D., and Katinger, H. W. D., Early screening for anti-plum pox virus monoclonal antibodies with different epitope specificities by means of gold-labelled immunosorbent electron microscopy, *J. Virol. Methods*, 22, 351, 1988.
49. Ohshima, K., Inoue, A. K., Ishikawa, Y., Shikata, E., and Hagita, T., Production and application of monoclonal antibodies specific to ordinary strain and necrotic strain of potato virus Y, *Ann. Phytopathol. Soc. Jpn.*, 56, 508, 1990 (in Japanese).
50. Kuwata, S., Imaizumi, S., and Kubo, S., Production of monoclonal antibodies to potato virus Y necrotic strain, *Ann. Phytopathol. Soc. Jpn.*, 52, 554, 1986 (Abstr. in Japanese).
51. Rose, D. G. and Hubbard, A. L., Production of monoclonal antibodies for the detection of potato virus Y, *Ann. Appl. Biol.*, 109, 317, 1986.
52. Diaco, R., Hill, J. H., and Durand, D. P., Purification of soybean mosaic virus by affinity chromatography using monoclonal antibodies, *J. Gen. Virol.*, 67, 345, 1986.

53. Hill, J. H., Benner, H. I., Permar, T. A., Bailey, T. B., Andrews, R. E., Jr., Durand, D. P., and Van Deusen, R. A., Differentiation of soybean mosaic virus isolates by one-dimensional trypsin peptide maps immunoblotted with monoclonal antibodies, *Phytopathology,* 79, 1261, 1989.
54. Slade, D. E., Johnston, R. E., and Dougherty, W. G., Generation and characterization of monoclonal antibodies reactive with the 49-kDa proteinase of tobacco etch virus, *Virology,* 173, 499, 1989.
55. Hsu, H. T., Franssen, J. M., van der Hulst, C. T. C., Derks, A. F. L. M., and Lawson, R. H., Factors affecting selection of epitope specificity of monoclonal antibodies to tulip breaking potyvirus, *Phytopathology,* 78, 1337, 1988.
56. Kantrong, S. and Sako, N., Characterization of epitopes recognized by monoclonal antibodies to aphid transmissible and non-transmissible strains of turnip mosaic virus, *Arch. Virol.*, 133, 11, 1993.
57. Somowiyarjo, S., Sako, N., and Nonaka, F., Production and characterization of monoclonal antibodies to watermelon mosaic virus 2, *Ann. Phytopathol. Soc. Jpn.,* 56, 541, 1990.
58. Somowiyarjo, S., Sako, N., and Nonaka, F., The use of monoclonal antibody for detecting zucchini yellow mosaic virus, *Ann. Phytopathol. Soc. Jpn.,* 54, 436, 1988.
59. Dietzgen, R. G. and Francki, R. I. B., Reducing agents interfere with the detection of lettuce necrotic yellows virus in infected plants by immunoblotting with monoclonal antibodies, *J. Virol. Methods,* 28, 199, 1990.
60. MacKenzie, D. J. and Tremaine, J. H., The use of a monoclonal antibody specific for the N-terminal region of southern bean mosaic virus as a probe of virus structure, *J. Gen. Virol.,* 67, 727, 1986.
61. Tremaine, J. H., MacKenzie, D. J., and Ronald, W. P., Monoclonal antibodies as structural probes of southern bean mosaic virus, *Virology,* 114, 80, 1985.
62. Al Moudallal, Z., Briand, J. P., and Van Regenmortel, M. H. V., Monoclonal antibodies as probes of the antigenic structure of tobacco mosaic virus, *EMBO J.,* 8, 1005, 1982.
63. Al Moudallal, Z., Briand, J. P., and Van Regenmortel, M. H. V., A major part of the polypeptide chain of tobacco mosaic virus protein is antigenic, *EMBO J.,* 4, 1231, 1985.
64. Altschuh, D., Al Moudallal, Z., Briand, J. P., and Van Regenmortel, M. H. V., Immunochemical studies of tobacco mosaic virus. VI. Attempts to localize viral epitopes with monoclonal antibodies, *Mol. Immunol.,* 22, 329, 1985.
65. Dore, I., Ruhlmann, C., Oudet, P., Cahoon, M., Caspar, D. L. D., and Van Regenmortel, M. H. V., Porarity of binding of monoclonal antibodies to tobacco mosaic virus rods and stacked disks, *Virology,* 176, 25, 1990.
66. Dore, I., Dekker, E. L., Porta, C., and Van Regenmortel, M. H. V., Detection by ELISA of two tobamoviruses in orchids using monoclonal antibodies, *J. Phytopathol.,* 20, 317, 1987.
67. Bahrani, Z., Sherwood, J. L., Sanborn, M. R., and Keyser, G. C., The use of monoclonal antibodies to detect wheat soil-borne mosaic virus, *J. Gen. Virol.,* 69, 1317, 1988.
68. Omura, T., Takahashi, Y., Shohara, K., Minobe, Y., Tsuchizaki, T., and Nozu, Y., Production of monoclonal antibodies against rice stripe virus for the detection of virus antigen in infected plant and viruliferous insects, *Ann. Phytopathol. Soc. Jpn.,* 52, 270, 1986.
69. de Ávila, A. C., Huguenot, C., Resende, R. de O., Kitajima, E. W., Goldbach, R. W., and Peters, D., Serological differentiation of 20 isolates of tomato spotted wilt virus, *J. Gen. Virol.,* 71, 2801, 1990.
70. Hsu, H. T., Wang, Y. C., Lawson, R. H., Wang, M., and Gonsalves, D., Splenocytes of mice with induced immunological tolerance to plant antigens for construction of hybridomas secreting tomato spotted wilt virus-specific antibodies, *Phytopathology,* 80, 158, 1990.
71. Huguenot, C., van den Dobbelsteen, G., de Haan, P., Wagemakers, C. A. M., Drost, G. A., Osterhaus, A. D. M. E., and Peters, D., Detection of tomato spotted wilt virus using monoclonal antibodies and riboprobes, *Arch. Virol.,* 110, 47, 1990.
72. Sherwood, J. L., Sanborn, M. R., Keyser, G. C., and Myers, L. D., Use of monoclonal antibodies in detection of tomato spotted wilt virus, *Phytopathology,* 79, 61, 1989.
73. Galfré, G. and Milstein, C., Preparation of monoclonal antibodies: strategies and procedures, Vol. 73, in *Methods in Enzymology,* Langone, J. J. and van Vunakis, H., Eds., Cambridge University Press, Cambridge, 1981, 3.
74. Campbell, A. M., Monoclonal antibody technology, in *Laboratory techniques in Biochemistry and Molecular Biology,* Vol. 13, Burdon, R. H. and van Knippenberg, P. H., Eds., Elsevier, Amsterdam, 1984.
75. Sander, E. and Dietzgen, R. G., Monoclonal antibodies against plant viruses, in *Advances in Virus Research,* Vol. 29, Maramorosch, K., Murphy, F. A., and Shatkin, A. J., Eds., Academic Press, Orlando, FL, 1984, 131.

76. Torrance, L., Serological methods to detect plant viruses: production and use of monoclonal antibodies, in *Techniques for the Rapid Detection of Plant Pathogens,* Duncan, J. M. and Torrance, L., Eds., Blackwell Scientific, Oxford, 1992, 7.
77. Spitz, M., Spitz, L., Thorpe, R., and Eugui, E., Intrasplenic primary immunization for the production of monoclonal antibodies, *J. Immunol. Methods,* 70, 39, 1984.
78. Reading, C. L., *In vitro* immunization for the production of antigen-specific lymphocyte hybridomas, in *Methods in Enzymology,* Vol. 121, Langone, J. J. and van Vunakis, H., Eds., Academic Press, London, 1986, 18.
79. Ohshima, K., Harjosudarmo, J., Ishikawa, Y., and Shikata, E., Relationship between hybridoma screening procedures and the characteristics of monoclonal antibodies for use in direct double antibody sandwich ELISA for the detection of plant viruses, *Ann. Phytopathol. Soc. Jpn.,* 56, 569, 1990.
80. Inoue, A. K., Studies on Potato Virus Y Strains in Hokkaido—Biological Properties, Serology and Molecular Characterization, Master's thesis, Faculty of Agriculture, Hokkaido University, Japan, 1992.
81. Jordan, R. and Hammond, J., Comparison and differentiation of potyvirus isolates and identification of strain-, virus-, subgroup-specific and potyvirus group-common epitopes using monoclonal antibodies, *J. Gen. Virol.,* 72, 25, 1991.
82. Shukla, D. D., Strike, P. M., Tracy, S. L., Gough, K. H., and Ward, C. W., The N and C termini of the coat proteins of potyviruses are surface-located and the N terminus contains the major virus-specific epitopes, *J. Gen. Virol.,* 69, 1497, 1988.
83. Ohshima, K., Nakaya, T., Inoue, A. K., Hataya, T., Hayashi, Y., and Shikata, E., Production and characteristics of strain common antibodies against a synthetic polypeptide corresponding to C-terminal region of potato virus Y, *J. Virol. Methods,* 40, 265, 1992.
84. Ohshima, K., Hataya, T., Sano, T., Inoue, A. K., and Shikata, E., Comparison of biological properties, serological characteristics and amino acid sequences of coat protein between potato virus Y ordinary strain and necrotic strain, *Ann. Phytopathol. Soc. Jpn.,* 57, 615, 1991 (in Japanese).
85. Benjamini, E., Immunochemistry of the tobacco mosaic virus protein, in *Immunochemistry of Proteins,* Vol. 2, Atassi, M. Z., Ed., Plenum Press, New York, 1977, 265.
86. Van Regenmortel, M. H. V., *Serology and Immunochemistry of Plant Viruses,* Academic Press, New York, 1982.
87. Hataya, T., Inoue, A. K., Ohshima, K., and Shikata, E., Amino acid sequence studies of the epitopes of potato virus Y ordinary and necrotic strains, *Ann. Phytopathol. Soc. Jpn.,* 57, 459, 1991 (Abstr. in Japanese).
88. Hu, J. S. and Rochow, W. F., Anti-idiotypic antibodies against an anti-barley yellow dwarf virus monoclonal antibody, *Phytopathology,* 78, 1302, 1988.
89. Mernaugh, R. L., Hill, J. H., Durand, D. P., Van Deusen, R. A., and Harris, K. F., Properties and potential applications of syngeneic monoclonal anti-idiotypic antibodies: use as a reference standard in enzyme-linked immunosorbent assay, *Phytopathology,* 81, 274, 1991.
90. Hiatt, A., Cafferkey, R., and Bowdish, K., Production of antibodies in transgenic plants, *Nature,* 342, 76, 1989.
91. Beachy, R. N., Coat protein mediated resistance in transgenic plants, in *Viral Genes and Plant Pathogenesis,* Pirone, T. P. and Shaw, J. G., Eds., Springer-Verlag, New York, 1990, 13.
92. Beachy, R. N., Loesch-Fries, S., and Tumer, N. E., Coat protein-mediated resistance against virus infection, in *Annual Review of Phytopathology,* Vol. 28, Gould, H. E., Ed., Annual Reviews, Palo Alto, CA, 1990, 451.

FURTHER READING

Hampton, R. O., Ball, E. M., and De Boer, S. H., *Serological Methods for Detection and Identification of Viral and Bacterial Plant Pathogens, a Laboratory Manual,* American Phytopathological Society, St. Paul, MN, 1990.

Honjo, T., Alt, F. W., and Rabbitts, T. H., *Immunoglobulin Genes,* Academic Press, San Diego, 1989.

Matthews, R. E. F., *Fundamentals of Plant Virology,* Academic Press, San Diego, 1992.

Sambrook, J., Fritsch, E. F., and Maniatis, T., *Molecular Cloning, a Laboratory Manual,* 2nd ed., Nolan, C., Ed. Cold Spring Harbor Laboratory, Cold Spring Harbor, NY, 1989.

Tijssen, P., Practice and theory of enzyme immunoassays, in *Laboratory Techniques in Biochemistry and Molecular Biology,* Burdon, R. H. and van Knippenberg, P. H., Eds., Elsevier Science Publishers, Amsterdam, 1985.

Van Regenmortel, M. H. V., Briand, J. P., Muller, S., and Plauè, S., Synthetic polypeptides as antigens, in *Laboratory Techniques in Biochemistry and Molecular Biology,* Burdon, R. H. and van Knippenberg, P. H., Eds., Elsevier Science Publishers, Amsterdam, 1988.
Wang, T. L., Immunology in plant science, in *Society for Experimental Biology,* Vol. 29, Cambridge University Press, Cambridge, 1985.
Watson, J. D., Hopkins, N. H., Roberts, J. W., Steitz, J. A., and Weiner, A. M., *Molecular Biology of the Gene,* 4th ed., The Benjamin/Cummings Publishing Company, USA, 1987.

Chapter 30

Monoclonal Antibodies: Nematodes

John M. S. Forrest

TABLE OF CONTENTS

I. INTRODUCTION

A. SEROLOGY AND THE ADVENT OF MONOCLONAL ANTIBODIES TO PLANT-PARASITIC NEMATODES

Application of serological methods as a means of solving problems in plant nematology developed more slowly than in some disciplines. Bird[1] was the first to demonstrate that secretions from *Meloidogyne javanica* were antigenic. Rabbits were immunized with live juveniles (J2) and immunoprecipitation was obtained around the excretory pores and oral apertures of J2 and adult *M. javanica.* The gelatinous matrix protecting the eggs and, to a lesser extent, the cuticle were also antigenic. His results suggested a closer physiological relationship between secretion, excretion, and the cuticle than had been previously suspected.

Much of the early effort was directed towards the separation of pathotypes or races of nematodes which are morphologically indistinct. Mabbott,[2,3] for example, used a polyclonal antibody raised in rabbits against adult females of potato cyst nematode for gel diffusion tests which demonstrated fixed recognizable patterns for populations of pathotype A (Boghall) and B (Duddingston) of *Heterodera rostochiensis* Woll = Dutch D. The results suggested that these pathotypes were actually separate species, so the work was abandoned as being unreliable. Stone,[4] 8 years later, finally separated *H. rostochiensis* into *Globodera rostochiensis* and *G. pallida* on morphological criteria.

Webster and Hooper[5] found that serological tests could differentiate species of *Ditylenchus* and *Heterodera* into broad groups. Antisera to extracts of these nematodes did not precipitate against extracts of host plants or nematodes from other genera. Nonimmune serum did not react with the nematodes.

These and other studies suggested that polyclonal antibodies were too cross reactive to be used reliably for differentiating between nematode species. However, antibodies raised by Davies and Lander[6] against soluble proteins from J2 of *M. incognita* race 2 were recently shown to recognize the cuticle surface of *M. incognita* races 1, 2, 3, and 4 and *M. javanica,* but not *M. arenaria* races 1 and 2. The antibodies bind to an 80-kDa protein and a number of 43-kDa proteins which are not recognized in *M. arenaria.* It remains to be seen whether or not these interesting results will lead to useful discrimination by means of polyclonal antibodies (see Section II.D).

0-87371-877-1/95/$0.00+$.50

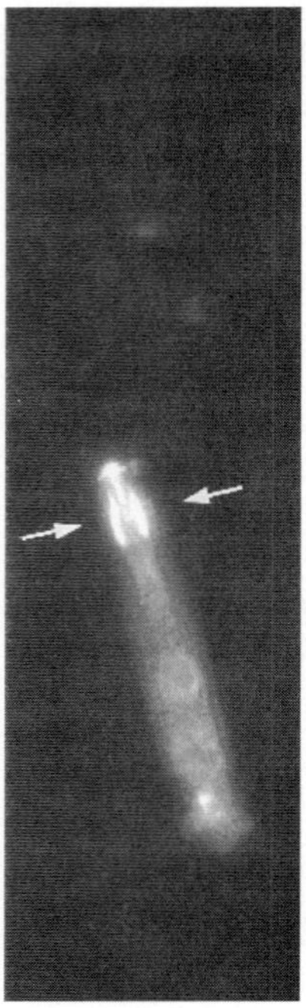

Figure 1 Binding of a mAb to amphidial pouches of J2 of *Heterodera glycines.* (After Atkinson et al.)

Nearly a quarter of a century after Bird's pioneering work, Atkinson et al.[7] first introduced monoclonal antibodies (mAbs) in studies of plant-parasitic nematodes. They did not seek to identify species, but to isolate salivary secretions of the soybean cyst nematode *Heterodera glycines,* which are involved in the modification of plant cells to induce the syncytia, without which the growth of females would be suboptimal. mAbs were produced not only to the pharyngeal glands, but also with specificity to many other tissues and organs (Figure 1).

Shortly afterwards, Jones *et al.*[8] raised two mAbs against *M. incognita* which distinguished it from *M. javanica, M. arenaria,* and other genera. Their aim was to apply these mAbs to identify *M. incognita* in infected plant tissue or soil, drastically reducing the time taken.

In the Netherlands, mAbs were raised to highly conserved, heat-stable proteins which were found to be representative of many populations of *G. rostochiensis* and *G. pallida* from Western Europe.[9,10] Each mAb had a preferred affinity for the species against which it was raised. An enzyme-linked immunosorbent assay (ELISA)[11] based on these mAbs is now marketed and can be used to determine the proportions of the two species in mixed populations. More recently in the U.K., two mAbs have been produced which may differentiate between these species even more clearly.[12]

The isolation of specific mAbs has also led to the creation of "plantibodies", Fv portions of mouse immunoglobulins (Ig) expressed *in planta,* capable of interfering with the action of nematode saliva. A major challenge will be to achieve correct folding of the proteins and delivery to the appropriate intracellular location.[13] mAbs to salivary components of nematodes are available in at least two other laboratories[7,14] (Figures 2 and 3). Thus, after a late beginning, there are several groups within the discipline of plant-parasitic nematology who are working close to the forefront in mAb technology and use. A summary of current applications is presented in Table 1.

II. FAVORED PROTOCOLS FOR THE PRODUCTION OF MONOCLONAL ANTIBODIES TO PLANT-PARASITIC NEMATODES

A. EXPERIMENTAL ANIMALS AND MYELOMAS

Female mice of the Balb/c strain have been the preferred subjects for immunization, since the readily available myeloma lines are also derived from Balb/c mice.[15] The resulting hybridomas can be multiplied by ascites tumors in this strain (see Chapter 25). However, as ascites production is now seriously discouraged in many countries, other commonly available mouse strains can serve as fusion partners for Balb/c myeloma lines. These strains may produce superior responses to certain antigens.[16] In our laboratory, New Zealand Black × Balb/c have been used almost exclusively.[17] The ability to produce a strong response to many antigens has been inherited from the New Zealand Black parent. The crosses are larger and more robust than the Balb/c and more difficult to handle, but their extra vigor is an asset in recovering from operations. Female mice are easier to handle because they are more docile than males. The significance of the age of 6 weeks is that many mice will still be free of natural infections. This is important to the characteristics of serum from nonimmune or immunized mice.

Different strains of mice also differ in their nonimmune response to nematode antigens. Hussey[14] found that nonimmune serum from Balb/c mice reacted strongly with secretory granules from the subventral esophageal glands of three species of *Meloidogyne.* Serum from C57 or C3H mice did not. We have encountered faint binding of nonimmune mouse and rabbit (New Zealand White) serum to the amphidial exudate of J2 of potato cyst nematodes.[17] This may be due either to nonspecific binding or specific binding to epitopes which are shared with animal pathogens.

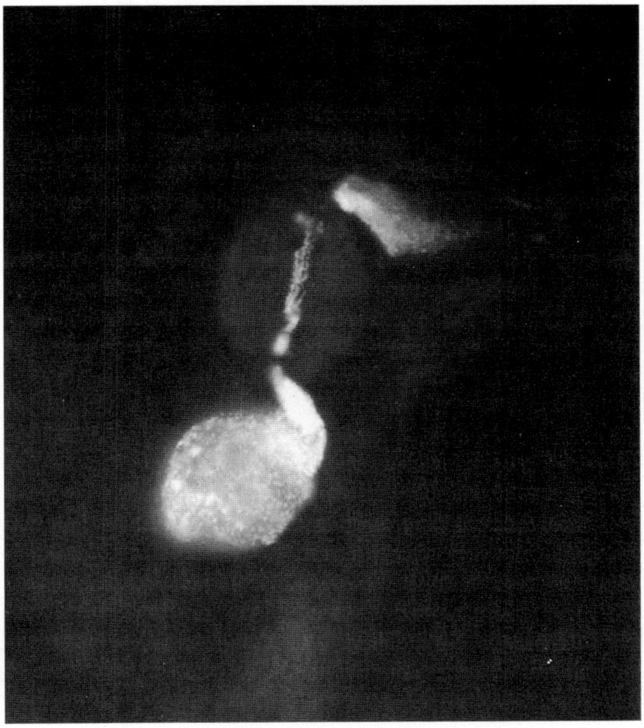

Figure 2 Binding of a mAb to secretory granules within the dorsal esophageal gland in a female of *Meloidogyne incognita.* (After Davis et al.)[24]

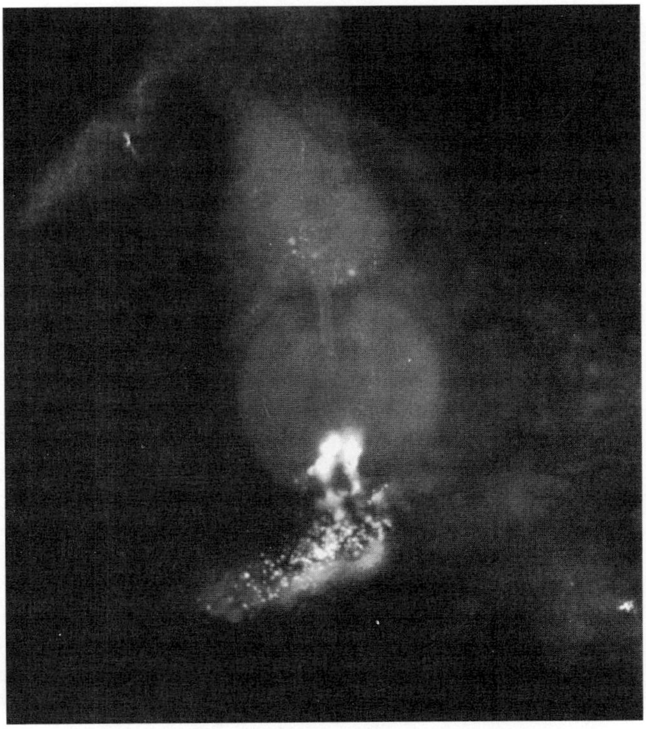

Figure 3 Binding of a mAb to secretory granules within the subventral gland extension and ampullae in a female of *M. incognita.* (After Davis et al.)[24]

Table 1 **Recent applications of monoclonal antibodies for research into plant-parasitic nematodes**

Application	Target	Nematode	Ref.
Identification, purification	Salivary secretions	*H. glycines*	7
		M. incognita	14,24
	Surface antigens	*D. dipsaci*	18
		G. pallida	58
	Amphidial exudate	*M. incognita*	52
Identification, quantification	Nematode proteins	*M. incognita*	8
		M. javanica	6
		M. arenaria	
		G. rostochiensis	11
		G. pallida	12
	RNA, esophageal bulb	*X. index*	59
Resistance screening	Nematode proteins	*M. incognita*	60
		G. pallida	61
Hyperparasitic adhesion	Surface proteins	*M. incognita*	51
		M. arenaria	
Synthetic resistance genes	Salivary secretions	*G. pallida*	13

Only two mouse myeloma lines have been used by plant nematologists, SP2/0-Ag14[6,10,14] and P3 × 63 Ag8.653.[7,8,18] Use of either line can be recommended as they do not secrete functional antibodies.[15]

The second report of raising specific mAbs to plant-parasitic nematodes[8] involved the use of Lou and 1AP rats because of certain advantages. Of growing fusions, 90% express spleen Igs as opposed to only 60% for mice. Reversion of parent lines to nonsecreting forms is lower than in the mouse cell lines available, under 10^{-4} cell/generation as opposed to 10^{-3} cell/generation. This is a valuable trait if retained in the fusion progeny. The fusion partners were rat myeloma cell lines IR 983F (nonsecreting) or Y3 Ag1.2.3. (secreting κ chain).[15]

B. PREPARATION AND PRESENTATION OF ANTIGENS

Antibody production is thought to be more successful with particulate than with soluble antigens.[20] Nematode antigens have been presented as whole nematodes, homogenates, or purified proteins in phosphate-buffered saline (PBS). Whole, unfixed nematodes have been injected into both mice[21] and rabbits.[1,21] Schots et al.[10] conjugated proteins to the carrier protein keyhole limpet hemocyanin (KLH). The proteins were spotted directly onto nitrocellulose or separated by sodium dodecyl sulfate-polyacrylamide gel electrophoresis (SDS-PAGE) and then blotted onto nitrocellulose; after converting to a slurry by adding 100 to 200 μl PBS, freezing in liquid nitrogen, and grinding, it was used to immunize mice intrasplenically or intraperitoneally. Antigens immobilized on small pieces of nitrocellulose can also be implanted[22] or shaped into pellets with small quantities of dimethyl sulfoxide prior to implantation.[23] Homogenates have been made in PBS with or without detergents and presented as crude homogenates or soluble protein extracts.[6–8,14,18] Jones et al.[8] found that extracts using detergent produced antibodies which displayed less cross reactivity.

Attempts have been made to enrich the antigen of interest prior to homogenization. Females of *H. glycines*[7] and *M. incognita*[24] were divided into anterior and posterior regions. Anterior portions were either freeze dried or fixed, ground, and centrifuged to remove coarse particles. Secretions from stylets were collected aseptically and deposited on nitrocellulose for intrasplenic implantation. Davis et al.[24] obtained mAbs to the secretory granules of *M. incognita* by using homogenates of females as well as purified secretory granules. The justification of the extra effort involved in extraction and purification of secretory granules is questioned.

The administration of an adjuvant should generate a high titer antibody response. Almost all the protocols published stipulate complete Freund's adjuvant (CFA) administered intraperitoneally in mice or intramuscularly in rats and rabbits, with subsequent booster injections with incomplete Freund's adjuvant (IFA). Booster injections given intravenously do not contain CFA and IFA. This schedule produces a high level of specific antibodies against the administered antigen as well as increasing the level of unrelated antibodies in the circulation, but it need not be the automatic choice. FA is difficult to emulsify and dispense because of its viscosity and incompatibility with polystyrene syringes. It produces a chronic inflammatory reaction at injection sites and can cause arthritis if accidentally injected into the operator. CFA has never been used in our laboratory and even IFA has now been replaced by the saponin Quil A (Superfos Biosector a/s Vedbaek, Denmark). At a recommended dose of 10 μg per injection for mice and 25 μg for rabbits, Quil A produced a measurable response after a single immunization, but IFA did not.[17] Schots et al.[10] used an adjuvant for humans, aluminum hydroxide. Their antigen conjugated to KLH was precipitated on the aluminum hydroxide, and a suspension of 50 to 100 μg of protein was injected intraperitoneally.

Kenney et al.[25] considered a number of criteria for assessing antibodies to chosen antigens, including titer, affinity, concentration, isotype, epitope specificity, and neutralizing activity of sera and culture supernatant fluids (CSF). The choice of adjuvant had a considerable effect on the end result. While FA was effective in inducing a high antibody titer, Quil A and aluminum hydroxide/threonine muramyldipeptide elicited the highest-affinity antibodies. The other adjuvants were superior to FA at producing antibodies to native rather than denatured antigen (human serum albumin). No adjuvant is necessary when antigen is administered as a deposit on nitrocellulose.[22]

C. ROUTES OF IMMUNIZATION

The majority of published protocols use intraperitoneal injections, which require no anesthetic, are easy to administer, and can accommodate a volume of 0.5 ml per injection. The final booster injection is often intravenous, via the tail vein in mice (≤0.2 ml), because it is reported to encourage the production of IgG. Considerable skill is required to locate the tail vein and administer the injection. Solids should not be injected intravenously, as they may cause embolisms. Footpad and subcutaneous injections in the inguinal and axial regions have been used successfully for immunization,[8] but are not recommended because they are thought to be too painful.

Intrasplenic immunization is the route of choice when only minute amounts of antigen are available. The technique is therefore ideal for raising antibodies to nematode products such as esophageal gland secretions which may only be available in nanogram quantities. Surgery must be carried out under anesthetic using either an inhalant such as isoflurane (WDC-10019-773-40, Anaquest, Madison, WI)[24] or an injection. We have used an anesthetic mixture of fentanyl citrate and fluanisone (Hypnovel, Roche Products Ltd., Welwyn Garden City, U.K.) and midazolam hydrochloride (Hypnorm, Janssen Pharmaceuticals Ltd., Oxford, U.K.)[23] Sterile distilled water, Hypnorm, and Hypnovel were mixed in the ratio 2:1:1 immediately prior to use to avoid precipitation. Administration is by a single intraperitoneal injection of 0.25 ml. For smaller mice (i.e., Balb/c of 6 weeks old or less) the dose can be reduced to 0.2 ml. The spleen is easily located midway down the left-hand side of the mouse, being visible through the pale skin of strains such as the Balb/c. After exposure, a volume of up to 100 μl of antigen solution/suspension can be introduced per injection. The spleen is then replaced and one or two small sutures are made in the wall of the peritoneum to prevent it adhering to the skin, which is then sealed with a further five to six stitches. We have found that silk sutures are immunogenic (W500 Mersilk, Ethicon Ltd., Edinburgh, U.K.). If the aim is to produce a monospecific polyclonal antibody in the mouse, then a non-immunogenic suture should be used (e.g., PDS II Polydioxanone W9777, Ethicon Ltd.).

In our laboratory, intrasplenic immunizations have been carried out with a mortality of 5%. The chief danger after the operation is hypothermia, and mice should be placed under an ordinary desk-top lamp to supply radiant heat until they have revived. There are clear danger signs in Balb/c mice as the color drains from the eyes. Healing of the wounds is always rapid and complete, and infections have never arisen as a result of these operations. With such a good recovery rate it is easy to notice the rare occasions when a toxic antigen is administered.[17]

D. MANIPULATION OF THE IMMUNE RESPONSE

The immunosuppressive drug cyclophosphamide (Sigma Ltd. Poole, U.K.) has been used to favor proliferation of lymphocytes which are specific for antigens in hatched J2 that were absent in unhatched J2,[7] and antigens present in anterior rather than posterior portions of females of *H. glycines*[7] and *M. incognita.*[24] Cyclophosphamide was injected intraperitoneally, at 50 μg/g mouse, 3 d after injection of "unwanted" antigens. This produced a significant reduction in mAbs with specificity for the subventral pharyngeal glands, but increased those against cell membranes of *H. glycines.* Davis et al.[24] also found that mAbs specific for the esophageal glands of *M. incognita* were eliminated by this procedure.

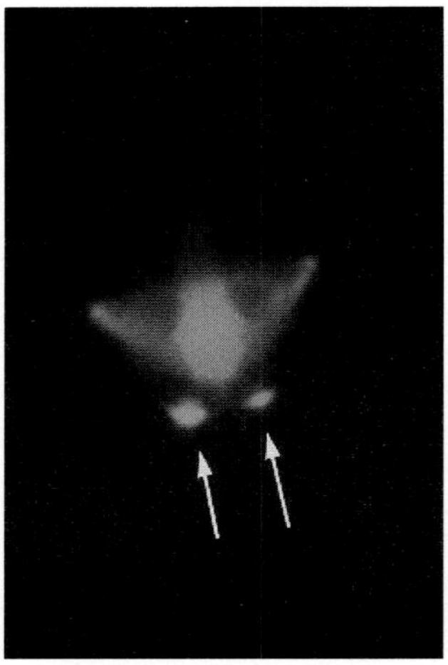

Figure 4 Binding of polyclonal antibodies of "non-IgG" fraction to amphidial exudate of live J2 of *Globodera rostochiensis.* (After Forrest et al.)[17]

Other immunization schedules designed to suppress idiotypes to shared epitopes reduced the overall polyclonal response to the surface of J2 of potato cyst nematodes, but did not enable discrimination between the species.[21] Part of the "non-IgG" fraction bound only to the amphidial exudate (Figure 4), and it may be that species- or organ-specific mAbs can be selected from within this fraction.

E. CELL FUSION AND HYBRIDOMA CULTURE

Most workers have used successfully protocols based on "long" (2.5-min) exposures of the cells to polyethylene glycol (PEG 1500 to 4000).[26,27] Efficiency of hybridoma production is relatively low, with an expectation of up to 1000 hybrid colonies from a single fusion of which 50 may be stable antibody producers. To obtain a worthwhile response to the antigen of interest, the mouse must be well immunized, typically showing a serum dilution titer of 1×10^5 measured by ELISA. Antibodies against antigens which are only weak immunogens are much more difficult to obtain under these conditions. Lane,[28] however, demonstrated that by reducing exposure to PEG from the standard 2.5 min to 30 to 60 s, the number of hybridomas of sp2/0 could be increased fivefold. The immunization schedule was short, with a second and final injection of antigen 14 d after the first and 3 d prior to the fusion (see also Table 2).

Dulbecco's modified Eagle's medium (DMEM) has been used almost exclusively for fusion, subsequent selection, and culture of hybridomas with only one citation for RPMI 1640.[19] We have also used α MEM, a medium recommended for *in vitro* immunization.[17,29] Both azaserine (HAZA) and aminopterin (HAT) have been used to select hybridomas from unfused myelomas. The use of HAZA is preferable to HAT, as aminopterin is toxic and slowly metabolized, and cells require weaning at the end of the process.[16]

To aid hybridoma growth, the fusion mixture is often supplemented with mouse peritoneal macrophages. This increases the risk of contamination at an advanced stage of the proceedings, and sterile,

Table 2 **Intraperitoneal and intrasplenic immunization schedules for monoclonal antibody production to plant-parasitic nematodes**

Immunization Route			
Intraperitoneal		Intrasplenic	
Day	Preparation*	Day	Preparation**
1	200 μl protein homogenate + FCA (200 μg)	1	100 μl protein homogenate + 0.1% Triton® X-100
22	As above + IFA	12	As above
42	As above: boost, via tail vein; no adjuvant	15	Carry out fusion
45	Carry out fusion		

*After Palmer et al.[18]; **After Davis et al.[24]

standardized, conditioned media can be used instead. These are available commercially from a number of suppliers; conditioned media are obtained when certain cell lines secrete growth factors into solution. After removal of cells the medium acts as an effective growth supplement. A third type of supplement is thymocyte conditioned medium prepared from the thymus glands of mice or rats.[30]

The majority of workers select and then expand their hybridomas before cloning by limiting dilution. Conversely, hybridomas have been selected and cloned simultaneously by the method of Davis et al.[24,31] The cells were plated at low density on petri dishes in semisolid medium which contains methylcellulose. The elimination of clonal competition and routine cloning achieved by this method may be essential for obtaining really stable hybridomas.[32]

F. SCREENING AND SELECTION FOR ANTIBODY AFFINITY

Purified antibodies or CSF should be screened by the method which will subsequently feature as part of an assay. Both indirect immunofluorescence and ELISA have been used for screening, sometimes in tandem. Where the aim is to isolate the products of minor organs, such as the pharyngeal glands, immunofluorescence is the method of choice. Davis et al.[24] found that ELISA was unsuitable for this purpose, because it yielded inconclusive results.

The immunofluorescence screen of Atkinson et al.,[7] later modified by Hussey,[14] has been the basis of selection of mAbs. The method is elaborate and involves considerable advance preparation, but it has been used successfully to detect many specific mAbs. J2 are fixed, dried, cut into pieces, and rendered permeable. After blocking nonspecific binding sites with serum they are exposed to antibodies, which are able to penetrate and bind to internal organs as well as to the cuticle surface. Similarly, the anterior portions of female *M. incognita* were rendered permeable and blocked prior to screening. Ten anterior portions were used to screen pooled CSF from six separate wells. Thus, the rescreening of only a few positive batches is necessary to locate the hybridoma of interest.

The technique allowed a throughput of up to 50 mAbs per operator per hour, because nematodes could be cut in large numbers and solutions rapidly exchanged by microfiltration under vacuum.[7,32]

Whole live juveniles and nematode pieces may be processed and screened in the wells of a Terasaki plate to detect antibodies to surface antigens. The plates fit the stage of a fluorescence microscope, and the wells can be scanned with a 10× objective, avoiding the need for mounting nematodes on slides.[17]

ELISA tests are very rapid, so many samples of CSF can be examined. Nematode homogenates and purified proteins were presented as plate-trapped antigens (PTA) at 0.05 to 0.1 μg per well. PTA from the posterior portions of female *M. incognita* were used to demonstrate a 100-fold reduction in mouse serum antibody after three injections of cyclophosphamide.[24] Plates were usually coated overnight or for 12 h at 4°C. Carbonate/hydrogen carbonate at pH 9.6 was the buffer of choice, although it has been shown that greater coating occurs at pH 7.8. At pH 9.6, the IgG molecule may be better presented with the Fc portion on the solid support.[33] A further refinement was the oxidation of oligosaccharide chains of PTA with periodate to gain information on the nature of the epitopes bound.[18] CSF, undiluted or diluted 1:5, was then placed in the wells for 1 h. Secondary antibodies, bound either to horseradish peroxidase or alkaline phosphatase, have been used for detection with *o*-phenylenediamine or *p*-nitrophenyl phosphate, respectively, as substrates. We are currently examining the ultrasensitive detection of a chemiluminescent signal generated by alkaline phosphatase from the substrate AMPPD (NBS Biologicals Ltd., Hatfield, U.K.).

Schots et al.[34] developed an ELISA test using antibodies to distinguish between *G. rostochiensis* and *G. pallida.* They selected the antibodies they required after using a direct ELISA to estimate the "binding constants" and "binding capacities". The former is a measure of intrinsic affinity, while the latter additionally takes into account the number of binding sites on the antigen to give the overall affinity. Their results were calculated from the quantitative direct ELISA using the computer program system LIGAND.[35]

G. BULKING AND PURIFICATION OF MONOCLONAL ANTIBODIES

Bulking of mAbs in ascitic fluid can be used to generate antibody concentrations of 1 to 10 mg/ml and is described in detail elsewhere (See Chapter 25). Such quantities may be in excess of the requirements of many research laboratories, and alternative methods should be considered which do not involve further procedures with animals. (see Section III.A).

Prior to the purification of an antibody it is important to establish the isotype and class, since this will dictate which procedures are subsequently undertaken. These can be determined either by Ouchterlony double-diffusion assays,[10] ELISA with isotype-specific antibodies, or an agglutination test with isotype and class-specific antibodies bound to red blood cells.[17]

Substantial losses of a mAb or its immunoreactivity may occur during purification. Precipitation with sodium or ammonium sulfate, for example, is dependent on pI of the antibody, and is commonly less than 50% efficient;[10,32] hydroxyapatite chromatography also has been replaced by hydrophobic interaction chromatography, which gives a higher yield of pure IgG.[10,32] Hussey et al.[34] used an anti-mouse IgM linked to an agarose affinity column for purification of mAbs specific for salivary secretions.

Caprylic (octanoic) acid precipitation is effective for purifying IgG_1 and IgG_2.[37] Large volumes of CSF can be processed rapidly and more economically than by affinity chromatography. IgG_3 and IgA, however, are irreversibly precipitated. The method has also caused an unacceptable reduction in the affinity of some mAbs, so a small trial separation should first be carried out.

Proteins A and G have been widely used for affinity separation of IgG. Protein G, an IgG-binding bacterial cell wall protein from streptococci, is the more avid binder of the two.[38] Protein A is ineffective for some subclasses of murine IgG and is not recommended for rat IgG.[15]

Desalting, removal of sodium azide, and exchange of buffers can be conveniently carried out with small volumes of antibody solution ($\leq$2.5 ml) on disposable PD10 columns which contain G-25 sephadex (Pharmacia Ltd.). Molecules larger than 25 kDa are excluded from the matrix of the column and are eluted in the equilibration buffer.[17]

mAbs can be concentrated by centrifugal evaporation (e.g., Uniscience Ltd.) or ultrafiltration in stirred cells, such as those of Flowgen Ltd. By using membranes with the appropriate MW cut-off points they can be purified from CSF;[39] the same principle should enable the separation of IgG and IgM.

III. FUTURE TRENDS

A. POSSIBLE IMPROVEMENTS IN STANDARD TECHNOLOGY

While mAbs are invaluable for investigating problems not previously possible by other methods, hybridoma technology is basically inefficient and expensive. A murine spleen contains approximately 10^8 B-cells, but only a few thousand of these are converted into hybridomas, and it is difficult to screen even 1000 clones. The argument for screening a library of tens of millions of antibody fragments generated by the phage display recombinant antibody system is therefore compelling. It may well be that hybridoma technology will be completely replaced, but for the purposes of this article we will assume it is not. The phage display recombinant antibody system is fully discussed in Chapter 25 and will only be mentioned briefly.

It is therefore possible to implement improvements which may reduce the scale and cost of a number of stages of hybridoma production. Possible improvements in standard technology are summarized in Figure 5.

An appropriate oligopeptide from the antigen of interest may be expressed in the Fv as an internal image idiotype, then presented as antigen.[40] The same synthetic peptide could also be used for *in vitro* immunization of B-cells to create a comprehensive idiotype library free from the *in vivo* constraints of the murine immune system.[29] The converse approach, of manipulating the *in vitro* response to a complex mixture of antigens, is also worth investigation. Subsequent screening by some form of "panning" or FACS to capture the important B-cells from a large population is essential. A few thousand "elite" B-cells must be fused with the myeloma cells with maximum efficiency, and this is best carried out by electroporation, perhaps in conjunction with prior enzyme treatment.[41] As a secondary screen and method for immunocytochemistry, antibody exchange on solid supports is very sensitive and economical.[42] Where possible, the use of ascites for bulking should be avoided. Simple culture in dialysis tubing can yield up to 5.4 mg/ml in 10 d,[43] while even greater yields are claimed for various culture systems using hollow-fiber cartridges. Thiophilic adsorption chromatography for IgG[44] and a modified single-step gel permeation for IgM[45] offer effective alternatives to most current purification methods.

B. THE CONTRIBUTION OF MONOCLONAL ANTIBODIES TO SOLVING PROBLEMS IN PLANT NEMATOLOGY

Many aspects of nematode biology depend on secretions. The saliva of plant-parasitic nematodes certainly controls the development and maintenance of syncytia and giant cells. Secretions synthesized in the subventral esophageal glands of *M. incognita* move anteriorly through the lumen of the esophagus and emerge from the stylet. This suggests a role in pathogenesis,[46] certainly in the development of the

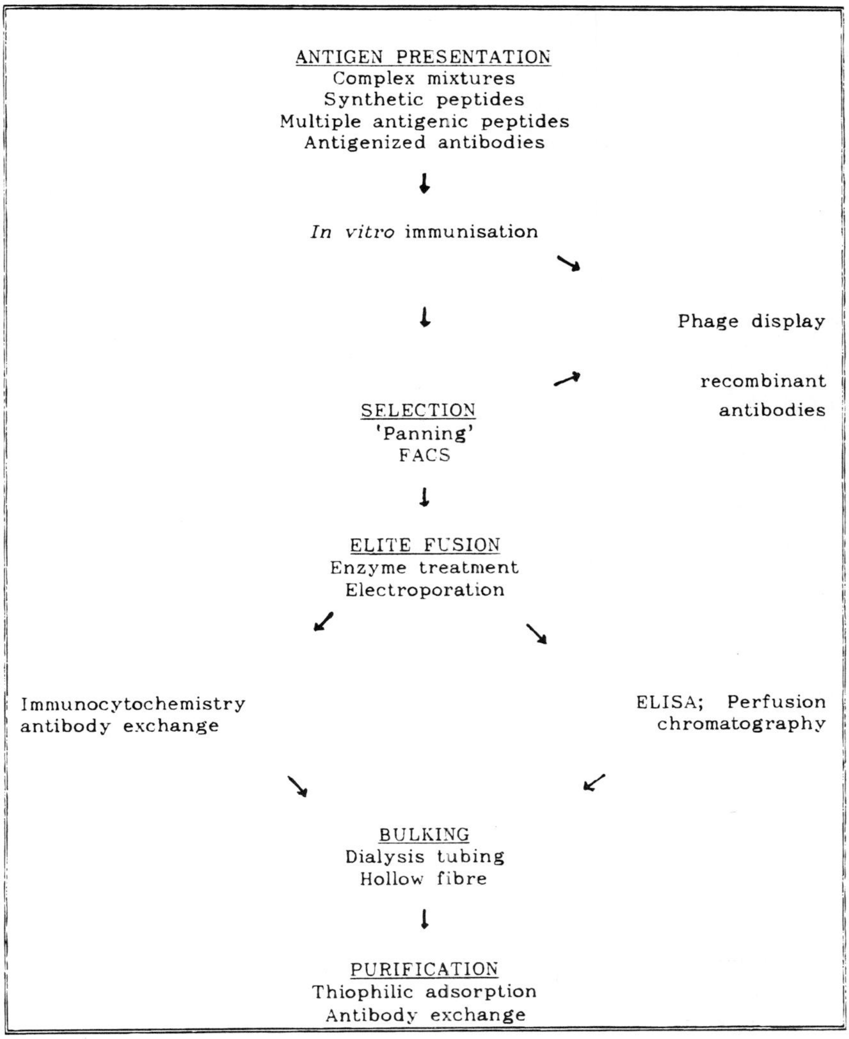

Figure 5 Possible improvements in standard technology.

compatible, and perhaps in the incompatible response. However, secreted cuticle surface proteins may in theory retard or promote elicitation of the incompatible response[47,48] (see also Figure 6). When the J2 has settled in the root and commenced feeding, amphidial exudate and cuticle surface proteins are contiguous or in close proximity to the plant cell wall and plasma membrane,[49] where receptors for nematode elicitors may exist. The area immediately adjacent to the head region of *H. glycines* has the highest accumulation of the phytoalexin glyceollin in the resistant soybean cv. Centennial.[50]

Outwith plant roots, surface proteins of nematodes are potential receptors for the adhesion of pathogenic microorganisms such as *Pasteuria penetrans*.[51] The amphids play a role in host finding, and the amphidial exudate may act as a receptor for attractant stimuli from the roots.[52] Specific mAbs are available for many of these secretions and therefore present a number of research opportunities. Firstly, they assist the collection of small quantities of protein for sequencing and identification. Specific peptide sequences will in turn offer alternative methods for distinguishing between different nematodes by raising new mAbs. If host recognition is based on post-translational modification of proteins, mAbs may be essential for identification relating to host preference.[53] To be useful, however, race-specific probes would have to be effective for all populations and suitable for all stages of nematode.[18] Secondly, immunoscreening of nematode DNA expression libraries will lead to the identification of genes coding

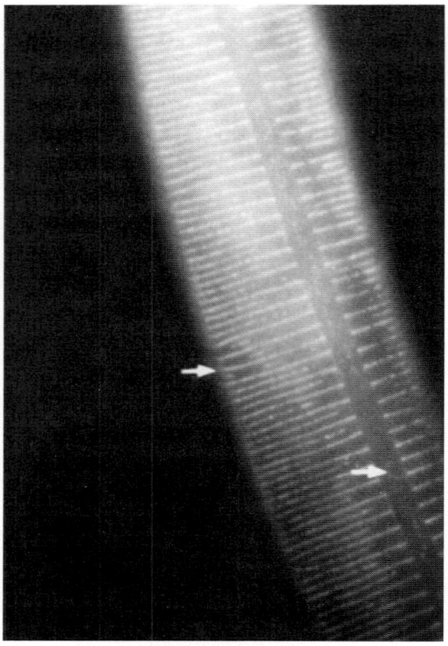

Figure 6 Binding of a mAb to surface antigen, on cut fragments of *D. dipsaci* (oat race). (After Palmer et al.)[16]

for secretions.[46,54] This is an alternative approach to the amplification of cDNA libraries derived from nematode mRNA isolated from syncytia.[55] Thirdly, it may be possible to use mAbs as competitive inhibitors of the binding of bacteria to the nematode surface or to inhibit host finding by J2.

The success of "plantibodies" may be enhanced by *in planta* studies using mAbs which bind to and inhibit the action of key plant enzymes.[56]

A final controversial approach to nematode interactions lies in the generation of internal image anti-idiotypic antibodies which are structural mimics of biological ligands and receptors. Such mimics have been substituted successfully for viral antigen in ELISA tests to demonstrate the efficacy of antiviral antibodies,[57] but could also play a role in studying nematode elicitors and plant receptors. The best chance of producing and isolating rare anti-idiotype antibodies is to guarantee the presentation of epitopes in a stable conformation[40] or to adopt a method such as the phage display recombinant antibody system which enables the screening of a very large number of clones.

ACKNOWLEDGMENTS

I am grateful to H. Atkinson, K. Davies, R. Hussey, M. Luc, T. Mabbott, H. Palmer, R. Perry, A. Schots, and D. Vaux for published or unpublished information included in this chapter. K. Davies, A. Schots, and L. Torrance kindly read the manuscript. Expert advice on all aspects of mAb production also came from M. Vernon, R. Burns, and L. Torrance. Finally, I thank my colleagues D. Stewart and J. Robb for the pleasure of shared ideas.

GLOSSARY

Idiotype. An idiotype is a set of one or more idiotopes (antigenic determinants) characteristic of the immunoglobulin produced by a clone or a few clones of cells and confined to the antigen binding sites. An anti-idiotypic antibody binds selectively to a particular idiotope of the antibody which has been used to immunize an animal.

Cyclophosphamide. Cyclophosphamide is an immunosuppressive drug which inhibits the division of lymphocytes, particularly B-lymphocytes, thereby inducing tolerance to a particular antigen, but not to unrelated antigens. It has been used to reduce the response to determinants which are immunodominant.

Fluorescence-activated cell sorting. Fluorescence-activated cell sorting is the electronic identification and isolation from a large population of a subset of cells labeled with fluorochrome-tagged antibodies.

Panning. Panning can be used to select lymphocytes secreting antibodies to the antigen of interest, which may be bound to a plastic surface such as the base of a petri dish. Nonadherent lymphocytes can be removed by washing.

Perfusion chromatography. Perfusion chromatography is a technique based on fluid mechanics, which allows a decrease in separation time of proteins and hence enables the rapid screening of a series of antibodies against an antigen on the column.

REFERENCES

1. Bird, A. F., Serological studies on the plant parasitic nematode *Meloidogyne javanica, Exp. Parasitol.,* 15, 350, 1964.
2. Mabbott, T. W., Serological techniques as an aid to potato-root eelworm strain identification, in *Rep. 8th Int. Symp. Nematol.,* Antibes, 1965, Brill, Leiden, Netherlands, 1968.
3. Mabbott, T. W., personal communication, 1992.
4. Stone, A. R., *Heterodera pallida* n.sp. (Nematoda: Heteroderidae), a second species of potato cyst nematode, *Nematologica,* 18, 591, 1973.
5. Webster, J. M. and Hooper, D. J., Serological and morphological studies on the inter- and intraspecific differences of the plant-parasitic nematodes *Heterodera* and *Ditylenchus, Parasitology,* 58, 879, 1968.
6. Davies, K. G. and Lander, E. B., Immunological differentiation of root-knot nematodes (*Meloidogyne* spp.) using monoclonal and polyclonal antibodies, *Nematologica,* 38, 353, 1992.
7. Atkinson, H. J., Harris, P. D., Halk, E. J., Novitski, C., Leighton-Sands, J., Nolan, P., and Fox, P. C., Monoclonal antibodies to the soy bean cyst nematode, *Heterodera glycines, Ann. Appl. Biol.,* 112, 459, 1988.
8. Jones, P., Ambler, D. J., and Robinson, M. P., The application of monoclonal antibodies to the diagnosis of plant pathogens and pests, *Brighton Crop Prot. Conf.,* 2, 767, 1988.
9. Bakker, J., Schots, A., Bouwman-Smits, L., and Gommers, F. J., Species specific and thermostable proteins from second-stage larvae of *Globodera rostochiensis* and *G. pallida, Phytopathology,* 78, 300, 1988.
10. Schots, A., Hermsen, T., Schouten, S., Gommers, F. J., and Egberts, E., Serological differentiation of the potato cyst nematodes *Globodera pallida* and *G. rostochiensis, Hybridoma,* 8, 401, 1989.
11. Schots, A., Gommers, F. J., and Egberts, E., Quantitative ELISA for the detection of potato cyst nematodes in soil samples, *Fundam. Appl. Nematol.,* 15, 55, 1992.
12. Robinson, M. P., Butcher, G., Curtis, R. H., Davies, K. G., and Evans, K., Characterisation of a 34 kD protein from potato cyst nematodes, using monoclonal antibodies with potential for species diagnosis, *Ann. Appl. Biol.,* 123, 337, 1993.
13. Schots, A., de Boer, J., Schouten, A., Roosien, J., Zilverentant, J. F., Pomp, H., Bouwman-Smits, L., Overmars, H., Gommers, F. J., Visser, B., Stiekema, W. J., and Bakker, J., 'Plantibodies': a flexible approach to design resistance against pathogens, *Neth. J. Plant Pathol.,* 98 (Suppl. 2), 183, 1992.
14. Hussey, R. S., Monoclonal antibodies to secretory granules in esophageal glands of *Meloidogyne* species, *J. Nematol.,* 21, 392, 1989.
15. Harlow, E. and Lane, D., *Antibodies: a Laboratory Manual,* Cold Spring Harbor Laboratory, Cold Spring Harbor, NY, 1988, 726 pp.
16. Torrance, L., Serological methods to detect plant viruses: production and use of monoclonal antibodies, in *Rapid Techniques for the Detection of Plant Pathogens,* Duncan, J. M. and Torrance, L., Eds., Blackwell Scientific, Oxford, 1991.
17. Forrest, J. M. S., unpublished information, 1993.
18. Palmer, H. M., Atkinson, H. J., and Perry, R. N., Monoclonal antibodies (mAbs) specific to surface expressed antigens of *Ditylenchus dipsaci, Fundam. Appl. Nematol.,* 15, 511, 1992.
19. Cordonnier, M.-M., Smith, C., Greppin, H., and Pratt, L. H., Production and purification of monoclonal antibodies to *Pisum* and *Avena* phytochrome, *Planta,* 158, 369, 1983.
20. Stähli, C., Staehelin, T., Miggiano, V., Schmidt J., and Häring, P., High frequencies of antigen-specific hybridomas: dependence on immunization parameters and prediction by spleen cell analysis, *J. Immunol. Methods,* 32, 297, 1980.
21. Forrest, J. M. S., Stewart, D., Macintosh, S. M., and Robb, J. M., The effect of different immunisation schedules on the specificity of polyclonal antibodies to *Globodera rostochiensis* and *G. pallida, Nematologica,* 38, 412, 1992.

22. Nilsson, B. O., Svalander, P. C., and Larsson, A., Immunization of mice and rabbits by intrasplenic deposition of nanogram quantities of protein attached to Sepharose beads or nitrocellulose paper strips, *J. Immunol. Methods,* 99, 67, 1987.
23. Forrest, J. M. S. and Ross, H. A., Splenic implantation of nitrocellulose pellets as a means of raising antibodies to nanogram quantities of antigen, *J. Immunol. Methods,* 159, 275, 1993.
24. Davis, E. L., Aron, L. M., Pratt, L. H., and Hussey, R. S., Novel immunization procedures used to develop monoclonal antibodies that bind to specific structures in *Meloidogyne* spp., *Phytopathology,* 82, 1244, 1992.
25. Kenney, J. S., Hughes, B. W., Masada, M. P., and Allison, A. C., Influence of adjuvants on the quantity, affinity, isotype and epitope specificity of murine antibodies, *J. Immunol. Methods,* 121, 157, 1989.
26. Fazekas de St. Groth, S. and Scheidegger, D., Production of monoclonal antibodies: strategy and tactics, *J. Immunol. Methods,* 35, 1, 1980.
27. Galfre, G. and Milstein, C., Preparation of monoclonal antibodies: strategies and procedures, *Methods Enzymol.,* 73, 1, 1981.
28. Lane, R. D., A short duration polyethylene glycol fusion technique for increasing production of monoclonal antibody-secreting hybridomas, *J. Immunol. Methods,* 81, 223, 1985.
29. Vaux, D. J. T., Helenius, A., and Mellman, I., Spike-nucleocapsid interaction in Semliki Forest virus reconstructed using network antibodies, *Nature (London),* 336, 36, 1988.
30. Micklem, L. R., McCann, M. C., and James, K., The use of rat mixed-thymocyte culture-conditioned medium for hybridoma production, cloning and revival. *J. Immunol. Methods,* 104, 81, 1987.
31. Davis, J. M., Pennington, J. E., Kubler, A. M., and Conscience, J-F., A simple, single-step technique for selecting and cloning hybridomas for the production of monoclonal antibodies, *J. Immunol. Methods,* 50, 161, 1982.
32. Schots, A. R., personal communication, 1992.
33. Sorensen, K. and Brodbeck, V., Assessment of coating efficiency in ELISA plates by direct protein determination, *J. Immunol. Methods,* 95, 291, 1986.
34. Schots, A., van der Leede, B. J., de Jongh, E., and Egberts, E., A method for the detection of antibody affinity using direct ELISA, *J. Immunol. Methods,* 109, 225, 1988.
35. Munson, P. J. and Rodbard, D., LIGAND: a versatile computerized approach for characterization of ligand-binding systems, *Anal. Biochem.,* 107, 220, 1980.
36. Hussey, R. S., Paguio, O. R., and Seabury, F., Localisation and purification of a secretory protein from the esophageal glands of *Meloidogyne incognita* with a monoclonal antibody, *Phytopathology,* 80, 709, 1990.
37. Temponi, M., Kageshita, T., Perosa, F., Ono, R., Okada, H., and Ferrone, S., Purification of murine IgG monoclonal antibodies by precipitation with caprylic acid: comparison with other methods of purification, *Hybridoma,* 8, 85, 1989.
38. Akerström, B., Brodin, T., Reis, K., and Björk, L., Protein G: a powerful tool for binding and detection of monoclonal and polyclonal antibodies, *J. Immunol.,* 135, 2589, 1985.
39. Saha, K., Case, R., and Wong, P. K. Y., A simple method of concentrating monoclonal antibodies from culture supernatant by ultrafiltration, *J. Immunol. Methods,* 151, 307, 1992.
40. Zanetti, M., Rossi, F., Lanza, P., Filaci, G., Lee, R. H., and Billetta, R., Theoretical and practical aspects of antigenized antibodies, *Immunol. Rev.,* 130, 125, 1992.
41. Igarashi, M. and Bando, Y., Enhanced efficiency of cell hybridisation by neuraminidase treatment, *J. Immunol. Methods,* 135, 91, 1990.
42. Hammarback, J. A. and Vallee, R. B., Antibody exchange immunochemistry, *J. Biol. Chem.,* 256, 12763, 1990.
43. Sjögren-Jansson, E. and Jansson, S., Large scale production of monoclonal antibodies in dialysis tubing, *J. Immunol. Methods,* 84, 359, 1985.
44. Lihme, A. and Heegaard, P. M. H., Thiophilic adsorption chromatography: the separation of serum proteins, *Anal. Biochem.,* 192, 64, 1991.
45. Bouvet, J-P. and Pirès, R., One-step purification of murine monoclonal antibodies of the IgM class, *J. Immunol. Methods,* 145, 263, 1991.
46. Davis, E. L. and Hussey, R. S., Monoclonal antibodies that bind to stylet secretions of *Meloidogyne incognita, J. Nematol.,* 24, 587, 1992.

47. Veech, J. A., Phytoalexins and their role in the resistance of plants to nematodes, *J. Nematol.,* 14, 2, 1982.
48. Forrest, J. M. S. and Robertson, W. M., Characterization and localization of saccharides on the head region of four populations of the potato cyst nematode, *Globodera rostochiensis* and *G. pallida, J. Nematol.,* 18, 23, 1986.
49. Forrest, J. M. S., Robertson, W. M., and Milne, E. W., Ultrastructure of syncytial cell walls and associated features at the head region of second stage juveniles of *G. rostochiensis* in susceptible and resistant potato roots, *Nematologia Mediterranea,* 21, 3, 1993.
50. Huang, J. S. and Barker, K. R., Glyceollin 1 in soybean-cyst nematode interactions, *Plant Physiol.,* 96, 1302, 1991.
51. Davies, K. G. and Danks, C., Interspecific differences in the nematode surface coat between *Meloidogyne incognita* and *M. arenaria* related to the adhesion of the bacterium *Pasteuria penetrans, Parasitology,* 105, 475, 1992.
52. Stewart, G. R., Perry, R. N., Alexander, J., and Wright, D. J., A glycoprotein specific to the amphids of *Meloidogyne* species, *Parasitology,* 106, 405, 1992.
53. Davies, K. G., A nematode case study focusing on the application of serology, in *Identification and Characterization of Pest Organisms,* D. L. Hawksworth, ed., CAB International, Wallingford, 1994.
54. Hussey, R. S., personal communication, 1993.
55. Harper, G., Gurr, S. J., Scollan, C., Glenn, M. P., McPherson, M. J., Atkinson, H. J., and Bowles, D. J., Gene expression during a plant nematode interaction, in *Mechanisms of Plant Defence Response,* Fritig, B. and Legrand, M., Kluwer Academic Publishers, Dordrecht, 340, 1993.
56. Maccarrone, M., Veldink, G. A., and Vliegenthart, J. F. G., Inhibition of lipoxygenase activity in lentil protoplasts by monoclonal antibodies introduced into cells via electroporation, *Eur. J. Biochem.,* 205, 995, 1992.
57. Mernaugh, R. L., Hill, J. H., Durand, D. P., van Deusen, R. A., and Harris, K. F., Properties and potential applications of syngeneic monoclonal anti-idiotype antibodies: use as a reference standard in enzyme-linked immunosorbent assay, *Phytopathology,* 81, 274, 1991.
58. Haslam, S. M., Atkinson, H. J., and Bowles, D. J., Characterization of proteins associated with the surface of *Globodera pallida, Nematologica,* 38, 415, 1992.
59. Robb, J. M. and Forrest, J. M. S., Staining of ribonucleic acid in the oesophageal bulb of *Xiphinema index* by bromination and microwave immunohistochemistry, *Fundam. Appl. Nematol.,* 15, 277, 1992.
60. Robinson, M. P., Quantification of soil and plant populations of *Meloidogyne* using immunoassay techniques, *J. Nematol.,* 21, 583, 1989.
61. Backett, K. D., Atkinson, H. J., and Forrest, J. M. S., Development of a rapid immunological screen for resistance in the potato to *Globodera pallida, Nematologica,* 36, 330, 1991.

FURTHER READING

Dresser, D. W., Immunization of experimental animals, in *Handbook of Experimental Immunology,* Vol. 1, Weir, D. M., Herzenberg, L. A., Blackwell, C., and Herzenberg, L. A., Eds., Blackwell Scientific, Oxford, 8.1, 1986.

Francotte, M. and Urbain, J., Induction of anti-tobacco mosaic virus antibodies in mice by rabbit anti-idiotypic antibodies, *J. Exp. Med.,* 160, 1485, 1984.

Greenspan, N. S. and Bona, C. A., Idiotypes: structure and immunogenicity, *FASEB J.,* 7, 437, 1993.

Hsu, H. T., Lee, I. M., Davis, R. E., and Wang, Y. C., Immunization for generation of hybridoma antibodies specifically reacting with plants infected with a mycoplasma like organism (MLO) and their use in detection of MLO antigens, *Phytopathology,* 80, 946, 1990.

King, S. W. and Morrow, K. J., Jr., Monoclonal antibodies produced against antigenic determinants present in complex mixtures of proteins, *BioTechniques,* 6, 856, 1988.

Vaux, D. and Fuller, S. D., The use of anti-idiotype antibodies for the characterisation of protein-protein interactions, *Methods Cell Biol.,* 34, 1, 1991.

Chapter 31

Immunochemical Methods for Detection of Toxins and Pesticides

Tsung-Che, Tseng

TABLE OF CONTENTS

I. INTRODUCTION

Landsteiner's[1] pioneering investigations on the antigenicity of small molecules, culminating in the development of the hapten concept, has provided the framework for the development of immunochemical assays for the detection of low-molecular weight compounds of both natural and synthetic origin. Immunoassays are detection methods based on a reaction between a target analyte and a specific antibody. Quantitation can be performed by monitoring a color change or by measuring radioactivity or fluorescence. Today, immunoassay is one of the most powerful analytical tools in biosciences, particularly long established in disciplines such as clinical chemistry and endocrinology, but still rather new to plant pathology.

Immunoassay has become increasingly important for the quantitation of low-molecular weight compounds such as pesticides or microbial toxins that are difficult to detect by conventional methodologies. The recent rapid growth in immunochemical methods is attributed in part to the availability of antibodies for a variety of compounds of environmental significance. For example, specific antibodies have been developed for many pesticides,[2–5] microbial toxins,[6,7] and various biological markers of exposure such as DNA adducts.[8,9] Another driving force in the development of immunoassays is attributed to the method requiring little sophisticated instrumentation, yet proving extreme versatility, sensitivity, specificity, and simplicity of operation. In addition, monoclonal antibody-based enzyme immunoassays are among the most attractive analytical tools available today.

In this chapter, a general introduction of the principles of immunoassay is given, followed by outlines for the practical realization of such assays. Applications of immunochemical technology for the determination of toxins with special emphasis on mycotoxins and pesticides will be addressed. Advantages and limitations of immunoassays will be discussed. Finally, it will include potential applications and future research areas for immunochemical methods.

0-87371-877-1/95/$0.00+$.50

II. PRINCIPLES

A. GENERAL PRINCIPLES

Immunochemical analytical methods are generally based on the principle of competition between an analyte and a labeled form of the analyte for a specific receptor. The specific receptor consists of an antibody protein synthesized in an animal in response to the injection of a suitable form of the analyte. Small molecules alone do not elicit an antibody response, but must be attached to a protein to form a derivative or conjugate which is termed the "immunogen". Thus, the initial task in method development is to couple the analyte to a suitable carrier protein.

B. IMMUNOGEN SYNTHESIS

Initiation of antibody production by lymphocytes in the spleen or lymph nodes requires that several reactive cell types specifically and simultaneously interact with an immunogen. The immunogen initiating the response should have a high molecular weight (> 2500 Da). Since toxin or mycotoxin and pesticide are low-molecular weight compounds (300 to 400 Da), they do not independently induce a hyperimmune response. In most cases, they must be derivatized and then conjugated to a carrier protein, such as bovine serum albumin, keyhole limpet hemocyanin, or ovalbumin to be rendered immunogenic. Conjugation of the mycotoxin to a protein molecule occurs by chemical reaction of functional groups present on the mycotoxin with groups present on the protein. Some mycotoxins, such as ochratoxin A (OTA), patulin, penicillic acid, and rubratoxin B, contain reactive groups which allow direct conjugation of the mycotoxin to a protein molecule. However, most mycotoxins, including the aflatoxins (AFs) and trichothecenes, contain no reactive groups for direct coupling, and a reactive group such as a carboxyl (-COOH) must first be introduced by chemical synthesis. Methods for the conjugation to protein carriers of AFB,[10–17] AFQ_1,[18] AFM_1,[19] AFB_1-DNA adduct,[20] Kojic acid,[21] or A,[22,23] rubratoxin B,[24] melanoic acid,[25] T-2 toxin,[26] zearalenone,[27] AFB_1 dihyrodiol,[28] diacetylscripenol,[29] sterigmatocystein,[30,31] fumonisin B1,[32] and 3-acetyl-deoxynivalenol[33] have been developed and employed in the production of polyclonal and monoclonal antibodies to the mycotoxins.

Pesticides of small size (< 1000 Da) must first be covalently linked to a larger molecule or carrier, usually a protein, to produce an immunogen.[1] The "perfect" hapten contains as much of the structure of the target compound as possible, plus a handle to facilitate recognition of the target structure by antibodies. This is usually three to six atoms long and contains a functional group (-NH2, -COOH, -OH, -SH) which can be covalently linked to a protein. If the target molecule has no reactive group at all, derivatization procedures are required to yield an appropriate spacer arm as well as a reactive functional group for conjugation to the protein.

The functional group of the hapten governs the selection of the method to be used to conjugate the hapten to the carrier. Two procedures routinely for conjugation of carboxyl-containing haptens to proteins are the mixed anhydride procedure, originally developed for peptide preparation,[34] and methods utilizing carbodiimides.[35,36] The mixed anhydride method has been used for benzoylphenylureas,[37] thiocarbamates,[38] and chlorinated biphenyls.[39] As examples of the water-soluble carbodiimide method, Newsome and Shields[40] coupled 2-succinamidobenzimidazole to human albumin at pH 7.0, and Wie and Hammock[37] reported the coupling of five benzoylphenylureas to several carrier proteins at pH 6.5. Wing et al.[41] prepared the active ester of *S*-bioallethrin hemisuccinate, which was coupled to several proteins and also to tyramine for radiosynthesis and structural proof of conjugation. This technique has also been used for triazines, dieldrin,[42] fenpropimorphic acid,[2] maleic hydrazide,[43] and endosulfan.[44]

Haptens containing amine groups can be conjugated by a simple diazotization as was done for molinate.[38] Hydroxyl-containing haptens can be conjugated to proteins directly after derivatization of the protein with succinic anhydride.[45] Sulfhydryl-containing haptens may be conjugated through homo- or heterobifunctional reagents.[46]

C. ANTIBODY PRODUCTION

After successful production and characterization of a mycotoxin-protein conjugate or pesticide-protein conjugate, suitable animal species are then immunized with the purified conjugate. If polyclonal antiserum is to be produced, then rabbits, goats, or sheep are normally used. However, if monoclonal antibodies are to be formed, then mice and possibly rats must be immunized because these are the only animals which have suitable tumor cell lines for the efficient fusion of plasma cells.

Immunization of animals is achieved by injection, usually intraperitoneally with a mixture of conjugate and Freund's adjuvant. Complete adjuvant is used for primary immunization and incomplete for secondary and subsequent injection. High-titer polyclonal antiserum normally can be produced within 5 to 7 weeks after the initial immunization.[11,23,26,30,47] With monoclonal antibody a final booster injection of conjugate alone is given 3 to 5 d prior to the fusion of spleen cells.[15]

Regardless of whether polyclonal sera or monoclonal antibodies are sought, the antibody response to a given antigen depends on the characterization of the conjugate, the immune system of the animal, and the immunization schedule and methods. Williams and Chase[48] and Vaitukaitis[49] describe a number of immunization procedures and schedules, many variations of which are widely used. For instance, in the production of polyclonal antibodies in rabbits, multiple intradermal injections are made along the back of the animal. An initial series of injections is followed by booster injections some weeks later. The animal is bled after each boost and the characteristics of the serum determined. One can either continue to collect and possibly pool sera following numerous booster injections or bleed the animal out. However, there is no standard protocol for immunization, and most approaches are largely empirical.

In general, monoclonal antibody production can be split into four major tasks: (a) immunization, (b) cell fusion and primary selection, (c) post-fusion cell management and secondary selection, and (d) expansion and scaled-up antibody production. Detailed manuals describing all steps of monoclonal antibody production have been addressed in the previous chapters, and also numerous descriptions of the general procedures for generating monoclonal antibodies have been published.[50] Optimal conditions, which can aid in the prediction of the success of production of the desired monoclonal antibody, are generally determined empirically.

III. IMMUNOCHEMICAL ASSAY FORMATS

Although a number of immunochemical methods have been used for the analysis of low-molecular weight compounds, two major methods, namely radioimmunoassays (RIA) and enzyme immunoassay (EIA), have been developed for the analysis of mycotoxins and pesticides.

A. RADIOIMMUNOASSAY

RAI is based on the competition between an isotopically labeled analyte (antigen) and an unlabeled analyte for a binding site on the antibody (receptor). Thus, the presence of large amounts of unlabeled analyte results in less radioactivity being bound to the antiserum. A comparison of the radioactivity of the bound to free labeled analyte with that obtained from a series of standard permits quantitation of unknown samples. The principle of the assay has been discussed by Hawker[51] and also in the previous chapter in this volume.

The requirement for high specific activity limits the choice of radioisotopes to the following (spec. activity in TBq/matom in brackets): ^{3}H(1.11), ^{35}S(56), ^{125}I(81), ^{32}P(340), ^{131}I(592), of which ^{3}H and ^{125}I are most widely used. The incorporation of isotope into the analyte may be carried out in various ways, ranging from the synthesis of the compound from a radioisotope precursor to synthesis of a tyramine derivative which is subsequently iodinated.[52]

Once equilibrium with the antibody has occurred, bound and free forms of the analyte are separated and quantitated. Ratcliffe[53] discussed the various methods available for separation. Selective adsorption of free analyte on dextran-coated charcoal remains a widely used technique because of its simplicity and low cost. Double antibody methods which employ a second antibody directed toward that bound to the analyte are less disruptive of the immune complex, but require additional equilibration time.

B. ENZYME IMMUNOASSAY

EIAs are similar in principle to RIA, but use an enzyme activity rather than radioactivity as the basis of quantitation. EIAs may be classified as either enzyme-linked immunosorbent assays (ELISA), which involve the use of a solid phase to effect separation of bound and free analyte, or enzyme-multiplied immunoassay techniques, which are conducted in solution and require no separation step. The latter methods use the alteration of an enzyme activity which may occur when the analyte binds to the antibody; although used extensively in drug and hormone analysis in human tissues and fluids, they have not yet been applied to toxin or pesticide analysis. Several types of ELISA have been developed and have been graphically described in reviews of the subject.[54] This paper is not intended to discuss the numerous formats of EIA that have been described. An introduction into this field should be found and more general treatments of EIA in References,[55,56] and also in the previous chapter. The enzyme most frequently used in the EIA are horseradish peroxidase (EC 1.11.1.7), alkalline phosphatase (EC

3.1.3.1), β-D-galactosidase (EC 3.2.1.23), glucoamylase (EC 3.2.1.3), and glucose oxidase (EC 1.1.3.4). However, alkaline phosphatase and horseradish peroxidase are preferable marker enzymes for small molecular analytes such as mycotoxin and pesticides,[57] and have been employed in a number of studies. The advantages of ELISA over RIA and conventional chemical methods include sensitivity, simplicity, ease of sample preparation, use of stable reagents, and absence of radiation hazard.

C. OTHER IMMUNOASSAYS

Alternatives to EIA and RIA have been developed, such as fluorescent[58] or luminescent immunoassays[59] and electroimmunoassay.[60] Metalloimmunoassay[61] makes use of organometallic labels which can be detected by atomic absorption spectrometry. In addition, differential pulse polarography or anodic stripping voltammetry can be used to detect low levels of haptens labeled with electroactive groups such as mercuric acetate.[62] In practice, some techniques are still at an experimental stage and sometimes lack sensitivity, are not easily carried out, or require expensive equipment. Fluorescent or luminescent labels are of special interest when designing an immunoassay for very low levels of haptens. In plant analysis, however, fluorescent labels may be of limited value due to the abundance of autofluorescent low-molecular weight compounds which may contribute to background noise. Luminescent immunoassays, on the other hand, do hold great promise for toxins and pesticides analysis, but this reservoir has not yet been tapped.

IV. IMMUNOCHEMICAL ASSAY FOR TOXINS

Although immunochemical techniques provide extremely specific and sensitive methods for studying the taxonomic, functional, and structural relationships of antigens derived from plant-pathogenic microorganisms or viruses, there have been few reports of these techniques having been applied to a plant pathological problem in terms of microbial toxin in plant disease. Until recently, naphthazarin toxins of *Fusarium solani* were detected by competitive ELISA analysis in xylem fluid or roots rotted by *F. solani* and symptomless scaffold roots and branches of healthy-appearing and diseased citrus trees in ridge and flatwoods Florida groves.[63] This study concentrated on blight, a wilt disease with an undefined cause and etiology, to determine if *F. solani* is a causal factor of the disease. Mycotoxins (toxic to animals) produced by plant pathogens and saprophytes have been found in various commodities, including fruit, grains, peanuts, and cottonseeds, and in forage plants and plant debris. Mycotoxins can be produced by fungi that grow in the living plant (field fungi), decaying plant materials (advanced decay fungi), and on stored plant material (storage fungi).[64] Contamination by mycotoxins may well occur before or during harvest of several agricultural products. The presence of mycotoxins in foods and feeds has been considered to be a potential hazard to human and animal health, and it is also a problem of great concern to plant pathologists, since the mycotoxin problem is difficult to avoid, and the most effective measure for their control depends on a rigorous detection of toxins present in the various commodities. Immunoassay has been developed against many mycotoxins such as aflatoxins, ochratoxin, trichothecences, and other mycotoxins and have been successfully applied to the analysis of corn, wheat, barley, cottonseed, peanuts, milk, and serum.[65–67,69,75] Table 1 summarizes reported immunoassays and applications for mycotoxins. Generally, ELISA has been found superior to RIA in terms of sensitivity, performance, and less time required for analysis.

A. IMMUNOASSAY FOR AFLATOXINS

Several different strategies have been reported for coupling AFB_1 to a carrier protein. The most common method to produce antibodies of the desired properties involves oxime formation prior to conjugation.[11,79,80] Thus, antibodies specific for AFB_1 or AFB_1 and B_2, or for AFB_1, B_2, G_1, and G_2 can be obtained. An alternative approach is to attach the AFB_1 to protein through the furan end of the hapten. Several possibilities have been tried through the 2,3-dihydro-2 hydroxy derivative, such as AFB_{2a}, which can be cross linked directly to protein with tetrazobenzidine,[81] which can be linked to protein after ester formation with glutaric anhydride.[82] Opening the furan ring of AFB_{2a} prior to conjugation has been an attractive proposition because of the potential for generating antibodies of very broad specificity for the AFs.[28,83,84] Another possibility is the conjugation to protein of the 2,3-dichloride derivative synthesized from AFB_1.[14] Antisera produced from conjugates through the furan ring will be unable to distinguish AFB_1 and B_2, and M_1 as well. Antibody preparations against AFG_1[85], AFQ_1[86] and AFB_2 have also been prepared.

Table 1 **Immunochemical methods reported in literature for mycotoxin analysis**

Toxin	Assay format	Detection limit	Applications	Ref.
1. Aflatoxins				
AFB_1	RIA	2.9 μg/kg	Peanut	65
AFB_1	ELISA	5 μg/kg	Corn	66
AFB_1	ELISA	0.1 ng/ml	Barley	6
AFB_1	ELISA	0.1 μg/kg	Ground nutmeal	68
AFM_1	RIA	0.5 μg/ml	Milk	69
2. Trichothecenes				
T-2	ELISA	2 ng/assay	Liposomes	70
T-2	ELISA	10 ng/ml	Barley	6
HT-2	ELISA	—	—	71
DON	RIA	0.1 ng/assay	Corn	72
3-ADN	ELISA	1 μg/kg	Rice	33
Roridin	ELISA	5 ng/ml	—	73
3. Other mycotoxins				
F-2	RIA	5 μg/ml	Human serum	74
OTA	ELISA	0.5 μg/kg	Wheat	75
OTA	RIA	2.5 μg/kg	Barley	77
ST	ELISA	—	Barley	31
FB	ELISA	50 ng/ml	Feed	76
TDP-1	ELISA	20 μg/kg	Barley	78
TDP-1	ELISA	20 μg/kg	Wheat	78

Note: AF = aflatoxin, T-2 = T-2 toxin, DON = deoxynivalenol, 3-ADN = 3-acetyl deoxynivalenol, HT-2 = HT-2 toxin, F-2 = zearalenone, OTA = ochratoxin A, ST = sterigmatocystin, FB = fumonisin, TDP-1 = fusarochromanone.

A number of monoclonal antibodies against AFs have been obtained in the past few years.[12,15,56,87–94] In general, the specificity of different monoclonal antibodies for AFs is also dependent on the type of conjugate used in the immunization. When the same immunogens are used for the preparation of either polyclonal or monoclonal antibody, most of the high-affinity monoclonal antibodies have almost the same specificity as that observed for the polyclonal antibodies. Although clones that elicited antibodies which have different cross reactivities have been found, these clones generally have low affinities. For example, Kawamura et al.[92] have recently found that some of the monoclonal antibodies for AFB_1 had higher cross reactivity with aflatoxicol, one of the AF metabolites. Nevertheless, the affinity of these antibodies to aflatoxicol was still lower than the affinity of the best monoclonal antibody to AFB.$_1$ For the determination of AFB_1 in corn, wheat, or peanut butter, ELISA has been found superior to RIA[95] in terms of performance and time required for analysis. The method requires extraction of the sample with organic solvents and compared well with other methods in International Agency for Research on Cancer check sample program.

B. IMMUNOASSAY FOR TRICHOTHECENES

Immunoassays for trichothecenes have received great attention recently, partly because of the difficulties experienced in analyzing these compounds by conventional chemical methods. Many reports have described the production of T-2 protein conjugated and synthesized after T-2 hemisuccinate formation. Such conjugation leads to subsequent production of very specific antisera. Comparatively, ELISAs are more sensitive than RIA for T-2 toxin analysis.[96,97]

Monoclonal antibodies against T-2 toxin[98] have performed poorly when used for analytical purposes compared to polyclonal antibodies. Several monoclonal antibodies were found to have specificities different from the polyclonal antibodies when the same type of immunogen was used. For example, immunization with T-2-HS-BSA conjugates resulted in monoclonal antibodies from different laboratories that each have different specificities. Some monoclonal antibodies have high specificity against HT-2 toxin, the others have specificity against 3′-OH-T-2 toxin.[99] Since trichothecenes have different side chains, the diversity of antibody specificity for these monoclonal antibodies might be due to the hydrolysis of either the immunogen in the immunization process or the hydrolysis of the T-2 toxin in the microtiter plate during selection of the specific clones.

Diacetoxyscripenol antibodies have been generated after derivatization through the hydroxyl function.[100,101] Again, the antibodies were of high specificity. The most sensitive of the assay, the indirect ELISA with polyclonal antibodies, was applied for determination of the toxin in wheat.

C. IMMUNOASSAY FOR OTHER MYCOTOXINS

Of other mycotoxins, most effort has been directed towards setting up assays for OTA, zearalenone, and sterigmatocystein. RIAs[102] and ELISAs[103,104] have been described for zearalenol determination, two reports of ELISAs for sterigmatocystin determination.[31,105] Many different immunoassays have been described for detection of OTA, all using antibodies generated against conjugate made by linking the toxin to protein through the carboxylic acid function. In general, RIAs, have been out-performed by ELISAs utilizing polyclonal[47,75] and monoclonal antibodies.[106] The assays have been applied to barley,[106] wheat,[75] and porcine kidney.[107] Immunoassay for rubratoxin,[108] fumonisin,[76] and fusarochromonone[78] have also been described.

Several new immunochemical techniques for the analysis of mycotoxins have been developed in recent years.

1. Immunoaffinity Chromatography

This unique separation technique has been used to isolate specific antibodies using immobilized analytes. Recent emphasis has been on the immobilization of antibodies onto stationary supports for the separation of small molecules. The antibody-coated support is packed into a column, and the solution containing the specific analyte is passed through. During this step, the immobilized antibody captures the analyte, separating it from other components in the solution. The captured analyte can then be removed from the adsorbent by dissociating the antibody-analyte complex with a buffer solution.

The immunoaffinity assay which traps the mycotoxins has been used for AFB_1, AFM_1, and OTA.[7,107,109] The toxin can be eluted from the column for subsequent analysis by ELISA or RIA. The affinity column serves as a specific cleanup and concentration tool for mycotoxin analysis, and it is particularly suited for concentrating polar compounds that are often difficult to concentrate from environmental samples using conventional resins.

2. Immunochromatography

Antibodies can be used as chemically selective detectors following HPLC separation of material from a complex mixture of chemicals. The use of ELISA as a post-column monitoring system for HPLC for the analysis of different group A trichothecenes was developed. Various group A trichothecenes were first separated on a C-18 reversed-phase column. Individual fractions eluted from the column were analyzed by ELISA using generic antibodies against group A trichothecenes. This approach not only can identify each individual group A trichothecene, but also can determine their concentration. The detection limit for T-2 toxin and related trichothecenes as well as their metabolites is as low as 2 ng. The combination of HPLC and immunoassay proved to be an efficient, sensitive, and specific method for the analysis of trichothecenes[110] and other mycotoxins. HPLC/Immunoassay (HPLC/IA) in some cases offers a 1000-fold increase in sensitivity over HPLC/ultraviolet absorption (HPLC/UV), and the sensitivity to quantify compounds present as only these fractions of the total. One current limitation of HPLC/IA is that there is no on-line immunodetector allowing real-time immunochemical sensing of the column effluent. Rather, fractions must be collected, the solvent exchanged to be compatible with the antibodies, and the immunoassays run on the individual fractions from the HPLC.

3. Novel Developed Assay

Several new methods for the analysis of mycotoxins have been developed. The "hit and run" assay was developed for T-2 toxin by Warden et al.[111] Briefly, an affinity column which was prepared by conjugating T-2 toxin to Sepharose gel was equilibrated with fluorescein isothiocynate (FITC)-labeled Fab fractions of immunoglobulin G (IgG) (anti-T-2 toxin). Samples containing T-2 toxin were applied to the column. The FITC-Fab which eluted together with the samples containing T-2 toxin was then determined in a standard flow-through fluorometer. The detection limit for T-2 toxin was found to be between 25 and 50 ng/ml. This assay is rapid and the column could be used many times, but it was very selective and required specific monoclonal antibodies. The sensitivity was also not very high.

Antibody against T-2 toxin has been conjugated by fiber optics which could be used in the future for the development of a biosensor.[112]

V. IMMUNOCHEMICAL ASSAY FOR PESTICIDES

The use of pesticides to control plant diseases and other pests has been increasing steadily at an annual rate of about 14% since the mid-1950s, and it is estimated that by the year 2000 more than 3 billion kg of pesticides will be used annually worldwide. There is little doubt that pesticide use has increased yields of crops in most cases. However, such huge amounts of poisonous substances damage our crop plants several times each year.

The existence of pesticides in almost all environmental components is a reality. Soil is a reservoir of these chemicals, from which other components of the environment can become contaminated.[113–115] Human food is the product of plants and animals, and can be the target of pesticide contamination through the food chain. Milk and dairy products are particularly susceptible to contamination with pesticides.[116] These health hazards have created an insatiable demand for increasing sensitivity in methods of analyses, since pesticides and their metabolites in general are toxic at low residual levels. Immunoassays have been recognized as versatile analytical tools in this field. The application of immunoassays for analysis of pesticide residue was first suggested by Ercegovich et al.[117] and its potential application in pesticide analysis was evaluated by Hammock and Mumma.[118] Table 2 lists the currently reported immunoassays for pesticides.

Relatively few studies have been conducted on the application of immunochemical techniques to pesticides in foods and agricultural commodities. Ercegovich et al.[117] developed an RIA method for the insectide parathion, which was capable of detecting 10 μg/kg in lettuce extract without cleanup. In the presence of pure standards, the only compound to compete significantly with parathion was amino parathion. Unfortunately, the presence of coextractives resulted in significant nonspecific inhibition of the binding of parathion by the antiserum. A similar nonspecific inhibition was observed in the ELISA method for paraoxon in serum[133] and for diflubenzuron in milk.[37] In the latter case, diflubenzuron could be determined without extraction at a level of 2 μg/kg. If lipid were removed from the extract by solvent partitioning, the interfering inhibition could be removed. A method for another insecticide, *S*-bioallethrin, has demonstrated promising sensitivity and specificity,[141] but has not been applied to food analysis.

An RIA method which determines benomyl as its degradation product methyl 2-benzimidazole carbamate was developed and applied to several commodities that had been spiked with benomyl.[40] No sample cleanup was required and recoveries correlated well by LC. The fungicides metalaxyl and triademefon have also been determined in food by ELISA. In the former case,[136] good recoveries were obtained as low as 0.1 mg/kg in several vegetables, but direct analysis of the methanol extract. A number of pesticides of similar structure to metalaxyl were investigated for cross reactivity, and it was found that the herbicides metolachlor and diethathyl ethyl and the fungicide furalaxyl were capable of some degree of interference if present. The method of triadimefon[127] was capable of quantitation at 0.5 mg/kg and above, again with no cleanup of the sample extract. The recoveries of both cases were verified by conventional GC analysis after purification by adsorption chromatography.

The herbicide diclofop methyl was determined in milk, wheat, soybean, and sugarbeets by enzyme immunoassay using horseradish peroxidase coupled to diclofop as a label and compared with a similar method using diclofop coupled to fluorescein amine.[134] The detection limit with either method varied greatly with the nature of the sample, ranging from 0.115 mg/kg in soybean to 9.0 mg/kg in wheat shoots. A large number of structurally related compounds were studied for cross reactivity, and it was observed that the (2,4-dichlorophenoxy) phenoxy moiety was necessary for binding to the antibody. Immunochemical assays for the herbicides 2,4-dichlorophenoxy acetic acid and trichlorophenoxy acetic acid,[135] paraquat,[137] terbutryn,[138] and chlorsulfuron[139] have been developed, and, although not applied to food analysis, should be adaptable for this purpose.

VI. EVALUATION AND PROSPECT OF IMMUNOCHEMICAL ASSAY

Immunochemical assays are finding increasing application in the chemical analysis of trace organic compounds. Applications include monitoring of residues in agricultural commodities, foods, in the environment, and in humans for both synthetic and naturally occurring toxins such as pesticides and

Table 2 **Immunochemical methods reported in literature for pesticide analysis**

Pesticide	Assay format	Detection limit	Application	Ref.
Alachlor	ELISA	1 ppb	Water	119
Aldicarb	ELISA	0.75 ppm	Citrus juices	120
		0.3 ppm	Water	
Bentazon	ELISA	2 ppb	Water	121
Benomyl	ELISA	0.35 ppm	Foods	122
Benomyl	RIA	0.5 mg/kg	Fruit	40
Bromacil	ELISA	0.25 ppb	—	143
Chlorsulfuron	ELISA	0.4–1.2 μg/kg	Soil	139
2,4-D and 2,4,5-T	RIA	100 pg	Water	123
Diflubenzuron	ELISA	1–40 ppb	Water	37,124
		1 ppb	Milk	
Diflubenzuron	ELISA	2 μg/kg	Milk	37
2,4-Dichloro-				135
phenoxyacetic acid	RIA	0.1 μg/kg	Water	
Diclofop methyl	EIA	0.1–1.2 mg/kg	Sugarbeet, wheat, soybean	134
Endosulfan	ELISA	3 ng/ml	Water	44
Fenitrothion	ELISA	1 ng	—	142
Fenpropimorph	ELISA	13 ng/l	Water	2
Molinate	ELISA	20 ppb	Water	125
Metalaxyl	ELISA	0.1 mg/kg	Fruit	136
Monolinuron	ELISA	14–22 ng/g	Foods	126
Diuron		0.1 μg/g		
Norflurazon	ELISA	1 ng/ml	Water	3
Desmethyl norflurazon		10 ng/ml		
Paraquat	ELISA	0.1–1 ng/ml	Air filters, clothing patches, hand rinses, biological fluids, foodstuffs	4
Paraquat	ELISA	0.8 μg/kg	Serum	137
Paraquat	RIA	0.6 μg/kg	Plasma	132
Terbutyrn	ELISA	25 μg/kg	Water	138
Parathion	RIA	10 μg/kg	Lettuce	117
Paraoxon	ELISA	0.26 μg/kg	Serum	133
Picloram and	ELISA	100 ng/ml	Water	123
2,4-D	RIA	50 ng/ml		
Triadimefon	ELISA	0.1 mg/kg	Fruit	127
Triadimefon	ELISA	0.5 ppm	Foods	127
Triazine herbicides	ELISA	0.1–10 μg/l	Water	5,128–130
		0.5–5 ppb	Soil	131
Warfarin	RIA	25 μg/kg	Plasma	140

mycotoxins. The primary motives for the development of these assays are their high sensitivity, selectivity, portability, short analysis time, low cost, and potential for parallel processing samples. It means that immunoassays are highly applicable to mass screening studies either for monitoring regulatory compliance or for epidemiological studies. Particularly, analytical approaches use antibodies in conjugation with other methods, for example, the use of immunoaffinity columns to concentrate and purify the analyte before measurement by conventional means. Immunoassays are of limited use in structural identification, but may help to screen for classes of chemicals suspected to be present in a sample. Their major disadvantage is that assay development is a multistep and somewhat lengthy procedure requiring immunological expertise. Sieber et al.[144] pointed out the barriers lying in adopting immunoassays to the pesticides analysis, including, first, the development time for the EIA method was unpredictable, ranging from several months to several years, but a new method could be developed in less than 1 month based on GC or HPLC. Second, the amount of information delivered by EIA was extremely limited, usually only a single piece of datum; nevertheless, GC, HPLC, or particularly GC/MS could provide dozens of analytes in a single run. Third, some inexplicable pitfalls in EIA were not easily identified instantly. Thus, it has reached a critical stage where the EIA could be an alternative or a complementary method for conventional assay methods in environmental toxic substance monitoring.

The development of nonisotopic immunoassays, based on monoclonal antibodies, however, is the method of choice for the future. Hybridoma technology provides unlimited amounts of highly pure antibodies of predefined characteristics. With the specific target of pesticide analysis in mind, perhaps the great benefit of using immunoassays is that such assays permit monitoring in great detail, pesticide uptake and translocation in treated organisms, in the environment, in worker exposure studies, and in toxicology studies in general. Immunochemical assays may furthermore prove very useful in helping the farmer optimize pesticide application programs, screen for residues in treated agricultural crops, etc. With the advent and further development of "dip-stick technology",[145,146] immunoassays may be carried out in the field and eventually become as simple as the "test-stick" analysis of many metabolites and pH values. There is also an increasing engagement of industry to develop immunoassay kits for pesticides, plant growth regulators, mycotoxins, and food and environmental contaminants. In addition, new application formats may be developed, such as the development of "immunosensors" to provide chemical detectors for passive monitoring.

Regarding mycotoxins, new methods have been developed rapidly in recent years. The most applicable techniques are simple, rapid, and inexpensive, employing fast screening methods such as minicolumn chromatography and immunoassays. Immunoassays, particularly ELISAs, have advantages over other techniques, allowing large-scale screening of samples for mycotoxins without the use of specialized expertise. However, the ELISA still requires rigorous testing by other workers, along with extensive collaborative studies, for a full assessment to be made. In addition, performance must be well documented and must address issues such as method detection limit, sensitivity, appropriate positive and negative controls, and interferences. Because immunoassays require little or no cleanup, time and the cost of an analysis are minimized. The performance of a method may be affected by interferences from the crude sample preparation. Thus, the most important area of research in immunoassay for detecting mycotoxins will be in the development and availability of high-affinity antibodies with the desired specificity. Hybridoma technology provides unlimited amounts of such highly pure antibodies of predefined characteristics.

Immunochemical methods for mycotoxin, including affinity chromatography as a cleanup tool and immunoassay as a post-column HPLC monitoring system, may be used in combination with other methodologies. For instance, a novel sequential injection immunoassay (SIIA) method[147] is described which utilizes immunomagnetic beads to detect short-time antibody binding. The SIIA system can be used to characterize antibodies which have been bound to the surface of the magnetic beads, as immunomagnetic bead reactor, for use in immunoassays. An antigen detection system, termed immuno-polymerase chain reaction (immuno-PCR), was developed in which a specific DNA molecule is used as the marker.[148] This immuno-PCR technology has a sensitivity greater than any existing antigen detection system and, in principle, could be applied to the detection of single antigen (analyte) molecules.

Further research effects, for mycotoxin analysis for example, multi-analyte analysis will be the focus of continuing research. This may be achieved by using different strips of immunoassay plate for each mycotoxin or by using a single well to detect all mycotoxins by placing various coating toxins in the well along with various specific antibodies. Greater emphasis may be focused on understanding the role of toxin or mycotoxin in diseased plants by an immunochemical method in order to elucidate the host-parasite interaction.[63] Other potential applications of immunochemical methods include immunocytochemistry with the antibody linked to a fluorescent enzyme label or immunogold label to allow visualization of toxin deposition in the producing organisms or infected hosts. The deposition of AFB_1 in the hyphae of *Aspergillus parasiticus* has been visualized by enzyme-linked immunocytochemistry.[149] The accumulation of AFB_1 in microsomes, mitochondria, and granual nucleoli in rat cells has been demonstrated by the immunogold labeling technique.[150] Other research efforts should be applied to the development of antibodies, particularly monoclonal antibodies, for other environmentally important mycotoxins such as patulin, cyclopiazonic acid, and fumonisin.

VII. CONCLUSIONS

Immunochemical methods, which are usually simple to perform, are powerful analytical techniques that can answer many questions concerning plant pathology and environmental contamination. The application of immunoaffinity chromatography to sample preparation, for instance, can minimize the use of organic solvents and subsequently reduce associated disposal costs. Furthermore, immunoassays are increasingly being recognized as one cost-effective alternative to chromatographic and spectroscopic procedures for analyzing environmental contaminants such as toxins and pesticides, and they can be used in the field for rapid screening. However, immunochemical methods will not replace existing analytical techniques, but they can augment current monitoring and measurement capabilities. The full potential of such methods has yet to be realized.

REFERENCES

1. Landsteiner, K., *The Specificity of Serological Reactions,* Harvard University Press, Boston, 1945.
2. Jung, F., Meyer, H. H. D., and Hamm, R. T., Development of a sensitive enzyme-linked immunosorbent assay for the fungicide fenpropimorph, *J. Agric. Food Chem.,* 37, 1183, 1989.
3. Riggle, B. and Dunban, B., Development of enzyme immunoassay for detection of the herbicide norflurazon, *J. Agric. Food Chem.,* 38, 1922, 1990.
4. Van Emon, J. M., Hammock, B., and Seiber, J. N., Enzyme-linked immunosorbent assay for paraquat and its application to exposure analysis, *Anal. Chem.,* 58, 1866, 1986.
5. Wittmann, C. and Hock, B., Development of an ELISA for the analysis of atrazine metabolites deethylatrazine and deisopropylatrazine, *J. Agric. Food Chem.,* 39, 1194, 1991.
6. Ramakrishna, N. and Lacey, J., Monoclonal antibody-based enzyme linked immunosorbent assay of aflatoxin B_1, T-2 toxin and ochratoxin A in barley, *J. Assoc. Off. Anal. Chem.,* 73, 71, 1990.
7. Pestka, J. J., Enhanced surveillance of foodborne mycotoxin by immunochemical assay, *J. Assoc. Off. Anal. Chem.,* 71, 1075, 1988.
8. Young, T. L., Habraken, Y., Ludlum, D. B., and Santella, R. M., Development of monoclonal antibodies recognizing 7-(2-hydroxyethyl)guanine and imidazole ring-open 7-(2-hydroxyethyl)guanine, *Carcinogenesis,* 11, 1685, 1990.
9. Foiles, P. G., Chung, F. L., and Hecht, S. S., Development of a monoclonal antibody-based immunoassay for cyclic DNA adducts resulting from exposure to crotonaldehyde, *Cancer Res.,* 360, 1987.
10. Bierman, V. D. and Terplan, G., Nachieviers von aflatoxin B_1 mittels ELISA, *Arch. Lebensmittelhyg.,* 31, 51, 1980.
11. Chu, F. S. and Ueno, I., Production of antibody against aflatoxin B_1., *Appl. Environ. Microbiol.,* 33, 1125, 1977.
12. Groopmen, J. D., Trudel, L. J., Donahue, P. R., Marshak-Rothstein, A., and Wogan, G. N., High affinity monoclonal antibodies for aflatoxins and their application to solid phase immunoassay, *Proc. Natl. Acad. Sci. U.S.A.,* 81, 7728, 1984.
13. Langone, J. J. and Van Vunakis, H., Aflatoxin B_1: specific antibodies and their use in radioimmunoassay, *J. Natl. Cancer Inst.,* 56, 596, 1986.
14. Sizaret, P., Malavaek, C., Montesand, R., and Frayisnet, C., Detection of aflatoxins and related metabolites by radioimmunoassay, *J. Natl. Cancer Inst.,* 6, 1375, 1983.
15. Candish, A. A. G., Stimson, W. H., and Smith, J. E., A monoclonal antibody to aflatoxin B_1: detection of the mycotoxin by enzyme immunoassay, *Lett. Appl. Microbiol.,* 1, 56, 1985.
16. Gaur, P. K., El-Nakib, and Chu, F. S., Comparison of antibody production against Aflatoxin B_1 in goats and rabbits, *Appl. Environ. Microbiol.,* 40, 678, 1980.
17. Lawellin, D. W., Grant, D. W., and Joyce, B. K., Enzyme-linked immunoabsorbent analysis of aflatoxin B_1, *Appl. Environ. Microbiol.,* 34, 94, 1977.
18. Lau, H. P., Fan, T., and Chu, S. F., Preparation and characterization of aflatoxin Q_1 hemisuccinate, *J. Assoc. Off. Anal. Chem.,* 64, 687, 1981.
19. Harder, W. O. and Chu, S. F., Production and characterization of antibody against aflatoxin M_1, *Experimentia,* 35, 1104, 1979.
20. Haugen, A., Groopman, J. D., Hsu, I. C., Goorrich, G. R., Wogan, G. N., and Harris, C. C., Monoclonal antibody to aflatoxin B_1-modified DNA detected by enzyme immunoassay, *Proc. Natl. Acad. Sci. U.S.A.,* 78, 4121, 1981.

21. Abdalla, A. E. and Grant, D. W., An immunoanalysis of kojic acid, *Sabouraudia,* 18, 191, 1981.
22. Aalund, O., Brundelt, I. K., Held, B., Krogh, P., and Poulsen, K., A radioimmunoassay for ochratoxin A: a preliminary investigation, *Acta Pathol. Microbiol., Scand. Sect. C.,* 83, 390, 1975.
23. Chu, S. F., Chang, F. C. C., and Hinsdill, R. D., Production of antibody against ochratoxin A, *Appl. Environ. Microbiol.,* 31, 831, 1976.
24. Davis, R. M. and Stone, S. S., Production of anti-rubratoxin antibody and its use in a radioimmunoassay for rubratoxin B, *Mycopathologia,* 67, 29, 1979.
25. Neucerne, J. N., Enzyme linked immunosorbent assay for secalonic acid D, *Mycopathologia,* 93, 39, 1986.
26. Chu, S. F., Groopman, S., Wei, R. D., and Mirocha, C. J., Production of antibody against T-2 toxin, *Appl. Environ. Microbiol.,* 37, 104, 1979.
27. Thouvenot, D. and Morfin, R. F., A radioimmunoassay for zearalenone and zearalenol in human serum: production, properties and use of porcine antibodies, *Appl. Environ. Microbiol.,* 45, 16, 1981.
28. Pestka, J. J. and Chu, F. S., Aflatoxin B_1 dihydrodiol antibody: production and specificity, *Appl. Environ. Microbiol.,* 47, 472, 1984.
29. Chu, F. S., Liang, M. Y., and Zhang, G. S., Production and characterisation of antibody against diacetoxyscirpenol, *Appl. Environ. Microbiol.,* 48, 777, 1984.
30. Li, Y. K. and Chu, S. F., Production and characterization of antibody against sterigmatocystin, *J. Food Safety,* 6, 119, 1984.
31. Morgan, M. R. A., Kang, A. J., and Chen, W. S., Production of antiserum against sterigmatocystin hemiacetal and its potential for use in an enzyme linked immunoabsorbent assay for sterigmatocystin in barley, *J. Sci. Food Agric.,* 37, 873, 1986.
32. Azcona-Olioera, J. I., Abouzied, M. M., Plattner, R. D., Norred, W. P., Pestka, J. J., Generation of antibodies reactive with fumonisins B_1, B_2, and B_3 by using cholera toxin as the carrier-adjuvant, *Appl. Environ. Microbiol.,* 58, 169, 1992.
33. Kemp, H. A., Mills, E. N. C., and Morgan, M. R. A., Enzyme-linked immunosorbent assay for 3-acetydeoxynivalenol applied to mice, *J. Sci. Food Agric.,* 37, 888, 1986.
34. Jung, F., Gee., S. J., Harrison, R. O., Goodrow, M. H., Karu, A. E., Braun, A. L., Li, Q. X., and Hammock, B. D., Use of immunochemical techniques for the analysis of pesticides, *Pestic. Sci.,* 26, 303, 1989.
35. Goodfriend, T. L., Levine, L., and Fasman, G. D., Antibodies to bradykinin and angiotensin: a use of carbodiimides in immunology, *Science,* 144, 1344, 1964.
36. Kurzer, F. and Zadeh, K. D., Advances in the chemistry of carbodiimides, *Chem. Rev.,* 67, 107, 1967.
37. Wie, S. J. and Hammock, B. D., Comparison of coating and immunizing antigen structure on the sensitivity and specificity of immunoassay for benzoylphenylurea insecticides, *J. Agric. Food Chem.,* 32, 1294, 1984.
38. Gee, S. J., Miyamato, T., Goodrow, M. H., Buster, D., and Hammock B. D., Development of an enzyme-linked immunosorbent assay for the analysis of the thiocarbamate herbicide molinate, *J. Agric. Food Chem.,* 36, 863, 1988.
39. Luster, M. I., Albro, P. W., Clark, G., Chae, K., Chaudhary, S. K., Lawson, L. D., Corbette, J. T., and McKinney, J. D., Effect of in utero exposure to diethylstilbestrol on the immuno response in mice, *Toxicol. Appl. Pharmacol.,* 47, 279, 1979.
40. Newsome, W. H. and Shields, J. B., A radioimmunoassay for benomyl and methyl 2-benzimidazolecarbamate on food crops, *J. Agric. Food Chem.,* 29, 220, 1981.
41. Wing, K. D., Hammock, B. D., and Wustner, D. A., Development of an S-bioallethrin specific antibody, *J. Agric. Food Chem.,* 26, 1328, 1978.
42. Langone, J. J. and Van Vunakis, H., Radioimmunoassay for dieldrin and aldrin, *Res. Commun. Chem. Pathol. Pharmacol.,* 10, 163, 1975.
43. Harrison, R. O., Brimfield, A. A., and Nelson, J. O., Development of a monoclonal antibody based enzyme immunoassay method for analysis of maleic hydrazide, *J. Agric. Food Chem.,* 37, 958, 1989.
44. Dreher, R. M. and Podratzki, B., Development of an enzyme immunoassay for endosulfan and its degradation products, *J. Agric. Food Chem.,* 36, 1072, 1988.
45. Hung, B. T., Benner, S. A., and Williams, C. M., Preparation of an ecdysone immunogen for radioimmunoassay work, *J. Biol. Chem.,* 255, 6047, 1980.
46. Tijssen, P., *Practice and Theory of Enzyme Immunoassay,* PH Knippenberg, Amsterdam, 1985, 221.

47. Morgan, M. R. A., McNerry, R., and Chan, H. W. S., Enzyme-linked immunosorbent assay of ochratoxin A in barley, *J. Assoc. Off. Anal. Chem.*, 66, 1480, 1983.
48. Williams, C. A. and Chase, M. W., *Methods in Immunology and Immunochemistry,* Academic Press, New York, 1967, 209.
49. Vaitukaitis, J. L., Production of antisera with small dose of immunogen: multiple interdermal injections, in *Methods in Enzymology: Immunochemical Techniques,* Part B, Langone, J. and Van Vunakis, H., Eds., Academic Press, New York, 1981, 46.
50. Westerwoudt, R. J., Factors affecting production of monoclonal antibodies in *Methods in Enzymology: Immunochemical Techniques,* Part I, Langane, J. J. and Van Vunakis, H., Eds., Academic Press, New York, 1986, 3.
51. Hawker, C. D., Radioimmunoassay and related methods, *Anal. Chem.,* 45, 878A, 1973.
52. Steiner, A. L., Assay of cyclic nucleotides by radioimmunoassay methods, in *Methods in Enzymology,* Hardman, J. G. and O'Malley, B. W., Eds., Academic Press, New York, 1974, 96.
53. Ratcliffe, J. G., Parathion techniques in saturation analysis, *Br. Med. Bull.,* 30, 32, 1974.
54. Monroe, D., Enzyme in immunoassay, *Anal. Chem.,* 56, 920A, 1984.
55. Maggio, E. T., *Enzyme-immunoassay,* CRC Press, Boca Raton, FL, 1980.
56. Woychik, N. A., Hinsdill, R. D., and Chu, S. F., Production and characterization of monoclonal antibodies against aflatoxin M_1, *Appl. Environ. Microbiol.,* 48, 1096, 1984.
57. Eberle, J., Arnscheidt, A., Klix, D., and Weiler, E. W., Monoclonal antibodies to plant growth regulator, *Plant Physiol.,* 81, 516, 1986.
58. Dandlinker, W. B., Dandlinker, J., Kelly, R. J., Farquhar, J., and Levin, J., Fluorescence polarization immunoassay theory and experimental method, *Immunochemistry,* 10, 219, 1973.
59. Weiter, E. W., Immunoassay for the quantitation of low molecular weight compounds in plant, in *Chemistry of Plant Protection,* Vol. 3, Blab, W. and Weiter, E. W., Eds., Springer-Verlag, Berlin, 1990, 147.
60. Heinemann, W. R. and Halsall, H. B., Strategies for electrochemical immunoassay, *Anal. Chem.,* 57, 1321A, 1985.
61. Cais, M., Dani, S., Eden, Y., Gandolfi, O., Horn, M., Isaacs, E. E., Josephy, Y., Saar, Y., Slovin, E., and Snarsky, L., Metalloimmunoassay, *Nature,* 270, 534, 1977.
62. Heinemann, W. R., Anderson, C. W., and Halsall, H. B., Immunoassay by differential pulse polarography, *Science,* 204, 865, 1979.
63. Nemec, S., Jabaji-Hare, S., and Charest, P. M., ELISA and immunocytochemical detection of *Fusarium soloni*—produced naphazarin toxins in citrus trees in Florida, *Phytopathology,* 81, 1497, 1991.
64. Hesseltine, C. W., Mycotoxins and other fungal related problems, in *Advances in Chemistry,* Series No. 149, Rodricks, J. V., Ed., Washington, 1976, 1.
65. El-Nakib, O., Pestka, J. J., and Chu, S. F., Determination of aflatoxin B_1 in corn, wheat and peanut butter by enzyme-linked immunosorbent assay and solid phase radioimmunoassay, *J. Assoc. Off. Anal. Chem.,* 64, 1077, 1981.
66. Schurrs, A. H. W. M. and Van Weoman, B. K., Enzyme immunoassay, *Clin. Chem. Acta,* 18, 1, 1977.
67. Fan, T. S. L., Zhang, G. S., and Chu, F. S., An indirect enzyme-linked immunosorbent assay for T-2 toxin in biological fluids, *J. Food Prot.,* 47, 964, 1984.
68. Bierman, V. A. and Terplan, G., Experience with a micro ELISA for the detection of aflatoxin B1 in food, *Arch. Lebensmittelhyg.,* 33, 17, 1982.
69. Pestka, J. J., Li, Y., Harder, W. O., and Chu, F. S., Comparison of radioimmunoassay and enzyme-linked immunosorbent assay for determining aflatoxin M1 in milk, *J. Assoc. Off. Anal. Chem.,* 64, 294, 1981.
70. Ligler, F. S., Bredehorst, R., Talebian, A., Shriver, L. C., Hammer, C. F., Sheriden, J. P., Vogel, C.-W., and Gaber, B. P., A homogeneous immunoassay for mycotoxin T-2 utilizing liposomes, monoclonal antibodies and complement, *Anal. Biochem.,* 163, 369, 1987.
71. Fan, T. S. L., Schubring, S. L., Wei, R. D., and Chu, S. F., Production and characterization of a monoclonal antibody cross-reactive with most group A trichothecenes, *Appl. Environ. Microbiol.,* 54, 2959, 1988.
72. Xu, Y-C., Zhang, G. S., and Chu, F. S., Radioimmunoassay of deoxynivalenol in wheat and corn, *J. Assoc. Off. Anal. Chem.,* 69, 967, 1986.

73. Martlbauer, E., Garies, M., and Terplan, G., Enzyme immunoassay for the macrocyclic trichothecene. Roridin A: production, properties, and use of rabbit antibodies, *Appl. Environ. Microbiol.*, 54, 225, 1988.
74. Pestka, J. J., Steinert, B. W., and Chu, S. F., Enzyme-linked immunosorbent assay for detection of ochratoxin A, *Appl. Environ. Microbiol.*, 41, 1472, 1981.
75. Lee, S. F. and Chu, F. S., Enzyme-linked immunosorbent assay for ochratoxin A in wheat, *J. Assoc. Off. Anal. Chem.*, 67, 45, 1984.
76. Azcona, J. I., Abouzied, M. M., Plathner R. D., and Pestka, J. J., Production of monoclonal antibodies to the mycotoxins fumonisins B_1, B_2, and B_3, *J. Agric. Food Chem.*, 40, 531, 1992.
77. Rousseau, D. M., Slegere, G. A., and Van Potegham, C. M., Radioimmunoassay of ochratoxin A in barley, *Appl. Environ. Microbiol.*, 50, 529, 1985.
78. Yu, J. and Chu, S. F., Immunochromatography of fusarochromanone mycotoxins, *J. Assoc. Off. Anal. Chem.*, 74, 655, 1991.
79. Morgan, M. R. A., Kang, A. S., and Chan, H. W.-S., Aflatoxin determination in peanut butter by enzyme-linked immunosorbent assay, *J. Sci. Food Agric.*, 37, 908, 1986.
80. Ram, B. P., Hart, L. P., Shotwell, O. L., and Pestka, J. J., Enzyme-linked immunosorbent assay of aflatoxin B_1 in naturally contaminated corn and cottonseed, *J. Assoc. Off. Anal. Chem.*, 69, 904, 1986.
81. Lawellin, D. W., Grant, D. W., and Joyce, B. K., Enzyme-linked immunosorbent analysis for aflatoxin B_1, *Appl. Environ. Microbiol.*, 34, 94, 1977.
82. Lau, H. P., Gaur, P. K., and Chu, F. S., Preparation and characterization of aflatoxin B2a-hemiglutarate and its use for the production of antibodies against aflatoxin B_1, *J. Food Safety*, 3, 1, 1981.
83. Gaur, P. K., Lau, H. P., Pestka, J. J., and Chu, F. S., Production and characterization of aflatoxin B_{2a} antiserum, *Appl. Environ. Microbiol.*, 41, 478, 1981.
84. Jackman, R., Determination of aflatoxins by enzyme-linked immunosorbent assay with special reference to aflatoxin M_1 in milk, *J. Sci. Food Agric.*, 36, 685, 1985.
85. Chu, F. S., Steinert, B. W., and Guar, P. K., Production and characterization of antibody against aflatoxin G_1, *J. Food Safety*, 7, 161, 1985.
86. Fan, T. S. L., Zhang, G. S., and Chu, F. S., Production and characterization of antibody against Aflatoxin Q_1, *Appl. Environ. Microbiol.*, 47, 526, 1984.
87. Garner, C., Ryder, R., and Montesano, R., Monitoring of aflatoxins in human body fluids and application to field studies, *Cancer Res.*, 45, 922, 1985.
88. Blankord, M. B. and Doerr, J. A., The development and characterization of monoclonal antibodies against aflatoxin, *Hybridoma*, 5, 57, 1986.
89. Candish, A. A. G., Stimson, W. H., and Smith, J. E., The detection of aflatoxin B1 in peanut kernels, butter and maize using a monoclonal antibody based enzyme immunoassay, *Food Microbiol*, 4, 147, 1987.
90. Dixon-Holland, D. E., Pestka, J. J., Bridigare, B. A., Casale, W. L., Warner, R. L., Ram, B. P., and Hart, L. P., Production of sensitive monoclonal antibodies to aflatoxin B_1 and aflatoxin M_1 and their application to ELISA of naturally contaminated food, *J. Food Prot.*, 51, 201, 1988.
91. Hastings, K. L., Tulis, J. J., and Dean, J. H., Production and characterization of a monoclonal antibody to aflatoxin B_2. *J. Agric. Food Chem.*, 36, 404, 1988.
92. Kawamura, O., Nagayama, S., Sato, S., Ohtani, K., Ueno, I., and Ueno, Y., A monoclonal antibody-based enzyme-linked immunosorbent assay of aflatoxin B1 in peanut products, *Mycotoxin Res.*, 4, 75, 1988.
93. Tseng, T. C. and Chung, C. S., A monoclonal antibody recognizing the specific epitope of aflatoxin analogs, *Bot. Bull. Acad. Sin.*, 31, 279, 1990.
94. Tseng, T. C., Li, I-shien, and Chao, Y. H., Preparation and characterization of a monoclonal antibody against aflatoxins B_1, B_2, G_1, and G_2, *Bot. Bull. Acad. Sin.*, 33, 369, 1992.
95. El-Nakib, O., Pestka, J. J., and Chu, F. S., Determination of aflatoxin B_1 in corn, wheat and peanut butter by enzyme-linked immunosorbent assay and solid phase radioimmunoassay, *J. Assoc. Off. Anal. Chem.*, 64, 1077, 1981.
96. Lee, S. and Chu, F. S., Radioimmunoassay of T-2 toxin in corn and wheat, *J. Assoc. Off. Anal. Chem.*, 64, 156, 1981.
97. Hewetson, J. F., Pace, J. G., and Beheler, J. E., Detection and quantitation of T-2 toxin mycotoxin in rat organ by radioimmunoassay, *J. Assoc. Off. Anal. Chem.*, 70, 654, 1987.

98. Hunter, K. W., Brimfield, A. A., Miller, M., Finkelman, F. D., and Chu, F. S., Preparation and characterization of monoclonal antibodies to the trichothecene mycotoxin T-2, *Appl. Environ. Microbiol.,* 49, 168, 1985.
99. Gendloff, E. H., Pestka, J. J., Dixon, D. E., and Hart, L. P., Production of a monoclonal antibody to T-2 toxin with strong cross-reactivity to T-2 metabolites, *Phytopathology,* 77, 57, 1987.
100. Chu, F. S., Immunochemical studies of mycotoxins, in *Toxigenic Fungi—Their Toxins and Health Hazard,* Kurata, H. and Meno, Y., Eds., Kodansha, Tokyo and Elsevier, Amsterdam, 1984, 234.
101. Schubring, S. L. and Chu, F. S., An indirect enzyme-linked immunosorbent assay for the detection of diacetoxyscirpenol in wheat and corn, *Mycotoxin Res.,* 3, 97, 1987.
102. Chu, F. S., Immunoassay for mycotoxins, in *Modern Methods in the Analysis and Structural Elucidation of Mycotoxins,* Cole, R. I., Ed., Academic Press, New York, 1986, 207.
103. Lin, M. T., Ram, B. P., Hart, L. P., and Pestka, J. J., Indirect enzyme-linked immunosorbent assay for the mycotoxin zearalenone, *Appl. Environ. Microbiol.,* 50, 332, 1985.
104. Dixon, D. E., Warxer, R. L., Ram, B. P., Hart, L. P., and Pestka, J. J., Hybridoma cell line production of a specific monoclonal antibody to the mycotoxins zearalenone and α-zearalenol, *J. Agric. Food Chem.,* 35, 122, 1987.
105. Candlish, A. A. G., The determination of mycotoxins in animal feeds by biological methods, in *Mycotoxins and Animal Foods,* Smith, J. E. and Henderson, R. S., Eds., CRC Press, London, 1991, 223.
106. Candish, A. A. G., Stimson, W. H., and Smith, J. E., Determination of ochratoxin A by monoclonal antibody-based enzyme immunoassay, *J. Assoc. Off. Anal. Chem.,* 71, 961, 1988.
107. Rousseau, D. M., Candlish, A. A. G., Slegers, G. A., Van Peteghem, C. H., Stimson W. H., and Smith, J. E., Detection of ochratoxin A in porcine kidneys by a monoclonal antibody-based radioimmunoassay, *Appl. Environ. Microbiol.,* 53, 514, 1987.
108. Davis, R. M. and Stone, S. S., Production of anti-rubratoxin antibody and its use in a radioimmunoassay for rubratoxin B, *Mycopathologia,* 67, 29, 1979.
109. Groopman, J. D. and Donahue, K. F., Aflatoxin, a human carcinogen: determination in foods and biological samples by monoclonal antibody affinity chromatography, *J. Assoc. Off. Anal. Chem.,* 71, 861, 1988.
110. Chu, F. S. and Lee, R. C., Immunochromatography of group A trichothecene mycotoxins, *Food Agric. Immunol.,* 1, 127, 1989.
111. Warden, B. A., Allam, K., Sentissi, A., Cecchini, D. J., and Giese, R. W., Repetitive hit-and-run fluoroimmunoassay for T-2 toxin, *Anal. Biochem.,* 162, 363, 1987.
112. Williamson, M. L., Atha, D. H., Reeder, D. J., and Sundaram, P. V., Anti-T-2 monoclonal antibody immobilization on quarts fibers: stability and recognition of T-2 mycotoxin, *Anal. Lett.,* 22, 803, 1989.
113. Al-Rubae, A. Y., *The Enzyme-Linked Immunosorbent Assay, a New Method for the Analysis of Pesticides Residues,* Doctor's thesis, Pennsylvania State University, Department of Entomology.
114. Estep, C. B., Menon, G. N., Williams, and Cole, H. C., Chlorinated hydrocarbon insecticide residues in Tennessee Honey and Beeswax, *Bull. Environ. Contam. Toxicol.,* 17, 168, 1977.
115. Harris, C. R. and Sans, W. W., Absorption of organochlorine insecticide residues from agricultural soils by crops used for animal feed, *Pest. Monit. J.,* 3, 182, 1969.
116. Brown, W. H., Witt, J. M., Whiting, F. M., and Stull, J. W., Secretion of DDT in fresh milk by cows, *Bull. Environ. Contam. Toxicol.,* 1, 21, 1966.
117. Ercegovich, C. D., Vallejo, R. D., Gettig, R. R., Woods, L., Bogus, E. R., and Mumma, R. O., Development of a radioimmunoassay for parathion, *J. Agric. Food Chem.,* 29, 559, 1981.
118. Hammock, B. D. and Mumma, R. O., Potential of immunochemical technology for pesticide analysis, in *Pesticide Analytical Methodology,* ACS Symposium Series 136, Zweig, G, Ed., American Chemical Society, Washington, D.C., 1980, 321.
119. Rittenburg, J. H., Grothaus, G. D., Fitzpatrick, D. A., and Lankow, R. K., in *Immunoassay for Trace Chemical Analysis,* ACS Symposium Series 451, Watkin, B. E. and Roberts, D. W., Eds., American Chemical Society, Washington, DC, 1990, chap. 3.
120. Brady, J. F., Fleeker, J. R., Wilson, R. A., and Mumma, R. O., in *Biomedical Monitoring for Pesticide Exposure,* ACS Symposium Series 382, Wang, R. G. M., Franklin, C. A., Honeywett, R. C., and Reinert, J. C., Eds., American Chemical Society, Washington, D.C., 1989, chap. 21.
121. Li, Q. X., Hammock, B. D., and Seiber, J. N., Development of an enzyme-linked immunosorbent assay for the herbicide bentazon, *J. Agric. Food Chem.,* 39, 1537, 1991.

122. Newsome, H. W. and Collins, P. G., Enzyme-linked immunosorbent assay of benomyl and thiabendazole in foods, *J. Assoc. Off. Anal. Chem.*, 70, 1025, 1987.
123. Hall, C. J., Deschamps, R. J. A., and Kreig, K. K., Immunoassays for detection of 2,4-D and picloram in river water, *J. Agric. Food Chem.*, 37, 981, 1989.
124. Wie, S. I. and Hammock, B. D., Development of enzyme-linked immunosorbent assay for residue analysis of diflubenzuron and BAY SIR 8514, *J. Agric. Food Chem.*, 30, 949, 1982.
125. Harrison, R. O., Braun, A. L., Gee, S. J., O'Brian, D. J., and Hammock, B. D., Evaluation of an ELISA for the direct analysis of molinate(Ordram) in rice field water, *Food Agric. Immunol.*, 1, 37, 1989.
126. Newson, W. H. and Collins, P. G., Immunochemical detection of aflatoxigenic potential of *Aspergillus* species with antisera prepared against enzymes specific to aflatoxin biosynthesis, *Food Agric. Immunol.*, 2, 75, 1990.
127. Newson, W. H., Development of an enzyme-linked immunosorbent assay for triadimefon in foods, *Bull. Environ. Contam. Toxicol.*, 36, 9, 1986.
128. Bushway, R. J., Perkins, B., Savage, S. A., Lekousi, S., and Ferguson, B. S., Determination of atrazine residues in water and soil by enzyme immunoassay, *Bull. Environ. Contam. Toxicol.*, 40, 647, 1988.
129. Giersch, T. and Hock, B., Production of monoclonal antibodies for the determination of S-triazines with enzyme immunoassay, *Food Agric. Immunol.*, 2, 85, 1990.
130. Schlaeppi, J. M., Fory, W., and Ramsteinner, K., Hydroxyatrazine and atrazine determination in solid and water by enzyme-linked immunosorbent assay using specific monoclonal antibodies, *J. Agric. Food Chem.*, 37, 1532, 1989.
131. Goh, K. S., Hernandez, J., Powell, S. J., Garretson, C., Troiano, J., Ray, M., and Greene, C. D., Enzyme immunoassay for determination of atrazine residues in soil, *Bull. Environ. Contam. Toxicol.*, 46, 30, 1991.
132. Newson, W. H., Potential and advantages of immunochemical methods for analysis of foods, *J. Assoc. Off. Anal. Chem.*, 69, 919, 1986.
133. Hunter, K. W., Jr. and Lenz, D. E., Detection and quantification of the organophosphate insecticide paraoxon by competitive inhibition enzyme immunoassay, *Life Sci.*, 30, 355, 1982.
134. Schwalbe, M., Dorn, E., and Beyerman, K., Enzyme immunoassay and fluoroimmunoassay for the herbicide diclofop-methyl, *J. Agric. Food Chem.*, 32, 734, 1984.
135. Rinder, D. F. and Fleeker, J. R., A radioimmunoassay to screen for 2,4-dichlorophenoxyacetic acid and 2,4,5-tricholorophenoxyacetic acid in surface water, *Bull. Environ. Contam. Toxicol.*, 26, 375, 1981.
136. Newson, W. H., An enzyme-linked immunosorbent assay for metalaxyl in foods, *J. Agric. Food Chem.*, 33, 528, 1985.
137. Niewola, Z., Hayward, C., Symington, B. A., and Robson, R. F., Quantitative estimation of paraquat by an enzyme linked immunosorbent assay using a monoclonal antibody, *Clin. Chim. Acta*, 148, 149, 1985.
138. Huber, S. J. and Hock, B., A solid-phase enzyme immunoassay for quantitative determination of herbicide terbutryn, *Z. Pflanzenkr. Pflanzenschutz*, 92, 147, 1985.
139. Kelley, M. M., Zahnow, E. W., Peterson, C., and Toy, S. T., Chlorsulfuron determination in extracts by enzyme immunoassay, *J. Agric. Food Chem.*, 33, 962, 1985.
140. Cook, C. E., Ballentine, N. H., Seltz, T. B., and Tallent, C. R., Warfarin enantiomer disposition: determination by stereoselective radioimmunoassay, *J. Pharmacol. Exp. Ther.*, 210, 391, 1979.
141. Wie, S. I., Sylwester, A. P., Wing, K. D., and Hammock, B. D., Synthesis of haptens and potential radioligand and development of antibodies to insect growth regulators diflubenzuron and BAY SIR 8514, *J. Agric. Food Chem.*, 30, 943, 1982.
142. McAdam, D. P., Hill, A. S., Beasley, H. L., and Skerritt, J. H., Mono- and polyclonal antibodies to the organophosphate fenitrothion 1. Approaches to hapten-protein conjugation, *J. Agric. Food Chem.*, 40, 1466, 1992.
143. Szurdoki, Ference, Bekleit, H. K. M., Marco, M. P., Goodrow, M. H., and Hammock, B. D., Synthesis of haptens and conjugates for an enzyme immunoassay for analysis of the herbicide bromacil, *J. Agric. Food Chem.*, 40, 1459, 1992.
144. Seiber, J. N., Xiao, Q., and Van Emon, J. M., *Barrier to Adopting Immunoassay in the Pesticide Analytical Laboratory*, ACS Symposium Series 442, American Chemical Society, Washington, D.C., 1990, 156–167.
145. Litman, D. J., in *Enzyme-Mediated Immunoassay*, Ngo, T. T. and Lenhoff, H. M., Eds., Plenum Press, New York, 1985, 155.

146. Zuk, R. F., Ginsberg, V. K., Houts, T., Rabbi, J., Merrick, H., Ullman, E. F., Fischer, M. M., Sizto, C. C., Stiso, S. N., and Litman, D. J., Enzyme immunochromatography—a quantitative immunoassay requiring no instrumentation, *Clin. Chem.,* 31, 1144, 1985.
147. Pollema, C. H., Ruzicka, J., and Christian, G. D., Sequential injection immunoassay utilizing immunomagnetic beads, *Anal. Chem.,* 64, 1356, 1992.
148. Sano, T., Smith, L., and Cantor, C. R., Immuno-PCR: very sensitive antigen detection by means of specific antibody DNA conjugates, *Science,* 258, 120, 1992.
149. Lawellini, D. W., Grant, D. W., and Joyce, B. K., Aflatoxin localization by enzyme-linked immunocytochemical techniques, *Appl. Environ. Microbiol.,* 34, 88, 1977.
150. Tseng, T. C., Lee, I-Shien, and Wang, J. J., High affinity monoclonal antibodies for aflatoxins and their application to localize aflatoxin B_1 in rat liver cell by immunogold labelling, *Toxicol. Lett., Suppl.,* 288, 1992.

FURTHER READING

Langane, J. J. and Van Vukanis, H., Eds., *Methods in Enzymology: Immunochemical Techniques,* Academic Press, New York, 1986.

Tijssen, P., *Practice and Theory of Enzyme Immunoassay,* PH Knippenberg, Amsterdam, 1985, 221.

Trione, E. J., The use of monoclonal antibodies in plant pathology, in *Experimental and Conceptual Plant Pathology,* Vol. I, *Techniques,* Singh, R. S., Singh, U. S., Hess, W. M., and Weber, D. J., Eds., Gordon and Breach Science Publishers, New York, 1988, 83.

Vanderlaan, M., Van Emon, J., Watkins, B., and Stanker, L., Monoclonal antibodies for the detection of trace chemicals, in *Pesticide Science and Biotechnology,* Greenhalgh, R. and Roberts, T. R., Eds., Blackwell Scientific, Oxford, 1987, 597.

Watkin, B. E. and Roberts, D. W., Eds., *Immunoassay for Trace Chemical Analysis,* ACA Symposium Series 451, American Chemical Society, Washington, D.C., 1990.

Chapter 32

Plant Disease Diagnosis: Biotechnological Approaches

Sally A. Miller

TABLE OF CONTENTS

I. INTRODUCTION

This chapter is an overview of some of the new approaches to plant disease diagnosis and pathogen detection that have come about in the last decade as a result of advances in biotechnology. It is focused on nucleic acid hybridization- and antibody-based techniques, and detailed descriptions of these and other methodologies that form the basis of modern plant disease diagnostics are to be found in other chapters. The reader is directed to these chapters to find in-depth treatments of these techniques. This chapter presents some of the techniques that have been applied to plant disease diagnostics and pathogen detection at the practical level, whether in diagnostic clinics or in the hands of growers or others involved in crop management, as well as those that are still primarily suitable for research laboratories, but have promise for future applications in practical diagnostics or studies of pathogen ecology and/or epidemiology.

II. TECHNOLOGIES

A. IMMUNOASSAYS

1. Enzyme-Linked Immunosorbent Assay

Modern serological techniques were introduced to plant pathology in the 1970s with ground-breaking papers by Voller et al.[1] and Clark and Adams,[2] describing the use of enzyme-linked immunosorbent assay (ELISA) to detect plant viruses. At the time, ELISA represented a significant advancement over other serological techniques, especially in terms of sensitivity, ability to quantify virus protein in plant extracts, and applicability to large-scale testing. It quickly became a standard technique for virus diagnostics, and antisera are routinely produced and widely available.[3] Over the years, many modifications of ELISA have been developed, and improvements in speed, accuracy, and instrumentation have been made. Other antibody-based technologies are available, including radioimmunoassay, immunofluorescence, and immunogold microscopy, but ELISA remains the dominant serological technique in plant pathology. Applications of ELISA have expanded well beyond virus detection to include a wide range of plant pathogens,[4] and include detection and quantitation of pathogens in plant tissue, soil, and water.[5]

There are numerous variations on the double-antibody sandwich (DAS-) ELISA format described by Clark and Adams.[2] Each format has characteristics that make it suitable for particular applications. In a typical DAS-ELISA, a specific capture antibody is immobilized onto a solid surface, such as the wells of a microtiter plate. The sample is added, and unbound material is washed away. Bound antigen is detected through the addition of a detecting antibody that has been conjugated with an enzyme,

0-87371-877-1/95/$0.00+$.50

typically horseradish peroxidase or alkaline phosphatase, and unbound material is again washed away. The presence of the detecting antibody is determined through the addition of a substrate for the enzyme. The amount of color that develops is proportional to the amount of antigen present in the sample (within the linear portion of the dose-response curve), and thus the assay is also quantitative. The intensity of the color is determined numerically through the use of automated equipment, although for qualitative uses, color change can be determined by eye. A modification of DAS-ELISA is indirect ELISA, in which the specific detecting antibody is not conjugated to an enzyme, but is detected with another enzyme-conjugated antibody specific to the Fc portion of the animal species in which the detecting antibody was made. While the direct methods usually require fewer steps to perform, the indirect methods are generally simpler to develop and may provide a broader range of reactivity.[6] The purification and enzyme conjugation steps required for direct ELISA make it impractical in instances where numerous specific detecting antibodies are used, e.g., when samples are being tested for many virus strains. On the other hand, the use of indirect ELISA limits the user to antigen capture methods, unless antibodies from two different source animals are used as capture and detecting antibodies (triple-antibody sandwich ELISA) or $F(ab')_2$ fragments are used to capture antigen.[7] Other variations on DAS-ELISA are too numerous to describe fully, but include the use of amplification systems such as avidin-biotin-enzyme complexes, which increase sensitivity,[8] and simultaneous incubation of the sample and enzyme-conjugated detecting antibody, which reduces the number of assay steps without sacrificing assay sensitivity.[9–11]

2. Immunofluorescence

Immunofluorescence is another antibody-based technique that has important applications in plant pathogen detection. While it has been used for detection of plant-pathogenic viruses[12] and fungi in plant tissue[13–15] and soil,[16,17] its most important application is in testing for the presence of plant-pathogenic bacteria in seed and other samples.[18] As in ELISA, both direct and indirect methods are used, although in practice indirect immunofluorescence is used most commonly. In the indirect technique, the primary antibody is detected by another antibody that has been conjugated with a fluorochrome such as fluorescein or rhodamine. Fluorescing cells are visualized by using a compound microscope equipped with fluorescence optics. Immunofluorescence may be used in combination with bacterial cell enrichment techniques, including the use of semiselective media and immunoisolation.[19] Concerns about cell viability can be addressed by using immunofluorescence colony staining, a procedure in which colonies embedded in agar medium are stained with fluorescein-conjugated antibodies.[19–21] The specificity of this technique has been vastly improved with the development of monoclonal antibodies to bacterial cell antigens. Immunofluorescence is also quite sensitive, with a limit of detection of about 10^3 to 10^4 cells per milliliter.[22] This technique allows the individual staining and counting of cells, as well as visual checks on cell morphology, which can be an advantage in seed testing and other programs. Immunofluorescence is considered to be somewhat more sensitive than ELISA for detection of bacteria,[21,23] although in some cases the sensitivities of the two assays have been shown to be roughly equivalent[24] or at least positively correlated.[25] The principle disadvantages of immunofluorescence techniques over ELISA are the absolute requirement for a good-quality microscope with epifluorescence optics, the fact that the assays can be tedious and time consuming, and technical problems resulting from plant and soil autofluorescence.

3. Dot Immunobinding

Dot immunobinding assays are immunoassays in which the immunochemical reactions are carried out on membranes made of nitrocellulose, nylon, or other materials (see Chapter 27 and References 26,27). Detecting antibodies are usually conjugated with enzymes, although colloidal gold has also been used.[28] These tests can be carried out in both direct and indirect formats. Where enzymes are used, the soluble substrate precipitates onto the solid surface after reacting with bound enzyme, instead of remaining in solution as in a microplate ELISA. Color intensity can be determined by reflectance using a calibrated meter, although for many applications a visual determination is made. Some of the commonly used configurations of dot immunobinding assays include dot-blot, slot-blot, and dipstick assays.[26,29–32] Membrane-based assays offer speed, convenience, and often increased sensitivity over microplate ELISAs.[32–36] However, quantitation is more difficult than in microplate ELISA.

An interesting variation on the dot immunobinding technique is the tissue print immunoblot, which has been shown to detect the endophyte *Acremonium coenophialum* in tall fescue seed and tillers,[37,38] and a variety of viruses as well as tomato big bud mycoplasma-like organism (MLO) in leaf tissue.[39] Plant tissue is pressed onto a nitrocellulose membrane, and antigens diffuse from the tissue and are bound onto the membranes. Bound antigens are detected by direct or indirect immunoassays using alkaline phosphatase as the enzyme marker. This type of assay was demonstrated to be as effective as a microplate ELISA and visual examination by light microscopy in detecting the tall fescue endophyte, and in fact was superior to ELISA in some instances where interference from seed proteins occurred. While tissue print immunoblot is not quantitative, its ability to localize antigens in tissues can be an advantage for some applications. It is also easy to perform and is suitable for large numbers of samples.[37,38]

4. Rapid Formats

For many routine diagnostic applications, simple rapid assays that require little equipment or formal training on the part of the user are needed. Commercial availability of standardized assays is critical for wide-scale routine application of ELISA for disease diagnosis, in diagnostic laboratories or directly by growers, consultants, and others in an advisory role in agriculture. Easy-to-use microplate ELISAs for detection of viruses and some bacteria have been available commercially to diagnostic laboratories and researchers for a number of years (e.g., products developed by AGDIA, Inc., Elkhart, IN). However, even simpler, more rapid ELISA formats have been developed in recent years, some of which are now commercially available. One such format is a "flow-through" ELISA that can be completed in 10 min.[40,41] This is a DAS-ELISA in which the primary antibody is immobilized onto beads or membranes on the surface of an absorbent plastic cylinder within a small well. The sample is macerated in a buffer, filtered, and added dropwise to the top of the cylinder where it is allowed to flow past the immobilized antibodies. The enzyme-conjugated second antibody is then added, followed by a rinse and addition of a precipitating substrate. Color intensity is recorded using a reflectance meter. Assays of this type have been developed to detect diseases caused by *Rhizoctonia solani, Pythium* spp., and *Sclerotinia homoeocarpa* in turf grass,[40,41] *Phytophthora* spp. in ornamental, vegetable, and other crops,[42] and *Septoria* in wheat.[43,44] The tests detect low levels of the pathogens in plant tissue and appear to correlate well with traditional isolation methods.[45] Similar products for the detection of the cereal foot rot pathogen,[46] *Botrytis cinerea* in grape extracts, and numerous plant protection chemicals have also been developed.[47] Another rapid assay designed for simplified testing is the "PIN ELISA", in which primary antibodies are coated onto pins attached to the cover of a microtiter plate (BIOREBA AG, Basel, Switzerland). The cover (with pins) is moved from plate to plate containing sample extract, conjugated antibody, and finally substrate, with tap water rinses in between. All reactions take place on the surface of the pins. The assay is slightly less sensitive than standard ELISA in detection of several plant viruses. Rapid ELISA tests are also available in standard microtiter plate formats for screening and quantitation of aflatoxin and other mycotoxins in grain samples (Neogen, Inc., East Lansing, MI).

5. Antibodies as Reagents

The backbone of ELISA and other immunoassays is the antibody component, and the availability of antibodies with the appropriate specificity and affinity for a particular pathogen in large part defines the quality of the assay. For many plant pathogens, particularly the less complex ones or those from which specific components (antigens) can be purified and characterized, polyclonal antisera can provide excellent detection capabilities. Specificity of an antiserum can also be improved by removing cross-reacting antibodies through cross absorption with heterologous antigens either through mixing and removal of precipitating antibody/antigen complexes or by affinity chromatography.[48] Polyclonal antisera are relatively easy and inexpensive to produce, and purification and conjugation with enzymes or other markers is usually straightforward. In some cases, a more broad specificity may be preferable, so that a wide range of strains of a particular pathogen can be detected. Polyclonal antisera have been particularly effective for plant virus diagnostics, and antisera are routinely produced and widely available through depositories such as the American Type Culture Collection (Rockville, MD) and through individual researchers. However, for more complex pathogens such as bacteria and fungi, polyclonal antisera often are not as specific for the target pathogen as needed. Cross reactivity cannot always be removed from antisera by cross absorption without removing too many homologous antibodies. Progress in development and application of serological techniques for these types of pathogens has been significantly slower than that for viruses, in large part as a result of the difficulty in obtaining highly specific antisera. However, the advent of monoclonal antibody technology has provided a means of overcoming this obstacle to the production of good-quality antibody reagents.

The monoclonal antibody concept was introduced in 1975,[49] and applications in plant pathology appeared in the early 1980s.[50] Since that time, monoclonal antibody technology has been applied widely to plant-pathogenic microorganisms[3,4,50,51] (see also Chapters by Wycoff and Brill, Jordan, Ohshima, and Forrest, this volume). It has been particularly useful for plant viruses that are difficult to purify, and eliminates the problem of plant background often observed in such cases with polyclonal antisera due to the presence of antibodies that react with plant components. For more complex pathogens, development of monoclonal antibodies may be the only means of overcoming cross reactivity with closely related, but nontarget microorganisms. Monoclonal antibodies capable of differentiating plant-pathogenic fungi at the species[43,52] and subspecies[31,53,54] levels have been developed in recent years. Similar results have been observed for plant-pathogenic bacteria[55–58] and nematodes.[59–60]

In addition to providing a high degree of specificity towards the target organism, monoclonal antibodies also provide a theoretically infinite source of quality reagent. Once a stable cell line has been established, it can be stored indefinitely and revived as needed for growth and antibody production. This is a significant advantage over polyclonal antisera, where batch to batch variation, either within breeds from the same animal or between different animals, can be a major problem. However, there are some disadvantages to the development and use of monoclonal antibodies compared to polyclonal antibodies. Production of monoclonal antibodies is a complicated, labor-intensive, time-consuming, and consequently expensive process. Monoclonal antibodies may be too specific for diagnostic applications, missing strains of the target pathogens. However, this can be overcome through the use of a cocktail of monoclonal antibodies to produce a "synthetic polyclonal" reagent that recognizes the complete range of variants of a given pathogen.[42] This requires extensive screening of the monoclonal antibodies against a large number of related and unrelated strains, again a time-consuming undertaking.

Monoclonal and polyclonal antibodies can be combined in immunoassays to provide both sensitivity and specificity towards a target pathogen. In indirect triple-antibody ELISA, the use of monoclonal antibodies derived from mice, and polyclonal antisera from rabbits, sheep, or other animals provides the two different source animal species required as well as the specificity of the monoclonal antibody.[61–63] For direct double-antibody ELISA, the combination of polyclonal and monoclonal antibodies may be better than either type used alone.[64,65] Whether the monoclonal antibody should be used as the capture antibody or enzyme-conjugated antibody should be determined empirically, since monoclonal antibodies may vary in their adaptability to particular uses.[65]

B. NUCLEIC ACID HYBRIDIZATION-BASED TECHNIQUES

The use of nucleic acid hybridization techniques to detect plant pathogens has been reviewed recently,[4,5,66,67] including a discourse on the comparative advantages and disadvantages of these and antibody-based techniques.[3] From a practical point of view, immunoassays are far more widely available and widely used than nucleic acid hybridization techniques for routine disease diagnosis and pathogen detection. While numerous immunoassays for this purpose are available commercially (see above), hybridization assays are, with a few exceptions (see below), still confined to research laboratories. There are also no nucleic acid hybridization assays that can compete with rapid immunoassays for speed and ease of use. However, nucleic acid hybridization techniques can provide specificity and/or sensitivity superior to that of immunoassays, particularly with the advent of polymerase chain reaction (PCR) technology. For example, it is now possible to detect point mutations in DNA through the use of PCR combined with allele-specific oligonucleotide analysis.[68] This technique has been used to detect point mutations in the β-tubulin gene of *Venturia inaequalis* that confers resistance to benomyl.[69]

1. Nucleic Acid Probes

Nucleic acid probes are sequences of nucleic acids that are labeled with a marker and used to detect complementary nucleic acid sequences. Techniques for cloning specific sequences are described (by Hadidi et al. and Young, this volume). Nucleic acid probes have been developed over the last decade for detection of a wide range of plant pathogens. Some examples are listed in Table 1.

Typically, the probes are used to detect nucleic acids of target organisms in dot-blot or squash hybridization assays. In these assays, small amounts of denatured nucleic acids are immobilized on a membrane, usually nitrocellulose or nylon, blocked, and probed with a specific nucleic acid probe. Hybridization between the sample and probe nucleic acids is detected by autoradiography if the marker is radioactive, or by colorimetric reactions if enzymatic systems are used. Plant samples can be extracted and applied to the membrane (dot-blot), or squashed directly onto the membrane (squash hybridization), then dried and stored before analysis. In this way, samples can be prepared in less than optimal conditions

Table 1 **Some examples of plant pathogens detected by nucleic acid probes**

Viroids	*Bacteria*
Avocado sunblotch[70]	*Erwinia caratovora*[87]
Potato spindle tuber[71]	*Xanthomonas campestris* pv. *oryzae*[88]
	X. campestris pv. *phaseoli*[89]
Viruses	*Clavibacter michiganense* subsp. *michiganense*[90]
Tomato spotted wilt[72,73]	*Pseudonomas syringae* pv. *tomato*[91]
Potato viruses S, X, Y, M[74–76]	*P. solanacearum*[92]
Tomato yellow leaf curl[77]	*E. herbicola*[93]
Barley yellow dwarf[78]	
Grapevine fanleaf[79]	*Fungi*
	Leptosphaeria korrae[94]
Mycoplasma-like Organisms	*Ophiosphaerella herpotricha*[95]
Western X-disease[80]	*Phytophthora parasitica*[96]
Aster yellows[81–83]	*P. megasperma* f. sp. *glycinea* (= *P. sojae*)[97]
Palm lethal yellowing[84]	*Gaeumannomyces graminis*[98]
Apple proliferation[85]	
Walnut witches'-broom[86]	*Nematodes*
	Globodera pallida[99]
	Heterodera glycines[100]

and mailed to laboratories for testing. This approach is being used for commercial testing of viroids, viruses, and MLOs (AGDIA, Inc., Elkhart, IN; Bresatec Ltd., Adelaide, Australia).

Perhaps the most significant roadblock to the widespread use of nucleic acid probes outside specialized laboratories has been the lack of easy-to-use, sensitive markers. While ^{32}P is a very sensitive marker, it poses restrictive handling and disposal problems. Fortunately, other marker systems are now available that may overcome some of the problems of radioactive labels. Nonradioactive labeling systems have been developed for the detection of numerous plant pathogens, including tomato yellow leaf curl virus,[77] tomato big bud, aster yellows, chrysanthemum yellows, and ash yellows MLOs,[101–104] *Pseudomonas syringae* pv. *tomato,*[91] pathogenic strains of *Erwinia herbicola,*[93] and the plant-parasitic nematode *Globodera pallida.*[99] Audy et al.[75] recently compared the sensitivity of four commercial labeling/detection systems to that of the standard ^{32}P label for detection of purified potato viruses. The use of sulfonated, biotinylated, and peroxidase-labeled cDNA probes resulted in about one tenth the sensitivity obtained using ^{32}P-labeled probes, while digoxigenin-labeled probes were equally as sensitive as the radioactive probes. The sulfonated and digoxigenin-labeled probes were detected colorimetrically, and the others were detected using X-ray films of chemiluminescence reactions. Although it was less sensitive than the ^{32}P and digoxigenin labeling systems, the peroxidase chemiluminescence system was considered to be the most convenient and adaptable to large-scale testing. However, none of these labeling systems were tested using infected plant tissue, so potential interference from components in plant sap, which has been seen in some other systems,[100] was not addressed.

2. Restriction Fragment Length Polymorphism (RFLP) Analysis

While RFLP analysis is generally too complex and time consuming to be used for routine diagnosis, it has become a useful tool in studying genetic relationships and variation within and between groups of plant pathogens. Genomic or organelle DNA is isolated from the microorganism of interest, purified, and digested with restriction endonucleases. Digested fragments of DNA are separated by gel electrophoresis and usually identified by hybridization with labeled DNA probes. The technique has been applied widely to plant pathogens (see recent reviews),[3] but has been particularly important in taxonomic studies of fungi. It has been used to differentiate isolates of *Phialophora gregata* from different hosts,[105] aggressive and nonaggressive isolates of *Ophiostoma ulmi*[106] and *Phoma lingam,*[107] and distinct genetic populations within *Colletotrichum gloeosporioides,*[108] as well as to clarify taxonomic issues in fungal genera including *Verticillium,*[109] *Fusarium,*[110,111] *Phytophthora,*[112–114] *Pythium,*[115,116] *Sclerotinia,*[117] and *Armillaria.*[118] Even finer distinctions among populations can be made by DNA fingerprinting, a technique in which restriction fragments of DNA are hybridized with simple repeat oligonucleotides or minisatellite

DNAs.[119,120] DNA fingerprinting has been used to distinguish isolates within populations of several plant-pathogenic fungi, including *Cryphonectria parasitica,*[121] *Mycosphaerella graminicola,*[122,123] and *Leptosphaeria maculans.*[124]

3. Polymerase Chain Reaction Techniques

A development that has generated considerable excitement in the last several years is the PCR, an *in vitro* method of amplifying sequences of DNA.[125] In PCR, sequences of DNA are amplified exponentially through repetitive cycles of DNA synthesis. Double-stranded DNA is converted to single strands at high temperatures, followed by annealing of oligonucleotide primers to target DNA and finally extension of the targeted DNA sequences through the action of a heat-stable DNA polymerase (see Hadidi et al., this volume, and Reference 126). The amplified product, which can represent a millionfold or higher amplification of the target sequence, can be detected by gel electrophoresis or dot-blot hybridization or used in RFLP analysis. PCR has tremendous potential in plant disease diagnosis and pathogen detection, and has already been applied widely among different pathogen groups. PCR will be especially useful in instances where high degrees of sensitivity and/or specificity are required. For example, Seal et al.[92] developed a highly specific DNA probe that routinely detected 10^5 to 10^6 cells of *Pseudomonas solanacearum.* However, when the probe was sequenced and oligonucleotide primers were produced, they could be used in PCR to detect as few as five bacterial cells. The authors noted that in the presence of excessive amounts of nontarget bacteria, sensitivity of the assay for the target was reduced. In another study, a limit of detection of approximately 100 cells of *E. amylovora* was demonstrated.[127] PCR technology has also been applied to the detection of plant-pathogenic viroids,[128,129] viruses,[130,131] MLOs,[132,133] bacteria,[134] fungi,[135–137] and nematodes.[138]

PCR is a fairly complicated technique that requires relatively expensive equipment and specialized technical support, and thus may not be immediately suitable for routine plant disease diagnosis outside specialized laboratories. Problems with interference from plant extracts and nontarget microorganisms must also be worked out reliably before PCR can find its way into routine diagnostic situations. However, it is an extremely promising technique, and research will undoubtedly continue in order to refine the technique for diagnostic applications.

III. CONCLUSION

In many ways, plant pathologists are faced with more difficult diagnostic problems than are their counterparts in human and veterinary medicine. Plant pathologists deal with many crop species and hundreds of pathogens ranging from viroids through parasitic plants, and have access to fewer products to assist in diagnosis. Agriculture lacks the extensive and highly developed infrastructure of the medical field for disease diagnosis, which includes a wide array of practitioners and supporting laboratories where sophisticated tests can be run. In medicine, these practitioners and laboratories provide an accessible market for diagnostic products, which encourages development of new technologies. While most states in the U.S. with significant agricultural production have at least one laboratory or clinic where diagnostic services are provided, by far the majority are in the public sector and often are underfunded and understaffed. In addition, the perceived value of a diagnosis for a plant disease is far less than that for a human or animal disease, and consumers are generally not willing to pay high prices for such services. These factors contribute overall to a relative lack of private sector involvement in development of diagnostic products for plant-related agriculture. Only a handful of companies worldwide have developed products for plant disease diagnosis. However, many new techniques have been brought forth during the last decade and are becoming available for applications in practical disease diagnosis. While immunoassays have led the way due to their relative economy, ease of use, and applicability, nucleic acid hybridization-based methods may also become more widely used, particularly as nonradioactive, user-friendly detection systems are developed and improved.

REFERENCES

1. Voller, A., Bartlett, A., Bidwell, D. E., Clark, M. F., and Adams, A. N., The detection of viruses by enzyme-linked immunosorbent assay (ELISA), *J. Gen. Virol.,* 33, 165, 1976.
2. Clark, M. F. and Adams, A. N., Characteristics of the microplate method of enzyme-linked immunosorbent assay for the detection of plant viruses, *J. Gen. Virol.,* 34, 475, 1977.

3. Chu, P. W. G., Waterhouse, P. M., Martin, R. R., and Gerlach, W. L., New approaches to the detection of microbial plant pathogens, *Biotechnol. Genet. Eng. Rev.*, 7, 45, 1989.
4. Miller, S. A. and Martin, R. R., Molecular diagnosis of plant disease, *Annu. Rev. Phytopathol.*, 26, 1988.
5. Miller, S. A. and Joaquim, T. R., Diagnostic techniques for plant pathogens, in *Biotechnological Prospects for Plant Disease Control,* Chet, I., Ed., Wiley-Liss, New York, 1993, 321.
6. van Regenmortel, M. H. V. and Burckard, J., Detection of a wide spectrum of tobacco mosaic virus strains by indirect enzyme-linked immunosorbent assays (ELISA), *Virology,* 106, 327, 1980.
7. Barbara, D. J. and Clark, M. F., A simple indirect ELISA using $F(ab')_2$ fragments of immunoglobulin, *J. Gen. Virol.*, 58, 315, 1982.
8. Shamsuddin, A. M. and Harris, C. C., Improved enzyme immunoassays using biotin-avidin-enzyme complex, *Arch. Pathol. Lab. Med.*, 107, 514, 1983.
9. van Vuurde, J. W. L. and Maat, D. Z., Enzyme-linked immunosorbent assay (ELISA) and disperse dye immuno assay (DIA): comparison of simultaneous and separate incubation of sample and conjugate for the routine detection of lettuce mosaic virus and pea early-browning virus in seeds, *Neth. J. Plant Pathol.*, 91, 3, 1985.
10. Flegg, C. L. and Clark, M. F., The detection of apple chlorotic leafspot virus by a modified procedure of enzyme-linked immunosorbent assay (ELISA), *Ann. Appl. Biol.*, 91, 61, 1979.
11. Resende, R., De O., De Avila, A. C., Goldbach, R. W., and Peters, D., Comparison of polyclonal antisera in the detection of tomato spotted wilt virus using the double antibody sandwich cocktail ELISA, *J. Phytopathol.*, 132, 46, 1991.
12. Gingery, R., Fluorescent antibody test, viruses, in *Serological Methods for Detection and Identification of Viral and Bacterial Plant Pathogens,* Hampton, R., Ball E., and De Boer, S., Eds., APS Press, St. Paul, MN, 1990, 287.
13. Nachmias, A. and Krikun, J., Diagnosis of *Verticillium dahliae* in potato seed tubers with immunoassay techniques, *Potato Res.*, 27, 423, 1984.
14. Kumar, S. U., Kalyanasundarum, R., and Kumar, V., The preparation and use of a fluorescent antibody reagent for the detection of *Fusarium vasinfectum, Can. J. Bot.*, 64, 1003, 1986.
15. Frankland, J. C., Bailey, A. D., Gray, T. R. G., and Holland, A. A., Development of an immunological technique for estimating mycelial biomass of *Mycena galopus* in leaf litter, *Soil Biol. Biochem.*, 13, 87, 1981.
16. Dewey, F. M., Barrett, D. K., Rose, I. R., and Lamb, C. J., Immunofluorescence microscopy for the detection and identification of propagules of *Phaeolus schweinitzii* in infested soil, *Phytopathology,* 74, 291, 1984.
17. MacDonald, J. D. and Duniway, J. M., Use of fluorescent antibodies to study the survival of *Phytophthora megasperma* and *P. cinnamomi* zoospores in soil, *Phytopathology,* 69, 436, 1979.
18. De Boer, S. H., Immunofluorescence for bacteria, in *Serological Methods for Detection and Identification of Viral and Bacterial Plant Pathogens,* Hampton, R., Ball E., and De Boer, S., Eds., APS Press, St. Paul, MN, 1990, 295.
19. van Vuurde, J. W. L., New approach in detecting phytopathogenic bacteria by combined immunoisolation and immunodetection assays, *EPPO Bull.*, 17, 139, 1987.
20. van Vuurde, J. W. L., Immunofluorescence colony staining, in *Serological Methods for Detection and Identification of Viral and Bacterial Plant Pathogens,* Hampton, R., Ball E., and De Boer, S., Eds., APS Press, St. Paul, MN, 1990, 299.
21. Franken, A. A. J. M. and van Vuurde, J. W. L., Problems and new approaches in the use of serology for seedborne bacteria, *Seed Sci. Technol.*, 18, 415, 1990.
22. De Boer, S. H., Current status and future prospects of bacterial ring rot testing, *Am. Potato J.*, 68, 107, 1991.
23. Wong, W. C., Methods for recovery and immunodetection of *Xanthomonas campestris* pv. *phaseoli* in navy bean seed, *J. Appl. Bacteriol.*, 71, 124, 1991.
24. Gudmestad, N. C., Baer, D., and Kurowski, C. J., Validating immunoassay test performance in the detection of *Corynebacterium sepedonicum* during the growing season, *Phytopathology,* 81, 475, 1991.
25. De Boer, S. H. and McCann, M., Determination of population densities of *Corynebacterium sepedonicum* in potato stems during the growing season, *Phytopathology,* 79, 946, 1989.
26. Hammond, J. and Jordan, R. L., Dot blots (viruses) and colony screening, in *Serological Methods for Detection and Identification of Viral and Bacterial Plant Pathogens,* Hampton, R., Ball E., and De Boer, S., Eds., APS Press, St. Paul, MN, 1990, 237.

27. Sherwood, J. L., Comparison of a filter paper immunobinding assay, Western blotting and an enzyme-linked immunosorbent assay for the detection of wheat streak mosaic virus, *J. Phytopathol.*, 118, 68, 1987.
28. Hsu, Y. H., Immunogold for detection of antigen on nitrocellulose paper, *Anal. Biochem.*, 142, 221, 1984.
29. Dewey, F. M., MacDonald, M. M., Phillips, S. I., and Priestly, R. A., Development of monoclonal-antibody-ELISA and -dip-stick immunoassays for *Penicillium islandicum* in rice grains, *J. Gen. Microbiol.*, 136, 753, 1990.
30. Mitchell, L. A., A sensitive dot immunoassay employing monoclonal antibodies for detection of *Sirococcus strobilinus* in spruce seed, *Plant Dis.*, 72, 664, 1988.
31. Dewey, F. M., Munday, C. J., and Brasier, C. M., Monoclonal antibodies to specific components of the Dutch elm disease pathogen *Ophiostoma ulmi*, *Plant Pathol.*, 38, 9, 1989.
32. Smith, F. D. and Banttari, E. E., Dot-ELISA on nitrocellulose membranes for detection of potato leafroll virus, *Plant Dis.*, 71, 795, 1987.
33. Banttari, E. E. and Goodwin, P. H., Detection of potato viruses S, X, and Y by enzyme-linked immunosorbent assay on nitrocellulose membranes (dot-ELISA), *Plant Dis.*, 69, 202, 1985.
34. Hsu, H. T. and Lawson, R. H., Direct tissue blotting for detection of tomato spotted wilt virus in *Impatiens*, *Plant Dis.*, 75, 292, 1991.
35. Parent, J.-G., Belanger, F., Desjardin, S., and Brisson, J. D., Dot-immunobinding for detection of tomato mosaic virus and potato virus X infecting greenhouse tomatoes, *Phytoprotection*, 66, 53, 1985.
36. Lazarovitz, G. B., The dot immunobinding assay (DIA)—bacteria, in *Serological Methods for Detection and Identification of Viral and Bacterial Plant Pathogens*, Hampton, R., Ball, E., and De Boer, S., Eds., APS Press, St. Paul, MN, 1990, 249.
37. Gwinn, K. D., Collins-Shepard, M. H., and Reddick, B. B., Detection of *Acremonium coenophialum* in tissue prints with immunoblot, *Tennessee Farm Home Sci.*, 158, 1991.
38. Gwinn, K. D., Collins-Shepard, M. H., and Reddick, B. B., Tissue print-immunoblot, an accurate method for the detection of *Acremonium coenophialum* in tall fescue, *Phytopathology*, 81, 747, 1991.
39. Lin, N. S., Hsu, Y. H., and Hsu, H. T., Immunological detection of plant viruses and a mycoplasma-like organism by direct tissue blotting on nitrocellulose membranes, *Phytopathology*, 80, 824, 1990.
40. Miller, S. A., Rittenburg, J. H., Petersen, F. P., and Grothaus, G. D., Application of rapid, field-usable immunoassays for the diagnosis and monitoring of fungal pathogens in plants, in *Brighton Crop Protection Conference—Pests and Diseases*, British Crop Protection Council, 1988, 795.
41. Miller, S. A., Rittenburg, J. H., Petersen, F. P., and Grothaus, G. D., From the research bench to the market place: development of commercial diagnostic kits, in *Techniques for the Rapid Detection of Plant Pathogens*, Duncan, J. M. and Torrance, L., Eds., Blackwell Scientific, Oxford, 1992, 208.
42. Miller, S. A., Rittenburg, J. H., Petersen, F. P., and Grothaus, G. D., Development of modern diagnostic tests and benefits to the farmer, in *Monoclonal Antibodies in Agriculture*, Schots, A., Ed., Pudoc, Wageningen, The Netherlands, 1990, 15.
43. Petersen, F. P., Rittenburg, J. H., Miller, S. A., and Grothaus, G. D., Development of monoclonal antibody-based immunoassays for detection and differentiation of *Septoria nodorum* and *S. tritici* in wheat, in *Brighton Crop Protection Conference—Pests and Diseases*, British Crop Protection Council, 1990, 751.
44. Mittermeier, L., Dercks, W., West, S. J. E., and Miller, S. A., Field results with a diagnostic system for the identification of *Septoria nodorum* and *Septoria tritici*, in *Brighton Crop Protection Conference—Pests and Diseases*, British Crop Protection Council, 1990, 757.
45. Benson, D. M., Detection of *Phytophthora cinnamomi* in azalea with commercial serological assay kits, *Plant Dis.*, 75, 478, 1991.
46. Lamison, C., Du Pont Advisor field test for cereal foot rot, *Phytopathology*, 82, 1113, 1992.
47. Charlton, R. R., Carlson, C. W., Kenyon, L. H., Carski, T. H., Smith, W. F., Steele, W. J., II, Liberatore, A., and Leung, K. S., Use of immunoassays to detect crop protection chemicals: applications in soil analysis and worker safety, *Phytopathology*, 82, 1168, 1992.
48. Ball, E. M., Hampton, R. O., De Boer, S. H., and Schaad, N. W., Polyclonal antibodies, in *Serological Methods for Detection and Identification of Viral and Bacterial Plant Pathogens*, Hampton, R., Ball E., and De Boer, S., Eds., APS Press, St. Paul, MN, 1990, 33.

49. Koehler, G. and Milstein, C., Continuous cultures of fused cells secreting antibody of predefined specificity, *Nature (London)*, 256, 495, 1975.
50. Halk, E. L. and De Boer, S. H., Monoclonal antibodies in plant-disease research, *Annu. Rev. Phytopathol.*, 23, 321, 1985.
51. Dewey, M., Evans, D., Coleman, J., Priestly, R., Hull, R., Horsley, D., and Hawes, C., Antibodies in plant science, *Acta Bot. Neerl.*, 40, 1, 1991.
52. Estrada-Garcia, M. T., Green, J. R., Booth, J. M., White, J. G., and Callow, J. A., Monoclonal antibodies to cell surface components of zoospores and cysts of the fungus *Pythium aphanidermatum* reveal species-specific antigens, *Exp. Mycol.*, 13, 348, 1989.
53. Hardham, A. R., Suzaki, E., and Perkin, J. L., Monoclonal antibodies to isolate-, species-, and genus-specific components on the surface of zoospores and cysts of the fungus *Phytophthora cinnamomi, Can. J. Bot.*, 64, 311, 1986.
54. Wong, W. C., White, M., and Wright, I. G., Production of monoclonal antibodies to *Fusarium oxysporum* f. sp. *cubense* race 4, *Lett. Appl. Microbiol.*, 6, 39, 1988.
55. Alvarez, A. M., Benedict, A. A., and Mizumoto, C. Y., Identification of Xanthomonads and grouping of strains of *Xanthomonas campestris* pv. *campestris* with monoclonal antibodies, *Phytopathology*, 75, 722, 1985.
56. Wingate, P. M., Norman, P. M., and Lamb, C. J., Analysis of the cell surface of *Pseudomonas syringae* pv. *glycinea* with monoclonal antibodies, *Mol. Plant-Microbe Interact.*, 3, 408, 1990.
57. Lamka, G. L., Hill, J. H., McGee, D. C., and Braun, E. J., Development of an immunosorbent assay for seedborne *Erwinia stewartii* in corn seeds, *Phytopathology*, 81, 839, 1991.
58. Benedict, A. A., Alvarez, A. M., Berestecky, J., Imanaka, W., Mizumoto, C. Y., Pollard, L. W., Mew, T. W., and Gonzalez, C. F., Pathovar-specific monoclonal antibodies for *Xanthomonas campestris* pv. *oryzicola, Phytopathology*, 79, 322, 1989.
59. Schots, A., Gommers, F. J., Bakker, J., and Egberts, E., Serological differentiation of plant-parasitic nematode species with polyclonal and monoclonal antibodies, *J. Nematol.*, 22, 16, 1990.
60. Jones, P., Ambler, D. J., and Robinson, M. P., The application of monoclonal antibodies to the diagnosis of plant pathogens and pests, in *Brighton Crop Protection Conference—Pests and Diseases*, British Crop Protection Council, 1988, 767.
61. Adam, G., Lesemann, D.-E., and Vetten, H. J., Monoclonal antibodies against tomato spotted wilt virus: characterization and application, *Ann. Appl. Biol.*, 118, 87, 1991.
62. Ellis, P. J. and Wieczorek, A., Production of monoclonal antibodies to beet western yellows virus and potato leafroll virus and their use in luteovirus detection, *Plant Dis.*, 76, 75, 1992.
63. D'Arcy, C. J., Hewings, A. D., and Eastman, C. E., Reliable detection of barley yellow dwarf viruses in field samples by monoclonal antibodies, *Plant Dis.*, 76, 273, 1992.
64. Haase, A., Richter, J., and Rabenstein, F., Monoclonal antibodies for detection and serotyping of cucumber mosaic virus, *J. Phytopathol.*, 127, 129, 1989.
65. Sherwood, J. L., Sanborn, M. R., and Keyser, G. C., Production of monoclonal antibodies to peanut mottle virus and their use in enzyme-linked immunosorbent assay and dot-immunobinding, *Phytopathology*, 77, 1158, 1987.
66. Salazar, L. F. and Querci, M., Detection of viroids and viruses by nucleic acid probes, in *Techniques for the Rapid Detection of Plant Pathogens*, Duncan, J. M. and Torrance, L., Eds., Blackwell Scientific, Oxford, 1992, 129.
67. Vivian, B., Identification of plant pathogenic bacteria using nucleic acid technology, in *Techniques for the Rapid Detection of Plant Pathogens*, Duncan, J. M. and Torrance, L., Eds., Blackwell Scientific, Oxford, 1992, 145.
68. Ala-Kokko, L., Baldwin, C. T., Moskowitz, R. W., and Prockop, D. J., Single base mutation in the type II procollagen gene (COL2A1) as a cause of primary osteoarthritis associated with a mild chondrodysplasia, *Proc. Natl. Acad. Sci. U.S.A.*, 87, 6565, 1990.
69. Koenraadt, H. and Jones, A. L., The use of allele-specific oligonucleotide probes to characterize resistance to benomyl in field strains of *Venturia inaequalis, Phytopathology*, 82, 1354, 1992.
70. Allen, R. N. and Dale, J. L., Application of rapid biochemical methods for detecting avocado sunblotch disease, *Ann. Appl. Biol.*, 98, 451, 1981.
71. Salazar, L. F., Owens, R. A., Smith, D. R., and Diener, T. O., Detection of potato spindle tuber viroid by nucleic acid spot hybridization: evaluation with tuber sprouts and true potato seed, *Am. Potato J.*, 60, 587, 1983.

72. Ronco, A. E., Bo, E. D., Ghiringhelli, P. D., Medrano, C., Romanowski, V., Sarachu, A. N., and Grau, O., Cloned cDNA probes for the detection of tomato spotted wilt virus, *Phytopathology,* 79, 1309, 1989.
73. Rice, D. J., German, T. L., Mau, R. F. L., and Fujimoto, F. M., Dot blot detection of tomato spotted wilt virus RNA in plant and thrips tissues by cDNA clones, *Plant Dis.,* 74, 274, 1990.
74. Weidemann, H. L. and Koenig, R., Differentiation of isolates of potato virus S which infect *Chenopodium quinoa* systematically by means of quantitative cDNA hybridization tests, *J. Plant Dis. Prot.,* 97, 323, 1990.
75. Audy, P., Parent, J.-G., and Asselin, A., A note on four nonradioactive labeling systems for dot hybridization detection of potato viruses, *Phytoprotection,* 72, 81, 1991.
76. Nokolaeva, O. V., Morozov, S. Yu., Zakhariev, V. M., and Skryabin, K. G., Improved dot-blot hybridization assay for large-scale detection of potato viruses in crude potato extracts, *J. Phytopathol.,* 129, 283, 1990.
77. Zilberstein, A., Navot, N., Ovadia, S., Reinhartz, A., Herzberg, M., and Czosnek, H., Field-usable assay for diagnosis of the tomato yellow leaf curl virus in squashes of plants and insects by hybridization with a chromogenic DNA probe, *Tech. J. Methods Cell Mol. Biol.,* 1, 118, 1989.
78. Fattouh, F. A., Ueng, P. P., Kawata, E. E., Barbara, D. J., Larkins, B. A., and Lister, R. M., Luteovirus relationships assessed by cDNA clones from barley yellow dwarf viruses, *Phytopathology,* 80, 913, 1990.
79. Fuchs, M., Pinck, M., Etienne, L., Pinck, L., and Walter, B., Characterization and detection of grapevine fanleaf virus by using cDNA probes, *Phytopathology,* 81, 559, 1991.
80. Kirkpatrick, B. C., Stenger, D. C., Morris, T. J., and Purcell, A. H., Cloning and detection of DNA from a nonculturable plant pathogenic mycoplasma-like organism, *Science,* 238, 197, 1987.
81. Kuske, C. R., Kirkpatrick, B. C., Davis, M. J., and Seemuller, E., DNA hybridization between western aster yellows mycoplasmalike organism plasmids and extrachromosomal DNA from other plant pathogenic mycoplasmalike organisms, *Mol. Plant-Microbe Interact.,* 4, 75, 1991.
82. Lee, I.-M. and Davis, R. E., Detection and investigation of genetic relatedness among aster yellows and other mycoplasmalike organisms by using cloned DNA and RNA probes, *Mol. Plant-Microbe Interact,* 1, 303, 1988.
83. Lee, I.-M., Davis, R. E., and Hiruki, C., Genetic interrelatedness among clover proliferation mycoplasmalike organisms (MLOs) and other MLOs investigated by nucleic acid hybridization and restriction fragment length polymorphism analysis, *Appl. Environ. Microbiol.,* 57, 3565, 1991.
84. Harrison, N. A., Bourne, C. M., Cox, R. L., Tsai, J. H., and Richardson, P. A., DNA probes for detection of mycoplasmalike organisms associated with lethal yellowing disease of palms in Florida, *Phytopathology,* 82, 216, 1992.
85. Bonnet, F., Saillard, C., Kollar, A., Seemuller, E., and Bove, J. M., Detection and differentiation of the mycoplasmalike organism associated with apple proliferation disease using cloned DNA probes, *Mol. Plant-Microbe Interact.,* 3, 438, 1990.
86. Chen, J., Chang, C. J., and Jarret, R. L., DNA probes as molecular markers to monitor the seasonal occurrence of walnut witches'-broom mycoplasmalike organism, *Plant Dis.,* 76, 1116, 1992.
87. Ward, L. J. and De Boer, S. H., A DNA probe specific for serologically diverse strains of *Erwinia caratovora, Phytopathology,* 80, 665, 1990.
88. Leach, J. E., White, F. F., Rhoads, M. L., and Leung, H., A repetitive DNA sequence differentiates *Xanthomonas campestris* pv. *oryzae* from other pathovars of *X. campestris, Mol. Plant-Microbe Interact.,* 3, 238, 1990.
89. Gilbertson, R. L., Maxwell, D. P., Hagedorn, D. J., and Leong, S. A., Development and application of a plasmid DNA probe for detection of bacteria causing common bacterial blight of bean, *Phytopathology,* 79, 518, 1989.
90. Thompson, E., Leary, J. V., and Chun, W. W. C., Specific detection of *Clavibacter michiganense* subsp. *michiganense* by a homologous DNA probe, *Phytopathology,* 79, 311, 1989.
91. Cuppels, D. A., Moore, R. A., and Morris, V. L., Construction and use of a nonradioactive DNA hybridization probe for detection of *Pseudomonas syringae* pv. *tomato* on tomato plants, *Appl. Environ. Microbiol.,* 56, 1743, 1990.
92. Seal, S. E., Jackson, L. A., and Daniels, M. J., Isolation of a *Pseudomonas solanacearum*-specific DNA probe by subtraction hybridization and construction of species-specific oligonucleotide primers for sensitive detection by the polymerase chain reaction, *Appl. Environ. Microbiol.,* 58, 3751, 1992.

93. Manulis, S., Evaluation of a DNA probe for the detection of *Erwinia herbicola* strains pathogenic on *Gypsophila paniculata, Plant Pathol.,* 41, 342, 1992.
94. Tisserat, N. A., Hulbert, S. H., and Nus, A., Identification of *Leptosphaeria korrae* by cloned DNA probes, *Phytopathology,* 81, 917, 1991.
95. Sauer, K. M., Hulbert, S. H., and Tisserat, N. A., Identification of *Ophiosphaerella herpotricha* by cloned DNA probes, *Phytopathology,* 83, 97, 1993.
96. Goodwin, P. H., Kirkpatrick, B. C., and Duniway, J. M., Identification of *Phytophthora citrophthora* with cloned DNA probes, *Appl. Environ. Microbiol.,* 56, 669, 1990.
97. Whisson, S. C., Maclean, D. J., Manners, J. M., and Irwin, J. A. G., Genetic relationships among Australian and North American isolates of *Phytophthora megasperma* f. sp. *glycinea* assessed by multicopy DNA probes, *Phytopathology,* 82, 863, 1992.
98. Henson, J. M., DNA probe for identification of the take-all fungus, *Gaeumannomyces graminis, Appl. Environ. Microbiol.,* 55, 284, 1989.
99. Burrows, P. R., The use of nucleic acid probes to identify plant parasitic nematodes, in *Brighton Crop Protection Conference—Pests and Diseases,* British Crop Protection Council, 1988, 811.
100. Besal, E. A., Powers, T. O., Radice, A. D., and Sandall, L. J., A DNA hybridization probe for detection of soybean cyst nematode, *Phytopathology,* 78, 1136, 1988.
101. Lee, I.-M., Davis, R. E., and DeWitt, N. D., Nonradioactive screening method for isolation of disease-specific probes to diagnose plant disease caused by mycoplasmalike organisms, *Appl. Environ. Microbiol.,* 56, 1471, 1990.
102. Davis, R. E., Sinclair, W. A., Lee, I.-M., and Dally, E. L., Cloned DNA probes specific for detection of a mycoplasmalike organism associated with ash yellows, *Mol. Plant-Microbe Interact.,* 5, 163, 1992.
103. Bertaccini, A., Davis, R. E., Lee, I.-M., Conti, M., Dally, E. L., and Douglas, S. M., Detection of chrysanthemum yellows mycoplasmalike organism by dot hybridization and Southern blot analysis, *Plant Dis.,* 74, 40, 1990.
104. Lee, I.-M. and Davis, R. E., Detection and investigation of genetic relatedness among aster yellows and other mycoplasmalike organisms by using cloned DNA and RNA probes, *Mol. Plant-Microbe Interact.,* 1, 303, 1988.
105. Gray, L. E. and Hepburn, A. G., Mitochondrial DNA restriction patterns of *Phialophora gregata* isolates from soybean and adzuki bean, *Phytopathology,* 82, 211, 1992.
106. Jeng, R. S., Duchesne, L. C., Sabourin, M., and Hubbes, M., Mitochondrial DNA restriction fragment length polymorphisms of aggressive and non-aggressive isolates of *Ophiostoma ulmi, Mycol. Res.,* 95, 537, 1991.
107. Hassan, A. K., Schulz, C., Sacristan, M. D., and Wostmeyer, J., Biochemical and molecular tools for the differentiation of aggressive and non-aggressive isolates of the oilseed rape pathogen, *Phoma lingam, J. Phytopathol.,* 131, 120, 1991.
108. Braithwaite, K. S., Irwin, J. A. G., and Manners, J. M., Restriction fragment length polymorphisms in *Colletotrichum gloeosporioides* infecting *Stylosanthes* spp. in Australia, *Mycol. Res.,* 94, 1129, 1990.
109. Carder, J. H. and Barbara, D. J., Molecular variation and restriction fragment length polymorphisms (RFLPs) within and between six species of *Verticillium, Mycol. Res.,* 95, 935, 1991.
110. Manicom, B. Q., Bar-Joseph, M., Rosner, A., Vigodsky-Haas, H., and Kotze, J. M., Potential applications of random DNA probes and restriction fragment length polymorphisms in the taxonomy of the Fusaria, *Phytopathology,* 77, 669, 1987.
111. Kistler, H. C., Momol, E. A., and Benny, U., Repetitive genomic sequences for determining relatedness among strains of *Fusarium oxysporum, Phytopathology,* 81, 331, 1991.
112. Forster, H., Kinscherf, T. G., Leong, S. A., and Maxwell, D. P., Restriction fragment length polymorphisms of the mitochondrial DNA of *Phytophthora megasperma* isolated from soybean, alfalfa, and fruit trees, *Can. J. Bot.,* 67, 529, 1989.
113. Forster, H. and Coffey, M. D., Molecular characterization of *Phytophthora* isolates with non-papillate sporangia causing root rot of raspberry using mtDNA restriction fragment length polymorphisms, *Mycol. Res.,* 96, 571, 1992.
114. Liew, E. C. Y., Maclean, D. J., Manners, J. M., Dawson, D., and Irwin, J. A. G., Use of restriction fragment length polymorphisms to study genetic relationships between Australian and Japanese isolates of *Phytophthora vignae, Aust. J. Bot.,* 39, 335, 1991.
115. Martin, F. N., Taxonomic classification of asexual isolates of *Pythium ultimum* based on cultural characteristics and mitochondrial DNA restriction patterns, *Exp. Mycol.,* 14, 47, 1990.

116. Martin, F. N. and Kistler, H. C., Species-specific banding patterns of restriction endonuclease-digested mitochondrial DNA from the genus *Pythium, Exp. Mycol.,* 14, 32, 1990.
117. Kohn, L. M., Petsche, D. M., Bailey, S. R., Novak, L. A., and Anderson, J. B., Restriction fragment length polymorphisms in nuclear and mitochondrial DNA of *Sclerotinia* species, *Phytopathology,* 78, 1047, 1988.
118. Anderson, J. B., Petsche, D. M., and Smith, M. L., Restriction fragment length polymorphisms in biological species of *Armillaria mellea, Mycologia,* 79, 69, 1987.
119. Jeffreys, A. J., Wilson, V., and Thein, S. L., Hypervariable 'minisatellite' regions in human DNA, *Nature,* 314, 67, 1985.
120. Jeffreys, A. J., Wilson, V., and Thein, S. L., Individual specific 'fingerprints' of human DNA, *Nature,* 316, 76, 1985.
121. Milgroom, M. G., Lipari, S. E., and Powell, W. A., DNA fingerprinting and analysis of population structure in the chestnut blight fungus, *Cryphonectria parasitica, Genetics,* 131, 297, 1992.
122. McDonald, B. A. and Martinez, J. P., DNA fingerprinting of the plant pathogenic fungus *Mycosphaerella graminicola* (anamorph *Septoria tritici*), *Exp. Mycol.,* 15, 146, 1991.
123. Zilberstein, A., Pnini-Cohen, S., Eyal, Z., Shuster, S., Hillel, J., and Lavi, U., Application of DNA fingerprinting for detecting genetical variation among isolates of the wheat pathogen *Mycosphaerella graminicola,* in *Biotechnological Prospects for Plant Disease Control,* Chet, I., Ed., Wiley-Liss, New York, 1993, 341.
124. Meyer, W., Lieckfeldt, E., Wostmeyer, J., and Borner, Th., DNA fingerprinting for differentiating aggressivity groups of the rapeseed pathogen *Leptosphaeria maculans, Mycol. Res.,* 96, 651, 1992.
125. Mullis, K. B. and Faloona, F. A., Specific synthesis of DNA in vitro via a polymerase-catalyzed chain reaction, *Methods Enzymol.,* 155, 335, 1987.
126. Stefan, R. J. and Atlas, R. M., Polymerase chain reaction: applications in environmental biology, *Annu. Rev. Microbiol.,* 45, 137, 1991.
127. Bereswill, S., Pahl, A., Bellemann, P., Zeller, W., and Geider, K., Sensitive and species-specific detection of *Erwinia amylovora* by polymerase chain reaction analysis, *Appl. Environ. Microbiol.,* 58, 3522, 1992.
128. Hadidi, A. and Yang, X., Polymerase chain reaction: molecular technology to enhance detection and diagnosis of pome fruit viroids, *Phytopathology,* 80, 982, 1990.
129. Levy, L., Hadidi, A., and Garnsey, S. M., Multiplex reverse transcription/polymerase chain reaction for the detection of mixed citrus viroids in a single reaction, *Phytopathology,* 81 (Abstr.) 1212, 1991.
130. Robertson, N. L., French, R., and Gray, S. M., Use of group-specific primers and the polymerase chain reaction for the detection and identification of luteoviruses, *J. Gen. Virol.,* 72, 1473, 1991.
131. Vunsh, R., Rosner, A., and Stein, A., The use of polymerase chain reaction (PCR) for the detection of bean yellow mosaic virus in gladiolus, *Ann. Appl. Biol.,* 117, 561, 1990.
132. Schaff, D., Lee, I.-M., and Davis, R. E., Sensitive detection and identification of mycoplasma-like organisms in plants by polymerase chain reactions, *Biochem. Biophys. Res. Commu.,* 186, 1503, 1992.
133. Ahrens, U. and Seemüller, E., Detection of DNA of plant pathogenic mycoplasmalike organisms by a polymerase chain reaction that amplifies a sequence of the 16S rRNA gene, *Phytopathology,* 82, 828, 1992.
134. Prossen, D., Hatziloukas, E., Panopoulos, N. J., and Schaad, N. W., Direct detection of the halo blight pathogen *Pseudomonas syringae* pv. *phaseolicola* in bean seeds by DNA amplification, *Phytopathology,* 81, 1159, 1991.
135. Schesser, K., Luder, A., and Henson, J. M., Use of the polymerase chain reaction to detect the take-all fungus, *Gaeumannomyces graminis,* in infected wheat plants, *Appl. Environ. Microbiol.,* 57, 553, 1991.
136. Henson, J. M., DNA hybridization and polymerase chain reaction (PCR) tests for identification of *Gaeumannomyces, Phialophora* and *Magnaporthe* isolates, *Mycol. Res.,* 96, 629, 1992.
137. Lee, S. B., White, T. J., and Taylor, J. W., Detection of *Phytophthora* species by oligonucleotide hybridization to amplified ribosomal DNA spacers, *Phytopathology,* 83, 177, 1993.
138. Harris, T. S., Sandall, L. J., and Powers, T. O., Identification of single *Meloidogyne* juveniles by polymerase chain reaction amplification of mitochondrial DNA, *J. Nematol.,* 22, 518, 1990.

FURTHER READING

Duncan, J. M. and Torrance, L., Eds., *Techniques for the Rapid Detection of Plant Pathogens,* Blackwell Scientific, Oxford, 1992, 235 pp.

Goding, J. W., *Monoclonal Antibodies: Principles and Practice,* Academic Press, London, 1983, 276 pp.

Hansen, M. A. and Wick, R. L., Plant disease diagnosis: present status and future prospects, in *Advances in Plant Pathology,* Vol. 10, Andrews, J. H., Ed., Academic Press, London, pp 65–126.

Hampton, R., Ball E., and De Boer, S., Eds., *Serological Methods for Detection and Identification of Viral and Bacterial Plant Pathogens,* APS Press, St. Paul, MN, 1990, 389 pp.

Harlow, E. and Lane, D., *Antibodies, a Laboratory Manual,* Cold Spring Harbor Laboratory, Cold Spring Harbor, NY, 1988, 726 pp.

Innis, M., Gelfand, D., Sninski, D., and White, T., *PCR Protocols: a Guide to Methods and Applications,* Academic Press, New York, 1990, 482 pp.

Saettler, A. W., Schaad, N. W., and Roth, D. A., Eds., *Detection of Bacteria in Seed and Other Planting Material,* APS Press, St. Paul, MN, 1989, 122 pp.

Sambrook, J., Fritsch, E. F., and Maniatis, T., *Molecular Cloning, a Laboratory Manual,* 2nd ed., Cold Spring Harbor Laboratory, Cold Spring Harbor, NY, 1989 (3 volumes).

Schotts, A., Ed., *Monoclonal Antibodies in Agriculture,* Pudoc, Wageningen, 1990, 92 pp.

Weir, D. M., Herzenberg, L. A., Blackwell, C., and Herzenberg, L. A., Eds., *Handbook of Experimental Immunology,* Vol. 1, *Immunochemistry,* 4th ed., Blackwell Scientific, Oxford, 1986.

Chapter 33

Monitoring of Biocontrol Agents and Genetically Engineered Microorganisms in the Environment: Biotechnological Approaches

Maarten H. Ryder

TABLE OF CONTENTS

0-87371-877-1/95/$0.00+$.50

I. INTRODUCTION

We need to detect biocontrol agents and genetically engineered microorganisms, both to study the ecology of these introduced organisms in the environment and for risk assessment. We may also want to monitor a biocontrol product in the environment, and the ability to do this may even be required in the not too distant future, for example, to comply with regulations.

The number of methods for detection of introduced organisms has increased dramatically in recent years, and there is now a variety of attractive possibilities for tracking organisms, especially since the application of recombinant DNA technology to this area. This review covers the methods currently available for specific detection of introduced organisms in the environment. It includes discussion of the advantages and disadvantages of the methods and examples of their use. Other reviews of this topic are referred to in "Further Reading" as well as in the reference section.

The recently developed technologies to detect introduced organisms in the environment offer a much greater variety of applications than traditional methods. Advantages of newer methods are that they can offer more specific and much greater ease of detection. They have also introduced exciting new possibilities in research. In some cases, we can determine not only the population present, but also the location and level of metabolic activity of the introduced microorganisms. We can follow genes and their transfer to other organisms, and we can enumerate organisms which become nonculturable (e.g., under starvation conditions). These are all important questions which have hitherto been difficult or impossible to answer. Most new methods have arisen out of advances in recombinant DNA technology, microbial genetics, and immunology (serology). However, the traditional methods can still be used very effectively in conjunction with newer methods.

For bacteria, the most common method of detection has traditionally been to use a selective medium for recovery of a spontaneous antibiotic-resistant derivative of the parent strain. This method is still often used in conjunction with newer detection methods and still has unique advantages. Because of its continued value, and its frequent use in conjunction with newer methods, this technique will also be discussed. There are far fewer examples of the specific detection of introduced fungi, but a large range of techniques, including those based on newer technologies, has been used.

For each group of methods there is a discussion of advantages, disadvantages, and important considerations. Examples are given, and in some cases there is also a section on ways to improve (lower) the detection limit.

In this review, selectivity refers to lack of background reaction or cross reaction. Thus a highly selective method allows detection without interference from other organisms. Sensitivity refers to detection limit. A highly sensitive method will allow the enumeration or detection of very low populations of an organism.

II. METHODOLOGY

Sections A to D describe methods for the detection of introduced bacteria. Following this, there is a section on tracking fungi and a brief mention of tracking viruses used in biological control. While many examples presented here refer to the detection of biological control agents and genetically engineered organisms, others are included because the methods are directly applicable to the subject of this review. For instance, some examples are taken from the plant pathology literature on the detection of pathogens, whereas others report the tracking of beneficial microorganisms other than biological control agents, especially *Rhizobium* and mycorrhizal fungi. There are also examples of detection of introduced microorganisms in aquatic environments. These are included because of the applicability of the methods to the terrestrial environment, and because risk assessment for release of genetically engineered organisms may require tests in aquatic microcosms.

A. SELECTIVE MEDIA AND SPONTANEOUS ANTIBIOTIC RESISTANCE

The bacterial strain of choice is marked by selecting a derivative that is resistant to an antibiotic that can be added to solidified agar media. Natural antibiotic resistance can also be used for detection of some organisms. The use of dilution plating techniques (spread plate, poured plate, droplet plate) on the particular selective medium then allows enumeration of microbial populations. Requirements are (a) that the organism be culturable, (b) that there be a low (or nondetectable) natural background of resistance to the chosen antibiotic in the environment into which the test strain is introduced, and (c) that the mutation to antibiotic resistance has not impaired the growth and function of the test strain.

There are many examples of the use of spontaneous antibiotic resistant derivatives to monitor bacteria, including biological control agents such as *Pseudomonas* spp.[1–3] and *Agrobacterium*[4] in the soil environment; *Erwinia herbicola* and *Enterobacter cloacae* in the phylloplane,[5] and *Escherichia coli* and *Pseudomonas* spp. in the aquatic environment.[6] Gram-negative bacteria have been most commonly detected in this way, but some Gram-positive organisms (*Bacillus* spp.[7] and *Clavibacter*[8] [= *Corynebacterium*]) have been tracked using spontaneous antibiotic resistance. The most popular antibiotics are rifampicin and nalidixic acid.

Virtually all genetically manipulated organisms so far released into the environment have been monitored. Antibiotic resistance is useful, especially where the genetic modification itself has not provided a selectable marker.[9] In other cases, the use of a spontaneous antibiotic resistant derivative has made the selective isolation of the genetically manipulated organism from the environmental sample much easier. For example, rifampicin resistance was used together with *lac*ZY marker genes to detect pseudomonads;[10] rifampicin resistance was combined with chromosomally inserted kanamycin resistance to monitor *Erwinia carotovora,*[11] and nalidixic acid and rifampicin resistance have been used together with plasmid-borne antibiotic resistances to detect *Pseudomonas* and *Escherichia coli* in water.[12]

Acea et al.[8] showed that for some soil bacteria, but not for others, the survival of antibiotic-resistant mutants in soil compared well to that of the parent strain. Compeau et al.[13] emphasized the point that some rifampicin-resistant mutants of *Pseudomonas* survived as well as the parent strain in soil, whereas others did not.

1. Advantages

The method is simple, relatively sensitive, and rapid. The data can be analyzed statistically, and the materials are normally relatively inexpensive. Assuming that resistance is on the chromosome, it allows the organism to be tracked rather than a particular gene.

Detection can be quite sensitive, as low as 10^3 colony forming units (CFU) per gram of sample. The limit depends upon the size of the sample that can be conveniently processed. For above-ground plant parts and for water samples, concentration of the sample prior to plating (e.g., by filtration) allows lower detection limits to be reached.

Sensitivity can be increased by enrichment and use of most probable number (MPN) enumeration.[14] The requirement here is that there should be strong selection against the growth of other microorganisms and a low frequency of appearance of new antibiotic resistance mutations in the environmental sample both during the enrichment phase and the subsequent selection phase, if performed.

2. Disadvantages

Use of this method is limited to environments where there is a low or nondetectable background of resistance to the particular antibiotic. A clean background is not always obtainable.[15] The use of a double-marked strain can overcome this problem, but this increases the risk of generating a derivative that does not survive or grow as well as the parent strain.

3. Important Considerations

When generating spontaneous antibiotic-resistant derivatives, the unintended alteration of bacterial properties such as growth rate and competitiveness in the environment is a serious problem. Normal practice should be to check that the antibiotic-resistant derivative behaves in the same way as the parent *in vitro* (growth rate in complex and also minimal medium if appropriate) and preferably also in the environment. Competition experiments and other tests, for example, biocontrol ability, can be used to check that the derivative behaves in the same way as the parent. The saturation constant (K_s) of the derivative could also be compared to that of the parent. This gives a measure of growth at less than optimal substrate concentration, which microorganisms are very likely to experience after release into the environment.

There is some uncertainty about the use of antibiotic resistance in longer-term environmental studies. Strains may not always be reisolated easily, and there is evidence for decreased antibiotic resistance in starved cells. Care should be taken when recovering bacteria from an environment where there are stresses such as starvation. Devanas et al.[16] showed that *E. coli* strains carrying plasmid-based antibiotic resistance were recovered poorly from soil unless they were first grown on a nonselective medium. Griffiths et al.[6] reported that *E. coli,* but not *P. fluorescens,* lost the ability to grow on media containing full-strength antibiotics after 49 d in starvation conditions in water. The cells were still alive and culturable as shown by their recovery on nonselective agar.

While there have been numerous studies of long-term survival of *Rhizobium* strains in soil, there have been few reports of the long-term stability (months or years) of antibiotic resistance markers in biological control agents. Glandorf et al.[17] demonstrated that rifampicin resistance was a suitable marker for *P. fluorescens* in field studies over a period of 4 months. Some combinations of organism and antibiotic resistance may be quite good, whereas others may be poor. We do not know enough to generalize, but we now can now cross check results obtained using antibiotic resistance markers with those obtained by other methods and thereby gain a better appreciation of the possibilities and the limitations of the method.

In summary, spontaneous antibiotic resistance markers can be useful, but the derivatives should be carefully tested before use in the environment. For long-term studies, recovery may be lowered when samples are plated on rich media with full concentration of antibiotics.

B. INTRODUCED MARKER GENES

The use of marker genes involves the addition of new DNA to an organism so that it can be uniquely identified and distinguished from other microorganisms in the environment into which it will be introduced. In this paper, marker genes are defined as genes that are introduced into an organism either on a plasmid or as an insertion into the chromosome and can be detected as a result of their expression in the new host organism.

A major consideration is whether the marker gene is located on a plasmid or on the chromosome. The method of choice is to use a chromosomal marker, because the chance of gene transfer to other organisms would normally be very much lower than for a plasmid-borne trait. Because of the relative ease of using plasmid-based markers, however, these have often preceded the development of chromosomal marker genes. The use of the *lac*ZY genes from *E. coli* to mark soil pseudomonads is an example.[10,18,19] A number of different types of marker genes have been used (Table 1).

1. Examples

a. New Metabolic Capability

Lactose utilization (*lac*ZY from *E. coli*) has been introduced into fluorescent pseudomonads as a plasmid-encoded or Tn*7*-based chromosomal marker.[10,19] The method has also been used for the nonfluorescent *P. corrugata.*[26a] Variations on the *lac*ZY theme include: (a) using *lac*Z from *Rhizobium* to mark *Pseudomonas*[27] and (b) a Mu d(*lac*) element to mark *P. aeruginosa.*[28] All of these methods rely on the use of 5-chloro-4-bromo-3-indolyl-β-D-galactopyranoside ("X-gal") or other synthetic β-galactosides as substrates to identify colonies of the marked strain on selective media. The β-galactosidase activity is constitutive in two of these examples.[19,28]

Bacteria carrying *xyl*E, a gene which codes for catechol-2,3-dioxygenase, can be readily identified by a yellow color reaction in colonies of the marked strain. This marker has been used to study the survival of *P. putida* in water and in soil[20,21,29] and *P. syringae* ice$^-$.[29a]

Apart from its use as a reporter gene, β-glucuronidase from *E. coli* can be used as a marker for detection of bacteria in the environment[22] with the substrate "X-gluc" in a selective medium.

Table 1 **Types of genetic markers used for detection of bacteria**

Marker	Example	Ref.
New metabolic capability	Lactose utilization (*lacZY*)	19
	Catechol dioxygenase(*xylE*)	20,21
	β-Glucuronidase (*gusA*)	22
Heavy metal resistance	Mercury resistance (*mer*)	23
	Arsenite resistance (*arsAB*)	23
Bioluminescence	*lux* operon, or part thereof	24,25
Herbicide resistance	Bialaphos resistance (*bar*)	23
Melanin gene	*mel* gene from *Streptomyces*	23
Transposon carrying antibiotic resistance	Tn*5*, Tn*903*	26,14

b. Heavy Metal Resistance

This category includes genes encoding resistance to mercury, organomercuric compounds, and to arsenite. Resistance to mercury is encoded by the *mer* operon of *Serratia marscescens.* Herrero et al.[23] developed methods for marking the chromosome of *P. putida* and a range of Gram-negative bacteria with part of the *mer* operon. While the method was developed for cloning, it can also be used to mark strains for ecological studies.

c. Bioluminescence

Genes that confer bioluminescence have been isolated from *Vibrio fischeri*[30] and *V. harveyi.*[31] Either the whole or parts of the *lux* operon have been used to monitor not only the survival and distribution of introduced bacteria,[25,32–36] but also the activity of the organism.[37,38] The *lux* marker genes have been used to monitor *Xanthomonas,*[25] *Pseudomonas* spp.,[33–35,37,39] and *Enterobacter.*[32] The addition of *lux*AB genes from *V. fischeri* to the chromosome of *P. fluorescens* strain 10586 did not impose a load on the organism.[39]

d. Herbicide Resistance

Resistance to the herbicide bialaphos (a tripeptide, phosphinotricin or "ptt", which is active against bacteria and plants) is coded for by the *bar* gene of *Streptomyces hygroscopicus.*[23] Ramos et al.[40] used the bialaphos resistance marker together with plasmid-coded *p*-ethylbenzoate degradation to specifically select a strain of *P. putida.*

e. Melanin Genes

A melanin gene from *S. antibioticus* was transferred to *Klebsiella pneumoniae,* giving a colony color that was easily recognizable.[23]

f. Transposons Carrying Antibiotic Resistance

Transposons which code for antibiotic resistance have been used frequently to monitor introduced organisms in soil, rhizosphere, and water. The most commonly used transposons are Tn*5* and Tn*903,* both of which encode kanamycin resistance. Bacteria that have been monitored in microcosms include *Azospirillum,*[41] *Pseudomonas* spp,[42–44] *Rhizobium,*[26] and *E. coli.*[45]

2. Advantages

The general advantages of using marker genes are similar to those for antibiotic-resistant mutants: the methods are simple, relatively inexpensive, quantitative, and statistics can be applied. As can be seen from Table 1, a wide variety of functions are available. In addition, cells can be visualized specifically *in situ* when bioluminescence genes are used. Another advantage is that the DNA sequences of the foreign genes are usually well known and therefore nucleic acid probes can also be used to monitor the organism (see Section II.C below).

Marker genes have the advantage that they can be specific and relatively sensitive where there is a low background. Sensitivity and specificity can be enhanced by combination with antibiotic resistance and enrichment can be performed to further increase sensitivity.

Bioluminescence genes have distinct advantages in microscope studies, enabling, for example, the detection of single cells in a soil slurry[34] as well as allowing enumeration of populations in soil suspensions by the use of a luminometer or other luminescence detector. Cells that are metabolically active, but not culturable on agar media, can be detected. The metabolic activity of the organism *in situ* may be assessed.[37,38]

Transposon-coded antibiotic resistance is simple to use, provided that the strain is amenable to insertion of a transposon. These markers can be used in the same way as spontaneous antibiotic resistance markers, but with the added advantage that specific DNA probes can be used for detection.[46] Increased sensitivity can be gained by enrichment and selective plating.[26]

3. Disadvantages

Some markers are not generally applicable: for example, the *lac*ZY genes can only be used for *lac*$^{-}$ organisms. Secondly, only culturable cells are recovered and enumerated. Marker genes can offer a greater specificity and sensitivity, but this is sometimes only achieved by the use of double-marked strains which also carry a spontaneous antibiotic resistance marker.[10]

The use of bioluminescence markers has the disadvantage that some types of equipment for detection of luminescence are expensive.

For transposons carrying antibiotic resistance genes, there are several disadvantages. Strains carrying these markers are not favored for environmental release because it is considered undesirable to release new antibiotic resistance determinants, especially resistance to medically important antibiotics. Plasmid-based transposons are even less favored, because plasmids are frequently conjugal or if not, may be mobilized to other bacteria, not necessarily of the same species or even genus. As a safeguard it is possible to use disarmed transposons, inserted in the chromosome. However, even here there is still a chance that the resistance can be transferred. Lastly, there may be problems with transposon instability.

4. Important Considerations

A suitable delivery system is required in order to mark the strain of choice, and the strains must be amenable to at least some types of genetic manipulation. The question of plasmid vs. chromosomal insertion is important because of the possibility of horizontal gene transfer.[47] The objectives of the work will determine which location is more appropriate. If the aim is to monitor a particular organism rather than a gene, chromosomal markers are preferred. If a plasmid-based marker gene is used, information on the transfer frequency *in vivo* would be desirable before field release.

5. Increased Sensitivity

This can be gained by enrichment coupled with MPN analysis.[14,26] Both selective plating and nucleic acid probes can be used for detection and enumeration of low populations. Scanferlato et al.[14] monitored a genetically manipulated *Erwinia carotovora* by incubating soil with a specific substrate (polypectate), and the antibiotics kanamycin, rifampicin, and cycloheximide. Enumeration was by MPN.

C. PROBES TO DNA AND RNA

Microorganisms can be detected by the use of specific probes to DNA or RNA sequences. A DNA probe is normally a short DNA sequence that matches and will bind uniquely to DNA of a particular organism or group of organisms, depending on the level of specificity desired. However, the labeled DNA that is used to detect an organism in an environmental sample may be based on anything from a whole genome to a plasmid, a part of a plasmid, a gene, or a short DNA sequence, as long as the required level of specificity is attained.

The sequences to which the probe binds in the genome of the introduced organism may be naturally occurring and unique amongst the microorganisms occurring in a particular environment. Alternatively, unique sequences can be produced or introduced via genetic manipulation. Within the latter category, probes can be made to detect introduced genes (e.g., Tn7-based *lac*ZY insertion).[48] The advantage of using nucleic acid probes to introduced marker genes is that the nature and the sequence of the foreign genes are usually well known. Thus, in many of the examples given in Section II.B above for use of marker genes, the organisms have been detected by nucleic acid probes as well as by selective plating.

Deletion of DNA can generate a junction that has a unique DNA sequence. Lindow and Panopoulos[9] detected an ice$^-$ derivative of *P. syringae* using a 21-bp probe sequence to the junction generated by deletion of part of the *ice* gene. It is also possible to insert a synthetic nucleotide sequence into the genome and to use a probe to that sequence to detect the organism.[49] Probes can be specifically targeted to rRNA sequences.[50]

DNA probes can be used to specifically detect organisms either by (1) hybridization to DNA from colonies that have been grown on culture media (colony hybridization)[51–55] or (2) hybridization with DNA extracted directly from soil or plant samples (direct detection).[56–61]

1. Examples

a. Colony Hybridization

Examples of colony hybridization are given in Table 2.

b. Direct Detection

Examples of direct detection are presented in Table 3.

Use of the direct detection methods requires isolation of nucleic acids from plant material, from soil, or from microorganisms extracted from soils or plants. The isolation of DNA from environmental samples has been reviewed recently.[62,63]

Table 2 **Detection of bacteria by colony hybridization**

Organism	Probe	Sample type	Ref.
Arthrobacter and *Pseudomonas putida*	Plasmid encoding 4-chlorobiphenyl degradation	Sediment	53
P. putida	*xylR*, for detection of TOL plasmid	Water	53
P. fluorescens	Part of Tn*5*	Soil	54
Escherichia coli	Plasmid encoding mercury resistance	Water	55

Table 3 **Direct detection of microorganisms using DNA hybridization**

Organism	Insertion/probe	Sample type	Ref.
Bradyrhizobium japonicum	Tn*5*/*nptII*	Soil	58
Pantoea agglomerans	*nif* sequence (3 kb)	Soil	60
Pseudomonas cepacia	Tn*1721*	Soil	61
Frankia sp.	Indigenous plasmid (8 kb)	Actinorhizal nodules	56

Probes can be based on whole chromosomal DNA, a specific insert in a cloning vector, whole plasmid DNA, part of a 16S rRNA sequence, or other specific oligonucleotide sequences. The type and size of the probe will depend on its specificity in the particular environment into which the organism is introduced. The background of cross reaction needs to be determined for each situation.

2. Advantages

The method can be very specific and sensitive, provided that the probe itself is specific. The direct detection of sequences of DNA from the environment allows measurement of nonculturable as well as culturable cells. DNA hybridization methodology is well developed, and quantitation is possible by using MPN-DNA hybridization. When used with a genetically engineered organism, the method can be used to follow the foreign gene, rather than the genome. Probes to mRNA could be developed to detect metabolic activity in both culturable and nonculturable cells. In some applications, such as slot-blot hybridization, many samples can be processed quickly, and routine analysis is possible.

The use of nucleic acid hybridization with fluorochrome-tagged nucleotides coupled with flow cytometry can allow very sensitive detection.[64]

3. Disadvantages

In the colony hybridization method, only culturable cells are detected, and when a nonselective medium is used, the usefulness of this method may be limited where there is a low frequency of positive colonies. The use of at least a semiselective medium will be advantageous, as demonstrated by Steffan et al.[52]

The methods are relatively expensive when compared to standard plating techniques, as access to a suitably equipped laboratory is required. In some applications, the method would not be suited to routine analysis.

When monitoring a gene rather than an organism, a high frequency of gene transfer to other organisms might not be detected. This could possibly be overcome by using the direct detection method in conjunction with colony hybridization. However, if different results were obtained using the two techniques, there is the difficulty that one might not distinguish whether cells of the introduced strain were becoming nonculturable or if the DNA was transferred to another type of organism which was not detectable by colony hybridization.

4. Important Consideration

The specificity of the nucleic acid probe needs to be checked thoroughly before it is used with samples collected from the environment.

5. Increased Sensitivity

This can be gained in two ways: (a) with amplification of the target DNA sequence by use of the polymerase chain reaction (PCR) and (b) by enrichment to increase the population of the target microorganism.

Table 4 **Detection of bacteria in soil and water using polymerase chain reaction (PCR)**

Organism	Amplified DNA/probe	Sample type	Ref.
Rhizobium leguminosarum	300 bp of Tn*5*/20 bp	Soil	67
Pseudomonas putida	Parts of *phoE* and colicin A from *E. coli*	*In vitro*	69
P. cepacia	1 kb of repeated sequence	Sediment	70
P. putida	72-bp synthetic DNA "numberplate"		49
Escherichia coli	0.3-kb plant DNA/17bp	Water	57

a. PCR Amplification

The use of amplification by PCR has been reviewed recently.[66] Amplification by PCR is used after extraction of DNA from environmental samples. Samples must be low in humic substances and any other extracted components that can interfere with the amplification step that is catalyzed by *Taq* polymerase. Methods of cleaning DNA for PCR are available. Recent improvements include separation of cells from soil material on a sucrose gradient combined with "double PCR"[67] and the use of cation exchange resin to release bacterial cells from soil particles.[61] Steps that are commonly taken include centrifugation in CsCl and "GeneClean" purification.[59]

Quantitative PCR has been reported by Picard et al.[68] for detection of *Agrobacterium tumefaciens* and for isolates of *Frankia*. Probes to the *vir* genes of *Agrobacterium* and 16S ribosomal RNA gene of *Frankia* were used.

Examples of the use of PCR to detect bacteria in soil and water are given in Table 4. A novel approach was taken by Amici et al.,[49] who inserted a 72-bp synthetic DNA sequence into *P. putida.* The central part of the DNA was an "identity sequence" which was flanked on each side by a 20-bp binding site for primers for PCR amplification. The novel *phoE/colA* gene used by Zaat et al.[69] to mark *P. putida* enabled detection by both PCR and immunological techniques.

When using PCR it is critical to check the level of specificity of the primer and the probe, and to optimize the conditions for PCR.

b. Enrichment

Enrichment to increase the population of the target microorganism, followed by dot-blot or slot-blot hybridization can be used together with MPN estimation to quantitate populations of organisms that are too low to be detectable by standard techniques.

Examples are

(a) Recovery of *Rhizobium,* labeled with Tn*5,* from soil using an enrichment medium containing the antibiotics rifampicin and kanamycin. Enumeration was by MPN.[26]
(b) Enrichment of soil for *P. fluorescens* carrying the Tn*7*::*lac*ZY marker by applying lactose to the soil, followed by selective plating and hybridization using part of Tn*7*.[48]
(c) Recovery of rifampicin-resistant *P. corrugata* carrying a Tn7-based chromosomal *lac*ZY insertion by incubating soil in a liquid medium containing rifampicin, cycloheximide, and lactose.[70a]

The presence and level of the marked organism was determined (a) by MPN/selective plating on solidified medium with glucose, X-gal, rifampicin, and cycloheximide, and (b) by MPN/slot-blot hybridization using labeled plasmid pMON7117[19] as a probe for Tn*7*::*lac*ZY. A mean of seven culturable cells per 100 g soil were detected 2 years after field release of the marked organism.[70a]

D. IMMUNOLOGICAL METHODS

The use of antibodies, including both polyclonal antisera and monoclonal antibodies, can be included among the traditional methods for finding introduced microorganisms in environmental samples. In recent years, however, there have been refinements that allow improved specificity and much greater sensitivity. These techniques are now quite useful for specific and sensitive detection. The improved methods include immunofluorescent colony staining (IFCS) and amplification steps that allow more sensitive detection in ELISA-based methods. The use of fluorescent tags on antibodies in combination with flow cytometry offers a further advance in technique.[64]

Polyclonal antisera are commonly raised against whole cells, so that the many antibodies will react with an antigenic component of the cell envelope. Foreign genes can also be inserted to give a novel

Table 5 **Detection of bacteria using antibodies**

Organism	Sample type	Ref.
Azospirillum brasilense	Rhizosphere, soil	71,72,75–77
Pseudomonas fluorescens with RP4::*pat*	Water	78
Rhizobium meliloti	Commercial inoculum, root nodules	74,79,80
R. leguminosarum with Tn*5*	Soil	26
Clavibacter michiganensis	Manure slurry	81

cell envelope protein that is detectable using an antibody.[69] Antibodies against an antigen in the cell envelope can in theory be used for light microscope and electron microscope studies, as well as for quantification of immunofluorescence (IF)-positive cells and for absorbance measurements in ELISA. For microscope studies, it is possible to use a labeled specific antibody or an unlabeled specific second antibody which is conjugated with either a fluorescent tag (e.g., fluorescein isothiocyanate or other fluorochrome),[71] or a gold particle tag.[72,73] The application of laser scanning confocal microscopy has the advantage that it reduces the autofluorescence of biological materials.[71] Antibodies against intracellular proteins, such as β-galactosidase, which can be found in some genetically manipulated organisms,[19] are probably only useful for transmission electron microscopy where the interior of the cell is exposed. Lack of specificity could sometimes be a problem unless an experimental system with a low background of cross reaction were chosen.

Quantification is normally achieved by direct counting of immunolabeled target cells or coupling the antibody-antigen reaction with a second reaction that is easily measured by standard biochemical techniques (e.g., ELISA). The latter methods commonly involve reaction of antibody-labeled alkaline phosphatase or horseradish peroxidase with a chromogenic substrate. An alternative to ELISA is immunoblot analysis on colonies that have been transferred onto nitrocellulose membrane.[74]

When antibodies are tagged with a fluorochrome, a flow cytometer can also be used to obtain quantitative and sensitive results.[64] Using a specific antibody, cells of the target strain can be sorted, counted, and can even be cultured afterwards, provided that they are in a culturable state. This combination of methods is simpler with aquatic samples, but can also be applied to soil samples.

1. Examples

Some examples for the detection of both genetically engineered and nonmodified bacteria are listed in Table 5.

Either polyclonal antisera or monoclonal antibodies may be used, and both Gram-negative and Gram-positive bacteria have been detected by immunological techniques. The soil bacterium *P. fluorescens* R2f with RP4::*pat* was detected using both IF and IFCS,[78] and the comparison of the two methods showed that nonculturable cells of the inoculated organism were present in environmental samples. Again, when *R. leguminosarum* carrying Tn*5* was monitored in soil with a fluorescent antibody,[26] a comparison of methods showed higher populations by the IF technique. This may have been due to the presence of nonculturable cells in the sample.

2. Advantages

Perhaps the greatest advantage of antibody-based detection is that it allows the visualization of cells *in situ,* by either light or electron microscopy, as well as being a quantitative technique. Recently, however, the use of bioluminescence genes and the fluorescent tagging of oligonucleotide probes have also permitted *in situ* visualization (see above). The ability to ascertain the precise location of an organism is of particular interest when bacteria are introduced into the rhizosphere. We may want to know the proportion of the root colonized by the introduced strain, and whether or not there is colonization of the interior of the root, for example.

Depending on the technique used, one can detect all cells bearing a recognizable antigen, whether the cell is live or dead, culturable or nonculturable. When IFCS is done on microscope slides and observations are made at high magnification, we can determine both the number of culturable and nonculturable cells in a sample.[82,83] IFCS allows isolation of viable bacteria directly from IFCS-positive colonies for confirmation of identity.[83]

3. Disadvantages

Nonspecific binding of antibodies can occur so that in some applications it is insensitive or difficult to achieve a satisfactory result. Sensitivity of ELISA is generally lower than for other methods; however, this is improving with novel ways to amplify the signal. Flow cytometry requires a very large initial investment of capital.

4. Increased Sensitivity

For quantification with enzyme reactions, sensitivity can be improved by using a second amplification step and by coupling the initial reaction with a chemical reaction such as chemoluminescence.[76,77] The latter gives a sensitivity 100 times greater than conventional ELISA.

The IFCS technique allows very sensitive detection. When the frequency of the target cells is low against a high background, a selective medium or at least a semiselective medium would be needed. The method has been used to find organisms *in situ* on roots[83] and *in planta* in sections.

Magnetic particles coated with a second antibody can be added to liquid samples and cells of the target strain can be retrieved using a super magnet.[84] This offers the possibility of enrichment and the retrieval of live cells.

E. DETECTION OF FUNGI AND VIRUSES

While there are many reports of the detection of introduced bacteria, there are relatively few that describe specific detection of introduced fungi and viruses. Nevertheless, a range of methods is potentially available for monitoring a fungal isolate after it has been introduced into the environment. The use of nucleic acid probes is the most common method to detect viruses.

1. Fungi

a. Antibiotic Resistance and Selective Media

This method has been used only rarely for fungi. Spontaneous resistance to cycloheximide at 75mg/l was used successfully by Chao et al.[85] to assess the colonization of soil and rhizosphere by isolates of *Trichoderma* spp. and by two other soil fungi, using both natural and sterilized soils.

Selective media are available for the isolation of some fungi from soil and rhizosphere. For example, *Talaromyces flavus*[86] and *Gliocladium* spp.[87] have been reisolated using such media. As an alternative, fungicide-resistant mutants can be used where the background of resistant organisms is low.

b. Marker Genes

The β-glucuronidase (*gusA*) gene of *Escherichia coli* has been used to label soilborne fungi. The *gus* marker gene enables specific detection using a synthetic β-glucuronide as a substrate. The pathogenic *Fusarium oxysporum* f.sp. *lini* was labeled with *gusA* as well as the selectable nitrate reductase marker (*nia*) from *Aspergillus nidulans*.[88] The tomato pathogen *Fulvia fulvia* (syn. *Cladosporium fulvum*) has also been marked with *gus*.[89]

c. Probes to DNA and RNA

Nucleic acid probes can be used to detect unmodified fungi and could also be used to monitor fungi that have been transformed with a plasmid.[90] Methods for extraction of DNA from fungi in environmental samples are not well developed; however, there are already methods for DNA extraction that allow PCR amplification of DNA from fungi in culture.[91] Genus-specific probes have been developed; for pathogens (e.g., *Gaeumannomyces*),[92] and the development of isolate-specific probes for introduced fungi appears to be quite feasible. Insertion of novel genes or DNA sequences into the fungal genome would make this method an attractive one.

Two species of *Verticillium* (*V. dahliae* and *V. albo-atrum*) have been distinguished using oligonucleotide probes that were based on minor differences between parts (the internal transcribed spacer regions) of their ribosomal RNA genes.[93] This type of technique could also be applied to fungal isolates that are deliberately introduced. Probes to rRNA sequences have also been used for the specific detection of *Glomus vesiculiferum* (a VA-mycorrhizal fungus) on the roots of leek plants.[94]

d. Immunological Methods

The VA-mycorrhizal fungus *Gigaspora margarita* has been specifically detected in soil and in the rhizosphere using a fluorescence-labeled polyclonal antiserum. The antiserum was useful only in soils where there was a low natural population of closely related fungi.[95] Hardham et al.[96] prepared a series of monoclonal antibodies to *Phytophthora cinnamomi* and some were found to be isolate specific, indicating the potential of the method for uniquely detecting introduced fungi.

e. General Comments

Most types of methods that have been used to monitor bacteria can also be used for the specific detection of fungi. Some degree of modification of the methods may be necessary. With our ability to manipulate soilborne fungi genetically,[88,97] methods for specific detection of introduced fungi should become more widely used.

2. Detection of Viruses

The detection of viruses is not covered in detail in this review. The topic is included in a review by Sayler and Layton,[98] where the use of DNA and RNA probes for detection of viruses is discussed. The biological control of chestnut blight by hypovirulence associated with the presence of double-stranded RNAs[99] is one example where a nucleic acid probe was used to track the spread of either a natural or an engineered virus in the fungal population.

III. COMPARISONS AND COMBINATIONS

The comments in this section refer to the detection of bacteria. Typical detection limits for various methods are shown in Table 6. Sensitivity can vary considerably. Generally speaking, the most sensitive methods, i.e., with lowest detection limit, are selective plating, with or without the use of marker genes, and nucleic acid probes. Immunological techniques have in the past been the least sensitive. However, combination with newer technologies such as flow cytometry, magnetic bead capture, and enrichment methods will probably overcome this deficiency.

Typical limits that can be obtained after enrichment or amplification steps are given in Table 7. Enrichment of samples using selective media and amplification of DNA sequences by PCR both allow lower detection limits to be reached, and quantification is possible via the use of MPN analysis.

Since a large range of monitoring techniques are available, more than one method can often be used for detection of an introduced organism in an environmental sample. This is potentially very powerful, partly because it allows us to cross check methods, so that our confidence in the results obtained by different methods is much increased. Several very useful studies have compared methods for the recovery and enumeration of the same organism in the same samples.[26,52,78] Variation in results between methods does occur, for example, higher counts have been obtained from immunological detection methods,[26,78] and this may partly be accounted for by the lack of culturability of a portion of the population of cells which, nevertheless, still react with antisera or antibodies. Variation in results between methods would be expected, as some will detect all cells, and others live cells only, or culturable cells only.

It can be advantageous to combine selective and diagnostic markers, and also to combine markers with other detection methods. There are many examples, viz., Tn-based antibiotic resistance + biolumi-

Table 6 **Typical detection limits for methods of monitoring bacteria**

Method	Detection limit (per g)	Example	Ref.
Selective plating (antibiotic resistance)	10^2–10^3, but can be 10 to 100	Numerous	
Antibody based	10^4–10^5, generally		
"CIA"[a]	10^3	*Azospirillum*	77
IFCS	1–5	*Erwinia*	82
IFCS + Rif	20	*Pseudomonas fluorescens*	100
DNA probes (soil DNA)	10^4	*Bradyrhizobium japonicum*	58
Marker genes			
lacZY	25 (can be < 10)	*P. fluorescens*	10
xylE	10	*P. putida*	21
luxAB	10^2–10^3	*Escherichia coli*	38
Tn antibiotic resistance	10–100	*R. leguminosarum, P. putida*	26

[a]Chemoluminescence-linked immunoassay

Table 7 **Detection limits after enrichment or amplification**

Method	Detection limit	Example	Ref.
Marker genes plus enrichment	1–20 cells/g (Hg^R, Tc^R)	*Escherichia coli*	16
	1 per 2 g (*lux*, Tc^R, Rif^R)	*Xanthomonas campestris*	25
Marker genes/Tn plus enrichment	1–10 per 10 g (Tn*903*, Rif^R)	*Erwinia carotovora*	14
	1–10 per 10 g (*lacZY*, Rif^R)	*Pseudomonas corrugata*)	100a
DNA probe, PCR, and MPN (total copies of sequence)	1 per g	*P. cepacia*	70
DNA probe plus enrichment and MPN	1–10 per 10 g (*lacZY*, Rif^R)	*P. corrugata*	100a

nescence;[101] Tn-based antibiotic resistance + DNA probe;[46,102] marker gene + spontaneous antibiotic resistance[17,40] marker gene + antibody detection; spontaneous antibiotic resistance + antibody detection.[100] It is noteworthy that in this context, spontaneous antibiotic resistance markers are still often used as selectable markers.

Advantages of combining methods in this way are that we can have (a) greater certainty about the unambiguous detection of the target strain; (b) greater ease of detection; (c) greater sensitivity.

IV. CONCLUSIONS

The diversity of methods available for tracking microorganisms that are introduced into the environment is much greater since the introduction of recombinant DNA technology and improved immunological methods, and the range of techniques will probably continue to expand. For biological control agents, the choice of methods for detection is more limited than for genetically engineered organisms. However, even with nonengineered organisms there is potential for the development of isolate-specific nucleic acid probes, for example, probes to ribosomal RNA sequences.

The choice of method will depend on the nature of the target organism, the background interference from other organisms that can be expected with the various methods, the availability and ease of use of genetic markers, DNA probes, antisera, or antibodies. Other considerations will be the cost and the number of samples that must be processed routinely.

Each method has its advantages and disadvantages, and there are scientific and financial considerations which will influence the final decision. For example, if the ability to find an organism *in situ* in cut sections of plant tissue is a high priority, this may limit the choice to immunological or bioluminescence marker techniques. If the organism cannot be easily manipulated genetically, then immunological methods may be the most appropriate.

The purpose of the monitoring program will also play a role in the selection of a method. Different techniques will allow us to count either all cells of a particular organism, all intact DNA sequences, culturable cells only, or live cells only. Some techniques will also permit measurement of gene expression or the metabolic state of the organism. In many cases, two or more detection methods can be combined, and this brings its own advantages, including greater sensitivity and selectivity, and greater ease of detection.

In addition to deciding upon the method of detection, attention should be given to a number of other issues which have not been addressed in this review. These include the reisolation of the organism or its DNA or RNA from soil, rhizosphere, above-ground plant parts, or water; the culturability of the organism, especially in long-term studies;[103] sampling strategy;[5] and comparison of results from microcosms with field results.[104–106]

Given that soil fungi are used frequently as biological control agents (e.g., *Trichoderma,*[107,108] *Pythium nunn,*[109] and isolates of binucleate *Rhizoctonia,*[110] and plant growth-promoting agents (VA- and ectomycorrhizal fungi), greater attention to tracking this group of microorganisms is warranted. At the same time, improved methods for quantifying plant pathogens, many of which are fungal, will allow us to study pathogens, biological control agents, and their interactions much more effectively.

There is a continued effort to develop improved and less expensive methods for routine extraction of DNA and RNA from soil and other environmental samples. Success in this area would open up opportunities for the wider use of nucleic acid probes and easier quantitation of results.

The need to monitor the fate of genetically engineered microorganisms in the environment is clear. Fortunately, the engineered organisms can usually be tracked readily, because they have been altered genetically and this alteration can be detected by using nucleic acid probes. The ability to track both genetically engineered microorganisms and biological control agents is necessary for both ecological studies of their fate after release in relation to their performance in the field, and for the accumulation of data for use in risk assessment.

ACKNOWLEDGMENTS

I thank Y. Couteaudier, F. Dane, J. D. van Elsas, P. G. Hartel, A. Hartmann, J. E. Hollebone, C. Jacobsen, K. Killham, J. W. Kloepper, S. E. Lindow, E. J. J. Lugtenberg, G. S. Sayler, and L. A. de Weger for providing materials prior to publication. I am grateful to C. E. Pankhurst, K. M. Ophel-Keller, J. van Vuurde and A. Hartmann for helpful discussions and suggestions on the manuscript.

REFERENCES

1. Weller, D. M., Colonization of wheat roots by a fluorescent pseudomonad suppressive to take-all, *Phytopathology,* 73, 1548, 1983.
2. Davies, K. G. and Whitbread, R., Factors affecting the colonisation of a root system by fluorescent pseudomonads: the effects of water, temperature and soil microflora, *Plant Soil,* 116, 247, 1989.
3. Dupler, M. and Baker, R., Survival of *Pseudomonas putida,* a biological control agent, in soil, *Phytopathology,* 74, 195, 1984.
4. Macrae, S., Thomson, J. A., and van Staden, J., Colonization of tomato plants by two agrocin-producing strains of *Agrobacterium tumefaciens, Appl. Environ. Microbiol.,* 54, 3133, 1988.
5. Donegan, D., Matyac, C., Seidler, R., and Porteous, A., Evaluation of methods for sampling, recovery, and enumeration of bacteria in the phylloplane, *Appl. Environ. Microbiol.,* 57, 51, 1991.
6. Griffiths, R. P., Moyer, C. L., Caldwell, B. A., Ye, C., and Morita, R. Y., Long-term starvation-induced loss of antibiotic resistance in bacteria, *Microb. Ecol.,* 19, 251, 1990.
7. Juhnke, M. E., Mathre, D. E., and Sands, D. C., Identification and characterization of rhizosphere-competent bacteria of wheat, *Appl. Environ. Microbiol.,* 53, 2793, 1987.
8. Acea, M. J., Moore, C. R., and Alexander, M., Survival and growth of bacteria introduced into soil, *Soil Biol. Biochem.,* 20, 509, 1988.
9. Lindow, S. E. and Panopoulos, N. J., Field tests of recombinant ice$^-$ *Pseudomonas syringae* for biological frost control in potato, in *Release of Genetically-Engineered Micro-organisms,* Sussman, M., Collins, C., Skinner, F., and Stewart-Tull, D., Eds., Academic Press, San Diego, 1988, 121.
10. Drahos, D. J., Hemming, B. C., and McPherson, S., Tracking recombinant organisms in the environment: β-galactosidase as a selectable non-antibiotic marker for fluorescent pseudomonads, *Bio/Technology,* 4, 439, 1986.
11. Scanferlato, V. S., Orvos, D. R., Cairns, J., Jr., and Lacy, G. H., Genetically engineered *Erwinia carotovora* in aquatic microcosms: survival and effects on functional groups of indigenous bacteria, *Appl. Environ. Microbiol.,* 55, 1477, 1989.
12. Genthner, F. J., Upadhyay, J., Campbell, R. P., and Sharak Genthner, B. R., Anomalies in the enumeration of starved bacteria on culture media containing nalidixic acid and tetracycline, *Microb. Ecol.,* 20, 283, 1990.
13. Compeau, G., Al-achi, B. J., Platsouka, E., and Levy, S. B., Survival of rifampin-resistant mutants of *Pseudomonas fluorescens* and *Pseudomonas putida* in soil systems, *Appl. Environ. Microbiol.,* 54, 2432, 1988.
14. Scanferlato, V. S., Orvos, D. R., Lacy, G. H., and Cairns, J., Jr., Enumerating low densities of genetically engineered *Erwinia carotovora* in soil, *Lett. Appl. Microbiol.,* 10, 55, 1990.
15. Schmidt, F. R. J., Rosien, J., and Brokamp, A., The role of soil bacteria in risk assessment analysis, in *Bacterial Genetics in Natural Environments,* Fry, J. C. and Day, M. J., Eds., Chapman and Hall, London, 1990, 207.
16. Devanas, M. A., Rafaoli-Eshkol, D., and Stotzky, G., Survival of plasmid-containing strains of *Escherichia coli* in soil: effect of plasmid size and nutrients on survival of hosts and maintenance of plasmids, *Curr. Microbiol.,* 13, 269, 1986.

17. Glandorf, D. C. M., Brand, I., Bakker, P. A. H. M., and Schippers, B., Stability of rifampicin resistance as a marker for root colonization studies of *Pseudomonas putida* in the field, *Plant Soil,* 147, 135, 1992.
18. Barry, G. F., Permanent insertion of foreign genes into the chromosomes of soil bacteria, *Bio/Technology,* 4, 446, 1986.
19. Barry, G. F., A broad-host-range shuttle system for gene insertion into the chromosomes of Gram-negative bacteria, *Gene,* 71, 75, 1988.
20. Winstanley, C., Morgan, J. A. W., Pickup, R. W., Jones, J. G., and Saunders, J. R., Differential regulation of lambda p_L and p_R promoters by a *c*I repressor in a broad-host-range thermoregulated plasmid marker system, *Appl. Environ. Microbiol.,* 55, 771, 1989.
21. Winstanley, C., Morgan, J. A. W., Pickup, R. W., and Saunders, J. R., Use of a *xyl*E marker gene to monitor survival of recombinant *Pseudomonas putida* populations in lake water by culture on nonselective media, *Appl. Environ. Microbiol.,* 57, 1905, 1991.
22. Jefferson, R. A., The GUS gene reporter system, *Nature,* 342, 837, 1989.
23. Herrero, M., de Lorenzo, V., and Timmis, K. N., Transposon vectors containing non-antibiotic resistance selection markers for cloning and stable chromosomal insertion of foreign genes in Gram-negative bacteria, *J. Bacteriol.,* 172, 6557, 1990.
24. Shaw, J. J. and Kado, C. I., Development of a *Vibrio* bioluminescence gene-set to monitor phytopathogenic bacteria during the ongoing disease process in a non-disruptive manner, *Bio/Technology,* 4, 560, 1986.
25. Shaw, J. J., Dane, F., Geiger, D., and Kloepper, J. W., Use of bioluminescence for detection of genetically engineered microorganisms released into the environment, *Appl. Environ. Microbiol.,* 58, 267, 1992.
26. Fredrickson, J. K., Bezdicek, D. F., Brockman, F. J., and Li, S. W., Enumeration of Tn*5* mutant bacteria in soil by using a most-probable-number-DNA hybridization procedure and antibiotic resistance, *Appl. Environ. Microbiol.,* 54, 446, 1988.
26a. Ryder, M. H. et al., unpublished, 1990.
27. O'Gara F., Boesten, B., and Fanning, S., The development and exploitation of 'marker genes' suitable for risk evaluation studies on the release of genetically engineered microorganisms in soil, in *Risk Assessment for Deliberate Releases,* Klingmüller, W., Ed., Springer-Verlag, Berlin, 1988, 50.
28. Höfte, M., Mergeay, M., and Verstraete, W., Marking the rhizopseudomonas strain 7NSK2 with a Mu d(*lac*) element for ecological studies, *Appl. Environ. Microbiol.,* 56, 1046, 1990.
29. MacNaughton, S. J., Rose, D. A., and O'Donnell, A. G., Persistence of a *xyl*E marker gene in *Pseudomonas putida* introduced into soils of differing texture, *J. Gen. Microbiol.,* 138, 667, 1192.
29a. Lindow, S. E., personal communication, 1992.
30. Engebrecht, J., Nealson, K., and Silverman, M., Bacterial bioluminescence: isolation and genetic analysis of functions from *Vibrio fischeri, Cell,* 32, 773, 1983.
31. Karp, M., Expression of bacterial luciferase genes from *Vibrio harveyi* in *Bacillus subtilis* and in *Escherichia coli, Biochim. Biophys. Acta,* 1007, 84, 1989.
32. Fravel, D. R., Lumsden, R. D., and Roberts, D. P., *In situ* visualization of the biocontrol rhizobacterium *Enterobacter cloacae* with bioluminescence, *Plant Soil,* 125, 233, 1990.
33. King, J. M. H., DiGrazia, P. M., Applegate, B., Burlage, R., Sanseverino, J., Dunbar, P., Larimer, F., and Sayler, G. S., Rapid, sensitive bioluminescent reporter technology for naphthalene exposure and biodegradation, *Science,* 249, 778, 1990.
34. Silcock, D. J., Waterhouse, R. N., Glover, L. A., Prosser, J. I., and Killham, K., Detection of a single genetically modified bacterial cell in soil by using charge coupled device-enhanced microscopy, *Appl. Environ. Microbiol.,* 58, 2444, 1992.
35. De Weger, L. A., Dunbar, P., Mahafee, W. F., Lugtenberg, E. J. J., and Sayler, G. S., Use of bioluminescence markers to detect *Pseudomonas* spp. in the rhizosphere, *Appl. Environ. Microbiol.,* 57, 3641, 1991.
36. Grant, F. A., Glover, L. A., Killham, K., and Prosser, J. I., Luminescence-based viable cell enumeration of *Erwinia carotovora* in the soil, *Soil Biol. Biochem.,* 23, 1021, 1991.
37. Meikle, A., Killham, K., Prosser, J. I., and Glover, L. A., Luminometric measurement of population activity of genetically modified *Pseudomonas fluorescens* in the soil, *FEMS Microbiol. Lett.,* 99, 217, 1992.

38. Rattray, E. A. S., Prosser, J. I., Glover, L. A., and Killham, K., Matric potential in relation to survival and activity of a genetically modified microbial inoculum in soil, *Soil Biol. Biochem.*, 24, 421, 1992.
39. Wright, D. A., Killham, K., Glover, L. A., and Prosser, J. I., The effect of location in soil on protozoal grazing of a genetically modified bacterial inoculum, *Geoderma,* 56, 633, 1993.
40. Ramos, J. L., Duque, E., and Ramos-Gonzalez, M-I., Survival in soils of an herbicide-resistant *Pseudomonas putida* strain bearing a recombinant TOL plasmid, *Appl. Environ. Microbiol.,* 57, 260, 1991.
41. Bentjen, S. A., Fredrickson, J. K., van Voris, P., and Li, S. W., Intact soil-core microcosms for evaluating the fate and ecological impact of the release of genetically engineered microorganisms, *Appl. Environ. Microbiol.,* 55, 198, 1989.
42. Trevors, J. T., van Elsas, J. D., van Overbeek, L. S., and Starodub, M-E., Transport of a genetically engineered *Pseudomonas fluorescens* strain through a soil microcosm, *Appl. Environ. Microbiol.,* 56, 401, 1990.
43. Hartel, P. G., Williamson, J. W., and Schell, M. A., Growth of genetically altered *Pseudomonas solanacearum* in soil and rhizosphere, *Soil Sci. Soc. Am. J.,* 54, 1021, 1990.
44. van Elsas, J. D., Trevors, J. T., and van Overbeek, L. S., Influence of soil properties on the vertical movement of genetically-marked *Pseudomonas fluorescens* through large soil microcosms, *Biol. Fertil. Soils,* 10, 249, 1991.
45. Recorbet, G., Givaudan, A., Steinberg, C., Bally, R., Normand, P., and Faurie, G., Tn*5* to assess soil fate of genetically marked bacteria: screening for aminoglycoside-resistance advantage and labelling specificity, *FEMS Microbiol. Ecol.,* 86, 187, 1992.
46. Pillai, S. D., Josephson, K. L., Bailey, R. L., Gerba, C. P., and Pepper, I. L., Rapid method for processing soil samples for polymerase chain reaction amplification of specific gene sequences, *Appl. Environ. Microbiol.,* 57, 2283, 1991.
47. Smit, E., van Elsas, J. D., and van Veen, J. A., Risks associated with the application of genetically modified microorganisms in terrestrial ecosystems, *FEMS Microbiol. Rev.,* 88, 263, 1992.
48. Kluepfel, D. A. and Tonkyn, D. W., The ecology of genetically altered bacteria in the rhizosphere, in *Biological Control of Plant Diseases,* Tjamos, E. C., Papavizas, G. C., and Cook, R. J., Eds., Plenum Press, New York, 1992, 407.
49. Amici, A., Bazzicalupo, M., Gallori, E., and Rollo, F., Monitoring a genetically engineered bacterium in a freshwater environment by rapid enzymatic amplification of a synthetic DNA "number-plate", *Appl. Microbiol. Biotechnol.,* 36, 222, 1991.
50. Amann, R. I., Binder, B. J., Olson, R. J., Chisholm, S. W., Devereux, R., and Stahl, D. A., Combination of 16S rRNA-targeted oligonucleotide probes with flow cytometry for analyzing mixed microbial populations, *App. Environ. Microbiol.,* 56, 1919, 1990.
51. Sayler, G. S., Harris, C., Pettigrew, C., Pacia, D., Breen, A., and Sirotkin, K. M., Evaluating the maintenance and effects of genetically engineered microorganisms, *Dev. Ind. Microbiol.,* 27, 135, 1987.
52. Steffan, R. J., Breen, A., Atlas, R. M., and Sayler, G. S., Application of gene probe methods for monitoring specific microbial populations in freshwater ecosystems, *Can. J. Microbiol.,* 35, 681, 1989.
53. Jain, R. K., Sayler, G. S., Wilson, J. T., Houston, L., and Pacia, D., Maintenance and stability of introduced genotypes in groundwater aquifer material, *Appl. Environ. Microbiol.,* 53, 996, 1987.
54. van Elsas, J. D., Trevors, J. T., Jain, D., Wolters, A. C., Heijnen, C. E., and van Overbeek, L. S., Survival of, and root colonization by, alginate-encapsulated *Pseudomonas fluorescens* cells following introduction into soil, *Biol. Fertil. Soils,* 14, 14, 1992.
55. Amy, P. S. and Hiatt, H. D., Survival and detection of bacteria in an aquatic environment, *Appl. Environ. Microbiol.,* 55, 788, 1989.
56. Simonet, P., Lee, N. T., du Cros, E. T., and Bardin, R., Identification of *Frankia* strains by direct DNA hybridization of crushed nodules, *Appl. Environ. Microbiol.,* 54, 2500, 1988.
57. Chaudhry, G. R., Toranzos, G. A., and Bhatti, A. R., Novel method for monitoring genetically engineered microorganisms in the environment, *Appl. Environ. Microbiol.,* 55, 1301, 1989.
58. Holben, W. E., Jansson, J. K., Chelm, B. K., and Tiedje, J. M., DNA probe method for the detection of specific microorganisms in the soil bacterial community, *Appl. Environ. Microbiol.,* 54, 703, 1988.
59. Smalla, K., Cresswell, N., Mendonca-Hagler, L. C., Wolters, A., and van Elsas, J. D., Rapid DNA extraction protocol from soil for polymerase chain reaction-mediated amplification, *J. Appl. Bacteriol.,* 74, 78, 1993.

60. Selenska, S. and Klingmüller, W., Direct detection of *nif*-gene sequences of *Enterobacter agglomerans* in soil, *FEMS Microbiol. Lett.*, 80, 243, 1991.
61. Jacobsen, C. S. and Rasmussen, O. F., Development and application of a new method to extract bacterial DNA from soil based on separation of bacteria from soil with cation-exchange resin, *Appl. Environ. Microbiol.*, 58, 2458, 1992.
62. Trevors, J. T., DNA extraction from soil, *Microb. Rel.*, 1, 3, 1992.
63. Holben, W. E., Isolation and purification of bacterial DNA from soil, in *Methods of Soil Analysis*, Agronomy Society of America, in press.
64. Pickup, R. W. and Saunders, J. R., Detection of genetically engineered traits among bacteria in the environment, *Trends Biotechnol.*, 8, 329, 1990.
65. Bentjen, S. A., Fredrickson, J. K., van Voris, P., and Li, S. W., Intact soil-core microcosms for evaluating the fate and ecological impact of the release of genetically engineered organisms, *Appl. Environ. Microbiol.*, 55, 198, 1989.
66. Steffan, R. J. and Atlas, R. M., Polymerase chain reaction: applications in environmental microbiology, *Annu. Rev. Microbiol.*, 45, 137, 1991.
67. Pillai, S. D., Josephson, K. L., Bailey, R. L., Gerba, C. P., and Pepper, I. L., Rapid method for processing soil samples for polymerase chain reaction amplification of specific gene sequences, *Appl. Environ. Microbiol.*, 57, 2283, 1991.
68. Picard, C., Ponsonnet, C., Paget, E., Nesme, X., and Simonet, P., Detection and enumeration of bacteria in soil by direct DNA extraction and polymerase chain reaction, *Appl. Environ. Microbiol.*, 58, 2717, 1992.
69. Zaat, S. A. J., Slegtenhorst-Eegdeman, K., Tommassen, J., Geli, V., Wijffelman, C. A., and Lugtenberg, E. J. J., personal communication, 1993.
70. Steffan, R. J. and Atlas, R. M., DNA amplification to enhance detection of genetically engineered bacteria in environmental samples, *Appl. Environ. Microbiol.*, 54, 2185, 1988.
70a. Ryder, M. H. and Ophel-Keller, K. M., unpublished data, 1993.
71. Schloter, M., Borlinghaus, R., Bode, W., and Hartmann, A., Direct identification and localization of *Azospirillum* in the rhizosphere of wheat with fluorescence labeled monoclonal antibodies and confocal laser scanning microscopy, *J. Microsc.*, 171, 173, 1993.
72. Levanony, H., Bashan, Y., Romano, B., and Klein, E., Ultrastructural localization and identification of *Azospirillum brasilense* Cd on and within wheat root by immuno-gold labeling, *Plant Soil,* 117, 207, 1989.
73. Underberg, H. and van Vuurde, J. W. L., *In situ* detection of *Erwinia chrysanthemi* on potato roots using immunofluorescence and immunogold staining, *Proc. 7th Int. Conf. on Plant Pathogenic Bacteria,* 1989, 937.
74. Olsen, P. E. and Rice, W. A., *Rhizobium* strain identification and quantification in commercial inoculants by immunoblot analysis, *Appl. Environ. Microbiol.*, 55, 520, 1989.
75. Levanony, H., Bashan, Y., and Kahana, Z. E., Enzyme-linked immunosorbent assay for specific identification and enumeration of *Azospirillum brasilense* Cd. in cereal roots, *Appl. Environ. Microbiol.*, 53, 358, 1987.
76. Schloter, M., Bode, W., and Hartmann, A., Characterization of monoclonal antibodies against cell surface structures of *Azospirillum brasilense* Sp7 using ELISA techniques, *Symbiosis,* 13, 37, 1992.
77. Schloter, M., Bode, W., Hartmann, A., and Beese, F., Sensitive chemoluminescence-based immunological quantification of bacteria in soil extracts with monoclonal antibodies, *Soil Biol. Biochem.*, 24, 399, 1992.
78. van Overbeek, L. S., van Elsas, J. A., Trevors, J. T., and Starodub, M. E., Long-term survival of and plasmid stability in *Pseudomonas* and *Klebsiella* species and appearance of nonculturable cells in agricultural drainage water, *Microb. Ecol.*, 19, 239, 1990.
79. Olsen, P. E. and Rice, W. A., Use of monoclonal antibodies in a colony immunoblot analysis of viable *Rhizobium* cell numbers in legume inoculants and on preinoculated seed, *Can. J. Microbiol.*, 37, 430, 1991.
80. Olsen, P., Collins, M., and Rice, W., Surface antigens present on vegetative *Rhizobium meliloti* cells may be diminished or absent when cells are in the bacteroid form, *Can. J. Microbiol.*, 38, 506, 1992.
81. Roozen, N. J. M. and van Vuurde, J. W. L., Development of a semi-selective medium and an immunofluorescence colony-staining procedure for the detection of *Clavibacter michiganensis* subsp. *sepedonicus* in cattle manure slurry, *Neth. J. Plant Pathol.*, 97, 324, 1991.

82. van Vuurde, J. W. L. and Roozen, N. J. M., Comparison of immunofluorescence colony-staining in media, selective isolation on pectate medium, ELISA and immunofluorescence cell staining for detection of *Erwinia carotovora* subsp. *atroseptica* and *E. chrysanthemi* in cattle manure slurry, *Neth. J. Plant Pathol.,* 96, 75, 1990.
83. van Vuurde, J. W. L., Immunofluorescence colony-staining (IFC) and immunofluorescence cell-staining as tools for the study of rhizosphere bacteria, in *Plant Growth-Promoting Rhizobacteria: Progress and Prospects,* Keel, C., Koller, B., and Défago, G., Eds., Intl. Org. for Biol. and Integrated Control of Noxious Animals and Plants, West Palaearctic Regional Section Bulletin XIV/8, 1991, 215.
84. Jones, J. B. and van Vuurde, J. W. L., Magnetic immunoisolation of *Xanthomonas campestris* pv. *pelargonii,* in *Proc. 7th Int. Conf. on Plant Pathogenic Bacteria,* 1989, 883.
85. Chao, W. L., Nelson, E. B., Harman, G. E., and Hoch, H. C., Colonization of the rhizosphere by biological control agents applied to seeds, *Phytopathology,* 76, 60, 1986.
86. Fravel, D. R. and Marois, J. J., Edaphic parameters associated with establishment of the biocontrol agent *Talaromyces flavus, Phytopathology,* 76, 643, 1986.
87. Park, Y., Stack, J. P., and Kenerly, C. M., Selective isolation and enumeration of *Gliocladium virens* and *G. roseum* from soil, *Plant Dis.,* 76, 230, 1992.
88. Couteaudier, Y., Daboussi, M.-J., Eparvier, A., Langin, T., and Orcival J., The GUS gene fusion system, a useful tool in studies of root colonization by *Fusarium oxysporum, Appl. Environ. Microbiol.,* 59, 1767, 1993.
89. Roberts, I. N., Oliver, R. P., Punt, P. J., and van den Hondel, C. A. M. J. J., Expression of the *Escherichia coli* β-glucuronidase gene in industrial and phytopathogenic fungi, *Curr. Gene.,* 15, 177, 1989.
90. Herrera-Estrella, A., Goldman, G. H., and van Montagu, M., High-efficiency transformation system for the biocontrol agents, *Trichoderma* spp., *Mol. Microbiol.,* 4, 839, 1990.
91. Cenis, J. L., Rapid extraction of fungal DNA for PCR amplification, *Nucleic Acids Res.,* 20, 2380, 1992.
92. Henson, J. M., DNA probe for the identification of the take-all fungus *Gaeumannomyces graminis, Appl. Environ. Microbiol.,* 55, 284, 1989.
93. Nazar, R. N., Hu, X., Schmidt, J., Culham, D., and Robb, J., Potential use of PCR-amplified ribosomal intergenic sequences in the detection and differentiation of verticillium wilt pathogens, *Physiol. Mol. Plant Pathol.,* 39, 1, 1991.
94. Simon, L., Lalonde, M., and Bruns, T. D., Specific amplification of 18S fungal ribosomal genes from vesicular-arbuscular endomycorrhizal fungi colonizing roots, *Appl. Environ. Microbiol.,* 58, 291, 1992.
95. Friese, C. F. and Allen, M. F., Tracking the fates of exotic and local VA mycorrhizal fungi: methods and patterns, *Agric. Ecosyst. Environ.,* 34, 87, 1991.
96. Hardham, A. R., Suzaki, E., and Perkin, J. L., Monoclonal antibodies to isolate-, species-, and genus-specific components on the surface of zoospores and cysts of the fungus *Phytophthora cinnamomi, Can. J. Bot.,* 64, 311, 1986.
97. Herrera-Estrella, A., Goldman, G. H., and van Montagu, M., High-efficiency transformation system for the biocontrol agents, *Trichoderma* spp., *Mol. Microbiol.,* 839, 1990.
98. Sayler, G. S., and Layton, A. C., Environmental application of nucleic acid hybridization, *Annu. Rev. Microbiol.,* 44, 625, 1990.
99. Nuss, D. L., Biological control of chestnut blight: an example of virus-mediated attenuation of fungal pathogenesis, *Microbiol. Rev.,* 56, 561, 1992.
100. Leeman, M., Raaijmakers, J. M., Bakker, P. A. H. M., and Schippers, B., Immunofluorescence colony staining for monitoring pseudomonads introduced into soil, in *Biotic Interactions and Soil-borne Diseases,* Beemster, A. B. R., Bollen, G. J., Gerlagh, M., Ruissen, M. A., Schippers, B., and Tempel, A., Eds., Elsevier, Amsterdam, 1991, 374.
100a. Ryder, M. H., Ophel-Keller, K., Dhindsa, H., and Veal, D., Detection and characterization of a genetically modified biocontrol bacterium in field sod, in *Improving Plant Productivity with Rhizosphere Bacteria,* Ryder, M. H., Stephens, P. M., and Bowen, G. D., Eds., CSIRO Division of Soils, Adelaide, 1994, 261.
101. Shaw, J. J., Settles, L. G., and Kado, C. I., Transposon Tn*4431* mutagenesis of *Xanthomonas campestris* pv. *campestris:* characterization of a nonpathogenic mutant and cloning of a locus for pathogenicity, *Mol. Plant-Microbe Interact.,* 1, 39, 1988.

102. Pillai, S. D. and Pepper, I. L., Transposon Tn*5* as an identifiable marker in rhizobia: survival and genetic stability of Tn*5* mutant bean rhizobia under temperature stressed conditions in desert soils, *Microb. Ecol.*, 21, 21, 1991.
103. Colwell, R., From counts to clones, *J. Appl. Bacteriol. Symp. Suppl.*, 1987, 1S.
104. Bolton, Jr., H., Fredrickson, J. K., Bentjen, S. A., Workman, D. J., Li, S. W., and Thomas, J. M., Field calibration of soil-core microcosms: fate of a genetically altered rhizobacterium, *Microb. Ecol.*, 21, 163, 1991.
105. Bolton, H., Jr., Fredrickson, J. K., Thomas, J. M., Li, S. W., Workman, D. J., Bentjen, S. A., and Smith, J. L., Field calibration of soil-core microcosms: ecosystem structural and functional comparisons, *Microb. Ecol.*, 21, 175, 1991.
106. Pedersen, J. C., Survival of *Enterobacter cloacae;* field validation of a soil/plant microcosm, *Microb. Rel.*, 1, 87, 1992.
107. Harman, G. E., Chet, I., and Baker, R., *Trichoderma hamatum* effects on seed and seedling disease induced in radish and pea by *Pythium* spp or *Rhizoctonia*, *Phytopathology*, 70, 1167, 1980.
108. Simon, A., Biological control of take-all of wheat by *Trichoderma koningii* under controlled environmental conditions, *Soil Biol. Biochem.*, 21, 323, 1989.
109. Paulitz, T. C. and Baker, R., Biological control of *Pythium* damping-off of cucumbers with *Pythium nunn:* influence of soil environment and organic amendments, *Phytopathology*, 77, 341, 1987.
110. Harris, A. R., Schisler, D. A., Rowden, R. G., and Ryder, M. H., Control of damping-off and growth promotion, in bedding plants, by soil bacteria and fungi, in *Plant Growth-Promoting Rhizobacteria—Progress and Prospects*, Keel, C., Koller, B., and Défago, G., Eds., Intl. Org. for Biol. and Integrated Control of Noxious Animals and Plants, West Palaearctic Regional Section Bulletin XIV/8, 1991, 63.

FURTHER READING

Edwards, C., Ed., *Monitoring Genetically Manipulated Microorganisms in the Environment*, John Wiley & Sons, Chichester, 1993.

Hampton, R., Ball, E., and de Boer, S., Eds., *Serological Methods for Detection and Identification of Viral and Bacterial Plant Pathogens: a Laboratory Manual*, APS Press, St Paul, MN, 1990.

Kluepfel, D. A., The behavior and tracking of bacteria in the rhizosphere. *Annu. Rev. Phytopathol.*, 31, 441, 1993.

MacKenzie, D. R. and Henry, S. C., *Biological Monitoring of Genetically Engineered Plants and Microbes*, Agric. Research Institute, Bethesda, MD, 1990.

Olson, B. H., Tracking and using genes in the environment, *Environ. Sci. Technol.*, 25, 604, 1991.

Pickup, R. W. and Saunders, J. R., Detection of genetically engineered traits among bacteria in the environment, *Trends Biotechnol.*, 8, 329, 1990.

Sayler, G. S. and Layton, A. C., Environmental application of nucleic acid hybridization, *Annu. Rev. Microbiol.*, 44, 625, 1990.

Steffan, R. J. and Atlas, R. M., Polymerase chain reaction: applications to environmental microbiology, *Annu. Rev. Microbiol.*, 45, 137, 1991.

Stotzky, G., Broder, M. W., Doyle, J. D., and Jones, R. A., Selected methods for the detection and assessment of ecological effects resulting from release of genetically engineered microorganisms to the terrestrial environments, *Adv. Appl. Microbiol.*, 3d, 2, 1993.

Tiedje, J. M., Colwell, R. K., Grossman, Y. L., Hodson, R. E., Lenski, R. E., Mack, R. N., and Regal, P. J., The planned introduction of genetically engineered organisms: ecological considerations and recommendations, *Ecology*, 70, 298, 1989.

van Elsas, J. D. and Waalwijk, C., Methods for the detection of specific bacteria and their genes in soil, *Agric. Ecosyst. Environ.*, 34, 97, 1991.

Chapter 34

Monitoring of Fungicide Resistance in Fungi: Biological to Biotechnological Approaches

Hideo Ishii

TABLE OF CONTENTS

I. INTRODUCTION

Since the late 1960s, fungicide resistance in phytopathogenic fungi has become one of the major problems worldwide in crop protection. In the case of benzimidazoles, for example, the problem of resistance occurred shortly after introduction of the fungicides, making disease control difficult with this class of fungicides.[1]

As mentioned by Delp and Dekker,[2] fungicide resistance can be defined as stable, inheritable adjustment by a fungus to a fungicide, resulting in a less than normal sensitivity to that fungicide. This definition is widely accepted, although resistance is not necessarily stable in some cases (e.g., resistance to sterol demethylation-inhibiting fungicides)[3] and the number of phytopathogenic fungi with which methods for crossing in culture have been established is still limited.

It is important to monitor fungicide resistance so that any shifts in fungicide sensitivity in fungal populations can be detected early and growers alerted to enable them to implement alternative disease-control strategies. Recognition of resistant strains must be made by comparison with wild-type sensitive strains.[4] Therefore, it is essential that the "baseline sensitivity" for the fungus-fungicide combination in question be established. General principles of monitoring fungicide resistance have been summarized elsewhere.[5,6]

When testing fungicide resistance, biological methods are usually employed. In this paper, conventional biological methods for monitoring resistance will be summarized. However, such methods are generally time-consuming, laborious, and allow relatively few strains to be monitored. Techniques based on DNA probe technology have been successful in the early diagnosis of human genetic disorders. Transfer of this technology to plant pathology should open the way to the rapid detection of fungicide resistance in pathogen populations.[7] Accordingly, biotechnological approaches to detect resistance based on the resistance mechanism will be further reviewed in the last parts of this paper.

II. BIOLOGICAL METHODS FOR MONITORING RESISTANCE

The action mechanism of fungicides differs from each other. Therefore, it is essential to develop the most appropriate testing methods in each combination of a fungus with a fungicide before the monitoring program starts. Standardized methods for monitoring resistance have been published by FAO[5] and by FRAC (Fungicide Resistance Action Committee).[8,9] Methodology and interpretation of results were described in detail.

0-87371-877-1/95/$0.00+$.50

A. MYCELIAL GROWTH TESTS

The most commonly used methods are mycelial growth tests on a culture medium supplemented with different concentrations of a fungicide. The minimum inhibitory concentration (MIC) of a fungicide necessary for the complete inhibition of mycelial growth of the fungal strain is determined. In *Venturia nashicola* (the cause of Japanese pear scab), for example, the procedure as below is followed to test benzimidazole resistance.

Each strain is previously cultured on potato dextrose agar (PDA) plates at 20°C for 45 d. Mycelial disks cut from the margins of the colonies are transferred onto PDA plates containing the fungicide carbendazim. After incubation at 20°C for 3 weeks, mycelial growth of the strains is observed, and the MIC values of carbendazim are determined. Based on the difference in sensitivity, strains are divided into four groups as follows: highly resistant, MIC >100 μg/ml; intermediately resistant, 100 μg/ml $\geq$ MIC $>$ 10 μg/ml; weakly resistant, 10 μg/ml $\geq$ MIC $>$ 1 μg/ml; sensitive, 1 μg/ml $\geq$ MIC. Genetic studies indicated that the occurrence of three different levels of benzimidazole resistance is due to three allelic mutations in a single gene, *Ben1,* and each level is controlled by one of the multiple alleles.[10,11]

The EC_{50} values are calculated by regressing the relative growth (colony diameter on a fungicide-amended medium divided by the diameter on unamended medium $\times$ 100) against the $\log_{10}$ fungicide concentration. Comparison of EC_{50} values is also frequently made for testing fungicide resistance. The strains of *Botrytis cinerea (Botryotinia fuckeliana)* are classified as sensitive (S), low-resistant (LR), or high-resistant (HR) to the dicarboximide fungicide vinclozolin on the basis of their EC_{50} values. *Daf1S, Daf1LR,* and *Daf1HR* alleles confer sensitivity, low, and high resistance to dicarboximides.[12]

These days, numerous commercial fungicides belong to sterol demethylation inhibitors (DMIs). When DMI sensitivity is examined in mycelial growth tests, comparisons are based on EC_{50} rather than MIC values in general,[13] then the resistance factor (the EC_{50} for a resistant strain divided by the EC_{50} for a sensitive strain) is calculated. However, this method is extremely laborious and time consuming for slow-growing fungi like *Venturia* in particular. Therefore, some alternative methods have been developed to reduce the labor and time.[3,14]

B. SPORE GERMINATION TESTS

A simple method named "Germ-Tube Septum Method" was developed to detect the benzimidazole-resistant strains of *V. nashicola.*[15] Conidia from scab lesions are directly transferred onto potato sucrose agar or PDA plates supplemented with a benzimidazole fungicide. After incubation at 15°C for 48 to 72 h, MIC of the fungicide is determined on the basis of septum formation in germ tubes. This method is often used in practice for "The Fungicide-Resistance Monitoring Program" in Japan.

Monitoring is especially important with fungi those quickly develop resistance, such as *B. cinerea.* A simple rapid method for detecting benzimidazole and dicarboximide resistance was proposed and adopted by the extension service.[16] The technique is based on the percent germination of conidia on a simple medium (glucose 10 g, agar 20 g, terramycin 50 g/l) amended with benomyl (5 mg/l) or vinclozolin (3 mg/l). Moreover, a diagnostic medium was developed for detection of strains resistant to vinclozolin and benomyl.[17] The medium contains 0.04% (w/v) bromocresol purple, 10% 0.1 *N* NaOH, and 2% agar. After autoclaving, filter-sterilized dextrose (4%) is added, then vinclozolin (40 mg/l) or benomyl (10 mg/l) and streptomycin sulfate are added. Germination and growth of resistant conidia cause a color change from red to yellow in 18 to 48 h after inoculation based on pH change of the medium. Comparisons between this method and other techniques showed excellent correlations. Other diagnostic media for identification of fungicide-resistant strains from the field are now under testing. Development of selective media is particularly useful for avoiding the contamination with nontarget microorganisms.

The antibiotic polyoxin causes swelling of mycelial tips and conidial germ tubes of *Alternaria alternata* Japanese pear pathotype (= *A. kikuchiana,* black spot fungus). Spore germination tests are most suitable for evaluating polyoxin sensitivity of this fungus. Conidia obtained from resistant isolates could grow normally at 1 μg/ml of polyoxin, whereas those from sensitive isolates formed bulbous germ tubes.[18] Polyoxin-resistant isolates varied considerably in their response: these variants were classified as intermediate in resistance and highly resistant. Not only highly, but intermediately resistant isolates reduced polyoxin efficacy on black spot control on detached pear leaves.[19]

C. INOCULATION TESTS ON HOST PLANTS

When field emergence of fungicide resistance is suspected, the following three steps must be fulfilled to evaluate the practical importance of the resistance: (1) isolation of the pathogen from the diseased plant materials from the field in question, (2) laboratory tests for resistance, and (3) reproduction of reduced control effects of the fungicide against the strains isolated.[18] The last procedure of these steps can be generally achieved by inoculation of the fungicide-treated plants with the strains.

In obligate parasites, such as downy mildew and powdery mildew, leaf disk methods are frequently employed for monitoring fungicide resistance. Disks of grapevine leaves were floated on a solution of the phenylamide fungicide metalaxyl in petri dishes. The disks were then inoculated with the sporangium suspension of *Plasmopara viticola.* After incubation in a moist chamber, the level of resistant strains in the pathogen population was assessed by comparing the sporulation on treated and untreated disks.[20] Leaf disks of cucumber were also used for testing the DMI sensitivity of *Sphaerotheca fuliginea.*[21] Inoculation was carried out by pressing mildew-free leaves onto leaves with the sporulating mildew isolate to be tested. The disks were placed onto fungicide solutions in petri dishes, then incubated to record mildew development.

Several methods are available for determination of the DMI sensitivity in populations of barley powdery mildew (*Erysiphe graminis* f. sp. *hordei*). For example, barley seed was dressed with triadimenol at different rates, and pieces of prophyll were mounted in square petri dishes containing water agar. The leaf pieces were inoculated with spores using a settling tower. The dishes were transferred to an optimal environment for powdery mildew development and assessed following the grading system.[22]

III. MECHANISMS OF RESISTANCE

The basic studies on biochemical genetics of fungicide resistance are indispensable for the application of biotechnological techniques to monitor the resistance in practice.

A. CLASSICAL GENETICS OF RESISTANCE

The genetics of fungicide resistance are closely related with development and stability of the resistance in the field. In general, it is believed that high levels of resistance can be acquired in one-step mutations in a major gene.[23] The best known example is provided by the benzimidazole resistance of *Venturia* species.[10,24] In *V. inaequalis,* five phenotypic responses to benomyl were identified as follows: sensitive, low resistance, medium resistance, high resistance, and very high resistance. These phenotypes were each governed by an allelic series in a single Mendelian gene.[24]

In case of a polygenic control, resistance development is expressed by a gradual shift of responses to the fungicide toward resistance. DMIs are often given as an example for this quantitative response of fungal populations.[25] However, the genetic mechanism of DMI resistance is in more complex situations than predicted before. In barley powdery mildew, the frequency distribution of triadimenol sensitivity of progeny from crosses was continuous, indicating that resistance is controlled not by one major gene, but probably by a complex genetic system.[26] In another case of this fungus, on the other hand, there was no evidence for polygenic control of response to triadimenol.[27] A similar scenario of discrepancy in the genetics of DMI resistance is the case of *V. inaequalis.* Stanis and Jones[28] concluded that reduced sensitivity to fenarimol of this fungus was determined by a single gene. However, recent studies indicated that more subtle genetic mechanisms operate in field strains, in addition to control by a single major gene.[29] Moreover, results from the progeny test suggested that triadimenol resistance of *Pyrenophoravteres* (the cause of barley net blotch) is controlled by a single, major genetic locus and by several additional loci.[30] These additional loci might modify the resistance phenotype of fungal isolates when recombined into a single isolate. They might also be significant in terms of fitness costs associated with resistance. The field data on the stability of DMI resistance are few. However, the decline of DMI resistance has recently been noticed in the populations of *V. inaequalis* after 3 years with no exposure to these fungicides.[31]

B. BIOCHEMICAL MECHANISM OF RESISTANCE

Recent advances in this area have already been reviewed by several researchers.[32,33] The mechanism of action of fungicides often relates to the resistance mechanism. The primary mode of action of benzimidazoles is considered to be the specific binding to the β-tubulin subunit of fungal tubulin and, consequently, an interference with microtubule assembly which is essential for a great number of cellular processes such as mitosis and meiosis.[34] Although several biochemical mechanisms of resistance have been proposed, a decreased binding of the fungicide to tubulin-like proteins was generally involved.[35] In *V. nashicola,* the binding of ^{14}C-carbendazim to tubulin-like proteins was lower in benzimidazole-resistant than in sensitive strains, suggesting that a decreased affinity of the fungicide to the target was closely related to resistance.[36] The mechanism of resistance was also elucidated in *B. cinerea.*[37] Benzimidazole-sensitive isolates showed high carbendazim binding, whereas binding was undetectable in highly resistant and intermediately resistant isolates. The different levels of resistance in these two fungal species are due to allelic mutations at the single chromosomal locus *Ben1*[10] and *Mbc1,*[12] respectively; each level of resistance is shown to be controlled by one of the multiple alleles. It is very likely that allelic mutations change the structure of the benzimidazole binding site at β-tubulin molecule.

The natural (inherent) resistance to benzimidazoles in some fungi, such as *A. brassicae* and *Pythium irregulare,* was explained by the less binding activity of tubulin-like proteins to the fungicide.[34] However, several other possibilities might also be concerned with the resistance. In *Colletotrichum acutatum,* one of the pathogens of strawberry anthracnose, high specific binding of cell-free extracts to ^{14}C-carbendazim has been found despite that this species revealed natural resistance to benzimidazole fungicides.[38] To understand the mechanism in more detail, it might be worthwile to characterize the β-tubulin gene of *C. acutatum,* since the β-tubulin genes have already been isolated from other *Colletotrichum* species and characterized.[39,40]

Ergosterol is now thought of as the dominant fungal sterol. Therefore, its biosynthesis is an important target for a great number of modern antifungal agents. The mechanism of action of DMIs is based on the interaction with a cytochrome P450-dependent 14α-demethylation enzyme which is important in the synthesis of ergosterol.[41] Several biochemical mechanisms of DMI resistance have been proposed.[33,42] De Waard and van Nistelrooy[43] have intensively studied the mechanism of resistance using laboratory-induced mutants of *Penicillium italicum.* The mechanism involved was ascribed to reduced accumulation of DMIs in fungal mycelia due to the energy-dependent efflux. It was further suggested that triadimenol resistance of *Rhynchosporium secalis* arose through alteration of the target sterol 14α-demethylase, since triadimenol sensitivity did not correlate with any differences in metabolism or uptake of this fungicide.[44] The third possible mechanism is the overproduction of the P450 14α-demethylase. In the haploid yeast *Candida glabrata,* it has been found that the azole antifungal agent fluconazole has the potency to induce an overproduction of P450.[45] The P450 content and ergosterol biosynthesis both decreased when the resistant isolate was subcultured repeatedly on drug-free media. Interestingly, decrease of the levels of DMI resistance on fungicide-free media was observed also in phytopathogenic fungi, such as *B. cinerea*[46] and *V. inaequalis.*[3] In the former fungus, the poly- and heterokaryotic nature might relate to the phenomenon. In *V. inaequalis,* moreover, strains maintained on PDA ammended with DMIs often showed reduced sensitivity to these fungicides.[29] Further studies will be carried out to examine if overproduction of the P450 is involved in the resistance.

A triadimenol-sensitive wild-type strain and a resistant mutant of *Ustilago avenae* were investigated with regard to their responses to triadimenol.[47] The expression of resistance was considered to be an induced response rather than a constitutive trait, although such reports had not been made previously for site-specific fungicides used in agriculture. The authors suspected that the induced response implies the activation of normally repressed "resistance genes" either induced by the fungicide or by the precursor sterols that initially accumulate. As reported for *C. albicans,* other possible mechanisms of DMI resistance will be a decreased binding affinity of the fungicides to their target site due to the mutation.[48]

IV. BIOTECHNOLOGICAL APPROACHES

Biotechnological approaches to detect fungicide resistance are only feasible where the resistance mechanism is elucidated at a molecular level. At present, these techniques are seen only as research tools. However, research in these areas is advancing quickly. Review articles on this subject have been written by Hollomon[49] and Hollomon and Butters.[7]

A. BIOCHEMICALLY BASED DIAGNOSTICS

Immunological techniques which are rapid, sensitive, and specific could be used for the rapid detection of the pathogens and diagnoses of the diseases.[50] A monoclonal antibody-based immunoassay has been developed and proved to be useful for the presymptomatic detection of fungal pathogens such as *Pseudocercosporella herpotrichoides,*[51] *Septoria nodorum,* and *S. tritici.*[52] Some diagnostic kits are now commercially available.

Table 1 **Amino acid substitutions in the β-tubulin gene for strains of nonplant-pathogenic fungi with resistance to benzimidazoles**

		Amino acids in position				
Fungal species	**Phenotype**[a]	**167**	**198**	**200**	**241**	**Ref.**
Neurospora crassa	S	Phe	Glu	Phe	Arg	54
	R	Tyr	Glu	Phe	Arg	54
	R	Phe	Gly	Phe	Arg	64
Aspergillus nidulans	S	Phe	Glu	Phe	Arg	68
	R	Phe	Asp	Phe	Arg	68
	R	Phe	Gln	Phe	Arg	68
	R	Phe	Lys	Phe	Arg	68
	R	Phe	Glu	Tyr	Arg	68

[a]Sensitive (S) and resistant (R) to benzimidazoles.

Based on the mechanism of benzimidazole resistance, attempts to isolate tubulins from fungi have been made. Although immunochemical detection of tubulins were performed for several fungi such as *B. cinerea,* isolation of tubulins to raise monoclonal antibodies have proved unsuccessful due to low yields and poor antigenicity.[53] The antigenicity of *B. cinerea* tubulin, and the epitopic availability of the carbendazim-binding site are uncertain yet. Orbach et al.[54] cloned the β-tubulin gene of *Neurospora crassa* from a benomyl-resistant strain and determined its nucleotide sequence. The mutation responsible for benomyl resistance was determined; it caused a phenylalanine-to-tyrosine change at position 167 (Table 1). Subsequently, Martin et al.[55] successfully synthesized peptides incorporating this region and have raised a specific monoclonal antibody capable of discerning the amino acid change. The antibody cross reacted well with tubulin from crude protein extracts of a carbendazim-resistant mutant of *N. crassa* on Western blots and was able to discriminate between this and tubulin from a sensitive strain. However, it was unable to detect tubulin from carbendazim-resistant strains of *B. cinerea.* This indicated that the tubulin conformation responsible for the resistance in *B. cinerea* is incongruous with that of *N. crassa.* It seems unlikely that rapid diagnostic methods for benzimidazole resistance can be developed using immunological methods, although the region around 198 rather than 167 is associated with the resistance in field strains as described below.[7]

B. DIAGNOSTICS BASED ON NUCLEIC ACID TECHNIQUES

DNA probe technology can be used to detect fungicide resistance once the mechanism of resistance has been determined at a molecular level. The mechanism of benzimidazole resistance has been well understood, and recently, sequencing of the β-tubulin genes has also been successfully achieved in phytopathogenic fungi. The carbendazim-resistant allele (*tubA*^{R}) of the *S. nodorum* β-tubulin gene was subcloned from a carbendazim-resistant mutant and transformation experiments demonstrated that the clone confered the resistance to this fungicide.[56,57] Subsequently, the wild-type allele of the β-tubulin gene (*tubA*$^+$) was cloned from amplified polymerase chain reaction (PCR) products using the genomic DNA of a carbendazim-sensitive strain as a template and the oligonucleotides to prime the reaction. The sequence data showed a single base difference between *tubA*$^+$ and *tubA*^{R}, resulting in an amino acid substitution from histidine to tyrosine.[58] The β-tubulin gene of *E. graminis* f. sp. *hordei* was isolated from a genomic library prepared from conidial DNA, using the *N. crassa* β-tubulin gene *tub-2* as a heterologous probe.[59] The amino acid sequence showed a high degree of homology to other fungal β-tubulins. Also for *C. graminicola,* the β-tubulin genes *TUB1* and *TUB2* have been cloned and characterized.[39]

The sequence data of wild-type sensitive and benzimidazole-resistant alleles of the β-tubulin gene from several fungi showed that resistance to benzimidazoles is attributed to single amino acid changes between 100 and 300 (Table 1). Accordingly, Martin et al.[60] attempted to use PCR for the diagnosis of carbendazim resistance in *B. cinerea* using conserved oligonucleotide primers encompassing this region. The β-tubulin gene segment from genomic DNA was amplified using primers β-101 and β-293 (Figure 1) in 100 μl reaction volumes containing 200 μ*M* of each deoxynucleotide triphosphate, 400 p*M* of each primer, 2.5 units of Taq polymerase, 50 m*M* KCl, 10 m*M* Tris-HCl (pH 8.3), 1.5 m*M* $MgCl_2$, and

Amplification of β-tubulin gene

Primers: β-101: TGG GCT AAA GGT CAC TAC AC
β-293: CAT TTG TTG TGT TGT TAA TTC TGG

Diagnosis of resistance using ASO primers

Primer S:	G AGA ACT CTG ACG A
Primer R:	G AGA ACT CTG ACG C
Tailed Primer S:	GCT GGC CAA CTG AGA ACT CTG ACG A
Tailed Primer R:	GCT GGC CAA CTG AGA ACT CTG ACG C

Southern blotting R-probe:
TCT GAC GCG ACC TTC TGT

Figure 1 Nucleotide sequences of primers and probe used for the diagnosis of carbendazim resistance in *Botrytis cinerea.* (From Martin, et al., in *Proc. 1992 Brighton Crop Protection Conf.—Pests and Diseases,* BCPC Publications, Surry, England, 1992, 207. With permission.)

0.01% gelatine. The condition was as follows: denaturation at 95°C for 2 min; annealing at 55°C for 2 min; extension at 72°C for 3 min for 30 cycles, with a final extension at 72°C for 7 min.

Amplification by PCR of genomic DNA from *B. cinerea* isolates resulted in a fragment 579 bp in size. These fragments were cloned into pBluescript and the nucleotide sequenced. It was concluded that a point mutation at amino acid 198, causing a change from glutamic acid to alanine, confered carbendazim resistance. Next, two allele-specific oligonucleotides (ASOs) were synthesized and used for determination of the point mutation for resistance and sensitivity. Reaction mixtures (50 μl) were prepared in duplicate, containing approximately 100 ng of the amplified product, primer β-293 and tailed primer R or tailed primer S (Figure 1). The samples were amplified according to the following procedure: denaturation at 95°C for 1.5 min, annealing at 62°C for 1.5 min, extension at 72°C for 3 min for two cycles, followed by the same condition with an annealing temperature of 67°C for 28 cycles. The assay was extremely successful. The resistant sequence was amplified by the R-primer, and amplification of the sensitive sequence occurred with the S-primer (Table 2). The ASO-PCR is remarkably sensitive, with the potential to amplify as little as 1 ng of target DNA within 48 to 72 h.

Koenraadt et al.[61] characterized mutations in the β-tubulin gene of benomyl-resistant field strains of numerous phytopathogenic fungi including *V. inaequalis, Monilinia fructicola, Penicillium digitatum,* etc. For rapid cloning of the β-tubulin gene, genomic DNA was prepared from *V. inaequalis* strains and subjected to PCR, in which a 22-mer oligonucleotide A (5′-CAAACCATCTCTGGCGAACACG) and a 22-mer oligonucleotide B (5′-TGGAGGACATCTTAAGACCACG) were used as primers. Further-

Table 2 **Screening of *Botrytis cinerea* isolates for carbendazim resistance**

Isolate	EC_{50} (μg/ml)	Phenotype	PCR primer[a] S	PCR primer[a] R	Southern blot R-probe
PC9385R	775	Resistant	−	+	+
K1145	733	Resistant	−	+	+
1805R(b)	425	Resistant	−	+	+
GB111/74	17.8	Intermediate	+	+	Not tested
PC9385S	0.066	Sensitive	+	−	−
A19	0.055	Sensitive	+	−	Not tested
B4	0.075	Sensitive	+	−	−

[a]+: Strong amplification by PCR and/or by Southern blot hybridization;
−: No visible amplification by PCR and/or by Southern blot hybridization.
From Martin et al., in *Proc. 1992 Brighton Crop Protection Conf.—Pests and Diseases,* BCPC Publications, Surry, England, 1992, 207. With permission.

Table 3 **Point mutations and deduced amino acid substitutions in the β-tubulin gene for strains of plant-pathogenic fungi with resistance to benzimidazoles**

		Amino acids in position		
Fungal species	**Phenotype**[a]	**198**	**200**	**Ref.**
Botrytis cinerea	S	Glu	Phe	60
	HR	Ala	Phe	60
Monilinia fructicola	S	Glu	Phe	61
	HR	Lys	Phe	61
Penicillium digitatum	S	Glu	Phe	61
	HR	Lys	Phe	61
P. italicum	S	Glu	Phe	61
	MR	Glu	Tyr	61
	HR	Lys	Phe	61
Rhynchosporium secalis	S	Glu	Phe	7
	R	Gly	Phe	7
	R	Lys	Phe	7
Venturia inaequalis	S	Glu	Phe	61
	LR	Glu	Phe	61
	MR	Gly	Phe	61
	MR	Glu	Tyr	61
	HR	Lys	Phe	61
	VHR	Ala	Phe	61
V. pirina	S	Glu	Phe	61
	MR	Glu	Tyr	61
	VHR	Ala	Phe	61

[a]Sensitive (S), low resistance (LR), moderate resistance (MR), high resistance (HR), and very high resistance (VHR) to benzimidazoles.

more, a 24-mer oligonucleotide C (5′-GAGGAATTCCCAGACCGTATGATG) and a 28-mer oligonucleotide D (5′-GCTGGATCCTATTCTTTGGGTCGAACAT) were chosen as generic β-tubulin primers for directional cloning of PCR products of other fungal species. All strains examined, except those with low resistance to benomyl, were found to contain a single base pair mutation in their β-tubulin genes, resulting in an amino acid substitution in β-tubulin (Table 3). In *V. inaequalis,* codon 198, which encodes glutamic acid in a sensitive strain, was converted to other amino acid in a strain with very high resistance, high resistance, or medium resistance to benomyl. Codon 200 for phenylalanine was converted to a codon for tyrosine in a second strain with medium resistance. Similarly, point mutations were found in codons 198 or 200 of the β-tubulin gene in benomyl-resistant strains of *Penicillium* species. It was indicated that a change in codons 198 or 200 of the β-tubulin gene confers resistance to benomyl and each level of resistance associated with a unique amino acid substitution.

Koenraadt and Jones[62] further developed a procedure for detecting point mutations in the β-tubulin gene of benomyl-resistant field strains of *V. inaequalis,* using PCR in combination with ASO analysis. PCR was used to amplify a specific 1191-bp DNA sequence of the β-tubulin gene in DNA extracts from pure fungal culture or apple scab lesions. The amplified DNA sequence was denatured, then applied to a nylon membrane in a dot-blot manifold. The dot-blots were probed with ^{32}P-labeled 18-mer ASO probes (Figure 2) specific for sensitive or for three benomyl-resistant phenotypes in fungal strains. This method could be employed to identify benzimidazole resistance in pathogens that are difficult to culture and in obligate parasites.

To cope with benzimidazole resistance, the *N*-phenylcarbamate fungicide diethofencarb has been introduced in several countries, since most of benzimidazole-resistant strains of *B. cinerea* showed increased sensitivity (negative cross resistance) to this fungicide.[63] The mechanism of increased sensitivity has recently been clarified in *N. crassa.* DNA sequencing of the β-tubulin gene revealed that a single amino acid substitution from glutamic acid to glycine at position 198 confers both carbendazim resistance and diethofencarb sensitivity.[64,65] Diethofencarb was bound in the 50,000-g supernatant of mycelial extracts from the benzimidazole-resistant strain, whereas binding was hardly observed in the supernatant

	198 200
ASO$^{S\text{-}LR}$	5′ C TCT GAC GAG ACA TTC TG 3′
ASO^{MR}	C GAG ACA TAC TGC ATT GA
ASO^{HR}	C TCT GAC AAG ACA TTC TG
ASO^{VHR}	C TCT GAC GCG ACA TTC TG

Figure 2 Sequences of allele-specific oligonucleotide (ASO) probes for *Venturia inaequalis*. The ASO probes for medium (MR), high (HR), and very high resistance (VHR) to benomyl differ from the β-tubulin DNA for sensitive (S) and low-resistant (LR) strains by one nucleotide (arrows). (From Koenraadt, H. and Jones, A. L., *Phytopathology*, 82, 1354, 1992. With permission.)

from the wild-type sensitive (diethofencarb-resistant) strain.[66] In phytopathogenic fungi, only those mutations in codon 198, converting the codon for glutamic acid in benomyl-sensitive strains to alanine or glycine in resistant strains, were associated with the increased sensitivity to diethofencarb.[61] Mutations in codon 200 from a phenylalanine to tyrosine were never associated with an altered sensitivity to diethofencarb.

V. FUTURE DIRECTIONS

The long time required for confirmation of fungicide resistance by traditional methods interferes with quick and accurate decision making to minimize the risk of control failure with the fungicide. An oligonucleotide probe based on the difference of sequence between resistant and sensitive strains will help the development of a new monitoring system. However, currently available nucleic acid hybridization assays use ^{32}P-labeled DNA probes which pose safety, disposal, and license requirement problems for the users. Therefore, other assays using a nonradioactive oligonucleotide probe ought to be developed.[67]

The mechanism of resistance to benzimidazoles has well been characterized in phytopathogenic fungi. Regarding DMIs, the mechanism of resistance is complex, thereby the report on biotechnological approaches to detect resistance is few despite occurrence of the resistance problem which is gradually increasing. Much fundamental research will be needed to elucidate the mechanism of resistance so that application of new technology can be considered.

At present, "sustainable development" is one of the keywords when discussing the global environment. Public concern about pesticide residues in food and the environment has forced policy changes in several countries. As indicated earlier, the "Multi-Year Crop Protection Plan" has been put into practice in the Netherlands to reduce the input of chemical crop protectants. It is also necessary to reduce fungicide applications so that resistance problems can be reduced. Increasing use of modern diagnostic tools to detect and quantify diseases and to identify fungicide resistance will play an important part in improving the rational use of fungicides within supervised crop protection systems.

REFERENCES

1. Smith, C. M., History of benzimidazole use and resistance, in *Fungicide Resistance in North America,* Delp, C. J., ed., APS Press, MN, 1988, 23.
2. Delp, C. J. and Dekker, J., Fungicide resistance: definitions and use of terms, *Bull. OEPP/EPPO Bull.,* 15, 333, 1985.
3. Köller, W., Smith, F. D., and Reynolds, K. L., Phenotypic instability of flusilazole sensitivity in *Venturia inaequalis, Plant Pathol.,* 40, 608, 1991.
4. Smith, F. D., Parker, D. M., and Köller, W., Sensitivity distribution of *Venturia inaequalis* to the sterol demethylation inhibitor flusilazole: baseline sensitivity and implications for resistance monitoring, *Phytopathology,* 81, 392, 1991.
5. FAO, Recommended methods for the detection and measurement of resistance of agricultural pests to pesticides, *FAO Plant Prot. Bull.,* 30, 36, 1982.
6. Brent, K., Monitoring fungicide resistance: purposes, procedures and progress, in *Resistance 91—Achievements and Developments in Combating Pesticide Resistance,* Denholm, I., Devonshire, A. L., and Hollomon, D. W., eds., Elsevier Applied Science, London, 1992, 1.

7. Hollomon, D. W. and Butters, J., New approaches to detecting fungicide resistance. *Agro-Food-Industry Hi-Tech,* 20, 1992.
8. FRAC, FRAC methods for monitoring fungicide resistance, *Bull. OEPP/EPPO Bull.,* 21, 291, 1991.
9. FRAC, FRAC methods for monitoring fungicide resistance, *Bull. OEPP/EPPO Bull.,* 22, 297, 1992.
10. Ishii, H., Yanase, H., and Dekker, J., Resistance of *Venturia nashicola* to benzimidazole fungicides, *Meded. Fac. Landbouwwet. Rijksuniv. Gent,* 49/2a, 163, 1984.
11. Ishii, H., Van Raak, M., Inoue, I., and Tomikawa, A., Limitations in the exploitation of *N*-phenylcarbamates and *N*-phenylformamidoximes to control benzimidazole-resistant *Venturia nashicola* on Japanese pear, *Plant Pathol.,* 41, 543, 1992.
12. Faretra, F. and Pollastro, S., Genetic basis of resistance to benzimidazole and dicarboximide fungicides in *Botryotinia fuckeliana (Botrytis cinerea), Mycol. Res.,* 95, 943, 1991.
13. Ishii, H., Takeda, H., Nagamatsu, Y., and Nakashima, H., Sensitivity of the pear scab fungus (*Venturia nashicola*) to three ergosterol biosynthesis-inhibiting fungicides, *Pestic. Sci.,* 30, 405, 1990.
14. Braun, P. G. and McRae, K. B., Composition of a population of *Venturia inaequalis* resistant to myclobutanil, *Can. J. Plant Pathol.,* 14, 215, 1992.
15. Umemoto, S. and Nagai, Y., Simple method for determining benomyl tolerance of *Venturia nashicola, Ann. Phytopathol. Soc. Jpn.,* 45, 430, 1979.
16. Gullino, M. L. and Garibaldi, A., Fungicide resistance monitoring as an aid to tomato grey mould management, in *Proc. 1986 British Crop Protection Conf.—Pests and Diseases,* BCPC Publications, Surry, England, 1986, 499.
17. Bardinelli, T. R., Butterfield, E. J., and Jones, T. L., Diagnostic media for the detection of fungi (*Botrytis cinerea*) resistant to vinclozolin and benomyl. *Phytopathology,* 79, 1212, 1989.
18. Nishimura, S., Kohmoto, K., and Udagawa, H., Field emergence of fungicide-tolerant strains in *Alternaria kikuchiana* Tanaka, *Rep. Tottori Mycol. Inst. (Japan),* 10, 677, 1973.
19. Nishimura, S., Kohmoto, K., and Udagawa, H., Tolerance to polyoxin in *Alternaria kikuchiana* Tanaka, causing black spot disease of Japanese pear, *Rev. Plant Prot. Res.,* 9, 47, 1976.
20. Clerjeau, M., Moreau, C., Piganeau, B., and Malato, G., Resistance of *Plasmopara viticola* to anilide-fungicides: evaluation of the problem in France. *Meded. Fac. Landbouwwet. Rijksuniv. Gent,* 49/2a, 179, 1984.
21. Schepers, H. T. A. M., Persistence of resistance to fungicides in *Sphaerotheca fuliginea, Neth. J. Plant Pathol.,* 90, 165, 1984.
22. Heaney, S. P., Humphreys, G. J., Hutt, R., Montiel, P., and Jegerings, P. M. F. E., Sensitivity of barley powdery mildew to systemic fungicides in the UK, in *Proc. 1984 British Crop Protection Conf.—Pests and Diseases,* BCPC Publications, Surry, England, 1984, 459.
23. Grindle, M., Genetic basis of fungicide resistance, in *Combating Resistance to Xenobiotics—Biological and Chemical Approaches,* Ford, M. G., ed., Ellis Horwood, Chichester, England, 1987, 74.
24. Katan, T., Shabi, E., and Gilpatrick, J. D., Genetics of resistance to benomyl in *Venturia inaequalis* isolates from Israel and New York, *Phytopathology,* 73, 600, 1983.
25. Georgopoulos, S. G., The genetics of fungicide resistance, in *Modern Selective Fungicides—Properties, Applications, Mechanisms of Action,* Lyn, H., ed., VEG Gustav Fischer Verlag, Jena, 1987, 53.
26. Hollomon, D. W., Butters, J., and Clark, J., Genetic control of triadimenol resistance in barley powdery mildew, in *Proc. 1984 British Crop Protection Conf.—Pests and Diseases,* BCPC Publications, Surry, England, 1984, 477.
27. Brown, J. K. M., Jessop, A. C., Thomas, S., and Rezanoor, H. N., Genetic control of the response of *Erysiphe graminis* f. sp. *hordei* to ethirimol and triadimenol, *Plant Pathol.,* 41, 126, 1992.
28. Stanis, V. F. and Jones, A. L., Reduced sensitivity to sterol-inhibiting fungicides in field isolates of *Venturia inaequalis, Phytopathology,* 75, 1098, 1985.
29. Ishii, H., Homma, F., Miura, T., Suzaki, H., and Van Raak, M., Resistance of *Venturia inaequalis* to DMIs—phenotypic instability and genetic control, in *Abstr. 6th Int. Congr. Plant Pathol.,* 1993, 92.
30. Peever, T. L. and Milgroom, M. G., Inheritance of triadimenol resistance in *Pyrenophora teres, Phytopathology,* 82, 821, 1992.
31. Braun, P. G., Development and decline of a population of *Venturia inaequalis* resistant to sterol-inhibiting fungicides, *Norw. J. Agric. Sci.,* in press.

32. Lyr, H., Advances in fungicide mode of action and resistance research, in *Pesticide Chemistry—Advances in International Research, Development, and Legislation.* Frehse, ed., VCH Publishers, Weinheim, Germany, 1991, 161.
33. Henry, M. J., Mechanisms of resistance to fungicides, in *Resistance 91—Achievements and Developments in Combating Pesticide Resistance,* Denholm, I., Devonshire, A. L., and Hollomon, D. W., Eds., Elsevier Applied Science, London, 1992, 218.
34. Davidse, L. C., Benzimidazole fungicides: mechanism of action and biological impact, *Annu. Rev. Phytopathol.,* 24, 43, 1986.
35. Ishii, H., Target sites of tubulin-binding fungicides, in *Target Sites of Fungicide Action,* Köller, W., ed., CRC Press, Boca Raton, FL, 1992, 43.
36. Ishii, H. and Davidse, L. C., Decreased binding of carbendazim to cellular protein from *Venturia nashicola* and its involvement in benzimidazole resistance, in *Proc. 1986 British Crop Protection Conf.—Pests and Diseases,* BCPC Publications, Surry, England, 1986, 567.
37. Groves, J. D., Fox. R. T. V., and Baldwin, B. C., Modes of action of carbendazim and ethyl *N*-(3, 5-dichlorophenyl)carbamate on field isolates of *Botrytis cinerea,* in *Proc. 1988 Brighton Crop Protection Conf.—Pests and Diseases,* BCPC Publications, Surry, England, 1988, 397.
38. Yamada, M., Ishii, H., and Nakazawa, Y., Mechanisms of carbendazim resistance of strawberry anthracnose fungi, in *Abstr. 6th Int. Congr. Plant Pathol.,* 1993, 89.
39. Panaccione, D. G. and Hanau, R. M., Characterization of two divergent β-tubulin genes from *Colletotrichum graminicola, Gene,* 86, 163, 1990.
40. Buhr, T. L. and Dickman, M. B., Isolation and characterization of a β-tubulin-encoding gene from *Colletotrichum gloeosporioides* f. sp. *aeschynomene, Gene,* 124, 121, 1993.
41. Vanden Bossche, H., Mode of action of pyridine, pyrimidine and azole antifungals, in *Sterol Biosynthesis Inhibitors—Pharmaceutical and Agrochemical Aspects,* Berg, D. and Plempel, M., Eds., Ellis Horwood, Chichester, England, 1988, 79.
42. Köller, W., Sterol demethylation inhibitors: mechanism of action and resistance, in *Fungicide Resistance in North America,* Delp, C. J., ed., APS Press, MN, 1988, 79.
43. De Waard, M. A. and Van Nistelrooy, J. G. M., Accumulation of SBI fungicides in wild-type and fenarimol-resistant isolates of *Penicillium italicum, Pestic. Sci.,* 22, 371, 1988.
44. Kendall, S. J. and Hollomon, D. W., DMI resistance and sterol 14α demethylation in *Rhynchosporium secalis,* in *Proc. 1990 Brighton Crop Protection Conf.—Pests and Diseases,* BCPC Publications, Surry, England, 1990, 1129.
45. Vanden Bossche, H., Marichal, P., Odds, F. C., Jeune, L. L., and Coene, M.-C., Characterization of an azole-resistant *Candida glabrata* isolate, *Antimicrob. Agents Chemother.,* 36, 2602, 1992.
46. De Waard, M. A. and Stehmann, C., Action of sterol biosynthesis inhibitors against *Botrytis cinerea,* in *Recent Advances in Botrytis Research,* Verhoeff, K., Malathrakis, N. E., and Williamson, B., eds., Pudoc Scientific Publishers, Wageningen, Germany, 1992, 223.
47. Smith, F. D. and Köller, W., The expression of resistance of *Ustilago avenae* to the sterol demethylation inhibitor triadimenol is an induced response, *Phytopathology,* 80, 584, 1990.
48. Vanden Bossche, H., Marichal, P., Gorrens, J., Bellens, D., Moereels, H., and Janssen, P. A. J., Mutation in cytochrome P-450-dependent 14α-demethylase results in decreased affinity for azole antifungals, *Biochem. Soc. Trans.,* 18, 56, 1990.
49. Hollomon, D. W., Molecular approaches to understanding the mechanisms of fungicide resistance, in *Proc. 1990 Brighton Crop Protection Conf.—Pests and Diseases,* BCPC Publications, Surry, England, 1990, 881.
50. Dewey, F. M., Detection of plant-invading fungi by monoclonal antibodies, in *Techniques for the Rapid Detection of Plant Pathogens,* Duncan, J. M. and Torrance, L., eds., Blackwell Scientific, Oxford, 1992, 47.
51. Collett, M., Clark, J. S. C., Kendall, S. J., and Hollomon, D. W., Field evaluation of an immunodiagnostic assay for cereal eyespot, in *Proc. 1992 Brighton Crop Protection Conf.—Pests and Diseases,* BCPC Publications, Surry, England, 1992, 697.
52. Joerger, M. C., Hirata, L. T., and Baxter, M. A., Research and development of enzyme-linked immunosorbent assays for the detection of the wheat pathogens *Septoria nodorum* and *Septoria tritici,* in *Proc. 1992 Brighton Crop Protection Conf.—Pests and Diseases,* BCPC Publications, Surry, England, 1992, 689.

53. Groves, J. D., Fox, R. T. V., and Baldwin, B. C., Tubulin from *Botrytis cinerea,* and the potential for development of an immunodiagnostic for benzimidazole resistance, in *Proc. 1988 Brighton Crop Protection Conf.—Pests and Diseases,* BCPC Publications, Surry, England, 1988, 415.
54. Orbach, M. J., Porro, E. B., and Yanofsky, C., Cloning and characterization of the gene for β-tubulin from a benomyl-resistant mutant of *Neurospora crassa* and its use as a dominant selectable marker, *Mol. Cell. Biol.,* 6, 2452, 1986.
55. Martin, L. A., Fox, R. T. V., and Baldwin, B. C., Rapid methods for detection of MBC-resistance in fungi: I. Immunological approach, in *Proc. 10th Int. Symp. on Systemic Fungicides and Antifungal Compounds,* Lyr, H. and Polter, C., eds., Eugen Ulmer, Stuttgart, Germany, 1992, 209.
56. Cooley, R. N. and Caten, C. E., Cloning and characterisation of the β-tubulin gene and determination of benomyl resistance in *Septoria nodorum, Found. Biotech. Ind. Fermentat. Res.,* 6, 207, 1989.
57. Cooley, R. N., Van Gorcom, R. F. M., Van den Hondel, C. A. M. J. J., and Caten, C. E., Isolation of a benomyl-resistant allele of the β-tubulin gene from *Septoria nodorum* and its use as a dominant selectable marker, *J. Gen. Microbiol.,* 137, 2085, 1991.
58. Cooley, R. N. and Caten, C. E., Molecular analysis of the *Septoria nodorum* β-tubulin gene and characterization of a benomyl-resistance mutation, *Mol. Gen. Genet.,* 237, 58, 1993.
59. Sherwood, J. E. and Somerville, S. C., Sequence of the *Erysiphe graminis* f. sp. *hordei* gene encoding β-tubulin, *Nucleic Acids Res.,* 18, 1052, 1990.
60. Martin, L. A., Fox, R. T. V., Baldwin, B. C., and Connerton. I. F., Use of polymerase chain reaction for the diagnosis of MBC resistance in *Botrytis cinerea,* in *Proc. 1992 Brighton Crop Protection Conf.—Pests and Diseases,* BCPC Publications, Surry, England, 1992, 207.
61. Koenraadt, H., Somerville, S. C., and Jones, A. L., Characterization of mutations in the beta-tubulin gene of benomyl-resistant field strains of *Venturia inaequalis* and other plant pathogenic fungi, *Phytopathology,* 82, 1348, 1992.
62. Koenraadt, H. and Jones, A. L., The use of allele-specific oligonucleotide probes to characterize resistance to benomyl in field strains of *Venturia inaequalis, Phytopathology,* 82, 1354, 1992.
63. Kato, T., Negative cross-resistance activity of MDPC and diethofencarb against benzimidazole-resistant fungi, in *Fungicide Resistance in North America,* Delp, C. J., Ed., APS Press, MN, 1988, 40.
64. Fujimura, M., Oeda, K., Inoue, H., and Kato, T., A single amino-acid substitution in the beta-tubulin gene of *Neurospora* confers both carbendazim resistance and diethofencarb sensitivity, *Curr. Genet.,* 21, 399, 1992.
65. Fujimura, M., Kamakura, T., Inoue, H., Inoue, S., and Yamaguchi, I., Sensitivity of *Neurospora crassa* to benzimidazoles and *N*-phenylcarbamates: effects of amino acid substitutions at position 198 in β-tubulin, *Pestic. Biochem. Physiol.,* 44, 165, 1992.
66. Fujimura, M., Kamakura, T., and Yamaguchi, I., Action mechanism of diethofencarb to a benzimidazole-resistant mutant in *Neurospora crassa, J. Pestic. Sci.,* 17, 237, 1992.
67. Kim, C., Swaminathan, B., Cassaday, P. K., Mayer, L. W., and Holloway, B. P., Rapid confirmation of *Listeria monocytogenes* isolated from foods by a colony blot assay using a digoxigenin-labeled synthetic oligonucleotide probe, *Appl. Environ. Microbiol.,* 57, 1609, 1991.
68. Jung, M. K., Wilder, I. B., and Oakley, B. R., Amino acid alterations in the *benA* (β-tubulin) gene of *Aspergillus nidulans* that confer benomyl resistance, *Cell Motil. Cytoskeleton*, 22, 170, 1992.

FURTHER READING

Chaube, H. S. and Singh, U. S., *Plant Disease Management: Principles and Practices,* CRC Press, Boca Raton, FL, 1991.

Denholm, I., Devonshire, A. L., and Hollomon, D. W., Eds., *Resistance 91—Achievements and Development in Combating Pesticide Resistance,* Elsevier Applied Science, London.

Duncan, J. M. and Torrance, L., eds., *Techniques for the Rapid Detection of Plant Pathogens,* Blackwell Scientific, Oxford.

INDEX

INDEX